Lecture Notes in Computer Science 7109

Commenced Publication in 1973
Founding and Former Series Editors:
Gerhard Goos, Juris Hartmanis, and Jan van Leeuwen

Antonio Fernández Anta
Giuseppe Lipari   Matthieu Roy (Eds.)

Principles of Distributed Systems

15th International Conference, OPODIS 2011
Toulouse, France, December 13-16, 2011
Proceedings

 Springer

Volume Editors

Antonio Fernández Anta
Institute IMDEA Networks
Avenida del Mar Mediterraneo, 22, 28918 Leganes, Madrid, Spain
E-mail: antonio.fernandez@imdea.org

Giuseppe Lipari
Scuola Superiore Sant'Anna
CEIICP, RETIS Lab
Via Moruzzi 1, 56127 Pisa, Italy
E-mail: g.lipari@sssup.it

Matthieu Roy
LAAS-CNRS
Dependability Group (TSF)
7 av du Colonel Roche, 31077 Toulouse, France
E-mail: roy@laas.fr

ISSN 0302-9743   e-ISSN 1611-3349
ISBN 978-3-642-25872-5   ISBN 978-3-642-25873-2 (eBook)
DOI 10.1007/978-3-642-25873-2
Springer Heidelberg Dordrecht London New York

Library of Congress Control Number: Applied for

CR Subject Classification (1998): C.2.4, C.2, F.2, D.2, I.2.11, G.2.2

LNCS Sublibrary: SL 1 – Theoretical Computer Science and General Issues

Typesetting: Camera-ready by author, data conversion by Scientific Publishing Services, Chennai, India

Printed on acid-free paper

Springer is part of Springer Science+Business Media (www.springer.com)

Preface

On behalf of the Technical Committee of the International Conference on Principles of Distributed Systems (OPODIS 2011), we are very pleased to present in this volume the proceedings of the 15th edition of the conference, which was held during December 13–16, and was hosted by LAAS-CNRS, Toulouse, France.

OPODIS is an international forum that attracts the best researchers and practitioners in the design, analysis and development of distributed and real-time systems.

In response to the call for papers, 96 complete submissions were received. After an accurate reviewing process that involved 32 Program Committee members, 36 papers were selected that, in our opinion, represent the current state of the art of the research in our field. We would like to thank all reviewers for their fundamental contribution in selecting the best papers.

The papers and associated presentations were grouped into 11 sessions. For this edition, we decided to generate randomly the order of presentations, and hence the content of sessions in order to encourage interactions. This volume also includes the abstract of the keynote speech, which was given by Marco Ajmone Marsan, from Politecnico di Torino, Italy, and Institute IMDEA Networks, Spain, on "From Energy-Efficient Networking to ZEN."

This edition also had two colocated workshops: the second annual *TORRENTS* workshop on Time-Oriented Reliable Embedded Networked Systems that featured an invited talk from Klaus Havelund (Nasa/JPL) and the First International Workshop on Dynamic Systems (*DYNAM*).

This event would not have been possible without the support of LAAS-CNRS, from the logistic and administrative support to its director Jean Arlat, and of the Midi-Pyrénées delegation of CNRS. We would like to express our gratitude to our sponsors and particularly to the RTRA STAE, the French Space and Aeronautic Sciences & Technologies foundation, that allowed us to provide insightful related workshops.

We hope you enjoy the proceedings.

December 2011

Antonio Fernández Anta
Giuseppe Lipari
Matthieu Roy

Organization

OPODIS 2011 was organized by LAAS-CNRS, Toulouse, France.

General Chair

Matthieu Roy   LAAS-CNRS, France

Program Co-chairs

Antonio Fernández Anta   Institute IMDEA Networks, Spain
Giuseppe Lipari   Scuola Superiore Sant'Anna, Italy

Steering Committee

Tarek Abdelzaher   University of Illinois at Urbana-Champaign,
                                   USA
Alain Bui   University of Versailles S.Q., France
Marc Bui   EPHE, France
Hacène Fouchal   University of Reims C.A., France
Roberto Gomez   ITESM-CEM, Mexico
Chanyang Lu   Washington University, USA
Toshimitsu Masuzawa   University of Osaka, Japan
Michel Raynal   IRISA, France
Nicola Santoro   Carleton University, Canada
Philippas Tsigas   Chalmers University of Technology, Sweden

Program Committee

Marcos K. Aguilera   Microsoft Research, USA
James H. Anderson   University of North Carolina at Chapel Hill,
                                   USA
Bjorn Andersson   Carnegie Mellon University, USA
Roberto Baldoni   Sapienza University of Rome, Italy
Olivier Beaumont   University of Bordeaux I, France
Thibault Bernard   University of Reims, France
Tommaso Cucinotta   Scuola Sant'Anna, Italy
Nathan Fisher   Wayne State University, USA
Laurent George   INRIA Rocquencourt, France
Chryssis Georgiou   University of Cyprus, Cyprus
Seth Gilbert   National University of Singapore, Singapore
Phuong H. Ha   University of Tromsø, Norway
Damir Isovic   Mälardalen University, Sweden

Shinpei Kato	Carnegie Mellon University, USA
Anis Koubaa	IMAMU, KSA and CISTER-ISEP, Portugal
Dariusz Kowalski	University of Liverpool, UK
Fabian Kuhn	University of Lugano, Switzerland
Adam Lackorzynski	University of Dresden, Germany
Zhiyong Liu	Institute of Computing Technology, China
Chenyang Lu	Washington University, USA
Toshimitsu Masuzawa	University of Osaka, Japan
Alberto Montresor	University of Trento, Italy
Mohamed Mosbah	University of Bordeaux 1, France
Miguel A. Mosteiro	Rutgers and Rey Juan Carlos University, USA and Spain
Boaz Patt-Shamir	Tel Aviv University, Israel
Luis Rodrigues	Universidade Tecnica de Lisboa, Portugal
Christian Scheideler	University of Paderborn, Germany
Maria J. Serna	Universitat Politecnica de Catalunya, Spain
Håkan Sundell	University of Boras, Sweden
Alex Shvartsman	University of Connecticut, USA
Sebastien Tixeuil	Pierre-Marie Curie University, France
Shmuel Zaks	Technion, Israel

External Referees

Maissa Ben	Mark Jelasity	Nesrine Ouled
Döbel Björn	Hirotsugu Kakugawa	Joao Paiva
Silvia Bonomi	Sebastian Kniesburges	Andrzej Pelc
Zohir Bouzid	Boris Koldehofe	Anna Philippou
Armando Castaneda	Andreas Koutsopoulos	Laurence Pilard
Adriano Cerocchi	Anissa Lamani	Nuno Preguica
Jérémie Chalopin	Derrick Lawrence	Giuseppe Prencipe
Octav Chipara	Joao Leitao	Paola Quaglia
Shantanu Das	Vincent Leroy	Hani Qusa
Sebastian Daum	Cong Liu	Michel Raynal
Seda Davtyan	Giorgia Lodi	Etienne Rivière
Oksana Denysyuk	Sofian Maabout	Paolo Romano
Swan Dubois	Bernard Mans	Abusayeed Saifullah
Björn Döbel	Fabien Mathieu	Devan Sohier
Donatella Firmani	Serge Midonnet	Julian Stecklina
Chien-Liang Fok	Alessia Milani	Céline Trapes
Allyx Fontaine	Alex Mills	Giovanni Viglietta
Yong Fu	Malcolm Mollison	Marcus Völp
Leszek Gasieniec	Achour Mostefaoui	Chengjie Wu
Cyril Gavoille	Peter Musial	Rafik Zitouni
Hugo Gimbert	Nicolas Nicolaou	
Fabiola Greve	Fukuhito Ooshita	
Martina Hüllmann	Rotem Oshman	

Related Workshops

TORRENTS Workshop Chairs

Claire Pagetti	ONERA, France
Christine Rochange	IRIT, France

DYNAM Workshop Committee

Lélia Blin (Chair)	Université d'Evry Val d'Essonne, LIP6, France
Yann Busnel (Chair)	Université de Nantes, LINA, France
Shantanu Das	Technion - Israel Institute of Technology, Israel
Fabíola Greve	Universidade Federal da Bahia, Brazil
David Ilcinkas	CNRS Bordeaux, France
Anne-Marie Kermarrec	INRIA Rennes Bretagne Atlantique, France
Mikel Larrea	University of the Basque Country, Spain
Leonardo Querzoni	Università La Sapienza di Roma, Italy
Etienne Rivière	University of Neuchâtel, Switzerland
Gilles Trédan	University of Cambridge, UK

Organizing Committee

Sonia De Sousa	LAAS-CNRS, France
Brigitte Ducrocq	LAAS-CNRS, France
Marc-Olivier Killijian	LAAS-CNRS, France
Nicolas Rivière	LAAS-CNRS, France
Matthieu Roy	LAAS-CNRS, France

Sponsoring Institutions

RTRA STAE
CNRS

Table of Contents

From Energy-Efficient Networking to ZEN*

Marco Ajmone Marsan[1,2]

[1] Politecnico di Torino, Italy
ajmone@polito.it
[2] Institute IMDEA Networks, Spain

1 Summary

Energy-efficiency has become a hot topic in networking research. Several large international research projects have been activated in recent years on this subject. Examples are GreenTouch [1], conceived at Alcatel Lucent Bell Labs, TREND, EARTH, ECONET, C2POWER, CHRON, STRONGEST, Fit4Green, COST IC 804 [2,3,4,5,6,7,8,9], all funded by the European Commission through its 7th Framework Programme, and many national research projects, such as COOL SILICON in Germany and EFFICIENT in Italy. In addition, most equipment manufacturers feature their own internal research projects on this topic, such as GREAT in Huawei. The objective of all these research efforts consists in the reduction of the energy consumption of data networks, but their targets vary, from the 20% saving in today's networks quite realistically claimed by TREND, to the reduction by a factor 1000 in future networks somewhat optimistically foreseen by GreenTouch.

In spite of all this interest in energy-efficient networking research, in reality, the power consumed by actual networks keeps growing at an alarming pace. While this is a significant concern for network operators in developed countries, because of the impact that energy costs have on the growth of OPEX, and of the consequent reduction of margins and profits, the energy-greedy attitude of today's networks, coupled with the lack of reliable power sources, remains one of the main obstacles (if not *the* obstacle) for the widespread diffusion of data networks in some developing countries.

This is the reason why we advocate the need for a paradigm shift in energy-efficient networking research, toward what we call Zero Electricity Networking (ZEN).

The ZEN concept is based on network elements (such as routers, base stations, etc.) that are not connected to a power grid, but can acquire limited amounts of energy from (probably intermittent) local generators exploiting renewable sources (solar, wind, etc.). During periods of sufficient energy production by the generator associated with a network element, energy is used to operate it, in a mode which is carefully chosen to balance performance and power consumption,

---

* The research leading to these results has received funding from the European Union Seventh Framework Programme (FP7/2007-2013) under grant agreement n. 257740 (Network of Excellence TREND).

A. Fernández Anta, G. Lipari, and M. Roy (Eds.): OPODIS 2011, LNCS 7109, pp. 1–3, 2011.

and any energy surplus is stored in a battery, so that the element can operate also in periods of low or no production, as long as energy is available, but it is forced to switch off when the battery is depleted. Typically, we can expect that some periodicity exists in the possibility of energy production at network elements, for example because of the day/night variation in solar energy production (similar periodicities also exist in traffic demands, and we can expect the correlation to be significant). This means that (most of) network elements may be fully operational only for fractions of time, until energy lasts, or that they must choose operation modes which correspond to less-than-desired capacity, but are compatible with the available energy.

Research on ZEN can build upon many studies performed under similar constraints, but with different objectives, in the field of ad hoc networks, sensor networks, and wireless mesh networks [10,11,12,13,14,15,16], as well as in delay tolerant networks, or in military networks, where the activity of network elements cannot be given for granted at all times [17,18,19,20,21,22,23].

The viability of the ZEN approach must be investigated under realistic assumptions as regards the quantity of energy that is necessary at network nodes, that can be obtained from renewable sources at limited cost, and that can be stored in reasonable size/cost batteries. The investigation must also consider realistic traffic patterns, and performance or QoS/QoE (Quality of Service / Quality of Experience) requirements, consistent with the needs of a network operator (this is a major difference with respect to research performed in ad hoc and sensor networks or in delay tolerant networks, and is also one of the main challenges in ZEN).

The assessment of the feasibility of the ZEN approach requires multidisciplinary research, including competences in energy generation and storage, in low-power networking systems and equipment, in energy-efficient distributed and adaptive algorithms, and, of course, in networking. The ZEN concept has the possibility of opening new opportunities for the development of modern data networks in regions where energy grids are inexistent, or unreliable, or temporarily unavailable (because of both structural problems, and exceptional events, such as earthquakes, wars, or terrorism), or simply where energy is too expensive for operators to provide services at reasonable cost (a risk that may be faced also by developed countries in the not-so-distant future).

References

1. http://greentouch.org/
2. http://www.fp7-trend.eu/
3. https://www.ict-earth.eu/
4. http://www.econet-project.eu/
5. http://www.ict-c2power.eu/
6. http://www.ict-chron.eu/
7. http://www.ict-strongest.eu/
8. http://www.fit4green.eu/
9. http://www.cost804.org/

10. Chang, J., Tassiluas, L.: Maximum lifetime routing in wireless sensor networks. IEEE/ACM Transactions on Networking 12(4) (August 2004)
11. Zussman, G., Segall, A.: Energy efficient routing in ad hoc disaster recovery networks. In: IEEE INFOCOM 2003, San Francisco, USA (March 2003)
12. Al-Hazmi, Y., de Meer, H., Hummel, K.A., Meyer, H., Meo, M., Remondo, D.: Energy-efficient wireless mesh infrastruture. IEEE Network 25(2) (March-April 2011)
13. Capone, A., Malandra, F., Sansò, B.: Energy savings in Wireless Mesh Networks in a time-variable context. ACM/Springer Mobile Networks and Applications (to appear, 2011)
14. Kar, K., Kodialam, M., Lakshman, T., Tassiulas, L.: Routing for network capacity maxi-mization in energy-constrained ad-hoc networks. In: IEEE INFOCOM 2003, San Francisco, USA (March 2003)
15. Melodia, T., Pompili, D., Akyildiz, I.: Optimal local topology knowledge for energy efficient geographical routing in sensor networks. In: IEEE INFOCOM 2004, Hong Kong (March 2004)
16. Luo, D., Zhu, X., Wu, X., Chen, G.: Maximizing lifetime for the shortest path aggregation tree in wireless sensor networks. In: IEEE INFOCOM 2011, Shanghai, China (April 2011)
17. Fall, K.: A delay-tolerant network architecture for challenged internets. In: ACM SIGCOMM 2003, Karlsruhe, Germany (August 2003)
18. Jain, S., Fall, K., Patra, R.: Routing in a delay tolerant network. In: ACM SIGCOMM 2004, Portland, Oregon, USA (August-September 2004)
19. Zhao, W., Ammar, M., Zegura, E.: A message ferrying approach for data delivery in sparse mobile ad hoc networks. In: ACM MobiHoc 2004, Tokyo, Japan (May 2004)
20. Jun, H., Ammar, M.H., Zegura, E.W.: Power management in delay tolerant networks: A framework and knowledge-based mechanisms. In: IEEE SECON 2005, Santa Clara, CA, USA (September 2005)
21. Small, T., Haas, Z.J.: Resource and performance tradeoffs in delay-tolerant wireless networks. In: ACM SIGCOMM Workshop on Delay-Tolerant Networking (WDTN 2005), Philadelphia, USA (August 2005)
22. Jain, S., Demmer, M., Patra, R., Fall, K.: Using redundancy to cope with failures in a delay tolerant network. In: ACM SIGCOMM 2005, Philadelphia, USA (August 2005)
23. http://www.dtnrg.org

Online Regenerator Placement[*]

George B. Mertzios[1], Mordechai Shalom[2],
Prudence W.H. Wong[3], and Shmuel Zaks[4]

[1] School of Engineering and Computing Sciences, Durham University, UK
george.mertzios@durham.ac.uk
[2] TelHai College, Upper Galilee, 12210, Israel
cmshalom@telhai.ac.il
[3] Department of Computer Science, University of Liverpool, Liverpool, UK
pwong@liverpool.ac.uk
[4] Department of Computer Science, Technion, Haifa, Israel
zaks@cs.technion.ac.il

Abstract. Connections between nodes in optical networks are realized by lightpaths. Due to the decay of the signal, a regenerator has to be placed on every lightpath after at most d hops, for some given positive integer d. A regenerator can serve only one lightpath. The placement of regenerators has become an active area of research during recent years, and various optimization problems have been studied. The first such problem is the Regeneration Location Problem (RLP), where the goal is to place the regenerators so as to minimize the total number of nodes containing them. We consider two extreme cases of online RLP regarding the value of d and the number k of regenerators that can be used in any single node. (1) d is arbitrary and k unbounded. In this case a feasible solution always exists. We show an $O(\log |X| \cdot \log d)$-competitive randomized algorithm for any network topology, where X is the set of paths of length d. The algorithm can be made deterministic in some cases. We show a deterministic lower bound of $\Omega \left(\frac{\log(|E|/d) \cdot \log d}{\log(\log(|E|/d) \cdot \log d)} \right)$, where E is the edge set. (2) $d = 2$ and $k = 1$. In this case there is not necessarily a solution for a given input. We distinguish between feasible inputs (for which there is a solution) and infeasible ones. In the latter case, the objective is to satisfy the maximum number of lightpaths. For a path topology we show a lower bound of $\sqrt{l}/2$ for the competitive ratio (where l is the number of internal nodes of the longest lightpath) on infeasible inputs, and a tight bound of 3 for the competitive ratio on feasible inputs.

Keywords: online algorithms, optical networks.

1 Introduction

Background. Optical wavelength-division multiplexing (WDM) is the most promising technology today that enables us to deal with the enormous growth of

[*] This work was supported in part by the Israel Science Foundation grant No. 1249/08 and British Council Grant UKTELHAI09.

A. Fernández Anta, G. Lipari, and M. Roy (Eds.): OPODIS 2011, LNCS 7109, pp. 4–17, 2011.

traffic in communication networks, like the Internet. Optical fibers using WDM technology can carry around 80 wavelengths (colors) in real networks and up to few hundreds in testbeds. As satisfactory solutions have been found for various coloring problems, the focus of studies shifts from the number of colors to the hardware cost. These new measures provide better understanding for designing and routing in optical networks.

A communication between a pair of nodes is done via a *lightpath*. The energy of the signal along a lightpath decreases and thus amplifiers are used every fixed distance. Yet, as the amplifiers introduce noise into the signal there is a need to place a regenerator every at most d hops.

There is a limit imposed by the technology on the number of regenerators that can be placed in a network node [3,5]. We denote this limit by k and refer to the case where this limit is not likely to be reached by any regenerator placement as $k = \infty$.

The problems. Given a network G, a set of lightpaths in G, and integers d and k, we need to place regenerators at the nodes of the network, such that a) for each lightpath there is a regenerator in at least one of each d consecutive internal nodes, and b) at most k regenerators are placed at any node. When $k = \infty$ we consider the *regenerator location problem* (RLP) where the objective is to minimize the number of nodes that are assigned regenerators. When k is bounded there are inputs for which there is no feasible regenerator placement that satify both conditions. For example, consider the case $d = 2$ and $k = 1$, and three identical lightpaths $u - v - w - x$. Each of these lightpaths must have a regenerator either at v or w, and this is clearly impossible). In this case we consider the *Path Maximization Problem* (PMP) that seeks for regenerator placements that serve as many lightpaths as possible. We consider online algorithms (see [2]) for these problems.

Online algorithms. In the online setting the lightpaths are given one at a time, the algorithm has to decide on the locations of the regenerators and cannot change the decision later. An algorithm is *c-competitive* for RLP if for every input the number of locations used is no more than c times the locations used by an optimal offline algorithm. An online algorithm is *c-competitive* for PMP if the number of lightpaths that it satisfies is at least $1/c$ times the number of lightpaths that could be satisfied by an optimal offline algorithm.

Related Work. Placement of regenerators in optical networks has become an active area in recent years. Most of the researches have focused on the technological aspects of the problems, heuristics and simulations in order to reduce the number of regenerators, (e.g., [3,4,7,9,10,11,12]). The regenerator location problem (RLP) was shown to be NP-complete in [3], followed by heuristics and simulations. In [5] theoretical results for the offline version of RLP are presented. The authors study four variants of the problem, depending on whether the number k of regenerators per node is bounded, and whether the routings of the requests are given. Regarding the complexity of the problem, they present polynomial-time algorithms and NP-completeness results for a variety of special cases.

We note that while considering the path topology, RLP has implications for the following scheduling problem: Assume a company has n cars and that car i needs to be serviced within every at most d days between day a_i and b_i. Furthermore, assume that the garage can serve at most k cars per day and charges a certain cost each time the garage is used. The objective is to service the cars in the fewest number of days and hence minimizing the number of times the garage is used.

Other objective functions have also been considered in the context of regenerator placement. E.g., in [8] the problem of minimizing the total number of regenerators is studied under other settings.

Our Contribution. In this paper we study the online version of the regenerator location problem, and consider two extreme cases regarding the value of d and the value k of the number of regenerators that can be used in any single node.

- RLP: $k = \infty$, G and d are arbitrary (in this case there is a solution for every input, and the measurement is the number of locations in which regenerators are placed). We show:
  - an $O(\log |X| \cdot \log d)$-competitive randomized algorithm for any network topology, that can be made deterministic (with the same competitive ratio) for some cases including tree topology networks, where X is the set of all paths of length d in G.
  - a deterministic lower bound of $\Omega\left(\frac{\log(|E|/d)\cdot \log d}{\log(\log(|E|/d)\cdot \log d)}\right)$, where E is the edge set of G.
- PMP: G is a path, $k = 1$ and $d = 2$ (in this case there is not necessarily a solution, and the measurement is the number of satisfied lightpaths). We distinguish between feasible inputs (for which there is a solution) and infeasible ones, on a path topology, and show:
  - a lower bound of $\sqrt{l}/2$ for the competitive ratio for general instances which may be infeasible (where l is the number of internal nodes of the longest lightpath).
  - a tight bound of 3 for the competitive ratio of deterministic online algorithms for feasible instances.

Organization of the paper. In Section 2 we present some preliminaries. In Section 3 we consider general topology and analyze the first extreme case (k unbounded). In Section 4 we analyze the other extreme case ($k = 1$) for a path topology. In Section 5 we present further research directions.

2   Preliminaries

Given an undirected underlying graph $G = (V, E)$ that corresponds to the network topology, a *lightpath* is a simple path in G. We are given a set $\mathcal{P} = \{P_1, P_2, ..., P_n\}$ of simple paths in G that represent the lightpaths. The *length* of a lightpath is the number of edges it contains. The *internal vertices* (resp. *edges*) of a path P are the vertices (resp. edges) in P except the first and the last ones.

A regenerator assignment is a function $reg : V \times \mathcal{P} \mapsto \{0,1\}$. For any $v \in V, P \in \mathcal{P}$, $reg(v, P) = 1$ if a regenerator is assigned to P at node v. Note that $reg(v, P) = 1$ only if v is an internal node of P. We denote by $reg(v)$ the number of regenerators located at node v, i.e., $reg(v) = \sum_{P \in \mathcal{P}} reg(v, P)$. Denote by $cost(reg)$ the cost of the assignment reg, measured by the total number of locations where regenerators have been placed. Let $R(reg) = \{v \in V | reg(v) \geq 1\}$, then $cost(reg) = |R(reg)|$.

Given an integer d, a lightpath P is *d-satisfied* by the regenerator assignment reg if it does not contain d consecutive internal vertices without a regenerator, in other words, for any d consecutive internal vertices of P, $v_1, v_2, \cdots, v_d$, $\sum_{i=1}^{d} reg(v_i, P) \geq 1$. A set of lightpaths is *d-satisfied* if each of its lightpaths is d-satisfied. Note that a path with at most d edges is d-satisfied regardless of reg, therefore we assume without loss of generality that every path $P \in \mathcal{P}$ has at least $d + 1$ edges. For the sake of the analysis we assume, without loss of generality, that every edge of the graph is used by at least one path $P \in \mathcal{P}$. We want to emphasize that this is not assumed by the online algorithms, (what would be a loss of generality).

The Regenerator Location Problem (RLP): given a graph $G = (V, E)$, a set $\mathcal{P}$ of paths in G, a distance $d \geq 1$, determine the smallest number of nodes $R \subseteq V$ to place regenerators so that all the paths in $\mathcal{P}$ are d-satisfied. Formally:

REGENERATOR LOCATION PROBLEM (RLP)[a]

Input: An undirected graph $G = (V, E)$, a set $\mathcal{P}$ of paths in G, $d \geq 1$
Output: A regenerator assignment reg such that every path $P \in \mathcal{P}$ is d-satisfied.
Objective: Minimize $cost(reg)$.

---
[a] The offline version of this problem is denoted as RPP/∞/+ in [5].

---

reg^* denotes an optimal regenerator assignment and $cost^*$ denotes its cost $cost(reg^*)$. We consider the online version of the problem in which G and d are given in advance and the paths $\mathcal{P} = \{P_1, P_2, \ldots, P_n\}$ arrive in an online manner, one at a time in this order. An online algorithm finds a regenerator assignment as the input arrives and once $reg(v, P)$ is set to 1 it cannot be reverted to 0. An online algorithm ALG for RLP is *c-competitive*, for $c \geq 1$, if its cost is at most $c \cdot cost^*$. Clearly, when $d = 1$, $cost(reg) = |V_I|$ for any regenerator assignment reg where V_I is the set of nodes that are internal nodes of some lightpaths, therefore any algorithm is 1-competitive. Hence we consider the case $d \geq 2$.

When k is finite, we study the Path Maximization Problem (PMP): given a graph $G = (V, E)$, a set $\mathcal{P}$ of paths in G, a distance $d \geq 1$, place regenerators so that the number of d-satisfied paths in $\mathcal{P}$ is maximized. Formally:

PATH MAXIMIZATION PROBLEM (PMP)

Input: An undirected graph $G = (V, E)$, a set $\mathcal{P}$ of paths in G, $d, k \geq 1$
Output: A regenerator assignment reg for which $reg(v) \leq k$ for every node $v \in V$.
Objective: Maximize the number of d-satisfied paths in $\mathcal{P}$.

An online algorithm ALG for PMP is c-*competitive*, for $c \geq 1$, if the number of paths it satisfies is at least $1/c$ times the number of paths satisfied by an optimal offline algorithm.

3   The Regenerator Location Problem

In this section we consider the case where the technological limit imposed on the number of regenerators at a node is unlikely to be reached by any regenerator assignment. In this case we can assume without loss of generality that whenever there is a node v and a path P with $reg(v, P) = 1$ then $reg(v, P') = 1$ for every other path $P' \in \mathcal{P}$, because this does not affect $cost(reg)$. In other words for any given node v and any two paths $P, P' \in \mathcal{P}$ we assume $reg(v, P) = reg(v, P')$, thus $reg(v) = \sum_{P \in \mathcal{P}} reg(v, P) \in \{0, |\mathcal{P}|\}$. In this section we divide the objective function by $|\mathcal{P}|$ and denote by $reg(v)$ the value $reg(v, P_1) = reg(v, P_2), = \cdots$.

3.1   Upper Bound for Path Topology

Lemma 1. *There is a 2-competitive deterministic online algorithm in path topologies for* RLP.

Proof. Let $V = \{v_1, v_2, \ldots, v_n\}$ be the nodes of the path and $E = \{\{v_i, v_{i+1}\} \,|\, 1 \leq i < n\}$ be its edge set. We set $R = \{v_d, v_{2d}, \ldots\} \subseteq V$ and start with the empty assignment, i.e. $reg(v) = 0$ for every node $v \in V$. When a path P is presented to the algorithm we set $reg(v) = 1$ for every $v \in R \cap P$. This strategy clearly d satisfies all the paths.

We show that this algorithm is 2-competitive. Consider the union $\cup \mathcal{P}$ of all the paths in the input. $\cup \mathcal{P}$ is a disjoint union of maximal sub-paths of G. Consider such a maximal sub-path, and let ℓ be its length. Clearly, our algorithm uses at most $\lceil \frac{\ell}{d} \rceil$ locations among the nodes of this sub-path. Note that $\ell > d$, because otherwise there is at least one path in the input with at most d edges. Using these fact one can show by induction on $m = \lceil \frac{\ell}{2d} \rceil$ that any solution uses at least m locations among the nodes of this sub-path. $\square$

3.2   Upper Bound for General Topologies

In this section we use the randomized algorithm presented in [1] for the online set-cover problem. For completeness, we provide brief descriptions of the problem and the algorithm.

An instance of the set cover problem is a pair $(X, \mathcal{S})$ where $X = \{x_1, x_2, \ldots\}$ is a ground set of elements, and $\mathcal{S} = \{S_1, S_2, \ldots\}$ is a collection of subsets of X. Given such an instance, one has to find a subset $\mathcal{C} \subseteq \mathcal{S}$ that covers X, i.e. $\cup_{S_i \in \mathcal{C}} S_i = X$. In [1] an online variant of the set cover problem is considered. An instance of the online set cover problem is a triple $(X, \mathcal{S}, X')$ where X and $\mathcal{S}$ are as before, and $X' \subseteq X$ is presented in an online manner, one element at a time. At any given time one has to provide a cover $\mathcal{C}' \subseteq \mathcal{S}$ of X', i.e. $X' \subseteq \cup_{S_i \in \mathcal{C}'} S_i$. Once a set is included in the cover $\mathcal{C}'$ this decision can not be changed when subsequent input is received. In other words, whenever an element is presented an online algorithm has to cover it by at least one set from $\mathcal{S}$ if it is not already covered. It is important to note that X and $\mathcal{S}$ are known in advance but X' is given online.

We proceed with a description of the online algorithm in [1]. We denote by $S^{(i)}$ the set of all sets containing x_i, i.e. $S^{(i)} \overset{def}{=} \{S_j \in \mathcal{S} | x_i \in S_j\}$. Let f an upper bound for the frequencies of the elements, i.e. $\forall x_i \in X, |S^{(i)}| \leq f$. The algorithm associates a weight w_j with each set S_j which is initiated to $1/f$. The weight $w^{(i)}$ of each element $x_i \in X$ is the sum of the weights of the sets containing it, i.e. $w^{(i)} = \sum_{S_j \in S^{(i)}} w_j$. See pseudo-code in Algorithm ONLINESETCOVER below for a description of the algorithm.

Algorithm 1. ONLINESETCOVER

1: When a non-covered element $x_i \in X$ is presented:
2: Find the smallest non-negative integer q such that $2^q \cdot w^{(i)} \geq 1$;
3: **for** each set $S_j \in S^{(i)}$ **do**
4:      $\delta_j = 2^q \cdot w_j - w_j$;
5:      $w_j += \delta_j$;
6: **end for**
7: **do** $4 \log |X|$ times
8:      Choose at most one set (from $S^{(i)}$) to the cover
9:          where each set S_j is chosen with probability $\delta_j/2$;

From an instance $(G, \mathcal{P}, d)$ of RLP we build an instance $(X, \mathcal{S}, X')$ of the online set cover problem. X is the set of all possible paths of length d in G and $|\mathcal{S}| = |V|$. Each set $S_j \in \mathcal{S}$ consists of all the paths in X containing the node v_j. For a path P, let $P^{(d)}$ be the set of all its sub-paths of length d. X' is $\cup_{P \in \mathcal{P}} P^{(d)}$. Now we observe that for any feasible regenerator assignment reg, $R(reg)$ is a set cover, and vice versa, i.e. any set cover $\mathcal{C}$ corresponds to a feasible regenerator assignment reg such that $R(reg) = \mathcal{C}$. Indeed, a path P is d-satisfied if and only if every path of $P^{(d)} \subseteq X'$ contains a node v_j with regenerators, that corresponds to a set $S_j \in \mathcal{C}$ containing this path. Therefore all the paths $P \in \mathcal{P}$ are d-satisfied if and only if $\mathcal{C}$ constitutes a set cover of X'. Moreover the cost of the set cover is equal to the number of regenerator locations, i.e. $|\mathcal{C}| = \sum_{v_j} reg(v_j) = cost(reg)$.

When a path P is presented, we present to ONLINESETCOVER all the paths of $P^{(d)}$ one at a time. For each set S_j added to the cover by ONLINESETCOVER, we set $reg(v_j) = 1$.

We first note that although the number of sets in X is exponential in terms of the input size of our problem, for every path P the set $P^{(d)}$ contains only a polynomial number of paths, therefore the first loop of Algorithm ONLINESET-COVER runs only a polynomial number of times. The second loop is executed $\log|X|$ times, which is also polynomial in terms of our input size.

Algorithm ONLINESETCOVER is proven to be $O(\log|X| \cdot \log f)$-competitive. Note that a path of length d contains $d + 1$ nodes, thus $f = d + 1$. As the cost of a cover is equal to the cost of a solution of $(G, \mathcal{P}, d)$ we conclude

Lemma 2. *There is an $O(\log|X| \cdot \log d)$-competitive polynomial-time randomized online algorithm for instances $(G, \mathcal{P}, d)$ of* RLP *where X is the set of all the paths of length d in G.*

In [1] algorithm ONLINESETCOVER is de-randomized using the method of conditional expectation. However in this method, in order to calculate the conditional expectancies, one has to consider all the elements of X. In our case X is the set of all paths of length d in G which is, in general, exponential in d, thus applying the technique in [1] directly to our case leads to an exponential algorithm. Although the definition of competitive ratio does not require polynomial running-time, for practical purposes we would like to have polynomial-time algorithms. The following theorem states some cases for which this condition is satisfied.

Theorem 1. *There is an $O(\log|X| \cdot \log d)$-competitive polynomial-time deterministic online algorithm for instances $(G, \mathcal{P}, d)$ of* RLP *in each one of the following cases where X is the set of all the paths of length d in G.*

- *Both d and the maximum degree $\Delta(G)$ of G are bounded by two constants.*
- *The number of cycles in G is bounded, in particular G is a ring.*
- *G has bounded treewidth, in particular G is a tree.*

3.3   Lower Bound for General Topologies

In this section we show a lower bound nearly matching the upper bound in the previous subsection, by using the online version of a reduction in [5] of set cover to RLP. Given an instance $(X, \mathcal{S}, X')$ of online set cover we build an instance $(G, \mathcal{P}, d)$ of RLP as follows (see Figure 1).

We set $d = |\mathcal{S}|$. The node set $V(G)$ of G is $\mathcal{S} \cup V_1 \cup V_2$ where $V_1 = \{s_i, t_i | 1 \leq i \leq |X|\}$ and $V_2 = \{v_{ij} | 1 \leq i \leq |X|, 1 \leq j \leq |\mathcal{S}|\}$. We proceed with a description of the paths $\mathcal{P}$. The edge set of G will be all the edges induced by the paths of $\mathcal{P}$. For each element x_i there is a path P_i in $\mathcal{P}$ between s_i and t_i. If $x_i \in S_j$ then $S_j \in V(G)$ is an internal node of the P_i, otherwise v_{ij} is an internal node of P_i. The internal nodes are ordered within the path P_i by their j index, i.e. the path x_i is of the form $(s_i - u_1 - u_2 - \cdots - u_{|\mathcal{S}|} - t_i)$ where u_j is either S_j or v_{ij} as described before.

By this construction every path x_i has exactly $|\mathcal{S}| = d$ internal nodes. Therefore a regenerator assignment is feasible if and only if it assigns at least one regenerator to one of the internal nodes of every path. Without loss of generality every element x_i is contained in at least one set S_j, otherwise no set

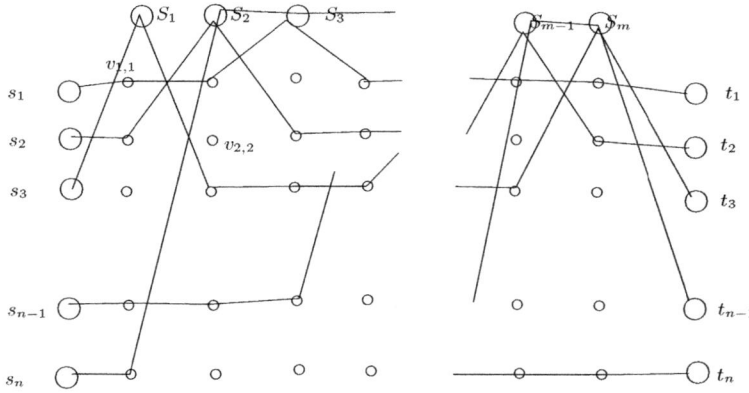

Fig. 1. Reduction from online set cover to RLP

cover exists. A feasible regenerator assignment *reg* corresponds to a set cover, in the following way. We first obtain a regenerator assignment *reg′* such that $reg'(v_{ij}) = 0$ for every $v_{ij} \in V_2$ and $cost(reg') \leq cost(reg)$. For every node with $reg(v_{ij}) = 1$ we set $reg'(v_{ij}) = 0$, and if P_i is not d-satisfied in *reg′* we choose arbitrarily a node S_j on P_i and set $reg'(S_j) = 1$. Now $R(reg') \subseteq \mathcal{S}$ is a set cover of cardinality at most $cost(reg)$.

Lemma 3. *There is no* $O(\frac{\log(|E|/d)\cdot\log d}{\log(\log(|E|/d)\cdot\log d)})$*-competitive online algorithm for* RLP.

Sketch of proof: Assume by contradiction that there is an $O(\frac{\log(|E|/d)\cdot\log d}{\log(\log(|E|/d)\cdot\log d)})$-competitive randomized algorithm ALG for RLP. From an instance $(X, \mathcal{S}, X')$ of online set cover we build an instance of RLP as described in the above discussion, and whenever we are presented an element $x_i \in X' \subseteq X$ we present the path P_i to ALG. We transform the regenerator assignment returned by ALG to a set cover $\mathcal{C}$ as described above. Note that the transformation does not exclude a set S_j from $\mathcal{C}$ if is was already in $\mathcal{C}$ before x_i was presented, thus $\mathcal{C}$ is an online set cover. We note that $|V| = \Theta(|X| \cdot |\mathcal{S}|), |E| = \Theta(|V|), d = \Theta(|\mathcal{S}|)$. This implies an $O(\frac{\log|X|\cdot\log|\mathcal{S}|}{\log(\log|X|\cdot\log|\mathcal{S}|)})$-competitive algorithm for the online set cover problem, which is proven to be impossible in [1]. $\qquad\square$

4   Path Maximization in Path Topology. ($k = 1, d = 2$)

In this section we consider possibly the simplest instances of the PMP problem, i.e. the case where the network is a path, and $k = 1, d = 2$.

We say that an instance is feasible, if there is a regenerator assignment that d-satisfies all the paths in $\mathcal{P}$, and infeasible otherwise. We first show in Section 4.1 that if the input instance is infeasible, no online algorithm (for PMP) has a small competitive ratio; precisely, we show that no online algorithm is better than $\sqrt{l}$-competitive, where l is the length of the longest path in the input. We then focus on feasible instances in Section 4.2.

4.1   Infeasible Instances

We show that there is a lower bound in terms of the length of the longest path if the input instance is infeasible, as follows:

Lemma 4. *Consider the path topology. For $k = 1$ and $d = 2$, any deterministic online algorithm for* PMP *has a competitive ratio at least $\sqrt{l}/2$, where l is the number of internal vertices of the longest path.*

Proof. The adversary first releases a path of length $l + 1$ with l internal vertices. The online algorithm has to satisfy this path, otherwise, the competitive ratio is unbounded. Then the adversary releases $\sqrt{l}$ paths along the first path each with $\sqrt{l}$ (disjoint) internal vertices. If the online algorithm does not satisfy any of these paths, the competitive ratio is at least $\sqrt{l}$ and we are done. Suppose x of these paths are satisfied. In order to make the first path and these x paths 2-satisfied, there is one regenerator placed in each node along these x paths. For each of these x paths P, the adversary releases $\sqrt{l}/2$ paths along P each with two (disjoint) internal vertices. The online algorithm is not able to satisfy any of these short paths and the total number of 2-satisfied paths is $x + 1$. On the other hand, the optimal offline algorithm satisfies all the paths except the first path of length l, i.e., $\sqrt{l} + x\sqrt{l}/2$ paths. As a result, the competitive ratio of the online algorithm is $\frac{(x+2)\sqrt{l}}{2(x+1)} > \sqrt{l}/2$.   □

4.2   Feasible Instances

We now consider feasible instances, that is, instances, where there exists a placement of regenerators such that all paths are satisfied. We will prove that, for feasible instances, there is a tight bound of 3 for the competitive ratio. That is, we provide an online algorithm Algorithm 2 with competitive ratio 3, and we show a lower bound of 3 for the competitive ratio of every deterministic online algorithm for feasible instances.

Algorithm 2 adopts a greedy approach and satisfies a newly presented path whenever possible. When a path P_i is presented, it checks whether there exist two consecutive internal vertices of P_i that are already assigned regenerators for previous paths. If yes, this means it is impossible (under the current assignment) to satisfy P_i. Otherwise, the algorithm satisfies P_i, as follows. There are two possible locations for the leftmost regenerator of P_i, namely, either its leftmost internal node, or the internal node adjacent to it. Among these two alternatives we choose the alternative that uses the smaller number of regenerators by trying the following regenerator allocation process. Suppose we put a regenerator at a certain internal node v of P_i. We check whether the node at distance 2 from v already has a regenerator; if no, we put a regenerator there and continue; if yes, we put a regenerator at the node at distance 1 from v^1. This continues until P_i is 2-satisfied.

Algorithm 2. Online algorithm for a path-topology, $k = 1$ and $d = 2$.

1: When the path P_i is presented:
2: **if** it is not possible to place regenerators to completely satisfy P_i **then**
3:     leave P_i unsatisfied;
4: **else**
5:     using the procedure described in the preamble of Algorithm 2, satisfy P_i using the smallest possible number of new regenerators
6: **end if**

Theorem 2. *Algorithm 2 is 3-competitive for* PMP *for feasible inputs in path topologies, when $k = 1$ and $d = 2$.*

Proof. Let S and U denote the sets of paths that have been satisfied and unsatisfied by the algorithm, respectively. We prove the theorem by showing that $|U| \leq 2|S|$. Then, the competitive ratio of Algorithm 2 is $\frac{|\mathcal{P}|}{|S|} = \frac{|U|+|S|}{|S|} \leq \frac{2|S|+|S|}{|S|} = 3$, i.e., Algorithm 2 is 3-competitive. In the sequel we prove that $|U| \leq 2|S|$ by associating with every path in U some paths of S, and showing that each path in S is associated with at most two paths in U.

Note that for $d = 2$ a feasible solution can be described as follows: Remove the first and last edges of every path $P \in \mathcal{P}$ presented, and return a vertex cover of the remaining edges. Therefore, in this proof, when we refer to a path P_i, we mean the path that the leftmost and rightmost edges have been removed.

Note also that, since the instance is assumed to be feasible, for every edge uv there exist *at most* two paths P_i, P_j, such that $uv \in P_i$ and $uv \in P_j$ (indeed, otherwise there would exist at least one path that is unsatisfied on the edge uv). Suppose that a path P_i presented at iteration i is unsatisfied, i.e., when P_i arrives, it cannot be satisfied by placing new regenerators. Then, there exists an edge $ab \in P_i$, where both a and b already have regenerators of paths that have been previously satisfied by the algorithm. We distinguish now two cases regarding the regenerators on vertices a and b.

Case 1: $reg(a, P_j) = reg(b, P_h) = 1$, with $j, h < i$ and $j \neq h$, where the paths P_j, P_h have been satisfied previously by the algorithm.

We first consider the cases where $ab \in P_j$ or $ab \in P_h$. Suppose that $ab \in P_j$. Then, since also $ab \in P_i$ by assumption, it follows that $ab \notin P_h$, since the instance is feasible. That is, b is an endpoint of P_h. In this case, associate the unsatisfied path P_i to the satisfied path P_h. Suppose now that $ab \in P_h$. Then it follows similarly that $ab \notin P_j$, and thus a is an endpoint of P_j. In this case, associate the unsatisfied path P_i to the satisfied path P_j.

---
[1] The node at distance 1 must have no regenerator, else there are two consecutive internal nodes with regenerators and the algorithm would have rejected the path.

Suppose now that $ab \notin P_j$ and $ab \notin P_h$, i.e., a is an endpoint of P_j and b is an endpoint of P_h. If there exists another path P_ℓ that is left unsatisfied by the algorithm, such that $ab \in P_\ell$, then associate the unsatisfied paths $\{P_i, P_\ell\}$ to the satisfied paths $\{P_j, P_h\}$. Otherwise, if no such path P_ℓ exists, then associate the path P_i to either P_j or P_h.

Case 2: $reg(a, P_j) = reg(b, P_j) = 1$, where $j < i$ and the path P_j has been satisfied previously by the algorithm.

The edge $ab \in P_j$. Furthermore, neither a nor b is an endpoint of path P_j, since otherwise Algorithm 2 would not place a regenerator on both vertices a and b of path P_j. That is, there exist two vertices d, c of P_j, such that (d, a, b, c) is a subpath of P_j. Moreover, since a and b are consecutive vertices of P_j, according to the algorithm there must exist two other satisfied paths P_h, P_ℓ, such that $reg(d, P_h) = reg(c, P_\ell) = 1$.[2] Note also that $ab \notin P_h$ and $ab \notin P_\ell$, since the instance is feasible, and since $ab \in P_i$ and $ab \in P_j$. That is, d or a is an endpoint of P_h, while b or c is an endpoint of P_ℓ.

We claim that there exist *at most* two different unsatisfied paths P_i and $P_{i'}$ that include at least one of the edges da, ab, bc. Suppose otherwise that there exist three such unsatisfied paths P_i, $P_{i'}$, $P_{i''}$. Recall that $ab \in P_i$ and that $da, ab, bc \in P_j$. Therefore, since the instance is assumed to be feasible, it follows that, either $da \in P_{i'}$ and $bc \in P_{i''}$, or $bc \in P_{i'}$ and $da \in P_{i''}$. Since these cases are symmetric, we assume without loss of generality that $da \in P_{i'}$ and $bc \in P_{i''}$. In any optimal (i.e., offline) solution, at least one of $\{a, b\}$ has a regenerator for path P_j; assume without loss of generality that $reg(b, P_j) = 1$ (the other case $reg(a, P_j) = 1$ is symmetric). Then, it follows that $reg(a, P_i) = 1$. Then, since the edge da must be satisfied for both paths P_j and $P_{i'}$, it follows that $reg(d, P_j) = reg(d, P_{i'}) = 1$. This is a contradiction, since every vertex can have at most one regenerator. Therefore there exist at most two different unsatisfied paths P_i, $P_{i'}$ that include at least one of the edges da, ab, bc.

In the case that P_i is the only unsatisfied path that includes at least one of the edges da, ab, bc, associate the unsatisfied path P_i to either the satisfied path P_h or to the satisfied path P_ℓ. Otherwise, if there exist two different unsatisfied paths P_i, $P_{i'}$ that include at least one of the edges da, ab, bc, associate the unsatisfied paths $\{P_i, P_{i'}\}$ to the satisfied paths $\{P_h, P_\ell\}$.

We observe that by the above associations of unsatisfied paths to satisfied ones, that at most two unsatisfied paths are associated to every satisfied path P (i.e., at most one to the left side and one to the right side of P, respectively). This gives $|U| \leq 2|S|$ and the theorem follows.   □

---

[2] Here we simplify the discussion slightly by assuming that the path P_i does not contain a chain of two internal edges that both do not belong to any other paths because the algorithm can simply assign regenerators to alternate internal nodes without conflicting any other paths and this would not affect the number of paths that can be satisfied by the algorithm.

Lemma 5. *Any deterministic online algorithm for* PMP *has a competitive ratio at least* 3 *even when the instance is restricted to feasible ones path topologies and* $k = 1, d = 2$.

Proof. We will prove that, for every $\varepsilon > 0$, there exists an input such that every algorithm has competitive ratio at least $3 - \varepsilon$. Choose n, such that $\frac{2}{n+1} < \varepsilon$. The adversary provides initially a path P_0 with $13n - 2$ edges. The algorithm must satisfy the path P_0, since otherwise the adversary stops and the competitive ratio is infinite. We divide P_0 into n subpaths P_i, $i = 1, 2, \ldots, n$, with 11 edges each, where between two consecutive subpaths there exist two edges.

Consider any such subpath P_i, $i = 1, 2, \ldots, n$. Suppose that there exist two edges ab and cd of P_i, where $\{a, b\} \cap \{c, d\} = \emptyset$, such that $reg(a, P_0) = reg(b, P_0) = 1$ and $reg(c, P_0) = reg(d, P_0) = 1$. Then the adversary provides next the paths $P_{i,1} = (a, b)$ and $P_{i,2} = (c, d)$. These two paths $P_{i,1}$ and $P_{i,2}$ can not be satisfied, since each of the vertices a, b, c, d has a regenerator for path P_0.

Suppose that there do not exist two such edges ab and cd of P_i. That is, there exist *at most* three consecutive vertices u_1, u_2, u_3 of P_i, such that $reg(u_1, P_0) = reg(u_2, P_0) = reg(u_3, P_0) = 1$, while for every other edge uu' of P_i, there exists a regenerator for P_0 either on vertex u or on vertex u'. Then, it is easy to check that there always exist five consecutive vertices v_1, v_2, v_3, v_4, v_5 of P_i, such that $reg(v_1, P_0) = reg(v_3, P_0) = reg(v_5, P_0) = 1$ and $reg(v_2, P_0) = reg(v_4, P_0) = 0$.

The adversary now provides the path $P_i' = (v_2, v_3, v_4)$. Thus, since $reg(v_3, P_0) = 1$ and $reg(v_2, P_0) = reg(v_4, P_0) = 0$, the only way that the algorithm can satisfy P_i' is to place regenerators for P_i' at the vertices v_2 and v_4 (that is, $reg(v_2, P_i') = reg(v_4, P_i') = 1$).

The adversary proceeds as follows. In the case where the algorithm chooses not to satisfy the path P_i', the adversary does not provide any other path that shares edges with P_i. Otherwise, if the algorithm satisfies P_i', then the adversary provides the paths $P_i'' = (v_1, v_2)$ and $P_i''' = (v_4, v_5)$. In this case, $reg(v_2, P_i') = reg(v_4, P_i') = 1$ and $reg(v_1, P_0) = reg(v_5, P_0) = 1$, and thus the paths P_i'' and P_i''' remain unsatisfied by the algorithm. In the sequel we show that the instance constructed in the proof is feasible. We prove that the instance delivered by the adversary is indeed a feasible instance. To this end, we provide a placement of the regenerators such that the path P_0, as well as all paths $P_{i,1}$, $P_{i,2}$, P_i', P_i'', and P_i''' are satisfied. First, we place a regenerator for P_0 on the vertex that lies between every two consecutive subpaths P_i and P_{i+1} of P_0. Inside the subpaths P_i of P_0, we place regenerators for P_0 on vertices with distance two between two regenerators. Then, we can assign appropriately regenerators to the paths $P_{i,1}$, $P_{i,2}$, P_i', P_i'', and P_i'''. In particular, for every subpath P_i of P_0, for which the opponent provides the path P_i', we have $reg(v_3, P_i') = 1$ and $reg(v_2, P_0) = reg(v_4, P_0) = 1$. Furthermore, for the subpaths P_i of P_0, for which the opponent provides also the paths P_i'' and P_i''', we have $reg(v_1, P_i'') = reg(v_5, P_i''') = 1$. Therefore, there exists a placement of regenerators on the vertices of the paths of the instance that the opponent delivers, such that all paths are satisfied. That is, the instance is feasible.

Denote now by h_1 the number of subpaths P_i, for which the algorithm adds the paths $P_{i,1}$ and $P_{i,2}$. Furthermore, denote by h_2 the number of subpaths P_i, for which the algorithm adds the path P_i', but not the paths P_i'' and P_i'''. Finally, denote by h_3 the number of subpaths P_i, for which the algorithm adds the three paths P_i', P_i'', and P_i'''. Clearly, $h_1 + h_2 + h_3 = n$. The total number of paths that the adversary provided equals $1 + 2h_1 + h_2 + 3h_3$, while the number of satisfied paths equals $1 + h_3$. That is, the competitive ratio of the algorithm is $\frac{1+2h_1+h_2+3h_3}{1+h_3} \geq \frac{1+h_1+h_2+3h_3}{1+h_3} = 3 + \frac{n-h_3-2}{1+h_3}$. Therefore, since $h_3 \geq n$, it follows that the competitive ratio of the algorithm is at least $3 - \frac{2}{1+n} > 3 - \varepsilon$. Since this holds for every $\varepsilon > 0$, it follows that any deterministic online algorithm has competitive ratio at least 3. This completes the proof of the lemma.   □

5   Future Work

We list some open problems and research directions:

– Close the gap between the bounds shown in this paper. In particular, we used in Section 3 a known approximation result of set cover and modified it for our problem. It might be of interest to improve the upper bound by developing a better algorithm for these special instances of the set cover problem. However we note that ONLINESETCOVER does not use the set of all potential elements but only its size. Therefore if the algorithm is supplied with an a priori information about the total length of the paths to be received, the algorithm can use it to get an upper bound which is logarithmic in terms of this bound, instead of the number of all possible paths of size d which can be much bigger.
– Extend the results for other values of the parameters d and k.
– Consider the regenerator location problem when also traffic grooming is allowed (that is, when up to g (the *grooming factor*) paths that share an edge can be assigned the same wavelength and can then share regenerators). In [6] optimizing the use of regenerators in the presence of traffic grooming is studied, but with two fundamental differences: (1) the cost function there is the number of locations where regenerators are used rather than the total number of regenerators suggested here, and (2) the authors consider the online case, where the requests for connection are not known a-priori, while here all requests are given in advance.
– Consider other objective functions (some of them are discussed in Section 1).

References

1. Alon, N., Awerbuch, B., Azar, Y., Buchbinder, N., Naor, S.: The online set cover problem. SIAM J. Computing 39(2), 361–370 (2009)
2. Borodin, A., El-Yaniv, R.: Online Computation and Competitive Analysis. Cambridge University Press, Cambridge (1998)
3. Chen, S., Ljubic, I., Raghavan, S.: The regenerator location problem. Networks 55(3), 205–220 (2010)

4. Fedrizzi, R., Galimberti, G.M., Gerstel, O., Martinelli, G., Salvadori, E., Saradhi, C.V., Tanzi, A., Zanardi, A.: Traffic independent heuristics for regenerator site selection for providing any-to-any optical connectivity. In: Proceedings of IEEE/OSA Conference on Optical Fiber Communications, OFC (2010)
5. Flammini, M., Marchetti-Spaccamela, A., Monaco, G., Moscardelli, L., Zaks, S.: On the complexity of the regenerator placement problem in optical networks. IEEE-TON 19(2), 498–511 (2011)
6. Flammini, M., Monaco, G., Moscardelli, L., Shalom, M., Zaks, S.: Optimizing Regenerator Cost in Traffic Grooming. In: Lu, C., Masuzawa, T., Mosbah, M. (eds.) OPODIS 2010. LNCS, vol. 6490, pp. 443–458. Springer, Heidelberg (2010)
7. Kim, S.W., Seo, S.W.: Regenerator placement algorithms for connection establishment in all-optical networks. IEE Proceedings Communications 148(1), 25–30 (2001)
8. Mertzios, G.B., Sau, I., Shalom, M., Zaks, S.: Placing Regenerators in Optical Networks to Satisfy Multiple Sets of Requests. In: Abramsky, S., Gavoille, C., Kirchner, C., Meyer auf der Heide, F., Spirakis, P.G. (eds.) ICALP 2010. LNCS, vol. 6199, pp. 333–344. Springer, Heidelberg (2010)
9. Pachnicke, S., Paschenda, T., Krummrich, P.M.: Physical impairment based regenerator placement and routing in translucent optical networks. In: Optical Fiber Communication Conference and Exposition and The National Fiber Optic Engineers Conference, page OWA2. Optical Society of America (2008)
10. Sriram, K., Griffith, D., Su, R., Golmie, N.: Static vs. dynamic regenerator assignment in optical switches: models and cost trade-offs. In: Proceedings of the IEEE Workshop on High Performance Switching and Routing (HPSR), pp. 151–155 (2004)
11. Yang, X., Ramamurthy, B.: Dynamic routing in translucent WDM optical networks. In: Proceedings of the IEEE International Conference on Communications (ICC), pp. 955–971 (2002)
12. Yang, X., Ramamurthy, B.: Sparse regeneration in translucent wavelength-routed optical networks: Architecture, network design and wavelength routing. Photonic Network Communications 10(1), 39–53 (2005)

A Quorum-Based Replication Framework for Distributed Software Transactional Memory

Bo Zhang and Binoy Ravindran

ECE Department, Virginia Tech
Blacksburg VA 24061, USA
{alexzbzb,binoy}@vt.edu

Abstract. Distributed software transactional memory (D-STM) promises to alleviate difficulties with lock-based (distributed) synchronization and object performance bottlenecks in distributed systems. Past single copy data-flow (SC) D-STM proposals keep only one writable copy of each object in the system and are not fault-tolerant in the presence of network node/link failures in large-scale distributed systems. In this paper, we propose a quorum-based replication (QR) D-STM model, which provides provable fault-tolerant property without incurring high communication overhead compared with SC model. QR model operates on an overlay tree constructed on a metric-space failure-prone network where communication cost between nodes forms a metric. QR model stores object replicas in a tree quorum system, where two quorums intersect if one of them is a write quorum, and ensures the consistency among replicas at commit-time. The communication cost of an operation in QR model is proportional to the communication cost from the requesting node to its closest read or write quorum. In the presence of node failures, QR model exhibits high availability and degrades gracefully when the number of failed nodes increases, with reasonable higher communication cost.

1 Introduction

Lock-based synchronization is non-scalable, non-composable, and inherently error-prone. Transactional memory (TM) is an alternative synchronization model for shared memory objects that promises to alleviate these difficulties. In addition to a simple programming model, TM provides performance comparable to highly concurrent fine-grained locking and is composable. TM for multiprocessors has been proposed in hardware, called HTM, in software, called STM, and in hardware/software combination [1].

Similar to multiprocessor TM, distributed STM (or D-STM) is motivated by the difficulties of lock-based distributed synchronization (e.g., distributed race conditions, composability). D-STM can be supported in any of the classical distributed execution models, including a) dataflow [2], where transactions are immobile, and objects are migrated to invoking transactions; b) control flow [3], where objects are immobile and transactions invoke object operations through

A. Fernández Anta, G. Lipari, and M. Roy (Eds.): OPODIS 2011, LNCS 7109, pp. 18–33, 2011.

RPCs; and c) hybrid models (e.g., [4]), where transactions or objects are migrated, based on access profiles, object size, or locality. The different models have their concomitant tradeoffs.

D-STM can be classified based on the system architecture: cache-coherent D-STM (cc D-STM) [2], where a number of nodes are interconnected using message-passing links, and a cluster model (cluster D-STM), where a group of linked computers works closely together to form a single computer ([4,5,6,7]). The most important difference between the two is communication cost. cc D-STM assumes a metric-space network, whereas cluster D-STM differentiates between local cluster memory and remote memory at other clusters.

In this paper, we focus on cc D-STM. The data-flow cc D-STM model is proposed by Herlihy and Sun [2]. In this model, only a single (writable) copy is kept in the system. Transactions run locally and objects move in the network to meet transactions' requests. When a node v_A initiates a transaction A that requests a read/write operation on object o, its TM proxy first checks whether o is in the local cache; if not, the TM proxy invokes a *cache-coherence* (*CC*) protocol to locate o in the network by sending a request $CC.locate(o)$. Assume that o is in use by a transaction B initiated by node v_B. When v_B receives the request $CC.locate(o)$ from v_A, its TM proxy checks whether o is in use by an active local transaction; if so, the TM proxy invokes a contention manager to handle the conflict between A and B. Based on the result of contention management, v_B's TM proxy decides whether to abort B immediately, or postpone A's request and let B proceed to commit. Eventually, CC moves o to v_A.

In the aforementioned single copy data-flow model (or SC model), the main responsibility of CC protocol is to locate and move objects in the network. A directory-based CC protocol is often adopted such that the latest location of the object is saved in the distributed directory and the cost to locate and move an object is bounded. Such CC protocols include Ballistic [2], Relay [8] and Combine [9].

Since SC model only keeps a single writable copy of each object, it is inherently vulnerable in the presence of node and link failures. If a node failure occurs, the objects held by the failed node will be simply lost and all following transactions requesting such objects would never commit. Hence, SC model cannot afford any node failures. Ballistic and Relay also assumes a reliable and fifo logical link between nodes, since they may not perform well when the message is reordered [10]. On the other hand, Combine can tolerate partial link failures and support non-fifo message delivery, as long as a logical link exists between any pair of nodes. However, similar to other directory-based CC protocols, Combine does not permit network partitioning incurred by link failures, which may make some objects inaccessible from outer transactions. In general, SC model is not suitable in a network environment with aforementioned node/link failures.

To achieve high availability in the presence of network failures, keeping only one copy of each object in the system is not sufficient. Inherited from database systems, replication is a promising approach to build fault-tolerant D-STM systems, where each object has multiple (writable) copies. However, only a few replicated D-STM solutions have been proposed for cluster-based D-STM ([4,5,6,7]). These solutions require some form of broadcasting to maintain consistency among replicas and assume a uniform communication cost across all pairs of nodes. As the result, we cannot directly apply these solutions for cc D-STM.

This paper presents QR model, a quorum-based replication cc D-STM model which provides provable fault-tolerance property in a failure-prone metric-space network, where communication cost between nodes forms a metric. To the best of our knowledge, this is the first replication cc D-STM proposal which provides provable fault-tolerant properties. In distributed systems, a quorum is a set of nodes such that the intersection of any two quorums is non-empty if one of them is a write quorum. By storing replicated copies of each object in an overlay tree quorum system motivated by the one in [11], QR model supports concurrent reads of transactions, and ensures the consistency among replicated copies at commit-time. Meanwhile, QR model exhibits a bounded communication cost of its operations, which is proportional to the communication cost from v to its closest read/write quorum, for any operation starting from node v. Compared with directory-based CC protocols, the communication cost of operations in QR model does not rely on the stretch of the underlying overlay tree (i.e., the worst-case ratio between the cost of direct communication between two nodes v and w and the cost of communication along the shortest tree path between v and w). Therefore QR model provides a more promising solution to support D-STM in the presence of network failures with communication cost comparable with SC model.

The rest of the paper is organized as follows. We introduce the system model and identify the limitations of SC model in Section 2. We present QR model and analyze its properties in Section 3. The paper concludes in Section 4.

2   Preliminaries

2.1   System Model

We consider a distributed system which consists of a set of distinct nodes that communicate with each other by message-passing links over a communication network. Similar to [2], we assume that the network contains n physical nodes scattered in a metric space of diameter D. The metric $d(u,v)$ is the distance between nodes u and v, which determines the communication cost of sending a message from u to v. Scale the metric so that 1 is the smallest distance between any two nodes.

We assume that nodes are *fail-stop* [12] and communication links may also fail to deliver messages. Further, node and link failures may occur concurrently and lead to network partitioning failures, where nodes in a partition may communicate with each other, but no communication can occur between nodes in different partitions. A node may become inaccessible due to node or partitioning failures.

We consider a set of *distributed transactions* $\mathcal{T} := \{T_1, T_2, \ldots\}$ sharing a set of objects $\mathcal{O} := \{o_1, o_2, \ldots\}$ distributed on the network. A transaction contains a sequence of requests, each of which is a read or write operation request to an individual object.

An execution of a transaction is a sequence of timed operations. An execution ends by either a commit (success) or an abort (failure). A transaction's status is one of the following three: *live, aborted*, or *committed*. A transaction is live after its first operation, and completes either by a commit or an abort operation. When a transaction aborts, it is restarted from its beginning immediately and may access a different set of shared objects. Two transactions are *concurrent* if they are both live at the same time. Suppose there are two live transactions T_j and T_k which request to access o_i and at least one of the access is a write. Then T_j and T_k are said to *conflict* at o_i, i.e., two live transactions conflict if they both access the same object and at least one of the accesses is a write. There are three types of conflicts: (1) Read-After-Write $(W \rightarrow R)$; (2) Write-After-Read $(R \rightarrow W)$; and (3) Write-After-Write $(W \rightarrow W)$. A *contention manager* is responsible for resolving the conflict, and does so by aborting or delaying (i.e., postponing) one of the conflicting transactions. Most contention managers do not allow two transactions to proceed (i.e., make progress) simultaneously. In other words, two operations from different transactions over the same object cannot be overlapped if one of them is a write. In this paper, we assume an underlying contention manager which has consistent policies to assign priorities to transactions. For example, the *Greedy* contention manager [13] always assigns higher priority to the transaction earlier timestamp.

Each node has a *TM proxy* that provides interfaces to the TM application and to proxies of other nodes. A transaction performs a read/write operation by first sending a read/write access request to its TM proxy. The TM proxy invokes a *CC* protocol to acquire a valid object copy in the network. For a read operation, the protocol returns a read-only copy of the object. For a write operation, the *CC* protocol returns a writable copy of the object. When there are multiple copies (or replicas) of an object existing in the network, the *CC* protocol is responsible to ensure the consistency over replicas such that multiple copies of an object must appear as a single logical object to the transactions, which is termed as one-copy equivalence [14].

2.2   Motivation: Limitations of SC D-STM Model

As mentioned in Section 1, SC model lacks the fault-tolerant property in the presence of network failures. SC model also suffers from some other limitations.

Limited support of concurrent reads. Although directory-based *CC* protocols for SC model allows multiple read-only copies of an object existing in the system, these protocols lacks the explanation on how they maintain the consistency over read-only and writable copies of objects. Consider two transactions A and B, where A contains operations $\{read(o_1), write(o_2)\}$ and B contains operations $\{read(o_2), write(o_1)\}$. In SC model, the operations of A and B could be interleaved, e.g., transaction A reads o_1 before B writes to o_1, and transaction B reads o_2 before A writes to o_2. Obviously, transactions A and

B conflict on both objects. In order to detect the conflict, each object needs to keep a record for any of its readers. When transaction A (or B) detects a conflict on object o_2 (or o_1), it does not know: i) the type of the conflicting transaction (read-only or read/write); and ii) the status of the conflicting transaction (live/aborted/committed). It is not possible for a contention manager to make distributed agreement without these knowledge (e.g., it is not necessary to resolve the conflict between a live transaction and an aborted/committed transaction). To keep each object updated with the knowledge of its readers, a transaction has to send messages to all objects in its readset once after its termination (commit or abort). Unfortunately, in SC model such mechanism incurs high communication and message overhead, and it is still possible that a contention manager may make a wrong decision if it detects a conflict between the time the conflicting transaction terminated and the time the conflicting object receives the updated information, due to the relatively high communication latency.

Due to the inherent difficulties in supporting concurrent read operations, practical implementations of directory-based CC protocols often do not differentiate between a read and write operation of a read/write transaction (i.e., if a transaction contains both read and write operations, all its operations are treated as write operations and all its requested objects have to be moved). Such over-generalization obviously limits the possible concurrency of transactions. For example, in the scenario where the workload is composed of late-write transactions [10], a directory-based CC protocol cannot perform better than a simple serialization schedule, while the optimal schedule maybe much shorter when concurrent reads are supported for read/write operations.

Limited locality. One major concern of directory-based CC protocols is to exploit locality in large-scale distributed systems, where remote access is often several orders of magnitude slower than local ones. Reducing communication cost and remote accesses is the key to achieving good performance for D-STM implementations. Existing CC protocols claim that the locality is preserved by their location-aware property: the cost to locate and move the objects between two nodes u and v is often proportional to the shortest path between u and v in the directory. In such a way directory-based CC protocols route transactions' requests efficiently: if two transactions requests an object at the same time, the transaction "closer" to the object in the directory will get the object first. The object will be first sent to the closer transaction, then to the further transaction.

Nevertheless, it is unrealistic to assume that all transactions start at the same time. Even if two transactions start at the same time, since a non-clairvoyant transaction may access a sequence of objects, it is possible that a closer transaction may request to access an object much later than a further transaction. In such cases, transactions' requests may not be routed efficiently by directory-based CC protocols. Consider two transactions A and B, where A is $\{\langle \text{some work} \rangle, write(o)\}$ and B is $\{write(o), \langle \text{some work} \rangle\}$. Object o is located at node v. Let $d(v, v_A) = 1$ and $d(v, v_B) = d(v_A, v_B) = D$, it is possible that

o first receives B's request of o. Assume that o is sent to B from v, then the directory of o points to v_B. Transaction A's request of o is forwarded to v_B and a conflict may occur at v_B. If B is aborted, the object o is moved to v_A from v_B. In this scenario, object o has to travel at least $3D$ distance to let two transactions A commit. On the other hand, when object o receives B's request at v, if we let o waits for time t_o to let A's request reach v, then o could be first moved to v_A and then to v_B. In this case, object o travels $t_o + D + 1$ distance to let two transactions commit. Obviously the second schedule may exploit more locality: as long as t_o is less than $2D - 1$ (which is a quite loose bound), the object is moved more quickly.

In practice, it is often impractical to predict t_o. As the result, directory-based CC protocols often overlook possible locality by simply keeping track of the single writable copy of each object. Such locality can be more exploited to reduce communication cost and improve performance.

3 Quorum-Based Replication Data-Flow D-STM Model

3.1 Overview

We present QR model, a quorum-based replication data-flow D-STM model, where multiple (writable) copies of each object are distributed at several nodes in the network. To perform a read or write operation, a transaction reads an object by reading object copies from a *read quorum*, and writes an object by writing copies to a *write quorum*. A quorum is assigned with the following restriction:

Definition 1 (Quorum Intersection Property). *A quorum is a collection of nodes. For any two quorums q_1 and q_2, where at least one of them is a write quorum, the two quorums must have a non-empty intersection: $q_1 \cap q_2 \neq \emptyset$.*

Generally, by constructing a quorum system over the network, QR model is able to keep multiple copies of each object. QR model provides 5 operations for a transaction: read, write, request-commit, commit and abort. Particularly, QR model provides a request-commit operation to validate the consistency of its readset and writeset before it commits. A transaction may request to commit if it is not aborted by other transactions before its last read/write operation. Concurrency control solely occurs during the request-commit operation: if a conflict is detected, the transaction may get aborted or abort the conflicting transaction. After collecting the response of the request-commit operation, a transaction may commit or abort.

We first present read and write operations of QR model in Algorithm 1. In following algorithms, notation "$msg \lhd v$" is interpreted as "receiving msg from node v", and notation "$msg \rhd v$" is interpreted as "sending msg to node v".

Read. When transaction T at node v starts a read operation, it sends a request message $req(T, read(o))$ to a selected read quorum q_r. The algorithm to find and select a read or write quorum will be elaborated in the next section. Node v', upon receiving $req(T, read(o))$, checks whether it has a copy of o. If not, it sends a null response to v.

In QR model, each object copy contain three fields: the value field, which is the value of the object; the version number field, starting from 0, and the *protected* field, a boolean value which records the status of the copy. The *protected* field is maintained and updated by request-commit, commit and abort operations. Each object copy o keeps a *potential readers list* $PR(o)$, which records the identities of the potential readers of o. Therefore, if v' has a copy of o, it adds T to $PR(o)$ and sends a response message $rsp(T, o)$ to v, which contains a copy of o.

Transaction T waits to collect responses until it receives all responses from a read quorum. Among all copies it receives, it selects the copy with the highest version number as the valid copy of o. The read operation finishes.

Algorithm 1. QR model: read and write

1 **procedure Read** (v, T, o)	14 **procedure Write** $(v, T, o, value)$
2 Local Phase:	15 Local Phase:
3 **ReadQuorum** $(v, req(T, read(o)))$;	16 **ReadQuorum** $(v, req(T, write(o)))$;
4 **wait until** $find(v) = true$;	17 **wait until** $find(v) = true$;
5 **foreach** $d \lhd v_i$ **do**	18 **foreach** $d \lhd v_i$ **do**
6 **if** $d.version > data(o).version$ **then**	19 **if** $d.version > data(o).version$ **then**
7 $data(o) \leftarrow d$;	20 $data(o) \leftarrow d$;
8 add o to $T.readset$;	21 $dataCopy(o) \leftarrow data(o)$;
9 Remote Phase:	22 $dataCopy(o).value \leftarrow value$;
10 Upon receiving $req(T, read(o)) \lhd v$;	23 $dataCopy(o).version \leftarrow data(o).version + 1$;
11 **if** $data(o)$ *exists* **then**	24 add o to $T.writeset$;
12 add T to $PR(o)$;	25 Remote Phase:
13 $rsp(T, o) \rhd v$;	26 Upon receiving $req(T, write(o)) \lhd v$;
	27 **if** $data(o)$ *exists* **then**
	28 add T to $PW(o)$;
	29 $rsp(T, o) \rhd v$;

Write. The write operation is similar to the read operation. Transaction T sends a request message $req(T, write(o))$ to a selected read quorum. Note that T does not need to send request to a write quorum because in this step it only needs to collect the latest copy of o. If a remote node v' has a copy of o, it adds T to o's *potential writers list* $PW(o)$ and sends a response message to T with a copy of o.

Transaction T selects the copy with with the highest version number among the responses from a read quorum. Then it creates a temporary local copy $(dataCopy(o))$ and updates it with the value it intends to write, and increases its version number by 1 compared with the selected copy.

Remarks: The read and write operations of QR model are simple: a transaction just has to fetch all latest copies of required objects and perform all computations locally. Unlike a directory-based CC protocol, there is no need to construct and update a directory for each shared object. In QR model a transaction can always query its "closest" read quorum to locate the latest copy of each object required. Therefore the locality is preserved.

If a transaction is not aborted (by any other transaction) during all its read and operations, the transaction can request to commit by requesting to propagate its changes to objects into the system. The concurrency control mechanism is needed when any non-consistent status of an object is detected. The request-commit operation is presented in Algorithm 2.

Algorithm 2. QR model: request-commit

---

1 **procedure Request-Commit** (v, T)
2 Local Phase:
3 **WriteQuorum** $(v, req\_cmt(T))$;
4 $AT(T) \leftarrow \emptyset$;
5 wait until $find(v) = true$;
6 **if** $\exists rsp\_cmt(T, abort)$ received **then**
7   **Abort** (v, T);
8 **else**
9   **foreach** $rsp\_cmt(T, cmt, CT(T))$ **do**
10          $AT(T) \leftarrow AT(T) \cup CT(T)$;
11   **Commit** (v, T);
12 Remote Phase:
13 Upon receiving $req\_cmt(T) \lhd v$;
14 $CT(T) \leftarrow \emptyset$;
15 $abort(T) \leftarrow false$;
16 **Conflict-Detect** (v, T);
17 **if** $abort(T) = false$ **then**
18   **if** $CT(T) = \emptyset$ **then**
19          $rsp\_cmt(T, cmt, CT(T)) \rhd v$;
20   **else**
21   **CM** $(T, CT(T))$;
22   **if** $CT(T) \neq \emptyset$ **then**
23              $rsp\_cmt(T, cmt, CT(T)) \rhd v$;
24 **if** $abort(T) = false$ **then**
25   **foreach** $o_T \in T.writeset$ **do**
26          $o_T.protected \leftarrow true$;
27 $\forall o$, remove T from $PR(o)$ and $PW(o)$;

28 **procedure Conflict-Detect** (v, T)
29 **foreach** $o_T \in T.readset \cup T.writeset$ of object o **do**
30   **if** $data(o).protected = true$ or
31      $data(o).version > o_T.version$ **then**
32          $abort(T) \leftarrow true$;
33          $rsp\_cmt(T, abort) \rhd v$;
34   **break**;
35   **if** $data(o).version = o_T.version$ **then**
36   **if** $data(o).value \neq o_T.value$ **then**
37              $abort(T) \leftarrow true$;
38              $rsp\_cmt(T, abort) \rhd v$;
39   **break**;
40   **else**
41   add $PW(o)$ to $CT(T)$;
42   **if** $o_T \in T.writeset$ **then**
43   add $PR(o)$ to $CT(T)$;

44 **procedure CM** $(T, CT(T))$
45 **foreach** $T' \in CT(T)$ **do**
46   **if** $T' \prec T$ **then**
47          $abort(T) \leftarrow true$;
48          $rsp\_cmt(T, abort) \rhd v$;
49          $CT(T) \leftarrow \emptyset$;
50   **break**;

Request-Commit. When transaction T requests to commit, it sends a message $req\_cmt(T)$ (which contains all information of its readset and writeset) to a write quorum q_w. Note that it is required that for each transaction T, and $\forall q_r, q_w$ selected by T, $q_r \subseteq q_w$.

In the remote phase, when node v' receives the message $req\_cmt(T)$, it immediately removes T from its potential read and write lists of all objects and creates an empty *conflicting transactions list* $CT(T)$ which records the

---

Algorithm 3. QR model: commit and abort

---

1 **procedure Commit** (v, T)	16 **procedure Abort** (v, T)
2 Local Phase:	17 Local Phase:
3 **foreach** *object* $o \in T.writeset$ **do**	18 **foreach** *object* $o \in T.writeset$ **do**
4 $data(o) \leftarrow dataCopy(o)$;	19 discard $dataCopy(o)$;
5 **foreach** $T' \in AT(T)$ **do**	20 **WriteQuorum** $(v, abort(T))$;
6 $req\_abt(T') \rhd T'$;	21 wait until $find(v) = true$;
7 **WriteQuorum** $(v, commit(T))$;	
8 wait until $find(v) = true$;	22 Remote Phase:
	23 Upon receiving $abort(T)$:
9 Remote Phase:	24 **foreach** $o_T \in T.writeset$ *of object* o
10 Upon receiving $commit(T)$:	**do**
11 **foreach** $o_T \in T.writeset$ *of object* o	25 $data(o).protected \leftarrow false$;
do	26 $\forall o$, remove T from $PR(o)$ and $PW(o)$;
12 $data(o) \leftarrow o_T$;	
13 $data(o).protected \leftarrow false$;	
14 Upon receiving $req\_abt(T')$:	
15 **Abort** (v', T')	

---

transactions conflicting with T. Node v' determines the conflicting transactions of T in the following manner:

1) if $o_T.protected = true$, then T must be aborted since o_T is waiting for a possible update;
2) if o_T is a copy read or written by T of object o, and the local copy of o at v' ($data(o)$) has the higher version than o_T, then T reads a stale version of o. In this case, T must be aborted.
3) if o_T is a copy read by T of object o, and the local copy of o at v' ($data(o)$) has the same version with o_T, then T conflicts with all transactions in $PW(o)$ (potential writers of object copy $data(o)$).
4) if o_T is a copy written by T of object o, and the local copy of o at v' ($data(o)$ has the same version with o_T, then T conflicts with all transactions in $PW(o) \cup PR(o)$ (potential readers and writers of $data(o)$).

The contention manager at v' compares priorities between T and its conflicting transactions (line 21). If $\forall T' \in CT(T)$, $T \prec T'$ (T has the higher priority than any of its conflicting transactions), T is allowed to commit by v'. Node v' sends a message $rsp\_cmt(T, cmt, CT(T))$ with $CT(T)$ to v and sets the status of $data(o)$ as $protected$, for any $o \in T.writeset$. If $\exists T' \in CT(T)$ such that $T' \prec T$, then T is aborted. Node v' sends $rsp\_cmt(T, abort)$ to v and resets $CT(T)$.

In the local phase, transaction T collects responses from all nodes in the write quorum. If any $rsp\_cmt(T, abort)$ message is received, T is aborted. If not, T can proceed to the commit operation. In this case, transaction T saves conflicting transactions from all responses into an aborted transactions list $AT(T)$.

Remarks: For each transaction T, its concurrency control mechanism is carried by the request-commit operation. Therefore, the request-commit operation must guarantee that all existing conflicts with T are detected. Note that a remote node makes this decision based on its potential read and write lists. Therefore, these lists must be efficiently updated: a terminated transaction must be removed from these lists to avoid an unnecessary conflict detected. By letting $q_r \subseteq q_w$ for all q_r and q_w selected by the same transaction T, QR model guarantees that all T's records in potential read and write lists are removed during T's request-commit operation.

On the other hand, if v' allows T proceed to commit, then v' needs to protect local object copies written by T from other accesses until T's changes to these objects propagate to v'. These objects copies become valid only after receiving T's commit or abort information. We describe T's commit and abort operations in Algorithm 3.

Commit. When T commits, it sends a message $commit(T)$ to each node in the same write quorum q_w as the one selected by the request-commit operation. Meanwhile, it sends a request-abort message $req\_abt(T')$ for any $T' \in AT(T)$. In the remote phase, when a node v' receives $commit(T)$, for any $o \in T.writeset$, it updates $data(o)$ with the new value and version number, and sets $data(o).protected = false$. If a transaction T' receives $req\_abt(T')$, it aborts immediately.

Abort. A transaction may abort in two cases: after the request-commit operation, or receives a request-abort message. When T aborts, it rolls back all its operations of local objects. Meanwhile, it sends a message $abort(T)$ to each node in the write quorum q_w (which is the same as the write quorum selected by the request-commit operation). Then transaction T restarts from the beginning. In the remote phase, when a node v' receives $abort(T)$, it removes T from any of its potential read and write list (if it has not done so), and sets $data(o).protected = false$ for any $o \in T.writeset$.

3.2   Quorum Construction: Flooding Protocol

One crucial part of QR model is the construction of a quorum system over the network. We adopt the hierarchical clustering structure similar to the one described in [2]. An overlay tree with depth L is constructed. Initially, all physical nodes are leaves of the tree. Starting from the leaf nodes at level $l = 0$, parent nodes at the immediate higher level $l + 1$ is elected recursively so that their children are all nodes at most at distance 2^l from them.

Our quorum system is motivated by the classic *tree quorum system* [11]. On the overlay tree, a quorum system is constructed by FLOODING protocol such that each constructed quorum is a valid tree quorum.

We present FLOODING protocol in Algorithm 4. For each node v, when the system starts, a *basic read quorum* $Q_r(v)$ and a *basic write quorum* $Q_w(v)$ are constructed by BASICQUORUMS method. The protocol tries to construct $Q_r(v)$ and $Q_w(v)$ by first putting $root$ into these quorums and setting a distance variable δ to $d(v, root)$. Starting from $level = L - 1$, the protocol recursively selects the majority of descendants $levelHead = closestMajority(v, parent, level)$ for each $parent$ selected in the previous level ($level + 1$), so that the distance

---

Algorithm 4. FLOODING protocol

---

1 **procedure BasicQuorums** $(v, root)$
2 $\delta \leftarrow d(v, root)$;
3 $Q_r(v) \leftarrow \{root\}$;
4 $Q_w(v) \leftarrow \{root\}$;
5 $Q_r(v).level \leftarrow L$;
6 $currentHead \leftarrow \{root\}$;
7 **for** $level = L - 1, L - 2, \ldots, 0$ **do**
8     $levelHead \leftarrow \emptyset$;
9   **foreach** $parent \in currentHead$ **do**
10         $new \leftarrow closestMajority(v, parent, level)$;
11   add new to $Q_w(v)$;
12   add new to $levelHead$;
13   **if** $d(v, levelHead) < \delta$ **then**
14             $Q_r(v) \leftarrow levelHead$;
15             $Q_r(v).level \leftarrow level$;
16             $\delta \leftarrow d(v, levelHead)$;
17     $currentHead \leftarrow levelHead$;

18 **procedure WriteQuorum** (v, msg)
19 $msg \triangleright Q_w(v)$;
20 **if** $v' \in Q_r(v)$ **is down then**
21     $find(v) \leftarrow false$;
22     $validAns(v) \leftarrow null$;
23     $validLevel(v) \leftarrow null$;
24   **for** $level = 1, \ldots, L$ **do**
25         $msg \triangleright ancestor(v, level)$;
26   **if** $ancestor(v, level)$ **is up then**
27             $validAns(v) \leftarrow ancestor(v, level)$;
28             $validLevel(v) \leftarrow level$;
29   **break**;
30   **if** $validAns(v) = null$ **then**
31   restart **WriteQuorum** (v, msg);
32   **if** $validLevel(v) > Q_r(v).level$ **then**
33         $msg \triangleright Q_r(v)$;
34   **DownProbe** $(validAns(v), validLevel(v), write)$;
35   **if** $find(v) = false$ **then**
36   restart **WriteQuorum** (v, msg);

37 **procedure DownProbe** $(v, validLevel, type)$
38 $curRdHead \leftarrow v$;
39 $curWrHead \leftarrow v$;
40 $noWriteQ \leftarrow false$;

41 **for** $level = [validLevel - 1, 0]$ **do**
42     $levelRdHead \leftarrow \emptyset$;
43     $levelWrHead \leftarrow \emptyset$;
44   **foreach** $parent \in curRdHead$ **do**
45         $msg \triangleright descend(parent, level) \cap Q_w(v)$;
46   **if** w **is down then**
47   add w to $levelRdHead$;
48   **if** $type = write$ **then**
49   **foreach** $parent \in curWrHead$ **do**
50   **if** $\exists newSet = closestMajority(v, parent, level)$ **then**
51                 $msg \triangleright newSet$;
52   add $newSet$ to $levelWrHead$;
53   **else**
54                 $noWriteQ \leftarrow true$;
55   **break**;
56   **if** $noWriteQ = true$ **then**
57   **break**;
58   **if** $levelRdHead = \emptyset$ and $type = read$ **then**
59         $find(v) \leftarrow true$;
60   **break**;
61   **else**
62         $curRdHead \leftarrow levelRdHead$;
63         $curWrHead \leftarrow levelWrHead$;
64 **if** $noWriteQ = false$ and $type = write$ **then**
65     $find(v) \leftarrow true$;

66 **procedure ReadQuorum** (v, msg)
67 $msg \triangleright Q_r(v)$;
68 $find(v) \leftarrow false$;
69 **if** $v' \in Q_r(v)$ **is down then**
70     $find(v) \leftarrow false$;
71   **if** $v' \neq root$ **then**
72   **for** $level = [Q_r(v).level + 1, L]$ **do**
73             $msg \triangleright ancestor(v, level)$;
74   **if** $ancestor(v, level)$ **is up then**
75                 $find(v) \leftarrow true$;
76   **break**;
77   **if** $find(v) = false$ **then**
78   **DownProbe** $(v', Q_r(v).level, read)$;
79   **if** $find(v) = false$ **then**
80   restart **ReadQuorum** (v, msg);

from v to $closestMajority(v, parent, level)$ is the minimum over all possible choices. Note that $closestMajority(v, parent, level)$ only contains $parent$'s descendants at level $level$. We define the distance from v to a quorum Q as: $d(v, Q) := \max_{v' \in Q} d(v, v')$. The basic write quorum $Q_w(v)$ is constructed by including all selected nodes.

At each $level$, after a set of nodes $levelHead$ has been selected, the protocol checks the distance from v to $levelHead$ ($d(v, levelHead)$). If $d(v, levelHead) <$ δ, then the protocol replaces $Q_r(v)$ with $levelHead$ and sets δ to $d(v, levelHead)$. If $d(v, levelHead) \geq \delta$, the protocol continues to the next level. At the end, $Q_r(v)$ contains a set of nodes from the same level, which is the $levelHead$ closest from v for all levels.

When node v requests to access a read quorum, the protocol invokes READQUORUM(v, msg) method. Initially, node v sends msg to every node in $Q_r(v)$. If all nodes in $Q_r(v)$ are accessible from v, then a live read quorum is found. If any node v' in $Q_r(v)$ is down, then the protocol needs to probe v''s substituting nodes $sub(v')$ such that $sub(v') \cup Q_r(v) \setminus v'$ still forms a read quorum.

The protocol first finds if there exists any v''s ancestor available. If so, v''s substituting node has been found. If not, the protocol probes downwards from v' to check if there exists v' substituting nodes such that a constructed read quorum is a subset of $Q_w(v)$ by calling DOWNPROBE method.

The protocol invokes WRITEQUORUM(v, msg) method when node v requests to access a write quorum. Similar to READQUORUM(v, msg), node v first sends msg to every node in $Q_w(v)$. If any node v' is down, then the protocol first finds if there is a live ancestor of v' ($validAns(v)$). Starting from $validAns(v)$, the protocol calls DOWNPROBE to probe downwards.

DOWNPROBE method works similarly as BASICQUORUMS by recursively probing an available closest majority set of descendants for each parent selected in the previous level. By adopting DOWNPROBE method, FLOODING protocol guarantees that READQUORUM and WRITEQUORUM can always probe an available quorum if at least one live read (or write) quorum exists in the network.

3.3   Analysis

We first analyze the properties of the quorum system constructed by FLOODING, then we prove the correctness and evaluate the performance of QR model.

Lemma 1. *Any read quorum q_r or write quorum q_w constructed by* FLOODING *is a classis tree quorum defined in [11].*

Proof. From the description of FLOODING, we know that for a tree of height $h + 1$,

$$q_r = \{root\} \vee \{\text{majority of read quorums for subtrees of height h}\},$$

$$q_w = \{root\} \cup \{\text{majority of write quorums for subtrees of height h}\}.$$

From Theorem 1 in [11], the lemma follows.

Then we immediately have the following lemma.

Lemma 2. *For any two quorums q_1 and q_2 constructed by* FLOODING, *where at least one of them is a write quorum, $q_r \cap q_w \neq \emptyset$.*

Lemma 3. *For any read quorum $q_r(v)$ and write quorum $q_w(v)$ constructed by* FLOODING *for node v, $q_r(v) \subseteq q_w(v)$.*

Proof. The theorem follows from the description of FLOODING. If no node fails, the theorem holds directly since $Q_r(v) \subseteq Q_w(v)$.

If a node $v' \notin Q_r(v)$ fails, then $q_r(v) = Q_r(v)$. If $v' \in Q_w(v)$, FLOODING detects that v' is not accessible when it calls WRITEQUORUM method. If $level(q) \geq Q_r(v).level$, then FLOODING adds $Q_r(v)$ to $q_w(v)$ and starts to probe v''s substituting nodes; if $level(v') < Q_r(v).level$, then the level of v''s substituting node is at most $Q_r(v).level$ and then the protocol starts to probe downwards. In either case, $q_r(v) \subset q_w(v)$.

If a node $v' \in Q_r(v)$ fails, then FLOODING detects that v' is not accessible when it calls READQUORUM or WRITEQUORUM method. Both methods starts to probe v''s substituting nodes from v'. When probing upwards, v''s ancestors are visited. If a live $ancestor(v')$ is found, then both methods add $ancestor(v')$ to the quorum. Then READQUORUM stops and WRITEQUORUM continues probing downwards from $ancestor(v')$. The theorem follows.

With the help of Lemmas 2, we have the following theorem.

Theorem 1. *QR model provides 1-copy equivalence for all objects.*

Proof. We first prove that for any object o, if at time t, no transaction requesting o is propagating its change to o (i.e., in the commit operation), then all transactions accessing o at t get the same copy of o.

Note that if any committed transaction writes to o before t, there exists a write quorum q_w such that $\{\forall v \in q_w\} \wedge \{\forall v' \notin q_w\}$, $data(o, v).version > data(o, v').version$. If any transaction T accesses o at time t, it collects a set of copies from a read quorum q_r. From Lemma 2, $\exists v \in \{q_w \cap q_r\}$ such that $data(o, v)$ is collected by T. Note that read and write operations select the object copy with the highest version number. Hence, for any transaction T, $data(o, v)$ is selected as the latest copy.

We now prove that for any object o, if at time t: 1) a transaction T is propagating its change to o; and 2) another transaction T' accesses a read quorum q_r before T's change propagates to q_r, then T' will never commit.

Note that in this case, T' reads a stale version $o_{T'}$ of o. When it requests to commit (if it is not aborted before that), it sends the request to a write quorum q_w. Then $\exists v \in q_w$, such that: 1) T's change of o still has not propagated to v and $data(o, v).protected = true$; or 2) T's change has been applied to $data(o, v)$ and $data(o, v).version > o_T.version$. In either case, T is aborted by CONFLICT-DETECT method.

As the result, at any time, the system exhibits that only one copy exists for any object and transactions observing an inconsistent state of object never commit. The theorem follows.

With the help of Lemma 3 and Theorem 1, we can prove that QR model provides one-copy serializability [14].

Theorem 2. *QR model implements one-copy serializability.*

QR model provides five operations and every operation incurs a remote communication cost. We now analyze the communication cost of each operation.

Theorem 3. *If a live read quorum $q_r(v)$ exists, the communication cost of a read or write operation that starts at node v is $O(k \cdot d(v, q_r(v)))$ for $k \geq 1$, where k is the number of nodes failed in the system. Specifically, if no node fails, the communication cost is $O(d(v, Q_r(v)))$.*

Proof. For a read or write operation, the transaction calls READQUORUM method to collect the latest value of the object from a read quorum. If no node fails, the communication cost is $2d(v, Q_r(v))$. If a node $v' \in Q_r(v)$ fails, the transaction needs to probe v''s substituting nodes to construct a new read quorum. The time for v to restart the probing is at most $2d(v, Q_r(v))$. Note that $\forall q_r(v), d(v, Q_r(v)) \leq d(v, q_r(v))$.

In the worst case, if k nodes fail and v detects only one failed node at each it accesses a read quorum, at most k rounds of probing are needed for v to detect a live read quorum. On the other hand, v always starts probing from the closest possible read quorum. Therefore for each round, the time for v to restart the probing is at most $2d(v, q_r(v))$. The theorem follows.

Similar to Theorem 3, the communication cost of other three operation can be proved in the same way.

Theorem 4. *If a live write quorum $q_r(v)$ exists, the communication cost of a request-commit, commit or abort operation that starts at node v is $O(k \cdot d(v, q_w(v)))$ for $k \geq 1$, where k is the number of nodes failed in the system. Specifically, if no node fails, the communication cost is $O(v, Q_w(v))$.*

Theorems 3 and 4 illustrate the advantage of exploiting locality for QR model. For read and write operations starting from v, the communication cost is only related to the distance from v to its closest read quorum. If no node fails, the communication cost is bounded by $2d(v, Q_r(v))$. Note that $d(v, Q_r(v)) \leq d(v, root)$ from the construction of $Q_r(v)$. On the other hand, the communication cost of other three operations is bounded by $O(v, Q_w(v))$. Since each transaction involves at most two operations from {request-commit, commit, abort}, when the number of read/write operations increases, the communication cost of a transaction only increases proportional to $d(v, Q_r(v))$. Compared with directory-based protocols, the communication cost of a operation in QR model is not related to the stretch provided by the underlying overlay tree.

When the number of failed nodes increases, the performance of each operation degrades linearly. In QR model, it is crucial to analyze the availability of the constructed quorum system. From the construction of the quorum system we know that if a live quorum exists, FLOODING protocol can always probe it. Let

p be the probability that node lives and R_h be the availability of a read quorum, i.e., at least one live read quorum exists in a tree of height h. Then we have the following theorem.

Theorem 5. *Assuming the degree of each node in the tree is at least $2d+1$, the availability of a read quorum is*

$$R_{h+1} \geq p + (1-p) \cdot \left[\binom{2d}{d+1}(R_h)^{d+1}(1-R_h)^d + \binom{2d}{d+2}(R_h)^{d+2}(1-R_h)^{d-1} \right.$$
$$\left. + \ldots + (R_h)^{2d}(1-R_h) \right]$$

Proof. From [11], we have

$$R_h = Prob\{\text{Root is up}\}$$
$$+ Prob\{\text{Root is down}\} \times [\text{Read Availability of Majority of Subtrees}].$$

Note that in our overlay tree, if a node v at level $h+1$ is down, then one of its descendants at h is also down for $h \geq 0$, because they are mapped to the same physical node. The theorem follows.

Similarly, let W_h be the availability of a write quorum in a tree of height h, then

Theorem 6.

$$W_{h+1} \geq p \cdot \left[\binom{2d}{d+1}(W_h)^{d+1}(1-W_h)^d + \binom{2d}{d+2}(W_h)^{d+2}(1-W_h)^{d-1} \right.$$
$$\left. + \ldots + (W_h)^{2d}(1-W_h) \right]$$

Initially, R_0 and W_0 is p (only the root exists). Theorems 5 and 6 provide the recurrence relations of R_h and W_h, which can be used to calculate specific tree configurations. As the result, FLOODING provides the availability similar to the classic tree quorum system in [11].

4   Conclusion

QR model requires that at least one read and one write quorums live in the system. If no live read (or write) quorum exists, FLOODING protocol cannot proceed after READQUORUM (or WRITEQUORUM) operation. In this case, a reconfiguration of the system is needed to rebuild a new overlay tree structure. Each node then runs FLOODING protocol to find their new basic read and write quorums.

QR model exhibits graceful degradation in a failure-prone network. In a failure-free network, the communication cost imposed by QR model is comparable with SC model. When failures occur, QR model continues executing operations with high probability and reasonable higher communication cost. Such property is especially desirable for large-scale distributed systems in the presence of failures.

Acknowledgement. This work is supported in part by NSF CNS 0915895, NSF CNS 1116190, NSF CNS 1130180, and US NSWC under Grant N00178-09-D-3017-0011.

References

1. Harris, T., Larus, J.R., Rajwar, R.: Transactional Memory, 2nd edn. Morgan and Claypool (2010)
2. Herlihy, M., Sun, Y.: Distributed transactional memory for metric-space networks. Distributed Computing 20(3), 195–208 (2007)
3. Arnold, K., Scheifler, R., Waldo, J., O'Sullivan, B., Wollrath, A.: Jini Specification. Addison-Wesley Longman Publishing Co., Inc., Boston (1999)
4. Bocchino, R.L., Adve, V.S., Chamberlain, B.L.: Software transactional memory for large scale clusters. In: PPoPP 2008: Proceedings of the 13th ACM SIGPLAN Symposium on Principles and Practice of Parallel Programming, pp. 247–258. ACM, New York (2008)
5. Manassiev, K., Mihailescu, M., Amza, C.: Exploiting distributed version concurrency in a transactional memory cluster. In: PPoPP 2006: Proceedings of the Eleventh ACM SIGPLAN Symposium on Principles and Practice of Parallel Programming, pp. 198–208. ACM, New York (2006)
6. Kotselidis, C., Ansari, M., Jarvis, K., Luján, M., Kirkham, C., Watson, I.: Distm: A software transactional memory framework for clusters. In: ICPP 2008: Proceedings of the 37th International Conference on Parallel Processing, pp. 51–58. IEEE Computer Society, Washington, DC (2008)
7. Couceiro, M., Romano, P., Carvalho, N., Rodrigues, L.: D2stm: Dependable distributed software transactional memory. In: Proceedings of the 2009 15th IEEE Pacific Rim International Symposium on Dependable Computing, PRDC 2009, pp. 307–313. IEEE Computer Society, Washington, DC (2009)
8. Zhang, B., Ravindran, B.: Brief announcement: Relay: A cache-coherence protocol for distributed transactional memory. In: Abdelzaher, T., Raynal, M., Santoro, N. (eds.) OPODIS 2009: Proceedings of the 13th International Conference on Principles of Distributed Systems. LNCS, vol. 5923, pp. 48–53. Springer, Heidelberg (2009)
9. Attiya, H., Gramoli, V., Milani, A.: A Provably Starvation-Free Distributed Directory Protocol. In: Dolev, S., Cobb, J., Fischer, M., Yung, M. (eds.) SSS 2010. LNCS, vol. 6366, pp. 405–419. Springer, Heidelberg (2010)
10. Attiya, H., Milani, A.: Transactional scheduling for read-dominated workloads. In: OPODIS 2009: Proceedings of the 13th International Conference on Principles of Distributed Systems. LNCS, vol. 5923, pp. 3–17. Springer, Heidelberg (2009)
11. Agrawal, D., El Abbadi, A.: The tree quorum protocol: an efficient approach for managing replicated data. In: Proceedings of the Sixteenth International Conference on Very Large Databases, pp. 243–254. Morgan Kaufmann Publishers Inc., San Francisco (1990)
12. Schlichting, R.D., Schneider, F.B.: Fail-stop processors: an approach to designing fault-tolerant computing systems. ACM Trans. Comput. Syst. 1, 222–238 (1983)
13. Guerraoui, R., Herlihy, M., Pochon, B.: Toward a theory of transactional contention managers. In: PODC 2005: Proceedings of the Twenty-Fourth Annual ACM Symposium on Principles of Distributed Computing, pp. 258–264. ACM, New York (2005)
14. Bernstein, P.A., Goodman, N.: Multiversion concurrency control - theory and algorithms. ACM Trans. Database Syst. 8, 465–483 (1983)

Error-free Multi-valued Broadcast and Byzantine Agreement with Optimal Communication Complexity

Arpita Patra[*]

Department of Computer Science, ETH Zurich, Switzerland
arpita.patra@inf.ethz.ch

Abstract. In this paper we present first ever *error-free, asynchronous* broadcast (called as A-cast) and Byzantine Agreement (called as ABA) protocols with optimal communication complexity and fault tolerance. Our protocols are multi-valued, meaning that they deal with ℓ bit input and achieve communication complexity of $\mathcal{O}(n\ell)$ bits for large enough ℓ for a set of $n \geq 3t + 1$ parties in which at most t can be Byzantine corrupted. Previously, Patra and Rangan (Latincrypt'10, ICITS'11) reported multi-valued, communication optimal A-cast and ABA protocols *that are only probabilistically correct.*

Following all the previous works on multi-valued protocols, we too follow reduction-based approach for our protocols, meaning that our protocols are designed given existing A-cast and ABA protocols for small message (possibly for single bit). Our reductions invoke less or equal number of instances of protocols for single bit in comparison to the reductions of Patra and Rangan. Furthermore, our reductions run in constant expected time, in contrast to $\mathcal{O}(n)$ of Patra and Rangan (ICITS'11). Also our reductions are much simpler and more elegant than their reductions.

By adapting our techniques from asynchronous settings, we present new *error-free*, communication optimal reduction-based protocols for broadcast (BC) and Byzantine Agreement (BA) in synchronous settings that are constant-round and call for only $\mathcal{O}(n^2)$ instances of protocols for single bit. Prior to this, communication optimality has been achieved by Fitzi and Hirt (PODC'06) who proposed *probabilistically correct* multi-valued BC and BA protocols with constant-round and $\mathcal{O}(n(n + \kappa))$ (κ is the error parameter) invocations to the single bit protocols. Recently, Liang and Vaidya (PODC'11) achieved the same *without* error probability. However, their reduction calls for round complexity and number of instances that are function of the message size, $\mathcal{O}(\sqrt{\ell} + n^2)$ and $\mathcal{O}(n^2\sqrt{\ell} + n^4)$, respectively where $\ell = \Omega(n^6)$.

Keywords: Multi-valued, Broadcast, Byzantine Agreement, A-cast, Asynchronous, Communication Complexity, Expected Running Time.

---

[*] The work was done when the author was a post-doctoral researcher at Department of Computer Science, Aarhus University, Denmark.

A. Fernández Anta, G. Lipari, and M. Roy (Eds.): OPODIS 2011, LNCS 7109, pp. 34–49, 2011.
© Springer-Verlag Berlin Heidelberg 2011

1   Introduction

The problem of Broadcast (BC) and Byzantine Agreement (BA) (also popularly known as consensus) were introduced in [PSL80] and since then they have been considered as the most fundamental problems in distributed computing. In brief, a BC protocol allows a special party among a set of parties, called sender, to send some message identically to all other parties. The challenge lies in achieving the above task despite the presence of some faulty parties (possibly including the sender), who may deviate from the protocol arbitrarily. The BA primitive is slightly different from BC. A BA protocol allows a set of parties, each holding some input bit, to agree on a common bit, even though some of the parties may act maliciously in order to make the honest parties disagree. The BC and BA primitives have been used as building blocks in several important secure distributed computing tasks such as Secure Multiparty Computation (MPC) [BOGW88, BKR94, RBO89], Verifiable Secret Sharing (VSS) [CGMA85, BOGW88, RBO89] etc.

An important, practically motivated variant of BC and BA problem are asynchronous broadcast (known as A-cast) and asynchronous BA (known as ABA) that study the conventional BC and BA problems in asynchronous network settings. It is well-known that asynchronous network setting is considered to be more realistic than synchronous network setting. The works of [BO83, Rab83, Bra84],[FM88, CR93, Can95, ADH08, PW92, PR11] have reported different A-cast and ABA protocols. In this paper, we focus on the communication complexity of *error-free* A-cast and ABA protocols and present first ever optimal protocols.

The Model. We follow the standard network model of [PSL80] for synchronous network and [CR93, Can95] for asynchronous network. Our A-cast, ABA, BC and BA protocols are carried out among a set of n parties, say $\mathcal{P} = \{P_1, \ldots, P_n\}$, where every two parties are directly connected by an authenticated and secure channel and at most t out of the n parties can be under the influence of a *computationally unbounded Byzantine adversary*, denoted as $\mathcal{A}_t$. The adversary corrupts the parties *adaptively* at any point during the course of the protocol execution and the choice may base on the information gathered so far by the adversary. We assume that $n = 3t + 1$ which is the minimum number of parties required to design *error-free* A-cast, ABA, BC and BA protocols [Lyn96, PSL80]. The parties not under the influence of $\mathcal{A}_t$ are called *honest or uncorrupted*.

We do not make any cryptographic assumptions such as public key infrastructure (PKI) etc in our protocols. All our protocols are *randomized*.

Definitions. We now define A-cast and ABA formally.

Definition 11 (A-cast [Can95]). *Let Π be an asynchronous protocol executed among the set of parties $\mathcal{P}$ and initiated by a special party called sender $S \in \mathcal{P}$, having input m (the message to be sent). Π is an A-cast protocol tolerating $\mathcal{A}_t$ if the following hold, for every behavior of $\mathcal{A}_t$ and every input m:*

- **Termination:** *If S is honest, then all honest parties in $\mathcal{P}$ will eventually terminate Π. If any honest party terminates Π, then all other honest parties will eventually terminate Π.*
- **Correctness:** *If the honest parties terminate Π, then they do so with a common output $m^\star$. Furthermore, if the sender S is honest then $m^\star = m$.*

Definition 12 (ABA [CR93]). *Let Π be an asynchronous protocol executed among the set of parties $\mathcal{P}$, with each party having a private binary input. We say that Π is an ABA protocol tolerating $\mathcal{A}_t$ if the following hold, for every possible behavior of $\mathcal{A}_t$ and every possible input:*

- **Termination***: All honest parties eventually terminate the protocol.*
- **Correctness***: All honest parties who have terminated the protocol hold identical outputs. Furthermore, if all honest parties had same input, say ρ, then all honest parties output ρ.*

The celebrated result of [FLP85] shows that any ABA protocol that never reaches disagreement must have some nonterminating executions. For a protocol that never reaches disagreement, the best we can hope for is that the set of nonterminating executions has probability zero. Such protocols are termed as *almost-surely terminating* by [ADH08]. In this work, we construct ABA protocol that is almost-surely terminating and has no error in correctness. The important complexity measures of A-cast and ABA protocol are: **Communication Complexity**: It is the total number of bits communicated by the *honest* parties in the protocol; **Expected Running Time**: Refer to [CR93, Can95] for a detailed definition of expected running time of a randomized asynchronous protocol.

While the basic definitions of A-cast and ABA consider message of single bit, *multi-valued* protocols allow message to be long string of bits and exploit the fact that the task is to be attained for the entire string and not bit by bit. This fact generally allows a *multi-valued* protocol to be considerably more efficient than many parallel executions of protocol for single bit.

Brief Literature. Error-free BC and BA protocol in synchronous network are possible if and only if $n \geq 3t + 1$ [PSL80, Lyn96]. The same bound holds for A-cast and ABA both with and without error probability [Lyn96]. The seminal result of [DR85] shows that any error-free BA or BC must communicate $\Omega(n^2)$ bits (which again carry over for the case of A-cast and ABA). Since the message must be at least single bit, the lower bound on the communication complexity for single bit is $\Omega(n^2)$ bits. However, communication complexity of $\mathcal{O}(n\ell)$ bits can be achieved for large enough value of ℓ (at least $\ell \geq n$ bits) as shown in [FH06]. Requiring large value for ℓ is practically motivated in many distributed computing applications, like reaching agreement on a large file in fault-tolerant distributed storage system, distributed voting where ballots containing gigabytes of data is to be handled, MPC where many broadcasts and agreements are invoked which can be combined into fewer executions of multi-valued protocols.

Following the approach of Turpin and Coan [TC84], all the subsequent multi-valued protocols apply reduction-based approach [FH06, LV11, PR11, PR10], meaning that they are constructed based on access to protocols for small message or single bit message. The reductions presented in [FH06, LV11] for synchronous settings and in [PR11, PR10] for asynchronous settings achieve optimal communication complexity of $\mathcal{O}(n\ell)$ bits. While the reduction presented in [FH06] involves error probability, the reduction of [LV11] is error-free. In the asynchronous settings, [PR11, PR10] reported multi-valued protocols with error probability.

Our Contribution. We achieve optimal complexity of $\mathcal{O}(n\ell)$ bits for *error-free* A-cast and ABA with optimal fault-tolerance of $t < n/3$. We too follow reduction-based approach of [TC84]. We now compare our reductions with that of [PR10, PR11] and show that our reductions are better in all the following aspects: (a) error-free, (b) running time and (c) number of invocations to protocols for single bit. All the protocols have optimal fault tolerance of $t < n/3$.

Ref.	Type	Running Time	# Invocations to single bit protocol
[PR10], A-cast	Probabilistic	constant	$\mathcal{O}(n^2 \log n)$ A-cast
[PR11], ABA	Probabilistic	$\mathcal{O}(n)$	$\mathcal{O}(n^3)$ ABA
This paper, A-cast	Error-free	constant	$\mathcal{O}(n^2 \log n)$ A-cast
This paper, ABA	Error-free	constant	$\mathcal{O}(n)$ ABA

We now compare our results with the current best *error-free* A-cast and ABA for single bit. The only error-free A-cast is due to [Bra84] that communicates $\mathcal{O}(n^2)$ bits and runs in constant time. Similarly, the only error-free ABA is due to [ADH08] that runs in $\mathcal{O}(n^2)$ time and requires communication of $\mathcal{O}(n^8 \log n)$ bits and A-cast of same number of bits. Our protocols in this paper show clear improvement over ℓ executions of these protocols for large enough ℓ.

Technically, our reductions are simple and are based on linear error correcting code (e.g. Reed-Solomon Code) and a graph theoretic algorithm for finding some special structure (called (n, t)-star; defined in Section 2) in undirected graph [CR93, Can95]. While the existing reductions for multi-valued protocols [PR11, LV11] are constructed in player-elimination [HMP00] or dispute control [BTH06] framework, our reductions do not require them and therefore they are more elegant. Finally, we note that the multi-valued A-cast protocol of [PR10] also employs the algorithm for finding (n, t)-star [CR93, Can95]. However, we mark an important and crucial observation about the outcome of the algorithm in our context that allows to construct our protocol in an error-free manner.

Finally, we discuss our results in synchronous settings. By adapting our techniques from asynchronous settings, we present new *error-free* reduction that is constant-round and calls for $\mathcal{O}(n^2)$ instances of protocols for single bit. We now compare our result with the communication optimal reductions of [FH06, LV11].

Ref.	Type	Fault Tolerance	Round Complexity	# Invocations to single bit protocol
[FH06]	Probabilistic	$t < n/2$	constant	$\mathcal{O}(n(n + \kappa))$
[LV11]	Error-free	$t < n/3$	$\mathcal{O}(\sqrt{\ell} + n^2)$	$\mathcal{O}(n^2 \sqrt{\ell} + n^4)$
This paper	Error-free	$t < n/3$	constant	$\mathcal{O}(n^2)$

Road-map. In section 2 and 3, we present our construction for A-cast and ABA respectively. We present our BA and BC protocols in Section 4 and then conclude in Section 5.

2   Error-free Communication Optimal A-cast

Here we present our A-cast protocol. We start with brief presentation of the tools that we use: (a) A-cast protocol of Bracha [Bra84]; (b) An algorithm for finding a graphical structure called (n,t)-star in an undirected graph; (c) Linear Error Correcting Code. We discuss them one by one.

Bracha's A-cast. The first ever protocol for A-cast is due to Bracha [Bra84] (a good description is available in [Can95]). The protocol is error-free, runs with $n \geq 3t+1$ in constant time and communicates $\mathcal{O}(n^2)$ bits for a *single bit* message.

Notation 21. *By saying that 'P_i A-casts M', we mean that P_i as a sender, initiates Bracha's A-cast protocol with M as the message. Similarly 'P_j receives M from the A-cast of P_i' will mean that P_j terminates the A-cast protocol initiated by P_i and outputs M. By the property of A-cast, if some honest party P_j terminates the A-cast of some sender P_i with M as the output, then every other honest party will eventually do so, irrespective of the behavior of the sender P_i.*

Finding (n,t)-star in an Undirected Graph. We now describe an existing solution for a graph theoretic problem, called finding (n,t)-star in an undirected graph $G = (V, E)$. Let G be an undirected graph with the n parties in $\mathcal{P}$ as its vertex set. A pair $(\mathcal{C}, \mathcal{D})$ of sets with $\mathcal{C} \subseteq \mathcal{D} \subseteq \mathcal{P}$ is an (n,t)-star [Can95, BOCG93] in G, if: (i) $|\mathcal{C}| \geq n - 2t$; (ii) $|\mathcal{D}| \geq n - t$; (iii) for every $P_j \in \mathcal{C}$ and every $P_k \in \mathcal{D}$ the edge (P_j, P_k) exists in G.

Following the idea of [GJ79], [BOCG93] presented an elegant and efficient algorithm for finding an (n,t)-star in a graph of n nodes, *provided that the graph contains a clique of size $n-t$*. Actually, the algorithm, called as Find-STAR takes the complementary graph $\overline{G}$ of G as input and tries to find (n,t)-$\overline{\text{star}}$ in $\overline{G}$, where (n,t)-$\overline{\text{star}}$ is a pair $(\mathcal{C}, \mathcal{D})$ of sets with $\mathcal{C} \subseteq \mathcal{D} \subseteq \mathcal{P}$, satisfying the following conditions: (a) $|\mathcal{C}| \geq n - 2t$; (b) $|\mathcal{D}| \geq n - t$; (c) There are no edges between the nodes in $\mathcal{C}$ and nodes in $\mathcal{C} \cup \mathcal{D}$ in $\overline{G}$. Clearly, a pair $(\mathcal{C}, \mathcal{D})$ representing an (n,t)-$\overline{\text{star}}$ in $\overline{G}$, is an (n,t)-star in G. Find-STAR outputs either an (n,t)-$\overline{\text{star}}$, or a message star-Not-Found. Whenever *the input graph $\overline{G}$ contains an independent set of size $n - t$*, Find-STAR always outputs an (n,t)-$\overline{\text{star}}$. For simple notation, we denote $\overline{G}$ by H. The algorithm Find-STAR is presented below:

Algorithm Find-STAR(H)

1. Find a maximum matching M in H. Let N be the set of matched nodes (namely, the endpoints of the edges in M), and let $\overline{N} = \mathcal{P} \setminus N$.
2. Compute output as follows:
   (a) Let $T = \{P_i \in \overline{N} | \exists P_j, P_k \text{ s.t } (P_j, P_k) \in M \text{ and } (P_i, P_j), (P_i, P_k) \in E\}$. T is called the set of triangle-heads. Let $\mathcal{C} = \overline{N} \setminus T$.
   (b) Let B be the set of matched nodes that have neighbors in $\mathcal{C}$. So $B = \{P_j \in N | \exists P_i \in \mathcal{C} \text{ s. t. } (P_i, P_j) \in E\}$. Let $\mathcal{D} = \mathcal{P} \setminus B$.
   (c) If $|\mathcal{C}| \geq n - 2t$ and $|\mathcal{D}| \geq n - t$, output $(\mathcal{C}, \mathcal{D})$. Otherwise, output star-Not-Found.

Linear Error Correcting Code. We use Reed-Solomon (RS) codes in our protocols. We consider an $(n, t+1)$ RS code in Galois Field $\mathbb{F} = GF(2^c)$, where $n \leq 2^c$. Each element of $\mathbb{F}$ is represented by c bits. An $(n, t+1)$ RS code encodes $t+1$ elements of $\mathbb{F}$ into a codeword consisting of n elements from $\mathbb{F}$. We denote the encoding function as ENC() and the corresponding decoding function as DEC(). Let $m_0, m_1, \ldots, m_t$ be the input to ENC, then ENC computes a codeword of length n, $(s_1, \ldots, s_n)$, as follows: It constructs a polynomial of degree-t, $f(x) = m_0 + m_1 x + \ldots + m_t x^t$. It then computes $s_i = f(i)$. We use the following syntax for ENC: $(s_1, s_2, \ldots, s_n) = \mathsf{ENC}(m_0, m_1, \ldots, m_t)$. Each element of the codeword is computed as a linear combination of the $t+1$ input data elements, such that every subset of $(t+1)$ elements from the codeword uniquely determine the input data elements. Similarly, knowledge of any $t+1$ elements from the codeword suffices to determine the remaining elements of the codeword.

The decoding function DEC can be applied as long as $t+1$ elements from a codeword are available. A RS code is capable of error correction and detection. The task of error correction is to find the error locations and error values in a received vector. On the other hand, error detection means an indication that errors have occurred, without attempting to correct them. We recall the following well known result from coding theory. DEC can correct up to c Byzantine error and simultaneously detect up to additional d Byzantine errors in a vector of length N (where $N \leq n$) if and only if $N - t - 1 \geq 2c + d$. In our protocols, we may invoke DEC on a vector of length $N \leq n$ with specific value of c and d. If c, d and N satisfy the above relation, then DEC returns back the correct data elements corresponding to the vector; otherwise DEC returns 'failure'.

2.1   Multi-valued A-cast Protocol

With the above tools, we are now ready to present our multi-valued A-cast protocol, called Multi-Valued-Acast. We assume that the sender S has a message m containing ℓ bits that he would like to communicate to all the parties in $\mathcal{P}$ identically. Our protocol is structured into two phases, (a) **S-dependent Phase** and (b) **S-independent Phase**. In the S-dependent phase, S proves that it has communicated the same message to at least a set of $2t+1$ parties, say $CORE$. The S-dependent phase, as the name suggests, demands S to perform some special roles. For an honest S, this phase will always be completed successfully. However, a corrupted S may choose not to perform his actions and therefore this phase may not be terminated for a corrupted S. The second phase, called S-independent phase is initiated upon completion of the first phase. If S successfully proves the existence of some $CORE$ in the first phase, then the parties in $CORE$ propagate their common message to the remaining parties without any help from S.

In the first phase, S communicates his message m to every party over private channel. Upon receiving a message from S, a party applies ENC on the message to get a codeword and communicates elements of the codeword to different party. Intuitively, the parties here check if they received the same message from S. They A-cast [Bra84] their responses. Based on the response of the parties, a consistency graph is constructed by the parties individually. S now finds a special structure in the graph, namely a quadruple $(\mathcal{C}, \mathcal{D}, \mathcal{F}, \mathcal{E})$ such that $(\mathcal{C}, \mathcal{D})$ is an (n, t)-star , $|\mathcal{F}| \geq 2t+1$ and every party in $\mathcal{F}$ has at least $t+1$ neighbors in $\mathcal{C}$, $|\mathcal{E}| \geq 2t+1$ and every party in $\mathcal{E}$ has at least $2t + 1$ neighbors in $\mathcal{F}$. Such a quadruple essentially proves that there is a set of at least $2t + 1$ parties, $CORE$ (same as $\mathcal{E}$), to whom S indeed communicated same message. On finding such a quadruple, S A-casts the same and all other parties can verify if indeed such quadruple exists in their individual graph. In this process, all the (honest) parties agree on $CORE$ and proceed to second phase. The algorithm for finding (n, t)-star and an important observation are combined intelligibly in order to find a quadruple in a graph. The observation is that *if S is honest then eventually, the set $\mathcal{C}$ of an (n, t)-star will contain at least $t + 1$ honest parties and when it happens, $\mathcal{F}$ and $\mathcal{E}$ can be computed such that a valid quadruple can be formed.* In the second phase, the parties use error correction and detection of RS code to compute and agree on the common message of the parties in $CORE$. We present the protocol in Figure 1 and Figure 2 and subsequently prove the properties.

Lemma 22. *The honest parties in $CORE$ hold same message of length ℓ. If S is honest then the message is S's message.*

Proof. The set $CORE$ is the $\mathcal{E}$ component of a quadruple $(\mathcal{C}, \mathcal{D}, \mathcal{F}, \mathcal{E})$. We start with proving that the honest parties in $\mathcal{C}$ hold the same message of length ℓ. We recall that $\mathcal{D}$ contains at least $t + 1$ honest parties and every $P_i \in \mathcal{C}$ is neighbor of every party in $\mathcal{D}$. Let $\{P_{i_1}, \ldots, P_{i_\alpha}\}$ be the set of α honest parties in $\mathcal{D}$, where $\alpha \geq t + 1$. Then for every P_i in $\mathcal{C}$, s_{ii_k} is same as $s_{i_k i_k}$ of all $k = [1, \alpha]$. Therefore the codewords corresponding to the messages of the honest parties in $\mathcal{C}$ are same at least at $t + 1$ locations corresponding to the identities of the honest parties in $\mathcal{D}$. Since the codewords belong to $(n, t + 1)$ RS code, the messages of the honest parties in $\mathcal{C}$ are same. Let the common message be m, $|m| = \ell$. Let $(s_1, \ldots, s_n) = \mathsf{ENC}(m_0, m_1, \ldots, m_t)$, where $m = m_0 | m_1 | \ldots, | m_t$. Now we show that every honest party $P_i \in \mathcal{F}$ holds s_i. Recall that P_i has at least $t + 1$ neighbors in $\mathcal{C}$ in which at least one is honest, say P_j. This implies that s_{ii} of P_i is same as s_{ji} of P_j. However, $s_{ji} = s_i$, since P_j holds m. Hence $s_{ii} = s_i$. Therefore every honest P_i in $\mathcal{F}$ holds s_i which is same as s_{ii}. Finally, we show that every honest $P_i \in \mathcal{E}$ holds m. Recall that P_i has at least $2t + 1$ neighbors in $\mathcal{F}$ in which at least $t + 1$ are honest. Let $\{P_{i_1}, \ldots, P_{i_\alpha}\}$ be the set of α honest parties in $\mathcal{F}$, where $\alpha \geq t + 1$. Then s_{ii_k} of P_i is same as $s_{i_k i_k}$ of every honest P_{i_k} for $k = [1, \alpha]$. Now $s_{i_k i_k}$ of P_{i_k} is same as s_{i_k}. Therefore the codeword corresponding to the message of $P_i \in \mathcal{E}$ matches with $(s_1, \ldots, s_n)$ at least at $t + 1$ locations corresponding to the identities of the honest parties in $\mathcal{F}$. This implies the codeword of P_i is identical to $(s_1, \ldots, s_n)$, since they belong to $(n, t + 1)$ RS code. Hence $P_i \in \mathcal{E}$ holds m. This completes the proof for the first part of the lemma. The second part of the lemma is easy to prove. □

To prove the lemma below, we will show that when S is honest then eventually an (n, t)-star can be found such that the set $\mathcal{C}$ will contain at least $t + 1$ *honest* parties. This observation is very crucial and is at the heart of our protocol.

Protocol Multi-Valued-Acast(S,m)

S-dependent Phase:

Code for S.
1. S sends his message m to every P_i.

Code for every P_i including S.
1. On receiving message m_i, divide the ℓ bit message m_i into $t + 1$ blocks, $m_{i0}, \ldots, m_{it}$, each containing $\frac{\ell}{t+1}$ (assume this to be an integer for simplicity) bits. Compute $(s_{i1}, \ldots, s_{in}) = \mathsf{ENC}(m_{i0}, \ldots, m_{it})$.
2. Send s_{ii} to every party. Send s_{ij} to P_j for $j = [1, n]$.
3. On receiving s_{jj} and s_{ji} from P_j, A-cast $\mathsf{OK}(P_i, P_j)$ if $s_{jj} = s_{ij}$ and $s_{ji} = s_{ii}$.
4. Construct a graph G_i with the parties in $\mathcal{P}$ as the vertices. Add an edge (P_j, P_k) in G_i if $\mathsf{OK}(P_j, P_k)$ and $\mathsf{OK}(P_k, P_j)$ are received from the A-cast of P_j and P_k respectively.

Code for S.
1. Upon every new receipt of some $\mathsf{OK}(*, *)$, update G_S. If a new edge is added to G_S, then execute Find-STAR$(\overline{G_S})$. Let there are $\alpha \geq 0$ distinct (n, t)-stars that are found in the past from different executions of Find-STAR$(\overline{G_S})$.
   (a) Now if an (n, t)-star is found from the current execution of Find-STAR$(\overline{G_S})$ that is distinct from all the α (n, t)-star obtained before, do the following:
      i. Call the new (n, t)-star as $(\mathcal{C}^{\alpha+1}, \mathcal{D}^{\alpha+1})$.
      ii. Create a list $\mathcal{F}^{\alpha+1}$ as follows: Add P_j to $\mathcal{F}^{\alpha+1}$ if P_j has at least $t + 1$ neighbors in $\mathcal{C}^{\alpha+1}$ in G_S.
      iii. Create a list $\mathcal{E}^{\alpha+1}$ as follows: Add P_j to $\mathcal{E}^{\alpha+1}$ if P_j has at least $2t + 1$ neighbors in $\mathcal{F}^{\alpha+1}$ in G_S.
      iv. For every γ, with $\gamma = 1, \ldots, \alpha$ update $\mathcal{F}^\gamma$ and $\mathcal{E}^\gamma$:
         A. Add P_j to $\mathcal{F}^\gamma$, if $P_j \notin \mathcal{F}^\gamma$ and P_j has at least $t + 1$ neighbors in $\mathcal{C}^\gamma$ in G_S.
         B. Add P_j to $\mathcal{E}^\gamma$, if $P_j \notin \mathcal{E}^\gamma$ and P_j has at least $2t + 1$ neighbors in $\mathcal{F}^\gamma$ in G_S.
   (b) If no (n, t)-star is found or an (n, t)-star that has been already found in the past is obtained, then update existing $\mathcal{F}^\gamma$'s and $\mathcal{E}^\gamma$'s.
   (c) Now let $(\mathcal{E}^\beta, \mathcal{F}^\beta)$ be the first pair such that $|\mathcal{E}^\beta| \geq 2t + 1$ and $|\mathcal{F}^\beta| \geq 2t + 1$. Assign $CORE = \mathcal{E}^\beta$ and A-cast $(\mathcal{C}^\beta, \mathcal{D}^\beta, \mathcal{E}^\beta, \mathcal{F}^\beta)$.

Code for P_i including S.
1. Assign $CORE = \mathcal{E}^\beta$, when all the following events occur: (a) $(\mathcal{C}^\beta, \mathcal{D}^\beta, \mathcal{E}^\beta, \mathcal{F}^\beta)$ is received from the A-cast of S; (b) $(\mathcal{C}^\beta, \mathcal{D}^\beta)$ becomes a valid (n, t)-star in G_i; (c) every party $P_j \in \mathcal{F}^\beta$ has at least $t + 1$ neighbors in $\mathcal{C}^\beta$ in G_i; and (d) every party $P_j \in \mathcal{E}^\beta$ has at least $2t + 1$ neighbors in $\mathcal{F}^\beta$ in G_i.

Fig. 1. Error-free Communication Optimal A-cast

Lemma 23. *If S is honest, then all the parties terminate* **S-dependent Phase,** *after agreeing on $CORE$.*

Proof. If S is honest, then he sends same message m to all the parties. Therefore, all honest parties generate same codeword, $(s_1, \ldots, s_n) = \mathsf{ENC}(m_0, \ldots, m_t)$, such

---

S-independent Phase:

Code for P_i including S.
1. If $CORE$ is constructed and $P_i \in CORE$, then assign $s_i = s_{ii}$.
2. If $CORE$ is constructed and $P_i \notin CORE$, then assign s_i to be the value s_{ji} that is received from at least $t+1$ P_j's in $CORE$.
3. Send s_i to all the parties.
4. On receiving $2t+1+r$, $r \geq 0$, s_j's apply DEC with $c = r$ and $d = t - r$, if DEC returns 'failure', then wait for more values. If DEC returns data elements $m_0, \ldots, m_t$, then output $m = m_0|m_1|\ldots|m_t$, where $|$ denotes concatenation.

Fig. 2. Error-free Communication Optimal A-cast

that $m = m_0|m_1|\ldots|m_t$. Therefore eventually there will be an edge between every pair of honest parties. This implies that there will be a clique of size at least $2t+1$ eventually. This guarantees that S will eventually find at least one (n,t)-star in G_S. Now we show that S will eventually find a quadruple $(\mathcal{C}, \mathcal{D}, \mathcal{F}, \mathcal{E})$ such that $(\mathcal{C}, \mathcal{D})$ is an (n,t)-star and every party in $\mathcal{F}$ has at least $t+1$ neighbors in $\mathcal{C}$ and every party in $\mathcal{E}$ has at least $2t+1$ neighbors in $\mathcal{F}$. To prove this we start with proving that an honest S will eventually find an (n,t)-star such that the set $\mathcal{C}$ will contain at least $t+1$ *honest* parties. For an honest S, eventually the edges between each pair of honest parties will vanish from the complementary graph $\overline{G_S}$. So the edges in $\overline{G_S}$ will be either (a) between an honest and a corrupted party OR (b) between two corrupted parties. Let β be the first index, such that (n,t)-star $(\mathcal{C}^\beta, \mathcal{D}^\beta)$ is generated in $\overline{G_S}$, when $\overline{G_S}$ contains edges of above two types only. Now, by construction of $\mathcal{C}^\beta$ (see Algorithm Find-STAR), it excludes the parties in N (set of parties that are endpoints of the edges of maximum matching M) and T (set of parties that are triangle-heads). An honest P_i belonging to N implies that $(P_i, P_j) \in M$ for some P_j and hence P_j is corrupted (as the current $\overline{G_S}$ does not have edge between two honest parties). Similarly, an honest party P_i belonging to T implies that there is some $(P_j, P_k) \in M$ such that (P_i, P_j) and (P_j, P_k) are edges in $\overline{G_S}$. This clearly implies that both P_j and P_k are surely corrupted. So for every honest P_i *not* in $\mathcal{C}^\beta$, at least one (if P_i belongs to N, then one; if P_i belongs to T, then two) corrupted party also remains outside $\mathcal{C}^\beta$. As there are at most t corrupted parties, $\mathcal{C}^\beta$ may exclude at most t honest parties. Still $\mathcal{C}^\beta$ is bound to contain at least $t+1$ honest parties.

Now all honest parties will be neighbors of the $t+1$ honest parties in $\mathcal{C}^\beta$ in G_S. Therefore $\mathcal{F}^\beta$ will eventually contain all the honest parties. Finally since all honest parties are neighbors of each other, $\mathcal{E}^\beta$ will contain all honest parties eventually and therefore it is guaranteed to contain at least $2t+1$ parties. Hence we proved that S can find a quadruple $(\mathcal{C}, \mathcal{D}, \mathcal{F}, \mathcal{E})$ with the required properties. S now A-casts the quadruple.

We now argue that every honest party will find $(\mathcal{C}, \mathcal{D}, \mathcal{F}, \mathcal{E})$ in their graphs and agree on the same. Though the graphs are constructed and maintained by parties individually in their local memory, it is always guaranteed that if an edge appears in the graph of an honest party, then the edge will eventually appear in the graphs of the other honest parties. This is ensured since the graphs are updated based on the responses of the parties that are A-casted. It now follows that if some honest party agree on $CORE$, then eventually all honest parties will also agree on the same. So we proved that all the honest parties will terminate **S-dependent Phase**, after agreeing on $CORE = \mathcal{E}$.   □

Lemma 24. *If the honest parties initiate* **S-independent Phase***, then they terminate the phase with the common message of the parties in $CORE$ as output.*

Proof. An honest party initiates **S-independent Phase**, if he agrees on $CORE$. By Lemma 22, all the honest parties in $CORE$ hold common message, say m of length ℓ and therefore same codeword $(s_1, \ldots, s_n) = \mathsf{ENC}(m_0, m_1, \ldots, m_t)$, where $m = m_0|m_1|\ldots, |m_t$. Then every honest P_i in $CORE$ already holds s_i, the ith element in the codeword. Every party P_i *not* in $CORE$ would receive s_i from the $t+1$ honest parties of $CORE$. Therefore every honest P_i will eventually hold the ith component of the codeword. Now every P_i send his s_i to every other party. Now on receiving at least $2t + 1 + r$, $0 \leq r \leq t$ s_j's, party P_i applies DEC with $c = r$ and $d = t - r$. Note that $c + d = t$, where t is the maximum number of corruption. Therefore if there are more than r wrong values (sent by Byzantine corrupted parties), DEC will return 'failure'. However for at least one value of r, $0 \leq r \leq t$, there will be at most r errors in the received vector and then the message can be reconstructed back successfully. This technique has been previously used in [CR93, Can95]. They call it as *Online Error Correction*.   □

Theorem 21. *Multi-Valued-Acast is an A-cast protocol satisfying Definition 11.*

Proof. We first consider the case of an honest S. By Lemma 23, for an honest S all the parties terminate **S-dependent Phase**, after agreeing on $CORE$. By Lemma 22, the honest parties in $CORE$ hold the message of S, i.e. m. By Lemma 24, all honest parties will terminate with the common message m.

For a corrupted S, all we need to show is that if some honest party terminates with message $m^\star$, then every other honest party do the same. Let P_i be the first honest party to terminate the protocol with $m^\star$ as output. Then P_i must have agreed on $CORE$ and the parties in $CORE$ holds $m^\star$. Then every other honest party will agree on the same $CORE$ and eventually terminate with $m^\star$ as the output (by Lemma 24).   □

Theorem 22. *Multi-Valued-Acast communicates $\mathcal{O}(n\ell)$ bits and invokes $\mathcal{O}(n^2 \log n)$ A-cast protocol for single bit.*

Proof. S communicates his message m, $|m| = \ell$ to all the parties. This requires $n\ell$ bits of communication. Every party P_i sends two values s_{ii} and s_{ij} to every other party P_j. The values are $\frac{\ell}{t+1}$ bits long each. Therefore in total there are $\frac{\ell}{t+1}\mathcal{O}(n^2) = \mathcal{O}(n\ell)$ bits of communication.

S A-casts $(\mathcal{C}, \mathcal{D}, \mathcal{F}, \mathcal{E})$. Each set in the quadruple can be represented by an n length bit vector. Therefore $4n$ invocations to A-cast protocol for single bit are required. Finally every party may A-cast OK signal for every other party. Each OK signal includes identities of two parties that can be represented by $2\log n$ bits. Therefore $\mathcal{O}(n^2\log n)$ invocations to A-cast for single bit are required. $\square$

We note that for an $(n, t+1)$ RS code, the field $\mathbb{F} = GF(2^c)$ in which the code is defined should satisfy $n \leq 2^c$ or $\log n \leq c$. In our case $c = \frac{\ell}{t+1}$ (recall that m is divided into $t+1$ parts each containing $\frac{\ell}{t+1}$ bits). Therefore $\frac{\ell}{t+1} \geq \log n \rightarrow \ell \geq (t+1)\log n$.

3   Error-free Communication Optimal ABA

In this section, we present our ABA protocol. We use our multi-valued A-cast protocol Multi-Valued-Acast from the previous section as one of the sub-protocols. Similar to Multi-Valued-Acast that uses A-cast protocol for single bit, our new ABA uses existing error-free ABA for single bit as another sub-protocol. In fact we use a very well-known asynchronous primitive called Agreement on Common Subset (ACS) introduced by [BKR94] that uses ABA for single bit as black box. We recall that the only error-free ABA is due to [ADH08]. We will use the following notation for invoking Multi-Valued-Acast.

Notation 31. *By saying that 'P_i Multi-casts M ', we mean that P_i as a sender, initiates Multi-Valued-Acast protocol with M as the message. Similarly 'P_j receives M from the Multi-cast of P_i' will mean that P_j terminates the execution of Multi-Valued-Acast protocol initiated by P_i and outputs M. By the property of Multi-Valued-Acast, if some honest party P_j terminates the Multi-Valued-Acast protocol of some sender P_i with M as the output, then every other honest party will eventually do so, irrespective of the behavior of the sender P_i.*

Agreement on a Common Subset (ACS). Consider the following scenario. The parties in $\mathcal{P}$ are asked to A-cast (or Multi-cast) some value. While the honest parties in $\mathcal{P}$ will eventually execute the A-cast (Multi-cast), the corrupted parties may or may not do the same. So the (honest) parties in $\mathcal{P}$ want to agree on a *common* set $\mathcal{T} \subset \mathcal{P}$, with $|\mathcal{T}| = 2t + 1$, such that A-cast (Multi-cast) of each party in $\mathcal{T}$ will be eventually terminated by the (honest) parties in $\mathcal{P}$. For this, the parties use ACS primitive presented in [BKR94]. The ACS protocol uses n instances of ABA for single bit.

3.1   Multi-valued ABA Protocol

Given the above sub-protocols, our ABA is very simple. Every party P_i on having a message m_i of length ℓ, computes an n length codeword $(s_{i1}, \ldots, s_{in}) =$

$\mathsf{ENC}(m_{i0}, \ldots, m_{it})$ where $m_{i0}| \ldots |m_{it}$. P_i Multi-casts s_{ii}. Using ACS, the parties then agree on some subset of $2t + 1$ parties, say $\mathcal{X}$ whose Multi-casts will be terminated eventually. Every party then verifies if the values Multi-casted by the parties in $\mathcal{X}$ match with their corresponding elements of the codeword and then A-cast their response. The parties again agree on some subset of $2t + 1$ parties using ACS, say $\mathcal{Y}$. Based on the responses of the parties in $\mathcal{Y}$ and the values Multi-casted by the parties in $\mathcal{X}$, the agreement is reached. Note that we use Multi-cast for the elements of the codewords (i.e. s_{ii}'s) and A-cast for the responses. The reason is that s_{ii}'s are message dependent and therefore can be arbitrarily large. Therefore by appropriately setting the value of ℓ, we can implement Multi-casting of s_{ii} values in $\mathcal{O}(n\ell)$ overall complexity. However, we will see from the protocol given below, the response vector will be always n length bit vector. Therefore, using Multi-cast for this case will worsen the complexity, as compared to the case when A-cast of Bracha is used for the same purpose. The protocol is now presented in Figure 3 and its properties are proved subsequently.

Protocol Multi-Valued-ABA()

Code for P_i.

1. On having message m_i, divide the ℓ bit message m_i into $t + 1$ blocks, $m_{i0}, \ldots, m_{it}$, each containing $\frac{\ell}{t+1}$ (assume this to be an integer) bits. Compute $(s_{i1}, \ldots, s_{in}) = \mathsf{ENC}(m_{i0}, \ldots, m_{it})$. Multi-cast s_{ii}.

2. Participate in an instance of ACS to agree on $\mathcal{X}$ containing $2t+1$ parties whose Multi-casts will be eventually terminated by all honest parties.

3. Construct a binary vector V_i of length n. Assign $V_i[j] = 1$, if $P_j \in \mathcal{X}$ and $s_{ij} = s_{jj}$ where s_{jj} is received from the Multi-cast of P_j. Otherwise assign $V_i[j] = 0$. A-cast V_i.

4. Participate in an instance of ACS to agree on $\mathcal{Y}$ containing $2t+1$ parties whose A-casts will be terminated eventually by all honest parties.

5. Check if there are at least $t + 1$ parties in $\mathcal{Y}$, whose vectors are identical and have at least $t + 1$ 1's. Let $\{i_1, \ldots, i_{i+1}\} \subseteq \mathcal{X}$ be the $t + 1$ minimum indices where they all have 1's. If there is no such set of $t + 1$ parties in $\mathcal{Y}$, then $\{i_1, \ldots, i_{i+1}\}$ be the $t + 1$ minimum indices in $\mathcal{X}$. Then apply DEC on $s_{i_1,i_1}, \ldots, s_{i_{t+1}i_{t+1}}$ and let $m_0, m_1, \ldots, m_t$ be the data returned by DEC. Then output $m = m_0| \ldots |m_t$.

Fig. 3. Error-free Communication Optimal ABA

Theorem 31. *Protocol Multi-Valued-ABA is an ABA protocol.*

Proof. The termination is guaranteed due to the termination properties of Multi-Valued-Acast, A-cast protocol of Bracha [Bra84] and ACS (the termination of ACS is guaranteed due to the termination of the underlying ABA for single bit). Since Multi-Valued-Acast initiated by the honest parties will eventually terminate, the set $\mathcal{X}$ will be agreed among the parties by the termination of ACS. Similarly, since A-cast (of Bracha) initiated by the honest parties will eventually terminate, the set $\mathcal{Y}$ will be agreed among the parties by the termination of ACS. Once $\mathcal{X}$ and $\mathcal{Y}$ are agreed, the rest is local computation. Therefore termination of Multi-Valued-ABA is guaranteed.

We now argue about the correctness of Multi-Valued-ABA. First we show that all the honest parties will agree on the same message. This follows from the fact that all the honest parties agree on $\{i_1, \ldots, i_{t+1}\} \subseteq \mathcal{X}$ in the last step of the protocol. Furthermore, by the correctness property of Multi-Valued-Acast, all the honest parties also agree on the values Multi-casted by the parties in $\{P_{i_1}, \ldots, P_{i_{t+1}}\}$. Our claim now follows trivially. We now consider the case when all the honest parties start with same input message m of length ℓ and argue that all honest parties will agree on m eventually. If all the honest parties start with m, then they generate $(s_1, \ldots, s_n) = \mathsf{ENC}(m_0, \ldots, m_t)$ locally, where $m = m_0 | \ldots | m_t$. Every honest party P_i then Multi-casts s_i. By the property of Multi-Valued-Acast, all the parties will receive the same value Multi-casted by the parties in $\mathcal{X}$. Therefore the honest parties in $\mathcal{Y}$ will have identical V_i vectors. Furthermore the V_i vectors will have 1's at least at $t+1$ locations corresponding to the parties in $\mathcal{X}$ who Multi-casted correct value from the codeword $(s_1, \ldots, s_n)$. So $\{i_1, \ldots, i_{t+1}\} \subseteq \mathcal{X}$ in the last step of the above protocol will be $t+1$ identities of the parties in $\mathcal{X}$ (having $t+1$ minimum indices) who Multi-casted correct value from codeword $(s_1, \ldots, s_n)$. So DEC when applied on the values Multi-casted by $\{P_{i_1}, \ldots, P_{i_{t+1}}\}$ will return $m_0, m_1, \ldots, m_t$ where $m = m_0 | \ldots | m_t$. □

Theorem 32. *Multi-Valued-ABA communicates $\mathcal{O}(n\ell)$ bits, invokes $\mathcal{O}(n^3 \log n)$ instances of A-cast for single bit and invokes $2n$ instances of ABA for single bit.*

Proof. Every party Multi-casts $\frac{\ell}{t+1}$ bits. This requires communication of $\mathcal{O}(n\ell)$ bits and $\mathcal{O}(n^3 \log n)$ invocations to A-cast for single bit. Then every party A-casts an n length bit vector. Therefore n^2 invocations to A-cast is required. Finally two invocations to ACS calls for $2n$ instances of ABA for single bit. □

To make the underlying protocol Multi-Valued-Acast work correctly, we require $\frac{\ell}{t+1} \geq (t+1) \log n$. Recall that when the input message size for Multi-Valued-Acast is ℓ, then we require that $\ell \geq (t+1) \log n$. In our ABA protocol, the input to Multi-valued-Acast is $\frac{\ell}{t+1}$. Therefore we have $\frac{\ell}{t+1} \geq (t+1) \log n \rightarrow \ell \geq (t+1)^2 \log n$.

4   Error-free Communication Optimal BA and BC

Here we present our new multi-valued BA and BC protocol. We first present a BA protocol. A BC protocol with same complexity of the BA protocol can be achieve by letting the sender send the message to all the parties and then running a BA to reach agreement. This is the standard reduction in synchronous settings from BA to BC [Lyn96]. Our BA protocol follows the idea of Multi-Valued-Acast. We use the BC protocol of [BGP09, CW92] for single bit that communicates $\mathcal{O}(n^2)$ bits. We now present the protocol in Figure 4.

Lemma 41. *The honest parties in CORE hold same message of length ℓ.*

The proof of Lemma 41 completely follows from the proof of Lemma 22.

Protocol Multi-Valued-BA()

Code for P_i.

1. On having message m_i, divide the ℓ bit message m_i into $t + 1$ blocks, $m_{i0}, \ldots, m_{it}$, each containing $\frac{\ell}{t+1}$ bits. Compute $(s_{i1}, \ldots, s_{in}) =$ ENC$(m_{i0}, \ldots, m_{it})$. Send s_{ii} to every party. Send s_{ij} to P_j for $j = [1, n]$.

2. Construct a binary vector V_i of length n. Assign $V_i[j] = 1$, if $s_{ij} = s_{jj}$ and $s_{ii} = s_{ji}$ where s_{jj} and s_{ji} are received from P_j. Otherwise assign $V_i[j] = 0$.

3. Broadcast V_i using BC protocol for single bit.

4. Construct graph G_i using parties in $\mathcal{P}$ as the vertices. Add edge (P_j, P_k) if $V_j[k] = 1$ and $V_k[j] = 1$. Execute Find-STAR$(\overline{G_i})$. If star-Not-Found is returned, then set $b_i = 0$. Else let $(\mathcal{C}_i, \mathcal{D}_i)$ be the (n, t)-star returned by Find-STAR. Let $\mathcal{F}_i$ be the set of parties who have at least $t + 1$ neighbors in $\mathcal{C}_i$ in graph G_i. Let $\mathcal{E}_i$ be the set of parties who have at least $2t + 1$ neighbors in $\mathcal{F}_i$ in graph G_i. If $|\mathcal{F}_i| \geq 2t + 1$ and $|\mathcal{E}_i| \geq 2t + 1$, then set $b_i = 1$, else set $b_i = 0$.

5. Broadcast only b_i when $b_i = 0$; else broadcast b_i and $(\mathcal{C}_i, \mathcal{D}_i, \mathcal{F}_i, \mathcal{E}_i)$ using BC protocol for single bit.

6. If $t+1$ b_j's are zero, then agree on some predefined message $m^\star$ of length ℓ. Else let α be the minimum index of the party where $b_\alpha = 1$ and $(\mathcal{C}_\alpha, \mathcal{D}_\alpha, \mathcal{F}_\alpha, \mathcal{E}_\alpha)$ be such that $(\mathcal{C}_\alpha, \mathcal{D}_\alpha)$ is an (n, t)-star in G_i, every party in $\mathcal{F}_\alpha$ has at least $t + 1$ neighbors in $\mathcal{C}_\alpha$ and every party in $\mathcal{E}_\alpha$ has at least $2t + 1$ neighbors in $\mathcal{F}_\alpha$. Assign $CORE = \mathcal{E}_\alpha$.

7. Assign s_i to be the value s_{ji} received from the majority of the parties in $CORE$. Send s_i to every party.

8. Let $(s_1, \ldots, s_n)$ be the vector where s_j is received from P_j. Apply DEC on $(s_1, \ldots, s_n)$ with $c = t$ and $d = 0$. Let $m_0, m_1, \ldots, m_t$ be the data returned by DEC. Output $m = m_0 | \ldots | m_t$.

Fig. 4. Error-free Multi-valued BA with Optimal Communication Complexity

Lemma 42. *If all honest parties start with same input m, then all the parties will agree on $CORE$, $|CORE| \geq 2t + 1$.*

Proof. The proof here follows from the proof of Lemma 23. Briefly, when all honest parties start with same input, every pair of honest parties will have edge between them. In other words, the edges in the complementary graph will be either (a) between an honest and a corrupted party OR (b) between two corrupted parties. Therefore following the argument given in Lemma 23, $\mathcal{C}$ component of an (n, t)-star will contain at least $t + 1$ honest parties, which subsequently will lead to the construction of $\mathcal{F}$ and $\mathcal{E}$ with size at least $2t + 1$. Although it is not guaranteed that all honest parties find same quadruple $(\mathcal{C}, \mathcal{D}, \mathcal{F}, \mathcal{E})$, but it is ensured that they will find some quadruple. So the honest parties never agree on predefined $m^\star$ in this case. Now since all the parties broadcast their quadruple, it is easy to reach agreement on a valid quadruple which the parties do by selecting the one broadcasted by the party with minimum index. Therefore all the parties will agree on $CORE$. $\qquad\square$

Lemma 43. *If CORE is agreed, all honest parties output the common message of the parties in CORE.*

Proof. By Lemma 41, all honest parties in $CORE$ hold same message, say m. The proof now follows from the proof of Lemma 24. $\qquad\square$

Theorem 41. *Multi-Valued-BA is a BA protocol.*

Proof. If $CORE$ is agreed, then all honest parties will output the common message of the parties in $CORE$ (by Lemma 43). If $CORE$ is not agreed, then there must be at least $t+1$ parties who broadcasted $b_i = 0$. Since b_i's are broadcasted, all honest parties will agree on predefined $m^\star$ of length. So agreement is always achieved at the end. Now if all the honest parties start with same m, then they will agree on $CORE$ (by Lemma 42) and output m (by Lemma 43). $\qquad\square$

Theorem 42. *Multi-valued-BA communicates $\mathcal{O}(n\ell)$ bits and invokes $\mathcal{O}(n^2)$ broadcast protocol for single bit.*

Proof. Every party P_i sends two values s_{ii} and s_{ij} to every other party P_j. The values are $\frac{\ell}{t+1}$ bits long each. Therefore in total there are $\frac{\ell}{t+1}\mathcal{O}(n^2) = \mathcal{O}(n\ell)$ bits of communication. Every party P_i broadcasts n-length binary vector V_i, a bit b_i and quadruple $(\mathcal{C}_i, \mathcal{D}_i, \mathcal{F}_i, \mathcal{E}_i)$. Each set in the quadruple can be represented by n-length bit vector. Therefore every party invokes $5n + 1$ instances of broadcast for single bit. This leads to total $\mathcal{O}(n^2)$ instances of broadcast for single bit. $\square$

The value of ℓ should be at least $(t + 1)\log n$ to make the underlying $(n, t + 1)$ RS code work (following the same logic as explained for Multi-Valued-Acast).

5　Open Problems

An important open question is to investigate whether multi-valued communication optimal protocols can be achieved with less number of invocations to protocols for single bit in comparison to what we provide in this paper.

References

[ADH08]　Abraham, I., Dolev, D., Halpern, J.Y.: An almost-surely terminating polynomial protocol for asynchronous Byzantine Agreement with optimal resilience. In: PODC, pp. 405–414. ACM Press (2008)

[BGP09]　Berman, P., Garay, G.A., Perry, K.J.: Bit optimal distributed consensus. Computer Science Research (2009)

[BKR94]　BenOr, M., Kelmer, B., Rabin, T.: Asynchronous secure computations with optimal resilience. In: PODC, pp. 183–192. ACM Press (1994)

[BO83]　Ben-Or, M.: Another advantage of free choice: Completely asynchronous agreement protocols. In: PODC, pp. 27–30. ACM Press (1983)

[BOCG93]　Ben-Or, M., Canetti, R., Goldreich, O.: Asynchronous Secure Computation. In: STOC, pp. 52–61. ACM Press (1993)

[BOGW88] Ben-Or, M., Goldwasser, S., Wigderson, A.: Completeness theorems for non-cryptographic fault-tolerant distributed computation (extended abstract). In: STOC, pp. 1–10. ACM Press (1988)

[Bra84] Bracha, G.: An asynchronous $\lfloor (n-1)/3 \rfloor$-resilient consensus protocol. In: PODC, pp. 154–162. ACM Press (1984)

[BTH06] Beerliová-Trubíniová, Z., Hirt, M.: Efficient Multi-Party Computation with Dispute Control. In: Halevi, S., Rabin, T. (eds.) TCC 2006. LNCS, vol. 3876, pp. 305–328. Springer, Heidelberg (2006)

[Can95] Canetti, R.: Studies in Secure Multiparty Computation and Applications. PhD thesis, Weizmann Institute, Israel (1995)

[CGMA85] Chor, B., Goldwasser, S., Micali, S., Awerbuch, B.: Verifiable secret sharing and achieving simultaneity in the presence of faults (extended abstract). In: STOC, pp. 383–395. ACM Press (1985)

[CR93] Canetti, R., Rabin, T.: Fast asynchronous Byzantine Agreement with optimal resilience. In: STOC, pp. 42–51. ACM Press (1993)

[CW92] Coan, B.A., Welch, J.L.: Modular construction of a Byzantine Agreement protocol with optimal message bit complexity. Information and Computation 97(1), 61–85 (1992)

[DR85] Dolev, D., Reischuk, R.: Bounds on information exchange for Byzantine Agreement. JACM 32(1), 191–204 (1985)

[FH06] Fitzi, M., Hirt, M.: Optimally Efficient Multi-valued Byzantine Agreement. In: PODC, pp. 163–168 (2006)

[FLP85] Fischer, M.J., Lynch, N.A., Paterson, M.: Impossibility of distributed consensus with one faulty process. JACM 32(2), 374–382 (1985)

[FM88] Feldman, P., Micali, S.: An Optimal Algorithm for Synchronous Byzantine Agreemet. In: STOC, pp. 639–648. ACM Press (1988)

[GJ79] Garey, M.R., Johnson, D.S.: Computers and Intractability: A Guide to the Theory of NP-Completeness. W. H. Freeman (1979)

[HMP00] Hirt, M., Maurer, U., Przydatek, B.: Efficient Secure Multi-party Computation. In: Okamoto, T. (ed.) ASIACRYPT 2000. LNCS, vol. 1976, pp. 143–161. Springer, Heidelberg (2000)

[LV11] Liang, G., Vaidya, N.H.: Error-Free Multi-Valued Consensus with Byzantine Failures. In: PODC, pp. 11–20. ACM Press (2011)

[Lyn96] Lynch, N.A.: Distributed Algorithms. Morgan Kaufmann (1996)

[PR10] Patra, A., Rangan, C.P.: Communication Optimal Multi-Valued Asynchronous Broadcast Protocol. In: Abdalla, M., Barreto, P.S.L.M. (eds.) LATINCRYPT 2010. LNCS, vol. 6212, pp. 162–177. Springer, Heidelberg (2010)

[PR11] Patra, A., Rangan, C.P.: Communication Optimal Multi-Valued Asynchronous Byzantine Agreement with Optimal Resilience. In: Fehr, S. (ed.) ICITS 2011. LNCS, vol. 6673, pp. 206–226. Springer, Heidelberg (2011)

[PSL80] Pease, M., Shostak, R.E., Lamport, L.: Reaching agreement in the presence of faults. JACM 27(2), 228–234 (1980)

[PW92] Pfitzmann, B., Waidner, M.: Unconditional Byzantine Agreement for Any Number of Faulty Processors. In: Finkel, A., Jantzen, M. (eds.) STACS 1992. LNCS, vol. 577, pp. 339–350. Springer, Heidelberg (1992)

[Rab83] Rabin, M.O.: Randomized Byzantine generals. In: FOCS, pp. 403–409. IEEE Computer Society (1983)

[RBO89] Rabin, T., Ben-Or, M.: Verifiable secret sharing and multiparty protocols with honest majority (extended abstract). In: STOC, pp. 73–85. ACM Press (1989)

[TC84] Turpin, R., Coan, B.A.: Extending binary Byzantine Agreement to multivalued Byzantine Agreement. IPL 18(2), 73–76 (1984)

Enhancing the Performance of High Availability Lightweight Live Migration

Peng Lu[1], Binoy Ravindran[1], and Changsoo Kim[2]

[1] ECE Department, Virginia Tech, USA
{lvpeng,binoy}@vt.edu
[2] ETRI, Daejeon, South Korea
cskim7@etri.re.kr

Abstract. Remus is one of the first systems which implemented whole virtual machine replication to achieve high availability (HA). Recently a fast, lightweight migration mechanism (LLM) was proposed to reduce the long network delay in Remus. However, these virtualized systems have the long downtime problem, which is a bottleneck to achieve HA. Based on LLM, in this paper, we describe a fine-grained block identification (or FGBI) mechanism to reduce the downtime in virtualized systems so as to achieve HA, with support for a block sharing mechanism and hybrid compression method. We implement the FGBI mechanism and evaluate it against LLM and Remus, using several benchmarks such as Apache, SPECweb, NPB and SPECsys. Our experimental results reveal that FGBI reduces the type I downtime over LLM and Remus by as much as 77% and 45% respectively, and reduces the type II downtime by more than 90% and more than 70%, compared with LLM and Remus respectively. Moreover, in all cases, the performance overhead of FGBI is less than 13%.

1 Introduction

High availability (HA) refers to a system and associated service implementation that is continuously operational for a long period of time. With respect to the clients, an ideal system never stops working, which also means the system will always respond to the clients' requests. Trying to achieve high availability is therefore one of the key concerns in modern cluster computing and failover systems. Whole-system replication is a conventional way to increase the system availability: once the primary machine fails, the running applications will be taken over by the backup machines. However, there are several limitations that make this method unattractive for deployment: it needs specialized hardware and software which are usually expensive. That the final system also requires complex customized configurations makes it hard to manage efficiently.

As virtualization becomes more and more prevalent, we can overcome these limitations by introducing the virtual machine (VM). In the virtual world, all the applications are running in the VM, so now it's possible to implement the whole-system replication in an easy and efficient way — by saving the copy of the whole VM running on the system. As VMs are totally hardware-independent,

A. Fernández Anta, G. Lipari, and M. Roy (Eds.): OPODIS 2011, LNCS 7109, pp. 50–64, 2011.

the cost is much lower compared to the hardware expenses in traditional HA solutions. Besides, virtualization technology can facilitate the management of multiple VMs on a single physical machine. With virtual machine monitors (VMM), the service applications are separated from physical machines, thus providing increased flexibility and improved performance.

Remus [6], built on top of the well-known Xen hypervisor [3], provides transparent, comprehensive high availability by using a checkpointing method under the Primary-Backup model (Figure 1). It checkpoints the running VM on the primary host, and transfers the latest checkpoint to the backup host as whole-system migration. Once the primary host fails, the backup host will take over the service based on the latest checkpoint. Remus proves that it is possible to create a general, fully transparent, high-availability solution entirely in software. However, checkpointing at high frequency will introduce significant overhead, since significant CPU and memory resources are consumed by the migration. Therefore, clients endure a long network delay.

Jiang *et. al.* [13] proposed an integrated live migration mechanism, called LLM, which integrates both whole-system checkpointing and input replay to reduce the network delay in Remus. The basic idea is that the primary host migrates the guest VM image (including CPU/memory status updates and new writes to the file system) to the backup host at low frequency. In the meanwhile, the service requests from network clients are migrated at high frequency. As its results show, LLM significantly outperforms Remus in terms of network delay by more than 90%.

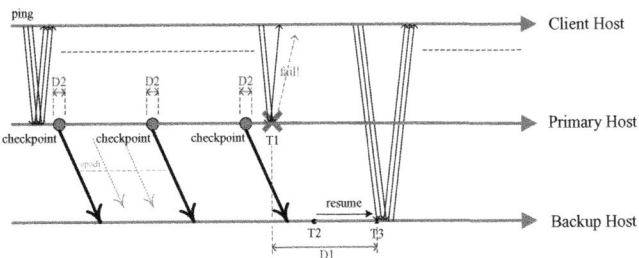

Fig. 1. Primary-Backup model and the downtime problem (T1: primary host crashes; T2: client host observes the primary host crashes; T3: VM resumes on backup host; D1 (T3 - T1): type I downtime; D2: type II downtime)

Downtime is the key factor for estimating the high availability of a system, since any long downtime experience for clients may result in loss of client loyalty and thus revenue loss. Under the Primary-Backup model, there are two types of downtime: I) the time from when the primary host crashes until the VM resumes from the last checkpointed state on the backup host and starts to handle client requests (shown as D1 in Figure 1); II) the time from when the VM pauses on the primary (to save for the checkpoint) until it resumes (shown as D2 in Figure 1). From Jiang's paper we observe that for memory-intensive workloads running on guest VMs (such as the HighSys workload [13]), LLM endures much longer type I downtime than Remus. This is because, these workloads update the guest memory at high frequency. On the other side, LLM migrates the guest VM image update (mostly from memory) at low frequency but uses input replay as an auxiliary. In this case, when a failure happens, a significant number of memory updates are needed in order to ensure synchronization between the primary and backup hosts. Therefore, it needs significantly more time for the input replay process in order to resume the VM on the backup host and begin handling client requests.

Regarding the type II downtime, there are several migration epochs between two checkpoints, and the newly updated memory data is copied to the backup host at each epoch. At the last epoch, the VM running on the primary host is suspended and the remaining memory states are transferred to the backup host. Thus, the type II downtime depends on the amount of memory that remains to be copied and transferred when pausing the VM on the primary host. If we reduce the dirty data which need to be transferred at the last epoch, we can reduce the type II downtime. Moreover, if we reduce the dirty data which needs to be transferred at each epoch, trying to synchronize the memory state between primary and backup host all the time, then at the last epoch, there will not be too many new memory updates that need to be transferred, so we can reduce the type I downtime as well.

Therefore, in order to achieve HA in these virtualized systems, especially to address the downtime problem under memory-intensive workloads, we propose a memory synchronization technique for tracking memory updates, called Fine-Grained Block Identification (or FGBI). Our main contributions include:

1) Based on LLM, we develop a novel, efficient and fine-grained approach called FGBI, to track and transfer the memory updates efficiently, by reducing the total number of dirty bytes which need to be transferred from primary to backup host. FGBI enhances LLM's performance by overcoming its downtime disadvantage, especially for applications with memory-intensive workloads.

2) We integrate memory block sharing support with FGBI to reduce the newly introduced memory and computation/comparison overheads. In addition, we also support a hybrid compression mechanism among the memory dirty blocks to further reduce the migration traffic in the transfer period.

3) We present a fully functional prototype implementation and demonstrate that it achieves comparable downtime compared with Remus/LLM. Our experimental results reveal that FGBI reduces the type I downtime over LLM and Remus by as much as 77% and 45% respectively, and reduces the type II downtime by more than 90% and more than 70%, compared with LLM and Remus respectively.

The rest of the paper is organized as follows. Section 2 discusses past and related work. Section 3 presents the design and implementation of the integrated FGBI mechanism. Section 4 reports our experimental environment, benchmarks, and the evaluation results. We conclude and discuss future work in Section 5.

2   Related Work

To achieve high availability, currently there exist many virtualization-based live migration techniques [12, 18, 24]. Two representatives are Xen live migration [4] and VMware VMotion [17], which share similar pre-copy strategies. During migration, physical memory pages are sent from the source (primary) host to the new destination (backup) host, while the VM continues running on the source host. Pages modified during the replication must be re-sent to ensure consistency. After a bounded iterative transferring phase, a very short stop-and-copy phase is executed, during which the VM is halted, the remaining memory pages are sent, and the destination VMM is signaled to resume the execution of the VM. However, these pre-copy methods incur significant VM downtimes, as the evaluation results in [8] show.

Remus [6] is now part of the official Xen repository. It achieves HA by maintaining an up-to-date copy of a running VM on the backup host, which automatically activates if the primary host fails. Remus (and also LLM [13]) copies over dirty data after memory update, and uses the memory page as the granularity for copying. However, the dirty data tracking method is not efficient, as shown in [16] (we also illustrate this inefficiency in Section 3.1). Thus, our goal in this paper is to further reduce the size of the memory transferred from the primary to the backup host, by introducing a fine-grained mechanism.

Lu *et. al.* [16] applied three memory state synchronization techniques to achieve HA in systems such as Remus: dirty block tracking, speculative state transferring and active backup. The first technology is similar to our proposed method, however, it incurs additional memory associated overhead. For example, when running the Exchange workload in their evaluation, the memory overhead is more than 60%. Since main memory is always a scarce resource, the high percentage overhead is a problem. Different from these authors' work, we reduce memory overhead incurred by FGBI by integrating a new memory blocks sharing mechanism, and a hybrid compression method when transferring the memory update.

To solve the memory overhead problems under Xen-based systems, there are several ways to harness memory redundancy in VMs, such as page sharing and patching. Past efforts showed the memory sharing potential in virtualization-based systems. Working set changes were examined in [4, 21], and their results showed that changes in memory were crucial for the migration of VMs from host to host. For a guest VM with 512 MB memory assigned, low loads changed roughly 20 MB, medium loads changed roughly 80 MB, and high loads changed roughly 200 MB. Thus, normal workloads are likely to occur between these extremes. The evaluation in [4, 21] also revealed the amount of memory changes (within minutes) in VMs running different light workloads. None of them changed more than 4 MB of memory within two minutes. The Content-Based Page Sharing (CBPS) method [22] also illustrated the sharing potential in memory. CBPS was based on the compare-by-hash technique introduced in [9, 10]. As claimed, CBPS was able to identify as much as 42.9% of all pages as sharable, and reclaimed 32.9% of the pages from ten instances of Windows NT doing real-world workloads. Nine VMs running Redhat Linux were able to find 29.2% of sharable pages and reclaimed 18.7%. When reduced to five VMs, the numbers were 10.0% and 7.2%, respectively.

To share memory pages efficiently, recently, the Copy-on-Write (CoW) sharing mechanism was widely exploited in the Xen VMM [19]. Unlike the sharing of pages within an OS that uses CoW in a traditional way, in virtualization, pages are shared between multiple VMs. Instead of using CoW to share pages in memory, we use the same idea in a more fine-grained manner, i.e., by sharing among smaller blocks. The Difference Engine project demonstrates the potential memory savings available from leveraging a combination of page sharing, patching, and in-core memory compression [8]. It shows the huge potential of harnessing memory redundancy in VMs. However, Difference Engine also suffers from complexity problems when applying the patching method. It needs additional modifications to Xen. We will present our corresponding mechanism and advantages over Difference Engine in Section 3.2.

Besides high availability systems such as Remus, LLM, and Kemari [20], which apply the pre-copy mechanism, there are also other related works that focus on migration optimization. Post-copy based migration [11] is proposed to address the drawbacks of pre-copy based migration. The experimental evaluation in [11] shows that the migration time using the post-copy method is less than the pre-copy method, under SPECweb2005 and Linux Kernel Compile benchmarks. However, its implementation only supports PV guests as the mechanism for trapping memory accesses and utilizes an in-memory pseudo-paging device in the guest OS. Since the post-copy mechanism needs to modify the guest OS, it is not so much widely used as the pre-copy mechanism.

3   Design and Implementation

We first overview the integrated FGBI design, including some necessary preliminaries about the memory saving potential. We then present the FGBI architecture, explain each component, and discuss the execution flow and other implementation details.

3.1   FGBI

Remus and LLM track memory updates by keeping evidence of the dirty pages at each migration epoch. Remus uses the same page size as Xen (for x86, this is 4KB), which is also the granularity for detecting memory changes. However, this mechanism is not efficient. For instance, no matter what changes an application makes to a memory page, even just modify a boolean variable, the whole page will still be marked dirty. Thus, instead of one byte, the whole page needs to be transferred at the end of each epoch. Therefore, it is logical to consider tracking the memory update at a finer granularity, like dividing the memory into smaller blocks.

We propose the FGBI mechanism which uses memory blocks (smaller than page sizes) as the granularity for detecting memory changes. FGBI calculates the hash value for each memory block at the beginning of each migration epoch. Then it uses the same mechanism as Remus to detect dirty pages. However, at the end of each epoch, instead of transferring the whole dirty page, FGBI computes new hash values for each block and compares them with the corresponding old values. Blocks are only modified if their corresponding hash values do not match. Therefore, FGBI marks such blocks as dirty and replaces the old hash values with the new ones. Afterwards, FGBI only transfers dirty blocks to the backup host.

However, because of using block granularity, FGBI introduces new overhead. If we want to accurately approximate the true dirty region, we need to set the block size as small as possible. For example, to obtain the highest accuracy, the best block size is one bit. That is impractical, because it requires storing an additional bit for each bit in memory, which means that we need to double the main memory. Thus, a smaller block size leads to a greater number of blocks and also requires more memory for storing the hash values. Based on these past efforts illustrating the memory saving potential (section 2), we present two supporting techniques: block sharing and hybrid compression. These are discussed in the subsections that follow.

3.2   Block Sharing and Hybrid Compression Support

From the memory saving results of related work (section 2), we observe that while running normal workloads on a guest VM, a large percentage of memory is usually not updated. For this static memory, there is a high probability that pages can be shared and compressed to reduce memory usage.

Block Sharing. Note that these past efforts [4, 9, 10, 21, 22] use the memory page as the sharing granularity. Thus, they still suffer from the "one byte differ, both pages cannot be shared" problem. Therefore, we consider using a smaller block in FGBI as the sharing granularity to reduce memory overhead.

The Difference Engine project [8] also illustrates the potential savings due to sub-page sharing, both within and across virtual machines, and achieves savings up to 77%. In order to share memory at the sub-page level, the authors construct patches to represent a page as the difference relative to a reference page. However, this patching method requires selected pages to be accessed infrequently, otherwise the overhead of compression/decompression outweighs the benefits. Their experimental evaluations reveal that patching incurs additional complexity and overhead when running memory-intensive workloads on guest VMs (from results for "Random Pages" workload in [8]).

Unlike Difference Engine, we apply a straightforward sharing technique to reduce the complexity. The goal of our sharing mechanism is to eliminate redundant copies of identical blocks. We share blocks and compare hash values in memory at runtime, by using a hash function to index the contents of every block. If the hash value of a block is found more than once in an epoch, there is a good probability that the current block is identical to the block that gave the same hash value. To ensure than these blocks are identical, they are compared bit by bit. If the blocks are identical, they are reduced to one block. If, later on, the shared block is modified, we need to decide which of the original constituent blocks has been updated and will be transferred.

Hybrid Compression. Compression techniques can be used to significantly improve the performance of live migration [14]. Compressed dirty data takes shorter time to be transferred through the network. In addition, network traffic due to migration is significantly reduced when much less data is transferred between primary and backup hosts. Therefore, for dirty blocks in memory, we consider compressing them to reduce the amount of transferred data.

Before transmitting a dirty block, we check for its presence in an address-indexed cache of previously transmitted blocks (through pages). If there is a cache hit, the whole page (including this memory block) is XORed with the previous version, and the differences are run-length encoded (RLE). At the end of each migration epoch, we send only the delta from a previous transmission of the same memory block, so as to reduce the amount of migration traffic in each epoch. Since smaller amount of data is transferred, the total migration time and downtime can both be decreased.

However, in the current migration epoch, there still may remain a significant fraction of blocks that is not present in the cache. In these cases, we find that Wilson *et. al.* [7] claims that there are a great number of zero bytes in the memory pages (so as in our smaller blocks). For this kind of block, we just scan the whole block and record the information about the offset and value of nonzero bytes. And for all other blocks with weak regularities, a universal algorithm with high compression ratio is appropriate. Here we apply a general-purpose and very fast compression technique, zlib [1], to achieve a higher degree of compression.

3.3   Architecture

We implement the FGBI mechanism integrated with sharing and compression support, as shown in Figure 2. In addition to LLM, which is labeled as "LLM Migration Manager" in the figure, we add a new component, shown as "FGBI", and deploy it at both Domain 0 and guest VM.

For easiness in presentation, we divide FGBI into three main components:

1) Dirty Identification: It uses the hash function to compute the hash value for each block, and identify the new update through the hash comparison at the end of migration epoch. It has three subcomponents:

Block Hashing: It creates a hash value for each memory block;

Hash Indexing: It maintains a hash table based on the hash values generated by the Block Hashing component. The entry in the content index is the hash value that reflects the content of a given block;

Block Comparison: It compares two blocks to check if they are bitwise identical.

2) Block Sharing Support: It handles sharing of bitwise identical blocks.

3) Block Compression: It compresses all the dirty blocks on the primary side, before transferring them to the backup host. On the backup side, after receiving

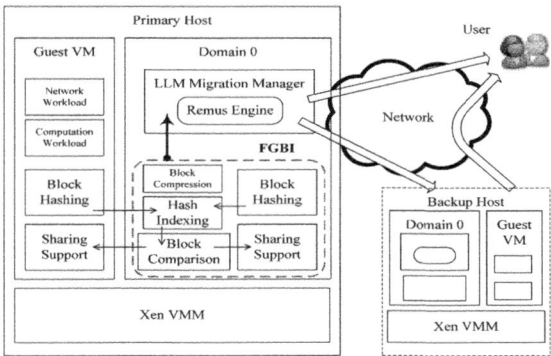

Fig. 2. The FGBI architecture with sharing and compression support

the compressed blocks, it decompresses them first before using them to resume the VM.

Basically, the Block Hashing component produces hash values for all blocks and delivers them to the Hash Indexing component. The Hash Indexing and Block Comparison components then check the hash table to determine whether there are any duplicate blocks. If so, the Hash Comparison component requests the Block Sharing Support components to update the shared blocks information. At the end of each epoch, the Block Compression component compresses all the dirty blocks (including both shared and not shared).

In this architecture, the components are divided between the privileged VM Domain 0 and the guest VMs. The VMs contain the Block Sharing Support components. We house the Block Sharing Support component in the guest VMs to avoid the overhead of using shadow page tables (SPTs). Each VM also contains a Block Hashing component, which means that it has the responsibility of hashing its address space. The Dirty Identification component is placed in the trusted and privileged Domain 0. It receives hash values of the hashed blocks generated by the Block Hashing component in the different VMs.

3.4   FGBI Execution Flow

Figure 3 describes the execution flow of the FGBI mechanism. The numbers on the arrows in the figure correspond to numbers in the enumerated list below:

1) Hashing: At the beginning of each epoch, the Block Hashing components at the different guest VMs compute the hash value for each block.

2) Storing: FGBI stores and delivers the hash key of the hashed block to the Hash Indexing component.

3) Index Lookup: It checks the content index for identical keys, to determine whether the block has been seen before. The lookup can have two different outcomes:

Key not seen before: Add it to the index and proceed to step 6.

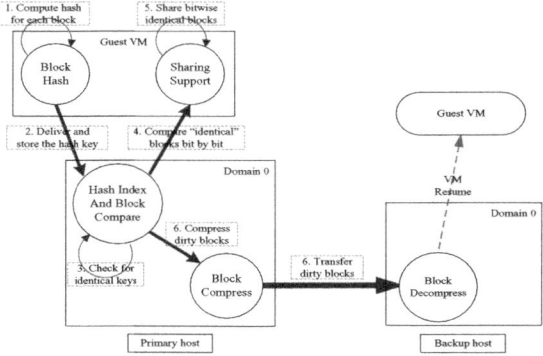

Fig. 3. Execution flow of FGBI mechanism

Key seen before: An opportunity to share, so request block comparison.

4) Block Comparison: Two blocks are shared if they are bitwise identical. Meanwhile, it notifies the Block Sharing Support Components on corresponding VMs that they have a block to be shared. If not, there is a hash collision, the blocks are not shared, and proceed to step 6.

5) Shared Block Update: If two blocks are bitwise identical, then store the same hash value for both blocks. Unless there is a write update to this shared block, it doesn't need to be compared at the end of the epoch.

6) Block Compression: Before transferring, compress all the dirty blocks.

7) Transferring: At the end of epoch, there are three different outcomes:

Block is not shared: FGBI computes the hash value again and compares with the corresponding old value. If they don't match, mark this block as dirty, compress and send it to the backup host. Repeat step 1 (which means begin the next migration epoch).

Block is shared but no write update: It means that either block is modified during this epoch. Thus, there is no need to compute hash values again for this shared block, and therefore, there is no need to make comparison, compression, or transfer either. Repeat step 1.

Block is shared and write update occurs: This means that one or both blocks have been modified during this epoch. Thus, FGBI needs to check which one is modified, and then compress and send the dirty one or both to the backup host. Repeat step 1.

4   Experimental Evaluation

We experimentally evaluated the performance of the proposed techniques (i.e., FGBI, sharing, and compression), which i simply referred to here as the FGBI mechanism. We measured downtime and overhead under FGBI, and compared the result with that under LLM and Remus.

4.1   Experimental Environment

Our experimental platform included two identical hosts (one as primary and the other as backup), each with an IA32 architecture processor (Intel Core 2 Duo Processor E6320, 1.86 GHz), and 3 GB RAM. We set up a 1 Gbps network connection between the two hosts, which is specifically used for migration. In addition, we used a separate machine as a network client to transmit service requests and examine the results based on the responses. We built Xen 3.4 from source [23], and let all the protected VMs run PV guests with Linux 2.6.18. The VMs were running CentOS Linux, with a minimum of services executing, e.g., sshd. We allocated 256 MB RAM for each guest VM, the file system of which is an image file of 3 GB shared by two machines using NFS. Domain 0 had a memory allocation of 1 GB, and the remaining memory was left free. The Remus patch we used was the nearest 0.9 version [5]. We compiled the LLM source code and installed its modules into Remus.

Our experiments used the following VM workloads under the Primary-Backup model:

Static web application: We used Apache 2.0.63 [15]. Both hosts were configured with 100 simultaneous connections, and repetitively downloaded a 256KB file from the web server. Thus, the network load will be high, but the system updates are not so significant.

Dynamic web application: SPECweb99 is a complex application-level benchmark for evaluating web servers and the systems that host them. This benchmark comprises a web server, serving a complex mix of static and dynamic page (e.g., CGI script) requests, among other features. Both hosts generate a load of 100 simultaneous connections to the web server [2].

Memory-intensive application: Since FGBI is proposed to solve the long downtime problem under LLM especially when running heavy computational workloads on the guest VM, we continued our evaluation by comparing FGBI with LLM/Remus under a set of industry-standard workloads, specifically NPB and SPECsys.

1. *NPB-EP:* This benchmark is derived from CFD codes, and is a standard measurement procedure used for evaluating parallel programs. We selected the Kernel EP program from the NPB benchmark [19], because the scale of this program set is moderate and its memory access style is representative. Therefore, this example involves high computational workloads on the guest VM.

2. *SPECsys:* This benchmark measures NFS (version 3) file server throughput and response time for an increasing load of NFS operations (lookup, read, write, and so on) against the server over file sizes ranging from 1 KB to 1 MB. The page modification rate when running SPECsfs has previously been reported as approximately 10,000 dirty pages/second [2], which is approximately 40% of the link capacity on a 1 Gbps network.

To ensure that our experiments are statistically significant, each data point is averaged from twenty sample values. The standard deviation computed from the samples is less than 7.6% of the mean value.

4.2 Downtime Evaluations

Type I Downtime. Figures 4a, 4b, 4c, and 4d show the type I downtime comparison among FGBI, LLM, and Remus mechanisms under Apache, NPB-EP, SPECweb, and SPECsys applications, respectively. The block size used in all experiments is 64 bytes. For Remus and FGBI, the checkpointing period is the time interval of system update migration, whereas for LLM, the checkpointing period represents the interval of network buffer migration. By configuring the same value for the checkpointing frequency of Remus/FGBI and the network buffer frequency of LLM, we ensure the fairness of the comparison. We observe that Figures 4a and 4b show a reverse relationship between FGBI and LLM. Under Apache (Figure 4a), the network load is high but system updates are rare. Therefore, LLM performs better than FGBI, since it uses a much higher frequency to migrate the network service requests. On the other hand, when running memory-intensive applications (Figures 4b and 4d), which involve high computational loads, LLM endures a much longer downtime than FGBI (even worse than Remus).

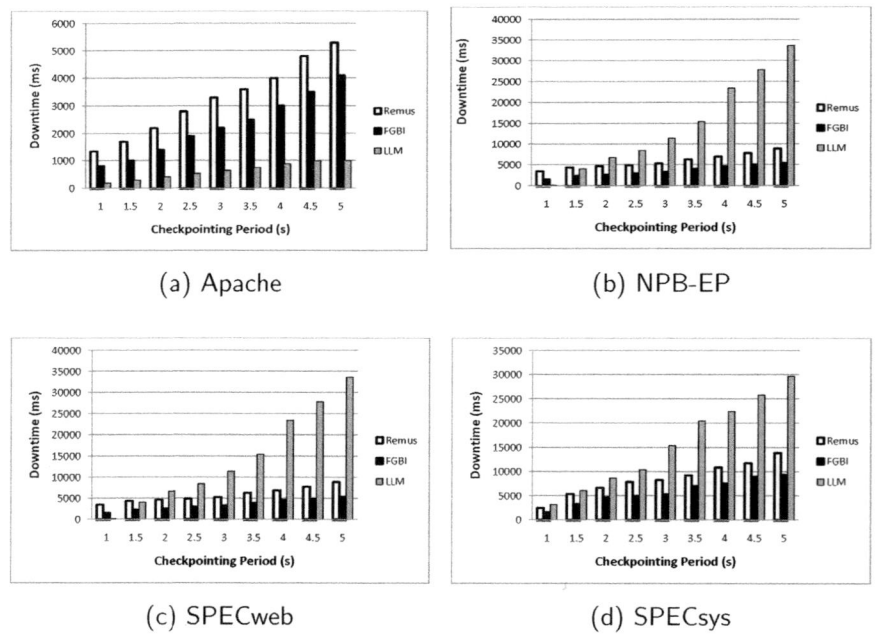

(a) Apache   (b) NPB-EP

(c) SPECweb   (d) SPECsys

Fig. 4. Type I downtime comparison under different benchmarks

Although SPECweb is a web workload, it still has a high page modification rate, which is approximately 12,000 pages/second [4]. In our experiment, the 1 Gbps migration link is capable of transferring approximately 25,000 pages/second. Thus, SPECweb is not a lightweight computational workload for these migration mechanisms. As a result, the relationship between FGBI and LLM in Figure 4c is more similar to that in Figure 4b (and also Figure 4d), rather than Figure 4a. In conclusion, compared with LLM, FGBI reduces the downtime by as much as 77%. Moreover, compared with Remus, FGBI yields a shorter downtime, by as much as 31% under Apache, 45% under NPB-EP, 39% under SPECweb, and 35% under SPECsys.

Type II Downtime. Table 1 shows the type II downtime comparison among Remus, LLM, and FGBI mechanisms under different applications. We have three main observations: (1) Their downtime results are very similar for the idle run. This is because, Remus is a fast checkpointing mechanism and both LLM and FGBI are based on it. Memory updates are rare during the "idle" run, so the type II downtime in all three mechanisms is short. (2) When running the NPB-EP application, the guest VM memory is updated at a high frequency. When saving the checkpoint, LLM takes much more time to save huge dirty data caused by its low memory transfer frequency. Therefore, in this case FGBI achieves a much lower downtime than Remus (more than 70% reduction) and LLM (more than 90% reduction). (3) When running the Apache application, the memory update is not so much as that when running NPB, but the memory update is definitely more than idle run. The downtime results shows that FGBI still outperforms both Remus and LLM.

Table 1. Type II downtime comparison

Application	Remus downtime	LLM downtime	FGBI downtime
idle	64 ms	69 ms	66 ms
Apache	1032 ms	687 ms	533 ms
NPB-EP	1254 ms	16683 ms	314 ms

4.3 Overhead Evaluations

Figure 5a shows the overhead during VM migration. The figure compares the applications' runtime with and without migration, under Apache, SPECweb, NPB-EP, and SPECsys, with the size of the fine-grained blocks varying from 64 bytes to 128 bytes and 256 bytes. We observe that in all cases the overhead is low, no more than 13% (Apache with 64 bytes block). As discussed in Section 3, the smaller the block size that FGBI chooses, greater is the memory overhead that it introduces. In our experiments, the smaller block size that we chose is 64 bytes, so this is the worst case overhead compared with the other block sizes. Even in this "worst" case, under all these benchmarks, the overhead is less than 8.21%, on average.

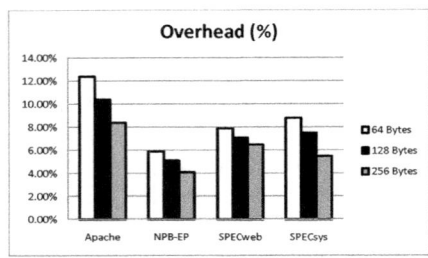 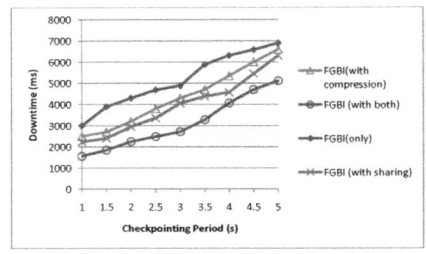

(a) Overhead under different block size.   (b) Comparison of proposed techniques.

Fig. 5. Overhead Measurements

In order to understand the respective contributions of the three proposed techniques (i.e., FGBI, sharing, and compression), Figure 5b shows the breakdown of the performance improvement among them under the NPB-EP benchmark. It compares the downtime between integrated FGBI (which we use for evaluation in this Section), FGBI with sharing but no compression support, FGBI with compression but no sharing support, and FGBI without sharing nor compression support, under the NPB-EP benchmark. As previously discussed, since NPB-EP is a memory-intensive workload, it should present a clear difference among the three techniques, all of which focus on reducing the memory-related overhead. We do not include the downtime of LLM here, since for this compute-intensive benchmark, LLM incurs a very long downtime, which is more than 10 times the downtime that FGBI incurs.

We observe from Figure 5b that if we just apply the FGBI mechanism without integrating sharing or compression support, the downtime is reduced, compared with that of Remus in Figure 4b, but it is not significant (reduction is no more than twenty percent). However, compared with FGBI with no support, after integrating hybrid compression, FGBI further reduces the downtime, by as much as 22%. We also obtain a similar benefit after adding the sharing support (downtime reduction is a further 26%). If we integrate both sharing and compression support, the downtime is reduced by as much as 33%, compared to FGBI without sharing or compression support.

5   Conclusions

One of the primary bottlenecks on achieving high availability in virtualized systems is downtime. We presented a novel fine-grained block identification mechanism, called FGBI, that reduces the downtime in lightweight migration systems. In addition, we developed a memory block sharing mechanism to reduce the memory and computational overheads due to FGBI. We also developed a dirty block compression support mechanism to reduce the network traffic at each migration epoch. We implemented FGBI with the sharing and compression mechanisms and integrated them with the LLM lightweight migration system. Our experimental evaluations reveal that FGBI overcomes the downtime disadvantage of LLM by more than 90%, and of Xen/Remus by more than 70%. In all cases, the performance overhead of FGBI is less than 13%.

Several directions for future work exist. It is possible to reduce the implementation complexity in the FGBI design. For instance, we can deploy some subcomponents (such as the Block Comparison part) in the VMM directly, and design a transparent solution by using the shadow page table mechanism. Moreover, compressing memory blocks that are unlikely to be accessed in the near future can further reduce the memory overhead.

Acknowledgment. We thank the anonymous reviewers for their very helpful comments. This work was supported by the IT R&D program of MKE/KEIT/ETRI, South Korea [K1001703, A Development of Cost Effective and Large Scale Global Internet Service Solution].

References

1. http://www.zlib.net
2. Akoush, S., Sohan, R., Rice, A., Moore, A.W., Hopper, A.: Predicting the performance of virtual machine migration. In: International Symposium on Modeling, Analysis, and Simulation of Computer Systems, pp. 37–46 (2010)
3. Barham, P., Dragovic, B., Fraser, K., Hand, S., Harris, T., Ho, A., Neugebauer, R., Pratt, I., Warfield, A.: Xen and the art of virtualization. In: Proceedings of the Nineteenth ACM Symposium on Operating Systems Principles, SOSP 2003, pp. 164–177. ACM, New York (2003)
4. Clark, C., Fraser, K., Hand, S., Hansen, J.G., Jul, E., Limpach, C., Pratt, I., Warfield, A.: Live migration of virtual machines. In: Proceedings of the 2nd Conference on Symposium on Networked Systems Design & Implementation, NSDI 2005, vol. 2, pp. 273–286. USENIX Association, Berkeley (2005)
5. Cully, B., Lefebvre, G., Hutchinson, N., Warfield, A.: Remus source code, http://dsg.cs.ubc.ca/remus/
6. Cully, B., Lefebvre, G., Meyer, D., Feeley, M., Hutchinson, N., Warfield, A.: Remus: high availability via asynchronous virtual machine replication. In: Proceedings of the 5th USENIX Symposium on Networked Systems Design and Implementation, NSDI 2008, pp. 161–174. USENIX Association, Berkeley (2008)
7. Ekman, M., Stenstrom, P.: A robust main-memory compression scheme. In: ISCA 2005: Proceedings of the 32nd Annual International Symposium on Computer Architecture, pp. 74–85. IEEE Computer Society, Washington, DC (2005)
8. Gupta, D., Lee, S., Vrable, M., Savage, S., Snoeren, A.C., Varghese, G., Voelker, G.M., Vahdat, A.: Difference engine: harnessing memory redundancy in virtual machines. Commun. ACM 53, 85–93 (2010)
9. Henson, V.: An analysis of compare-by-hash. In: Proceedings of the 9th Conference on Hot Topics in Operating Systems, vol. 9, pp. 3–3 (2003)
10. Henson, V., Henderson, R.: Guidelines for using compare-by-hash, http://infohost.nmt.edu/~val/review/hash2.pdf
11. Hines, M.R., Gopalan, K.: Post-copy based live virtual machine migration using adaptive pre-paging and dynamic self-ballooning. In: Proceedings of the 2009 ACM SIGPLAN/SIGOPS International Conference on Virtual Execution Environments, VEE 2009, pp. 51–60. ACM, New York (2009)

12. Huang, W., Gao, Q., Liu, J., Panda, D.K.: High performance virtual machine migration with RDMA over modern interconnects. In: CLUSTER 2007: Proceedings of the 2007 IEEE International Conference on Cluster Computing, pp. 11–20. IEEE Computer Society, Washington, DC (2007)
13. Jiang, B., Ravindran, B., Kim, C.: Lightweight Live Migration for High Availability Cluster Service. In: Dolev, S., Cobb, J., Fischer, M., Yung, M. (eds.) SSS 2010. LNCS, vol. 6366, pp. 420–434. Springer, Heidelberg (2010)
14. Jin, H., Deng, L., Pan, X.: Live virtual machine migration with adaptive, memory compression. In: 2009 IEEE International Conference on Cluster Computing and Workshops, pp. 1–10 (2009)
15. Liu, H., Jin, H., Liao, X., Hu, L., Yu, C.: Live migration of virtual machine based on full system trace and replay. In: HPDC 2009: Proceedings of the 18th ACM International Symposium on High Performance Distributed Computing, pp. 101–110. ACM, New York (2009)
16. Lu, M., Chiueh, T.C.: Fast memory state synchronization for virtualization-based fault tolerance. In: Dependable Systems Networks, pp. 534–543 (2009)
17. Nelson, M., Lim, B.H., Hutchins, G.: Fast transparent migration for virtual machines. In: ATEC 2005: Proceedings of the Annual Conference on USENIX Annual Technical Conference, pp. 25–25. USENIX Association, Berkeley (2005)
18. Bradford, A.F.R., Kotsovinos, E., Schioeberg, H.: Live wide-area migration of virtual machines including local persistent state. In: VEE 2007: Proceedings of the Third International Conference on Virtual Execution Environments, pp. 169–179. ACM Press, San Diego (2007)
19. Sun, Y., Luo, Y., Wang, X., Wang, Z., Zhang, B., Chen, H., Li, X.: Fast live cloning of virtual machine based on Xen. In: Proceedings of the 2009 11th IEEE International Conference on High Performance Computing and Communications, pp. 392–399. IEEE Computer Society, Washington, DC (2009)
20. Tamura, Y., Sato, K., Kihara, S., Moriai, S.: Kemari: Virtual machine synchronization for fault tolerance using DomT (technical report). http://wiki.xen.org/xenwiki/Open_Topics_For_Discussion? action=AttachFile&do=get&target=Kemari_08.pdf (June 2008)
21. Vrable, M., Ma, J., Chen, J., Moore, D., Vandekieft, E., Snoeren, A.C., Voelker, G.M., Savage, S.: Scalability, fidelity, and containment in the potemkin virtual honeyfarm. In: Proceedings of the Twentieth ACM Symposium on Operating Systems Principles, SOSP 2005, pp. 148–162. ACM, New York (2005)
22. Waldspurger, C.A.: Memory resource management in vmware esx server. SIGOPS Oper. Syst. Rev. 36, 181–194 (2002)
23. XenCommunity. Xen unstable source, http://xenbits.xensource.com/xen-unstable.hg
24. Zhao, M., Figueiredo, R.J.: Experimental study of virtual machine migration in support of reservation of cluster resources. In: VTDC 2007: Proceedings of the 2nd International Workshop on Virtualization Technology in Distributed Computing, pp. 5:1–5:8. ACM, New York (2007)

Towards Consistency Oblivious Programming

Yehuda Afek[1], Hillel Avni[1], and Nir Shavit[1,2]

[1] Tel-Aviv University, Tel-Aviv 69978, Israel
[2] MIT, Cambridge MA 02139, USA
hillel.avni@gmail.com

Abstract. It is well known that guaranteeing program consistency when accessing shared data comes at the price of degraded performance and scalability.

This paper initiates the investigation of consistency oblivious programming (COP). In COP, sections of concurrent code that meet certain criteria are executed without checking for consistency. However, checkpoints are added before any shared data modification to verify the algorithm was on the right track, and if not, it is re-executed in a more conservative and expensive consistent way. We show empirically that the COP approach can enhance a software transactional memory (STM) framework to deliver more efficient concurrent data structures from serial source code. In some cases the COP code delivers performance comparable to that of more complex fine-grained structures.

1   Introduction and Related Work

The need to maintain consistency when accessing data has been a major source of overhead in concurrent software and a great limitation on its scalability with respect to its matching sequential code.

There are in the literature concurrent algorithms that reduce the consistency overhead by traversing the data structure while ignoring all locks and meta-data. Examples are the concurrent lazy list [1] and skip list [2] algorithms that allow traversal operations to execute while ignoring the locks that are taken by threads modifying the structure. The traversal correctness is derived from properties of the structure and post validation. Another example is Lee's Java hash table [3], where a thread first traverses the bucket unsafely, and only if the key is not found, takes locks to guarantee consistency and then re-traverses it.

Our goal in this paper is to generalize this approach to derive a broader class of algorithms that execute without verifying consistency. We provide a methodology for designing concurrent algorithms in which we optimistically make a first attempt at executing an operation in a completely un-instrumented way. We will call this approach *consistency oblivious programming* (COP). This paper makes a first attempt at formulating COP and providing examples of its usefulness.

We base our COP generalization in part on using the software transactional memory (STM) programming paradigm. Transactional memory is a leading technique for simplifying concurrent programming. Software transactional memory

A. Fernández Anta, G. Lipari, and M. Roy (Eds.): OPODIS 2011, LNCS 7109, pp. 65–79, 2011.

systems suffer from major overheads because they maintain, in some form or another, a level of consistency among concurrently executing transactions. In [4] Shavit and Touitou allow a transaction to see an inconsistent state but from that point on the operation is considered a zombie, and will not complete successfully. Most modern STM algorithms [5,6,7,8] conservatively abort a transaction as soon as the possibility of inconsistency is detected. Others [5,7,9] force consistency by having even read-only operations check locks and global clocks or by maintaining multiple versions per address. Consistency, as many authors argue, simplifies the interaction of the STM algorithm with its environment, and allows simplified correctness proofs, yet it comes at a cost. Each and every access of a transaction to a shared variable or object must be instrumented in some way or another. This instrumentation is a major source of modern STM overhead, which in some cases can be reduced by algorithms such as the NORec STM [6]. These schemes avoid per object meta data, which reduces instrumentation overhead at the price of reduced scalability.

Our approach here is to provide a set of criteria that will allow a programmer to determine if a given data structure can be converted to work in the COP framework. If the conversion is possible, COP allows us to optimistically execute various operations on the data structure in native code, without any consistency checks, then test the outcome, and either retry or resort to a traditional STM based consistent execution if the earlier consistency oblivious ones failed.

1.1   COP and Acceptability Oriented Programming

Our approach follows along the lines of Rinard's acceptability oriented programming [10], where he introduces the idea of allowing programs to execute while making errors, attempting to recover from them only eventually.

In concurrent programs we consider sections of a thread's code (such as, but not only, transactional memory transactions) that can be viewed as having two types of sequential segments. A given section starts with a segment of code that does not affect the system, consisting usually (but not always) of reads, followed by a second segment that writes and updates the memory in addition to further reads. An example could be an insert operation in a data structure where it is first traversed to find the insert location and then writes to memory take place in order to implement the actual insertion of a new item.

In COP we add a third segment of code, called a Validating White-Box, between the first part and the updating part. The idea is to execute the first part, called the Black-Box, in a concurrent environment, as fast as possible without worrying about consistency of any form. Then, in the Validating White-Box, the values returned from the first part are checked to make sure they are in a consistent state with respect to the values returned from the first part. If the validation fails, then the entire operation re-starts.

We allow a transformed section of code to encompass multiple black and white boxes, as long as each black box is followed by its corresponding white box. It is interesting that a black box may include writes, as long as each write, by itself, is a valid addition to the system. We show an example for such writes later, in a union-find compression algorithm.

We say that the output of a black-box is *acceptable* if there exists a consistent execution of the black-box in which exactly the same outputs are generated by the black-box. As we do not check the consistency of the execution, we must extract other properties of the output that imply it is consistent with the system. We call these properties *acceptability properties*.

The programmer must first determine the acceptability properties of the output of the black-boxes and then devise tests that verify that these properties are satisfied given a set of outputs. If any of these tests fail, it means the output of the black-box is not acceptable. We call these tests *acceptability rules*.

We define our approach following Rinard and in the same way. Several activities characterize this approach:

- **Acceptability Property Identification:** The programmer identifies what in the output of the black-box marks it valid, i.e., what are the acceptability properties of the output. These properties are specific to each algorithm and implementation.
- **Enforcement:** Minimal set of acceptability rules is constructed to verify that the output of the black-box is acceptable. Once a rule is violated, it means an acceptability property is violated, and an action which fixes it is taken.
- **Monitoring:** The programmer produces components that enable the acceptability rules testing. These components must indicate whether a rule is violated in the output.

As mentioned, we split the concurrent code to boxes. The boxes terminology is borrowed again from Rinard, into COP, and relates to the above activities:

- **Black Box:** A native, sequential code section. We know the code does not crash or corrupt the system, yet we cannot trust its output.
- **White Box:** The code of this box monitors and enforces the acceptability rules on the outputs of the black box. It must enforce continuous correctness and undo any error that may occur.
- **Gray Box:** The code is not modified, but it is recompiled with synchronization, such as STM. Gray boxes are self contained as they are synchronized and are thus concurrency safe.

The performance advantage of COP is that the black boxes run without any synchronization. Our goal is to have as much code as possible in the black boxes, and as little of it as possible in the white and gray ones. The size of a box is its execution time.

The remainder of the paper is structured as follows. In Section 2 we present the conditions and justifications for the use of COP, and we use an example to demonstrate where it can be most affective. Then in section 3 we show several COP algorithms. In section 4 we evaluate the performance of the COP, and conclude in Section 5.

2   When Can We Use COP

To determine if one can use the COP approach, we must check if a breakup into the three boxes is possible. When checking a sequential section of code, to see if it can be in a black box, we need to examine each shared variable access in that function. If all accesses are reads If the access is a write, we must make sure that it is:

− Based only on committed data, and
− any single write is a valid modification, so we do not need synchronization. If a write is tentative, it must be reversible, and can not be visible, thus it must be instrumented.

The COP un-instrumented write accesses are immediately published and can not be undone. Thus, if the written values are calculated according to un-instrumented reads from transactional addresses, we have to use a commit time locking STM in the system. This means when a value is recorded in a transactional address, the value will be final and ready to use. We also need to see that any subset of the un-instrumented writes is a valid addition to the current state.

An un-instrumented read can be added if its encapsulating data structure precludes:

− permanent loops, even in deleted objects.
− un-initialized pointers, even for an object that is not yet fully connected.

Since COP does not validate consistency, an infinite loop may go undetected, unless it is caught by the serial algorithm. We notice that the second condition applies partially to any STM: this is the privatization problem of [11]. If transaction T_1 reached a node N during traversal and went to sleep, and then another transaction, T_2 freed N, then T_1 could wake up and read N's next pointer. If N is actually freed by T_2, T_1 may prompt an exception. Currently this problem is best solved by quiescence barriers [12].

When inspecting a data structure to see if it is fit for COP, the challenging part is to identify the set of acceptability checks that it requires. In the next section we show useful and common data structures that benefit from COP.

Any COP code has the layout of Algorithm 1 below. It executes a black box (BB) which generates some output V. Then a white box (WB) is used to verify V is acceptable, and finally it executes a gray box (GB), which is compiled with full synchronization using V to complete the operation. Multiple such operations may reside in the same transactions.

The gray box may be empty, but if we have a black box we must have a white box following it. If there are no black-boxes in a transaction it is not related to COP.

Algorithm 1. COP

1: $V \leftarrow$ BB
2: **if** failed(WB(V)) **then**
3:    **restart**
4: **end if**
5: **if** exists(GB(V)) **then**
6:    GB(V)
7: **end if**

3 COP Algorithms

This section shows the viability of the COP approach by way of a series of examples, presenting COP versions of a linked list data structure, a union-find data structure, and a linked, bottom balanced variant of a red-black tree data structure. The latter structure is a fundamental data structure [13] not known to be parallelizable before.

3.1 COP Linked List

In Algorithm 2 we show the delete function of the COP linked list. In the white boxes we assume the existence of an STM and use **txld**(address) and **txst**(address, value) which are respectively a transactional load and a transactional store. The transactional operations are affecting the read / write set, are validated, and if necessary, aborted, according to the used transactional memory algorithm.

Note that this list supports the standard single node insert, delete, and lookup operations. When we reach a node which has the desired value or greater, we

Algorithm 2. COP WBLookup, which deletes a node from a linked list.

1: $val, n, prev \leftarrow$ BBLookup(ValToDelete))
2: //n is the last traversed node, val is its value and $prev$ is its predecessor
3: **if** **txld**(n.val)$\neq$ val **then**
4:    Abort
5: **end if**
6: **if** **txld**($prev.next$) $\neq$n **then**
7:    Abort
8: **end if**
9: **if** **txld**($prev.val$) $\geq$ValToDelete **then**
10:    Abort
11: **end if**
12: **if** **txld**($n.val$) $= ValToDelete$ **then**
13:    **txst**(n.val, MAXVAL)
14:    **txst**(n.next, n)
15:    GBDelete(n)
16: **end if**

return that node together with its value and its predecessor at line 2. The addition we make to the serial algorithm is the setting of the node's deleted value to MAXVAL in line 13 to prevent BBLookup from continuing looping in the deleted node. Deleted nodes are pointing to themselves, and if their value are less than the looked up one, the code would get trapped in them.

The acceptance rules are that a successfully found node has the value it is supposed to have in line 5 and is still pointed-to by its predecessor in line 8. The predecessor, in turn, must have a lower value than the one to be deleted, as checked in line 11. If we would not have modified the deleted node to point to itself, it would pass an STM load as consistent, and no white box will be able to notice the node is deleted. If the deleted value would not be set to MAXVAL, the lookup would hit an infinite loop when a lookup is trapped in a concurrently deleted node.

3.2   COP Union Find

The union-find algorithm maintains disjoint sets under union. Each set is represented by a rooted tree whose nodes are the elements of the set. The root is called the representative of the set. The representative may change when the tree is updated by a union operation. The data structure provides two operations: find, which follows the path from the element to the representative and returns it, and union, which links two set representatives by making one point to the other.

The union function calls find for the two elements to be unified and then uses symmetry breaking to decide the direction of the link to be installed. We will look in the find. Actually it is called the find-compress function as it also compresses the path from the element to the representative.

The way FindCompress works is that it follows the next pointers of nodes to their parents until it hits a representative. If a node points to itself, then this node is the representative of its set. Otherwise, if the parent is not the representative, a compression occurs. In compression, the nodes' next pointer is rewritten to point to the grandparent of the node.

Algorithm 3 is searching for the representative of x and compresses the path of x to its representative. Forest is an array of nodes where each node has a

---

Algorithm 3. Transactional FindCompress in union-find.

```
1:  next = txld(forest[x].next)
2:  while x ≠ next do
3:      t = next
4:      t_next=txld(forest[next].next)
5:      txst(forest[x].next, forest[t].next)
6:      x=t
7:      next=txld(forest[t].next)
8:  end while
9:  return x
```

next field which is the index of its representative, or a member from its set that is closer to the representative. The interesting thing is that as long as all data read is valid, the write in line 5 does not need any instrumentation, because no matter what value is written by a correct reader, it will perform some helpful compression. However, if transaction T_1 wrote a link and that link was used by T_2 in compression, and then T_1 is aborted, the compression performed will be wrong. Thus, for algorithm 4, we use a commit time STM, which keeps written values in a buffer. BBFindCompress is the transactional Algorithm 3, replacing each instrumented access with an uninstrumented one.

Algorithm 4. COP WBFindCompress in union-find.

```
1: repeat
2:     x← BBFindCompress(x)
3:     y←txld(forest[x].next)
4:     if x≠y then
5:         x←y
6:         continue
7:     end if
8: until x=y
9: return  x
```

Transaction T which calls WBFindCompress(x) gets the representative of x, but then verifies it is the representative by reading the next link transactionally in line 3. Note that there is no read-after write hazard, as any transactional write will replace a self pointing pointer, and a self pointing pointer is always read transactionally eventually.

Now, either x is still pointing to itself, or it was changed within T, or T read it transactionally for the first time, and it is different. This will keep the function running, or it will be read transactionally for the second time by T. Now if x was changed by another transaction, T will abort.

As we will show empirically, the COP version of union-find performs well when there are many unions which introduce memory contention, because in that situation it saves cycles. When there are mostly finds, the path length to a representative becomes one, and there the COP and STM have the same overhead and performance. However, in applications, usually there is a burst of unions when a new network structure is constructed and then the structure becomes read only, where instrumentation can be eliminated by a barrier [14]. Thus, the high contention scenario seems more important.

3.3   COP Red-Black Tree

The next example for COP is a balanced binary red-black (RB) search tree, which is balanced bottom up. Balancing the tree from the bottom, makes the balancing effect unpredictable, so no locking mechanism, except global lock, can be used to parallelize it. The algorithm is taken from [13], where the interface is:

- **Lookup**(K, Tree) which searches the tree and returns the node with key K, or its successor if it is missing as a left son, or its predecessor in case K is missing as a right son.
- **Insert**(K,Tree) which runs Lookup and inserts the node with key K in case it is missing.
- **Delete**(K, Tree) which runs Lookup and deletes K in case it is present.

As can be seen, Lookup is the main part of all functions. Luckily, we can fit it into a black box. We verify that in the serial implementation of the RB tree, no uninitialized pointers are visible and no permanent loops are created. The challenge is then to add the appropriate white box with fitting acceptance rules.

During balancing there are two problematic states that should be addressed. A node can be temporarily detached and missed, or a successfully found node can be deleted by a concurrently executing transaction.

To solve the first issue we exploit the fact that a binary search tree has a total order on its keys and that a lookup in any binary search tree always arrives to the target node from the node with the predecessor or the successor key, depending on whether the key is the left or right son of its parent. In the white box we connect all nodes in a doubly linked list of successor - predecessor. Than we add the acceptance rule that if a node is found missing, then its predecessor must be connected to its successor or vice versa. We maintain the list with synchronization so we know it is correct. The list is not changed during balancing, and takes only a few accesses to maintain. We also have to verify that the node is missing in the parent we found. To determine this we verify the place where the node was supposed to be is still null. The problem of successfully found deleted nodes is solved with a transactionally maintained live mark, which is set by Insert and reset by Delete operations.

Algorithm 5 is the code for the Lookup operation which uses the original RB tree Lookup as a black box and adds the acceptance rules in order to make it always return a node that is valid, and which can be used safely by the other operations. In the code, we first check that the node was not removed by reading its live indication in line 8. Then in lines 14, 25, 25 and 28 we check the predecessor or the successor of the node, to verify that the requested key falls in the gap between them and is thus definitely missing. In lines 20 and 31 we verify the node is not only missing, but is missing in the found location.

In Figure 1 we see a lookup for key 26 that starts when the tree is unbalanced, and goes from the root which is 20 to 30 and to 27, then the tree is balanced, and the lookup continues through 27 back to 20, then to 25 and finally to 26. During the lookup we go through the right link of 20 to two different nodes. Before balancing we go to 30 and after we go to 25. This operation would abort on all known single version STM algorithms. The only known STM that will endure it is the multi version [9] which will introduce high overhead. As said, no existing locking protocol, including the new DL [15], will tolerate such a rebalancing move during lookup either.

In Algorithms 6 and 7, we see that Delete and Insert in the RB tree, which use the WB lookup and then insert or delete the node in the list in line 4 and set or unset its mark in line 5. Though it is a subjective matter, we found that the amount of work required to convert the bottom balancing serial RB tree we started from, into an efficient concurrent algorithm, is small and the correctness of it is easy to verify. If we wanted to keep all these features in a fully hand crafted algorithm we would end up spending a lot of work on and would involve a complex proof. Thus another good side of COP is that it is engineer friendly and saves work as well as inevitable bugs.

4   Evaluation

We first evaluated COP for the union-find algorithm. We started with an implementation of the classical union-find serial algorithm from [13] and transactified it using the TL2 STM [5]. We then created the COP version of the union-find according to the description in section 3.2. We compare the above to an optimized implementation of the parallel union-find algorithm from [16].

Next we examined the performance of COP for RB tree. We started with the classical red-black tree from [13] as it was transactified by Dave Dice. We derived the COP RB tree as described in 3.3. We compared the COP to the TL2-Enc [5] and NORec [6] STM, early release elastic transactions [17], and to Hanke's concurrent RB tree algorithm [18]. Later we will focus on comparing the COP RB tree to STMs and elastic transactions, because Hanke's tree is relaxed and top-down balanced, so its comparison to the bottom up and perfectly balanced RB tree from [13] is misleading.

Experimental setup: We collected results on two hardware platforms: a Sun SPARC T5240 and an Intel Core i7. The Sun is a dual-chip machine, powered by two UltraSPARC T2 Plus (Niagara II) chips. The Niagara II is a chip multithreading (CMT) processor, with 8 1.165 GHz in-order cores with 8 hardware strands per core, for a total of 64 hardware strands per chip. The Core i7 920 processor in the Intel machine holds four 2.67GHz cores that each multiplex 2 hardware threads.

4.1   Union Find

As explained in section 3.2, each representative lookup compresses the path from the element to the representative, so after a certain level of find operations, almost all paths are of length one. In this situation the COP and the transactional algorithm will perform the same number of instrumentations and their performance will be the same. However, usually the union-find is used to build a network and then is used in read only mode, so the practically important scenario is when the amount of unions is high.

Algorithm 5. COP WBLookup in RB tree.

1: $n \leftarrow$ BBLookup(K, Tree)
2: **if** **txld**(n.live)$\neq$ TRUE **then**
3:     Abort
4: **end if**
5: **if** $n.key = K$ **then**
6:     **if** **txld**($n.key$) $\neq K$ **then**
7:         Abort
8:     **end if**
9:     **return**
10: **else**
11:     **if** $n.key < K$ **then**
12:         **if** **txld**($n.key$) $> K$ **then**
13:             Abort
14:         **end if**
15:         **if** **txld**(**txld**($n.successor$).key) $< K$ **then**
16:             Abort
17:         **end if**
18:         **if** **txld**($n.right$) $\neq \perp$ **then**
19:             Abort
20:         **end if**
21:     **end if**
22: **else**
23:     **if** **txld**($n.key$) $< K$ **then**
24:         Abort
25:     **end if**
26:     **if** **txld**(**txld**($n.predecessor$).key) $> K$ **then**
27:         Abort
28:     **end if**
29:     **if** **txld**($n.left$) $\neq \perp$ **then**
30:         Abort
31:     **end if**
32: **end if**
33: **return** n

Algorithm 6. Insert in RB Tree.

1: $n \leftarrow$ WBLookup(key)
2: GBAddToTree(new_node)
3: GBBalanceTree
4: WBInsertToList(new_node)
5: **txst**($new\_node.live$, **true**)

Algorithm 7. Delete in RB Tree.

1: $n \leftarrow$ WBLookup(key)
2: GBRemoveFromTree(new_node)
3: GBBalanceTree
4: WBRemoveFromList(new_node)
5: **txst**($new\_node.live$, **false**)

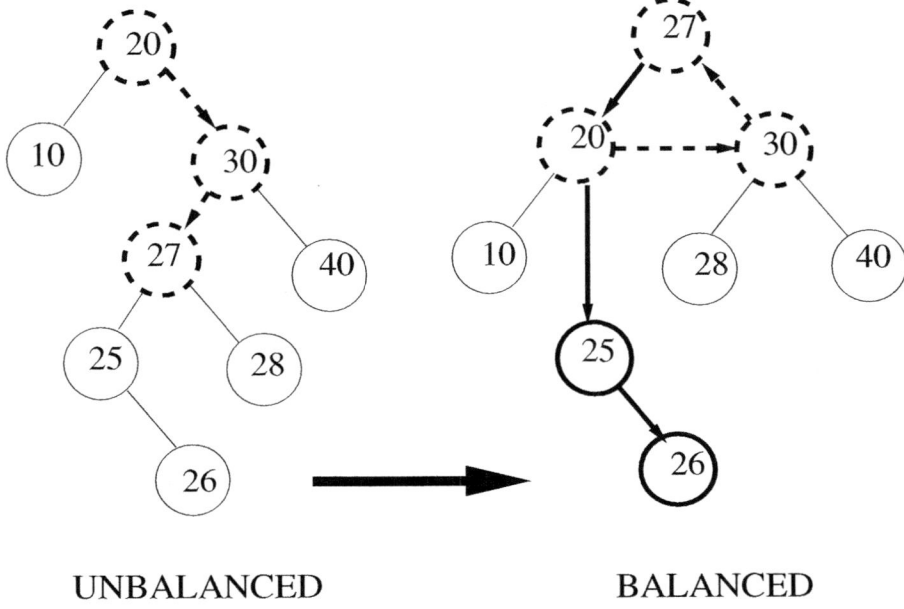

UNBALANCED   BALANCED

Fig. 1. COP RB Tree is finding a key 26 during rebalancing

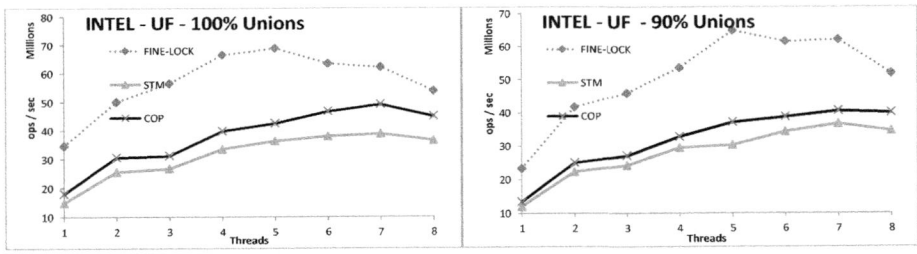

Fig. 2. Union Find with different amount of unions, benchmark results on Intel Nehalem machine with 4 cores, each multiplexing 2 hardware threads

In Figure 2, most 90-100 percents of the operations are the mutating union. As seen, when the amount of union operations is high, the COP performance is closer to the optimized hand crafted concurrent algorithm than to the STM version. We note that the COP is simple to develop and is composable and may participate in STM transactions.

4.2   Red Black Tree

Our bottom up and perfectly balanced serial algorithm does not have a concurrent, non-transactional version. Thus, we chose the top down, locally balanced Hanke algorithm as an upper bound for the performance of the concurrent red-black tree.

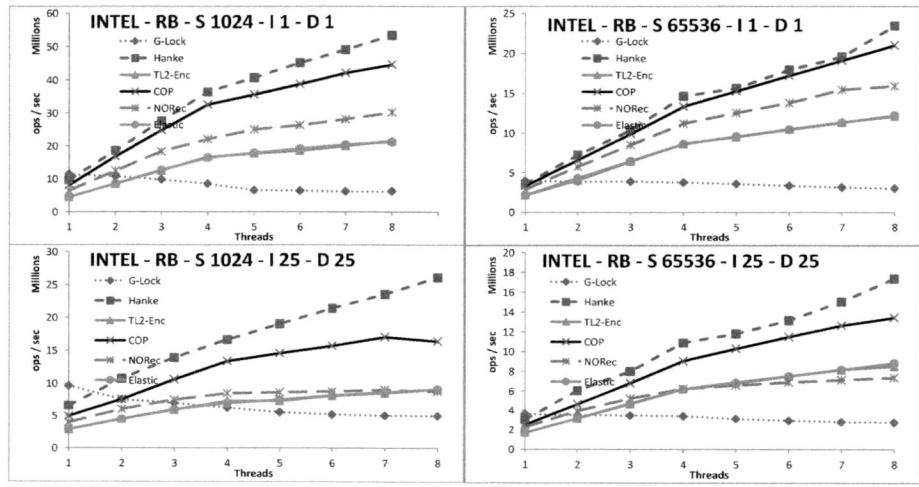

Fig. 3. RB Tree benchmark results for 1024 and 65536 nodes on Intel Nehalem machine. The benchmarks have 50% lookups with 25% inserts and 25% deletes or 98% lookups with 1% inserts and 1% deletes.

In Figure 3 we show the COP RB Tree throughput compared to Hanke and a global lock as upper and lower performance bounds. In addition, we compare to the NORec, TL2-Enc and Elastic STM algorithms. The results show that COP consistently outperforms all other transactional variations, for all tested combinations of tree size and contention level. However, it is not as good as the more loosely balanced Hanke algorithm.

In Figure 4 we compare throughput in the first row and cycles per operation in the second row, and see a perfect match. The lower the count, the higher the throughput. In this case the COP algorithm has the lowest count and respectively has about twice the throughput.

As our COP RB tree contains links, which can be useful for range queries, we present its comparison to the STM version of RB tree with links, i.e., trees whose nodes are chained in a predecessor - successor doubly linked list, in Figure 5. As expected, the COP wins by a large margin.

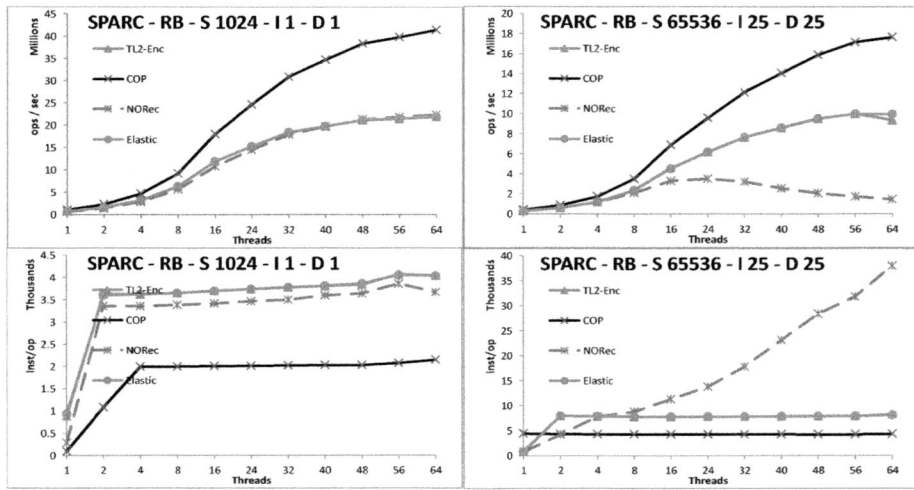

Fig. 4. RB Tree of 1024 and 65536 nodes. Graphs show rate of completed operations per second and hardware performance instruction counter per operation. The benchmarks have 50% lookups with 25% inserts and 25% deletes or 98% lookups with 1% inserts and 1% deletes.

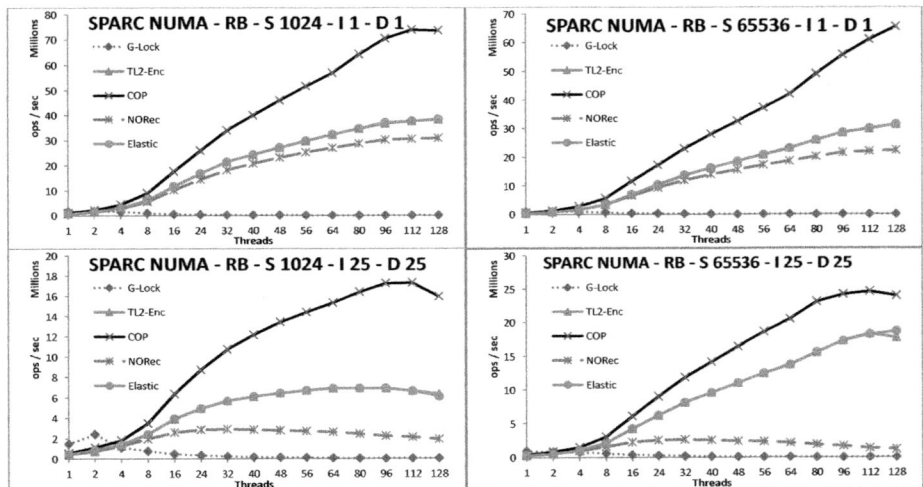

Fig. 5. RB Tree of 1024 and 65536 nodes with 50% lookups and 25% inserts and 25% deletes or 98% lookups with 1% inserts and 1% deletes when using both processors of the Sun machine (i.e., NUMA configuration). Default OS policy is to place threads on chips in round robin order.

5 Conclusion

In this paper we introduced consistency oblivious programming, an approach for efficient parallelization of serial code. It manages to reduce significantly the footprint of synchronization compared to automatic transactification or coarse locking.

The reasoning needed to parallelize an RB tree using COP is negligible, while its performance compares, in some cases, to that of the celebrated concurrent algorithm of Hanke [18]. In contrast, locking protocols such as hand over hand or DL [15] are not as flexible as COP and are not applicable, for example, to the RB tree we converted.

All the COP algorithms we presented are composable and have controllable number of transactional accesses, characteristics which make COP HTM friendly.

References

1. Heller, S., Herlihy, M., Luchangco, V., Moir, M., Scherer III, W.N., Shavit, N.: A lazy concurrent list-based set algorithm. Parallel Processing Letters 17(4), 411–424 (2007)
2. Herlihy, M.P., Lev, Y., Luchangco, V., Shavit, N.N.: A Simple Optimistic Skiplist Algorithm. In: Prencipe, G., Zaks, S. (eds.) SIROCCO 2007. LNCS, vol. 4474, pp. 124–138. Springer, Heidelberg (2007)
3. Lea, D.: The java.util.concurrent synchronizer framework. Science of Computer Programming 58(3), 293–309 (2005); Special Issue on Concurrency and synchonization in Java programs
4. Shavit, N., Touitou, D.: Software transactional memory. Distributed Computing 10(2), 99–116 (1997)
5. Dice, D., Shalev, O., Shavit, N.: Transactional Locking II. In: Dolev, S. (ed.) DISC 2006. LNCS, vol. 4167, pp. 194–208. Springer, Heidelberg (2006)
6. Dalessandro, L., Spear, M.F., Scott, M.L.: Norec: streamlining stm by abolishing ownership records. In: PPOPP, pp. 67–78 (2010)
7. Riegel, T., Fetzer, C., Felber, P.: Time-based transactional memory with scalable time bases. In: SPAA, pp. 221–228 (2007)
8. Herlihy, M., Luchangco, V., Moir, M., Scherer III, W.N.: Software transactional memory for dynamic-sized data structures. In: Proceedings of the Twenty-Second Annual Symposium on Principles of Distributed Computing, PODC 2003, pp. 92–101. ACM, New York (2003)
9. Afek, Y., Morrison, A., Tzafrir, M.: Brief announcement: view transactions: transactional model with relaxed consistency checks. In: PODC, pp. 65–66 (2010)
10. Rinard, M.C.: Acceptability-oriented computing. In: OOPSLA Companion, pp. 221–239 (2003)
11. Shpeisman, T., Menon, V., Adl-Tabatabai, A.R., Balensiefer, S., Grossman, D., Hudson, R.L., Moore, K.F., Saha, B.: Enforcing isolation and ordering in stm. SIGPLAN Not. 42, 78–88 (2007)
12. Matveev, A., Shavit, N.: Implicit privatization using private transaction. Transact. (2010)
13. Cormen, T.H., Leiserson, C.E., Rivest, R.L., Stein, C.: Introduction to Algorithms, 2nd edn. The MIT Press and McGraw-Hill Book Company (2001)

14. Scott, M.L., Spear, M.F., Dalessandro, L., Marathe, V.J.: Delaunay triangulation with transactions and barriers. In: Proceedings of the 2007 IEEE 10th International Symposium on Workload Characterization, IISWC 2007, pp. 107–113. IEEE Computer Society, Washington, DC (2007)
15. Golan-Gueta, G., Bronson, N., Aiken, A., Ramalingam, G., Sagiv, M., Yahav, E.: Automatic fine-grain locking using shape properties. In: SPLASH, pp. 194–208 (2011)
16. Schrijvers, T., Frühwirth, T.: Optimal union-find in constraint handling rules. Theory Pract. Log. Program. 6, 213–224 (2006)
17. Felber, P., Gramoli, V., Guerraoui, R.: Elastic Transactions. In: Keidar, I. (ed.) DISC 2009. LNCS, vol. 5805, pp. 93–107. Springer, Heidelberg (2009)
18. Hanke, S.: The Performance of Concurrent Red-Black Tree Agorithms. In: Vitter, J.S., Zaroliagis, C.D. (eds.) WAE 1999. LNCS, vol. 1668, p. 287. Springer, Heidelberg (1999)

Regret Freedom Isn't Free

Edmund L. Wong[1], Isaac Levy[1], Lorenzo Alvisi[1],
Allen Clement[2], and Mike Dahlin[1]

[1] Department of Computer Science, The University of Texas at Austin
[2] Max Planck Institute for Software Systems
{elwong,isaac,lorenzo,dahlin}@cs.utexas.edu, aclement@mpi-sws.org

Abstract. Cooperative, peer-to-peer (P2P) services—distributed systems consisting of participants from multiple administrative domains (MAD)—must deal with the threat of arbitrary (Byzantine) failures while incentivizing the cooperation of potentially selfish (rational) nodes that such services rely on to function. This paper investigates how to specify conditions (*i.e.*, a solution concept) for rational cooperation in an environment that also contains Byzantine and obedient peers. We find that *regret-free* approaches—which, inspired by traditional Byzantine fault tolerance, condition rational cooperation on identifying a strategy that proves a best response regardless of how Byzantine failures occur—are unattainable in many fault-tolerant distributed systems. We suggest an alternative *regret-braving* approach, in which rational nodes aim to best respond to their *expectations* regarding Byzantine failures: the chosen strategy guarantees no regret only to the extent that such expectations prove correct. While work on regret-braving solution concepts is just beginning, our preliminary results show that these solution concepts are not subject to the fundamental limitations inherent to regret freedom.

1 Introduction

Traditional fault-tolerant distributed computing relies on the assumption that nodes can be cleanly categorized as correct or faulty; the former can be counted on to run protocols that guarantee that systems will continue to provide desirable functionalities despite a limited number of the latter. The rise of cooperative, peer-to-peer (P2P) systems spanning multiple administrative domains (MAD) complicates this simple picture. Much evidence suggests that a large number of peers in MAD services will free-ride (*e.g.*,[5,24,36]) or deviate from the assigned protocol if it is in their interest to do so (*e.g.*, [1,36]). To maintain the service, it is essential to give these peers sufficient incentives to cooperate, and informal common-sense reasoning about incentives may still leave systems vulnerable to strategic attacks (*e.g.*, [27,30,35]). But what should be the basis for a rigorous treatment of MAD systems?

There is little controversy about the failure model. It is clear that one cannot simply assume that every peer will be rational, as in standard game theory: like

A. Fernández Anta, G. Lipari, and M. Roy (Eds.): OPODIS 2011, LNCS 7109, pp. 80–95, 2011.

other distributed systems, P2P services are susceptible to arbitrary failures.[1] And, of course, some peers may simply be happy to run whatever protocol is assigned to them—similar to correct nodes in traditional distributed systems. P2P services should hence be designed to function in environments consisting of a mix of Byzantine, acquiescent,[2] and rational (or *BAR*) participants.

Building a BAR-tolerant system then involves two steps: 1) designing a Byzantine fault-tolerant protocol and 2) proving that rational peers will cooperate and follow the prescribed protocol. But how does one specify the conditions, *i.e.*, the *solution concept* [17], under which rational peers will be willing to cooperate?

A natural approach is to draw inspiration from traditional Byzantine fault-tolerant (BFT) computing. In threshold-based BFT, as long as the number of Byzantine nodes does not exceed a threshold t, the system is guaranteed to provide its safety properties *independent of who the t Byzantine nodes are and how they behave*. Similarly, it is appealing to aim for a notion of equilibrium in which rational nodes—either unilaterally or as a part of a coalition—cannot improve their utility by deviating *independent of who the t Byzantine nodes are and of how they behave*. This approach, elegantly formalized in the notion of (k,t)-robustness [3,4], is in principle very attractive: at equilibrium, peers will never have reason to regret their chosen strategy, which is guaranteed to prove a best response to any Byzantine strategy, independent of the identities of Byzantine nodes.

The main result of this paper is to show that, despite its appeal, a solution concept that guarantees *regret freedom* is fundamentally unable to yield non-trivial equilibria in games (which we name *communication games*) that capture three key characteristics of many practical fault-tolerant distributed systems: (a) to achieve some desired functionality, some nodes need to communicate; (b) bandwidth is not free; and (c) the desired functionality can be achieved despite t Byzantine failures.

More, we find that weakening (k,t)-robustness, even considerably, seems unlikely to help. For example, suppose that, magically, all rational nodes in a communication game knew precisely the identity of all Byzantine nodes (but not their strategy); or, alternatively, that they knew their strategy (but not their identities). We find that in both cases a regret-free equilibrium can be achieved only under very limited circumstances.

These results are not interesting because of their proofs, which are straightforward, but because they show that in fault-tolerant distributed systems, conditioning rational cooperation on the expectation of regret freedom may be fundamentally too much to ask. Furthermore, the limitations of this approach

---

[1] Of course, arbitrarily faulty peers too can be modeled as rational peers who follow an unknown utility function. Unfortunately, doing so does not simplify the problem.

[2] We originally named these nodes *altruistic* [6] but have since been made aware [2] of the risk of confusing such peers (whose irrational generosity is only driven by obedience to the given protocol) with peers who are irrationally generous for arbitrary reasons. We believe that "acquiescent" better captures our original intentions.

appear hard to fix, since they are rooted in the universal quantifiers (*e.g.*, "for all strategies" or "for all sets of t Byzantine nodes") that are at the very essence of regret freedom.

The second part of the paper points to a promising research agenda to overcome this impasse, an approach we call *regret braving*. Regret braving is motivated by the observation that rational agents that operate under uncertainty about the strategy of other players (as is the case when players are Byzantine) are often willing to cooperate without requiring absolute regret freedom. For instance, when stock traders buy or sell shares, they are well aware of the possibility of regretting their actions. Nonetheless, they follow a particular strategy as long as they cannot improve their utility with respect to their expectation about their environment—the worth of the traded asset, their comfort with risk, and what they believe will be the trends in the market—by deviating. Similarly, we consider solution concepts in which rational nodes aim to best respond to their *expectations* regarding Byzantine failures: the chosen strategy guarantees no regret only to the extent that such expectations prove correct.

We find that regret-braving solution concepts admit simple and intuitive equilibria for communication games where even the weakened versions of (k, t)-robustness could not. In particular, we consider two solution concepts: in the first, rational nodes play a maximin strategy that guarantees the best worst-case outcome despite any possible Byzantine failure; in the other, rational nodes assign probabilities to various possible faulty behaviors and aim for a Bayesian equilibrium. We do *not* suggest that these solution concepts are the "right" ones or that they can be directly applied to every BAR-tolerant system; in fact, we believe that an exciting research opportunity lies in identifying increasingly realistic models for Byzantine failure expectations. What these preliminary results do show, however, is that regret-braving solution concepts are not subject to the fundamental limitations inherent to regret freedom.

The paper proceeds as follows. Section 2 formalizes how we model players and introduces the communication game that we use to compare solution concepts. Section 3 explores the land of the (regret) free, showing why equilibria that base rational cooperation on regret freedom are fundamentally hard to achieve. Section 4 describes instead the home of the (regret) brave: we discuss two models of rational beliefs that admit useful equilibria in an instantiation of the communication game. Section 5 discusses related work, and Section 6 concludes.

2   Model

A *communication game* models any fault-tolerant system in which communication is not free and at least some nodes need to communicate in order to achieve the desired functionality.

Definition 1. *A communication game consists of some set of nodes $N = \{1, \ldots, n\}$ in which*

- *Communication incurs some cost and does not generate immediate benefit to the sender,*

- *Communication incurs some cost to the receiver, and*
- *Benefit is obtained from functionality that (a) can be achieved in the presence of up to t < n Byzantine failures and (b) requires communication between some pair of nodes.*

For simplicity, we use the same communication cost γ for both sending and receiving, and we assume that messages are never lost.

Protocols are strategies played in the communication game, and strategies involve actions drawn from a non-empty, finite set. We refer to the service-assigned protocol as the *assigned strategy*. A *strategy profile* $\sigma = (\sigma_x)_{x \in N}$ assigns a strategy σ_x to each node x, and Σ denotes the space of all possible strategy profiles σ that nodes may use. Every strategy profile σ results in some utility $U_x(\sigma)$ for every node x. Following common game theory notation, we use (σ_x', σ_{-x}) to denote the strategy profile in which x plays σ_x' and everyone else plays their component in σ (we also do this for sets of players, e.g., (σ_K', σ_{-K})), and we drop redundant parentheses when using a strategy profile as a parameter to a utility function, e.g., $U_x(\sigma_x', \sigma_{-x})$ vs. $U_x((\sigma_x', \sigma_{-x}))$. We primarily focus on *non-trivial* strategy profiles, in which some positive utility is expected for at least one node; this implies that some communication must occur.

We are interested in systems that include Byzantine, rational, and (optionally) acquiescent nodes; each node x belongs to a *type* θ_x that falls into one of these groups. For simplicity, we assume that all rational nodes are of the same type R, and we ignore acquiescent nodes (who would anyway follow any strategy assigned to them). These assumptions do not affect our impossibility results, and they simplify the analysis for the positive results in regret braving—which, as in any game-theoretic analysis, depend on the types of players and solution concept. Because a Byzantine node may potentially play one of many different strategies, it is convenient to denote the node's type using the strategy it plays. Formally, if some Byzantine node z plays some strategy τ_z, then we say that $\theta_z = \tau_z$; the type space Θ then consists of $\Sigma \cup \{R\}$.

We focus on environments in which neither trusted hardware nor trusted third-parties are used to monitor communication. Although a trusted mediator is useful [10,23,38], it is often impractical or even infeasible to provide one, and in practice few cooperative systems leverage trusted hardware to prove communication. We express this reality in the following assumption:

Assumption 2. it A node that sent a message m cannot unilaterally prove that it sent m.

3   Byzantine Regret Freedom in Communication Games

In BFT systems, safety properties hold regardless of how Byzantine failures occur. Ideally, one would like rational cooperation to be achieved under similarly strong guarantees. (k, t)-robustness [3,4] is an elegant solution concept that captures this attractive intuition. A (k, t)-robust equilibrium is completely impervious to the actions of Byzantine nodes: rational nodes will never have to second-guess their decision even if the identities and strategies of the Byzantine nodes become known. Specifically, (k, t)-robustness offers two key properties. The first, *t-immunity* [3], captures the intuition that nodes following a strategy profile should not be adversely affected by up to t Byzantine failures.

Definition 3. *A strategy profile σ is t-immune if, for all $T \subseteq N$ such that $|T| \leq t$, all strategy profiles τ, and $x \notin T$,*

$$U_x(\sigma_{-T}, \tau_T) \geq U_x(\sigma)$$

Note that t-immunity is *not* equivalent to Byzantine fault-tolerance, as t-immunity does not specify that a strategy profile σ must provide any sort of desirable safety or liveness properties despite t faults. In fact, any σ, fault-tolerant or not, is t-immune if it specifies actions so bad that Byzantine nodes, playing anything other than σ, cannot hurt a player's utility.

The second, *k-resilience* [3], addresses the possibility of collusion: a k-resilient strategy guarantees that a coalition of size at most k cannot deviate in a way that benefits every member.[3]

Definition 4. *A strategy profile σ^* is k-resilient if, for all $K \subseteq N$ such that $|K| \leq k$, there exists no alternate strategy profile σ' such that for all $x \in K$,*

$$U_x(\sigma'_K, \sigma^*_{-K}) > U_x(\sigma^*)$$

The (k, t)-robustness solution concept is the combination of t-immunity, k-resilience, *and regret freedom with respect to Byzantine failure*: regardless of how Byzantine failures occurs, (k, t)-robustness guarantees that no coalition of at most k nodes can ever do better than following the equilibrium strategy.

Definition 5. *A strategy profile σ^* is a (k, t)-robust equilibrium if σ^* is t-immune and, for all (a) $K, T \subseteq N$ such that $K \cap T = \emptyset$, $|K| \leq k$, and $|T| \leq t$, and (b) strategy profiles τ, there does not exist an alternate strategy profile σ' such that for all $x \in K$,*

$$U_x(\sigma'_K, \tau_T, \sigma^*_{-\{K \cup T\}}) > U_x(\sigma^*_{-T}, \tau_T)$$

3.1 (k, t)-robustness Is Infeasible in Communication Games

We show that the very property that makes (k, t)-robustness so appealing—regret freedom regardless of how Byzantine failures occur—makes it infeasible in many real-world systems. The reason, fundamentally, is that *communication always incurs cost but could potentially yield no benefit if one is communicating with a Byzantine node.* In other words, a rational node may realize in hindsight

---

[3] Abraham et al. also define a strong version of collusion resilience in which there must not exist a deviation in which even *one* coalition member can do better [3,4]. We focus on the weak version as Abraham et al. do in [4]. Since any strongly k-resilient equilibria is (weakly) k-resilient, our impossibility results hold in both versions.

that it could have reduced its costs without affecting its benefits by avoiding all communication with Byzantine nodes, thus improving its utility. As any node can be Byzantine, this implies that the only possible (k,t)-robust equilibrium is one in which no node communicates.

Theorem 6. *There exist no non-trivial (k,t)-robust equilibria in any communication game.*

Proof. Consider some non-trivial (k,t)-robust strategy σ^*. There must exist some node x which, with positive probability α under σ^*, sends a message to some other node z before receiving any other messages. Suppose that z is Byzantine. Since σ^* is (k,t)-robust, x must not be able to do better with some alternate strategy, regardless of who has failed and what a failed node will do. In particular, for all alternate strategies σ'_x for x and Byzantine strategies τ_z for z, it must be that

$$U_x(\sigma^*_{-z}, \tau_z) \geq U_x(\sigma'_x, \tau_z, \sigma^*_{-\{x,z\}}) \tag{1}$$

Suppose τ_z is the strategy in which z "crashes" immediately, *i.e.*, z never sends any messages. Let σ'_x be the strategy in which x does everything in σ^*_x, except x sends nothing to z. By Assumption 2, x cannot prove that it communicated with z; it thus follows that $(\sigma'_x, \tau_z, \sigma^*_{-\{x,z\}})$ has the same functionality as (σ^*_{-z}, τ_z) and is indistinguishable to any node in $N \setminus \{x,z\}$. Clearly, if z follows τ_z, x can do better by never communicating with z; x's outcome will not change (since z never communicates with anyone), and x's communication costs are lower. Formally,

$$U_x(\sigma'_x, \tau_z, \sigma^*_{-\{x,z\}}) = U_x(\sigma^*_{-z}, \tau_z) + \alpha\gamma > U_x(\sigma^*_{-z}, \tau_z)$$

which directly contradicts inequality (1). □

More broadly, Theorem 6 suggests that it may be hard to build non-trivial (k,t)-robust equilibria for any game where a player's actions incur cost. Indeed, in all the games for which Abraham et al. derive (k,t)-robust equilibria [3,4], a node's utility depends only on the game's outcome (*e.g.*, in a secret-sharing game based on Shamir's scheme, utility depends on whether a node can learn the secret) and is independent of how much communication is required to reach that outcome.

Discussion. (k,t)-robustness promises regret freedom simultaneously along two axes: *who* the Byzantine nodes are and *how* they behave. Theorem 6 suggests that this may be too strong to require in practice. But what if we only require regret freedom along only one axis? If we know exactly who the Byzantine nodes are, but not how they will behave, can we achieve regret freedom in communication games? What if we do not know who is Byzantine, but we know their strategy?

3.2   What If We Know Who Is Byzantine?

Let us assume that we know *exactly* who all the Byzantine players are before the game begins. This may already appear a strong assumption, but it is necessary, since if the identity of even one Byzantine node were unknown, Theorem 6 would still apply. We show that, even with this strong assumption, a solution concept that is regret-free with respect to the strategies of Byzantine nodes is possible only to the extent that it defines away the problem: the only possible equilibria are those in which rational nodes communicate only among themselves, completely excluding Byzantine nodes from the system. Furthermore, we show that many interesting communication games do not yield a regret-free equilibrium even if one takes the drastic step of excluding Byzantine nodes: specifically, communication games in which Byzantine nodes may take actions that can *affect* a rational node's utility by more than the cost of sending a *single* message have no regret-free equilibrium, even if the identity of all Byzantine nodes are known a priori.

We first define the equivalent of t-immunity (Definition 3) and (k, t)-robustness (Definition 5) for a fixed set T of Byzantine nodes.

Definition 7. *A strategy profile σ is T-strategy-immune if for all strategy profiles τ and $x \notin T$,*

$$U_x(\sigma_{-T}, \tau_T) \geq U_x(\sigma)$$

Definition 8. *A strategy profile σ^* is (k, T)-strategy-robust with respect to $T \subseteq N$ iff σ^* is T-strategy-immune and for all $K \subseteq N \setminus T$ such that $|K| \leq k$ and all Byzantine strategies τ, there does not exist some σ' such that for all $x \in K$,*

$$U_x(\sigma'_K, \tau_T, \sigma^*_{-(K \cup T)}) > U_x(\sigma^*_{-T}, \tau_T)$$

A (k, T)-*strategy*-robust equilibrium need only be a best response to the specified set T of Byzantine nodes. The following theorem shows that no (k, T)-*strategy*-robust equilibrium is possible unless rational nodes "blacklist" all nodes in T.

Theorem 9. *In a communication game, there does not exist any (k, T)-strategy-robust equilibrium σ^* where any $x \notin T$ communicates with any $z \in T$.*

Proof. Similar to proof of Theorem 6 (see [42]). □

Although Theorem 9 does not rule out all (k, T)-*strategy*-robust equilibria, Theorem 10 proves that these equilibria, which must be regret-free for *any* Byzantine strategy, only exist in limited circumstances.

Theorem 10. *No communication game can yield a (k, T)-strategy-robust equilibrium for any set $T \subseteq N$ of Byzantine nodes if for some $x \notin T$ and some $z \in T$, (a) x has at least one opportunity to send a message to z and (b) for any strategy profile σ, there exist two Byzantine strategies τ_z and τ'_z such that τ_z and τ'_z are the same until x's first opportunity to communicate with z and*

$$U_x(\sigma_{-z}, \tau_z) - U_x(\sigma_{-z}, \tau'_z) > \gamma$$

We omit the straightforward proof for lack of space (see [42]): in essence, if there exists a Byzantine strategy in which a rational node may *gain* by interacting with Byzantine nodes, then ignoring Byzantine players may not prove, in hindsight, an optimal strategy.

Theorem 10—unlike Theorem 6—provides conditions under which no (k, t)-*strategy*-robust equilibria exist, whether trivial or not. Since (k, t)-*strategy*-robust equilibria are a superset of (k, t)-robust equilibria, it naturally follows from Theorem 10 that no (k, t)-robust equilibria exist under the same conditions.

3.3   What If We Know How Byzantine Nodes Behave?

Let us now consider a solution concept that assumes that the strategy played by every Byzantine node is known a priori and yields equilibria that are regret-free with respect to who the Byzantine nodes are.

Definition 11. *The strategy profile σ^* is a (k, t, τ)-type-robust equilibrium iff σ^* is t-immune and for all $K, T \subseteq N$ such that $K \cap T = \emptyset$, $|K| \leq k$, and $|T| \leq t$, there does not exist some σ' such that for all $x \in K$,*

$$U_x(\sigma'_K, \tau_T, \sigma^*_{-(K \cup T)}) > U_x(\sigma^*_{-T}, \tau_T)$$

Despite the strong assumption on which they rely, (k, t, τ)-*type*-robust equilibria are impossible to achieve for many Byzantine behaviors. In particular, it follows immediately from Theorem 6 that no such equilibrium is possible if the known Byzantine strategy calls for any Byzantine node to crash at the very beginning of the game.

Theorem 12. *There exist no non-trivial (k, t, τ)-type-robust equilibria in the communication game in which a Byzantine node z, following τ_z, crashes at the beginning of the game.*

Proof. Same as proof of Theorem 6.   □

In general, it is possible to show (see [42]) that non-trivial (k, t, τ)-*type*-robust equilibria are impossible whenever there is a point in the known Byzantine strategy after which a Byzantine node becomes "unresponsive," *i.e.*, the node's behavior becomes independent of how the game has been played so far (*e.g.*, the node crashes or starts flooding all other nodes with messages).

4   Dealing with Byzantine Failures through Regret Bravery

Finding a single strategy that is a best response against all possible Byzantine strategies or all possible t-sized subsets of Byzantine nodes (or both) appears fundamentally hard: regret-free solution concepts, for which rational cooperation depends on finding such a strategy, seem unlikely to provide a viable theoretical framework for many BAR-tolerant systems.

Regret bravery, the alternative we explore in this section, explicitly forgoes seeking a "universal" best response. Instead, it makes rational cooperation dependent on identifying a strategy that is a best response to the Byzantine behavior that rational nodes *expect* to be exposed to. Before we proceed to look at examples of regret-braving equilibria, we answer some natural questions.

Is aiming for a best response towards only a subset of all possible Byzantine behaviors in effect abdicating the general claims (and benefits) of Byzantine fault tolerance? No. Any BAR-tolerant protocol, independent of the underlying solution concept, must be a strategy that guarantees Byzantine fault tolerance. The choice of a solution concept is not about fault tolerance; rather, it specifies under which conditions rational nodes will be willing to follow a given strategy, fault-tolerant or not. Regret-braving solution concepts are motivated by the observation that rational nodes may be willing to cooperate even without the guarantee that the considered strategy will, in *all* circumstances, prove to be a best response.

Do regret-braving solution concepts limit how Byzantine node can behave? No more than a threshold t on the number of Byzantine faults limits a system to experience, in reality, more than t faults. Regret braving asks rational nodes to build a model of expected Byzantine behavior, but of course Byzantine nodes are in no way bound to follow that model. If Byzantine behavior does not match the expectation of rational nodes, then a regret-braving equilibrium strategy may not, in hindsight, prove to be a best response.

What is the right set of expectations when it comes to Byzantine behavior? It all depends on the application being considered. We discuss below two concrete examples inspired by approaches (maximin and Bayes equilibria) that have been extensively studied in the economics literature, but we do not claim that these solution concepts model "realistic" expectations for all distributed systems. For example, the maximin approach produces a best response to the expectation that the system always includes exactly t Byzantine nodes, when it may instead often be reasonable to expect that the actual number of Byzantine faults will be lower.[4] Indeed, we believe that the challenge of finding equilibrium strategies under more flexible solution concepts is an extremely exciting research opportunity.

Regret Braving the *Quorum* Communication Game. To show the viability of regret-brave solution concepts in a communication game, we consider a concrete communication game: a *quorum game*, which models protocols, such as secret-sharing [38], replicated state machines [26] and terminating reliable broadcast [21] in which functionality is achieved if and only if some subset of nodes (a *quorum*) work together.

Definition 13. *A (synchronous) quorum game is an infinitely-repeated communication game where*

---

[4] A worst-case attitude is actually not uncommon when designing fault-tolerant systems, even for benign failures. For instance, non-early stopping protocols for synchronous terminating reliable broadcast always run for $t + 1$ rounds, even in executions that experience no failures.

- *There are at least 3 nodes ($n \geq 3$).*
- *The game repeats indefinitely. In every round, for each $y \in N$, a node $x \neq y$ decides whether to send a message ("contribute") or not ("snub") to y.*
- *At the end of the round, every $x \in N$ simultaneously (1) observes who contributed to it and (2) receives its payoff.[5] x incurs a cost of γ for each node x contributes to and for each node that contributes to x; x incurs no cost for snubbing or being snubbed. x realizes a positive benefit of $b > 2n\gamma$ in any round where q other nodes (a quorum) contribute to x.[6]*
- *The total payoff is the δ-discounted sum of each individual round's payoff, where $0 < \delta < 1$.*

δ-discounting is a commonly-accepted way of handling utility in infinite-horizon games [17]. This models the reality that earning benefit (incurring cost) now is better (worse) than doing so later.[7]

We consider two concrete regret-braving solution concepts for the quorum game. In the first, rational nodes best-respond to fearing the worst, *i.e.*, they follow a maximin strategy with respect to Byzantine failures.

Definition 14. *The strategy profile σ^* is a k-resilient t-maximin equilibrium iff for any coalition $K \subseteq N$ such that $|K| \leq k$, there does not exist an alternate strategy profile σ' such that for all $x \in K$,*

$$\min_{\substack{T \subseteq N \backslash K: \\ |T| \leq t}} \min_\tau U_x(\sigma'_K, \tau_T, \sigma^*_{-(K \cup T)}) \geq \min_{\substack{T \subseteq N \backslash K: \\ |T| \leq t}} \min_\tau U_x(\sigma^*_{-T}, \tau_T)$$

and for some $y \in K$, the inequality is strict.

In the second, rational nodes weigh the probabilities of various Byzantine failures; an equilibrium is thus these probabilities—known as *beliefs* in game theory parlance—and the strategy profile that is an expected best response given these beliefs. A set of beliefs $\mu = \{\mu_x\}_{x \in N}$ is, for each node, a probability distribution over sets of nodes and their types—whether they are rational, or Byzantine and playing a particular strategy. We use $\mu_x((R_{-T}, \tau_T)|R_K)$ to denote a rational node x's belief that all nodes $z \in T$ are Byzantine and of type (*i.e.*, playing strategy) τ_z and all nodes $w \notin T$ are rational (*i.e.*, of type R), given that there is some K (the coalition) in which $x \in K$ and all $y \in K$ are rational.

Definition 15. *The strategy profile/belief tuple (σ^*, μ^*) is a k-resilient Bayes equilibrium iff for all $K \subseteq N$ such that $|K| \leq k$, there does not exist an alternate*

---

[5] In game theory parlance, the game is a simultaneous game; in distributed systems, synchronous.

[6] Technically, the quorum size is $q + 1$: q other nodes and the node itself (we assume that it costs nothing for a node to contribute to itself). For simplicity, we will simply say that the quorum size is q.

[7] For example, it is often preferable to have a dollar now rather than later, since money can be invested and can earn interest in the meantime.

strategy profile σ' such that for all $x \in K$,

$$\sum_{T \subseteq N \setminus K} \sum_{\tau} \mu_x^*((R_{-T}, \tau_T)|R_K)U_x(\sigma'_K, \tau_T, \sigma^*_{-(K \cup T)})$$

$$\geq \sum_{T \subseteq N \setminus K} \sum_{\tau} \mu_x^*((R_{-T}, \tau_T)|R_K)U_x(\sigma^*_{-T}, \tau_T)$$

and for some $y \in K$, the inequality is strict.

In both definitions, we extend previous work that uses regret-brave solution concepts [6,28,29,41] by explicitly considering collusion, which prior work avoided by either considering collusion a Byzantine failure or making informal arguments on the basis of experimental results. For simplicity, we use k-resilience (Definition 4); however, we could have used any notion of collusion resilience, as this choice is orthogonal to how rational participants view Byzantine peers.

An example of a t-maximin equilibrium. We prove a k-resilient t-maximin equilibrium in the quorum game. Although we argue that communication always has cost and the quorum game does not explicitly model communication that coalition members may perform to coordinate, our proof implicitly assumes that the coalition can coordinate its actions. Thus, our results hold even if we augmented the game to allow coalition members to coordinate via cheap talk [12,16].

Theorem 16. *Let the strategy profile σ^* be defined as follows: any $x \in N$ following σ^*_x contributes to some $y \neq x$ iff x and y have always contributed to each other in the past and x has been snubbed by at most t different nodes. σ^* is a k-resilient t-maximin equilibrium if $q = n - t - 1$, $k \leq q$, and*

$$\frac{b}{\gamma} \geq \max\left(\frac{1 + \delta^2}{\delta^2}(n-1), \frac{1}{1-\delta}(t+k) + 1 \right) \tag{2}$$

Proof. (Sketch)[8] Since $q = n - t - 1$, a rational node needs the cooperation of all other rational nodes to achieve a quorum; as $k \leq q$, a coalition cannot achieve quorum by itself.[9] Consider some coalition K of size at most k. It can be easily verified that, given the conditions above, a coalition member $x \in K$ never snubs a cooperative, non-coalition node $y \notin K$ following σ^*. Intuitively, suppose x snubs y in some round r and y is not Byzantine. If t Byzantine nodes snub every node at least once by round r, y, having observed $t + 1$ snubs, will then snub every node in round $r + 1$. This causes all non-coalition nodes to follow suit and snub in round $r + 2$. It follows that all members of K, including x, will only receive up to $k - 1 < q$ other contributions for the remainder of the game starting from round $r + 2$. As this is not enough to achieve quorum, such a deviation results in the loss of benefit for the remainder of the game and is thus not worthwhile for K given the above conditions.

---

[8] See [42] for the full details.

[9] Recall that a node needs q other nodes to contribute in order to achieve quorum.

However, coalition members have an additional possible deviation: they may choose to help each other save on receiving extraneous contributions (stemming from the fault-tolerant nature of the quorum game, nodes typically send and receive contributions from more than q members) by "snubbing" one another without threat of punishment.

Suppose that nodes in K play such an alternate strategy σ'_K in which some nodes in K snub, for the first time, some $x \in K$ in round r. Then Byzantine nodes may also snub x in order to cause x to lose quorum in round r. Specifically, by deviating, x may

- lose the benefit b it would have normally gained from playing σ^*,
- save at most $(t+1)\gamma$ from not receiving contributions from $t+1$ members (the reason why x did not achieve quorum and lost benefit), and
- save at most $k\gamma$ from not contributing to other coalition members.

Therefore, as compared to σ^*_K, σ'_K loses x at least $b - (t+k+1)\gamma$ in utility. However, in all subsequent rounds, x could save on contributing to

- Byzantine nodes that snubbed x in round r, saving at most $t\gamma$ per round (in the worst case, the Byzantine nodes still continue to contribute to x), and
- coalition members, saving at most $k\gamma$ per round.

This implies that x saves at most $\delta/(1-\delta)(t+k)\gamma$ in utility over all subsequent rounds.

Thus, in order for σ'_K to be worthwhile for x, it must be the case that

$$-b + (t+k+1)\gamma + \frac{\delta}{1-\delta}(t+k)\gamma > 0$$

which is never satisfied given inequality (2).   □

An example of a Bayesian equilibrium. One advantage of using the t-maximin solution concept is its simplicity: because we need only consider the worst possible case, t-maximin equilibria are simple to analyze. Unfortunately, although a rational node playing a t-maximin equilibrium may receive a safe, steady amount of utility, Byzantine failures are unlikely to always occur in the worst possible way, and a rational node willing to take a risk and deviate from the prescribed strategy may be able to do better in expectation.

In the remainder of this section, we demonstrate that the Bayesian approach provides flexibility in how Byzantine nodes are modeled by rational nodes by demonstrating a simple example of a k-resilient Bayes equilibrium. Our goal is to simply illustrate the existence of Bayesian equilibria, not to derive tight bounds for when these equilibria exist. Thus, for simplicity of exposition, we use simple beliefs, optimistic bounds about the utility earned by deviating, and pessimistic bounds about the utility earned by cooperating.

Theorem 17. *Define the strategy profile σ^* such that any $x \in N$, following σ_x^*, contributes to any $y \neq x$ iff x and y have always contributed to each other in the past and x has been snubbed by at most t peers, where t is some constant.*

Let τ be defined as the random t-crash strategy: in any given round, a node z playing τ_z has some positive probability ρ of crashing. Define the set of beliefs μ^* such that for all subsets $K \subseteq N$ such that $|K| \leq k$ and all $y \in K$,

- *$\mu_y^*((R_{-T}, \tau_T)|R_K) = 0$ for any T such that $|T| \neq t$, and*
- *$\mu_y^*((R_{-T_1}, \tau_{T_1})|R_K) = \mu_y^*((R_{-T_2}, \tau_{T_2})|R_K) > 0$ for any $T_1, T_2 \subseteq N \setminus K$ such that $|T_1| = |T_2| = t$.*

Then (σ^*, μ^*) is a k-resilient Bayes equilibrium if $n > t + k$, $k \leq q$, and

$$\frac{b}{\gamma} \geq \frac{n+t-1}{\rho^t \delta^2 (1-\delta)} \frac{n-k}{n-k-t} + n - t - 1 \tag{3}$$

Proof. Fix some rational node x and some coalition K, where $x \in K$ and $|K| \leq k$. We optimistically assume a rational node that deviates in round r only loses utility if t nodes crash on or before round r, which occurs with probability at least ρ^t.

It can be easily verified that by following σ^*, each member of K, including x, earns no less than

$$\frac{1}{1-\delta}(b - 2(n-1)\gamma) \tag{4}$$

in utility, since x can achieve quorum even if every Byzantine node crashes, so the "worst" that happens is x achieves quorum in every round while incurring cost from communication from everyone.

Suppose that x snubs some node $y \notin K$. Since the probability that a node is rational is uniform across all (non-coalition) nodes, y is rational with probability at least $1 - t/(n-k)$, and with probability at least ρ^t, y will observe t other nodes snub it by round r. y then snubs everyone starting in round $r+1$, all non-coalition nodes snub everyone starting in round $r+2$, and x earns at most 0 in every round starting from round $r+2$. Otherwise, we assume x earns the maximum round payoff $b - q\gamma$. Thus, deviating is worthwhile only if

$$\rho^t \left(1 - \frac{t}{n-k}\right)(1+\delta)(b-q\gamma) + \left(1 - \rho^t\left(1 - \frac{t}{n-k}\right)\right)\frac{1}{1-\delta}(b-q\gamma)$$

exceeds expression (4). This never holds given inequality (3).

Otherwise, suppose that $x \in K$ "snubs" its peer $y \in K$ to save on y's communication costs. Again, y, with probability at least ρ^t, will not achieve quorum if all t nodes crash on or before round r. However, unlike before, y only loses quorum for one round; we otherwise assume that it achieves the maximum round payoff $b - q\gamma$. Thus, deviating as a coalition is worthwhile only if

$$\rho^t \frac{\delta}{1-\delta}(b-q\gamma) + (1-\rho^t)\frac{1}{1-\delta}(b-q\gamma) > \frac{1}{1-\delta}(b - 2(n-1)\gamma)$$

which never holds given inequality (3). □

5   Related Work

Outside of (k,t)-robustness [3,4], Eliaz [15] also defined a solution concept which is effectively $(1,t)$-robustness. Gradwohl [19] explored regret-free equilibria with t arbitrary or colluding nodes in leader election and random sampling games. Our results still apply to the solution concepts used in these papers. Moscibroda et al. [33] use an approach similar to t-maximin to consider worst-case Byzantine behavior in the context of a computer virus propagation model.

Coalitions have been studied in depth in the game theory literature. Aumann [40] proposed a notion of collusion resilience which is the basis for k-resilience. Berheim et al. [11], Moreno et al. [32], Einy et al. [14], among others, have proposed weaker solution concepts that only consider deviations that are self-enforcing, meaning that there does not exist an even more profitable deviation for a sub-coalition within the coalition. All of these notions are complementary to regret-brave equilibria and can be used as a part of a regret-brave solution concept.

Our results are similar in spirit to previous work in mechanism design [13,18,20,25,34,37] where mechanisms that incentivize nodes to reveal their true preferences or types for every possible realization of types are found to be often impossible or heavily restricted. Others [13,34] found positive results by using Bayesian solution concepts instead of dominant ones. Mookherjee et al. [31] define conditions in which Bayesian incentive-compatible mechanisms can be replaced by equivalent dominant-strategy mechanisms.

Maximin strategies have been previously explored in conjunction with adversarial or possibly irrational agents. Alon et al. [7] quantify how, in a two-player zero-sum game, the payoff of playing a mixed maximin strategy is affected by an adversary who can choose its actions based on some information about its peer's realized strategy. Tennenholtz [39], extending the work of Aumann et al. [8,9], explores how maximin strategies can approximate the payoff of a Nash equilibrium when a rational node may not want to rely on the rationality of its peers.

6   Conclusion

Distributed systems that span multiple administrative domains must tolerate the possibility that nodes may be Byzantine, rational, and (possibly) acquiescent. To formally reason about such services, we need a solution concept that provides rigorous guarantees for rational cooperation without sacrificing real-world applicability. This paper argues that solution concepts based on regret freedom, despite their intuitive correspondence to the traditional guarantees of fault-tolerant distributed computing, are unlikely to provide the basis for a viable theoretical framework for real-world systems. In particular, we believe that any practical solution concept should be able to admit equilibria in games where a rational node's payoff is not based simply on the outcome but also on the cost of the actions required to achieve said outcome. While our discussion here has focused on communication costs, other costs should be included, such as the computational costs discussed in the recent work of Halpern and Pass [22]. We believe that regret-brave solution concepts provide a rigorous and realistic framework for games that account for these costs.

References

1. Kazaa Lite, http://en.wikipedia.org/wiki/Kazaa_Lite
2. Abraham, I., Dolev, D., Halpern, J.Y.: Private communication
3. Abraham, I., Dolev, D., Gonen, R., Halpern, J.: Distributed computing meets game theory: Robust mechanisms for rational secret sharing and multiparty computation. In: PODC 2006 (2006)
4. Abraham, I., Dolev, D., Halpern, J.Y.: Lower Bounds on Implementing Robust and Resilient Mediators. In: Canetti, R. (ed.) TCC 2008. LNCS, vol. 4948, pp. 302–319. Springer, Heidelberg (2008), http://portal.acm.org/citation.cfm?id=1802614.1802638
5. Adar, E., Huberman, B.A.: Free riding on Gnutella. First Monday 5(10), 2–13 (2000), http://www.firstmonday.org/issues/issue5_10/adar/index.html
6. Aiyer, A.S., Alvisi, L., Clement, A., Dahlin, M., Martin, J.P., Porth, C.: BAR fault tolerance for cooperative services. In: SOSP 2005 (2005)
7. Alon, N., Emek, Y., Feldman, M., Tennenholtz, M.: Adversarial leakage in games. In: ICS 2010 (2010)
8. Aumann, R.J., Maschler, M.: Some thoughts on the minimax principle. Management Science 18(5), P54–P63 (1972)
9. Aumman, R.J.: On the non-transferable utility value: A comment on the Roth-Shaper examples. Econometrica 53(3), 667–677 (1985)
10. Ben-Porath, E.: Cheap talk in games with incomplete information. Journal of Economic Theory 108 (2003)
11. Bernheim, B.D., Peleg, B., Whinston, M.D.: Coalition-proof nash equilibria i. concepts. Journal of Economic Theory 42(1), 1–12 (1987)
12. Crawford, V.P., Sobel, J.: Strategic information transmission. Econometrica 50(6), 1431–1451 (1982)
13. d'Aspremont, C., Gerard-Varet, L.A.: Incentives and incomplete information. Journal of Public Economics 11(1), 25–45 (1979)
14. Einy, E., Peleg, B.: Coalition-proof communication equilibria. In: EISET 1995 (1995)
15. Eliaz, K.: Fault tolerant implementation. Rev. of Econ. Studies 69, 589–610 (2002)
16. Farrell, J., Rabin, M.: Cheap talk. The Journal of Economic Perspectives 10(3), 103–118 (1996)
17. Fudenberg, D., Tirole, J.: Game Theory. MIT Press (August 1991)
18. Gibbard, A.: Manipulation of voting schemes: A general result. Econometrica 41(4), 587–601 (1973)
19. Gradwohl, R.: Rationality in the full-information model. In: Theory of Cryptography (2010)
20. Green, J., Laffont, J.J.: On coalition incentive compatibility. The Review of Economic Studies 46(2), 243–254 (1979)
21. Hadzilacos, V., Toueg, S.: Fault-tolerant broadcasts and related problems (1993)
22. Halpern, J., Pass, R.: Game theory with costly computation (2010)

23. Halpern, J., Teague, V.: Rational secret sharing and multiparty computation. In: STOC 2004 (2004)
24. Hughes, D., Coulson, G., Walkerdine, J.: Free riding on Gnutella revisited: the bell tolls? IEEE Distributed Systems Online 6(6) (June 2005)
25. Jehiel, P., Meyer-ter Vehn, M., Moldovanu, B., Zame, W.R.: The limits of ex post implementation. Econometrica 74(3), 585–610 (2006), http://dx.doi.org/10.1111/j.1468-0262.2006.00675.x
26. Lamport, L.: Time, clocks, and the ordering of events in a distributed system. CACM (July 1978)
27. Levin, D., LaCurts, K., Spring, N., Bhattacharjee, B.: BitTorrent is an auction: analyzing and improving BitTorrent's incentives. SIGCOMM Comput. Commun. Rev. 38(4), 243–254 (2008)
28. Li, H., Clement, A., Marchetti, M., Kapritsos, M., Robinson, L., Alvisi, L., Dahlin, M.: FlightPath: Obedience vs choice in cooperative services. In: OSDI 2008 (2008)
29. Li, H.C., Clement, A., Wong, E., Napper, J., Roy, I., Alvisi, L., Dahlin, M.: BAR Gossip. In: OSDI 2006 (2006)
30. Locher, T., Moor, P., Schmid, S., Wattenhofer, R.: Free riding in bittorrent is cheap. In: HotNets 2006 (2006)
31. Mookherjee, D., Reichelstein, S.: Dominant strategy implementation of Bayesian incentive compatible allocation rules. Journal of Economic Theory 56(2), 378–399 (1992)
32. Moreno, D., Wooders, J.: Coalition-proof equilibrium. Games and Economic Behavior 17, 80–112 (1996)
33. Moscibroda, T., Schmid, S., Wattenhofer, R.: When selfish meets evil: Byzantine players in a virus inoculation game. In: PODC 2006 (2006)
34. Myerson, R.B., Satterthwaite, M.A.: Efficient mechanisms for bilateral trading. Journal of Economic Theory 29(2), 265–281 (1983)
35. Piatek, M., Isdal, T., Anderson, T., Krishnamurthy, A., Venkataramani, A.: Do incentives build robustness in BitTorrent? In: NSDI 2007, pp. 1–14 (April 2007)
36. Saroiu, S., Gummadi, K.P., Gribble, S.D.: A Measurement Study of Peer-to-Peer File Sharing Systems (January 2002)
37. Satterthwaite, M.A.: Strategy-proofness and arrow's conditions: Existence and correspondence theorems for voting procedures and social welfare functions. Journal of Economic Theory 10(2), 187–217 (1975)
38. Shamir, A.: How to share a secret. Comm. ACM 22(11), 612–613 (1979)
39. Tennenholtz, M.: Competitive safety analysis: robust decision-making in multi-agent systems. J. Artif. Int. Res. 17, 363–378 (2002), http://portal.acm.org/citation.cfm?id=1622810.1622822
40. Tucker, A.W.: Acceptable points in general cooperative n-person games. Annals of Mathematics Study 40(4), 287–324 (1959)
41. Wong, E.L., Leners, J.B., Alvisi, L.: It's on Me! the Benefit of Altruism in BAR Environments. In: Lynch, N.A., Shvartsman, A.A. (eds.) DISC 2010. LNCS, vol. 6343, pp. 406–420. Springer, Heidelberg (2010)
42. Wong, E.L., Levy, I., Alvisi, L., Clement, A., Dahlin, M.: Regret freedom isn't free, http://www.cs.utexas.edu/ elwong/research/ publications/bar-no-regret-tr.pdf

On the Complexity of the Regenerator Cost Problem in General Networks with Traffic Grooming*

Michele Flammini[1], Gianpiero Monaco[1], Luca Moscardelli[2],
Mordechai Shalom[3], and Shmuel Zaks[4]

[1] Department of Computer Science, University of L'Aquila, L'Aquila, Italy
{flammini,gianpiero.monaco}@di.univaq.it
[2] Department of Science, University of Chieti-Pescara, Pescara, Italy
moscardelli@sci.unich.it
[3] Tel Hai Academic College, Upper Galilee, 12210, Israel
cmshalom@telhai.ac.il
[4] Department of Computer Science, Technion, Haifa, Israel
zaks@cs.technion.ac.il

Abstract. We consider the problem of minimizing the number of regenerators in optical networks with traffic grooming. In this problem we are given a network with an underlying topology of a graph G, a set of requests that correspond to paths in G and two positive integers g and d. There is a need to put a regenerator every d edges of every path, because of a degradation in the quality of the signal. Each regenerator can be shared by at most g paths, g being the grooming factor. On the one hand, we show that even in the case of $d = 1$ the problem is $APX - hard$, i.e. a polynomial time approximation scheme for it does not exist (unless $P = NP$). On the other hand, we solve such a problem for general G and any d and g, by providing an $O(\log g)$-approximation algorithm and thus extending previous results holding only for specific topologies and specific values of d or g.

Keywords: Optical Networks, Wavelength Division Multiplexing (WDM), Regenerators, Traffic Grooming, Approximation Algorithms and Complexity.

1 Introduction

In modern optical networks, high-speed signals are sent through optical fibers using WDM (Wavelength Division Multiplexing) technology. The decrease in the energy of the signal with the traveled distance necessitates optical amplification at every (almost) fixed distance. However this amplification introduces

* This work was partially supported by the Israel Science Foundation grant No. 1249/08, and by the PRIN 2008 research project COGENT (COmputational and GamE-theoretic aspects of uncoordinated NeTworks), funded by the Italian Ministry of University and Research.

A. Fernández Anta, G. Lipari, and M. Roy (Eds.): OPODIS 2011, LNCS 7109, pp. 96–111, 2011.

noise into the signal, so that it has to be regenerated after a certain number of amplifications. The signal is regenerated by first using a ROADM (Reconfigurable Optical Add-Drop Multiplexer) to extract a set of wavelengths from the optical fiber. Then, for each extracted wavelength, an optical regenerator regenerates the signal carried by that wavelength. That is, at a given optical node, one needs one ROADM if any regeneration will take place, and as one regenerators per wavelength to be regenerated.

The dominant part of the regeneration cost is the cost of the regenerators, because they are (a) expensive and (b) needed one per wavelength. Therefore the *total* number of regenerators is an important cost parameter to be minimized [15].

A logical path formed by a signal traveling from its source to its destination using a unique wavelength is termed a *lightpath*. Let d be the maximum number of hops a lightpath can make without meeting a regenerator. Then an optimal solution can be found by simply placing one regenerator for every d consecutive vertices of each lightpath ℓ. However the problem becomes harder when the *traffic grooming* enters the picture.

Traffic grooming: The network usually supports traffic that is at rates lower than the full wavelength capacity. Therefore the network operator puts together (= grooms) low-capacity connection requests into high capacity lightpaths. In graph-theoretic terms, we associate a path in the graph with each connection, and the problem can viewed as assigning wavelengths to these paths so that at most g of them using the same wavelength (g being the *grooming factor*) can share one edge. Thus, all paths (i.e. connections) that get the same color (i.e., the same wavelength) and form a connected subgraph correspond to grooming of these connections into one lightpath. The optical signal is routed in the intermediate nodes, based on wavelength only, therefore connection requests assigned the same wavelength can not *split* from each other, i.e. they might not induce a graph with a node with degree 3 or higher. In other words a set of path assigned the same wavelength induces a graph with maximum degree two.

1.1   Related Work

Various variants of regenerator placement problems were studied in [3,8,16,19]. Most of these results concentrate in heuristics and simulations and do not consider traffic grooming.

In the literature, two different scenarios have been studied, depending on whether or not it is allowed to split the paths in order, for instance, to reduce the number of used wavelengths or the cost of hardware components. In particular, [5,7] assume that no splitting is allowed, while [6] allows to split paths and [14] considers both scenarios. In this work, we focus on the case in which splitting lightpaths is not allowed.

In [9] theoretical results (upper bounds and lower bounds) are presented for a family of related problems. The objective in that work is to minimize the number of regenerator locations (as opposed to the total number of regenerators), and traffic grooming is not considered. On the other hand, [15] consider the same cost measure as this work but still does not consider traffic grooming.

The problem we study is shown to be NP-hard in other contexts such as fiber minimization in [18] and its NP-hardness is also implied by the proof of a similar result in [11] holding even for path topology and $g = 2$.

When the underlying graph is a path the problem is equivalent to a machine scheduling problem studied in [10], where several approximation algorithms are presented for it and for some of its special cases. In [12] and [13] these results have been extended to the tree topology, and also an algorithm for general networks has been provided. Unfortunately, the worst case approximation ratio of that algorithm is very high for general topologies, namely as the order of the number of lightpaths.

1.2   Our Contribution

In this work we consider the problem of minimizing the number of regenerators in optical networks with traffic grooming, extending the results in [10,12,13] for general settings.

We first show that even in the case of $d = 1$, G being a bipartite graph, the problem is $APX - hard$ for any $g \geq 2$, i.e. a polynomial time approximation scheme (PTAS) for it does not exist (unless $P = NP$). We then provide an $O(\log g)$-approximation algorithm for the most general version of this problem in which general topologies are admitted and both d and g can be arbitrary.

The paper is organized as follows. In Section 2 we define our problem. On the one hand, in Section 3 we show that our problem is APX-Hard even in the case of $d = 1$, G being a bipartite graph and $g \geq 2$. On the other hand, in Section 4 we provide a polynomial time approximation algorithm solving the problem for general topologies and any value of g and d, with an approximation ratio logarithmic in g. We conclude by suggesting open research directions in Section 5.

2   Definitions and Problem Statement

We consider instances $(G, \mathcal{P}, g, d)$ where $G = (V, E)$ is a graph modeling the optical network, $\mathcal{P}$ is a set of simple paths in G, $g \in \mathbb{N}^+$ is the grooming factor and d is the maximum number of hops a lightpath can travel without meeting a regenerator.

A coloring (or wavelength assignment) of $(G, \mathcal{P})$ is a function $w : \mathcal{P} \mapsto \mathbb{N}$. For a coloring w and color λ, $\mathcal{P}_\lambda^w$ denotes the subset of paths from $\mathcal{P}$ colored λ by w, i.e. $\mathcal{P}_\lambda^w \overset{def}{=} \{P \in \mathcal{P}|w(P) = \lambda\}$. When there is no ambiguity on the coloring w under consideration, we omit the superscript w and use $\mathcal{P}_\lambda$.

For a node v, $\mathcal{P}_v$ denotes the subset of paths of $\mathcal{P}$ having v as an intermediate node, and similarly for an edge e, $\mathcal{P}_e$ denotes the subset of paths of $\mathcal{P}$ using the edge e. For every $e \in E$ we define $load(\mathcal{P}, e) \overset{def}{=} |\mathcal{P}_e|$ and $load(\mathcal{P}) \overset{def}{=} \max_{e \in E} load(\mathcal{P}, e)$.

A set of paths is called a *no-split instance* or shortly an NSI if the union of its paths (as sets of edges) induces a graph of maximum degree at most 2. In particular, if also the minimum degree of such a graph is 2 (i.e., such a graph is a vertex disjoint union of rings), we call every connected component of it a *ring*-NSI, otherwise we call every connected component a *path*-NSI.

In this work we assume (as in [5,7,14]) that splitting of paths is not allowed, i.e. paths using the same wavelength and going through the same edge of the network can be routed only to another unique edge. Moreover, since we do not consider bounds on the number of colors and our goal is independent of it, without loss of generality we assume that, in any solution, paths belonging to different NSIs are assigned different colors, and therefore every set of paths with the same color has to be an NSI.

A *valid* coloring (or wavelength assignment) w of $(G, \mathcal{P}, g, d)$ is a coloring of $\mathcal{P}$ such that for every λ, P_λ^w satisfies the following two conditions:

- *The load condition*: For any edge e at most g paths using e are colored with λ, i.e. $load(\mathcal{P}_\lambda^w) \leq g$.
- *The no-splitting condition*: P_λ^w is an NSI.

Given a valid coloring w of $(G, \mathcal{P}, g, d)$, a regenerator assignment is a boolean function $r_w : V \times \mathbb{N} \mapsto \{0, 1\}$; in particular, $r_w(u, \lambda) = 1$ if and only if a regenerator operating at wavelength λ is placed at node u.

We are now ready to give a formal definition of our problem.

TOTAL REGENERATORS WITH GROOMING (TRG)

Input: A quadruple $(G, \mathcal{P}, g, d)$, where $G = (V, E)$ is a graph, $\mathcal{P} = \{P_1, P_2, ..., P_n\}$ is a set of simple paths in G, g is an integer, namely the grooming factor, and d is the maximum number of hops a lightpath can go through without needing a regenerator.
Output: A valid coloring $w : \mathcal{P} \mapsto \mathbb{N}$ and a regenerator assignment r_w such that, r_w satisfies the constraint that every lightpath has a regenerator every at most d hops (we will refer to this condition as the *regeneration condition* through this work).
Objective: The cost of a solution is given by the total number of regenerators $REG^w \stackrel{def}{=} \sum_\lambda \sum_{u \in V} r_w(u, \lambda)$. The goal is to minimize the total number of regenerators REG^w.

$OPT(G, \mathcal{P}, g, d)$ denotes the cost of any optimal coloring.

3   Hardness of Approximation

In this section we show that the problem TRG is *APX*-hard even if restricted to instances $(G, \mathcal{P}, g, 1)$, with g at least 2.

Notice that, if $d = 1$, the coloring w univocally identifies the regenerator assignment r_w; in fact, given an NSI $\mathcal{N}$ colored λ by w, a regenerator is needed at each node being an internal node of some path in $\mathcal{N}$, i.e., $r_w(u, \lambda) = 1$ if and only if u is an internal node of some path in $\mathcal{N}$.

We first define a problem that will be used in our proof.

B-BOUNDED EDGE PARTITION INTO TRIANGLES AND MINIMUM PATHS (MEPTP-B):
Input: A graph $G = (V, E)$ with $\Delta(G) \leq B$, where $\Delta(G)$ is the maximum degree of a node of G.
Output: A partition of E into connected graphs with at most 3 edges.
Measure of a solution: The number of paths of the returned partition.
Objective: Minimizing the measure of the returned solution.

By exploiting a reduction similar to the one used in [1] we will prove that this problem is APX-Hard. Finally we will reduce this problem to the $(G, \mathcal{P}, g, 1)$ problem with $g \geq 2$ in order to show the APX-Hardness of the latter.

Definition 1. *Given a tripartite graph $G = (V_0 \cup V_1 \cup V_2, E)$ we can obtain a directed graph, by directing the edges from nodes of V_i nodes of $V_{(i+1) \mod 3}$. We will say that G is directed Eulerian if the directed graph obtained in this way is directed Eulerian.*

Lemma 1. *The MEPTP-B problem is APX-Hard for any fixed $B \geq 12$ even when the graph is directed Eulerian tripartite and the optimum is at least $|E|/10$.*

Theorem 1. *The set of $(G, \mathcal{P}, g, 1)$ instances of the TRG problem is APX-hard for any $g \geq 2$ and even when G is a bipartite graph.*

Proof. We will give an approximation ratio preserving reduction from the MEPTP-B problem in graphs satisfying the conditions of Lemma 1 to the $(G, \mathcal{P}, g, 1)$ instances of TRG, with $g \geq 2$.

Let $G' = (V_0' \cup V_1' \cup V_2', E')$ be an instance of MEPTP-B. We build an instance $(G = (V, E), \mathcal{P}, g, 1)$ of our problem as follows (see Figure 1): For each $i \in \{1, 2, 3\}$ and for every node $v \in V_i'$, G contains a path with three nodes v^-, v, v^+ and two edges $(v^-, v), (v, v^+)$. G contains 3 special nodes u_{01}, u_{12}, u_{20}. Node u_{ij} is connected to all the v^+ nodes corresponding to any $v \in V_i'$ and to all the v^- nodes corresponding to any $v \in V_j'$.

For each edge $(v, w) \in E'$ where $v \in V_i'$ and $w \in V_j'$ ($j \equiv (i + 1) \mod 3$), $\mathcal{P}$ contains the path $(v^-, v, v^+, u_{ij}, w^-, w, w^+)$.

The constructed instance has the following properties:

- Any two distinct edges of G' both connecting nodes from V_i' and V_j' correspond to two paths in $\mathcal{P}$ that induce a graph with degree 3 or 4 at node u_{ij}. Therefore these two paths cannot be part of an NSI. We conclude an NSI can contain at most 3 paths, i.e., one corresponding to an edge of G' from V_0' to V_1', another one corresponding to an edge of G' from V_1' to V_2', and another one corresponding to an edge of G' from V_2' to V_0'.
- The two paths corresponding to any pair of adjacent edges in G' are compatible (i.e., form together an NSI) and have one intermediate node in common.

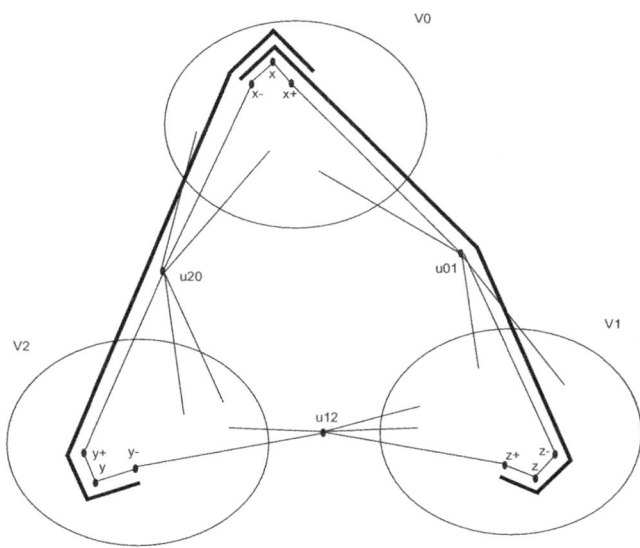

Fig. 1. The bipartite graph in the proof of Theorem 1

We conclude that the edges in G' corresponding to the paths of an NSI form either a triangle, or a path with 1, 2 or 3 edges, and conversely, every such subgraph of G' corresponds to an NSI. Let T be the number of NSIs corresponding to triangles of G', and l_k be the number of NSIs corresponding to paths of length k of G' for $k \in \{1, 2, 3\}$. Note that each path of $\mathcal{P}$ has 5 intermediate nodes. Therefore the number Reg of regenerators used by such a solution is $Reg = 5|E'| - 3T - 2l_3 - l_2$. As G' is directed Eulerian tripartite, for any T there is a solution with $l_2 = l_1 = 0$, with cost $Reg = 5|E'| - 3T - 2l_3 = 5|E'| - (3T + 2l_3) = 5|E'| - (|E'| - l_3) = 4|E'| + l_3$. Therefore the minimum number of regenerators is obtained at the optimum of the MEPTP-B instance. Let Reg^* be the optimum of instance $(G, \mathcal{P}, g, 1)$, and consider a ρ-approximate solution of it with $Reg = \rho \cdot Reg^*$. Moreover, let l_k^*, for $k \in \{1, 2, 3\}$, be the number of paths of length k in an optimal solution of the corresponding instance of the MEPTP-B problem. Then

$$l_3 = Reg - 4|E'| = \rho \cdot Reg^* - 4|E'| = \rho \cdot (l_3^* + 4|E'|) - 4|E'|$$
$$= \rho \cdot l_3^* + 4(\rho - 1)|E'| \le \rho \cdot |l_3^*| + 40(\rho - 1)l_3^* \qquad (1)$$
$$= (\rho + 40(\rho - 1))l_3^*,$$

where 1 holds because by Lemma 1 we can assume that the optimum of G' is at least $\frac{|E'|}{10}$.

Assume that our problem admits a PTAS. For any $\epsilon > 0$ we run the PTAS with the parameter $\epsilon' = \epsilon/41$ to obtain a $\rho = 1 + \epsilon/41$ approximated solution.

This corresponds to a solution of the MEPTP-B with $l_3 \leq (1 + \epsilon)l_3^*$. A contradiction to the fact that MEPTP-B does not admit a PTAS unless P=NP.

$\square$

4 Approximation Algorithm

In this section we provide an approximation algorithm for the TRG problem for general topologies, guaranteeing an $O(\log g)$ approximation ratio in polynomial time.

A *proper* set $\bar{\mathcal{P}}$ of paths, is a set of paths that constitute and independent set with respect to inclusion. In other words no paths of $\bar{\mathcal{P}}$ is included in another. An instance is said to be proper if its set of paths $\mathcal{P}$ is proper.

This section is organized as follows: We first provide an $O(\log g)$-approximation algorithm for the case of proper instances with $d = 1$. We then extend this result to the more general case in which $d = 1$ but the instance is not necessarily proper. Finally we extend the result to any value of d. Each time we extend the previous result, we lose only a constant factor in the approximation ratio, therefore achieving an $O(\log g)$-approximation ratio for the general case.

We introduce some definitions that will be useful in the proofs contained in this section. We denote by $INT(P)$ the set of intermediate nodes, i.e. of all the nodes not being endpoints, of a path P in G, and $int(P) \overset{def}{=} |INT(P)|$. For a set $\mathcal{P}$ of paths we define:

$$SPAN(\mathcal{P}) \overset{def}{=} \bigcup_{P \in \mathcal{P}} INT(P), \; span(\mathcal{P}) \overset{def}{=} |SPAN(\mathcal{P})|, \; len(\mathcal{P}) \overset{def}{=} \sum_{P \in \mathcal{P}} int(P).$$

Notice that, if $d = 1$, the number of regenerators operating at wavelength λ is $span(\mathcal{P}_\lambda^w)$; in fact, at each node being an intermediate node of some path in $\mathcal{P}_\lambda^w$ a regenerator operating at this wavelength is needed. Moreover, when $d = 1$, we have the following trivial lower bound (the grooming bound) for the cost of any coloring w (in particular for an optimal coloring), holding because a regenerator can be used by at most g intermediate nodes of paths: $REG^w \geq \frac{len(\mathcal{P})}{g}$.

4.1 Proper Instances with $d = 1$

In this section, we focus on the case $d = 1$, i.e., a regenerator is needed at every internal node of a path, and the set of paths constitute a proper set. In particular, we provide Algorithm 2 working for $(\mathcal{G}, \bar{\mathcal{P}}, g, 1)$ instances, with $\bar{\mathcal{P}}$ being a proper set of paths. It exploits the greedy set cover approximation algorithm $GreedySetCover$ for the minimum weight set cover problem presented in [4]. Such an algorithm guarantees an H_k approximation ratio, where k is the maximum cardinality of a subset in the input and H_k is the k-th harmonic number $\sum_{i=1}^{k} \frac{1}{i}$.

More formally, the Set Cover problem is defined as follows:

MINIMUM WEIGHTED SET COVER

Output: A subcollection $\mathcal{SC} \subseteq \mathcal{S}$ of subsets covering the elements in A, i.e. such that $\cup \mathcal{SC} = A$.

Measure of a solution: $\sum_{S \in \mathcal{SC}} weight[S]$, i.e. the sum of the weights of the selected subsets.

Objective: Minimizing the measure of the returned solution.

We present here the *GreedySetCover* Algorithm of [4] because we will slightly modify it in the sequel, to improve the time complexity of our algorithm.

Algorithm 1. [4] *GreedySetCover*$(A, \mathcal{S}, weight)$

1: $\mathcal{SC} \leftarrow \emptyset$
2: $Covered \leftarrow \emptyset$
3: **while** $Covered \neq A$ **do**
4:     **for** $i = 1$ to m **do**
5:         $eff[S_i] \leftarrow \frac{weight[S_i]}{|S_i \setminus Covered|}$
6:     **end for**
7:     $min \leftarrow \operatorname{argmin}_{i=1}^{m} eff[S_i]$
8:     $\mathcal{SC} \leftarrow \mathcal{SC} \cup \{S_{min}\}$
9:     $Covered \leftarrow Covered \cup S_{min}$
10: **end while**
11: **return** $\mathcal{SC}$

Definition 2. *Given a set $\mathcal{Q}$ of paths, and a path $P \in \mathcal{Q}$, we say that P dominates $\mathcal{Q}$ if $\forall P' \in \mathcal{Q}$, $E(P') \cap E(P) \neq \emptyset$. A set $\mathcal{Q}$ of paths is said to be dominated if there exists a path $P \in \mathcal{Q}$ that dominates $\mathcal{Q}$.*

We term an NSI $\mathcal{N}$ such that $load(\mathcal{N}) \leq g$ as a g-NSI. Our algorithm is based on the following basic lemma.

Lemma 2. *Let $\mathcal{N}$ be a proper g-NSI. $\mathcal{N}$ can be covered with proper, dominated g-NSI's $\mathcal{Q}_0^{\mathcal{N}}, \mathcal{Q}_1^{\mathcal{N}}, \ldots$, such that $\sum_i span(\mathcal{Q}_i^{\mathcal{N}}) \leq 2 \cdot span(\mathcal{N})$.*

Proof. Let $\widehat{\mathcal{N}} \subseteq \mathcal{N}$ be a maximal subset of pairwise edge-disjoint paths in $\mathcal{N}$. It follows from the maximality, that every path $P \in \mathcal{N}$ edge-intersects with at least one path of $\widehat{\mathcal{N}}$. Let $\widehat{\mathcal{N}} = \{P_0, P_1, \ldots, P_{|\widehat{\mathcal{N}}|-1}\}$; if $\mathcal{N}$ is a path-NSI we assume without loss of generality that $P_0, P_1, \ldots$ are ordered from left to right, otherwise (i.e., if $\mathcal{N}$ is a ring-NSI) we assume that they are ordered clockwise. In the following the terms *before* and *after* refer to this order, *right* and *clockwise* are used interchangeably, and when $\mathcal{N}$ is a ring-NSI index arithmetic is done modulo $|\widehat{\mathcal{N}}|$.

We observe that a path $P \in \mathcal{N}$ intersects either exactly one path P_i, or two consecutive paths P_i, P_{i+1}, because otherwise there would exist a path P_j included in P, contradicting the properness of $\mathcal{N}$. We partition $\mathcal{N}$ into $\mathcal{Q}_0^{\mathcal{N}}, \ldots, \mathcal{Q}_{|\widehat{\mathcal{N}}|-1}^{\mathcal{N}}$ such that, for $i = 0, \ldots, |\widehat{\mathcal{N}}| - 1$, $\mathcal{Q}_i$ consists of the paths of

$\mathcal{N}$ intersecting only P_i, or both P_i and P_{i+1}. Clearly each $\mathcal{Q}_i^{\mathcal{N}}$ is a proper g-NSI dominated by P_i. It remains to show that the last condition in the statement of the Lemma holds. We will show that every node $v \in SPAN(\mathcal{N})$ is in at most two sets $SPAN(\mathcal{Q}_i^{\mathcal{N}})$ and $SPAN(\mathcal{Q}_{i+1}^{\mathcal{N}})$.

Clearly, the claim holds when $\left|\widehat{\mathcal{N}}\right| \leq 2$. Assume that $\left|\widehat{\mathcal{N}}\right| \geq 3$ and that there is a node $v \in SPAN(\mathcal{Q}_i^{\mathcal{N}}) \cap SPAN(\mathcal{Q}_{i+j}^{\mathcal{N}})$ where $j > 1$. If v is not before the right endpoint of P_{i+1} then there is a path $P \in \mathcal{Q}_i^{\mathcal{N}}$ whose right endpoint is not before the right endpoint of P_{i+1}, thus including P_{i+1}, contradicting the properness of $\mathcal{N}$. If v is before the right endpoint of P_{i+1}, then there is a path $P \in \mathcal{Q}_{i+j}^{\mathcal{N}}$ that intersects P_{i+1} contradicting the way we partitioned $\mathcal{N}$.   □

Algorithm 2. $(\mathcal{G}, \bar{\mathcal{P}}, g, 1)$

 1:   ▷ Prepare the input for $GreedySetCover$
 2: $\mathcal{S} \leftarrow \emptyset$
 3: **for** each $\mathcal{Q} \subseteq \bar{\mathcal{P}}$ such that $|\mathcal{Q}| \leq 2g - 1$ **do**
 4:     **if** $load(\mathcal{Q}) \leq g$ **and** $\mathcal{Q}$ is dominated **and** $\mathcal{Q}$ is an NSI **then**
 5:         $\mathcal{S} \leftarrow \mathcal{S} \cup \{\mathcal{Q}\}$
 6:         $weight[\mathcal{Q}] \leftarrow span(\mathcal{Q})$
 7:     **end if**
 8: **end for**
 9: $\mathcal{SC} \leftarrow GreedySetCover(\bar{\mathcal{P}}, \mathcal{S}, weight)$
10:   ▷ Eliminate inclusions
11: **while** there exist $S, S' \in \mathcal{SC}$ such that $S \cap S' \neq \emptyset$ **do**
12:     $S \leftarrow S \setminus S'$
13: **end while**
14:   ▷ Assign colors to paths
15: **for** each $S_i \in \mathcal{SC}$ **do**
16:     **for** each $P \in S$ **do**
17:         $w(P) \leftarrow i$
18:     **end for**
19: **end for**
20: **return** w

Theorem 2. *Algorithm 2 is a $2H_{2g-1}$-approximation algorithm for $(\mathcal{G}, \bar{\mathcal{P}}, g, 1)$ instances, where $\bar{\mathcal{P}}$ is a proper set of paths. Its running time is not polynomial in g.*

Proof. The cost of the solution is at most the cost of the set cover returned by the greedy algorithm, because the elimination of inclusions (lines 11–13 of Algorithm 2) can only reduce the cost of the cover.

Since the maximum cardinality of a subset in the collection $\mathcal{S}$ given in input to $GreedySetCover$ is $2g - 1$ and therefore, by [4], $GreedySetCover$ guarantees an H_{2g-1} approximation ratio, in order to prove the claim, it remains to show that there exists a subcollection $\overline{SC} \subseteq S$ such that $\sum_{S \in \overline{SC}} weight[S] \leq 2 \cdot OPT(\mathcal{G}, \bar{\mathcal{P}}, g, 1)$.

Let $\mathcal{N}_1^*, \mathcal{N}_2^*, \ldots, \mathcal{N}_{W^*}^*$ be the NSIs of an optimal solution. For each $1 \leq i \leq W^*$, let $\mathcal{Q}_0^{\mathcal{N}_i^*}, \ldots, \mathcal{Q}_{|\widehat{\mathcal{N}_i^*}|-1}^{\mathcal{N}_i^*}$ be the proper, dominated g-NSIs whose existence is guaranteed by Lemma 2, and let $\overline{SC} = \left\{ \mathcal{Q}_j^{\mathcal{N}_i^*} \mid 1 \leq i \leq W^*, 0 \leq j \leq |\widehat{\mathcal{N}_i^*}| - 1 \right\}$. It holds that:

$$\sum_{S \in \overline{SC}} weight[S] = \sum_{S \in \overline{SC}} span(S) = \sum_{i=1}^{W^*} \sum_{j=0}^{|\widehat{\mathcal{N}_i^*}|-1} span(\mathcal{Q}_j^{\mathcal{N}_i^*})$$

$$\leq \sum_{i=1}^{W^*} 2 \cdot span(\mathcal{N}_i^*) = 2 \cdot OPT(\mathcal{G}, \bar{\mathcal{P}}, g, 1).$$

To conclude the proof we note that a proper, dominated g-NSI contains at most $2g-1$ paths. This is because each path of such a set must use one of the extremal edges of the dominating path, otherwise such a path either does not intersect the dominating path, or it is included in it, both of which contradict the definition of dominated, proper set. There can be at most $g-1$ paths except the dominating path using each extremal edge, therefore at most $2(g-1) + 1 = 2g - 1$ paths. We thus conclude $\overline{SC} \subseteq S$. □

Though Algorithm 2 is not polynomial in g (as it considers all subset of paths of cardinality at most $2g-1$) we are able to provide a polynomial time algorithm preserving the same approximation ratio up to a constant factor. First of all, we relax the greedy choice (line 7) of algorithm $GreedySetCover$ as the following algorithm does.

Algorithm 3. $GreedySetCover_2(A, S, weight, \rho)$

1: $SC \leftarrow \emptyset$
2: $Covered \leftarrow \emptyset$
3: **while** $Covered \neq A$ **do**
4:      **for** $i = 1$ to m **do**
5:          $eff[S_i] \leftarrow \frac{weight[S_i]}{|S_i \setminus Covered|}$
6:      **end for**
7:      Let j such that $eff[S_j] \leq \rho \cdot \min_{i=1}^{m} eff[S_i]$
8:      $SC \leftarrow SC \cup \{S_j\}$
9:      $Covered \leftarrow Covered \cup S_j$
10: **end while**
11: **return** SC

The following lemma states a well known fundamental result on the approximation ratio guaranteed by Algorithm $GreedySetCover_2$.

Lemma 3. *Algorithm $GreedySetCover_2$ guarantees a $(\rho \cdot H_k)$-approximation for the set cover problem, where k is the size of the biggest set in the input.*

An immediate consequence of the above lemma and Theorem 2 is

Corollary 1. *Algorithm 2 in which, at line 9, $GreedySetCover_2$ is invoked with $\rho = 2$ instead of $GreedySetCover$ is a $4H_{2g-1}$-approximation algorithm for $(\mathcal{G}, \bar{\mathcal{P}}, g, 1)$ instances, where $\bar{\mathcal{P}}$ is a proper set of paths. Its running time is not polynomial in g.*

We are now ready to provide and analyze the following algorithm.

Algorithm 4. $(\mathcal{G}, \bar{\mathcal{P}}, g, 1)$

1: $\mathcal{SC} \leftarrow \emptyset$
2: $Covered \leftarrow \emptyset$
3: **while** $Covered \neq \bar{\mathcal{P}}$ **do**
4:     **for each** $\mathcal{Q}_i \subseteq \bar{\mathcal{P}} \setminus Covered$ such that $|\mathcal{Q}_i| \leq 3$ **and** $\mathcal{Q}_i$ is a dominated NSI **do**
5:         **for each** $P \in \bar{\mathcal{P}} \setminus Covered$ **do**
6:             **if** $\mathcal{Q}_i \cup \{P\}$ is a dominated g-NSI **and** $SPAN(\mathcal{Q}_i \cup \{P\}) = SPAN(\mathcal{Q}_i)$
      then
7:                $\mathcal{Q}_i \leftarrow \mathcal{Q}_i \cup \{P\}$
8:             **end if**
9:         **end for**
10:         $eff[\mathcal{Q}_i] \leftarrow \frac{span(\mathcal{Q}_i)}{|\mathcal{Q}_i|}$
11:     **end for**
12:     Choose $\mathcal{Q}_{min}$ that minimizes $eff[\mathcal{Q}_{min}]$
13:     $\mathcal{SC} \leftarrow \mathcal{SC} \cup \{\mathcal{Q}_{min}\}$
14:     $Covered \leftarrow Covered \cup \mathcal{Q}_{min}$
15: **end while**
16:                                   ▷ Assign colors to paths
17: **for each** $\mathcal{Q}_i \in \mathcal{SC}$ **do**
18:     **for each** $P \in \mathcal{Q}_i$ **do**
19:         $w(P) \leftarrow i$
20:     **end for**
21: **end for**
22: **return** w

Lemma 4. *Algorithm 4 is a polynomial time $4H_{2g-1}$-approximation algorithm for $(\mathcal{G}, \bar{\mathcal{P}}, g, 1)$ instances, with $\bar{\mathcal{P}}$ being a proper set of paths.*

Proof. In this proof Algorithm 2 refers to the variant in which, at line 9, instead of $GreedySetCover$, $GreedySetCover_2$ is invoked with $\rho = 2$. Recall that Corollary 1 holds for this variant. Actually Algorithm 4 is equivalent to Algorithm 2 in the following sense: Instead of preparing an exponential number of sets (lines 2–8 of Algorithm 2) and passing it to $GreedySetCover_2$ (line 9 of Algorithm 2), it actually simulates it, and each time calculates the greedy choice of $GreedySetCover_2$, by iterating over a polynomial number of sets.

In particular, in the following we show that the subcollection $\mathcal{SC}$ computed at the end of line 15 of Algorithm 4 is one of the possible subcollections that Algorithm $GreedySetCover_2$, executed at line 9 of Algorithm 2 on input $(\bar{\mathcal{P}}, \mathcal{S}, weight)$ (where $\mathcal{S}$ and $weight$ are those computed at lines 2–8 of Algorithm 2), could return as output.

Let without loss of generality the subcollection $\mathcal{SC}$ computed at lines 1–15 of Algorithm 4 be $\{\mathcal{B}_1, \mathcal{B}_2, \ldots\}$ in the order they are chosen in the *while loop*. We prove that $\mathcal{SC}$ is a possible subcollection that can be returned by the $GreedySetCover_2$ algorithm invoked at line 9 of Algorithm 2. It is enough to show that for every k, $\mathcal{B}_k$ is a possible outcome of iteration k of the while loop of $GreedySetCover_2$. In the following discussion we confine ourselves to the k-th iteration of both algorithms and to the values of $\mathcal{SC}$ and $Covered$ in the beginning of this iteration. Specifically we have to show that for every $\mathcal{D} \in \mathcal{S}$, $eff[\mathcal{B}_k] \leq 2 \cdot eff[\mathcal{D}]$ that means that $\mathcal{B}_k$ can be chosen by $GreedySetCover_2$ algorithm at iteration k. Consider an arbitrary set $\mathcal{D} \in \mathcal{S}$. Since $\mathcal{D}$ is a dominated, proper g-NSI then $|\mathcal{D}| \leq 2g - 1$ and there exists a set of three[1] paths $P_1, P_2, P_3 \in \mathcal{D}$ dominated by P_1, such that $SPAN(\mathcal{D}) = SPAN(\{P_1, P_2, P_3\})$. The set $\{P_1, P_2, P_3\}$ is considered by Algorithm 4 at line 4 and therefore a corresponding set of paths $\mathcal{Q}_i$ with at least $\min(g, |\mathcal{D}|)$ paths is built, such that $SPAN(\mathcal{Q}_i) = SPAN(\{P_1, P_2, P_3\})$. We get $|\mathcal{Q}_i| \geq \frac{|\mathcal{D}|}{2}$ and therefore

$$eff[\mathcal{Q}_i] = \frac{SPAN(\{P_1, P_2, P_3\})}{|\mathcal{Q}_i|} = \frac{SPAN(\mathcal{D})}{|\mathcal{Q}_i|} \leq \frac{2 \cdot SPAN(\mathcal{D})}{|\mathcal{D}|} = 2 \cdot eff[\mathcal{D}].$$

Since $\mathcal{B}_k$ is chosen as the subset of paths with minimum cost-effectiveness at line 12 of Algorithm 4, $eff[\mathcal{B}_k] \leq eff[\mathcal{Q}_i] \leq 2 \cdot eff[\mathcal{D}]$, as required.

As a final remark, notice that the algorithm puts each path in exactly one subset of the subcollection $\mathcal{SC}$, and therefore there is no need to eliminate inclusions as in lines 11–13 of Algorithm 2.   □

4.2   Case $d = 1$

In this section we deal with the case $d = 1$ and general instances, by reducing the problem to the special case of proper instances.

In order to show such a reduction, we exploit Algorithm "FirstFit" of [13] and we also need a lemma of [13].

Algorithm $FirstFit$ colors the paths greedily by considering them one after the other, from longest to shortest. Each path is assigned the lowest possible color for it. $FirstFit$ uses colors starting from λ_{start}.

Lemma 5. *[13] Let w be the coloring returned by $FirstFit$ and $W \geq 1$ be the number of colors used by w; for any $\lambda_{start} < \lambda \leq \lambda_{start} + W$, $len(\mathcal{P}_{\lambda-1}) \geq \frac{g}{3} span(\mathcal{P}_\lambda)$.*

We are now ready to prove the following lemma.

---

[1] Actually, the number of such paths could be less than three, but the proof easily extends to these cases.

Algorithm 5. [13] $FirstFit(G, \mathcal{P}, g, \lambda_{start})$ where G is a path or a ring

1: Sort the paths in non-increasing order of length, i.e., $int(P_1) \geq int(P_2) \geq \ldots \geq int(P_n)$.
2: Consider the paths by the above order and, for any path P_j, $j \in \{1, \ldots, n\}$, let $w(P_j)$ be the first possible color $\lambda \geq \lambda_{start}$ that will not violate the load condition. Namely, find the minimum value $\lambda \geq \lambda_{start}$ such that, for every edge e of P_j, $load(\mathcal{P}_\lambda, e) \leq g - 1$ and set $w(P_j) \leftarrow \lambda$.
3: **return** w

Lemma 6. *Given a polynomial time ρ-approximation algorithm $\mathcal{A}$ for $(\mathcal{G}, \bar{\mathcal{P}}, g, 1)$ such that $\bar{\mathcal{P}}$ is a proper set of paths, it is possible to obtain a polynomial time $(2\rho + 3)$-approximation algorithm $\mathcal{A}'$ for $(\mathcal{G}, \mathcal{P}, g, 1)$, where $\mathcal{P}$ is not necessarily proper.*

Proof. As a first step we calculate a maximal subset $\bar{\mathcal{P}} \subseteq \mathcal{P}$ of proper paths, by picking up all the paths $P \in \mathcal{P}$ and adding P to $\bar{\mathcal{P}}$ as long as P is not included in any other path of $\bar{\mathcal{P}}$. Clearly, $\bar{\mathcal{P}}$ is proper and every path $P \in \mathcal{P} \setminus \bar{\mathcal{P}}$ is included in a path of $\bar{\mathcal{P}}$. Consider Algorithm 6 executed on input $((\mathcal{G}, \mathcal{P} = \mathcal{P}' \cup \bar{\mathcal{P}}, g, 1), \bar{w})$, where $\bar{w} = \mathcal{A}(\mathcal{G}, \bar{\mathcal{P}}, g, 1)$. For any $1 \leq \lambda \leq \bar{W}$, let $first_\lambda$ and $last_\lambda$ be the minimum and the maximum color used by w_λ, respectively. By Lemma 5, we obtain

$$\sum_{\lambda'=first_\lambda+1}^{last_\lambda} span(\mathcal{P}_{\lambda'}) = \sum_{\lambda'=first_\lambda}^{last_\lambda-1} span(\mathcal{P}_{\lambda'+1})$$

$$\leq \frac{3}{g} \sum_{\lambda'=first_\lambda}^{last_\lambda-1} len(\mathcal{P}_{\lambda'}) < \frac{3}{g} len(P'_\lambda). \quad (2)$$

Moreover, Algorithm 6 guarantees that $span(\mathcal{P}_{first_\lambda}) \leq span(\bar{\mathcal{N}}_\lambda)$, because all the paths in $\mathcal{P}_{first_\lambda}$ are included in some path of $\bar{\mathcal{N}}_\lambda$.

The number of regenerators used by the coloring w returned by Algorithm 6 is

$$\sum_{\lambda=1}^{\bar{W}} span(\bar{\mathcal{N}}_\lambda) + \sum_{\lambda=1}^{\bar{W}} \sum_{\lambda'=first_\lambda}^{last_\lambda} span(\mathcal{P}_{\lambda'})$$

$$= \sum_{\lambda=1}^{\bar{W}} span(\bar{\mathcal{N}}_\lambda) + \sum_{\lambda=1}^{\bar{W}} \left(span(\mathcal{P}_{first_\lambda}) + \sum_{\lambda'=first_\lambda+1}^{last_\lambda} span(\mathcal{P}_{\lambda'}) \right)$$

$$\leq \sum_{\lambda=1}^{\bar{W}} span(\bar{\mathcal{N}}_\lambda) + \sum_{\lambda=1}^{\bar{W}} \left(span(\bar{\mathcal{N}}_\lambda) + \frac{3}{g} \cdot len(P'_\lambda) \right) \quad (3)$$

$$= 2 \sum_{\lambda=1}^{\bar{W}} span(\bar{\mathcal{N}}_\lambda) + \frac{3}{g} \sum_{\lambda=1}^{\bar{W}} len(P'_\lambda)$$

$$\leq 2\rho \cdot OPT(\mathcal{G}, \bar{\mathcal{P}}, g, 1) + 3 \cdot OPT(\mathcal{G}, P', g, 1) \tag{4}$$

$$\leq (2\rho + 3) \cdot OPT(\mathcal{G}, P' \cup \bar{\mathcal{P}}, g, 1), \tag{5}$$

where inequality 3 holds by (2); inequality 4 holds because A is ρ-approximation algorithm for $(\mathcal{G}, \bar{\mathcal{P}}, g, 1)$ and by the grooming bound; finally, inequality 5 holds because both $OPT(\mathcal{G}, \bar{\mathcal{P}}, g, 1)$ and $OPT(\mathcal{G}, P', g, 1)$ are at most $OPT(\mathcal{G}, P' \cup \bar{\mathcal{P}}, g, 1)$. □

Algorithm 6. $((\mathcal{G}, \mathcal{P}, g, 1), \bar{w})$, where $\bar{w}$ is a valid coloring for the instance $(\mathcal{G}, \bar{\mathcal{P}}, g, 1)$ with $\bar{\mathcal{P}} = \mathcal{P} \setminus \mathcal{P}'$ being a maximal proper set of paths

1: Let $1, 2, \ldots, \bar{W}$ be the colors used by $\bar{w}$
2: $\lambda_{new} \leftarrow \bar{W} + 1$
3: **for** $\lambda = 1$ to $\bar{W}$ **do**
4:     $\bar{\mathcal{N}}_\lambda = \bar{\mathcal{P}}_\lambda$ ▷ $\bar{\mathcal{P}}_\lambda$ is the set of paths in $\bar{\mathcal{P}}$ colored λ by $\bar{w}$
5:     $\mathcal{P}'_\lambda \leftarrow \emptyset$
6:     **for each** $P \in \mathcal{P}'$ such that P is included in some path of $\bar{\mathcal{N}}_\lambda$ **do**
7:         $\mathcal{P}'_\lambda \leftarrow \mathcal{P}'_\lambda \cup \{P\}$
8:     **end for**
9:     $w_\lambda \leftarrow FirstFit(\bar{\mathcal{N}}_\lambda, \mathcal{P}'_\lambda, g, \lambda_{new})$
10:     $\lambda_{new} \leftarrow 1 +$ the maximum color used in w_λ
11: **end for**
12: **return** $w = \cup_{\lambda=1}^{\bar{W}} w_\lambda$.

4.3   The General Case

The following lemma of [13] shows that, given a ρ-approximation algorithm for $(G, \mathcal{P}, g, 1)$, it is possible to obtain a 4ρ-approximation algorithm for $(G, \mathcal{P}, g, d)$.

Lemma 7 ([13]). *Given a polynomial time ρ-approximation algorithm $\mathcal{A}$ for $(G, \mathcal{P}, g, 1)$, for any $d > 1$, it is possible to obtain a polynomial time algorithm $\mathcal{A}'$ guaranteeing a $(4 \cdot \rho)$-approximation for $(G, \mathcal{P}, g, d)$.*

Given an NSI $\mathcal{N}$, let $G(\mathcal{N}) = (V(\mathcal{N}), E(\mathcal{N}))$ be the graph corresponding to the NSI $\mathcal{N}$, i.e. such that $V(\mathcal{N}) = \bigcup_{P \in \mathcal{N}} V(P)$ and $E(\mathcal{N}) = \bigcup_{P \in \mathcal{N}} E(P)$. We now provide a polynomial time algorithm (Algorithm 7) whose existence is shown in Lemma 7, transforming a feasible solution for $(G, \mathcal{P}, g, 1)$ into a feasible solution for $(G, \mathcal{P}, g, d)$.

By combining Algorithms 4, 6 and 7, we are finally able to provide a polynomial time approximation algorithm (Algorithm 8) working for any $(G, \mathcal{P}, g, d)$ instance of the general problem.

By exploiting Lemmata 7, 6 and 4, we obtain the following theorem.

Theorem 3. *Algorithm 8 is a $(32H_{2g-1} + 12)$-approximation polynomial time algorithm for general instances.*

Algorithm 7. $((G, \mathcal{P}, g, d), w)$, w being a valid coloring such that $\mathcal{P}_\lambda$ is an NSI for any λ

1: Let $1, 2, \ldots, W$ be the colors used by w
2: **for** $\lambda = 1$ to W **do**
3:     $\mathcal{N}_\lambda = \mathcal{P}_\lambda$
4:     **for** each connected component G' in the subgraph of $G(\mathcal{N}_\lambda)$ induced by the nodes in $SPAN(\mathcal{N}_\lambda)$ **do**   ▷ G' is either a path or a cycle
5:         Build r_w such that in G' there is a regenerator every d nodes
6:     **end for**
7: **end for**
8: **return** (w, r_w)

Algorithm 8. $(\mathcal{G}, \mathcal{P}, g, d)$

1: $\mathcal{P}' \leftarrow \emptyset$ ▷ Partition $\mathcal{P}$ into $\bar{\mathcal{P}}$ and $\mathcal{P}'$ such that $\bar{\mathcal{P}}$ is a maximal proper set of paths.
2: $\bar{\mathcal{P}} \leftarrow \mathcal{P}$
3: **while** there exist $P, \bar{P} \in \bar{\mathcal{P}}$ such that P is included in $\bar{P}$ **do**
4:     $\mathcal{P}' \leftarrow \mathcal{P}' \cup \{P\}$
5:     $\bar{\mathcal{P}} \leftarrow \bar{\mathcal{P}} \setminus \{P\}$
6: **end while**
7: $w' \leftarrow$ **Algorithm 4** $(\mathcal{G}, \bar{\mathcal{P}}, g, 1)$
8: $w'' \leftarrow$ **Algorithm 6** $((\mathcal{G}, \mathcal{P}, g, 1), w')$
9: $w \leftarrow$ **Algorithm 7** $((\mathcal{G}, \mathcal{P}, g, d), w'')$
10: **return** w

---

5 Open Problems

The main open problem is that of closing the gap between the hardness result of Section 3 and the approximation ratio guaranteed by the Algorithm 8 provided in Section 4. In particular, determining whether the problem is in APX constitutes a very interesting research direction.

Interesting research directions are that of modeling the network by means of a edge-weighted graph and also that of considering lightpaths requiring a bandwidth being $\frac{b}{g}$, with $1 \leq b \leq g$; notice that in this paper we have dealt with the case $b = 1$.

It would be also interesting to extend our result by considering more involved cost functions taking into account other switching parameters (e.g., the ADMs - Add-Drop-Multiplexers - used at the endpoints of the lightpath) or the possibility of splitting paths.

References

1. Amini, O., Pérennes, S., Sau, I.: Hardness and approximation of traffic grooming. Theor. Comput. Sci. 410(38-40), 3751–3760 (2009)
2. Ausiello, G., Crescenzi, P., Gambosi, G., Kann, V., Marchetti-Spaccamela, A., Protasi, M.: Complexity and Approximation, Combinatorial Optimization Problems and Their Approximability Properties. Springer, Heidelberg (1999)

3. Chen, S., Ljubic, I., Raghavan, S.: The regenerator location problem, vol. 55, pp. 205–220 (2010)
4. Chvátal, V.: A greedy heuristic for the set covering problem. Mathematics of Operation Research 4, 233–235 (1979)
5. Călinescu, G., Frieder, O., Wan, P.-J.: Minimizing electronic line terminals for automatic ring protection in general wdm optical networks. IEEE Journal of Selected Area on Communications 20(1), 183–189 (2002)
6. Călinescu, G., Wan, P.-J.: Splitable traffic partition in wdm/sonet rings to minimize sonet adms. Theoretical Computer Science 276(1-2), 33–50 (2002)
7. Călinescu, G., Wan, P.-J.: Traffic partition in wdm/sonet rings to minimize sonet adms. Journal of Combinatorial Optimization 6(4), 425–453 (2002)
8. Fedrizzi, R., Galimberti, G.M., Gerstel, O., Martinelli, G., Salvadori, E., Saradhi, C.V., Tanzi, A., Zanardi, A.: A Framework for Regenerator Site Selection Based on Multiple Paths. In: Prooceedings of IEEE/OSA Conference on Optical Fiber Communications, OFC (2010)
9. Flammini, M., Marchetti-Spaccamela, A., Monaco, G., Moscardelli, L., Zaks, S.: On the complexity of the regenerator placement problem in optical networks. IEEE/ACM Transactions on Networking 19(2), 498–511 (2011)
10. Flammini, M., Monaco, G., Moscardelli, L., Shachnai, H., Shalom, M., Tamir, T., Zaks, S.: Minimizing total busy time in parallel scheduling with application to optical networks. Theor. Comput. Sci. 411(40-42), 3553–3562 (2010)
11. Flammini, M., Monaco, G., Moscardelli, L., Shalom, M., Zaks, S.: Approximating the Traffic Grooming Problem with Respect to ADMs and OADMs. In: Luque, E., Margalef, T., Benítez, D. (eds.) Euro-Par 2008. LNCS, vol. 5168, pp. 920–929. Springer, Heidelberg (2008)
12. Flammini, M., Monaco, G., Moscardelli, L., Shalom, M., Zaks, S.: Optimizing Regenerator Cost in Traffic Grooming (extended abstract). In: Lu, C., Masuzawa, T., Mosbah, M. (eds.) OPODIS 2010. LNCS, vol. 6490, pp. 443–458. Springer, Heidelberg (2010)
13. Flammini, M., Monaco, G., Moscardelli, L., Shalom, M., Zaks, S.: Optimizing regenerator cost in traffic grooming. Technical Report CS-2011-07, Technion, Department of Computer Science (2011)
14. Gerstel, O., Lin, P., Sasaki, G.: Wavelength assignment in a wdm ring to minimize cost of embedded sonet rings. In: INFOCOM 1998, Seventeenth Annual Joint Conference of the IEEE Computer and Communications Societies, pp. 69–77 (1998)
15. Mertzios, G.B., Sau, I., Shalom, M., Zaks, S.: Placing Regenerators in Optical Networks to Satisfy Multiple Sets of Requests. In: Abramsky, S., Gavoille, C., Kirchner, C., Meyer auf der Heide, F., Spirakis, P.G. (eds.) ICALP 2010. LNCS, vol. 6199, pp. 333–344. Springer, Heidelberg (2010)
16. Sriram, K., Griffith, D., Su, R., Golmie, N.: Static vs. dynamic regenerator assignment in optical switches: models and cost trade-offs. In: Workshop on High Performance Switching and Routing (HPSR), pp. 151–155 (2004)
17. Vazirani, V.V.: Approximation Algorithms. Springer, Heidelberg (2004)
18. Winkler, P., Zhang, L.: Wavelength assignment and generalized interval graph coloring. In: SODA, pp. 830–831 (2003)
19. Yang, X., Ramamurthy, B.: Sparse Regeneration in Translucent Wavelength-Routed Optical Networks: Architecture, Network Design and Wavelength Routing. Photonic Network Communications 10(1), 39–53 (2005)

On the Cost of Concurrency
in Transactional Memory[*]

Petr Kuznetsov and Srivatsan Ravi

TU Berlin/Deutsche Telekom Laboratories

Abstract. The promise of software transactional memory (STM) is to combine an easy-to-use programming interface with an efficient utilization of the concurrent-computing abilities provided by modern machines. But does this combination come with an inherent cost?

We evaluate the cost of concurrency by measuring the amount of expensive synchronization that must be employed in an STM implementation that ensures *positive* concurrency, i.e., allows for concurrent transaction processing in some executions. We focus on two popular progress conditions that provide positive concurrency: *progressiveness* and *permissiveness*.

We show that in permissive STMs, providing a very high degree of concurrency, a transaction may perform a linear number of expensive synchronization patterns with respect to its read-set size. In contrast, progressive STMs provide a very small degree of concurrency but, as we demonstrate, can be implemented using at most one expensive synchronization pattern per transaction. However, we show that even in progressive STMs, a transaction has to "protect" (e.g., by using locks or strong synchronization primitives) a linear amount of data with respect to its write-set size. Our results suggest that achieving high degrees of concurrency in STM implementations may bring a considerable synchronization cost.

1 Introduction

The software transactional memory (STM) paradigm promises to efficiently exploit the concurrency provided by modern computers while offering an easy-to-use programming interface. It allows a programmer to write a concurrent program as a sequence of *transactions*. A transaction is a series of read and write operations on *transactional objects* (or *t-objects*). An STM implementation turns this series into a sequence of accesses to underlying *base objects* and exports "all-or-nothing" semantics: a transaction either *commits* in which case all its operations instantaneously "take effect", or *aborts* in which case the transaction does not affect any other transaction. In this paper, the default STM correctness property is *opacity* [13,15] that, informally, requires that in every execution,

---

[*] The research leading to these results has received funding from the European Union Seventh Framework Programme (FP7/2007-2013) under grant agreement N 238639, ITN project TRANSFORM.

A. Fernández Anta, G. Lipari, and M. Roy (Eds.): OPODIS 2011, LNCS 7109, pp. 112–127, 2011.

there is a total order on all transactions, including aborted ones, where every read operation returns the argument of the last committed write operation on the read t-object.

An STM implementation that aborts every transaction is trivially correct but useless. Therefore, we need to specify a *progress condition* that captures the execution scenarios in which a transaction should commit. Consider, for example, a simple non-trivial progress condition that only requires a transaction to commit if it does not run concurrently with any other transaction. This condition can be implemented using a single lock that is acquired at the beginning of a transaction and released at its end. The resulting "single-lock" STM will be running one transaction at a time, thus ignoring the potential benefits of multiprocessing. Similarly, an *obstruction-free* STM [12] that only requires a transaction to commit if it eventually runs with no contention allows for no concurrency at all. But to exploit the power of modern multiprocessor machines, an STM implementation must allow at least *some* transactions to make progress concurrently. If this is the case, we say that the implementation provides *positive* concurrency, in contrast to zero concurrency provided by "single-lock" and obstruction-free STMs.

In this paper, we try to understand the inherent costs of allowing multiple concurrent transactions to commit. Therefore, we focus on progress conditions that provide positive concurrency: *progressiveness* [14] and *permissiveness* [11]. Informally, a progressive STM [14] provides a very small degree of concurrency by only enforcing a transaction T to commit if it encounters no concurrent *conflicting* transaction T': T and T' conflict on a t-object X if they concurrently access X and one of the transactions tries to update X. A stronger variant of progressiveness, called *strong progressiveness*, additionally requires that in case a set of transactions conflict on at most one t-object, at least one transaction commits. A much more demanding permissive STM [11] stipulates that a transaction must commit, unless committing it violates correctness, which, intuitively, provides the highest possible degree of concurrency.

To understand the inherent cost of positive concurrency in STM implementations, we first consider the number of $RAW/AWAR$ synchronization patterns [6] that must be performed by a process in the course of a transaction. A *read-after-write* (RAW) pattern consists of a write to a (shared) base object x followed by a read from a different base object y (without a write to y in between). An *atomic write-after-read* (AWAR) pattern consists of an atomic (indivisible) execution of a read of a base object followed by a write on (possibly the same) base object. Accounting for RAW/AWAR patterns is important since most modern processor architectures use relaxed memory models, where maintaining the order of operations in a RAW requires a *memory fence* [22] and each AWAR is manifested as an atomic instruction such as Compare-and-Swap (CAS). In most architectures, memory fences and atomic instructions are believed to be considerably slower than regular shared-memory accesses [1, 20, 22, 21].

We show that every permissive and opaque STM implementation has, for any $m \in \mathbb{N}$, an execution in which a transaction with a read set of size m

incurs $\Omega(m)$ consecutive RAW/AWAR patterns. This contrasts with a single-lock STM that uses only one such pattern, since a successful lock acquisition can be implemented using only one (multi-) RAW [19] or (multi-)AWAR [4].[1] We show that one RAW/AWAR is in fact optimal for single lock STMs. Moreover, we present implementations of *progressive* STMs that employ just a single multi-RAW or multi-AWAR pattern per transaction. Additionally, we describe a *strongly progressive* space-bounded STM implementation that incurs four RAWs per transaction.

These implementations suggest that the (multi-)RAW/AWAR metric is too coarse-grained to evaluate the complexity of progressive STMs: for example, a multi-AWAR such as $mCAS$ may be very hard to implement in practice. Therefore, we introduce a new metric called *protected data size* that, intuitively, captures the amount of data that a transaction must exclusively control at some point of its execution. All progressive STM implementations we are aware of (see, e.g., an overview in [14]) use locks or timing assumptions to give an updating transaction exclusive access to all objects in its write set at some point of its execution. E.g., lock-based progressive implementations require that a transaction grabs all locks on its write set before updating the corresponding base objects. Our results show that this is an inherent price to pay for providing progressive concurrency: every committed transaction in a progressive and *disjoint-access-parallel*[2] STM implementation must, at some point of its execution, protect every object in its write set. Interestingly, as our progressive implementations show, the transaction's read set does not need to be protected.

In brief, our results imply that providing high degrees of concurrency in opaque STM implementations incurs a considerable synchronization cost. Permissive STMs, while providing the best possible concurrency in theory, require a strong synchronization primitive or a memory fence per read operation, which may result in excessively slow execution times. Progressive STMs provide only basic concurrency but perform considerably better in this respect: we present progressive implementations that incur constant RAW/AWAR complexity. Does this mean that maximizing the ability of processing multiple transactions in parallel should not be an important factor in STM design? Should we rather assume little positive concurrency provided by progressiveness or even focus on speculative single-lock solutions á la *flat combining* [16]? Difficult to say affirmatively, but our results suggest that the question makes sense.

The rest of the paper is organized as follows. Section 2 briefly introduces our system model and defines the correctness criteria of STM implementations. Section 3 recalls the definitions of the progress conditions. Sections 4 presents a

---

[1] A multi-RAW consists of a series of writes followed by a series of reads from distinct locations. Maintaining the multi-RAW order can be achieved with a single memory fence. A multi-AWAR (e.g., multi-CAS) performs an atomic write-after-read on multiple base objects.

[2] A disjoint-access-parallel STM implementation [17,8] guarantees that concurrent transactions accessing disjoint sets of transactional objects are executed independently of each other, i.e., without conflicting on the base objects.

linear lower bound on the number of RAW/AWAR patterns executed by a transaction in a permissive STM. Section 5 describes our progressive STM implementations that perform constant RAWs or AWARs per transaction and presents a lower bound on the amount of data to be protected by a transaction in a progressive STM. Section 6 summarizes some related work and Section 7 concludes the paper. The full version of this paper [18] contains detailed correctness proofs for our implementations along with some extensions and side results.

2   Model

Our STM model, while keeping the spirit of the original definitions of [13, 15], introduces some refinements that are instrumental for our results.

Transactions. Transactional memory provides the ability of reading and writing to a set of *transactional* objects, or *t-objects* using atomic *transactions*. A transaction is a sequence of accesses (reads or writes) to t-objects. We assume that every transaction T_k has a unique identifier k. Formally, STM exports the following operations (called *tm-operations* in the paper): (1) $read_k(X)$ that returns a value in a set V or a special value $A_k \notin V$ (*abort*); (2) $write_k(X, v)$ that returns ok_k or A_k; (3) $tryC_k$ that returns $C_k \notin V$ (*commit*)or A_k and (4) $tryA_k$ that returns A_k.

A *history* H is a sequence of invocations and responses of tm-operations. A history H is *sequential* if every invocation is either the last event in H or is immediately followed by a matching response. $H|k$ denotes the subsequence of H restricted to events with index k. If $H|k$ is non-empty we say that T_k *participates* in H, and $parts(H)$ denotes the set of transactions that participate in H. A history is *well-formed* if for all T_k, $H|k$ is sequential and contains no events that appear after A_k or C_k. Throughout this paper, we assume that all histories are well-formed, i.e., the user of transactional memory never invokes a new operation before receiving a response from the current one and does not invoke any operation op_k after T_k has returned C_k or A_k. A history H is *complete* if for every $T_k \in parts(H)$, $H|k$ ends with a response event. A transaction $T_k \in parts(H)$ is *live* in H if $H|k$ does not end with A_k or C_k. Otherwise, T_k is called *complete*. A history is *t-complete* if $parts(H)$ contains only complete transactions. A transaction $T_k \in parts(H)$ is *forcefully aborted* in H if some operation $op_k \neq tryA_k$ returns A_k. Two histories H and H' are *equivalent* if for every transaction T_k, $H|k = H'|k$.

The *read set* (resp., the *write set*) of a transaction $T_k \in parts(H)$, denoted $Rset(T_k)$ (resp., $Wset(T_k)$), is the set of t-objects that T_k reads (resp., writes to) in H. $Dset(T_k) = Rset(T_k) \cup Wset(T_k)$ is called the *data set* of T_k. A transaction T_k is called *read-only* if $Wset(T_k) = \emptyset$, otherwise, it is called *updating*.

Real-time and deferred-update orders. For $T_k, T_m \in parts(H)$, we say that T_k *precedes* T_m in the *real-time order* in H, and we write $T_k \prec_H T_m$, if T_k is committed or aborted and the last event of T_k precedes the first event of T_m

in H. If neither $T_k \prec_H T_m$ nor $T_m \prec_H T_k$, then we say that T_k and T_m are *concurrent* in H. A transaction $T_k \in parts(H)$ which is not concurrent with any other transaction in H is called *uncontended* in H. A history H is *t-sequential* if no two transactions are concurrent in H.

For $T_k, T_m \in parts(H)$, we say that T_k *precedes* T_m in the *deferred-update order*, and we write $T_k \prec_H^{DU} T_m$ if there exists $X \in Rset(T_k) \cap Wset(T_m)$, T_m has committed, such that the response of $read_k(X)$ precedes the invocation of $tryC_m()$ in H. For $T_k, T_m \in parts(H)$, we write $T_k \overset{X}{\prec_H} T_m$, if T_k has committed and the response of $read_m(X)$, $X \in Rset(T_m) \cap Wset(T_k)$ returns v, the value of X updated in $write_k(X, v)$.

Legal histories. Let H be a complete t-sequential history. For every operation $read_k(X)$ in H that reads a t-object X, we define the *latest written value* of X as follows: (1) If T_k contains a $write_k(X, v)$ preceding $read_k(X)$ then the latest written value of X is the value of the latest such write. (2) Otherwise, if H contains a $write_m(X, v)$ such that $m \neq k$, T_m precedes T_k, and T_m commits in H, then the latest written value of X is the value of the latest such write in H. (3) Otherwise, the latest written value of X is the initial value of X. Without loss of generality, we assume that H starts with a fictitious initializing transaction T_0 that writes 0 to every t-object. We say that a complete t-sequential history H is *legal* if for every t-object X, every read of X in H returns the latest written value of X.

Opacity. Let H be any complete sequential history. Now $\bar{H}$ denotes a history constructed from H as follows: (1) For every live transaction T_k in H, we insert $tryC_k \cdot A_k$ immediately after the last event of T_k in H and (2) For every aborted transaction T_k in H, we remove all write operations in T_k with the matching responses.

Definition 1. *A complete sequential history H is* opaque *if there exists a legal complete t-sequential history S such that (1) $\bar{H}$ and S are equivalent and (2) S respects $\prec_H$ and $\prec_H^{DU}$.*

We call such a legal complete t-sequential history S a *serialization* of H. A weaker property, called *strict serializability* [23], guarantees opacity with respect to committed transactions in H. Obviously, every opaque history is also strictly serializable.

Implementations. We consider an asynchronous shared-memory system in which processes $p_1, \ldots p_N$ communicate by executing atomic operations on shared *base objects*.

An STM *implementation* provides the processes with algorithms for operations $read_k$, $write_k$, $tryC_k$ and $tryA_k$. Without loss of generality, we assume that base objects are accessed with atomic read-write operations, but we allow the programmer to aggregate a sequence of operations on base objects using clearly demarcated *atomic sections*: the operations within an atomic section are to be executed sequentially. The atomic-section construct is general enough to implement various strong synchronization primitives, such as *test-and-set* (TAS) or *compare-and-swap* (CAS). We assume that atomic sections may only contain a bounded number of base-object operations.

An *execution* of an implementation M is a sequence of atomic accesses to base objects (*base-object events*), and invocation and responses of the TM operations (*TM-events*). If a base-object event is a write or an atomic-section that contains a write (in one of its execution paths), we say that the event is *non-trivial*.

A *configuration* of M (after some execution E) is determined by the states of all base objects and the states of the processes. An *initial state* of M is determined by the initial states of base objects and t-objects. We assume that each base object and each t-object is initialized to 0. A *history* of an execution E, denoted by $E|_{TM}$ is the subsequence of E restricted to TM-events. $E|_{TM,p_i}$ denotes the subsequence of $E|_{TM}$ restricted to events issued by process p_i.

The *interval of a transaction* T_k in E is the fragment of E that starts with the first event of T_k in E and ends with the completing event of T_k (A_k or C_k) in E, or, if T_k has not completed in E, with the last event of E. A tm-operation op_1 *precedes* op_2 in H if the invocation of op_2 appears after the response of op_1 in H. An execution E is *well-formed* if every atomic section is executed sequentially in E, $E|_{TM,p_i}$ is t-sequential for each p_i, and no event on behalf of a transaction T_k is taking place outside of an interval between invocation and response of some TM-operation in T_k. We assume here that a TM implementation generates only well-formed executions.

A *completion* of H is a history constructed from H by removing some pending invocations and adding responses to the remaining pending invocations to the end of H. To account for initial values of t-objects, we add to the beginning of H a (fictitious) transaction T_0 that writes 0 to every t-object and commits. A complete sequential history H' is a *linearization* of H if there exists a history H'', a completion of H, such that (1) H' respects the precedence order of H, and (2) H' and H'' are equivalent.

Definition 2. *An STM implementation M is* opaque *if for every execution E of M, there exists an opaque linearization of $E|_{TM}$.*

RAW/AWAR complexity. Let M be an STM implementation. Let π be a fragment of an execution of M and let π_i denote the i-th event in π ($i = 0, \ldots, |\pi|-1$). We say that process p performs a *RAW* (read-after-write) in π if $\exists i, j; 0 \leq i < j < |\pi|$ such that (1) π_i is a write to a base object x by process p, (2) π_j is a read of a base object $y \neq x$ by process p and (3) there is no π_k such that $i < k < j$ and π_k is a write to y by p. We say that two RAWs by process p *overlap* in an execution E with the read event of the first RAW occurs after the write event of the second RAW. A *multi-RAW* consists of series of writes to a set of base objects, followed by a series of reads from different base objects.

We say a process p performs an *AWAR* (atomic-write-after-read) in π if $\exists i, j, 0 \leq i < j < |\pi|$ such that (1) π_i is a read of a base object x by process p, (2) π_j is a write to a base object y by process p and (3) π_i and π_j belong to the same atomic section. A multi-AWAR performs an atomic write-after-read on multiple base objects. Examples of AWAR and multi-AWAR are *CAS* and *mCAS* [5] respectively.

In the rest of the paper we simply say RAW/AWAR instead of multi-RAW/AWAR.

Disjoint-access parallelism. Let I be a fragment of an execution E. Following [17, 8], we first define a *conflict graph* which relates transactions that are live in I. Vertices of the graph represent t-objects. The vertices representing distinct t-objects X and Y are related with an edge if and only if there is a transaction T such that $\{X, Y\} \subseteq Dset(T)$ and the interval of T overlaps with I in E.

Two transactions T_i and T_j are *disjoint-access in E* if there is no path between an item in $Dset(T_i)$ and an item in $Dset(T_j)$ in the conflict graph of the minimal execution interval containing the intervals of T_i and T_j.

Two processes *concurrently contend on a base object x* in a given configuration if they have pending events on x in the configuration and one of these events is non-trivial (contains a write to x).

Definition 3. *An STM implementation M is* disjoint-access parallel *(DAP) if, for all executions E of M, two processes executing T_i and T_j concurrently contend on the same base object in E only if T_i and T_j are not disjoint-access.*

Lemma 1. *[8] Let E be an execution of a DAP STM implementation M in which a complete execution of T_1 is immediately followed by a (possibly incomplete) execution of T_2 such that T_1 and T_2 are disjoint-access. Then there does not exist a base object x such that both processes executing T_1 and T_2 access x in E and one of the accessing events is non-trivial.*

3   Liveness and Progress

To describe the conditions under which a TM implementation does something useful, we need to address two orthogonal dimensions. First, we need to give a *tm-liveness* property [3] that determines the conditions under which an individual tm-operation must return. Second, we need to give a *progress condition* that describes the conditions under which a transaction must commit.

TM-liveness properties. A TM implementation M is *wait-free* if in every infinite execution of M, each tm-operation returns in a finite number of its own steps, regardless of the behavior of concurrent transactions. In other words, a wait-free individual tm-operation (tm-read, tm-write, tryC or tryA) cannot be delayed because of a concurrent operation. The property can be very beneficial if executions of transactions are subject to unpredictable delays or failures.

In this paper, we do not assume failures: every operation is expected to take steps until it terminates. Moreover, we are interested in deriving inherent costs of implementing non-trivial concurrency in TM. Therefore, we assume a weaker default tm-liveness guarantee, that we call *starvation-freedom*. A TM implementation M is starvation-free in every infinite execution of M, each tm-operation eventually returns, assuming that no concurrent tm-operation stops indefinitely before returning. Starvation-freedom allows a tm-operation to be delayed only by a concurrent tm-operation.

Progress conditions. A progress condition determines the scenarios in which a transaction is allowed to abort. Technically, unlike tm-liveness, a progress condition is a safety property [3], since it can be violated in a finite execution. The simplest non-trivial progress property we consider in this paper is *single-lock progressiveness* that says that a transaction can only abort if there is a concurrent transaction. Clearly, an opaque single-lock TM can be implemented using any mutual exclusion algorithm [24] with one critical section per transaction. Stronger progress conditions allow some transactions to progress concurrently in some scenarios implying *positive concurrency*.

Progressiveness allows an implementation to abort a transaction only in case of a conflict. Transactions T_i, T_j *conflict* in a history H on a t-object X if T_i and T_j are concurrent in H, $X \in Dset(T_i) \cap Dset(T_j)$, and $X \in Wset(T_i) \cup Wset(T_j)$.

Definition 4. *A TM implementation M is* (weakly) progressive *if for every history H of M and every transaction $T_i \in parts(H)$ that is forcefully aborted, there exists a prefix H' of H and a transaction $T_k \in parts(H')$ that is live in H', such that T_k and T_i conflict in H'.*

The *strong progressiveness* property [14] additionally requires that, in case of a conflict on at most one t-object, at least one transaction commits. A formal definition can be found in [15].

Let C be any correctness property, i.e., any safety property on TM histories [3]. The following property guarantees that no transaction is forcefully aborted if there is a chance of committing the transaction and preserving C.

Definition 5. *A TM implementation M is* permissive with respect to C *if for every history H of M such that H ends with a response r_k and replacing r_k with some $r_k \neq A_k$ gives a history that satisfies C, we have $r_k \neq A_k$.*

In this paper, we consider TM implementations that are permissive with respect to opacity. Clearly, permissiveness with respect to opacity is strictly stronger than progressiveness: every opaque and permissive with respect to opacity implementation is also opaque and progressive.

4   RAW/AWAR Cost of Permissive STMs

In this section, we show that an execution of a transaction in a permissive STM implementation may require to perform at least one RAW/AWAR pattern *per* tm-read.

Let M be a permissive, opaque TM implementation. Consider an execution E of M with a history H consisting of transactions T_1, T_2, T_3 as shown in Figure 1:

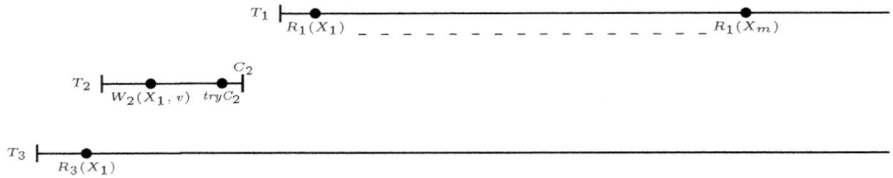

Fig. 1. Execution E of a permissive, opaque STM: T_2 and T_3 force T_1 to perform a RAW/AWAR in each $R_1(X_k)$, $2 \leq k \leq m$

T_3 performs a read of X_1, then T_2 performs a write on X_1 and commits, and finally T_1 performs a series of reads from objects $X_1, \ldots, X_m$. Here, $R_k(X)$, $W_k(X, v)$ denote complete executions of $read_k(X)$ and $write_k(X, v)$ respectively. Since the implementation is permissive, no transaction can be forcefully aborted in E, and the only valid serialization of this execution is T_3, T_2, T_1. Note also that the execution generates a sequential history: each invocation of a tm-operation is immediately followed by a matching response in H. Thus, since we assume starvation-freedom as a liveness property, such an execution exists.

Imagine that we modify the execution E as follows. Immediately after $R_1(X_k)$ executed by T_1 we add $W_3(X, v)$, and $tryC_3$ executed by T_3 (let $TC_3(X_k)$ denote the complete execution of $W_3(X_k, v)$ followed by $tryC_3$). Obviously, $TC_3(X_k)$ must return abort: neither T_3 can be serialized before T_1 nor T_1 can be serialized before T_3. On the other hand if $TC_3(X_k)$ takes place just before $R_1(X_k)$, then $TC_3(X_k)$ must return commit but $R_1(X_k)$ must return the value written by T_3. In other words, $R_1(X_k)$ and $TC_3(X_k)$ are *strongly non-commutative* [6]: both of them see the difference when ordered differently. As a result, intuitively, $R_1(X_k)$ needs to perform a RAW or AWAR to make sure that the order of these two "conflicting" operations is properly maintained. A formal proof follows.

Theorem 1. *Let M be a permissive opaque STM implementation. Then, for any $m \in \mathbb{N}$, M has an execution in which some transaction performs m tm-reads such that the execution of each tm-read contains at least one RAW or AWAR.*

Proof. We consider $R_1(X_k)$, $2 \leq k \leq m$ in execution E.

Imagine a modification E' of E, in which T_3 performs $W_3(X_k)$ immediately after $R_1(X_k)$ and then tries to commit. A serialization of $H' = E'|_{TM}$ should obey $T_3 \prec_{H'}^{DU} T_2$ and $T_2 \prec_{H'} T_1$. The execution of $R_1(X_k)$ does not modify base objects, hence, T_3 does not observe $R_1(X_k)$ in E'. Since M is permissive, T_3 must commit in E'. But since T_1 performs $R_1(X_k)$ before T_3 commits and T_3 updates X_k, we also have $T_1 \prec_{H'}^{DU} T_3$. Thus, T_3 cannot precede T_1 in any serialization—contradiction. Consequently, each $R_1(X_k)$ must perform a write to a base object.

Let π be a fragment of E that represents the complete execution of $R_1(X_k)$. Clearly, π contains a write to a base object. Let π_j be the first write to a base object in π and π_w, the shortest fragment of π that contains the atomic section to which π_j belongs, else if π_j is not part of an atomic section, $\pi_w = \pi_j$. Thus, π can be represented as $\pi_s \cdot \pi_w \cdot \pi_f$.

Suppose that π does not contain a RAW or AWAR. Since π_w does not contain an AWAR, there are no read events in π_w that precede π_j. Thus, π_j is the first base object event in π_w. Consider the execution fragment $\pi_s \cdot \rho$, where ρ is the complete execution of $TC_3(X_k)$ by T_3. Such an execution exists since π_s does not perform any base object write, hence, $\pi_s \cdot \rho$ is indistinguishable to T_3 from ρ.

Since, by our assumption, $\pi_w \cdot \pi_f$ contains no RAW, any read performed in $\pi_w \cdot \pi_f$ can only be applied to base objects previously written in $\pi_w \cdot \pi_f$. Thus, there exists an execution $\pi_s \cdot \rho \cdot \pi_w \cdot \pi_f$ that is indistinguishable to T_1 from π. In $\pi_s \cdot \rho \cdot \pi_w \cdot \pi_f$, T_3 commits (as in ρ) but T_1 ignores the value written by T_3 to X_k. But T_3, T_2, T_1 is the only valid serialization for $E|_{TM}$—contradiction. Thus, each $R_1(X_k)$, $2 \leq k \leq m$ must contain a RAW/AWAR.

Note that since all tm-reads of T_1 are executed sequentially, all these RAW/AWAR patterns are pairwise non-overlapping.

5   RAW/AWAR Cost and Protected Data in Progressive STMs

In this section, we first describe our progressive STM implementations that perform at most one RAW/AWAR per transaction. Then we present a lower bound on the amount of data to be protected by a transaction in a progressive STM.

5.1   Constant RAW/AWAR Implementations for Progressive STM

We show first that even a *single-lock* progressive STM cannot avoid performing one RAW/AWARs per transaction in some executions (the proof is a simple variation of the arguments of [6], where a single RAW/AWAR is shown to be necessary to acquire an indivisible lock).

Theorem 2. *Let M be a single-lock progressive opaque STM implementation. Then every execution of M in which an uncontended transaction performs at least one read, at least one write, and commits, must contain a RAW/AWAR pattern.*

Since every progressive or permissive STM implementation is also single-lock progressive, the RAW/AWAR lower bound of Theorem 2 also holds for progressive and permissive STM implementations. The lower bound is actually tight, and we sketch two progressive opaque implementations (complete proofs of correctness can be found in [18]). Both implementations are *strict data-partitioned* [15] (split the set of base objects used into disjoint subsets, each subset storing information of only a single t-object) and *single-version* (maintain exactly one copy of a t-object's state at a time). They also use *invisible reads*, i.e., no execution of a tm-read operation performs a write to a base object.

Our first implementation uses a simple *multi-trylock* primitive which can be implemented with a single RAW. The *multi-trylock* primitive exports operations $acquire(W)$, $release(W)$ and $isContended(X)$, for all sets of t-objects W and all t-objects X. Informally, if there is no contention on the locks on objects in W, then $acquire(W)$ returns *true* which means that exclusive locks on all objects in W are acquired. Otherwise, $acquire(W)$ returns *false* which means that no locks on objects in W are acquired. Operation $release(W)$ releases the acquired locks on objects in W and $isContended(X)$ returns *true* *iff* a lock on X is currently held by any other process. The implementation of $acquire(W)$ first writes to a series of base objects and then reads a series of base objects incurring a single RAW, while operations $release(W)$ and $isContended(X)$ incur no RAW.

Algorithm 1. Progressive STM with one multi-RAW: the implementation of T_k executed by p_i

---

1: **Shared variables:**
2:     v_j, for each t-object X_j
3:     L, a multi-trylock object

4: $read_k(X_j)$:
5:     $ov_j := read(v_j)$
6:     $Rset(T_k) := Rset(T_k) \cup \{X_j\}$
7:     **if** isAbortable() **then**
8:         **return** A_k
9:     **return** the value of ov_j

10: $write_k(X_j, v)$:
11:     **if** $X_j \notin Wset(T_k)$ **then**
12:         $nv_j := v$
13:         $Wset(T_k) := Wset(T_k) \cup \{X_j\}$
14:     **return** ok_k

15: $tryA_k()$:
16:     **return** A_k

17: $tryC_k()$:
18:     **if** $|Wset(T_k)| = \emptyset$ **then**
19:         **return** C_k
20:     $locked := L.acquire(Wset(T_k))$
21:     **if** not locked **then**
22:         **return** A_k
23:     **if** isAbortable() **then**
24:         $L.release(Wset(T_k))$
25:         **return** A_k
26:     **for all** $X_j \in Wset(T_k)$ **do**
27:         $write(v_j, (nv_j, k))$
28:     $L.release(Wset(T_k))$
29:     **return** C_k

30: **Function: isAbortable():**
31:     **if** $\exists X_j \in Rset(T_k):L.isContended(X_j)$ **then**
32:         **return** *true*
33:     **if** isInvalid() **then**
34:         **return** *true*
35:     **return** *false*

36: **Function: isInvalid():**
37:     **if** $\exists X_j \in Rset(T_k):ov_j \neq read(v_j)$ **then**
38:         **return** *true*
39:     **return** *false*

Algorithm 1 describes the implementation of a progressive STM incurring a single RAW per updating transaction. Every t-object X_i is associated with a distinct base object v_i that stores the "most recent" value of X_i together with the *id* of the transaction that was the last to update X_i. Each time a transaction T_k performs a read of a t-object X_i, it reads v_i, adds X_i to its read set and checks if the t-objects in the current read set of T_k have not been updated since T_k has read them. Additionally, the implementation checks if no object in the current read set is *locked* by an updating transaction. If some object in the read set has been modified or is locked, the transaction is forcefully aborted. Otherwise, T_k returns the value read in v_i. Each time T_k performs a write to a t-object X_i, it adds X_i to its write set and returns *ok*. For every updating transaction T_k, $tryC_k()$ invokes $acquire(Wset(T_k))$. If it returns *true*, $tryC_k()$ returns C_k, otherwise it returns A_k. Read-only transactions simply returns C_k.

Theorem 3. *There exists a progressive opaque STM implementation with wait-free operations that employs a single RAW per transaction. Moreover, no RAWs are performed in read-only transactions.*

Alternatively, a single AWAR implementation can be trivially obtained using a (theoretical) *mCAS* primitive [5].

In [18], we also derive a *strongly progressive* STM using only reads and writes that incurs at most four RAWs per updating transaction and uses a finite number of bounded registers. This implementation uses a *starvation-free multi-trylock* primitive inspired by the Black-White Bakery Algorithm [25], a bounded version of the Bakery Algorithm [19]. Informally, if no concurrent process contends infinitely long on some $X \in W$, then the $acquire(W)$ operation of the starvation-free multi-trylock eventually returns *true* which means that exclusive locks on all objects in W are acquired. The implementation of $acquire(W)$ incurs three RAWs, while operation $release(W)$ performs a single RAW.

Implementations of tm-reads and tm-writes are identical to the ones in Algorithm 1. For every updating transaction T_k, $tryC_k()$ invokes the *acquire* operation of the starvation-free multi-trylock over $Wset(T_k)$. Note that this always returns *true* and a transaction T_k with $Rset_k = \emptyset$ eventually returns C_k. Read-only transactions simply returns C_k. Consequently, the implementation incurs four RAWs per updating transaction.

Theorem 4. *There exists a strongly progressive single-version opaque STM implementation with starvation-free operations that uses invisible reads and employs four RAWs per transaction. Moreover, no RAWs are performed in read-only transactions.*

Since we only assume that that transactional operations are stravation-free, our implementation does not violate the impossibility result of Guerraoui and Kapalka [15] who proved that a strongly progressive opaque STM cannot be implemented using only reads and writes if tm-operations are required to be *wait-free*.

5.2   Protected Data

Let M be a progressive STM implementation. Intuitively, a t-object X_j is protected at the end of some finite execution π of M if some transaction T_0 is about to atomically change the value of X_j in its next step (e.g., by performing a CAS operation) or does not allow any concurrent transaction to read X_j (e.g., by holding a lock on X_j).

Formally, let $\alpha \cdot \pi$ be an execution of M such that π is an uncontended complete execution of a transaction T_0, where $Wset(T_0) = \{X_1, \ldots, X_m\}$. Let u_j ($j = 1, \ldots, m$) denote the value written by T_0 to t-object X_j in π. We say that π' is a *proper prefix* of π if π' is a prefix of π and every atomic section is complete in π'. In this section, let π^t denote the t-th shortest proper prefix of π. Let π^0 denote the empty prefix. (Recall that an atomic event is either a tm-event, a read or write on a base object, or an atomic section.)

For any $X_j \in Wset(T_0)$, let T_j denote a transaction that tries to read X_j and commit. Let $E_j^t = \alpha \cdot \pi^t \cdot \rho_j^t$ denote the extension of $\alpha \cdot \pi^t$ in which T_j runs solo until it completes. Note that, since we only require the implementation to be starvation-free, ρ_j^t can be infinite.

We say that $\alpha \cdot \pi^t$ is $(1,j)$-valent if the read operation performed by T_j in $\alpha \cdot \pi^t \cdot \rho_j^t$ returns u_j (the value written by T_0 to X_j). We say that $\alpha \cdot \pi^t$ is $(0,j)$-valent if the read operation performed by T_j in $\alpha \cdot \pi^t \cdot \rho_j^t$ does not abort and returns an "old" value $u \neq u_j$. Otherwise, if the read operation of T_j aborts or never returns in $\alpha \cdot \pi^t \cdot \rho_j^t$, we say that $\alpha \cdot \pi^t$ is $(\bot, j)$-valent.

Definition 6. *We say that T_0 protects an object X_j in $\alpha \cdot \pi^t$, where π^t is the t-th shortest proper prefix of π ($t > 0$) if one of the following conditions holds: (1) $\alpha \cdot \pi^t$ is $(0,j)$-valent and $\alpha \cdot \pi^{t+1}$ is $(1,j)$-valent, or (2) $\alpha \cdot \pi^t$ or $\alpha \cdot \pi^{t+1}$ is $(\bot, j)$-valent.*

For *disjoint-access parallel* (DAP) progressive STM, we show that every uncontended transaction must protect every object in its write set at some point of its execution.

Theorem 5. *Let M be a progressive, opaque and disjoint-access-parallel STM implementation. Let $\alpha \cdot \pi$ be an execution of M, where π is an uncontended complete execution of a transaction T_0. Then there exists π^t, a proper prefix of π, such that T_0 protects $|Wset(T_0)|$ t-objects in $\alpha \cdot \pi^t$.*

The lower bound of Theorem 5 is tight: it is matched by all progressive implementations we are aware of, including ones in Section 5.1. Note that any DAP single-lock STM implementation automatically provides a stronger progress condition than just single-lock progressiveness. A transaction T in a DAP single-lock STM can only be forcefully aborted if it observes a concurrent transaction T' such that $Dset(T) \cap Dset(T') \neq \emptyset$. This is not very far from progressiveness, where T may abort only if T and T' experience a write-write or write-read conflict on a t-object. Thus, in the realm of DAP STM implementations, progressiveness is very close to the weakest non-trivial progress condition.

6 Related Work

Crain et al. [9] proved that a permissive opaque TM implementation cannot maintain invisible reads, which inspired the derivation of our lower bound on RAW/AWAR complexity in Section 4.

The RAW/AWAR complexity for concurrent implementations was recently introduced in [6]. It was shown in [6] that at least one RAW/AWAR is needed to perform a strongly non-commutative linearizable operation (an operation that cannot be commuted with any other operation so that at least one of them does not see the difference). The proofs of Theorems 1 and 2 extend the arguments used in [6] to the STM context, where a transaction may consist of multiple such operations.

A related paper by Attiya et al. [8] showed that every permissive strictly serializable and DAP TM in which every read-only transaction must commit in a wait-free manner has an execution in which some read-only transaction T_k performs at least $|Dset(T_k)|$-1 base-object writes. In this paper, we do not assume that a read operation must be wait-free and we do not require disjoint-access parallelism. Also, we focus the number of RAW/AWAR patterns and not only base-object writes. On the other hand, we consider a stronger correctness property (opacity). Therefore, our lower bound in Section 4 incomparable with the one of [8].

To establish the lower bound on t-objects that must be "protected" in an opaque, progressive TM (Section 5.2), we use the definition of disjoint-access parallelism introduced in [8]. Guerraoui and Kapalka [15] considered a stronger version of DAP called *strict data-partitioning* to prove a linear lower bound on the number of steps performed by a successful read operation in a progressive, opaque TM that uses invisible reads. Interestingly, the constant RAW/AWAR implementations of progressive, opaque TMs sketched in Section 5 are strict data-partitioned.

7   Concluding Remarks

In this paper, we derived inherent costs of implementing STMs with non-trivial concurrency guarantees. At a high level, our results suggest that providing high degrees of concurrency in STM may incur considerable unavoidable costs. Our results give rise to many intriguing questions, and we list some of them below.

In this paper, we focused on progress conditions that provide positive concurrency, progressiveness and permissiveness. The results do not apply to *obstruction-free* STMs [12] that only guarantee that a transaction commits if it eventually runs without contention. Effectively, an obstruction-free STM provides zero concurrency, since progress is guaranteed only when one transaction is active at a time. However, unlike single-lock implementations, it does allow overlapping transactions to make progress (one at a time). Does this incur higher RAW/AWAR complexity?

We cannot expect the lower bound of Theorem 5 (the protected-data size) to apply to non-DAP STMs, including trivial ones that allow storing the state of the whole STM in one base object. One way to generalize our result and avoid trivialities is to assume that a base object can store information only about a constant number of t-objects (the *constant-size information* property in [13]) which can potentially give asymptotically close lower bounds.

We focused on implementations that allow a tm-operation to be delayed only by concurrent operations performed by other transactions. Does relaxing the tm-liveness property by allowing a read operation to wait until a concurrent transaction terminates [7] improve the RAW/AWAR complexity with respect to permissive implementations? It is easy to see that the proof of our permissive lower bound (Theorem 1) does not work for this case. But it is unclear a priori how this may affect the cost of progressive implementations.

Last but not least, the results of this paper assume opacity as a correctness property. Recently, multiple relaxations of opacity were proposed [10, 2, 9, 8]. It would be very interesting to understand the concurrency benefits gained by such relaxed consistency conditions.

Acknowledgements. The authors are grateful to Michel Raynal and Rachid Guerraoui for inspiring discussions on the properties and costs of STM and Damien Imbs for valuable comments on the previous drafts. The comments and suggestions of anonymous reviewers on an earlier version of this paper are also gratefully acknowledged.

References

1. Adve, S.V., Gharachorloo, K.: Shared memory consistency models: A tutorial. IEEE Computer 29(12), 66–76 (1996)
2. Afek, Y., Morrison, A., Tzafrir, M.: View transactions: Transactional model with relaxed consistency checks. In: PODC 2010: Proceedings of the 29th Annual ACM SIGACT-SIGOPS Symposium on Principles of Distributed Computing (2010)
3. Alpern, B., Schneider, F.B.: Defining liveness. Information Processing Letters 21(4), 181–185 (1985)
4. Anderson, T.E.: The performance of spin lock alternatives for shared-memory multiprocessors. IEEE Trans. Parallel Distrib. Syst. 1(1), 6–16 (1990)
5. Attiya, H., Hendler, D.: Time and space lower bounds for implementations using k-cas. IEEE Transactions on Parallel and Distributed Systems 21(2), 162–173 (2010)
6. Attiya, H., Guerraoui, R., Hendler, D., Kuznetsov, P., Michael, M.V.M.: Laws of order: Expensive synchronization in concurrent algorithms cannot be eliminated. In: POPL (2011)
7. Attiya, H., Hillel, E.: Single-Version STMs Can Be Multi-Version Permissive (Extended Abstract). In: Aguilera, M.K., Yu, H., Vaidya, N.H., Srinivasan, V., Choudhury, R.R. (eds.) ICDCN 2011. LNCS, vol. 6522, pp. 83–94. Springer, Heidelberg (2011)
8. Attiya, H., Hillel, E., Milani, A.: Inherent limitations on disjoint-access parallel implementations of transactional memory. In: Proceedings of the Twenty-First Annual Symposium on Parallelism in Algorithms and Architectures, SPAA 2009, pp. 69–78. ACM, New York (2009)
9. Crain, T., Imbs, D., Raynal, M.: Read invisibility, virtual world consistency and permissiveness are compatible. Research Report, ASAP - INRIA - IRISA - CNRS: UMR6074 - INRIA - Institut National des Sciences Appliquées de Rennes - Université de Rennes I (November 2010)
10. Felber, P., Gramoli, V., Guerraoui, R.: Elastic Transactions. In: Keidar, I. (ed.) DISC 2009. LNCS, vol. 5805, pp. 93–107. Springer, Heidelberg (2009)

11. Guerraoui, R., Henzinger, T.A., Singh, V.: Permissiveness in Transactional Memories. In: Taubenfeld, G. (ed.) DISC 2008. LNCS, vol. 5218, pp. 305–319. Springer, Heidelberg (2008)
12. Guerraoui, R., Kapalka, M.: On obstruction-free transactions. In: Proceedings of the Twentieth Annual Symposium on Parallelism in Algorithms and Architectures, SPAA 2008, pp. 304–313. ACM, New York (2008)
13. Guerraoui, R., Kapalka, M.: On the correctness of transactional memory. In: PPOPP, pp. 175–184 (2008)
14. Guerraoui, R., Kapalka, M.: The semantics of progress in lock-based transactional memory. In: POPL, pp. 404–415 (2009)
15. Guerraoui, R., Kapalka, M.: Principles of Transactional Memory, Synthesis Lectures on Distributed Computing Theory. Morgan and Claypool (2010)
16. Hendler, D., Incze, I., Shavit, N., Tzafrir, M.: Flat combining and the synchronization-parallelism tradeoff. In: SPAA, pp. 355–364 (2010)
17. Israeli, A., Rappoport, L.: Disjoint-access-parallel implementations of strong shared memory primitives. In: Proceedings of the Thirteenth Annual ACM Symposium on Principles of Distributed Computing, PODC 1994, pp. 151–160. ACM, New York (1994)
18. Kuznetsov, P., Ravi, S.: On the cost of concurrency in transactional memory. CoRR, abs/1103.1302 (2011)
19. Lamport, L.: A New Solution of Dijkstra's Concurrent Programming Problem. Commun. ACM 17(8), 453–455 (1974)
20. Lee, J.: Compilation Techniques for Explicitly Parallel Programs. PhD thesis, Department of Computer Science, University of Illinois at Urbana-Champaign (1999)
21. McKenney, P.: Concurrent code and expensive instructions. Linux Weekly News (January 2011), http://lwn.net/Articles/423994/
22. McKenney, P.E.: Memory barriers: a hardware view for software hackers. Linux Technology Center, IBM Beaverton (June 2010)
23. Papadimitriou, C.H.: The serializability of concurrent database updates. J. ACM 26, 631–653 (1979)
24. Raynal, M.: Algorithms for Mutual Exclusion. MIT Press (1986)
25. Taubenfeld, G.: The Black-White Bakery Algorithm and Related Bounded-Space, Adaptive, Local-Spinning and FIFO Algorithms. In: Liu, H. (ed.) DISC 2004. LNCS, vol. 3274, pp. 56–70. Springer, Heidelberg (2004)

Response Time Bounds for G-EDF without Intra-Task Precedence Constraints

Jeremy P. Erickson and James H. Anderson

University of North Carolina at Chapel Hill

Abstract. Prior work has provided bounds on the deadline tardiness that a set of sporadic real-time tasks may incur when scheduled using the global earliest-deadline-first (G-EDF) scheduling algorithm. Under the sporadic task model, it is necessary that no individual task overutilize a single processor and that the set of all tasks does not overutilize the set of all processors. In this work we generalize the task model by allowing jobs within a single task to run concurrently. In doing so we remove the requirement that no task overutilize a single processor. We also provide tardiness bounds that are better than those available with the standard sporadic task model.

1 Introduction

Multicore processors have been shown to be useful for supporting traditional soft real-time (SRT) workloads when bounded deadline tardiness is acceptable [1, 2, 3, 4]. In this paper we extend these works to a broader class of workloads in which jobs are independent of each other and can be executed in parallel, such as servers handling independent requests. For both types of SRT workloads, temporal correctness requires that *tardiness bounds* exist, i.e., for each task, there exists an upper bound on the amount of time between the deadline of any job of that task and its actual completion time. Prior work has shown that the global earliest-deadline-first (G-EDF) algorithm is a good candidate scheduler when bounded tardiness is desired, as its use allows all available processing capacity to be utilized.

In most previous analysis of G-EDF, successive jobs (i.e. invocations) of each task are required to execute in sequence. This constraint arises naturally when jobs correspond to separate invocations of the same code segment. However, in some settings, jobs are released as separate threads in response to interrupts, in which case, successive jobs of the same task may execute concurrently. In prior hard real-time analysis of G-EDF [5], the impact of such concurrently-executing jobs has been considered, but to our knowledge, no such analysis exists for SRT systems for which bounded deadline tardiness is acceptable. Such analysis is the focus of this paper.

The task model considered in this paper is based on the widely-studied sporadic model, but differs from the usual specification of that model in two ways. First, as implied by the discussion above, successive jobs of the same task are

A. Fernández Anta, G. Lipari, and M. Roy (Eds.): OPODIS 2011, LNCS 7109, pp. 128–142, 2011.

allowed to execute in parallel. Second, *early release* behavior [6] is allowed: a job may have an *actual release time* (or, *a-release time*) that is earlier than its *scheduler release time* (or, *s-release time*) A job's deadline is defined based on its s-release time, and constructive s-releases of each task τ_i are constrained to be no closer than T_i time units apart, where T_i is the minimum separation parameter of τ_i. However, a job may begin execution as early as its a-release time. These changes to the traditional sporadic model allow us to support general event models, as the following example illustrates.

Example In high-frequency trading systems, short response times are critical to minimize risk [7]. Consider such a system that responds to data from the market about two stocks. One stock is highly critical and should receive new information every 2 ms (but due to network uncertainty may not be timed precisely.) It may take up to 3 ms to process and should be processed as quickly as possible, so its deadline is 3 ms. Observe that this stock overutilizes a single processor and could not be supported using the traditional sporadic task model, even on a multiprocessor. However, it can be supported using the methodology provided in this paper. A second stock is less critical, should receive new information every 4 ms, and can take up to 2 ms to process. One possible execution on two processors is depicted in Fig. 1. Observe that the a-release times sometimes do occur before the s-release times (because incoming packets can arrive early or late) and that some jobs do miss deadlines.

The main contribution of this paper is to show that tardiness under G-EDF is greatly lessened if jobs of the same task are not constrained to execute in sequence. We show this by deriving per-job response-time bounds, from which tardiness bounds can be deduced. After deriving such bounds, we compare them experimentally to prior bounds, which were derived assuming no intra-task parallelism. We begin in the next section by more fully describing our system model.

2   System Model

We consider a system τ of n arbitrary-deadline sporadic tasks $\{\tau_1, \tau_2, \ldots, \tau_n\}$ running on m processors, with each task τ_i characterized by a worst-case execution time C_i, a minimum separation time (between s-releases) T_i, and a relative deadline D_i. No job may run concurrently with itself, but distinct jobs within the same task may run concurrently. In addition, we define a task's *utilization* $U_i \stackrel{\text{def}}{=} \frac{C_i}{T_i}$, the *task system utilization* $U(\tau) \stackrel{\text{def}}{=} \sum_{\tau_i \in \tau} U_i$, and $m^+ \stackrel{\text{def}}{=} \lceil U(\tau) \rceil$.

Under the traditional task model with implicit precedence constraints, providing bounded response time required that no τ_i had $U_i > 1$ and that $U(\tau) \leq m$ [3]. However, under the task model considered here, a job with $U_i > 1$ can have bounded response time if subsequent invocations run on separate processors, as depicted for τ_1 in Fig. 1. $U(\tau) \leq m$ remains necessary so that the entire system is not overutilized. In this work we demonstrate that $U(\tau) \leq m$ is also a sufficient condition for bounded response times and provide response-time bounds relative to the s-release time of each job.

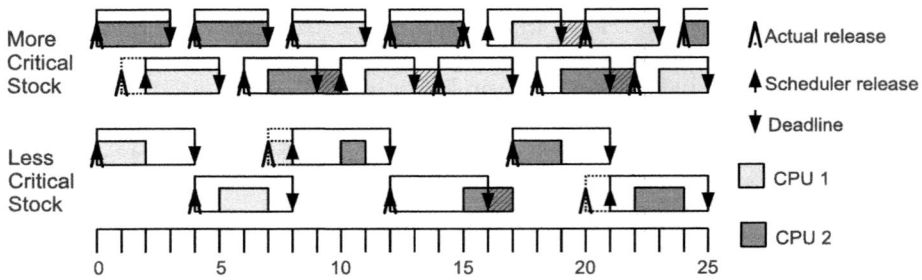

Fig. 1. Example high-frequency trading system

3   Response Time Characterization

Over an interval of any given length t, the total amount of work from jobs of τ_i (with both s-release times and deadlines inside the interval) is bounded. This bound, called the *demand bound function*, was first defined in [8]:

$$\text{DBF}(\tau_i, t) \stackrel{\text{def}}{=} C_i \cdot \max\left\{0, \left\lfloor \frac{t - D_i}{T_i} \right\rfloor + 1\right\}. \tag{1}$$

(1) is defined by counting the number of possible jobs of τ_i having both s-release times and deadlines in an interval of length t, and multiplying that number by the worst-case execution time C_i. Because we still assume a sporadic s-release pattern for jobs, and deadlines are based solely on s-release times, allowing multiple jobs within a task to execute at the same time does not invalidate (1). An early release of a job can only reduce the demand as compared to that predicted in (1).

Lem. 1 of [3] used (1) to demonstrate that for all τ_i and $t \geq 0$,

$$\text{DBF}(\tau_i, t) \leq U_i t + S_i, \tag{2}$$

where $S_i \stackrel{\text{def}}{=} C_i \cdot \max\{0, 1 - C_i/D_i\}$. Essentially, S_i accounts for the extra demand that can be created by a job with a short deadline.

For an n-task system τ, we wish to define a vector of non-negative real numbers $\langle x_1, x_2, \ldots x_n \rangle$ such that the response time of each task τ_i, $1 \leq i \leq n$, is at most $x_i + C_i$ when τ is scheduled using G-EDF on m unit-speed processors. Each x_i value depends upon the other x_i values. Therefore, we initially define the vectors using an implicit criterion, and as in [2, 3] we define the notion of a *compliant vector* as one that meets this criterion.

Definition 1. *For each task τ_i, non-negative integer $p < m^+ - 1$, and non-negative real number x_i, let*

$$l(\tau_i, x_i, p) = \min\{C_i, \max\{0, x_i + C_i - pT_i\}\}. \tag{3}$$

For any $\boldsymbol{x} = \langle x_1, x_2, \ldots, x_n \rangle$, *an ordered list of* n *non-negative real numbers, let*

$$L(\boldsymbol{x}) \stackrel{def}{=} \sum_{m^+ - 1 \text{ largest}} l(\tau_i, x_i, p), \tag{4}$$

$$S(\tau) \stackrel{def}{=} \sum_{\tau_i \in \tau} S_i. \tag{5}$$

We define $\boldsymbol{x}$ *as a **compliant vector** if and only if*

$$\frac{L(\boldsymbol{x}) + S(\tau) + U(\tau)D_i - C_i}{m} \leq x_i \tag{6}$$

is satisfied for all i, $1 \leq i \leq n$.

Our definition differs from that of [3] both in the definition of $L(\boldsymbol{x})$ and in the additional $U(\tau)D_i$ term.

We now derive a response-time bound by considering a compliant vector $\boldsymbol{x} = \langle x_1, x_2, \ldots, x_n \rangle$ and an arbitrary collection I' of jobs generated by τ. We order jobs by deadline with ties broken arbitrarily (as per the standard G-EDF algorithm). We analyze the response time of an arbitrary job J_k with s-release time r_k and deadline d_k, assuming that each job of each τ_i ordered prior to J_k completes within $(C_i + x_i)$ units of its s-release time. We denote as I the set of all jobs ordered at or before J_k, which (by the definition of G-EDF) contains all jobs that affect the scheduling of J_k. We also denote $I_c \stackrel{def}{=} I \setminus \{J_k\}$ (i.e., the work *competing* with J_k).

Without loss of generality, we assume that the earliest s-release time for any job in I is 0. We denote as $W_i(t)$ the remaining execution for jobs in I of task τ_i at time t, and let $W(t) \stackrel{def}{=} \sum_{\tau_i \in \tau} W_i(t)$.

Lemma 1. *If* $\boldsymbol{x}$ *is a compliant vector and for all* $i, 1 \leq i \leq N$, *the response time of each job of* τ_i *in* I_c *is at most* $x_i + C_i$, *then*

$$W(t) \leq U(\tau)(d_k - t) + L(\boldsymbol{x}) + S(\tau). \tag{7}$$

Proof. We will define an interval as *busy* if at least m^+ processors are executing work throughout the interval, and *nonbusy* if fewer than m^+ processors are executing work. We will consider a set of time instants $\{t_0, t_1, \ldots t_k\}$, $t_0 = 0$, $t_k = r_k$, such that each $[t_i, t_{i+1})$ is either all busy or all nonbusy. We will prove the lemma by induction.

Base Case *($t_0 = 0$).* All jobs with both s-release times and deadlines within $[0, d_k]$ contribute to $W(0)$. By (2) $W_i(0) \leq U_i d_k + S_i$, and therefore, summing over all $W_i(0)$ values, $W(0) \leq U(\tau)d_k + S(\tau) \leq U(\tau)d_k + S(\tau) + L(\boldsymbol{x})$.

Induction Step. Suppose the lemma is true for t_i. We will consider two subcases, based on whether $[t_i, t_{i+1})$ is busy or nonbusy.

Case A. Suppose $[t_i, t_{i+1})$ is busy. Then,

$$W(t_{i+1})$$
$$\leq \{\text{Since at least } m^+(t_{i+1} - t_i) \text{ work is completed}\}$$
$$W(t_i) - m^+(t_{i+1} - t_i)$$
$$\leq \{\text{By the inductive assumption}\}$$
$$U(\tau)(d_k - t_i) + S(\tau) + L(\boldsymbol{x}) - m^+(t_{i+1} - t_i)$$
$$\leq \{\text{Since } U(\tau) \leq m^+\}$$
$$U(\tau)(d_k - t_i) + S(\tau) + L(\boldsymbol{x}) - U(\tau)(t_{i+1} - t_i)$$
$$= \{\text{Simplifying}\}$$
$$U(\tau)(d_k - t_{i+1}) + S(\tau) + L(\boldsymbol{x}),$$

so the lemma is true for t_{i+1} as well.

Case B. Suppose $[t_i, t_{i+1})$ is nonbusy. We will say that a job J is "executing at time instant t_{i+1}^-" if there is an ϵ greater than 0 such that J is executing over the entire interval $[t_{i+1} - \epsilon, t_{i+1})$. In [3], the presence of an idle CPU implied that at most $m^+ - 1$ *tasks* have work available for execution at time instant t_{i+1}^-, whereas here the same condition implies that at most $m^+ - 1$ *jobs* are available for execution. In [3] it was necessary to account for released jobs that were not running due to a precedence constraint, despite the presence of an idle CPU. In order to do so, assuming that $U_i \leq 1$ for each τ_i was necessary. Here we do not need to account for such a case, but do need to account for the fact that several jobs running in a non-busy interval could be from the same task. The assumption that $U_i \leq 1$ is no longer necessary.

We now consider two cases for jobs that may contribute to $W(t_{i+1})$: jobs that are executing at time instant t_{i+1}^- (Case B.1) and jobs that have s-release time at or after t_{i+1} (Case B.2).

B.1 In total, there may be at most $m^+ - 1$ jobs executing at time instant t_{i+1}^-. We ignore early-released jobs that have s-releases at or after t_{i+1}^-, as these are accounted for in Case B.2. We consider the jobs of each task τ_j that has jobs executing at time instant t_{i+1}^-. We will use p to index each executing job relative to the job with the most recent s-release within τ_j: $p = 0$ indicates the job with the most recent s-release, $p = 1$ the next most recent s-release, etc. By the assumption of the lemma, if $p > 0$ for job $J \in \tau_j$, then J must complete by $x_j + C_j$ units after its s-release time, and must be have a s-release time before $t_{i+1} - pT_j$. Therefore, J must complete by time $t_{i+1} + x_j + C_j - pT_j$, and its contribution to $W_j(t_{i+1})$ is at most $\min\{C_j, \max\{0, x_j + C_j - pT_j\}\} \overset{\text{By (3)}}{=} l(\tau_j, x_j, p)$.

When $p = 0$ for $J \in \tau_j$, $x_j + C_j - pT_j \geq C_j$. Therefore $l(\tau_j, x_j, p) = C_j$ by (3), so J's contribution to $W_j(t_{i+1})$ is also at most $l(\tau_j, x_j, p)$.

B.2 We now consider jobs with s-release time at or after t_{i+1}. By (2), each task τ_j contributes at most $U_j(d_k - t_{i+1}) + S_j$ units of work over $[t_{i+1}, d_k)$.

Cumulatively, all tasks contribute at most $U(\tau)(d_k - t_{i+1}) + S(\tau)$ units of work over $[t_{i+1}, d_k)$.

Total $W(t_{i+1})$ contains at most $m^+ - 1$ jobs from Case B.1, in addition to all jobs from Case B.2, so $W(t_{i+1}) \leq U(\tau)(d_k - t_{i+1}) + S(\tau) + L(\boldsymbol{x})$.
   Thus the lemma is true for t_{i+1}.

We now use the previous lemma to bound the response time of a job under the same assumptions.

Lemma 2. *If $\boldsymbol{x}$ is a compliant vector and for all $i, 1 \leq i \leq N$, the response time of each job of τ_i in I_c is at most $x_i + C_i$, then the response time of J_k is at most $x_k + C_k$.*

Proof. Recall that r_k is the s-release time of J_k, and d_k is its deadline. By Lem. 1,

$$W(r_k) \leq U(\tau)(d_k - r_k) + S(\tau) + L(\boldsymbol{x}). \tag{8}$$

After r_k, J_k is continuously running until it is finished, except when all other CPUs are occupied by jobs from I_c. Recall that, by definition, $W(r_k)$ is the total remaining work after time r_k for jobs in I. We define $W_c(r_k)$ as the total amount of remaining work after time r_k for jobs in I_c. Because the upper bound in (8) assumes that all jobs (including J_k) run for their full worst-case execution times, (8) implies

$$W_c(r_k) \leq U(\tau)(d_k - r_k) + S(\tau) + L(\boldsymbol{x}) - C_k. \tag{9}$$

The total amount of time after r_k during which m CPUs are busy with work from I_c can be at most

$$\frac{W_c(r_k)}{m} \leq \{\text{By (9)}\}$$
$$\frac{L(\boldsymbol{x}) + S(\tau) + U(\tau)D_k - C_k}{m}$$
$$\leq \{\text{By (6)}\}$$
$$x_k.$$

Thus, J_k is prevented from executing after its s-release time for at most x_k time units, so its response time is at most $x_k + C_k$.

This lemma leads directly to the main result of this section:

Theorem 1. *If $\boldsymbol{x}$ is a compliant vector then $\forall i, 1 \leq i \leq N$, each job of τ_i completes within $x_i + C_i$ units of its s-release time.*

Proof. By inducting over the jobs of I' using Lem. 2.

4   The Minimum Compliant Vector

Thm. 1 uses compliant vectors to express response-time bounds. Our objective is to compute response-time bounds that are as small as possible. We show that for any arbitrary-deadline sporadic task system τ without implicit precedence constraints there exists a unique *minimum* compliant vector. This proof closely follows a similar one provided in [3], and some lemmas have nearly identical proofs. For space reasons, proofs of such lemmas are ommitted. We also provide an Appendix with an algorithm for computing the minimum compliant vector in polynomial time.

We first characterize the behavior of $L(\boldsymbol{x})$. We consider two vectors $\boldsymbol{x}$ and $\boldsymbol{y}$ differing by a constant for some of their values, and are the same elsewhere. For example, $\boldsymbol{x} = \langle 1, 2, 3 \rangle$ and $\boldsymbol{y} = \langle 2, 2, 4 \rangle$ differ by exactly 1 in two places (the first and third) and are the same in the second; Lem. 3 would apply to $\boldsymbol{x}$ and $\boldsymbol{y}$ with $k = 2$. The reasoning for Lem. 3 is identical to Lem. 4 of [3].

Lemma 3. *Suppose length-n vectors $\boldsymbol{x}$ and $\boldsymbol{y}$ differ at exactly k values, and for these values $y_i = x_i + \delta$, where δ is a positive constant. Denote $w = \min\{k, m^+ - 1\}$. Then, the following inequality holds:*

$$L(\boldsymbol{x}) \le L(\boldsymbol{y}) \le L(\boldsymbol{x}) + \delta \cdot w. \tag{10}$$

We say that length-n $\boldsymbol{x}$ is *strictly smaller* than length-n $\boldsymbol{y}$ if for all $i, x_i \le y_i$ and there exists a j such that $x_j < y_j$. Clearly $\boldsymbol{y}$ cannot be considered "minimum" if there exists such an $\boldsymbol{x}$. We next use Lem. 3 to characterize the minimum compliant vector, with logic identical to Lem. 5 of [3].

Lemma 4. *If $\boldsymbol{y}$ is compliant and there is a j such that $y_j > (L(\boldsymbol{y}) + S(\tau) + U(\tau)D_i - C_i)/m$, then there exists a strictly smaller vector $\boldsymbol{x}$ that is also compliant.*

Lem. 4 demonstrates that each inequality in (6) should actually be an equality, or the vector cannot be the minimum. A minimum compliant vector must therefore be of the form

$$x_i = \frac{L(\boldsymbol{x}) + S(\tau) + U(\tau)D_i - C_i}{m} \quad \forall i. \tag{11}$$

Because $L(\boldsymbol{x})$ does not depend on i, there must exist a real number

$$s = \frac{L(\boldsymbol{x})}{m} \tag{12}$$

such that

$$x_i = s + \frac{S(\tau) + U(\tau)D_i - C_i}{m} \quad \forall i. \tag{13}$$

We define some functions:

$$\boldsymbol{v}(s) \stackrel{\text{def}}{=} \boldsymbol{x} \text{ such that (13) holds} \tag{14}$$

$$L(s) \stackrel{\text{def}}{=} L(\boldsymbol{v}(s)) \tag{15}$$

$$M(s) \stackrel{\text{def}}{=} L(s) - ms. \tag{16}$$

By (13), any minimum compliant vector must be $\boldsymbol{v}(s)$ for some s. Furthermore, $L(s)$ must equal ms, by (12). Therefore, $M(s) = 0$ if and only if $\boldsymbol{v}(s)$ is a compliant vector in the form of (11), and thus the minimum compliant vector. We are now ready to prove this section's main result:

Theorem 2. *For any given task set τ, there exists a unique minimum compliant vector.*

Proof. We wish to demonstrate that exactly one real s exists such that $M(s) = 0$. We will use the Intermediate Value Theorem from calculus.

A necessary precondition for the Intermediate Value Theorem is that $M(s)$ is a continuous function. The following lemma is essentially Lem. 21 of [3] and leads to the desired result as a corollary.

Lemma 21. $L(s)$ *is continuous over* $\mathbb{R}$

Let $C_{\max}$ denote the largest C_i value in τ. We now show that $M(0) > 0$ and $M(C_{\max}) < 0$, completing the preconditions for the Intermediate Value Theorem.

Lemma 22. $M(0) > 0$

Proof. Let $1 \leq i \leq N$ be arbitrary. Then:

$$
\begin{aligned}
&M(0) \\
&= \{\text{By (16) with } s = 0\} \\
&L(0) \\
&= \{\text{By (15) and (4)}\} \\
&\sum_{m^+ - 1 \text{ largest}} l(\tau_i, v_i(0), p) \\
&\geq \{\text{Since, by (3), } l(\tau_i, v_i(0), p) \text{ can't be negative}\} \\
&l(\tau_i, v_i(0), 0) \\
&= \{\text{By (3) and (14), with } s = 0 \text{ and } p = 0\} \\
&\min\left\{C_i, \max\left\{0, \frac{S(\tau) + U(\tau)D_i - C_i}{m} + C_i\right\}\right\} \\
&= \{\text{Simplifying}\} \\
&= \min\left\{C_i, \max\left\{0, \frac{S(\tau) + U(\tau)D_i + (m-1)C_i}{m}\right\}\right\} \\
&> 0.
\end{aligned}
$$

Lemma 23. $M(C_{\max}) < 0$.

Proof. By (3), $l(\tau_i, x_i, p) \leq C_i$ for any i and p. Therefore, for any i and p,

$$
l(\tau_i, x_i, p) \leq C_{\max}. \tag{17}
$$

Therefore,

$$
\begin{aligned}
&M(C_{\max}) \\
&= \{\text{By (16) with } s = C_{\max}\} \\
&\quad L(C_{\max}) - mC_{\max} \\
&\leq \{\text{By (15), (4), and (17)}\} \\
&\quad (m^+ - 1)C_{\max} - mC_{\max} \\
&\leq \{\text{Since } m^+ \leq m\} \\
&\quad - C_{\max} \\
&< 0.
\end{aligned}
$$

Lemma 24. *There is an s in $(0, C_{\max})$ such that $M(s) = 0$.*

Proof. By Lem. 21, Lem. 22, Lem. 23, and the Intermediate Value Theorem.

We now verify that the s value of Lem. 24 is unique, using the following lemma, with identical reasoning to Lem. 22 of [3].

Lemma 25. $s_1 \neq s_2$ *implies* $M(s_1) \neq M(s_2)$

Lem. 25 demonstrates that $s_1 \neq s_2$ and $M(s_1) = 0$ imply $M(s_2) \neq 0$, so the value of s characterized in Lem. 24 is unique.

In Lem. 24 we have a substantial improvement compared to [3], where the upper bound was given as the sum of the $m^+ - 1$ largest values of C_i. This improvement leads to Thm. 3, which provides a response-time bound that can be quickly calculated.

Theorem 3. *The response time of any job of any task τ_i cannot exceed $C_{\max} + \frac{S(\tau) + U(\tau)D_i - C_i}{m} + C_i$.*

Proof. Follows from Lem. 24, (13), and Thm. 1.

5   Evaluation

This work allows smaller response-time bounds than are possible using prior work. In particular, these results are especially competitive for implicit-deadline sporadic task systems. By Thm. 3, combined with $U(\tau) \leq m$ (a necessary condition), and the fact that, for implicit-deadline systems, $S(\tau) = 0$, the response time of any job of any task τ_i must be upper-bounded by $C_{\max} + D_i + \frac{m-1}{m}C_i$. Therefore, the *tardiness* of any job of τ_i must be no greater than $C_{\max} + \frac{m-1}{m}C_i$.

In order to evaluate the improvement to the bounds we obtain by eliminating implicit precedence constraints, we compared our results to the best available analysis for implicit-deadline sporadic tasks, found in [2]. For the experiments in this paper, we compared the best results of our work to the best bounds attainable using [2].

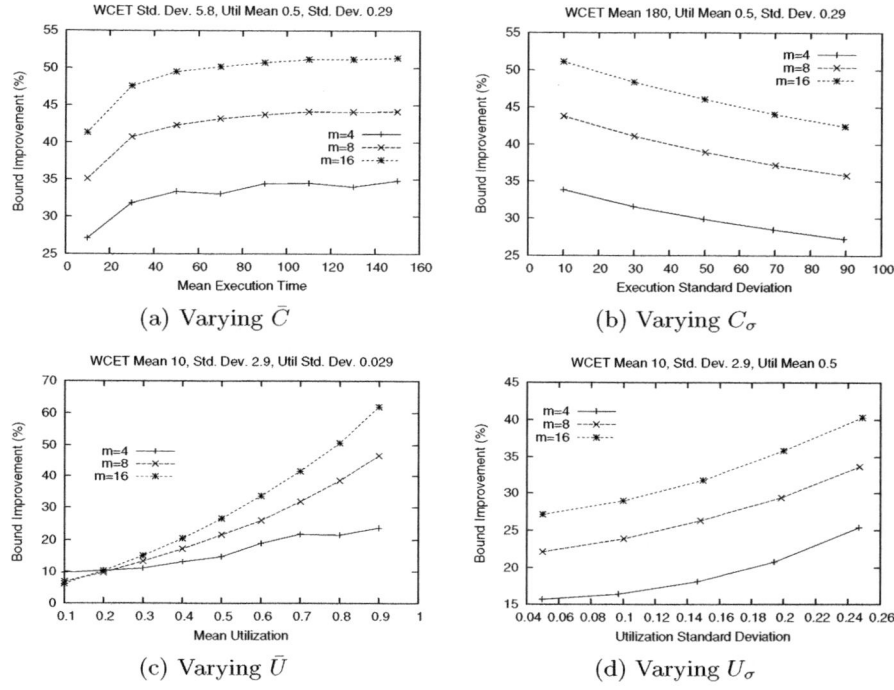

Fig. 2. Results of experiments

Our experimental methodology is inspired by the tests in [2]. All experiments were done with processor counts of 4, 8, and 16. We used uniform distributions for the task worst-case execution times and utilizations, and we determined the effects of varying each of four parameters: mean worst-case execution time ($\bar{C}$), standard deviation of worst-case execution time (C_σ), mean utilization ($\bar{U}$), and standard deviation of utilization (U_σ). For mean x and standard deviation σ, values were chosen uniformly over $(x - \sigma\sqrt{3}, x + \sigma\sqrt{3})$.

In each experiment, the processor count m and three of the four parameters above were fixed, and the remaining parameter was varied. For each value of the varied parameter, we generated 1000 task sets. For each individual task set, we generated tasks until a task was generated that would cause $U(\tau)$ to exceed m. For each task set we computed the mean tardiness bound with respect to [2], δ, and with respect to our work, δ'. For each set of 1000 task sets we computed $\bar{\delta}$ (the mean value of δ) and $\bar{\delta}'$ (the mean value of δ'). The *absolute improvement* for each set of sets is defined as $\bar{\delta} - \bar{\delta}'$, and the *relative improvement* for each set of sets is defined as $(\bar{\delta} - \bar{\delta}')/\bar{\delta}$.

Results are in Fig. 2. We see that the improvement to tardiness is quite substantial, particularly with large execution times, small variance in execution times, large utilizations, and large variance in utilizations. More significant improvement occurs with larger processor counts because the bounds of [2] increase significantly with m, while our bounds are upper-bounded by $C_{\max} + \frac{m-1}{m}C_i$. This improvement is possible even when per-task utilization is restricted to be less than one to make our results comparable to prior work. We do not have results comparing our work to previous results when per-task utilization may exceed one, because prior work is not applicable in this case.

6 Conclusion

G-EDF scheduling has already proven useful for traditional SRT workloads in which jobs of the same task have implicit precedence constraints. Here we have demonstrated that G-EDF scheduling may be even more useful for SRT workloads in which jobs may be released as separate threads that can safely run concurrently. We have shown that doing so not only improves response times compared to prior work, but enables new workloads where a single task may overutilize a single processor.

For future work, allowing critical sections would be useful, so that tasks that write shared data but do not have precedence constraints could be handled. Supporting integrated workloads where some tasks have internal precedence constraints and some do not would also be interesting to consider. Furthermore, in past work on slack reclaiming, precedence constraints between jobs have prevented slack produced by a job from being reclaimed after its successor is released. Using the methods in this paper, we may be able to overcome this limitation.

References

1. Devi, U.C., Anderson, J.H.: Tardiness bounds under global EDF scheduling on a multiprocessor. The Journal of Real-Time Systems 38(2), 133–189 (2008)
2. Erickson, J.P., Devi, U.C., Baruah, S.K.: Improved tardiness bounds for global EDF. In: ECRTS, pp. 14–23 (2010)
3. Erickson, J.P., Guan, N., Baruah, S.K.: Tardiness Bounds for Global EDF with Deadlines Different from Periods. In: Lu, C., Masuzawa, T., Mosbah, M. (eds.) OPODIS 2010. LNCS, vol. 6490, pp. 286–301. Springer, Heidelberg (2010)
4. Leontyev, H., Anderson, J.H.: Generalized tardiness bounds for global multiprocessor scheduling. The Journal of Real-Time Systems 44(1), 26–71 (2010)
5. Baker, T., Baruah, S.K.: An analysis of global EDF schedulability for arbitrary-deadline sporadic task systems. The Journal of Real-Time Systems 43(1), 3–24 (2009)
6. Anderson, J.H., Srinivasan, A.: Early-release fair scheduling. In: ECRTS, pp. 35–43 (2000)
7. Durbin, M.: All About High-Frequency Trading, 1st edn. McGraw-Hill (2010)
8. Baruah, S.K., Mok, A.K., Rosier, L.E.: Preemptively scheduling hard-real-time sporadic tasks on one processor. In: RTSS, pp. 182–190 (1990)

Appendix

Computation Algorithm

We now show how to compute the minimum compliant vector for a task system τ in time polynomial to the size of τ and the number of processors. $L(s)$ as defined in (15) is a piecewise linear function; our algorithm works by tracing $L(s)$ until we find a fixed point $L(s) = ms$.

In order to assist the reader's understanding of this algorithm, we provide an example task system in Table 1.[1] Simple calculations reveal that, for this system, $S(\tau) = 0$ and $U(\tau) = 2$. Furthermore, in a two-CPU system, by Def. 1, we only need to consider $p = 0$. A graph of the relevant $l(\tau_i, v_i(s), 0)$ functions with respect to s is provided in Fig. 3.

Table 1. 2 CPU task system example for Sec. 6

	C_i	T_i	D_i
τ_1	6	10	10
τ_2	12	10	10
τ_3	4	20	20

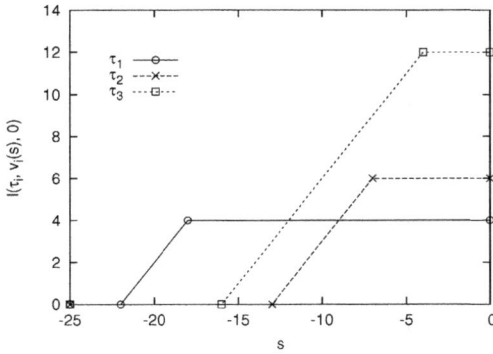

Fig. 3. l functions for the system in Table 1

We define the *slope* at point s of a piecewise linear function $f(s)$ to be $lim_{\epsilon \to 0+} \frac{f(s+\epsilon)-f(s)}{\epsilon}$. This definition differs from the common notion of derivative

---

[1] In this system, the worst-case execution time of τ_2 exceeds its deadline, so it appears that it is impossible for τ_2 to meet its deadline. However, because execution times given are worst-case rather than exact, it is actually possible for this job to complete before its deadline. Furthermore, here we are interested in response-time bounds rather than hard deadlines.

in that its limit is taken from the right; it is thus defined for all real s. For example, $l(\tau_1, v_1(s), 0)$ in Fig. 3 has a slope of 1 at $s = -22$, but is not differentiable at $s = -22$.

For each value of s we will define $l(\tau_i, v_i(s), p)$ as being in one of three states, depending on the value of $v_i(s) + C_i - pT_i$:

- If $v_i(s) + C_i - pT_i < 0$, then $l(\tau_i, v_i(s), p)$ is in state 0, is equal to 0, and has a slope of 0. $l(\tau_1, v_1(s), 0)$ in Fig. 3 is in state 0 in the interval $(-\infty, -22)$.
- If $0 \le v_i(s) + C_i - pT_i < C_i$, then $l(\tau_i, v_i(s), p)$ is in state 1, is equal to $v_i(s) + C_i - pT_i$, and has a slope of 1. $l(\tau_1, v_1(s), 0)$ in Fig. 3 is in state 1 in the interval $[-22, -18)$.
- If $C_i \le v_i(s) + C_i - pT_i$, then $l(\tau_i, v_i(s), p)$ is in state 2, is equal to C_i, and has a slope of 0. $l(\tau_1, v_1(s), 0)$ in Fig. 3 is in state 1 in the interval $[-18, \infty)$.

In order to analyze the piecewise linear function $L(s)$, we will need to determine where the slope changes. To do so, we need to determine which $l(\tau_i, v_i(s), p)$ components contribute to $L(s)$ for various intervals. For some intervals, the choice is arbitrary. For example, the task system in Fig. 3 has only one $l(\tau_i, v_i(s), p)$ component contributing to $L(s)$, because $m - 1 = 2 - 1 = 1$. However, for $s < -22$ all $l(\tau_i, v_i(s), p)$ components equal zero. We provide a sufficient solution by arbitrarily tracking some valid set of $l(\tau_i, v_i(s), p)$ components.

We will create a set $Points$ of tuples, one for each possible change in the slope of $L(s)$. (Each will have an associated s value, but there could be multiple possible changes at the same s value.) Each tuple will identify a point where some $l(\tau_{i_0}, v_{i_0}(s), p_0)$ in state h_0 is replaced by some $l(\tau_{i_1}, v_{i_1}(s), p_1)$ in state h_1. Such a tuple will be of the form $\{s, i_0, p_0, h_0, i_1, p_1, h_1\}$. In some cases, more than one old component may be appropriate. To handle these cases efficiently, any of i_0, p_0, or h_0 may be set to $*$, which is defined as matching any value of the relevant parameter. For example, the tuple $\{s, *, *, 0, i_1, p_1, 1\}$ indicates that any arbitrary $l(\tau_{i_0}, v_{i_0}(s), p_0)$ in state 0 should be replaced by $l(\tau_{i_1}, v_{i_1}(s), p_1)$ in state 1.

The slope of $L(s)$ may change in any of the following cases:

1. Some $l(\tau_i, v_i(s), p)$ changes from state 0 to state 1. This occurs where $v_i(s) + C_i - pT_i = 0$. The resulting tuple will be $\{s, *, *, 0, i, p, 1\}$, as we can view $l(\tau_i, v_i(s), p_i)$ as replacing any $l(\tau_j, v_j(s), p_j)$ in state 0 in the system—they all have value 0. This change occurs exactly once per $l(\tau_i, v_i(s), p)$ and therefore $m - 1$ times per task (once per value of p), for a total of $O(mn)$ times for the system. In Fig. 3, this state change occurs for $l(\tau_1, v_1(s), 0)$ at $s = -22$, for $l(\tau_2, v_2(s), 0)$ at $s = -13$, and for $l(\tau_3, v_3(s), 0)$ at $s = -16$.
2. Some $l(\tau_i, v_i(s), p)$ changes from state 1 to state 2. This occurs where $v_i(s) + C_i - pT_i = C_i$ (so $v_i(s) = pT_i$). The resulting tuple will be $\{s, i, p, 1, i, p, 2\}$. As above, this change occurs $O(mn)$ times for the system. In Fig. 3, this state change occurs for $l(\tau_1, v_1(s), 0)$ at $s = -18$, for $l(\tau_2, v_2(s), 0)$ at $s = -7$, and for $l(\tau_3, v_3(s), 0)$ at $s = -4$.
3. Some $l(\tau_i, v_i(s), p_i)$ is in state 1 and crosses C_j, and thus potentially crosses $l(\tau_j, v_j(s), p_j)$ (for some p_j) where the latter is in state 2. This occurs when

$C_i > C_j$ and $v_i(s) + C_i - p_i T_i = C_j$. The resulting tuple will be $\{s, j, *, 2, i, p, 1\}$. This point may exist at most $n - 1$ times per $l(\tau_i, v_i(s), p)$ (in the worst case, $l(\tau_i, v_i(s), p)$ crosses one $l(\tau_j, v_j(s), p_j)$ for each other τ_j), so occurs at most $O(mn^2)$ times for the system. In Fig. 3, this point does not occur for τ_1 (as C_1 is the smallest value in the system), occurs for $l(\tau_2, v_2(s), 0)$ with τ_1 at $s = -9$, and occurs for $l(\tau_3, v_3(s), 0)$ with τ_1 at $s = -12$ and with τ_2 at $s = -10$. (Although $l(\tau_3, v_3(s), 0)$ does not actually cross $l(\tau_2, v_2(s), 0)$ at $s = -10$, our algorithm nonetheless records the point where $l(\tau_3, v_3(s), 0)$ crosses C_2.)

In order to track $L(s)$, we order the tuples in $Points$ by s value, breaking ties in favor of tuples indicating a change in state for a particular $l(\tau_i, v_i(s), p)$ component. We create a list $Active$ containing tuples $\{i, p, h\}$, each representing the corresponding $l(\tau_i, v_i(s), p)$ in state h that contributes its value to $L(s)$. For s smaller than the smallest in $Points$, we may arbitrarily make $m^+ - 1$ choices of $l(\tau_i, v_i(s), p)$ components, each in state 0. Therefore, we initialize $Active$ to an arbitrary choice of $m^+ - 1$ tuples of the form $\{i, p, 0\}$.

The appropriate s value is computed using Algorithm 1, which works by tracing the piecewise linear function and checking for $L(s) = ms$ (as per (12), (14), and (15)) in each segment.

Algorithm 1.   Compute s value

---

tuple set $Active$, $Points$, described in text
integer $slope$, $current$ **initially** 0
real s, s_2
for all $\{s_1, i_1, p_1, h_1, i_2, p_2, h_2\} \in Points$ **do**
   if $\{i_1, p_1, h_1\}$ matches some $\{i, p, h\}$ in $Active$ **then**
      Replace $\{i, p, h\}$ with $\{i_2, p_2, h_2\}$
      if $h_2 = 1$ **then**
         {Changing to state 1 means slope increases}
         $slope := slope + 1$
      else
         {Must be changing away from state 1 or $\{s_1, i_1, p_1, h_1, i_2, p_2, h_2\}$ wouldn't be in $Points$}
         $slope := slope - 1$
      end if
      $s_2 :=$ next s value from $Points$, or $C_{\max}$ if there is no such value
      $s := \frac{current - slope \cdot s_1}{m - slope}$
      if $s \in [s_1, s_2)$ **then**
         return s
      end if
      $current := current + slope \cdot (s_2 - s_1)$
   end if
end for

As an example, suppose $Active$ is initialized to $\{\{3, 0, 0\}\}$, which represents $l(\tau_3, v_3(s), 0)$ in state 0. The first tuple in $Points$ is $\{-22, *, 0, 0, 1, 0, 1\}$, representing the leftmost slope change in Fig. 3. This tuple will match the single tuple in *Active*, so *Active* will become $\{1, 0, 1\}$. *slope* is used to track the slope between s_1 and the next s value in *Points* (which is called s_2). *current* is used to represent the correct value of $L(s_1)$. In this case, the current interval of interest is $-22 \leq s < -18$. The new state h_2 is 1, so the *slope* (which was initially 0) will be incremented by 1, resulting in a new *slope* of 1. We now know the slope $slope = 1$ of $L(s)$ over $[-22, -18)$ and its value $L(s_1) = current = 0$ at $s_1 = -22$. We therefore compute the point where $L(s) = ms$ would hold, assuming a linear function that is equal to the correct piecewise linear function over the interval of interest. In this case, s is assigned the value $\frac{0-(-22)}{2-1} = 22$, which is not in $[-22, -18)$, so the desired value of s for the algorithm is not in the current interval of interest. We do not return, so we update the value *current* to match the value of $L(s_2)$ at the end of the current interval of interest (and thus in the next iteration the correct value of $L(s_1)$). In this case, *current* will be assigned to $0 + 1 \cdot 4 = 4$.

Points is of size $O(mn^2)$ and *Active* of size $O(m)$, so checking for matches will require $O(m^2n^2)$ operations over the execution of the algorithm. Each match requires $O(1)$ time to process, so the complexity of Algorithm 1 is $O(m^2n^2)$. Computing *Points* requires $O(mn^2)$ time, and sorting requires $O(mn^2 \log(mn))$ time, so the complexity of computing s is $O(mn^2 log(mn) + m^2n^2)$. Once an s value has been computed using Algorithm 1, the correct minimum compliant vector is simply $\boldsymbol{v}(s)$, which can be computed in $O(n)$ time.

Node-Disjoint Multipath Spanners and Their Relationship with Fault-Tolerant Spanners

Cyril Gavoille[1,*], Quentin Godfroy[1,*], and Laurent Viennot[2,**]

[1] University of Bordeaux, LaBRI
[2] INRIA, University Paris 7, LIAFA

Abstract. Motivated by multipath routing, we introduce a multi-connected variant of spanners. For that purpose we introduce the p-multipath cost between two nodes u and v as the minimum weight of a collection of p internally vertex-disjoint paths between u and v. Given a weighted graph G, a subgraph H is a p-multipath s-spanner if for all u, v, the p-multipath cost between u and v in H is at most s times the p-multipath cost in G. The s factor is called the stretch.

Building upon recent results on fault-tolerant spanners, we show how to build p-multipath spanners of constant stretch and of $\tilde{O}(n^{1+1/k})$ edges[1], for fixed parameters p and k, n being the number of nodes of the graph. Such spanners can be constructed by a distributed algorithm running in $O(k)$ rounds.

Additionally, we give an improved construction for the case $p = k = 2$. Our spanner H has $O(n^{3/2})$ edges and the p-multipath cost in H between any two node is at most twice the corresponding one in G plus $O(W)$, W being the maximum edge weight.

1 Introduction

It is well-known [2] that, for each integer $k \geqslant 1$, every n-vertex weighted graph G has a subgraph H, called *spanner*, with $O(n^{1+1/k})$ edges and such that for all pairs u, v of vertices of G, $d_H(u, v) \leqslant (2k - 1) \cdot d_G(u, v)$. Here $d_G(u, v)$ denotes the distance between u and v in G, i.e., the length of a minimum cost path joining u to v. In other words, there is a trade-off between the size of H and its *stretch*, defined here by the factor $2k - 1$. Such trade-off has been extensively used in several contexts. For instance, this can be the first step for the design of a Distance Oracle, a compact data structure supporting approximate distance query while using sub-quadratic space [25,4,5]. It is also a key ingredient for several distributed algorithms to quickly compute a sparse skeleton of a connected graph, namely a connected spanning subgraph with only $O(n)$ edges. This can

---

* Supported by the European project "EULER", the ANR-project "ALADDIN", the équipe-projet INRIA "CÉPAGE". The first author is Member of the "Insitut Universitaire de France".
** Supported by the European project "EULER", the ANR-project "ALADDIN", and the équipe-projet INRIA "GANG".
[1] Tilde-O notation is similar to Big-O up to poly-logarithmic factors in n.

A. Fernández Anta, G. Lipari, and M. Roy (Eds.): OPODIS 2011, LNCS 7109, pp. 143–158, 2011.
© Springer-Verlag Berlin Heidelberg 2011

be done by choosing $k = O(\log n)$. The target distributed algorithm can then be run on the remaining skeleton [3]. The skeleton construction can be done in $O(k)$ rounds, whereas computing a spanning tree requires diameter rounds in general. We refer the reader to [21] for an overview on graph spanner constructions.

However, it is also proved in [25] that if G is directed, then it may have no sub-digraph H having $o(n^2)$ edges and constant stretch, the stretch being defined analogously by the maximum ratio between the *one-way* distance from u to v in H and the one-way distance from u to v in G. Nevertheless, a size/stretch trade-off exists for the *round-trip* distance, defined as the sum of a minimum cost of a dipath from u to v, and a minimum cost dipath from v to u (see [7,23]). Similar trade-offs exist if we consider the *p-edge-disjoint multipath* distance (in undirected graphs) for each $p \geqslant 1$, that is the minimum sum of p edge-disjoint paths joining u and v, see [10].

1.1   Trade-Offs for Non-increasing Graph Metric

More generally, we are interested in size/stretch trade-offs for graphs (or digraphs) for some non-increasing graph metric. A *non-increasing* graph metric δ associates with each pair of vertices u, v some non-negative cost that can only decrease when adding edges. In other words, $\delta_G(u, v) \leqslant \delta_H(u, v)$ for all vertices u, v and spanning subgraphs H of G. Moreover, if $\delta_H(u, v) \leqslant \alpha \cdot \delta_G(u, v) + \beta$, then we say that H is an (α, β)-spanner and that its *stretch* (w.r.t. the graph metric δ) is at most (α, β). We simply say that H is an α-spanner if $\beta = 0$. The *size* of a spanner is the number of its edges.

In the previous discussion we saw that every graph or digraph has a spanner H of size $o(n^2)$ and with bounded stretch for graph metrics δ such as round-trip, p-edge-disjoint multipath, and the usual graph distance. However, it does not hold for one-way distance. A fundamental task is to determine which graph metrics δ support such size/stretch trade-off. We observe that the three former graph metrics cited above have the triangle inequality property, whereas the one-way metric does not.

This paper deals with the construction of spanners for the vertex-disjoint multipath metric. A *p-multipath* between u and v is a subgraph composed of the union of p pairwise internally vertex-disjoint paths joining u and v. The *cost* of a p-multipath between u and v is the sum of the weight of the edges it contains. Given an undirected positively weighted graph G, define $\delta_G^p(u, v)$ as the minimum cost of a p-multipath between u and v if it exists, and ∞ otherwise. A *p-multipath s-spanner* is a spanner H of G with stretch at most s w.r.t. the graph metric δ^p. In other words, for all vertices u, v of G, $\delta_H^p(u, v) \leqslant s \cdot \delta_G^p(u, v)$, or $\delta_H^p(u, v) \leqslant \alpha \cdot \delta_G^p(u, v) + \beta$ if $s = (\alpha, \beta)$. It generalizes classical spanners as $d_G(u, v) = \delta_G^p(u, v)$ for $p = 1$.

1.2   Motivations

Our interest in the node-disjoint multipath graph metric stems from the need for multipath routing in networks. Using multiple paths between a pair of nodes is an obvious way to aggregate bandwidth. Additionally, a classical approach to quickly overcome link failures consists in pre-computing fail-over paths which are disjoint from primary paths [14,19,18]. Multipath routing can be used for traffic load balancing and for minimizing delays. It has been extensively studied in ad hoc networks for load balancing, fault-tolerance, higher aggregate bandwidth, diversity coding, minimizing energy consumption (see [17] for a quick overview). Considering only a subset of links is a practical concern in link state routing in ad hoc networks [13]. This raises the problem of computing spanners for the multipath graph metric, a first step towards constructing compact multipath routing schemes.

1.3   Our Contributions

Our main contribution is to show that sparse p-multipath spanners of constant stretch do exist for each $p \geqslant 1$. Moreover, they can be constructed *locally* in a constant number of rounds. More precisely, we show that:

1. Every weighted graph with n vertices has a p-multipath $kp \cdot O(1 + p/k)^{2k-1}$-spanner of size $\tilde{O}(p^2 \cdot n^{1+1/k})$, where k and p are integral parameters $\geqslant 1$. Moreover, such a multipath spanner can be constructed distributively in $O(k)$ rounds.
2. For $p = k = 2$, we improve this construction whose stretch is 18. Our algorithm provides a 2-multipath $(2, O(W))$-spanner of size $O(n^{3/2})$ where W is the largest edge weight of the input graph.

Distributed algorithms are given in the classical $\mathcal{LOCAL}$ model of computations (cf. [20]), a.k.a. the free model [15]. In this model nodes operate in synchronous discrete rounds (nodes are also assumed to wake up simultaneously). At each round, a node can send and/or receive messages of unbounded capacity to/from its neighbors and can perform any amount of local computations. Hence, each round costs one time unit. Also, nodes have unique identifiers that can be used for breaking symmetry. As long as we are concerned with running time (number of rounds) and not with the cost of communication, synchronous and asynchronous message passing models are equivalent.

1.4   Overview

Multipath spanners have some flavors of fault-tolerant spanners, notion introduced in [6] for general graphs. A subgraph H is an r-*fault tolerant s-spanner* of G if for any set F of at most $r \geqslant 0$ faulty vertices, and for any pair u, v of vertices outside F, $d_{H \setminus F}(u, v) \leqslant s \cdot d_{G \setminus F}(u, v)$.

At first glance, r-fault tolerant spanners seem related to $(r + 1)$-multipath spanners. (Note that both notions coincide to usual spanners if $r = 0$.) This is motivated by the fact that, if for an edge uv of G that is not in H, and if, for each set F of r vertices, u and v are connected in $H \setminus F$, then by Menger's Theorem H must contain some p-multipath between u and v. If the connectivity condition fulfills, there is no guarantee however on the cost of the p-multipath in H compared to the optimal one in G. Actually, as presented on Fig. 1, there are 1-fault tolerant s-spanners that are 2-multipath but with arbitrarily large stretch.

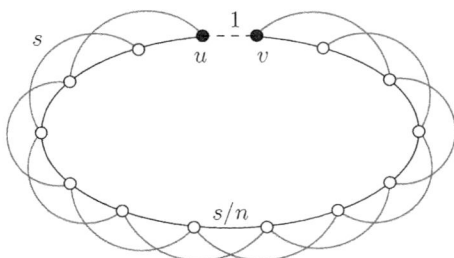

Fig. 1. A weighted graph G composed of a cycle of $n+1$ vertices plus $n-1$ extra edges, and a spanner $H = G \setminus \{uv\}$. Edge uv has weight 1, non-cycle edges have weight s, and cycle edges weight s/n so that $d_H(u,v) = s$. Removing any vertex $z \notin \{u,v\}$ implies $d_{G \setminus \{z\}}(u,v) = 1$ and $d_{H \setminus \{z\}}(u,v) = 2s(1 - 1/n)$. For other pairs of vertices x, y, $d_{H \setminus \{z\}}(x,y)/d_{G \setminus \{z\}}(x,y) < 2s$. Thus, H is a 1-fault tolerant $2s$-spanner. However $\delta_H^2(u,v)/\delta_G^2(u,v) \geqslant sn/s$. Thus, H is a 2-multipath spanner with stretch at least n.

Nevertheless, a relationship can be established between p-mutlipath spanners and *some* r-fault tolerant spanners. In fact, we prove in Section 2.4 that every r-fault tolerant s-spanner that is b-*hop* is a $(r+1)$-multipath spanner with stretch bounded by a function of b, r and s. Informally, a b-hop spanner H must replace every edge uv of G not in H by a path simultaneously of low cost and composed of at most b edges. We observe that many classical spanner constructions (including the greedy one) do not provide bounded-hop spanners, although such spanners exist as proved in Section 2.1. Some variant presented in [6] of the Thorup-Zwick constructions [25] are also bounded-hop (Section 2.2). Combining these specific spanners with the generic construction of fault tolerant spanners of [9], we show in Section 2.3 how to obtained a $\mathcal{LOCAL}$ distributed algorithm for computing a p-mutlipath spanner of bounded stretch. A maybe surprising fact is that the number of rounds is independent of p and n. We stress that the distributed algorithm that we obtain has significantly better running time than the original one presented in [9] that was $\Omega(p^3 \log n)$.

For instance for $p = 2$, our construction can produce a 2-multipath 18-spanner with $O(n^{3/2} \log^{3/2} n)$ edges. For this particular case we improve the general construction in Section 3 with a completely different approach providing a low multiplicative stretch, namely 2, at the cost of an additive term depending of the largest edge weight.

We note that the graph metric δ^p does not respect the triangle inequality for $p > 1$. For $p = 2$, a cycle from u to w and a cycle from w to v does not imply the existence of a cycle from u to v. The lack of this property introduces many complications for our second result. Basically, there are $\Omega(n^2)$ pairs u, v of vertices, each one possibly defining a minimum cycle $C_{u,v}$ of cost $\delta_G^2(u, v)$. If we want to create a spanner H with $o(n^2)$ edges, we cannot keep $C_{u,v}$ for all pairs u, v. Selecting some vertex w as pivot for going from u to v is usually a solution of save edges (in particular at least one between u and v). One pivot can indeed serve for many other pairs. However, without the triangle inequality, $C_{u,w}$ and $C_{w,v}$ do not give any cost guarantee on $\delta_H^2(u, v)$.

2   Main Construction

In this section, we prove the following result:

Theorem 1. *Let G be a weighted graph with n vertices, and p, k be integral parameters $\geqslant 1$. Then, G has a p-multipath $kp \cdot O(1 + p/k)^{2k-1}$-spanner of size $O(kp^{2-1/k}n^{1+1/k}\log^{2-1/k}n)$ that can be constructed w.h.p. by a randomized distributed algorithm in $O(k)$ rounds.*

Theorem 1 is proved by combining several constructions presented now.

2.1   Spanners with Few Hops

An s-spanner H of a weighted graph G is b-hop if for every edge uv of G, there is a path in H between u and v composed of at most b edges and of cost at most $s \cdot \omega(uv)$ (where $\omega(uv)$ denotes the cost of edge uv). An s-hop spanner is simply an s-hop s-spanner.

If G is unweighted (or the edge-cost weights are uniform), the concepts of s-hop spanner and s-spanner coincide. However, not all s-spanners are s-hop. In particular, the $(2k-1)$-spanners produced by the greedy[2] algorithm [2] are not.

For instance, consider a weighted cycle of $n + 1$ vertices and any stretch s such that $1 < s < n$. All edges of the cycle have unit weight, but one, say the edge uv, which has weight $\omega(uv) = n/s$. Note that $d_G(u, v) = \omega(uv) > 1$. The greedy algorithm adds the n unit cost edges but the edge uv to H because $d_H(u, v) = n \leqslant s \cdot \omega(uv)$ (recall that uv is added only if $d_H(u, v) > s \cdot d_G(u, v)$). Therefore, H is an s-spanner but it is only an n-hop spanner.

However, we have:

Proposition 1. *For each integer $k \geqslant 1$, every weighted graph with n vertices has a $(2k-1)$-hop spanner with less than $n^{1+1/k}$ edges.*

Proof. Consider a weighted graph G with edge-cost function ω. We construct the willing spanner H of G thanks to the following algorithm which can be seen as the dual of the classical greedy algorithm, till a variant of Kruskal's algorithm:

---

[2] For each edge uv in non-decreasing order of their weights, add it to the spanner if $d_H(u, v) > s \cdot d_G(u, v)$.

(1) Initialize H with $V(H) := V(G)$ and $E(H) := \varnothing$;
(2) Visit all the edges of G in non-decreasing order of their weights, and add the edge uv to H only if every path between u and v in H has more than $2k - 1$ edges.

Consider an edge uv of G. If uv is not in H then there must exist a path P in H from u to v such that P has at most $2k - 1$ edges. We have $d_H(u, v) \leqslant \omega(P)$. Let e be an edge of P with maximum weight. We can bound $\omega(P) \leqslant (2k - 1) \cdot \omega(e)$. Since e has been considered before the edge uv, $\omega(e) \leqslant \omega(uv)$. It follows that $\omega(P) \leqslant (2k - 1) \cdot \omega(uv)$, and thus $d_H(u, v) \leqslant (2k - 1) \cdot \omega(uv)$. Obviously, if uv belongs to H, $d_H(u, v) = \omega(uv) \leqslant (2k - 1) \cdot \omega(uv)$ as well. Therefore, H is $(2k - 1)$-hop.

The fact that H is sparse comes from the fact that there is no cycle of length $\leqslant 2k$ in H: whenever an edge is added to H, any path linking its endpoints has more than $2k - 1$ edges, i.e., at least $2k$.

We observe that H is simple even if G is not. It has been proved in [1] that every simple n-vertex m-edge graph where every cycle is of length at least $2k + 1$ (i.e., of girth at least $2k + 1$), must verify the Moore bound:

$$n \geqslant 1 + d \sum_{i=0}^{k-1} (d - 1)^i > (d - 1)^k$$

where $d = 2m/n$ is the average degree of the graph. This implies that $m < \frac{1}{2}(n^{1+1/k} + n) < n^{1+1/k}$.

Therefore, H is a $(2k - 1)$-hop spanner with at most $n^{1+1/k}$ edges.   □

2.2   Distributed Bounded Hop Spanners

There are distributed constructions that provide s-hop spanners, at the cost of a small (poly-logarithmic in n) increase of the size of the spanner compared to Proposition 1.

If we restrict our attention to deterministic algorithms, [8] provides for unweighted graphs a $(2k - 1)$-hop spanner of size $O(kn^{1+1/k})$. It runs in $3k - 2$ rounds without any prior knowledge on the graph, and optimally in k rounds if n is available at each vertex.

Proposition 2. *There is a distributed randomized algorithm that, for every weighted graph G with n vertices, computes w.h.p. a $(2k - 1)$-hop spanner of $O(kn^{1+1/k} \log^{1-1/k} n)$ edges in $O(k)$ rounds.*

Proof. The algorithm is a distributed version of the spanner algorithm used in [6], which is based on the sampling technique of [25]. We make the observation that this algorithm can run in $O(k)$ rounds. Let us briefly recall the construction of [6, p. 3415].

To each vertex w of G is associated a tree rooted at w spanning the *cluster* of w, a particular subset of vertices denoted by $C(w)$. The construction of $C(w)$ is a refinement over the one given in [25]. The main difference is that the clusters' depth is no more than k edges. The spanner is composed of the union of all such cluster spanning trees. The total number of edges is $O(kn^{1+1/k}\log^{1-1/k}n)$. It is proved in [6] that for every edge uv of G, there is a cluster $C(w)$ containing u and v. The path of the tree from w to one of the end-point has at most $k-1$ edges and cost $\leqslant (k-1)\cdot\omega(uv)$, and the path from w to the other end-point has at most k edges and cost $\leqslant k\cdot\omega(uv)$. This is therefore a $(2k-1)$-hop spanner.

The random sampling of [25] can be done without any round of communications, each vertex randomly select a level independently of the other vertices. Once the sampling is performed, the clusters and the trees can be constructed in $O(k)$ rounds as their the depth is at most k. □

2.3   Fault Tolerant Spanners

The algorithm of [9] for constructing fault tolerant spanners is randomized and generic. It takes as inputs a weighted graph G with n vertices, a parameter $r\geqslant 0$, and any algorithm **A** computing an s-spanner of $m(\nu)$ edges for any ν-vertex subgraph of G. With high probability, it constructs for G an r-fault tolerant s-spanner of size $O(r^3\cdot m(2n/r)\cdot\log n)$. It works as follows: Set $H:=\varnothing$, and repeat independently $O(r^3\log n)$ times:

(1) Compute a set S of vertices built by selecting each vertex with probability $1-1/(r+1)$;
(2) $H:=H\cup\mathbf{A}(G\setminus S)$.

Then, they show that for every fault set $F\subset V(G)$ of size at most r, and every edge uv, there exists with high probability a set S as computed in Step (1) for which $u,v\notin S$ and $F\subseteq S$. As a consequence, routine $\mathbf{A}(G\setminus S)$ provides a path between u and v in $G\setminus S$ (and thus also in $G\setminus F$) of cost $\leqslant s\cdot\omega(uv)$. If uv lies on a shortest path of $G\setminus F$, then this cost is $\leqslant s\cdot d_{G\setminus F}(u,v)$. From their construction, we have:

Proposition 3. *If $\mathbf{A}$ is a distributed algorithm constructing an s-hop spanner in t rounds, then algorithm [9] provides a randomized distributed algorithm that in t rounds constructs w.h.p. an s-hop r-fault tolerant spanner of size $O(r^3\cdot m(2n/r)\cdot\log n)$.*

Proof. The resulting spanner H is s-hop since either the edge uv of G is also in H, or a path between u and v approximating $\omega(uv)$ exists in some s-hop spanner given by algorithm **A**. This path has no more than s edges and cost $\leqslant s\cdot\omega(uv)$.

Observe that the algorithm [9] consists of running in parallel $q=O(r^3\log n)$ times independent runs of algorithm **A** on different subgraphs of G, each one using t rounds. Round i of all these q runs can be done into a single round of communication, so that the total number of rounds is bounded by t, not by q.

More precisely, each vertex first selects a q-bit vector, each bit set with probability $1-1/(r+1)$, its jth bit indicating whether it participates to the jth run of **A**. Then, q instances of algorithm **A** are run in parallel simultaneously by all the vertices, and whenever the algorithms perform their ith communication round, a single message concatenating the q messages is sent. Upon reception, a vertex expands the q messages and run the jth instance of algorithm **A** only if the jth bit of its vector is set.

The number of rounds is no more than t. $\qquad\qquad\qquad\qquad\qquad\qquad$ □

2.4   From Fault Tolerant to Multipath Spanner

Theorem 2. *Let H be a s-hop $(p-1)$-fault tolerant spanner of a weighted graph G. Then, H is also a p-multipath $\varphi(s,p)$-spanner of G where $\varphi(s,p) = sp \cdot O(1 + p/s)^s$ and $\varphi(3,p) = 9p$.*

To prove Theorem 2, we need the following intermediate result, assuming that H and G satisfy the statement of Theorem 2.

Lemma 1. *Let uv be an edge of G of weight $\omega(uv)$ that is not in H. Then, H contains a p-multipath connecting u to v of cost at most $\varphi(s,p) \cdot \omega(uv)$ where $\varphi(s,p) = sp \cdot O(1 + p/s)^s$ and $\varphi(3,p) = 9p$.*

Proof. From Menger's Theorem, the number of pairwise vertex-disjoint paths between two non-adjacent vertices x and y equals the minimum number of vertices whose removal disconnects x and y.

By definition of H, $H \setminus F$ contains a path P_F of at most s edges between u and v for each set F of at most $p - 1$ vertices (excluding u and v). This is because u and v are always connected in $G \setminus F$, precisely by a single edge path of cost $\omega(uv)$. Consider P_H the subgraph of H composed of the union of all such P_F paths (so from u to v in $H \setminus F$ – see Fig. 2 for an example with $p = 2$ and $s = 5$).

Vertices u and v are non-adjacent in P_H. Thus by Menger's Theorem, P_H has to contain a p-multipath between u and v. Ideally, we would like to show that this multipath has low cost. Unfortunately, Menger's Theorem cannot help us in this task.

Let $\kappa_s(u,v)$ be the minimum number of vertices in P_H whose deletion destroys all paths of at most s edges between u and v, and let $\mu_s(u,v)$ denote the maximum number of internally vertex-disjoint paths of at most s edges between u and v. Obviously, $\kappa_s(u,v) \geqslant \mu_s(u,v)$, and equality holds by Menger's Theorem if $s = n - 1$. Equality does not hold in general as presented in Fig. 2. However, equality holds if s is the minimum number of edges of a path between u and v, and for $s = 2, 3, 4$ (cf. [16]).

Since not every path of at most s edges between u and v is destroyed after removing $p - 1$ vertices in P_H, we have that $\kappa_s(u,v) \geqslant p$. Let us bound the total number of edges in a p-multipath Q of minimum size between u and v in P_H. Let r be the least number such that $\mu_r(u,v) \geqslant p$ subject to $\kappa_s(u,v) \geqslant p$. The total number of edges in Q is therefore no more than pr.

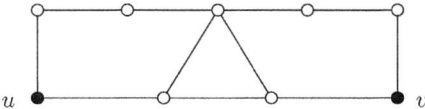

Fig. 2. A subgraph P_H constructed by adding paths between u and v with at most $s = 5$ edges and with $p = 2$. Removing any vertex leaves a path of at most 5 edges, so $\kappa_5(u, v) > 1$. However, there aren't two vertex-disjoint paths from u to v of at most 5 edges, so $\kappa_5(u, v) > \mu_5(u, v)$. Observe that $\mu_6(u, v) = \kappa_5(u, v) = 2$.

By construction of P_H, each edge of P_H comes from a path in $H \setminus F$ of cost $w(P_F) \leqslant s \cdot d_{G \setminus F}(u, v) \leqslant s \cdot w(uv)$. In particular, each edge of Q has weight at most $s \cdot w(uv)$. Therefore, the cost of Q is $w(Q) \leqslant prs \cdot w(uv)$.

It has been proved in [22] that r can be upper bounded by a function $r(s, p) < \binom{p+s-2}{s-2} + \binom{p+s-3}{s-2} = O(1 + p/s)^s$ for integers s, p, and $r(3, p) = 3$ since as seen earlier $\kappa_3(u, v) = \mu_3(u, v)$. It follows that H contains a p-multipath Q between u and v of cost $w(Q) \leqslant sp \cdot O(1 + p/s)^s \cdot w(uv)$ as claimed. $\qquad\square$

Proof of Theorem 2. Let x, y be any two vertices of a graph G with edge-cost function w. We want to show $\delta_H^p(x, y) \leqslant \varphi(s, p) \cdot \delta_G^p(x, y)$. If $\delta_G^p(x, y) = \infty$, then we are done. So, assume that $\delta_G^p(x, y) = w(P_G)$ for some minimum cost p-multipath P_G between x and y in G. Note that $w(P_G) = \sum_{uv \in E(P_G)} w(uv)$.

We construct a subgraph P_H between x and y in H by adding: (1) all the edges of P_G that are in H; and (2) for each edge uv of P_G that is not in H, the p-multipath Q_{uv} connecting u and v in H as defined by Lemma 1.

The cost of P_H is therefore:

$$w(P_H) = \sum_{uv \in E(P_H)} w(uv) = \left(\sum_{uv \in E(P_G) \cap E(H)} w(uv) \right) + \left(\sum_{uv \in E(P_G) \setminus E(H)} w(Q_{uv}) \right).$$

By Lemma 1, $w(Q_{uv}) \leqslant \varphi(s, p) \cdot w(uv)$. It follows that:

$$w(P_H) \leqslant \varphi(s, p) \cdot \sum_{uv \in E(P_G)} w(uv) = \varphi(s, p) \cdot w(P_G) = \varphi(s, p) \cdot \delta_G^p(x, y)$$

as $\varphi(s, p) \geqslant 1$ and by definition of P_G.

Clearly, all edges of P_H are in H. Let us show now that P_H contains a p-multipath between x and y. We first assume x and y are non-adjacent in P_H. By Menger's Theorem applied between x and y in P_H, if the removal of every set of at most $p - 1$ vertices in P_H does not disconnect x and y, then P_H has to contain a p-multipath between x and y.

Let S be any set of less than $p - 1$ faults in G. Since P_G is a p-multipath, P_G contains at least one path between x and y avoiding S. Let's call this path Q. For each edge uv of Q not in H, Q_{uv} is a p-multipath, so it contains one path avoiding S. Note that Q_{uv} may intersect Q_{wz} for different edges uv and wz of

Q. If it is the case then there is a path in $Q_{uv} \cup Q_{wz}$ from u to z (avoiding v and w), assuming that u, v, w, z are encountered in this order when traversing Q. Overall there must be a path connecting x to y and avoiding S in the subgraph $(Q \cap H) \cup \bigcup_{uv \in Q \setminus H} Q_{uv}$. By Menger's Theorem, P_H contains a p-multipath between x and y.

If x and y are adjacent in P_H, then we can subdivide the edge xy into the edges xz and zy by adding a new vertex z. Denote by P'_H this new subgraph. Clearly, if P'_H contains a p-multipath between x and y, then P_H too: a path using vertex z in P'_H necessarily uses the edges xz and zy. Now, P'_H contains a p-multipath by Menger's Theorem applied on P'_H between x and y that are non-adjacent.

We have therefore constructed a p-multipath between x and y in H of cost at most $\omega(P_H) \leqslant \varphi(s, p) \cdot \delta_G^p(x, y)$. It follows that $\delta_H^p(x, y) \leqslant \varphi(s, p) \cdot \delta_G^p(x, y)$ as claimed.   □

Theorem 1 is proved by applying Theorem 2 to the construction of Proposition 3, which is based on the distributed construction of s-hop spanners given by Proposition 2. Observe that the number of edges of the spanner is bounded by $O(kp^3 \cdot m(2n/p) \cdot \log n) = O(kp^{2-1/k} n^{1+1/k} \log^{2-1/k} n)$.

3   Bi-path Spanners

In this section we concentrate our attention on the case $p = 2$, i.e., 2-multipath spanners or *bi-path* spanners for short. Observe that for $p = k = 2$ the stretch is $\varphi(3, 2) = 18$ using our first construction (cf. Theorems 1 and 2). We provide in this section the following improvement on the stretch and on the number of edges.

Theorem 3. *Every weighted graph with n vertices and maximum edge-weight W has a 2-multipath $(2, O(W))$-spanner of size $O(n^{3/2})$ that can be constructed in $O(n^4)$ time.*

While the construction shown earlier was essentially working on edges, the approach taken here is more global. Moreover, this construction essentially yields an additive stretch whereas the previous one is only multiplicative. Note that a 2-multipath between two nodes u and v corresponds to an elementary cycle. We will thus focus on cycles in this section.

An algorithm is presented in Section 3.1. Its running time and the size of the spanner are analyzed in Section 3.2, and the stretch in Section 3.3. Due to space limitation some proofs of these sections appear in the long version [11].

3.1   Construction

Classical spanner algorithms combines the use of trees, balls, and clusters. These standard structures are not suitable to the graph metric δ^2 since, for instance, two nodes belonging to a ball centered in a single vertex can be in two different bi-components[3] and therefore be at an infinite cost from each other. We will adapt theses standard notions to structures centered on edges rather than vertices.

Consider a weighted graph G and with an edge uv that is not a cut-edge[4]. Let us denote by $G[uv]$ the bi-component of G containing uv, and by $\delta_H^2(uv, w)$ the minimum cost of a cycle in subgraph H passing through the edge uv and vertex w, if it exists and ∞ otherwise.

We define a 2-*path spanning tree of root* uv as a minimal subgraph T of G such that every vertex w of $G[uv]$ belongs to a cycle of T containing uv. Such definition is motivated by the following important property (see Property 1 in Section 3.3): for all vertices a, b in $G[uv] \setminus \{u, v\}$, $\delta_G^2(a, b) \leqslant \delta_T^2(uv, a) + \delta_T^2(uv, b)$. This can be seen as a triangle inequality like property.

If $\delta_T^2(uv, w) = \delta_G^2(uv, w)$ for every vertex w of $G[uv]$, T is called a *shortest 2-path spanning tree*. An important point, proved in Lemma 2 in Section 3.2, is that such T always exists and contains $O(\nu)$ edges, ν being the number of vertices of $G[uv]$.

In the following we denote by $B_G^2(uv, r) = \{w : \delta_G^2(uv, w) \leqslant r\}$ and $B_G(u, r) = \{w : d_G(u, w) \leqslant r\}$ the 2-*ball* (resp. 1-*ball*) of G centered at edge uv (at vertex u) and of radius r. We denote by $N_G(u)$ the set of neighbors of u in G. We denote by BFS(u, r) any shortest path spanning tree of root u and of depth r (not counting the edge weights). Finally, we denote by SPST$^2{}_G(uv)$ any shortest 2-path spanning tree of root uv in $G[uv]$.

The spanner H is constructed with Algorithm 1 from any weighted graph G having n vertices and maximum edge weight W. Essentially, the main loop of the algorithm selects an edge uv from the current graph lying at the center of a dense bi-component, adds the spanner H shortest 2-path spanning tree rooted at uv, and then destroys the neigborhood of uv.

$F := G$, $H := (\varnothing, \varnothing)$;
while $\exists uv \in E(G)$, $|B_G^2(uv, 4W) \cap (N_G(u) \cup N_G(v))| > \sqrt{n}$ **do**
$\quad | \quad H := H \cup \text{SPST}^2{}_F(uv) \cup \text{BFS}_G(u, 2) \cup \text{BFS}_G(v, 2)$;
$\quad \lfloor \quad G := G \setminus (B_G^2(uv, 4W) \cap (N_G(u) \cup N_G(v)))$
$H := H \cup G$

Algorithm 1. Construction of H

3.2   Size Analysis

The proof of the spanner's size is done in two steps, thanks to the two next lemmas.

First, Lemma 2 shows that the while loop does not add too much edges: a shortest 2-path spanning tree with linear size always exists. It is built upon the algorithm of Suurballe-Tarjan [24] for finding shortest pairs of edge-disjoint paths in weighted digraphs.

---

[3] A short for 2-vertex-connected components.
[4] A cut-edge is an edge that does not belong to a cycle.

Lemma 2. *For every weighted graph G and for every non cut-edge uv of G, there is a shortest 2-path spanning tree of root uv having $O(\nu)$ edges where ν is the number of vertices of $G[uv]$. It can be computed in time $O(n^2)$ where n is the number of vertices of G.*

Secondly, Lemma 3 shows that the graph G remaining after the while loop has only $O(n^{3/2})$ edges. For that, G is transformed as an unweighted graph (edge weights are set to one) and we apply Lemma 3 with $k = 2$. The result we present is actually more general and interesting in its own right. Indeed, it gives an alternative proof of the well-known fact that graphs with no cycles of length $\leqslant 2k$ have $O(n^{1+1/k})$ edges since $B_G^2(uv, 2k) = \varnothing$ in that case.

Lemma 3. *Let G be an unweighted graph with n vertices, and $k \geqslant 1$ be an integer. If for every edge uv of G, $|B_G^2(uv, 2k) \cap N_G(u)| \leqslant n^{1/k}$, then G has at most $2 \cdot n^{1+1/k}$ edges.*

Combining these two lemmas we have:

Lemma 4. *Algorithm 1 creates a spanner of size $O(n^{3/2})$ in time $O(n^4)$.*

Proof. Each step of the while loop adds $O(n)$ edges from Lemma 2, and as it removes at least $\sqrt{n}$ vertices from the graph this can continue at most $\sqrt{n}$ times. In total the while loop adds $O(n^{3/2})$ edges to H.

After the while loop, the graph G is left with every $B_G^2(uv, 4W) \cap (N_G(u) \cup N_G(v))$ smaller than $\sqrt{n}$. If we change all edges weights to 1, it is obvious that every $B_G^2(uv, 4) \cap (N_G(u) \cup N_G(v))$ is also smaller than $\sqrt{n}$. Then as $B_G^2(uv, 4) \cap N_G(u)$ is always smaller than $B_G^2(uv, 4) \cap (N_G(u) \cup N_G(v))$ we can apply Lemma 3 for $k = 2$, and therefore bound the number of edges added in the last step of Algorithm 1.

The total number of edges of H is $O(n^{3/2})$.

The costly steps of the algorithm are the search of suitable edges uv and the cost of construction of SPST^2.

The search of suitable edges is bounded by the number of edges as an edge e which is not suitable can be discarded for the next search: removing edges from the graph cannot improve $B_G^2(e, 4W)$. Then for each edge a breadth first search of depth 3 must be computed, whose cost is bounded by the number of edges of G. So in the end the search costs at most $O(n^4)$.

The cost of building a SPST^2 is bounded by the running time of [24], which at worst costs $O(n^2)$ (the reduction is essentially in $O(m + n)$). Since the loop is executed at most $\sqrt{n}$ times, the total cost is $O(n^{7/2})$.

So the total running time is $O(n^4)$.   □

3.3   Stretch Analysis

The proof for the stretch is done as follows: we consider a, b two vertices such that $\delta_F^2(a, b) = \ell$ is finite (if it is infinite there is nothing to prove). We need to prove

that the spanner construction is such that at the end, $\delta_H^2(a,b) \leqslant 2\ell + O(W)$.
To this effect, we define $P_F = P_F^1 \cup P_F^2$ as a cycle composed of two disjoint paths
(P_F^1 and P_F^2) going from a to b such that its weight sums to $\delta_F^2(a,b)$.

Proving the stretch amounts to show that there exists a cycle $P_H = P_H^1 \cup P_H^2$
joining a and b in the final H, with cost at most $2\ell + O(W)$. Observe that if the
cycle P_F has all its edges in H then one candidate for P_H is P_F and we are done.
If not, then there is at least one 2-ball whose deletion provokes actual deletion
of edges from P_F (that is edges of P_F missing in the final H).

In the following, let uv be the root edge of the first 2-ball whose removal
deletes edges from P_F (that is they are not added in H neither during the while
loop nor the last step of the algorithm). Let G_i be the graph G just before the
removal of $B_G^2(uv, 4W) \cap (N_G(u) \cup N_G(v))$, and G_{i+1} the one just after.

The rest of the discussion is done in G_i otherwise noted.

The proof is done as follows: we first show in Lemma 5 that any endpoint of
a deleted edge (of P_F) belongs to an elementary cycle comprising the edge uv
and of cost at most $6W$. We then show in Lemma 6 that we can construct cycles
using a and/or b passing through the edge uv, effectively bounding $\delta_H^2(uv, a)$
and $\delta_H^2(uv, b)$ due to the addition of the shortest 2-path spanning tree rooted at
uv. Finally we show in Lemma 7 that the union of a cycle passing through uv
and a and another one passing through uv and b contains an elementary cycle
joining a to b, its cost being at most the sums of the costs of the two original
cycles.

Lemma 5. *Let $e = wt$ be an edge of $(G_i \setminus G_{i+1}) \setminus H$. Then in G_i both w and t
are connected to uv by a cycle of cost at most $6W$.*

We now show that we can use this lemma to exhibit cycles going from a to uv
and from b to uv.

From the vertices belonging to both $B_{G_i}^2(uv, 6W)$ and P_F we choose the ones
which are the closest from a and b (we know that at least two of them exist
because one edge was removed from P_F during step i of the loop). There are
at maximum four of them (a_1, a_2, b_1, b_2), one for each sub-path P_F^i and each
extremity $\{a, b\}$. Note that each extremity is connected to the root edge by an
elementary cycle of cost at most $6W$. Two cases are possible (the placement of
the vertices is shown on Fig. 3):

Case 1: There are only two extremities (then they belong to the same sub-
path) and their cycles which connect them to uv do not intersect the second
subpath (w.l.o.g we can suppose it is a_1 and b_1).

Case 2: There are more than two extremities: either some edges of the second
path were removed or one of the cycles going from one of the extremities a_1
or b_1 to uv intersects the second path.

We show next that we can bound $\delta_H^2(uv, a)$ and $\delta_H^2(uv, b)$ with the help of the
cycles connecting the endpoints and the path P_G. This is done with the two next
lemmas.

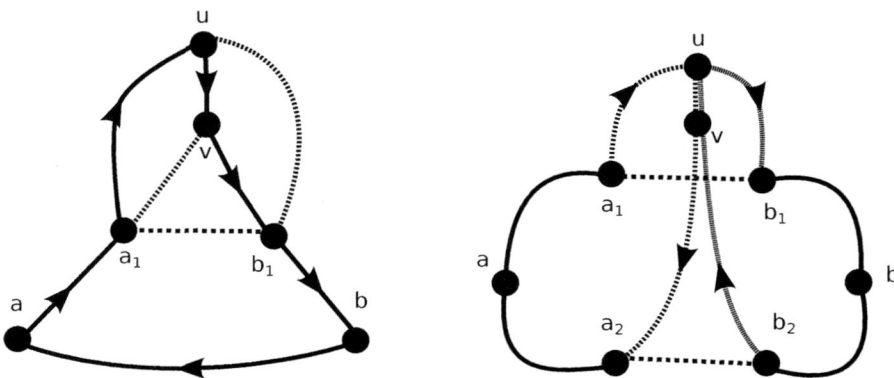

Fig. 3. Proof of Lemma 7: the two cases for the simple paths

Lemma 6. *For any two vertices joined to the same edge uv by elementary cycles there is a simple path of cost at most the sum of the cycles' costs and passing through the edge uv.*

Lemma 7. *Let a, b be two vertices such that an elementary cycle of cost $\delta^2(a, b)$ has common vertices with some $B^2(uv, 6W)$. Then $\delta^2(a, uv)$ and $\delta^2(b, uv)$ are bounded by $\delta^2(a, b) + 12W$*

Property 1. Let uv be a non cut-edge of G and T be any 2-path spanning tree rooted at uv. Then, for all vertices a, b in $G[uv] \setminus \{u, v\}$, $\delta^2_G(a, b) \leqslant \delta^2_T(uv, a) + \delta^2_T(uv, b) - \omega(uv)$.

Proof. There is in T a cycle joining a to uv of cost $\delta^2_T(uv, a)$, and another one joining b to uv of cost $\delta^2_T(uv, b)$. Consider the subgraph P containing only the edges from these two cycles. The cost of P is $\omega(P) \leqslant \delta^2_T(uv, a) + \delta^2_T(uv, b) - \omega(uv)$ as edge uv is counted twice. It remains to show that P contains an elementary cycle between a and b. Note that since $a \notin \{u, v\}$, a has in P two vertex-disjoint paths leaving a and excluding edge uv: one is going to u, and one to v. Similarly for vertex b.

W.l.o.g. we can assume that a and b are not adjacent in P. Otherwise we can subdivide edge ab to obtain a new subgraph P'. Clearly, if P' contains an elementary cycle between a and b, then P too. Consider that one vertex z, outside a and b, is removed in P. From the remark above, in $P \setminus \{z\}$, there must exists a path leaving a and joining some vertex $w_a \in \{u, v\} \setminus \{z\}$ and one path leaving b and joining some vertex $w_b \in \{u, v\} \setminus \{z\}$. If $w_a = w_b$, then a and b are connected in $P \setminus \{z\}$. If $w_a \neq w_b$, then edge uv belongs to $P \setminus \{z\}$ since in this case $z \notin \{u, v\}$, and thus a path connected a to b in $P \setminus \{z\}$. By Menger's Theorem, P contains a 2-multipath between a and b. □

Lemma 8. *H is a 2-multipath $(2, 24W)$-spanner.*

Proof. If there is in F a path of cost $\delta^2(a, b)$ such that every edge of it is in H, then there is nothing to prove. If there is some removed edge, then we can identify the loop order i which removed the first edge, and we can associate the graph G_i just before the deletion performed in the second step of the loop (so P_F still completely exist in G_i). By virtue of Lemma 5 we can identify some root-edge uv and we know that there are some vertices of P_F linked to uv by an elementary cycle of length at most $6W$. Lemma 7 can then be applied, and so in G_i, $\delta^2_{G_i}(a, uv)$ and $\delta^2_{G_i}(b, uv)$ are both bounded by $\delta^2_{G_i}(a, b) + 12W$. As the loop's first step is to build a shortest 2-path spanning tree rooted in uv we know that in H

$$\delta^2_H(a, uv) \leqslant \delta^2_{G_i}(a, uv) \leqslant \delta^2_{G_i}(a, b) + 12W$$

and the same for b. Property 1 can then be used in the 2-path spanning tree, to bound $\delta^2_H(a, b)$:

$$\delta^2_H(a, b) \leqslant \delta^2_H(a, uv) + \delta^2_H(b, uv) \leqslant 2 \cdot \delta^2_{G_i}(a, b) + 24W$$

Finally, as in G_i P_F still exists completely, we have that $\delta^2_{G_i}(a, b) = \delta^2_F(a, b)$, so

$$\delta^2_H(a, b) \leqslant 2 \cdot \delta^2_F(a, b) + 24W$$

4   Conclusion   □

We have introduced a natural generalization of spanner, the vertex-disjoint path spanners. We proved that there exists for multipath spanners a size-stretch trade-off similar to classical spanners. We also have presented a $O(k)$ round distributed algorithm to construct p-multipath $kp \cdot O(1 + p/k)^{2k-1}$-spanners of size $\tilde{O}(p^2 n^{1+1/k})$, showing that the problem is *local*: it does not require communication between distant vertices.

Our construction is based on fault tolerant spanner. An interesting question is to know if better construction (in term of stretch) exists as suggested by our alternative construction for $p = 2$.

The most challenging question is to explicitly construct the p vertex-disjoint paths in the p-multipath spanner. This is probably as hard as constructing efficient routing algorithm from sparse spanner. We stress that there is a significant difference between proving the existence of short routes in a graph (or subgraph), and constructing and explicitly describing such short routes. For instance it is known (see [12]) that sparse spanners may exist whereas routing in the spanner can be difficult (in term of space memory and stretch of the routes).

References

1. Alon, N., Hoory, S., Linial, N.: The Moore bound for irregular graphs. Graphs and Combinatorics 18, 53–57 (2002)
2. Althöfer, I., Das, G., Dobkin, D.P., Joseph, D.A., Soares, J.: On sparse spanners of weighted graphs. Discr. & Comp. Geometry 9, 81–100 (1993)
3. Barenboim, L., Elkin, M.: Deterministic distributed vertex coloring in polylogarithmic time. In: 29th ACM Symp. PODC, pp. 410–419 (2010)

4. Baswana, S., Gaur, A., Sen, S., Upadhyay, J.: Distance Oracles for Unweighted Graphs: Breaking the Quadratic Barrier with Constant Additive Error. In: Aceto, L., Damgård, I., Goldberg, L.A., Halldórsson, M.M., Ingólfsdóttir, A., Walukiewicz, I. (eds.) ICALP 2008, Part I. LNCS, vol. 5125, pp. 609–621. Springer, Heidelberg (2008)
5. Baswana, S., Kavitha, T.: Faster algorithms for approximate distance oracles and all-pairs small stretch paths. In: 47th Annual IEEE Symp. on Foundations of Computer Science (FOCS), pp. 591–602. IEEE Comp. Soc. Press (October 2006)
6. Chechik, S., Langberg, M., Peleg, D., Roditty, L.: Fault tolerant spanners for general graphs. SIAM Journal on Computing 39, 3403–3423 (2010)
7. Cowen, L.J., Wagner, C.: Compact roundtrip routing in directed networks. In: 19th ACM Symp. PODC, pp. 51–59 (2000)
8. Derbel, B., Gavoille, C., Peleg, D., Viennot, L.: On the locality of distributed sparse spanner construction. In: 27th ACM Symp. PODC, p. 273 (2008)
9. Dinitz, M., Krauthgamer, R.: Fault-tolerant spanners: Better and simpler, Tech. Rep. 1101.5753v1 [cs.DS], arXiv (January 2011)
10. Gavoille, C., Godfroy, Q., Viennot, L.: Multipath Spanners. In: Patt-Shamir, B., Ekim, T. (eds.) SIROCCO 2010. LNCS, vol. 6058, pp. 211–223. Springer, Heidelberg (2010)
11. Gavoille, C., Godfroy, Q., Viennot, L.: Node-Disjoint Multipath Spanners and their Relationship with Fault-Tolerant Spanners, HAL-00622915 (September 2011)
12. Gavoille, C., Sommer, C.: Sparse spanners vs. compact routing. In: 23rd ACM Symp. SPAA, pp. 225–234 (June 2011)
13. Jacquet, P., Viennot, L.: Remote spanners: what to know beyond neighbors. In: 23rd IEEE International Parallel & Distributed Processing Symp. (IPDPS). IEEE Computer Society Press (May 2009)
14. Kushman, N., Kandula, S., Katabi, D., Maggs, B.M.: R-bgp: Staying connected in a connected world. In: 4th Symp. on NSDI (2007)
15. Linial, N.: Locality in distributed graphs algorithms. SIAM Journal on Computing 21, 193–201 (1992)
16. Lovász, L., Neumann-Lara, V., Plummer, M.D.: Mengerian theorems for paths of bounded length. Periodica Mathematica Hungarica 9, 269–276 (1978)
17. Mueller, S., Tsang, R.P., Ghosal, D.: Multipath Routing in Mobile Ad Hoc Networks: Issues and Challenges. In: Calzarossa, M.C., Gelenbe, E. (eds.) MASCOTS 2003. LNCS, vol. 2965, pp. 209–234. Springer, Heidelberg (2004)
18. Nasipuri, A., Castañeda, R., Das, S.R.: Performance of multipath routing for on-demand protocols in mobile ad hoc networks. Mobile Networks and Applications 6, 339–349 (2001)
19. Pan, P., Swallow, G., Atlas, A.: Fast Reroute Extensions to RSVP-TE for LSP Tunnels. RFC 4090 (Proposed Standard) (2005)
20. Peleg, D.: Distributed Computing: A Locality-Sensitive Approach. SIAM Monographs on Discrete Mathematics and Applications (2000)
21. Pettie, S.: Low Distortion Spanners. In: Arge, L., Cachin, C., Jurdziński, T., Tarlecki, A. (eds.) ICALP 2007. LNCS, vol. 4596, pp. 78–89. Springer, Heidelberg (2007)
22. Pyber, L., Tuza, Z.: Menger-type theorems with restrictions on path lengths. Discrete Mathematics 120, 161–174 (1993)
23. Roditty, L., Thorup, M., Zwick, U.: Roundtrip spanners and roundtrip routing in directed graphs. ACM Transactions on Algorithms 3, Article 29 (2008)
24. Suurballe, J.W., Tarjan, R.E.: A quick method for finding shortest pairs of disjoint paths. Networks 14, 325–336 (1984)
25. Thorup, M., Zwick, U.: Approximate distance oracles. Journal of the ACM 52, 1–24 (2005)

The First Fully Polynomial Stabilizing Algorithm for BFS Tree Construction⋆

Alain Cournier[1], Stéphane Rovedakis[2], and Vincent Villain[1]

[1] Laboratoire MIS, Université de Picardie, 33 Rue St Leu, 80039 Amiens Cedex 1, France
{alain.cournier,vincent.villain}@u-picardie.fr
[2] Laboratoire CEDRIC, CNAM, 292 Rue St Martin, 75141 Paris Cedex 03, France
stephane.rovedakis@cnam.fr

Abstract. The construction of a spanning tree is a fundamental task in distributed systems which allows to resolve other tasks (i.e., routing, mutual exclusion, network reset). In this paper, we are interested in the problem of constructing a *Breadth First Search* (BFS) tree. *Stabilization* is a versatile technique which ensures that the system recover a correct behavior from an arbitrary global state resulting from transient faults.

A *fully polynomial* algorithm has a round complexity in $O(d^a)$ and a step complexity in $O(n^b)$ where d and n are the diameter and the number of nodes of the network and a and b are constants. We present the first fully polynomial stabilizing algorithm constructing a BFS tree under a distributed daemon. Moreover, as far as we know, it is also the first fully polynomial stabilizing algorithm for spanning tree construction. Its round complexity is in $O(d^2)$ and its step complexity is in $O(n^6)$.

To our knowledge, since in general the diameter of a network is much smaller than the number of nodes ($\log(n)$ in average instead of n), this algorithm reaches the best compromise of the literature between the complexities in terms of rounds and in terms of steps.

Keywords: Distributed systems, Fault-tolerance, Stabilization, Spanning tree construction.

1 Introduction

The construction of spanning trees is a fundamental problem in the field of distributed systems. A spanning tree is a virtual structure which contains no cycle and interconnects all the nodes of a network. In distributed systems, the construction of a spanning tree is commonly used to design algorithms resolving other distributed tasks, like routing, token circulation or message broadcasting in a network. Spanning trees are also used to obtain algorithms resolving a particular distributed problem with a better time complexity compared to algorithms for the same problem which do not use this structure. There are many different spanning tree construction problems guaranteeing various properties, e.g., the construction of a depth first search (DFS) tree, a spanning tree of minimum weight or a spanning tree of minimum diameter. A crucial class of

---

⋆ This work has been supported in part by the ANR project SPADES (08-ANR-SEGI-025).

A. Fernández Anta, G. Lipari, and M. Roy (Eds.): OPODIS 2011, LNCS 7109, pp. 159–174, 2011.

spanning trees is the construction of a Breadth First Search (BFS) tree, which contains shortest paths (in hops) from every node to the root of the tree. This structure is mainly used in networks to quickly broadcast information from a source node. When a cost is associated to communication links, this problem is known as the construction of a Shortest Path tree.

Self-stabilization introduced first by Dijkstra in [16] and later publicized by several books [17,23] is one of the most versatile techniques to handle transient faults arising in distributed systems. A distributed algorithm is self-stabilizing if starting from any arbitrary global state (due to faults or attacks) the system is able to recover from this catastrophic situation in finite time without external (e.g., human) intervention. As self-stabilization makes no hypothesis about the nature or the extent of the faults, this paradigm can also be used to handle dynamic changes on the network topology since these modifications are seen as faults by the system. Another kind of stabilization was introduced by Bui *et al* [4], called *snap-stabilization*. These algorithms have the ability to always guarantee a correct system behavior according to the specifications of the problem to be solved, starting from any arbitrary global state.

Related work. Due to the importance of the construction of spanning trees, there are a lot of works which study this task. A survey on several self-stabilizing tree constructions can be found in [19]. Moreover, Table 1 summarizes the time complexities (round and step complexities) of some self-stabilizing tree construction algorithms. The number of steps required to compute a solution is an important criterion since it reflects the number of messages exchanged by an algorithm; especially for a self-stabilizing algorithm for which each node has to send messages to its neighbors in order to inform them that its state has been changed. However, few works give a step complexity analysis as can be seen in Table 1, except for [22,11,10,13] and [9] which presented an algorithm improving the step complexity to $\Theta(n^2)$ steps for the construction of an arbitrary spanning tree (with n the number of nodes of the network). Another essential criterion concerns the round complexity of a distributed algorithm, that is to have a round complexity only function of the network diameter (which is much smaller than the size of network for most network topologies). Some of the algorithms cited in Table 1 are optimal in terms of rounds for the construction of an arbitrary spanning tree or a BFS tree.

From the above discussion, we give a characterization for self-stabilizing distributed algorithms having an efficient complexity to solve a task, called fully polynomial algorithms. A *fully polynomial* algorithm has a round complexity in $O(d^a)$ and a step complexity in $O(n^b)$ where d and n are the diameter and the number of nodes of the network and a and b are constants. As presented in Table 1, the existing self-stabilizing spanning tree construction algorithms with a polynomial step complexity requires $\Omega(n)$ rounds, or a round complexity of $\Theta(\max(d^2, n))$ for the construction of a BFS tree. To our knowledge, no fully polynomial stabilizing algorithm was given for the construction of a spanning tree. Therefore, a legitimate question can be the following: Is it possible to construct in a self-stabilizing manner a spanning tree with a polynomial step complexity and a round complexity lower than $\Theta(\max(d^2, n))$?

Contributions. In this paper, we present the first fully polynomial stabilizing algorithm for the construction of a spanning tree with a round complexity lower than $\Theta(\max(d^2, n))$. Notice that the algorithm presented in [13] does not satisfy the

Table 1. Distributed stabilizing algorithms for the construction of spanning trees. n, d and Δ are respectively the number of nodes, the diameter and the maximum degree in the network, while N is an upper bound of n and Max is the maximum height value in the tree of a node in the initial configuration. The *silent* property for a self-stabilizing algorithm is to guarantee that when a legitimate configuration is reached the values stored in the registers do not change anymore.

	References	Round complexity	Step complexity	Memory complexity	Silent property
BFS	[2]	$O(N^2)$	Undetermined	$O(\log(n))$	Yes
	[18]	$O(d)$	Undetermined	$O(\Delta\log(n))$	Yes
	[1]	$O(n^2)$	Undetermined	$O(\log(n))$	Yes
	[20]	$\Theta(d)$	$O(n(Max+d)^n)^4$	$O(\log(n))$	Yes
	[3]	$O(d)$	Undetermined	$O(\log^2(n))$	Yes
	[21]	$\Omega(d^2)$	Undetermined	$O(\log(\Delta))$	No
	[5]	$O(d)$	Undetermined	$O(\log^2(n))$	Yes
	[15]	$O(n)$	Undetermined	$O(\log(n))$	Yes
	[13]	$\Theta(d^2+n)$	$O(\Delta n^3)$	$O(\log(n))$	No
	This paper	$O(d^2)$	$O(n^6)$	$O(\log(n))$	Yes
Any	[6]	$O(n)$	$\Omega(2^n)^4$	$O(\log(n))$	Yes
	[22]	$O(n)$	$\Theta(n^2d)$	$O(\log(n))$	Yes
	[9]	$\Theta(n)$	$\Theta(n^2)$	$O(\log(n))$	Yes
DFS	[7]	$O(dn\Delta)$	Undetermined	$O(n\log(\Delta))$	Yes
	[11]	$O(n^2)$	$O(n^3)$	$O(\log(n))$	Yes
	[10]	$O(n)$	$O(n^2)$	$O(n\log(n))$	Yes
	[13]	$O(n)$	$O(\Delta n^3)$	$O(\log(\Delta+n))$	No

definition of a fully polynomial algorithm since it has a round complexity which is related with the network size. Our algorithm computes a BFS tree in $O(d^2)$ rounds with a polynomial number of steps in $O(n^6)$ (the step complexity is $O(mn^4)$ and $m << n^2$) under a distributed daemon without any fairness assumptions, with d the diameter, m the number of edges and n the number of nodes in the network. To our knowledge, since in general the diameter of a network is much smaller than the number of nodes ($\log(n)$ in average instead of n), this algorithm reaches the best compromise of the literature between the complexities in terms of rounds and in terms of steps. Moreover, this BFS tree construction is based on a snap-stabilizing algorithm given in this paper resolving the Question-Answer problem, in which each node requests a permission (delivered by a subset of network nodes) in order to perform a defined computation, which is of independent interest.

Outline of the paper. The paper is organized as follows. In Section 2 we present the model assumed in this paper. We then present a fully polynomial stabilizing algorithm to construct a BFS tree in Section 3, based on a snap-stabilizing algorithm to the Question-Answer problem given in Section 4. We describe in Section 5 how these stabilizing algorithms are composed together and give an explanation about the time complexity to solve the BFS tree problem. Finally, we conclude in the last section.[1]

---

[1] This is detailed in the analysis given in [8].

2 Model

Notations. We consider a network as an undirected connected graph $G = (V,E)$ where V is a set of nodes (or *processors*) and E is the set of *bidirectional asynchronous communication links*. We state that n is the size of G ($|V| = n$). We assume that the network is *rooted*, i.e., among the processors, we distinguish a particular one, r, which is called the *root* of the network. In the network, p and q are neighbors if and only if a communication link (p,q) exists (i.e., $(p,q) \in E$). Every processor p can distinguish all its links. To simplify the presentation, we refer to a link (p,q) of a processor p by the *label* q. We assume that the labels of p, stored in the set $Neig_p$, are locally ordered by $\prec_p$. We also assume that $Neig_p$ is a constant input from the system. Δ is the maximum degree of the network (i.e., the maximal value among the local degrees of the processors). A tree $T = (V_T, E_T)$ is an acyclic connected subgraph such that $V_T \subseteq V$ and $E_T \subseteq E$, where the root of tree T is noted by $root(T)$. Moreover, any processor has a *parent* in a tree T which is the neighbor on the path leading to $root(T)$. A processor $p \in V_T$ with at least two neighbors in tree T is called an *internal* processor and a *leaf* processor otherwise.

Programs. In our model, protocols are *semi-uniform*, i.e., each processor executes the same program except r. We consider the local shared memory model of computation. In this model, the program of every processor consists in a set of *variables* and an *ordered finite set of actions* inducing a *priority*. This priority follows the order of appearance of the actions into the text of the protocol. A processor can write to its own variable only, and read its own variables and that of its neighbors. Each action is constituted as follows: $< label > :: < guard > \rightarrow < statement >$. The guard of an action in the program of p is a boolean expression involving variables of p and its neighbors. The statement of an action of p updates one or more variables of p. An action can be executed only if its guard is satisfied. The *state* of a processor is defined by the value of its variables. The *state* of a system is the product of the states of all processors. We will refer to the state of a processor and the system as a (*local*) *state* and (*global*) *configuration*, respectively. We note $\mathcal{C}$ the set of all possible configuration of the system. Let $\gamma \in \mathcal{C}$ and A an action of p ($p \in V$). A is said to be *enabled* at p in γ if and only if the guard of A is satisfied by p in γ. Processor p is said to be *enabled* in γ if and only if at least one action is enabled at p in γ. When several actions are enabled simultaneously at a processor p: only the priority enabled action can be activated.

Let a distributed protocol P be a collection of binary transition relations denoted by $\mapsto$, on $\mathcal{C}$. A *computation* of a protocol P is a *maximal* sequence of configurations $e = (\gamma_0,\gamma_1,...,\gamma_i,\gamma_{i+1},...)$ such that, $\forall i \geq 0$, $\gamma_i \mapsto \gamma_{i+1}$ (called a *step*) if γ_{i+1} exists, else γ_i is a terminal configuration. *Maximality* means that the sequence is either finite (and no action of P is enabled in the terminal configuration) or infinite. All computations considered here are assumed to be maximal. $\mathcal{E}$ is the set of all possible computations of P.

As we already said, each execution is decomposed into steps. Each step is shared into three sequential phases atomically executed: (i) every processor evaluates its guards, (ii) a *daemon* (also called *scheduler*) chooses some enabled processors, (iii) each chosen processor executes its priority enabled action. When the three phases are done, the next step begins.

A *daemon* can be defined in terms of *fairness* and *distributivity*. In this paper, we use the notion of *unfairness*: the *unfair* daemon can forever prevent a processor from executing an action except if it is the only enabled processor. Concerning the *distributivity*, we assume that the daemon is *distributed* meaning that, at each step, if one or more processors are enabled, then the daemon chooses at least one of these processors to execute an action.

We consider that any processor p executed a *disabling action* in the computation step $\gamma_i \mapsto \gamma_{i+1}$ if p was *enabled* in γ_i and not enabled in γ_{i+1}, but did not execute any protocol action in $\gamma_i \mapsto \gamma_{i+1}$. The disabling action represents the following situation: at least one neighbor of p changes its state in $\gamma_i \mapsto \gamma_{i+1}$, and this change effectively made the guard of all actions of p false in γ_{i+1}.

To compute the time complexity, we use the definition of *round*. This definition captures the execution rate of the slowest processor in any computation. Given a computation e ($e \in \mathcal{E}$), the *first round* of e (let us call it e') is the minimal prefix of e containing the execution of one action (an action of the protocol or a disabling action) of every enabled processor from the initial configuration. Let e'' be the suffix of e such that $e = e'e''$. The *second round* of e is the first round of e'', and so on.

3   Spanning Tree Construction

In this section, we are interested in the problem of constructing a tree spanning all the processors of the network. We consider there is a particular *root* processor, noted r, which is used to construct a spanning tree. More precisely, we consider the construction of a *Breadth First Search* (BFS) tree rooted at processor r. We can define a BFS tree as in Definition 1.

Definition 1 (BFS Tree). *Let $G = (V, E)$ be a network and r a node called the* root. *A graph $T = (V_T, E_T)$ of G is called a* Breadth First Search *tree if the following conditions are satisfied:*

1. *$V_T = V$ and $E_T \subseteq E$, and*
2. *T is a connected graph (i.e., there exists a path in T between any pair of nodes $x, y \in V_T$) and $|E_T| = |V| - 1$, and*
3. *For each node $p \in V_T$, there exists no shorter path (in hops) between p and r in G than the path between p and r in T.*

We give a formal specification to the problem of constructing a BFS tree, stated in Specification 1.

Specification 1 (Tree Construction). *Let $\mathcal{C}$ the set of all possible configurations of the system. An algorithm $\mathcal{A}_{\mathcal{BFS}}$ solving the problem of constructing a stabilizing BFS tree satisfies the following conditions:*

[TC1] *Algorithm $\mathcal{A}_{\mathcal{BFS}}$ reaches a set of terminal configurations $\mathcal{T} \subseteq \mathcal{C}$ in finite time, and*

[TC2] *Every configuration $\gamma \in \mathcal{T}$ satisfies Definition 1.*

3.1 Breadth First Search Tree Algorithm

In this section, we present a snap-stabilizing algorithm, called $\mathcal{BFS}$, to construct a BFS tree. Algorithm $\mathcal{BFS}$ is a semi-uniform algorithm, this means that exactly one of the processors, called the *root* and denoted r, is distinguished. This distinguished processor is used in Algorithm $\mathcal{BFS}$ as the root of the spanning tree.

Algorithm $\mathcal{BFS}$ is a composition of two algorithms. Algorithm 1 is based on the fact that a processor has to choose a neighbor with the minimal distance to the root as its parent in the tree. It is well known that this common idea is enough to get a round complexity in $O(d)$, but does not ensure a step complexity in $O(n^b)^2$ So we allow a processor to connect to a neighbor only if this neighbor is in the tree rooted at r and in the shortest path to r. The detection of such neighbors is assigned to Algorithm 2 (see Section 4) which can be seen as an oracle by Algorithm 1. The second role of Algorithm 1 is to remove the abnormal trees, i.e., those that are not rooted at r.

Variables. We define below the different variables used by Algorithm 1. For Algorithm 1, we characterize r by the predicate *Allowed* (i.e., $Allowed(p) \equiv (p = r)$, $\forall p \in V$).

Shared variable. Each processor $p \in V$ has a local shared variable $p.Req$ which is used by Algorithm 1 to monitor Algorithm 2 at p. This shared variable can take four values: $ASK, WAIT, REP$, and OUT. By setting the shared variable $p.Req$ to ASK, Algorithm 1 informs Algorithm 2 that a permission from the root of the tree that p belongs to is needed at p. In this case, Algorithm 2 tries to send a request and to obtain a permission for p if it is possible (i.e., if p belongs to an allowed tree and this request has the highest priority during enough time). If a permission is delivered to processor p, then Algorithm 2 sets this shared variable to REP in order to inform Algorithm 1. Then, every neighbor of p can execute Algorithm 1 to join the tree that p belongs to. When there is no neighbor of p to connect, then Algorithm 1 sets $p.Req$ to OUT which allows to Algorithm 1 to request another permission through Algorithm 2 if needed.

Local variables. Each processor $p \in V$ maintains three local variables:

- $p.P$: it gives the parent of p in the tree it belongs to, $p.P = \perp$ for processor $p = r$.
- $p.L$: it stores the level (or height) of p in the tree it belongs to, $p.L = 0$ for processor $p = r$.
- $p.S$: it defines the status of processor p. It can take two values: E if p does not belong to a tree rooted to a processor x satisfying Predicate $Allowed(x)$, C otherwise. We have $p.S = C$ for processor $p = r$.

Algorithm Description. As described before, we consider a forest $\mathcal{F}$ of trees and a distinguished processor r which is the only processor authorized to deliver permissions in the network (i.e., $Allowed(p) \equiv (p = r)$ for every processor $p \in V$). We can notice

---

[2] Indeed, this approach is used in [20] to construct a BFS tree with a round complexity in $\Theta(d)$ but with a step complexity in $\Omega(Max \times n^2)$, as demonstrated in [8]. However, Max is an upper bound of n and can be arbitrary high with respect to n so the step complexity can be at least exponential. Note that the gap between the lower and the upper bound (see Table 1) of the step complexity lead us to think that the lower bound in [8] is not tight.

that in a tree there is a strong constraint between the level of a processor and the level of its parent in the tree: For any processor $p \neq r$, the level of p's parent must be equal to p's level minus 1. Therefore, the root of a tree in forest $\mathcal{F}$ is either (i) processor r, or (ii) a processor $p \neq r$ such that $p.L \leq (p.P).L$ (it is used to detect cycles in the network). Since we want to construct a spanning tree, in case (ii) we say that processor p is an *abnormal root*. Moreover, any processor $p \neq r$ in a tree in $\mathcal{F}$ rooted at an abnormal root belongs to an *abnormal tree*. Every processor $p \in V$ in an abnormal tree can execute *E-action* to change its Status to E (i.e., $p.S = E$) and to inform its descendants in the tree (see the formal description of Algorithm 1). Note that to reduce the number of moves executed by Algorithm $\mathcal{BFS}$, a processor $p \in V$ in an abnormal tree does not ask any permission. Processor p waits until a neighbor q in the tree rooted at r authorizes p to connect to q.

Algorithm 1.   Spanning Tree Construction for any $p \in V$

Inputs: $Neig_p$: set of (locally) ordered neighbors of p;
Shared variable: $p.Req \in \{ASK, WAIT, REP, OUT\}$;

. . . .

Macros:
$$Child(p) \quad = \{q \in Neig_p :: q.P = p \wedge q.L = p.L + 1\}$$
$$Parent(p) \quad = p.P$$
$$Height(p) \quad = p.L$$
$$ChPar(p) \quad = \{q \in Neig_p \backslash Child(p) :: q.S = C\}$$
$$MinChPar(p) = \min\{q \in ChPar(p) :: \forall t \in ChPar(p), q.L \leq t.L\}$$

Global Predicates:
$$GoodT(p) \quad \equiv p.S \neq E \wedge (p \neq r \Rightarrow p.L = (p.P).L + 1)$$
$$GoodL(p) \quad \equiv (\forall q \in Neig_p :: |p.L - q.L| > 1 \Rightarrow (p.L < q.L \vee q.S = E))$$
$$GP\text{-}REP(p) \equiv (\exists q \in Neig_p :: q.S = E \vee q.L - p.L > 1)$$
$$Start(p) \quad \equiv p.Req = OUT \wedge GP\text{-}REP(p)$$
$$End(p) \quad \equiv p.Req = REP \wedge \neg GP\text{-}REP(p)$$

Algorithm for $p = r$:
Constants: $p.S = C; p.P = \bot; p.L = 0;$
Predicates:
$\quad Allowed(p) \equiv true$
Actions:
$\quad$ A-action :: $Start(p) \rightarrow p.Req := ASK;$
$\quad$ O-action :: $End(p) \quad \rightarrow p.Req := OUT;$
Algorithm for $p \neq r$:
Variables: $p.S \in \{C, E\}; p.P \in Neig_p; p.L \in \mathbb{N};$
Predicates:
$\quad Allowed(p) \qquad \equiv false$
$\quad AbnormalTree(p) \equiv p.S = C \wedge ((p.P).S = E \vee (p.P).L \geq p.L)$
$\quad Connect(p) \qquad \equiv (\exists q \in Neig_p :: q.Req = REP \wedge q = MinChPar(p)$
$\qquad\qquad\qquad\qquad \wedge (p.S = C \Rightarrow p.L - q.L > 1))$
Actions:
$\quad$ E-action :: $AbnormalTree(p) \rightarrow p.S := E;$
$\quad$ C-action :: $Connect(p) \qquad\quad \rightarrow p.S := C; p.P := MinChPar(p); p.L := (p.P).L + 1;$
$\qquad\qquad\qquad\qquad\qquad\qquad\qquad p.Req := OUT;$
$\quad$ A-action :: $Start(p) \qquad\qquad \rightarrow p.Req := ASK;$
$\quad$ O-action :: $End(p) \qquad\qquad\; \rightarrow p.Req := OUT;$

When a BFS tree is constructed, the following property is verified at each processor $p \in V, p \neq r$: The level of p's parent is equal to p's level minus 1 (i.e., $(p \neq r) \Rightarrow (p.L = (p.P).L + 1)$). For processor r, we have the following constant values: r has no parent and a level equal to zero (i.e., $(p = r) \Rightarrow (p.P = \bot \wedge p.L = 0)$). Moreover, according to Claim 3 of Definition 1 we must have that the deviation on the level values between any processor $p \in V$ and its neighbors does not exceed one (i.e., $\forall q \in Neig_p, |q.L - p.L| < 1$). If one of these above constraints are not verified then a BFS tree is not constructed. Therefore, we have either at least one abnormal tree in $\mathcal{F}$ or there is a processor $p \in V$ with a neighbor q such that $q.L - p.L > 1$ (i.e., Predicate $GP\text{-}REP(p)$ is satisfied at p). In these cases, processor p executes $A\text{-}action$ to set the shared variable $p.Req$ to ASK in order to ask the permission to allow q to connect to p, if p is not already asking a permission (i.e., we have $p.Req = OUT$). To this end, Algorithm 2 sends a request to the root of the tree.

Inputs for Algorithm 2. In order to allows Algorithm 2 to send a request the following inputs are given at processor p: (i) $Child(p)$ is the set of children of p in the tree (i.e., $Child(p) \equiv \{q \in Neig_p : q.P = p\}$), (ii) $Parent(p)$ is the parent of p in the tree (i.e., $Parent(p) \equiv p.P$), (iii) $Height(p)$ is the height in the tree of the requesting processor p, and (iv) $Allowed(p)$ is a predicate which notifies if p can deliver permissions (i.e., $Allowed(p) \equiv (p = r)$). Remind that $Allowed(p)$ must be satisfied only at processor $p = r$ in Algorithm 2 to allow that eventually every processor joins the tree rooted at r, since eventually the processors cannot join another tree in forest $\mathcal{F}$.

In the case a permission is delivered at processor p (i.e., we have $p.Req = REP$), then each neighbor q of p can execute $C\text{-}action$ to connect to p. However to construct a BFS tree without an overcost on moves, processor q waits for until its neighbor x with the smallest level in a normal tree gives its authorization to q to connect by executing $C\text{-}action$ (i.e., we have $x.Req = REP \wedge x = MinChPar(q)$). When processor q executes $C\text{-}action$ then it sets its variables $p.P$ and $p.L$ according to its new parent in the tree, and it changes its status to Status C and its shared variable $p.Req$ to OUT. Finally, if there is no neighbor for which processor p needs a permission (i.e., Predicate $GP\text{-}REP(p)$ is no more satisfied at p), then p executes $O\text{-}action$ to set its shared variable $p.Req$ to OUT. This informs Algorithm 2 that the permission can be removed at p, then this allows p to ask a new permission later.

4   Question-Answer problem

In this section, we present a snap-stabilizing algorithm to implement the oracle used by the BFS tree construction given in Section 3. Formally, this oracle has to solve the Question-Answer problem which can be stated as following, a formal specification is given in Specification 2.

Given a static forest $\mathcal{F}$ of trees in a network $G = (V, E)$, a set of processors $De \subseteq V$ requesting a permission to make a defined computation and a set of processors $AP \subset V$ authorized to deliver permissions. Each $p \in AP$ is a root of a tree $T \in \mathcal{F}$. The *Question-Answer* problem is to deliver a permission (or *acknowledgement*) to a processor p in a tree $T \in \mathcal{F}$ if and only if the root q of T is in AP.

Specification 2 (Question-Answer). *Let* $G = (V, E)$ *be a network and* $\mathcal{F}$ *the static forest of trees in* G. *Let a tree* $T \in \mathcal{F}$ *and* $root(T)$ *the root of* T. T *is an* allowed *tree if* $root(T) \in AP$ *and* not allowed *otherwise. A protocol* P *which resolves the Question-Answer problem satisfies:*

[Liveness 1] *During an infinite computation, if a processor has to send infinitely often a request and it cannot send its request in an allowed tree, then there exist an infinite number of requests which were sent.*

[Liveness 2] *For every computation suffix, if a processor in an allowed tree has sent a request at time t, then there exist at least one processor in the same tree which receives an acknowledgement to its own sent request at time $t' > t$.*

[Safety 1] *Every processor which has sent a request receives at most one acknowledgement causally related to its sent request.*

[Safety 2] *Every processor in a not allowed tree which has sent a request never receives an acknowledgement.*

Remark that only semi-algorithms can satisfy Specification 2, that is no acknowledgement is sent to processors in a not allowed tree, from Property [Safety 2] of Specification 2.

4.1 Question-Answer Algorithm

In this section, we present a snap-stabilizing algorithm for the Question-Answer problem, a formal description is given by Algorithm 2. This is a non-uniform algorithm because some rules are only executed by a subset of processors $p \in V$ satisfying a local Predicate $Allowed(p)$ (i.e., p can deliver a permission or not).

Variables. We define below the different variables used by Algorithm 2.

Shared variable. Each processor $p \in V$ has a local shared variable $p.Req$ which allows an external algorithm to require the Question-Answer algorithm at p. This shared variable can take four values: ASK, $WAIT$, REP, and OUT. By setting the shared variable $p.Req$ to ASK in the external algorithm, p requests a permission through the Question-Answer algorithm to its root of the tree. To this end, Question-Answer algorithm tries to send a request to the root of the tree and sets the shared variable $p.Req$ to $WAIT$. At least the request of a requesting processor with the lowest level (or height) in the tree will reach the root and then receive a permission (an *acknowledgement*). When p receives an acknowledgement, it sets $p.Req$ to REP. Finally, the external algorithm must set $p.Req$ to OUT to request another permission through Question-Answer algorithm.

Local variables. Each processor $p \in V$ maintains two local variables:

- $p.Q$: it defines the status of the Question-Answer algorithm at processor p. There are three distinct status: R, W, and A. Status R notifies that p transmits a request to the root of the tree, whereas Status W indicates that p waits for an acknowledgement from the root for the transmitted request. The third status, Status A, indicates that p has received an acknowledgement from the root.
- $p.HQ$: it stores at p the height of the processor which has sent the request.

Algorithm Description. To simplify the presentation of the algorithm, consider a forest of allowed trees (i.e., trees rooted at nodes p satisfying Predicate $Allowed(p)$) and a fixed set of requests. In the following, we explain the way our algorithm handles requests focusing on a single tree T of the forest, but this is the same for other trees since

---

Algorithm 2.   Question-Answer algorithm for any $p \in V$

Inputs: $Neig_p$: set of (locally) ordered neighbors of p;
    $Child(p)$: set of neighbors considered as children of p in the tree;
    $Allowed(p)$: predicate which indicates if p is able to acknowledge to a request;
    $Parent(p)$: parent of p in the tree, equal to a processor $q \in Neig_p$ if $\neg Allowed(p)$ or equal to $\perp$ otherwise ;
    $Height(p)$: height of p in the tree;
Shared variable: $p.Req \in \{ASK, WAIT, REP, OUT\}$;
Variables: $p.Q \in \{R, W, A\}$; $p.HQ \in \mathbb{N}$;

....

Macros:
$RC(p)$ $= \{q \in Child(p) :: q.Q \in \{R, W\}\}$
$PrioRC(p) = \{q \in RC(p) :: \forall t \in RC(p), q.HQ \le t.HQ\}$
Ch_p $= \min\{q \in PrioRC(p)\}$

....

Global Predicates:
$Transmit(p)$ $\equiv p.Q = A \wedge (\forall q \in Child(p) :: q.Q = W \Rightarrow q.HQ \ne p.HQ)$
$Retransmit(p) \equiv p.Q = W \wedge (\exists q \in Child(p) :: q.Q = R \wedge q.HQ = p.HQ)$
$Error(p)$ $\equiv p.Q \ne A \wedge [(p.Req \notin \{ASK, WAIT\} \wedge p.HQ = Height(p)) \vee (p.HQ \ne Height(p)$
$\wedge (p.Req \ne REP \Rightarrow (\forall q \in Child(p) :: q.HQ = p.HQ \Rightarrow q.Q = A)))]$
$Request(p)$ $\equiv p.Req = ASK \wedge (|PrioRC(p)| > 0 \Rightarrow Height(p) \le (Ch_p).HQ)$
$RequestT(p)$ $\equiv p.Req \ne REP$
$\wedge |PrioRC(p)| > 0 \wedge [((Ch_p).HQ \ge p.HQ \Rightarrow Transmit(p)) \vee Retransmit(p)]$

....

Algorithm:
Predicates:
$Wait(p)$ $\equiv (Allowed(p) \wedge p.Q = R \wedge (\forall q \in Child(p) :: q.HQ = p.HQ \Rightarrow q.Q = W)) \vee$
    $(\neg Allowed(p) \wedge Parent(p).Q = R \wedge p.Q = R \wedge Parent(p).HQ = p.HQ$
    $\wedge (\forall q \in Child(p) :: q.HQ = p.HQ \Rightarrow q.Q = W))$
$Answer(p) \equiv (Allowed(p) \wedge p.Q = W) \vee$
    $(\neg Allowed(p) \wedge Parent(p).Q = A \wedge p.Q = W \wedge Parent(p).HQ = p.HQ)$

Actions:
$QE\text{-}action$ :: $Error(p)$ $\rightarrow p.Q := A; p.HQ := Height(p)$;
$QR\text{-}action$ :: $Request(p)$ $\rightarrow p.Q := R; p.HQ := Height(p); p.Req = WAIT$;
$QRC\text{-}action$ :: $RequestT(p)$ $\rightarrow p.Q := R; p.HQ := (Ch_p).HQ$;
    if $p.HQ < Height(p) \wedge p.Req = WAIT$ then $p.Req := ASK$; fi
$QW\text{-}action$ :: $Wait(p)$ $\rightarrow p.Q := W$;
$QA\text{-}action$ :: $Answer(p)$ $\rightarrow p.Q := A$;
    if $p.Req = WAIT$ then $p.Req := REP$; fi

---

the requests in each tree are handled independently. In the algorithm, the requests sent by nodes of lowest height in the tree are handled in priority.

When a processor p has a *local request* requested by the external algorithm (i.e., $p.Req = ASK$), p can execute $QR\text{-}action$ to set its variables $p.Req, p.Q$, and $p.HQ$ to $WAIT$, R, and to $Height(p)$ respectively, in order to send its request to the root of the tree it belongs to. The external algorithm is informed that the request is sent since $p.Req = WAIT$. Otherwise, an internal processor p in the tree with no local request (i.e., $p.Req \ne REP$) could have to transmit requests from its children (the request from a requesting descendant of lowest height first) in the following cases:

– a child of p is sending a request with a highest priority (i.e., $(Ch_p).HQ < p.HQ$);
– the acknowledgement received for the transmitted request is no more needed at p (all its children waiting it have transfered the acknowledgment, see Predicate $Transmit(p)$);
– p is waiting for an acknowledgement for a request and a new request is transmitted by a child of p with the same height (see Predicate $Retransmit(p)$).

In all these above cases, p executes $QRC\text{-}action$ to set $p.Q$ to R and $p.HQ$ to the lowest height among requesting descendant of p (i.e., $p.HQ = (Ch_p).HQ$).

A processor p waits for an acknowledgement for a current request when its parent has transmitted the request (see Predicate $Wait(p)$). Moreover, all p's children transmitting the same request (i.e., with the same height) have to wait for an acknowledgement. Hence, Status W allows to remove bad requests due to an incorrect initial configuration and to synchronize request transmissions of same priority. In this case, p sets its variable $p.Q$ to W using QW-$action$.

When the root $root(T)$ of the tree T has no local request and is waiting for an acknowledgement for requesting descendant(s) (see Predicate $Answer$), then it executes QA-$action$ to set its variable $root(T).Q$ to A. This permission is propagated down in the tree to the requesting descendant(s) following the path(s) used to transmit the request. Finally, a processor p waiting for an acknowledgement to a local request (i.e., $p.Q = W$ and $p.Req = WAIT$) executes QA-$action$ to receive the acknowledgement and sets the shared variable $p.Req$ to REP to notify to the external algorithm of the delivered permission. Note that as soon as a received acknowledgement is no more needed at a processor p (i.e., $p.Req$ is setted to OUT by the external algorithm), then another request transmitted by a child of p can be transmitted by p up in the tree.

However, a processor must be able to detect *wrong* requests due to an incorrect initial configuration. A request treated by a processor p is a *wrong request* in the following cases (see Predicate $Error(p)$):

- p is sending a local request whereas it has no local request (i.e., $p.Q \neq A \wedge p.Req \notin \{ASK, WAIT\}$ and $p.HQ = Height(p)$);
- p is transmitting a request from a child, however no child of p has a request with the same height (i.e., $p.Q \neq A \wedge p.HQ \neq Height(p) \wedge (\forall q \in Child(p), q.HQ = p.HQ \Rightarrow q.Q = A)$).

When a processor p detects a wrong request, then p executes QE-$action$. This action has the highest priority among the actions at p, and it reinitiates p's state like if an acknowledgement to a local request was received, i.e., to set $p.Q$ to A and $p.HQ$ to $Height(p)$ (without changing the state of the shared variable $p.Req$).

A questioning mechanism close to the mechanism presented here was used in [12] to design a snap-stabilizing solution to the problem of Propagation of Information with Feedback (PIF) with a round complexity in $O(n)$ and a step complexity in $O(\Delta n^3)$. However, solving the PIF problem involves a strong synchronization in the network to insure that all the nodes in the network belong to the same broadcast tree before to initiate the feedback phase. Indeed, each time a node is added to the broadcast tree the questioning mechanism is reset leading to a $O(n)$ round complexity. Contrary to this questioning mechanism, here our mechanism needs a weakest synchronization to resolve the Question-Answer problem. Let De the set of requesting nodes and h_{min} the height of closest requesting nodes from the root in T. The first requests acknowledged by $root(T)$ are the requests from nodes at height h_{min}. Then, if the set of requests is static then the requests at height $h_{min} + 1$ are acknowledged by $root(T)$ (if any) and so on. In fact, only a synchronization for the requests of requesting nodes at height h_{min} (whose requests are of highest priority) in tree T is required leading to a round complexity function of the height of T. The transmission of a request requires $O(n)$ steps, however this transmission can be interrupted only by a requesting node with the same height in T, that is at most $|De|$ times.

The following lemma summarizes the above discussion:

Lemma 1. *Let T an allowed tree and h_{min} the height of closest requesting nodes in De from the root in T, in $O(h_{min})$ rounds and $O(n|De|)$ steps at least one requesting node in De receive an acknowledgement from $root(T)$ to its request.*

5 Composition and Complexities

Algorithm $\mathcal{BFS}$ is obtained by composition of Algorithm 2 and Algorithm 1. These two algorithms are composed together at each processor $p \in V$ with a conditional composition (first introduced in [14]): Algorithm 1 $\circ$ $|_{Cond(p)}$ Algorithm 2, where each guard g of the actions of Algorithm 2 at each processor $p \in V$ has the form $Cond(p) \wedge g$ with Predicate $Cond(p)$ defined below (see Algorithm 1 for the description of predicates): $Cond(p) \equiv GoodT(p) \wedge GoodL(p)$.

Using this composition, each processor $p \in V$ can execute Algorithm 2: (i) to transmit requests and acknowledgements only if the tree containing p is locally correct (i.e., Predicate $GoodT(p)$ is satisfied), and (ii) to ask a permission if needed (i.e., Predicate $GoodL(p)$ is satisfied). Indeed, actions in Algorithm 2 can be locked to avoid processors belonging to a tree not rooted at r (abnormal tree) to transmit useless requests since no acknowledgement can be received (only r can deliver acknowledgements). Therefore, processors in abnormal trees can only execute actions in Algorithm 1 to hook on to another tree in the forest via a neighbor with a permission (acknowledgement delivered by Algorithm 2). Moreover, actions of Algorithm 2 and Algorithm 1 can be enabled at p simultaneously. In this case, Algorithm 2 is executed before Algorithm 1 at p.

Algorithm $\mathcal{BFS}$ uses Algorithm 2 which can be viewed as a synchronizer allowing the BFS tree construction of T rooted at r layer by layer, the addition of any new layer of processors depending of a permission request. The requesting processors closest to r at height k in T receive an acknowledgement to their request from r in $O(k)$ rounds (Lemma 1) which allows their neighbors to hook on to T. The same argument holds for the addition of each new layer of T. Moreover, the height of a BFS tree is lower than or equal to the network diameter. Therefore, summing up the round complexity associated to each layer we obtain a round complexity $O(d^2)$ to construct a BFS tree, with d the network diameter. In another hand, the mechanism we use for deleting the abnormal trees is obviously in $O(n)$ rounds, since the height of such a tree can be in $O(n)$. But any processor in an abnormal tree far from the root of this tree will become the neighbor of at least a processor of the normal BFS tree in $O(d^2)$ rounds and will hook to it even if the abnormal tree is not yet deleted. So the global round complexity is still $O(d^2)$ as stated in the following lemma.

Lemma 2. *From any configuration, in $O(d^2)$ rounds Algorithm $\mathcal{BFS}$ reaches a configuration $\gamma \in \mathcal{C}$ satisfying Definition 1, with d the diameter of the network.*

We discuss above the ideas leading to the round complexity of Algorithm $\mathcal{BFS}$. We give below the main arguments allowing to show that Algorithm $\mathcal{BFS}$ has a step complexity in $O(mn^4)$. We define a *topological change* as follows: Given a forest $\mathcal{F}$ of trees in a configuration $\gamma \in \mathcal{C}$, a *topological change* in $\mathcal{F}$ is obtained by the execution of E-action or C-action at a processor $p \in V$ in step $\gamma \mapsto \gamma'$. We first consider the step complexity of Algorithm 1. A processor can hook on to several abnormal trees until

belonging to the tree rooted at r. First of all, we establish the number of connections to an abnormal tree that any processor $p \in V$ can make until belonging to the tree rooted at r. In the reminder, the tree rooted at r is noted $Tree(r)$ and $root(T)$ describes the root node of a tree T.

Proposition 1. *Every processor $p \in V$ is hooked on to the neighbor q such that $\forall s \in Neig_p, q.L \le s.L$.*

Proof. According to the formal description of Algorithm 1, a processor hooks on to a neighbor using C-$action$. Assume, by the contradiction, that there is a processor $p \in V$ such that $\exists s \in Neig_p, (p.P).L > s.L$. We must consider two cases: s is in an abnormal tree or not. If s is in an abnormal tree then either $s.S = E$ then $s \notin MinChPar(p) \Rightarrow \neg Connect(p)$ a contradiction, or $s.S = C$ then by Property [Safety 2] of Specification 2 s never receives an acknowledgement and we have that $s.Req \ne REP \Rightarrow \neg Connect(p)$, otherwise C-$action$ is enabled at p, a contradiction. If s is in a normal tree then by Property [Liveness 2] of Specification 2 we have that $s.Req = REP$ and C-$action$ is enabled at p, a contradiction.   □

Lemma 3. *Let any abnormal tree $T \in \mathcal{F}$ and the set of processors $B = \{p \in V : p \notin T \wedge (\exists q \in Neig_p :: q \in T)\}$. In any execution, only processors in B can hook on to T.*

Proof. Consider any abnormal tree $T \in \mathcal{F}$ in configuration $\gamma \in \mathcal{C}$. According to the formal description of Algorithm 1, a processor p must execute C-$action$ to hook on to a tree, i.e., there is a neighbor q such that $q.Req = REP$. Suppose that every processor $q \in B$ executes C-$action$ and they are hooked on to T in configuration γ_k. Note that after executing C-$action$, we have $q.Req = OUT$ at every processor $q \in B$. Assume, by the contradiction, that there is a processor $p \notin T$ in configuration γ_k which hooks on to T in step $\gamma_k \mapsto \gamma_{k+j}, j > 0$. This implies that p hooks on to a neighbor $q \in B$ (by definition of B) such that $q.Req = REP$, a contradiction by Property [Safety 2] of Specification 2 because q cannot receive an acknowledgement from $root(T)$ since T is an abnormal tree.   □

Proposition 2. *Let a processor $p \in V$ which hooks on to a tree T in configuration $\gamma_i \in \mathcal{C}$. If another processor $q \in V$ hooks on to T by p in $\gamma_{i+j}, j > 0$, then T is a normal tree.*

Proof. According to Lemma 3, the expansion of an abnormal tree T' is limited at distance one from T'. After p hooks on to T, to allow the processor q to hook on to T by p then p receives an acknowledgement from $root(T)$. Therefore, T is a normal tree by Specification 2.   □

Lemma 4. *Let any abnormal tree $T \in \mathcal{F}$. A processor $p \in V$ can hook on to T at most once by the same neighbor $q \in T$.*

Proof. Assume, by the contradiction, that there is a configuration $\gamma_k \in \mathcal{C}$ such that there is a processor $p \in V$ which hooks on to T by the same neighbor $q \in T$ a second time. To hook on to T, p must execute C-$action$, i.e., there is a neighbor $x \in T$ of p such that $x.S = C$ and $x.Req = REP$. According to Proposition 1, p hooks on to the neighbor $x \in V$ such that $x.S = C \wedge (\forall s \in Neig_p, x.L \le s.L)$. Suppose that

p hooks on to T by the neighbor q a first time in step $\gamma_{i-1} \mapsto \gamma_i \in \mathcal{C}$, then p hooks on to another neighbor s of p, $s \neq q$, in step $\gamma_{j-1} \mapsto \gamma_j \in \mathcal{C}, j > i$. Now, we must consider several cases in configuration $\gamma_k, i < j < k$. If p is hooked on to s in γ_j because $q.S = E$ and $s.Req = REP$ in γ_i then since $q \in T$ we have $q.S = E$ in γ_k and $q \notin MinChPar(p) \Rightarrow \neg Connect(p)$, a contradiction. Otherwise $s.S = q.S = C$ and p is hooked on to s in $\gamma_j, i < j < k$, because $s.L < q.L$ and $s.Req = REP$. When p hooks on to q the first time in step $\gamma_{i-1} \mapsto \gamma_i$, we have $s.S = E$ or $s.L > q.L$. Since we have $s.S = C \wedge s.L < q.L \wedge s.Req = REP$ and p hooks on to s in step $\gamma_{j-1} \mapsto \gamma_j$, this implies that s is in a normal tree in γ_j according to Proposition 2. Thus, we have $s.S = C \wedge s.L < q.L$ in γ_k and $q \notin MinChPar(p) \Rightarrow \neg Connect(p)$, a contradiction.   □

Lemma 5. *In any execution, every processor* $p \in V \backslash \{r\}$ *produces at most* 2Δ *topological changes in forest* $\mathcal{F}$ *while* $p \notin Tree(r)$, *with* Δ *the maximum degree of a processor in the network.*

Proof. To hook on to a tree, a processor $p \in V$ must execute C-action. According to Lemma 4, p cannot hook on to an abnormal tree $T \in \mathcal{F}$ twice by the same neighbor q of p. Since a processor can have at most Δ neighbors, p can hook on at most Δ times to an abnormal tree. Observe that E-action has a higher priority than C-action and E-action can be executed between two executions of C-action, i.e., at most Δ times while $p \notin Tree(r)$. Therefore, by the definition of a topological change the lemma follows.   □

After giving a bound for the number of connections to abnormal trees, we provide below an upper bound for the number of connections that a processor can make in the normal tree.

Remark 1. *For every processor* $p \in Tree(r)$, E-action *is disabled at* p.

Lemma 6. *In any execution, every processor* $p \in V \backslash \{r\}$ *produces at most* n *topological changes in forest* $\mathcal{F}$ *while* $p \in Tree(r)$, *with* n *the number of processors in the network.*

Proof. Observe that for every processor $p \in Tree(r)$ we have $p.S = C$. Moreover, by Remark 1, for every processor $p \in Tree(r)$, E-action is disabled. So, by definition the only topological change in $\mathcal{F}$ that a processor $p \in Tree(r)$ can produce is to execute C-action in order to reduce its level in $Tree(r)$. Thus, by Proposition 1 each execution of C-action by a processor $p \in Tree(r)$ in step $\gamma_i \mapsto \gamma_{i+1}$ implies that p hooks on to the neighbor with the lowest level in γ_{i+1} and $p.L$ in γ_i is higher than $p.L$ in γ_{i+1}. Therefore, since the size of $Tree(r)$ is bounded by n then any processor p can hook on to at most $n - 1$ processors by executing C-action while $p \in Tree(r)$.   □

From the above lemmas, each processor can hook on to at most $2\Delta + n$ times until reaching its correct position in the final BFS tree. Thus, there are at most $2\Delta n + n^2$ topological changes in the forest until a BFS tree is reached. Moreover, each topological change yields at most Δ requests in the network, so to construct a BFS tree at most

$2\Delta m + mn$ requests are generated by Algorithm 1. We now consider the step complexity of Algorithm 2. According to Lemma 1, a processor receives an acknowledgement with Algorithm 2 in $O(n^2)$ steps (since $|De| \leq n$). So, in $O(n^3)$ steps every requesting processor receives an acknowledgement (there are at most n requests in the network).

Given an upper bound on the number of requests generated by Algorithm 1, we have to multiply this amount by the number of steps needed by Algorithm 2 to acknowledge these requests in order to obtain an upper bound to the total step complexity of Algorithm $\mathcal{BFS}$. This is stated by the following lemma.

Lemma 7. *From any configuration, $O(\Delta mn^3 + mn^4)$ steps are needed by Algorithm $\mathcal{BFS}$ to reach a terminal configuration.*

Notice that in one hand using a questioning mechanism allows us to save steps by avoiding the transmission of useless requests, but in the other hand we obtain a higher round complexity ($O(d^2)$ instead of $O(d)$ with standard algorithms for BFS trees) due to the fact that permissions must be delivered before the add of new nodes to the constructed tree. Moreover, the step complexity established in Lemma 7 is not related with any initial value of a variable and it holds under any fairness assumptions.

Lemma 7 implies that Algorithm $\mathcal{BFS}$ always satisfies Property [TC1] of Specification 1 and is silent. We now consider any terminal configuration. Since the configuration is terminal, no action is enabled in Algorithm $\mathcal{BFS}$. It is trivial to verify by induction on the distance of a processor to r that every processor is in $Tree(r)$ and at the right level as stated in the following lemma.

Lemma 8. *Every terminal configuration reached by Algorithm $\mathcal{BFS}$ satisfies Definition 1.*

According to Lemmas 7 and 8, Algorithm $\mathcal{BFS}$ always satisfies respectively Properties [TC1] and [TC2] of Specification 1. Therefore, we can state the following theorem.

Theorem 1. *Algorithm $\mathcal{BFS}$ is a silent snap-stabilizing algorithm to construct a BFS tree.*

6 Conclusion

In this paper a silent snap-stabilizing algorithm resolving the Question-Answer problem has been given, in which each node requests a permission (delivered by a subset of network nodes) in order to perform a defined computation. Based on this algorithm, the first fully polynomial stabilizing algorithm for the construction of a spanning tree has been presented. A Breadth First Search tree is constructed in $O(d^2)$ rounds and in $O(n^6)$ steps, with d the diameter and n the number of nodes in the network. Moreover, a distributed daemon without any fairness assumptions is considered.

One crucial open question is the following: Is it possible to design a fully polynomial self-stabilizing algorithm to construct a spanning tree in $O(d)$ rounds with a polynomial step complexity?

References

1. Afek, Y., Kutten, S., Yung, M.: Memory-Efficient Self-Stabilizing Protocols for General Networks. In: van Leeuwen, J., Santoro, N. (eds.) WDAG 1990. LNCS, vol. 486, pp. 15–28. Springer, Heidelberg (1991)

2. Arora, A., Gouda, M.: Distributed reset (extended abstract). In: Veni Madhavan, C.E., Nori, K.V. (eds.) FSTTCS 1990. LNCS, vol. 472, Springer, Heidelberg (1990)
3. Awerbuch, B., Kutten, S., Mansour, Y., Patt-Shamir, B., Varghese, G.: Time optimal self-stabilizing synchronization. In: 25th Annual ACM Symposium on Theory of Computing (STOC), pp. 652–661 (1993)
4. Bui, A., Datta, A.K., Petit, F., Villain, V.: State-optimal snap-stabilizing pif in tree networks. In: Workshop on Self-stabilizing Systems (WSS), pp. 78–85. IEEE Computer Society (1999)
5. Burman, J., Kutten, S.: Time Optimal Asynchronous Self-Stabilizing Spanning Tree. In: Pelc, A. (ed.) DISC 2007. LNCS, vol. 4731, pp. 92–107. Springer, Heidelberg (2007)
6. Chen, N.-S., Yu, H.-P., Huang, S.-T.: A self-stabilizing algorithm for constructing spanning trees. Inf. Process. Lett. 39(3), 147–151 (1991)
7. Collin, Z., Dolev, S.: Self-stabilizing depth-first search. Inf. Process. Lett. 49(6), 297–301 (1994)
8. Cournier, A.: Mémoire d'Habilitation à Diriger les Recherches: Graphes et algorithmique distribuée stabilisante. Université de Picardie Jules Verne (2009)
9. Cournier, A.: A New Polynomial Silent Stabilizing Spanning-Tree Construction Algorithm. In: Kutten, S., Žerovnik, J. (eds.) SIROCCO 2009. LNCS, vol. 5869, pp. 141–153. Springer, Heidelberg (2010)
10. Cournier, A., Devismes, S., Petit, F., Villain, V.: Snap-stabilizing depth-first search on arbitrary networks. The Computer Journal 49(3), 268–280 (2006)
11. Cournier, A., Devismes, S., Villain, V.: A Snap-Stabilizing DFS with a Lower Space Requirement. In: Tixeuil, S., Herman, T. (eds.) SSS 2005. LNCS, vol. 3764, pp. 33–47. Springer, Heidelberg (2005)
12. Cournier, A., Devismes, S., Villain, V.: Snap-stabilizing pif and useless computations. In: 12th International Conference on Parallel and Distributed Systems (ICPADS), pp. 39–48. IEEE Computer Society (2006)
13. Cournier, A., Devismes, S., Villain, V.: Light enabling snap-stabilization of fundamental protocols. ACM Transactions on Autonomous and Adaptive Systems (TAAS) 4(1) (2009)
14. Datta, A.K., Gurumurthy, S., Petit, F., Villain, V.: Self-stabilizing network orientation algorithms in arbitrary rooted networks. Stud. Inform. Univ. 1(1), 1–22 (2001)
15. Datta, A.K., Larmore, L.L., Vemula, P.: Self-Stabilizing Leader Election in Optimal Space. In: Kulkarni, S., Schiper, A. (eds.) SSS 2008. LNCS, vol. 5340, pp. 109–123. Springer, Heidelberg (2008)
16. Dijkstra, E.W.: Self-stabilizing systems in spite of distributed control. Commun. ACM 17(11), 643–644 (1974)
17. Dolev, S.: Self-Stabilization. MIT Press (2000)
18. Dolev, S., Israeli, A., Moran, S.: Self-stabilization of dynamic systems assuming only read/write atomicity. In: 9th ACM Symposium on Principles of Distributed Computing (PODC), pp. 103–117 (1990)
19. Gärtner, F.: A survey of self-stabilizing spanning-tree construction algorithms. Tech. rep., EPFL (October 2003)
20. Huang, S.-T., Chen, N.-S.: A self-stabilizing algorithm for constructing breadth-first trees. Inf. Process. Lett. 41(2), 109–117 (1992)
21. Johnen, C.: Memory-efficient self-stabilizing algorithm to construct bfs spanning trees. In: 3rd Workshop on Self-stabilizing Systems (WSS), pp. 125–140 (1997)
22. Kosowski, A., Kuszner, Ł.: A Self-Stabilizing Algorithm for Finding a Spanning Tree in a Polynomial Number of Moves. In: Wyrzykowski, R., Dongarra, J., Meyer, N., Waśniewski, J. (eds.) PPAM 2005. LNCS, vol. 3911, pp. 75–82. Springer, Heidelberg (2006)
23. Tel, G.: Introduction to distributed algorithm, 2nd edn. Cambridge University Press (2000)

Anonymous Agreement: The Janus Algorithm

Zohir Bouzid[1,*], Pierre Sutra[1,**], and Corentin Travers[2,***]

[1] University Pierre et Marie Curie - Paris 6, LIP6-CNRS 7606, France
name.surname@lip6.fr
[2] LaBRI University Bordeaux 1
name.surname@labri.fr

Abstract. We consider the consensus problem in an n-process shared-memory distributed system when processes are anonymous, i.e., they have no identities and are programmed identically.

We present Janus, a new anonymous consensus algorithm that reaches decision after $O(\sqrt{n})$ writes in every solo execution. The set of values that can be proposed is unbounded and the algorithm tolerates an arbitrary number of crash failures. The algorithm relies on an anonymous eventual leader election mechanism. Furthermore, during solo executions in which a non-faulty process is elected since the beginning, the individual step complexity of Janus is $O(n)$, matching a recent lower bound by Aspnes and Ellen (SPAA 2011).

The algorithm is then extended to the case of *homonymous* system in which c, $1 \leq c \leq n$, identities are available. In every solo execution, the modified algorithm achieves $O(\sqrt{n-c+1} + \frac{\log c}{\log \log c})$ individual write complexity and $O(n - c + \frac{\log c}{\log \log c})$ individual step complexity.

Keywords: Anonymity, asynchronous shared memory, consensus, failure detectors, homonym processes, indulgent algorithms.

1 Introduction

In a typical distributed system, processes are *eponymous*, i.e., they have unique identities. On the other hand, in *anonymous* systems, processes have no identity and are programmed identically. When provided with the same input, processes in such systems are indistinguishable. Anonymity adds a new, challenging, difficulty to distributed computing.

From a practical point of view, anonymity is sometimes unavoidable. For example, consider a system composed of many tiny nodes, e.g., sensors networks. Sensors nodes might have limited storage and computational capability, and might not have been provided with unique identifiers [2]. Some other systems, like peer-to-peer file sharing applications [13], might require users to remain anonymous as a prerequisite to ensure privacy. See [19] for more details regarding anonymous computing and privacy.

---

* Supported by DIGITEO project PACTOLE.
** Supported in part by the ANR projects PROSE and CONCORDANT.
*** Supported in part by the ANR project DISPLEXITY.

A. Fernández Anta, G. Lipari, and M. Roy (Eds.): OPODIS 2011, LNCS 7109, pp. 175–190, 2011.

Recently, several papers [4,5,14,22,24,29] have addressed the question of the computational power of anonymous systems, with an emphasis on the consensus problem. In particular, Aspnes and Ellen [4] have shown that, when the number of proposed values is unbounded, the solo step complexity of consensus is $\Theta(n)$ in an n-process system. This paper presents a new, efficient, consensus algorithm for anonymous system.

The consensus problem. Consensus is a fundamental problem in fault-tolerant distributed computing. Informally, n processes, each starting with a private value, are required to agree on one value chosen among their initial values. For shared memory systems, it is well known that *asynchronous fault tolerant* consensus is impossible as soon as at least one process may fail by *crashing* [28]. Trivially, consensus is thus impossible in anonymous, asynchronous and failure-prone shared memory. The same impossibility holds for non-anonymous message passing asynchronous systems [20].

Since the publication of this result, several approaches have been identified to overcome this impossibility, including randomization (e.g., [6]), strengthening the model with timing assumptions (e.g., [18]) or failure detectors (e.g., [12]) and strong synchronization primitives [25]. Similarly, in anonymous systems, randomization [10], failure detectors [7,14], as well as additional synchrony assumptions [16] have been investigated to solve consensus.

A *failure detector* is a distributed device which provides processes with possibly unreliable information about failures. Unreliable failure detectors, and more generally system assumptions which are not guaranteed to always hold, have motivated the study of *indulgent* algorithms [23]. Informally, an algorithm is indulgent if it is always *safe*, i.e., it never violates the safety part of the problem it is supposed to solve, and *converges to a decision* when the failure detector matches its eventual property. In this line of research, the key question is determining how fast indulgent algorithms converge when the eventual property of the failure detector is satisfied [17].

Contributions of the paper. This paper investigates the consensus problems in an anonymous, crash prone and asynchronous shared memory systems. In particular, we are interested in the individual *write* step complexity of anonymous consensus. Typically, shared memory systems use caching techniques to improve performances. When a write is performed, the system has to ensure that every cached copy is updated, which is costly. Differently, repeatedly reading a shared location may be a local operation. The paper presents the following two main results:

- The first result is a consensus algorithm. The set of input values that processes might propose is unbounded. The algorithm relies on a failure detector of the class $A\Omega$ [8] and tolerates up to $n - 1$ process crashes. The "anonymous leader" class $A\Omega$ is the anonymous counterpart of the class Ω, which is the weakest failure detector for solving consensus [11] in the eponymous settings. Informally, when queried, a failure detector of the class $A\Omega$ returns a boolean. Eventually, each query, except the queries issued by some non-faulty process, returns false. If no failure detector is available, we note that our algorithm can easily be made obstruction-free [26] by simply removing failure detector invocations. The algorithm is write-efficient in the following sense : a process executing solo decides after performing $O(\sqrt{n})$ write operations and $O(n)$ shared memory operations in total.

— The second result is a generalization of our consensus algorithm to the case of *homonymous* systems recently introduced by Delporte-Gallet et al [15], in which a small number c, $1 \leq c \leq n$ of identities is available. The system is no longer totally anonymous since processes have identities. However, when the number of ids is smaller than n, several processes may share the same id. The generalized algorithm achieves $O(\sqrt{n - c + 1} + \log c / \log \log c)$ individual write complexity and $O(n - c + \log c / \log \log c)$ step complexity in solo execution. As in the case of anonymous systems, the algorithm relies on a failure detector of the class $A\Omega$ and the set of values that can be proposed is unbounded.

Roadmap. The paper is composed of 6 sections. Section 2 describes the anonymous shared memory model and the failure detector class $A\Omega$. An anonymous consensus is presented in Section 3. Its generalization to the case of systems with homonym processes follows (Section 4). Section 5 surveys related work and Section 6 concludes the paper.

2   System Model

Anonymous shared memory model. We consider a system Π of $n \geq 2$ deterministic processes. Processes are anonymous: they do not have identifiers, and they execute identical algorithms. The total number of processes n is however known by the processes. The system is asynchronous, in the sense that each process runs at its own speed, independently of the other processes.

Processes communicate with each other by reading and writing *atomic* shared registers (they are linearizable [27]). Registers are multi-writer and multi-reader: every register can be written in, or read from, by every process. In the pseudo-code we use to describe our algorithm, shared objects are denoted by upper-case letters, while lower-case identifiers are reserved for processes' local variables.

Failures and failure detectors. Processes may *crash*. A process is *correct* in an execution if it never crashes in this execution; otherwise it is *faulty*. We make no assumption on the number of crashes that may occur during a run.

As noted in the Introduction, a *failure detector* is a distributed oracle that provides processes with possibly unreliable information about failures [12]. Several classes of failure detectors suited to anonymous systems have been defined [8]. The failure detector we consider is anonymous Ω, denoted hereafter $A\Omega$. Each process is provided with a primitive $A\Omega.query()$, which returns *true* or *false*.

The following property, termed *eventual leadership* is ensured: there exists some correct process p_0 such that eventually every $A\Omega.query()$ always returns *true* at p_0, and *false* at every other process.

Consensus. Consensus is a distributed task which consists in a single operation *propose(v)* that takes as input a value v in some (possibly unbounded) set $\mathbb{V}$, and returns a value v' in $\mathbb{V}$. When a process p invokes *propose(v)*, we say that p *proposes* v. Similarly, when *propose(v)* returns a value v, we say that p *decides* v. Consensus requires that in every run: (Agreement) two processes cannot decide different values; (Validity) if a process decides some value v, then v was proposed before; and (Termination) every correct process eventually decides.

Time complexity. Consider an algorithm $\mathcal{A}$ that solves consensus in an asynchronous system equipped with an eventual failure detector such as $A\Omega$. In every execution, a correct leader process eventually emerges, but there is no bound on the time at which a correct process is elected. Obviously, the worst-case number of reads, or writes, performed by a process is unbounded. Thus, we measure the time complexity of asynchronous consensus algorithms in solo executions. Specifically, the *individual write complexity* (respectively the *individual step complexity*) is the worst-case number of write operations (respectively the total number of read aand write operation) that occur in solo executions in which only one process participates and this process is the leader output by the failure detector from the beginning of the execution.

3 The Janus[1] Algorithm

3.1 Description of Janus

The Janus algorithm solves consensus among n asynchronous and anonymous processes. Its pseudo-code is depicted in Figure 1. Janus relies on a failure detector of the class $A\Omega$ and tolerates up to $n-1$ process failures. No knowledge of the set of values that can be proposed is required. In particular, this set might be unbounded.

A process p initiates its algorithm by invoking propose(v), where v is the input value of p. Process p then launches two tasks **T1** and **T2** that run in parallel (line 1). In task **T2**, p monitors a shared register *decision D*, which is initialized to $\perp$[2]. If p reads a non-$\perp$ value d in D, p decides that value (line 21) and terminates.

In task **T1**, the execution proceeds in asynchronous rounds. Process p maintains an *estimate* (stored in the local variable *est*), which is the value it currently

---

[1] In Roman religion and mythology, Janus is the god of gates. Most often he is depicted as having two heads, facing opposite directions (Wikipedia). The choice of the name is explained by the fact that each process in our algorithm has to look in two directions: forward to check if another process has started a new round, and back to check if another process concurrently executed the $\mathcal{K}$ past rounds.

[2] $\perp$ is a special value that is never proposed by the processes.

shared variables

      $\forall r > 0 : T[r]$ is a multivalued MWMR atomic register, initially $\bot$

      $\forall r > 0 : C[r]$ is a binary MWMR atomic register, initially *false*

      D is a multivalued MWMR atomic register, initially $\bot$

propose(v)

  (1) $est \leftarrow v$; $rnd \leftarrow 0$; start **T1**; start **T2**;

task T1 :

  (2)   **while** (*true*) **do**

  (3)   **if** ($A\Omega$-QUERY()) **then**

  (4) $rnd \leftarrow rnd + 1$

          % Look for an estimate with higher priority %

  (5)   **if** ($T[rnd] \neq \bot$) **then** let $r \leftarrow \min\{r' > rnd \mid T[r'] = \bot\}$;

  (6) $est \leftarrow T[r-1]$; $rnd \leftarrow r - 1$

  (7)   **else** $T[rnd] \leftarrow est$

  (8)   **end if**

          % Look for conflicting estimates in the last $\mathcal{K}$ rounds %

  (9)   **for each** $i : 0 \leq i < \min(rnd, \mathcal{K})$ **do**

 (10)   **if** ($T[rnd - i] \neq est$) **then** $C[rnd - i] \leftarrow true$ **end if**

 (11)   **end for**

          % Check if no conflict occurs in the last $\mathcal{K}$ rounds %

 (12) $can\_decide \leftarrow true$;

 (13)   **if** ($rnd \geq \mathcal{K}$) **then**

 (14)   **for each** $i : 0 \leq i < \mathcal{K}$ **do**

 (15)   **if** ($C[rnd - i] = true$) $\vee$ ($T[rnd - i] \neq est$) **then** $can\_decide \leftarrow false$ **endif**

 (16)   **end for end if**

 (17)   **if** (can_decide) **then** $D \leftarrow est$ **endif**

 (18)   **end if**

 (19) **end while**

task T2 :

 (20)   **repeat** $d \leftarrow D$ **until** $d \neq \bot$

 (21)   **stop T1**; decide(d)

Fig. 1. The Janus algorithm, $\mathcal{K} = 2\lceil \sqrt{n} \rceil + 1$

favors. During each round to which it participates, p tries to *commit* its estimate by writing it in the decision register D (line 17). The algorithm ensures that (1) no two distinct values are committed and (2) at least one process eventually commits its estimate. To that end, each round r is associated with two multi-writer/multi-reader shared registers: the *value* register $T[r]$ and the *conflict* register $C[r]$. Intuitively, $T[r]$ stores a value that some process is willing to commit in round r, while $C[r]$, when set to *true*, indicates that two or more processes try to commit distinct values in round r.

A process p entering round r first checks whether a value has already been written in $T[r]$ (line 5). If this happens, p immediately enters round $r' \geq r$, where r' is the greatest round for which a value has been written to the associated register $T[r']$, thus possibly skipping rounds $r, \ldots, r' - 1$. In addition, p adopts the value currently stored in $T[r']$ as its new estimate. Otherwise, i.e., when $T[r]$ equals $\bot$, p writes its estimate in $T[r]$.

Writing/reading value v to/from the value register $T[r]$ is however not sufficient to allow this value to be committed. Several processes may be performing write operations concurrently on $T[r]$ and thus, assuming that v is committed, a process entering round r later might adopt a value $v' \neq v$ and commits this value. Therefore, before committing its estimate v (that is, writing v in D, line 17), process p first checks that no conflicts have been detected in the last $\mathcal{K}$ rounds and that the registers $T[r], T[r-1], \ldots, T[r-\mathcal{K}+1]$ still store v (lines 14–16). For large enough values of $\mathcal{K}$, these two conditions prevent any other value different from v from being written in $T[r]$. We show in the proof (Lemma 6) that for $\mathcal{K} \geq \lceil 2\sqrt{n} \rceil + 1$ this property is ensured.

Conflicts are detected at lines 9–11. A process p with estimate v executing round r performs a read operation in every register $T[r'], r - \mathcal{K} + 1 \leq r' \leq r$. Whenever a value different from v is returned, the corresponding conflict register $C[r']$ is updated to *true*.

Finally, the progress of Janus relies on the underlying failure detector $A\Omega$. A process is allowed to enter round r only if it considers itself as a leader. In more details, before entering round r, each process queries its local failure detector module (line 3). Only if this query returns *true*, the process starts round r. Eventually, a unique non-faulty process is elected by the failure detector. This process eventually executes rounds alone, and eventually decides (See Lemma 3). When a failure detector is not available, we note that Janus is easily made obstruction-free by removing the query to the failure detector at line 3.

3.2   Proof of the Janus Algorithm

Fix some execution of the algorithm. Since the shared objects (i.e. the registers) are atomic the execution (as an interleaved sequence of reads and writes operation of the processes) is linearizable [27]. As a consequence, we may consider σ a linearization of the reads and writes operations. We shall say that an operation in σ on some register *occurs at time* τ if τ is the linearization point of that operation. As usual, we shall note var_p the local variable var of process p. The *execution of the (asynchronous) round r by p* is the interval during which $rnd_p = r$. More precisely, it is the sequence of steps applied by p when $rnd_p = r$. Missing proofs of Lemmata 1 and 1 can be found in the full version [9].

A process, executing round r, writes its estimate v in $T[r]$, provided it observes that no value has been previously written in $T[r]$ (line 7). The following Lemma implies that if this occurs, v has been previously written to $T[1], \ldots, T[r-1]$.

Lemma 1. *Let $r > 1$. Suppose that a write operation op with parameter v is performed on $T[r]$. Then a write operation op$'$ of value v to $T[r-1]$ occurs before op.*

It then follows from the previous Lemma that algorithm 1 satisfies the validity requirement of consensus.

Lemma 2 (Validity). *Every decided value is a proposed value.*

Termination then followed from the eventual leadership property of the failure detector $A\Omega$.

Lemma 3 (Termination). *Every correct process eventually decides.*

Proof. Assume for contradiction that some correct process q never decides. As, (1) only non-$\bot$ values can be written in D, and (2) q reads D infinitely many times and never decides, no value $v \neq \bot$ is written in D. As a process may decide only if it reads a value different from $\bot$ in D, this implies that no process decides.

By the eventual leadership property of the failure detector class $A\Omega$, there is a correct process p and a time τ such that each $A\Omega$-QUERY() performed after τ returns *true* if and only if the invoking process is p. At time τ, let R be the largest round such that $T[R-1] = \bot$. Clearly, p is the only process that can execute rounds $R+1, R+2, \ldots$ (line 3). Moreover by Lemma 1, for all $i > 0$, we have that $T[R+i] = \bot$.

As p is correct, it never decides, and for all $i > 0$ we have that $T[R+i] = \bot$, p eventually executes rounds $R+1, R+2, \ldots$ As p is the only process that executes those rounds, it follows from the code (lines 5–7) that p writes in each register $T[R+i]$ for all $i > 0$. Besides, it is not difficult to observe that the same value, say v, is written by p in each register $T[R+i]$.

As no process except p executes rounds $R+i, i > 0$, no process except p performs write operations on registers $T[R+i], i > 0$. Therefore it holds forever that $C[R+i] = \textit{false}$ and $T[R+i] = v$, once p has written v in $T[R+i]$. Consider the execution of round $R+\mathcal{K}$ by p. Process p first writes v in $T[R+\mathcal{K}]$ (line 7). After this occurs, we have $C[R+i] = \textit{false}$ and $T[R+i] = v$ for each $i, 0 < i \leq \mathcal{K}$. Hence, $can\_decide_p = \textit{true}$ after the execution of the **for each** loop at lines 14–16. We conclude that p writes v in D (line 17), and decides by the code of task **T2**: contradiction. $\qquad\square$

Proof of agreement. We divide the execution in *epochs* as follows. Epoch e_i is an interval that starts with the first write (according to the linearization σ) to register $T[i]$ and ends immediately before the first write (if any) performed to register $T[i+1]$. Given a read, or write, operation op, we say that *op occurs in epoch e_i*, or equivalently, that *op is performed in e_i*, if op is linearized in the interval e_i. Clearly, if a write to $T[j]$ occurs in e_i, then $j \leq i$. The next lemma directly follows from the code of Janus (lines 5 and 7).

Lemma 4. *Suppose that p performs a write operation op on $T[i]$. The last operation preceding op performed by p is a read on $T[i]$, and the value returned by that operation is $\bot$.*

Suppose that process p performs a write operation on register $T[j]$ in epoch e_i. When this operation terminates, a value has already been written in $T[i]$ by definition of e_i. Lemma 4 then implies that the next write operation by p (if any) is performed on some register $T[j']$ such that $j' > i$. Lemma 5 bellow captures precisely this observation.

Lemma 5. *Denote by op, op' two write operations performed by the same process p. Suppose that: (1) op occurs in e_i, (2) op' is a write on register $T[j]$ with $j \neq i$, and (3) op precedes op'. Then, $j > i$.*

Proof. By Lemma 4, p reads from $T[j]$ immediately before executing op', and this read operation returns $\perp$. Let op'' denote that operation. It follows from the third condition of the Lemma that op'' occurs after op, which in turn occurs after some non-$\perp$ value has been written in $T[i']$ for each $i' \leq i$ (By definition of e_i, and the fact op occurs in e_i.). Since the read operation op'' performed on $T[j]$ returns $\perp$, we conclude that $j > i$. $\qquad\square$

Consider a round number r, and a value v. We say that value v is *committed at round* r if there exists a process p that writes v in D (line 17) while it is executing round r. Observe that in such a case, v is the estimate of p, and v has been written in $T[r]$ (by p itself or some other process). Note moreover that for each decided value v, there exists a round during which v is committed.

The following lemma is central to the proof of the agreement property. Informally, this lemma says that if some process writes a value v in the decision register D while executing round r, no other value than v can be written to $T[r]$.

Lemma 6. *Let v be a value, and R be a round number such that v is committed at round R. For every value v' written in $T[R]$, it holds that $v' = v$.*

The agreement property then follows by combining Lemma 1 and Lemma 6, and observing that every decided value has been committed.

Lemma 7 (Agreement). *No two process decide different values.*

Proof. Let v and v' be two decided values (at line 21). By the code of Algorithm 1, v and v' have been previously written in D (at line 17). Hence, v and v' are committed at some round, say, r and r' respectively. Without loss of generality, assume that $r \leq r'$. Let p' be a process that writes v' in D in round r'. Observe that v' is the estimate of p' in round r'. Therefore, v' has been written in $T[r']$, either by p' (at line 7) or by some other process (in the latter case, v' was read by p' at line 6). As $r \leq r'$, it follows from Lemma 1 that v' is written in $T[r]$ as well. Since v is committed at round r, we conclude by Lemma 6 that $v = v'$. $\qquad\square$

The rest of this section is devoted to the proof of Lemma 6. We proceed by contradiction. We name H the following assumption:

> There exists a round R such that two write operations with parameters $u \neq v$ are performed on $T[R + \mathcal{K}]$ and v is committed in round $R + \mathcal{K}$.

In the following, we show that to satisfy assumption H the system must consist of at least $n + 1$ processes.

Denote by R the round number appearing in assumption H. For each $i, j, 1 \leq i, j \leq \mathcal{K}$, note W^i_j the set of processes that perform a write operation to register $T[R + j]$ during epoch e_i. More precisely, a process p belongs to W^i_j if and only if there exists a write operation to $T[R + j]$ by p which occurs in e_i. By the definition of epochs, we know that if $j > i$, then $W^i_j = \emptyset$. The three lemmata below further precise how the sizes of the W^i_j's and the round numbers are related.

Lemma 8. *If assumption H holds, then:* $\forall i, 1 \leq i < \mathcal{K}, |W^i_i| \geq 2$.

Proof. By assumption H, at least two values v and u are written in $T[R + \mathcal{K}]$. It follows from Lemma 1 that v and u must have been written in $T[R + i]$ for each i such that $1 \leq i < \mathcal{K}$. It remains to show that such a write operation with parameter v (resp. u) occurs in e_i.

Let us consider the first write of v in $T[R + i]$. Clearly, this operation occurs in epoch $e_{R+i'}$, for some $i' \geq i$. Suppose for the sake of contradiction that $i' > i$. Hence, the first time v is written in $T[R+i]$, a value has already been written in $T[R+i+1]$. Let p be the process that performs this first write of v in $T[R+i+1]$. As v is written to $T[R + \mathcal{K}]$, p must exist by Lemma 1. Denote $w_p(R+i+1)$ the write operation of p. According to the code of Janus we know that: (1) p performs that operation while it is executing round $R + i + 1$ (line 7), (2) $w_p(R + i + 1)$ is preceded by a read operation of $T[R + i + 1]$ (denoted $r_p(R + i + 1)$) by p that returns $\perp$, and (3) in round $R + i$, there is a read operation from $T[R+i]$ that returns v or a write of v by p to $T[R + i]$. Denote by $op_p(R + i)$ this last operation, and $op_p(R+i), r_p(R+i+1), w_p(R+i+1)$ the operations that occur in this order. Moreover, $op_p(R + i)$, which reads or writes v in $T[R + i]$ occurs in epoch $e_{R+i''}$ for some $i'' \geq i'$, since the write of v in $T[R + i]$ occurs in $e_{R+i'}$. Therefore, operation $r_p(R+i+1)$ occurs after a write in $T[R+i+1]$, from which we conclude that $r_p(R + i + 1)$ returns a non-$\perp$ value. It thus follows by Lemma 4 that p does not write in $T[R + i + 1]$: a contradiction.

We have shown that a write of v in $T[R + i]$ occurs in epoch e_i. A similar argument applied to value u yields that a write of u in $T[R + i]$ occurs in e_i. Since each process does not write twice in the same register, $|W^i_i| \geq 2$. $\square$

Lemma 9. *If assumption H holds, then* : $\forall i, j : 1 \leq i < \mathcal{K}$ *and* $1 \leq j < i$, $|W^i_j| \geq 1$.

Proof. We start by establishing that two read operations that return v and u respectively occur in e_i.

As v is written in $T[R + \mathcal{K}]$, v is also written in $T[R + i + 1]$ (Lemma 1). Let p the process that performs the first write of v in $T[R + i + 1]$. By the code, p executes round $R + i$ before performing that write operation, and v is the estimate of p in that round. At the beginning of round $R + i$, p either reads v in $T[R + i]$ or writes v in $T[R + i]$. Moreover, the read operation on $T[R + i + 1]$ performed by p at the beginning of round $R + i + 1$ returns $\perp$ (Otherwise p does not perform a write operation on $T[R + i + 1]$). Therefore, every operation performed by p while it is executing round $R + i$ occurs in epoch e_{R+i}.

In particular, the read of $T[R+j]$ performed by p at line 10 occurs in e_{R+i}. This read must return v. Otherwise, p writes $true$ in $C[R+i]$, and this operation occurs in e_{R+i}. As no process ever writes $false$ in $C[R+i]$, every read operation performed on $C[R+i]$ that occurs in later epochs return $true$. Consider a process p' executing round $R+\mathcal{K}$. p' reads $C[R+i]$ at line 15. This read operation occurs after a write operation has been performed on $T[R+\mathcal{K}]$, so it occurs after the end of epoch e_{R+i}. Hence, that operation returns $true$ and thus p' cannot write in D in that round. Therefore, no value is committed in round $R+\mathcal{K}$, contradicting assumption H.

Similarly, by considering the process that performs the first write of u in $T[R+i+1]$, we get that a read operation of $T[R+j]$ that returns u occurs in e_{R+i}.

Finally, as there are two read operations of $T[R+j]$ returning two different values occur in e_i, there must exist a write operation on $T[R+j]$ that occurs in e_i. We thus conclude that $W_j^i \neq \emptyset$. $\qquad\square$

Lemma 10. *Suppose that assumption H holds. Let i, i', j, j' such that $1 \leq i \leq i' < \mathcal{K}$ and $1 \leq j < i$, $1 \leq j' < i'$. $W_j^i \cap W_{j'}^{i'} \neq \emptyset \Rightarrow (i = i' \wedge j = j') \vee (i < j')$*

Proof. Let $p \in W_j^i \cap W_{j'}^{i'}$. By definition, a write operation by p occurs in e_i and $e_{i'}$. Either $i = i'$ and $j = j'$ or, by Lemma 5, $i < j'$. $\qquad\square$

Proof of Lemma 6. Assume for the sake of contradiction that assumption H is satisfied, and consider the following set:

$$S = \left\{ (i,j) : \left\lceil \frac{\mathcal{K}-1}{2} \right\rceil \leq i \leq \mathcal{K}-1, 1 \leq j \leq \left\lceil \frac{\mathcal{K}-1}{2} \right\rceil \right\}$$

In what follows, we count the total number of processes that appear in the union of the sets W_j^i, where $(i,j) \in S$, then we show that this union includes at least $n + 1$ distinct processes.

Let $(i,j) \neq (i',j') \in S$ such that $i \leq i'$. By definition of S, $i \geq j'$ and thus it follows from Lemma 10 that $W_j^i \cap W_{j'}^{i'} = \emptyset$. Hence,

$$\left| \bigcup_{(i,j)\in S} W_j^i \right| = \sum_{(i,j)\in S} |W_j^i|$$

Moreover, It follows from Lemmas 8 and 9 that $|W_j^i| \geq 1$ for each $(i,j) \in S$ and $|W_i^i| \geq 2$ for each $(i,i) \in S$. Therefore,

$$\left| \bigcup_{(i,j)\in S} W_j^i \right| \geq \left\lceil \frac{\mathcal{K}-1}{2} \right\rceil \cdot \left\lceil \frac{\mathcal{K}-1}{2} \right\rceil + 1$$

Finally, as $\mathcal{K} = 2 \cdot \lceil \sqrt{n} \rceil + 1$, we get $\left| \bigcup_{(i,j)\in S} W_j^i \right| \geq n + 1$. Therefore, assuming that H is satisfied, we have exhibited a set of $n+1$ distinct processes : a contradiction. Consequently, H cannot be satisfied, from which we conclude that no value different from v is written in $T[R]$, as desired. $\qquad\square$

Theorem 1. *The Janus algorithm described in Figure 1, when instantiated with a failure detector of the class $A\Omega$ solves consensus in an n-processes, anonymous shared memory system.*

Proof. Immediately follows from Lemmas 2, 3 and 7.   □

The following theorem proves that the step complexity of Janus is $O(n)$, which is optimal [4], and that its write complexity equals to $O(\sqrt{n})$.

Theorem 2. *The Janus algorithm has a step complexity of $O(n)$, and a write complexity of $O(\sqrt{n})$.*

Proof. Consider a solo execution of some process p. During this execution, p executes $\mathcal{K} = 2\lceil\sqrt{n}\rceil + 1$ rounds, then decides. Name $\{1, \ldots, \mathcal{K}\}$ the rounds executed by p, and consider some round i. According to the code of Algorithm 1, during round i process p executes a single write (line 7), and reads $3i + 1$ shared registers (lines 5, 9 to 11, and 14 to 16). As a consequence, the step complexity of Janus is $O(n)$, and its write complexity equals $O(\sqrt{n})$.   □

4   The Case of Homonymous Systems

In an homonymous system, $c, 1 \le c \le n$ identities are available [15,30]. Each process has an identifier in the range $\{1, \ldots, c\}$. Processes that share the same identifier are said to be homonym, and for each $i \in \{1, \ldots, c\}$, there is at least one process with id i (and thus at most $n - c + 1$).

In this section we present a consensus algorithm for homonymous shared-memory systems that tolerates up to $n - 1$ process failures. As in the case of anonymous systems, the algorithm relies on a failure detector of the class $A\Omega$ and the set of values that can be proposed is unbounded. The algorithm is built in a modular way from several copies of the Janus algorithm and an efficient implementation of m-valued adopt-commit objects due to Aspnes and Ellen [4].

Adopt-commit. An adopt-commit object [21] is a shared object that supports a single operation denoted $propose(v)$ where v is a value taken from some set $\mathbb{V}$. Every invocation of $propose(\cdot)$ returns a response of the form (b, v') where $b \in \{commit, adopt\}$ and $v' \in \mathbb{V}$ such that the following properties hold: (Termination) Every invocation of $propose(\cdot)$ by a correct process terminates; (Validity) If (b, v) is returned, then some process previously invoked $propose(v)$; (Agreement) If $(commit, v)$ is returned, then every decision has the form $(*, v)$; (Convergence) If every process proposes the same value v, then $(commit, v)$ is the only possible decision.

An efficient crash-tolerant asynchronous implementation of m-valued adopt-commit objects from multi-reader multi-writer registers in anonymous system is presented by Aspnes and Ellen in [4]. The algorithm achieves $O(\frac{\log m}{\log \log m})$ individual step-complexity provided that the set $\mathbb{V}$ from which proposed values are taken is a priori known contains at most m values.

Overview of the algorithm. The algorithm, described in Figure 2, proceeds in asynchronous rounds. Each round is divided in two phases, an *agreement* phase in which each group of homonym processes agree on a common value, and a *conciliation* phase in which processes check whether every group agrees on the same value.

The agreement phase of round r is implemented by c instances of the Janus algorithm that we note $J[r][1], \ldots, J[r][c]$. As in the Janus algorithm, each process maintains an estimate stored in the local variable *est*. Processes with identity *id* propose their estimate to the same instance of Janus $J[r][id]$ (line 4). The array $V[r][1..c]$ is then used to store the decisions that occur (if any) in each of the c instances $J[r][1..c]$ (line 5). This completes the agreement phase of round r.

Note that each instance of Janus is implemented with its own collection of registers. Processes however share a single failure detector $A\Omega$. This means that a given instance of Janus might not progress if no process participating in this instance is elected by the failure detector. Nevertheless, if every correct process participates in at least one of the Janus instances of round r, termination is ensured in at least one instance, namely the instance $J[r][id]$, where *id* is the identity of the eventual leader. The conciliation phase of round r is implemented by a single adopt-commit object denoted $AC[r]$. A process p with identity *id* that has previously obtained a decision d from the instance of Janus $J[r][id]$

---

shared objects
    $\forall r > 0 : J[r][1..c]$ is an array of c copies of Janus
    $\forall r > 0 : AC[r]$ is an adopt-commit object
    $\forall r > 0 : V[r][1..c]$ is an array of c MRMW registers, initially $\perp$
    DD is a multivalued MWMR atomic register, initially $\perp$

propose(v)
  (1) $est \leftarrow v$; $rnd \leftarrow 0$; start **T1**; start **T2**;

task T1 :
  (2)   **while** (*true*) **do**
  (3) $rnd \leftarrow rnd + 1$;
  (4) $est \leftarrow J[rnd][id].\text{propose}(est)$;
  (5) $V[rnd][id] \leftarrow est$;
  (6) $(b, id') \leftarrow AC[rnd].\text{propose}(id)$;
  (7) $est \leftarrow V[rnd][id']$;
  (8)   **if** $b = commit$ **then** $DD \leftarrow est$ **endif**
  (9)   **end while**

task T2 :
 (10) **repeat** $d \leftarrow DD$ **until** $d \neq \perp$
 (11) stop **T1**; *decide*(d)

Fig. 2. Consensus with homonyms, code for processes with identity *id*

and has written this value to the register $V[r][id]$ checks whether it is safe to decide this value. To do so, it proposes its identity to the adopt-commit object $AC[r]$ (line 6). Let (b, id') denote the response of the object obtained by p. p first adopts the value it reads from $V[r][id']$ as its new estimate (line 7). Note that the read operation of $V[r][id']$ returns a non-$\perp$ value. This is because by the validity property of adopt-commit, a process p' with identity id' must have proposed its identity to $AC[r]$ before p obtains the response (b, id'). In addition, before accessing $AC[r]$, p' must have written a value to $V[r][id']$.

Second, if $b = commit$, p then writes its estimate in the shared register DD, indicating that this value can be safely decided. Indeed, by the agreement property of adopt-commit, every propose() operation to $AC[r]$ returns $(adopt, id')$ or $(commit, id')$. Hence, as a unique value v is written in $V[r][id']$, the estimate of each process that completes round r is equal to v. It thus follows that v is the only value that may be written to DD in round r and any subsequent round.

Termination relies on the underlying failure detector $A\Omega$. The eventual leadership property ensures that after some time τ, a single correct process considers itself as a leader. Let id denote the identity of this eventual leader. Observe that, by the code of Janus, a process participating in the execution of an instance of Janus does not take write steps unless it considers itself as a leader (Figure 1, line 3). Therefore, no decisions occur in every instance $J[r][id']$ that starts after τ if $id' \neq id$. On the other hand, every instance $J[r'][id]$, $r' \geq 1$ eventually produces a decision because the set of processes that participate in these instances includes the eventual leader. Consequently, if each round r instance of Janus starts after τ, only process with identity id may access the object $AC[r]$. Since they all propose the same value, namely id, it follows from the convergence property of adopt-commit that they get back $(adopt, id)$. This implies that a value is eventually written to the decision register DD, and termination follows.

Complexity. Since at most $n-c+1$ processes participate in each instance of Janus ($n - c + 1$ is the maximal size of a group of homonym processes), the parameter $\mathcal{K}$ is set to $2\sqrt{n - c + 1} + 1$ in each instance. Values proposed to objects $AC[r]$ are always taken from the set of available identities $\{1, \ldots, c\}$. Each adopt-commit object is thus implemented by the optimal algorithm by Aspnes and Ellen [4]. A process executing solo, and elected leader by the failure detector from the beginning of the execution, decides after participating in one instance of Janus, and performing one propose() operation on an adopt-commit object. In addition, it performs two write operations (at lines 5 and 8). Therefore, in solo executions, the individual write complexity equals to $O(\sqrt{n - c + 1} + \frac{\log c}{\log \log c})$ and the individual step complexity equals to $O(n - c + 1 + \frac{\log c}{\log \log c})$.

Proof. The correctness proof of the algorithm described in Figure 2 is presented in a companion technical report [9].

5   Related Work

Attiya et al. [5] characterized *failure-free* tasks that are solvable using registers when the number of processes n is unknown. In particular, the authors show, using bivalence and covering arguments, that consensus in such an environment requires more than $\Omega(\log n)$ atomic registers, and at least $\Omega(\log n)$ total work. Recently, Aspnes and Ellen [4] proved that the individual step complexity of adopt-commit object in anonymous shared-memory is $\Theta(min\ (\frac{\log m}{\log\log m}, n))$, where m is the number of different values that might be proposed to the object. Because consensus satisfies the specification of an adopt-commit object [21], this lower bound also holds for the consensus object.

Guerraoui and Ruppert [24] studied the computational power of shared memory distributed systems in the presence of both anonymity and failures. They propose constructions for several fundamental abstractions: wait-free timestamping and snapshots, and obstruction-free consensus. In particular, the authors depict an anonymous *binary* consensus algorithm having a step complexity of $O(1)$. *When m is known*, this algorithm solves anonymous consensus in $O(\log m)$ write operations and $O(\log m)$ individual work. Delporte-Gallet and Fauconnier [14] proposed an anonymous consensus which relies on failure detector $A\Omega$ and a weak set abstraction. If m is known, this algorithm solves consensus in $O(\log m)$ individual work and $O(1)$ writes.

Abrahamson [1] studied binary consensus in the probabilistic-write model with eponymous processes, when identities are only used to label registers. Recently, Aspnes [3] proposed a consensus algorithm for the probabilistic-write anonymous model which solves consensus in $O(\log m)$ individual work. The algorithm is based on the decomposition of consensus into two distinct components: an adopt-commit object which detects agreement, and a conciliator, which ensure agreement with some probability. Aside from their lower bound result, the authors of [4] proposed two asymptotically optimal implementations of adopt-commit objects: a $O(\frac{\log m}{\log\log m})$ solution which requires that m is known, and a $O(n)$ solution which solves the problem without any assumptions over m. During a solo execution, the latter algorithm writes in $O(n)$ different registers.

The Janus algorithm we depicted in Section 3 solves anonymous consensus in $O(n)$ individual work, and $O(\sqrt{n})$ write operations, a result which matches the lower bound of [3] and further improves the write complexity of anonymous consensus.

The notion of partial anonymity in which some processes may share the same identifier was first introduced by Yamashita et al. [30] in the context of the leader election problem. The term homonyms was coined recently by Delporte et al. [15]. In this work, the authors study the Byzantine consensus problem in message passing systems when a limited number of identities is available.

6   Conclusion

This paper has presented two efficient consensus algorithms for anonymous and partially anonymous asynchronous shared memory systems. Both algorithms do not impose restrictions on the set $\mathbb{V}$ from which proposed values are taken. The complexity depends solely on the number of processes n and the number of available identifiers c in the partially anonymous case. To the best of our knowledge, the generalized algorithm presented in Section 4 is the first non-trivial consensus implementation for shared memory homonymous systems.

Of note, by limiting the Janus algorithm to its first $\mathcal{K}$ rounds and removing the queries to the failure detector, we obtain an anonymous adopt-commit implementation whose individual write complexity is $O(\sqrt{n})$, while retaining an optimal $O(n)$ individual work. With respect to the write complexity, this is an improvement over existing implementations.

This paper focuses on consensus algorithms for which the set of input values is not restricted. A direction for future research is to investigate the interplay between the size of the input set m, the number of available identifiers c, the number of processes n, and the number of distinct values k that can be decided.

References

1. Abrahamson, K.: On achieving consensus using a shared memory. In: Proc. of the 17th Symp. on Principles of Distributed Computing (PODC), pp. 291–302. ACM (1988)
2. Angluin, D., Aspnes, J., Diamadi, Z., Fischer, M.J., Peralta, R.: Computation in networks of passively mobile finite-state sensors. Distributed Computing 18(4), 235–253 (2006)
3. Aspnes, J.: A modular approach to shared-memory consensus, with applications to the probabilistic-write model. In: Proc. of the 29th Symp. on Principles of Distributed Computing (PODC), pp. 460–467. ACM (2010)
4. Aspnes, J., Ellen, F.: Tight bounds for anonymous adopt-commit objects. In: Proc. of the 23rd Symp. on Parallelism in Algorithms and Architectures (SPAA), pp. 317–324. ACM (2011)
5. Attiya, H., Gorbach, A., Moran, S.: Computing in totally anonymous asynchronous shared memory systems. Inf. Comput. 173, 162–183 (2002)
6. Ben-Or, M.: Another advantage of free choice: Completely asynchronous agreement protocols (extended abstract). In: Proc. of the 2nd Symp. on Principles of Distributed Computing (PODC), pp. 27–30. ACM (1983)
7. Bonnet, F., Raynal, M.: The Price of Anonymity: Optimal Consensus Despite Asynchrony, Crash and Anonymity. In: Keidar, I. (ed.) DISC 2009. LNCS, vol. 5805, pp. 341–355. Springer, Heidelberg (2009)
8. Bonnet, F., Raynal, M.: Anonymous Asynchronous Systems: The Case of Failure Detectors. In: Lynch, N.A., Shvartsman, A.A. (eds.) DISC 2010. LNCS, vol. 6343, pp. 206–220. Springer, Heidelberg (2010)
9. Bouzid, Z., Sutra, P., Travers, C.: Anonymous Agreement: The Janus Algorithm. Technical report, http://hal.inria.fr/inria-00625704/en/
10. Buhrman, H., Panconesi, A., Silvestri, R., Vitányi, P.M.B.: On the importance of having an identity or, is consensus really universal? Distributed Computing 18(3), 167–176 (2006)
11. Chandra, T.D., Hadzilacos, V., Toueg, S.: The weakest failure detector for solving consensus. J. ACM 43(4), 685–722 (1996)

12. Chandra, T.D., Toueg, S.: Unreliable failure detectors for reliable distributed systems. J. ACM 43(2), 225–267 (1996)
13. Chothia, T., Chatzikokolakis, K.: A Survey of Anonymous Peer-to-Peer File-Sharing. In: Enokido, T., Yan, L., Xiao, B., Kim, D.Y., Dai, Y.-S., Yang, L.T. (eds.) EUC-WS 2005. LNCS, vol. 3823, pp. 744–755. Springer, Heidelberg (2005)
14. Delporte-Gallet, C., Fauconnier, H.: Two Consensus Algorithms with Atomic Registers and Failure Detector Ω. In: Garg, V., Wattenhofer, R., Kothapalli, K. (eds.) ICDCN 2009. LNCS, vol. 5408, pp. 251–262. Springer, Heidelberg (2008)
15. Delporte-Gallet, C., Fauconnier, H., Guerraoui, R., Kermarrec, A.M., Ruppert, E., Tran-The, H.: Byzantine agreement with homonyms. In: Proc. of the 30th Symp. on Principles of Distributed Computing (PODC), pp. 21–30. ACM (2011)
16. Delporte-Gallet, C., Fauconnier, H., Tielmann, A.: Fault-tolerant consensus in unknown and anonymous networks. In: Proc. of the 29th Int'l Conference on Distributed Computing Systems (ICDCS), pp. 368–375. IEEE (2009)
17. Dutta, P., Guerraoui, R.: The inherent price of indulgence. Distributed Computing 18(1), 85–98 (2005)
18. Dwork, C., Lynch, N., Stockmeyer, L.: Consensus in the presence of partial synchrony. J. ACM 35(2), 288–323 (1988)
19. Federrath, H. (ed.): Designing Privacy Enhancing Technologies. LNCS, vol. 2009. Springer, Heidelberg (2001)
20. Fischer, M.J., Lynch, N.A., Paterson, M.: Impossibility of distributed consensus with one faulty process. J. ACM 32(2), 374–382 (1985)
21. Gafni, E.: Round-by-round fault detectors: unifying synchrony and asynchrony. In: Proc. of the 17th Symp. on Principles of Distributed Computing (PODC), pp. 143–152. ACM (1998)
22. Guerraoui, R., Ruppert, E.: What can be implemented anonymously? In: Fraigniaud, P. (ed.) DISC 2005. LNCS, vol. 3724, pp. 244–259. Springer, Heidelberg (2005)
23. Guerraoui, R., Lynch, N.A.: A general characterization of indulgence. TAAS 3(4) (2008)
24. Guerraoui, R., Ruppert, E.: Anonymous and fault-tolerant shared-memory computing. Distributed Computing 20(3), 165–177 (2007)
25. Herlihy, M.: Wait-free synchronization. ACM Trans. Program. Lang. Syst. 13(1), 124–149 (1991)
26. Herlihy, M., Luchangco, V., Moir, M.: Obstruction-free synchronization: Double-ended queues as an example. In: Proc. of the 23rd Int'l Conference on Distributed Computing Systems (ICDCS), pp. 522–529. IEEE (2003)
27. Herlihy, M., Wing, J.: Linearizability: a correctness condition for concurrent objects. ACM Trans. on Prog. Lang. 12(3), 463–492 (1990)
28. Loui, M., Abu-Amara, H.: Memory requirements for agreement among unreliable asynchronous processes. Advances in Computing Research 4, 163–183 (1987)
29. Ruppert, E.: The Anonymous Consensus Hierarchy and Naming Problems. In: Tovar, E., Tsigas, P., Fouchal, H. (eds.) OPODIS 2007. LNCS, vol. 4878, pp. 386–400. Springer, Heidelberg (2007)
30. Yamashita, M., Kameda, T.: Leader election problem on networks in which processor identity numbers are not distinct. IEEE Transactions on Parallel and Distributed Systems 10(9), 878–887 (1999)

Communication Complexity of Consensus in Anonymous Message Passing Systems*

Emanuele G. Fusco[1],** and Andrzej Pelc[2],***

[1] Computer Science Department, Sapienza, University of Rome, 00198 Rome, Italy
fusco@di.uniroma1.it
[2] Département d'informatique, Université du Québec en Outaouais,
Gatineau, Québec J8X 3X7, Canada
pelc@uqo.ca

Abstract. We consider the message complexity of achieving consensus in synchronous anonymous message passing systems. Unlabeled processors (nodes) communicate through links of a network. In each round every processor can exchange messages with all neighbors and the duration of each transmission is one round. An adversary wakes up some subset of processors at possibly different times and assigns them arbitrary numerical input values. All other processors are dormant and do not have input values. Any message wakes up a dormant processor. The goal of consensus is to wake up all processors and have them agree on one of the input values. We seek deterministic consensus algorithms using as few messages as possible. As opposed to most of the literature on consensus, the difficulty of our scenario are not faults (we assume that the network is fault-free) but the arbitrary network topology combined with the anonymity of nodes.

For unknown n-node networks we show a consensus algorithm using $O(n^2)$ messages; this complexity is optimal for this class. We show that if the network is known, then the complexity of consensus decreases significantly. Our main contribution is an algorithm that uses $O(n^{3/2} \log^2 n)$ messages on any n-node network and we show that some networks require $\Omega(n \log n)$ messages to achieve consensus. We also observe that availability of distinct labels of nodes helps to improve complexity of consensus for known networks but has no effect for the class of unknown networks. Indeed, even with labeled nodes, $\Omega(n^2)$ messages are sometimes necessary if the network is unknown but for known labeled networks consensus can be always achieved with $O(n)$ messages.

Keywords: algorithm, consensus, anonymous network.

---

* This work was done during the visit of Emanuele G. Fusco at the Research Chair in Distributed Computing of the Université du Québec en Outaouais.
** Partially supported by MIUR of Italy under project AlgoDEEP prot. 2008TFBWL4.
*** Partially supported by NSERC discovery grant and by the Research Chair in Distributed Computing at the Université du Québec en Outaouais.

A. Fernández Anta, G. Lipari, and M. Roy (Eds.): OPODIS 2011, LNCS 7109, pp. 191–206, 2011.

1   Introduction

1.1   The Problem and the Model

Consensus is one of the fundamental problems in distributed computing. In this paper we consider a generalized version of consensus in synchronous anonymous message passing systems. Processors communicate through links of a network, modeled as an undirected connected graph. Nodes of this network represent processors, and we use terms "processor" and "node" as synonyms. We assume that nodes are unlabeled. It is desirable to be able to achieve consensus without relying on labels of processors because the latter may refrain from revealing their identities, due to privacy or security reasons. On the other hand, ports at each node of degree δ are labeled $1, \ldots, \delta$, but no coherence between these labelings is assumed. Nodes are equipped with local clocks that tick at the same rate, in synchronous rounds. An adversary wakes up some subset of nodes at possibly different rounds and assigns them arbitrary numerical input values. All other nodes are dormant and do not have input values. The local clock of every awake node is initialized to 0 at its wake-up round. In each round every awake node can exchange messages with all neighbors, and the duration of each transmission is one round. Any message wakes up a dormant node. The goal of consensus is to wake up all nodes and have them agree on exactly one of the input values in the same round; all nodes have to be aware when this is done. As opposed to most of the literature on consensus, the difficulty of our scenario are not faults (we assume that the network is fault-free) but the arbitrary network topology combined with the anonymity of nodes.

Note that, in a fault-free environment, this version of consensus is more general and has a stronger requirement than the usual formulation. First, the adversary can wake up only some processors and may do this at different times, whereas in the classic version [21] all processors are active from the beginning. Second, we require that consensus be made on one of the input values, whereas the classic validity condition only stipulates that this be the case if all input values are identical. Third, consensus has to be achieved by all processors in the same round, which is not required in the classic version. As will be seen, all our positive results concern this stronger, more general version, while our negative results are valid even for the classic, weaker version.

We consider two scenarios: that of ad-hoc (i.e., unknown) and that of known networks. In the scenario of unknown networks the only knowledge that nodes have about the network is a linear upper bound on the number of its nodes. Note that without knowing any bound on the size of the network, consensus is impossible even in an oriented ring. Indeed, due to anonymity, nodes cannot distinguish if they are in a small or in a large ring. At some point each node must make a decision and it is easy to construct an instance with a large ring, where two remote groups of nodes make incompatible decisions before communicating. In the scenario of known networks, every node is provided with a map of the network, which is an isomorphic copy of it containing all port numbers, with the location of the given node marked in the map. Note that, due to the lack of node labels, if the network has non-trivial automorphisms preserving port numbers, then it is impossible to distinguish between isomorphic nodes in the map.

A problem related to consensus is that of *establishing global time*. In our setting, clocks of all awake nodes tick at the same rate but they do not necessarily show the same round number; instead, the clock at each node shows the number of rounds since the wake-up of this node. This is sometimes called *local synchronization* [17]. Establishing global time (or achieving *global synchronization*) consists in waking up all nodes and having all their clocks show the same round number. All nodes must be aware when this happens. These two levels of synchrony of the system have been previously studied in various contexts and it turns out that global synchronization is much more powerful than local synchronization.

1.2   Our Results

We seek deterministic consensus algorithms using as few messages as possible. For unknown n-node networks we show a consensus algorithm using $O(n^2)$ messages. This complexity is optimal for this class: indeed, if the network is unknown to nodes, some networks require $\Omega(n^2)$ messages for consensus. We show that if the network is known, then the complexity of consensus decreases significantly. Our main contribution is an algorithm that uses $O(n^{3/2} \log^2 n)$ messages on any n-node network and we show that some networks require $\Omega(n \log n)$ messages to achieve consensus. We also observe that the availability of distinct labels of nodes helps to improve complexity of consensus for known networks but has no effect for the class of unknown networks. Indeed, even with labeled nodes, $\Omega(n^2)$ messages are sometimes necessary if the network is unknown but for known labeled networks consensus can be always achieved with $O(n)$ messages.

The main challenge in achieving low message complexity of consensus in anonymous systems is that nodes woken up by the adversary behave identically in highly symmetric networks and thus may collectively send many messages before communicating and coordinating their actions. Our algorithmic techniques for known networks use careful pruning of the subnetworks informed by some nodes and stopping the growth of subnetworks informed by others, depending on their age in the network and on their input value.

Our results also imply the same complexity bounds for the problem of global synchronization in locally synchronized systems with arbitrary wake up times. To get the upper bounds, our algorithms can be transformed as follows. Since our requirement for consensus stipulates that all nodes must be awake and agree on one of the input values *in the same round*, we can run the respective consensus algorithm with all input values 0 and use the round when consensus is achieved to reset all clocks to 0, thus achieving global synchronization. For the lower bounds, we indicate in each case how the argument should be modified to work for the global synchronization problem.

Due to lack of space, proofs of several results are omitted and will appear in the full version of the paper.

1.3   Related Work

Consensus is a classic problem in distributed computing, mostly studied assuming that processes communicate by shared variables or through message passing networks [4,21]. Most of the literature on consensus concerns the presence of processor faults, that can be either crash or Byzantine, starting from the seminal paper [22]; see the recent book [23] for a comprehensive survey. In [18] the authors showed a randomized consensus for crash faults with optimal communication complexity. In [11], feasibility and complexity of consensus in a multiple access channel (MAC) with simultaneous wake-up and crash failures were studied in the context of different collision detectors. Consensus (without faults) in a MAC with different wake-up times was studied in [15]. The authors also investigated the impact of global synchronization on the time efficiency of consensus. It should be noted that communication through a MAC significantly differs from our setting. First, the underlying topology of the MAC is a complete graph, unlike in our case where the topology is arbitrary. Second, one of the main problems in a MAC are message collisions that do not occur in message passing systems (cf. [6]) for which we investigate consensus. Consensus in the quantum setting has been studied, e.g., in [10].

The differences between local and global synchronization for the wake-up problem were first studied in [17] and then in [8,9,14]. The communication model used in these papers was that of radio networks in which the main challenge are collisions between simultaneously received messages. Global synchronization is often used in the study of broadcasting in radio networks (cf. [7,14]).

Computability in anonymous networks and feasibility of various distributed tasks performed using message exchange in anonymous networks have been studied, e.g., in [1,5,13,19,20,24,25,26] for arbitrary network topologies, and in [2,3,16] for rings. To the best of our knowledge, the present paper is the first to study communication complexity of consensus in arbitrary anonymous networks.

2   Unknown Networks

In this section we assume that nodes of the network do not know its topology but only have a linear bound N on the total number n of nodes. We first show a consensus algorithm that uses $O(n^2)$ messages.

2.1   Algorithm Flooding-with-Delays

Messages circulating in the network have *signatures* which are pairs $(age, value)$, where *age* is a counter set to 0 when the node initially sending the message is woken up, and incremented by 1 in each round; *value* is the input value of the node initially sending the message. Signatures are ordered lexicographically. (In each round the signature of a message changes, as its age is incremented.) Any node of degree δ woken up by the adversary creates a message μ with signature $(0, val)$, where *val* is its input value. It also switches on its *termination counter* initialized at 0, that increments by 1 in each round. The node waits δ rounds. If during this waiting time it does not obtain any message of larger signature than the *current* signature of μ, it sends μ with its current signature to all its neighbors.

A node of degree δ that obtains a message ν whose signature is larger than *current* signatures of all messages it has seen previously, resets its termination counter to 0 and switches on its *delay counter* initialized to 0. Recall that both counters increment by 1 at each round. The node waits δ rounds and if it does not obtain any message of larger signature than the current signature of ν during this waiting time, it relays message ν on all incident links. If during the waiting period some message of larger signature arrives, the delay counter and the termination counter are reset to 0 and waiting δ rounds for relay restarts. In the round when the termination counter gets to $4N - a$, where a is the age in the largest signature of any received message at its reception time, the node terminates executing the algorithm and decides on the value in this signature.   ∎

In order to prove the correctness and analyze the complexity of Algorithm Flooding-with-Delays we will use the following well known combinatorial lemma.

Lemma 1. *For any n-node graph G and for any nodes u and v in G, there exists a path $(v_1, \ldots, v_k)$ in the graph, such that $u = v_1$, $v = v_k$, and $\sum_{i=1}^{k} \delta_i \leq 3n$, where δ_i is the degree of v_i.*

Theorem 1. *Algorithm Flooding-with-Delays is correct and uses $O(n^2)$ messages.*

Proof. Consider a message μ having the largest signature and let t be the time when nodes creating message μ are woken-up by the adversary. Message μ is received by all nodes in the network within time $t + 4n$. Indeed, since no message with a signature larger than the one of μ exists in the network, any node of degree δ that receives message μ at time t' for the first time forwards it to all its neighbors at time $t' + \delta$. By Lemma 1 the sum of the degrees of nodes in a shortest path between any two nodes u and v is bounded by $3n$ and any simple path between two nodes is bounded by $n - 1$.

The correctness of the algorithm follows from the fact that all nodes agree on the value in the largest signature of a message they have ever seen and that by the time of the decision each node has seen this message. The way nodes use their termination counter guarantees that the decision is made by all nodes in the same round.

It remains to estimate the message complexity of the algorithm. A node of degree δ sends at most δ messages in any segment of δ rounds. Hence the amortized number of messages per round sent by any node is at most 1. The duration of the entire algorithm is at most $4n$ rounds, hence the total number of messages is at most $4n^2$.

It should be noted that waiting periods while flooding are a crucial tool to decrease message complexity in our algorithm. The following example shows that simple flooding without waiting can result in message complexity $\Omega(n^3)$ for some networks. Let $n = 4x$ and consider the n-node network composed of a path $(v_1, w_1, v_2, w_2, \ldots, v_x, w_x)$ of $2x$ nodes, whose extremity w_x is adjacent to x nodes forming the set S, each of which is in turn adjacent to x nodes forming the set T (the graph induced by S and T is complete bipartite). If the adversary wakes up node v_i in round i of some global time (unknown to nodes), for $i = 1, \ldots, x$, then node w_x would relay all the messages initiated at these nodes, one after another, which would result in all nodes from S relaying all these messages to all nodes from T. The total number of messages would then be $\Omega(n^3)$.

We now show that it is impossible to improve the complexity $O(n^2)$ of Algorithm Flooding-with-Delays for the entire class of n-node networks, if the topology is unknown to nodes.

Proposition 1. *For any consensus algorithm working correctly for all n-node connected networks without knowledge of topology, there exists an n-node connected network for any positive integer n, on which this algorithm requires $\Omega(n^2)$ messages.*

Proof. For simplicity we assume that n is divisible by 4. We give a proof of this proposition that holds also for the classic (weaker) version of the consensus problem. In fact, we prove that $\Omega(n^2)$ messages are needed for the clique, even if nodes know that they are in the clique, as long as the arrangement of port numbers is unknown to nodes.

Consider any consensus algorithm A for the n-node clique. The port numbers will be assigned by the adversary in such a way that for every edge the port numbers at both endpoints of the edge are equal. We will call this common port number the *color* of the edge.

Without loss of generality we may assume that any node that sends a message according to algorithm A appends to it its entire history, consisting of its input value and of the sequence of all messages received in each round since its wake-up (some of the messages could be empty), together with the color of the edge on which each message was received. Suppose that all nodes are woken up simultaneously by the adversary and each node is given value 0 or 1. The edge color on which each node sends its first message, before it got any message, depends only on its input value. Suppose that this color is i_1 for nodes with input value 0. Now suppose that integers $i_1, \ldots, i_r$ have been defined. We define the integer i_{r+1} as the $(r + 1)$-st edge color on which a node v with input value 0 sends a message, if it has the following history: in every round in which it sent a message on edge color i, it received a message from its neighbor w on this edge color and the history of w before this round is exactly the same as the history of v. Similarly we define by induction integers $j_1, j_2 \ldots$, with the only difference that in the definition input value 0 is replaced by 1. (If many new colors are used by a node in a round, they are added to the sequence in the order of increasing numbers.)

Let $n = 4m$. Partition all nodes of the clique into subsets $X = \{a_0, \ldots, a_{m-1}, b_0, \ldots, b_{m-1}\}$ and $Y = \{c_0, \ldots, c_{m-1}, d_0, \ldots, d_{m-1}\}$. Consider the scenario σ in which all nodes in X have input value 0 and all nodes in Y have input value 1. In this scenario the adversary assigns colors as follows (additions of indices are modulo m): pairs of nodes a_i and b_i, for $0 \leq i \leq m - 1$, are joined by an edge of color i_1, pairs of nodes a_i and b_{i+1} are joined by an edge of color i_2, and in general, pairs of nodes a_i and b_{i+c} are joined by an edge of color i_{c+1}. A similar coloring is done in the set Y using colors $j_1, j_2 \ldots$ instead of $i_1, i_2 \ldots$, and then the adversary colors all other edges arbitrarily. (Notice that the subgraphs induced by the node sets X and Y are fully symmetric.) We claim that in this scenario at least m^2 messages have to be sent. Suppose not. By induction on the round number, all nodes in X have identical history until the round in which they send a message on an edge of color i_m , and all nodes in Y have identical history until the round in which they send a message on an edge of color j_m. Hence, by the definition of numbers $i_1, i_2 \ldots$ and $j_1, j_2 \ldots$, the order of colors on which nodes from X and Y send messages (disregarding repetitions of already used colors) is $i_1, i_2 \ldots$ and $j_1, j_2 \ldots$, respectively. Since we assumed that fewer than m^2 messages were sent, none of the nodes in X could get to sending a message on an edge of color i_m and none of the nodes in Y could get to sending a message on an edge of color j_m. It follows that no communication occurred between any node of X and any node of Y. Suppose that algorithm A reaches consensus on input value 1 in scenario σ (the case of input value 0 is symmetric and thus omitted from the proof). Now consider the scenario σ_0 in which all nodes have input value 0. In this scenario, the adversary assigns edge colors according to the sequence $i_1, i_2, \ldots$ to both sets X and Y. Nodes in X and Y, however, have the exact same history in scenario σ_0 as nodes in X in scenario σ, up to the round when they have to reach consensus on input value 1. Hence, all nodes would reach consensus on input value 1 in scenario σ_0 as well, which contradicts validity, as all nodes in scenario σ_0 have input value 0.

In the case when n is not divisible by 4, the proof can be easily adapted using an $n - (n \mod 4)$ node clique as before, using the remaining $(n \mod 4)$ nodes as *dummy* nodes that are left dormant by the adversary. Each node in partitions X and Y of the clique is connected to the dummy nodes by ports that are not in $\{i_1, \ldots, i_m\} \cup \{j_1, \ldots, j_m\}$. Port numbers at the dummy nodes are arbitrary.

The following simple modification of the above proof allows to get the same lower bound for global synchronization. The adversary wakes up simultaneously only nodes of the set X. The same argument shows that $\Omega(n^2)$ messages have to be sent before any node in Y is woken up.

3   Known Networks

In this section we assume that each node is provided with a map of the network, which is an isomorphic copy of it containing all port numbers, with the location of the given node marked in the map. We will show that in this setting the complexity of consensus decreases significantly with respect to the scenario of unknown networks. We start considering fully symmetric networks, and we later extend our algorithm to handle arbitrary networks.

3.1   Fully Symmetric Networks

Algorithm Span-and-Prune, presented below, achieves consensus in fully symmetric networks. In any such network G, for any pair of nodes u, v, there exists a port-preserving automorphism of G that carries u to v. Hence different behavior of nodes can only occur if the histories of these nodes are different, which may be caused by different wake-up times or input values. The overall idea of our algorithm is to first find a sparse spanner of the network, using the available map, and then grow a spanning forest of this spanner, where each tree is rooted at a node awaken by the adversary. The reason why we grow a forest instead of a single tree is that two trees rooted at nodes having the same value and the same wake-up time may be perfectly symmetric, making it impossible to chose which one should be killed and which one should survive. At the end, every node is in some tree, such that all roots have the same input value and the same wake-up time. All nodes terminate in the same round, agreeing on this input value. We start with the following lemma.

Lemma 2. *Let G be a fully symmetric connected n-node graph. Then there exists a fully symmetric connected spanning subgraph G' of G with at most $n\lfloor \log n \rfloor$ edges.*

Lemma 2 implies that it is always possible to fix a permutation π of port numbers that results in the construction of a fully symmetric, connected spanning subgraph G' of G, when the first $\lfloor \log n \rfloor$ ports are selected by each node. Hence from now on we can assume that nodes of G have degree bounded by $\lfloor \log n \rfloor$.

Similarly as in Algorithm Flooding-with-Delays, messages sent by nodes during the execution of Algorithm Span-and-Prune have *signatures* which are pairs $(age, value)$. As before, *age* is a counter set to 0 when the node initially sending the message is woken up, and incremented by 1 in each round, and *value* is the input value of the node initially sending the message. Signatures of different messages are compared lexicographically based on their *current* age.

Define the *code* of a path $P = (v_1, v_2, \ldots, v_k)$ in the graph G as the sequence $(p_1, p_2, \ldots, p_k)$, where p_i is the port number at node v_i, corresponding to the edge (v_i, v_{i+1}) in P.

Define a spanning tree T_d of G, rooted at a node u, according to the following rule.

Rule 1. *Let P be the path in T_d connecting node u to a node v, and let $(p_1, p_2, \ldots, p_k)$ be the code of P. Then P is a shortest path between u and v in G and $(p_1, p_2, \ldots, p_k)$ is the lexicographically smallest code of any shortest path connecting u to v.*

Notice that Rule 1 defines a unique spanning tree rooted at u, for any node u, i.e., the spanning tree corresponding to a breadth-first search performed following the increasing order of port numbers.

Each node v in the tree T_d, rooted at node u, is assigned a unique number $r_u(v)$, in $[0, n-1]$, according to the following rule.

Rule 2. $r_u(v) < r_u(v')$, *if and only if, the path P connecting u to v in T_d is shorter than the path P' connecting u to v' in T_d, or P and P' have the same length but the code of P is lexicographically smaller than the code of P'.*

The assignment of ranks to nodes in T_d is performed according to a breadth-first visit of the tree, visiting children in increasing order of port numbers. Notice that neither the removal of edges that bounds node degrees in G' by $\lfloor \log n \rfloor$ nor the design of the spanning trees and local numbering of nodes require any message exchange, as both tasks can be performed independently by each node using the map.

Algorithm Span-and-Prune

Any node u woken up by the adversary computes the permutation π, designs the spanning tree T_d rooted at u, computes ranks of all nodes in T_d, and initializes its age counter to 0.

Algorithm Span-and-Prune proceeds in phases. Each phase lasts $2n^2$ rounds and is divided in two parts: Part 1 uses rounds from 1 to n^2 and Part 2 uses rounds from n^2+1 to $2n^2$. In phase i, the root u having a *designed* tree T_d tries to *conquer* all nodes in T_d having ranks in $[2^{i-1}, 2^i-1]$. When a node v is conquered, it becomes part of the *conquered* tree T_c of u. If node u is unable to conquer some node v during a phase, either its designed tree T_d is pruned or it is completely destroyed. Pruning is done when the conquest of node v failed because v has been already included in a tree T'_c of another node u', whose messages have *the same* signature as those of u. After pruning, the whole subtree rooted in v is removed from T_d. The tree T_d is destroyed when it meets another tree rooted at a node whose messages have *larger* signatures. Note that different nodes in the network could be running different phases (or different rounds of the same phase), due to different activation times.

Part 1. Rounds from 1 to n^2 of phase i are used to try to conquer nodes having ranks in $[2^{i-1}, 2^i - 1]$. In particular, if $r_u(v) = 2^{i-1} + j$ and the parent w of v in T_d is in T_c, then a *conquer* message is sent from w to v in round $nj + 1$, unless node w received a message with a larger signature than those originated by tree T_c in some previous round. Conquer messages contain the subtree of T_d rooted at the node to be conquered, annotated with node ranks. Node v is included in T_c if the signature of the message it receives from w is the largest it has ever received. If by the time when it receives the message from w, node v already received a message with a larger signature (i.e., either it received it in some previous round, or in the same round as the message from w), it notifies w that tree T_c must be destroyed, by sending it a *kill* message. If node v already belongs to a tree T'_c whose messages have the same signature as those from tree T_c, it notifies w that the conquest of v by tree T_c failed. This is done by sending a *prune* message from v to w. Both the sender and the receiver of a prune message save the edge connecting them in memory, calling it a *connecting edge*. As a consequence of the failed conquest of node v, the tree T_d is pruned by removing the entire subtree rooted at v. Here we are not specifying how to give precedence to conquer messages coming from trees with the same signature in the same round. The omission is legitimate since we can prove that this event is impossible.

If a node v in a tree T_c is subsequently conquered by a tree T'_c (whose messages have a larger signature), node v reports this event in the following round by sending a kill message to all its neighbors in T_c and through all its connecting edges. Then it removes the connecting edges from its memory. A node that receives for the first time in some round a kill message having a signature larger than the one of its tree T_c, relays this message in the following round to all its neighbors in T_c and to all its connecting edges. Then it removes the connecting edges from its memory.

In Part 2 of each phase only kill messages are transmitted.

A node terminates executing the algorithm and decides on the value from the signature of messages coming from the root of the tree to which it belongs, when the age in the signature of these messages reaches $2n^2 \lceil \log n \rceil$.   ■

In order to prove the correctness of Algorithm Span-and-Prune we will use the following lemma.

Lemma 3. *Let M be the set of nodes generating messages with the largest signature s in G. Let v be a node in G that does not belong to M. Let m be the node in M that minimizes $r_{m'}(v)$, for $m' \in M$. Finally, let T_c be the conquered tree of m at the end of the execution of algorithm Span-and-Prune. Then v belongs to T_c.*

Theorem 2. *Algorithm Span-and-Prune reaches consensus, in an arbitrary fully symmetric n-node network G, exactly after $2n^2 \lceil \log n \rceil$ rounds.*

Proof. By Lemma 3, after $2n^2 \lceil \log n \rceil$ rounds, each node belongs to a tree T_c rooted at a node that generates messages with the largest signature. Hence all nodes terminate in the same round, and agree on the input value in this signature.

The rest of this subsection is devoted to the analysis of the communication complexity of Algorithm Span-and-Prune.

Consider a round t on the global clock of the adversary (t is unknown to the nodes). Two conquered trees T_c and T'_c *met* if one tried to conquer a node already conquered by the other in a round $t' \leq t$. A *component* is a forest of conquered trees. Two trees T_c and T'_c belong to the same component, if and only if, their roots have the same signature s and there exists a sequence $(T_c, T_1, \ldots, T'_c)$ of trees with signature s such that consecutive trees in this sequence met. Two components are *separated*, if no tree inside one component met a tree inside the other. The set of edges of a component is the union of all edges of the trees inside the component, together with connecting edges (traversed by messages that made pair of trees inside the component meet). We say that a component is *alive* in round t, if none of its trees met a tree whose messages have a larger signature. A component is *dying* in round t, if one or more of its trees met a tree whose messages have a larger signature. A component is *dead* when all its nodes have received a kill message. Since kill messages flood a component in at most n rounds, a component that is dying in a round i becomes dead by round $i + n$.

Lemma 4. *The total number of messages sent in an execution of Algorithm Span-and-Prune in an n-node fully symmetric network G is in $O(n^{3/2} \log^2 n)$.*

Proof. We will show that, for some constant c and an arbitrary round t (on the global clock of the adversary), at most $cn^{3/2} \log n$ messages are sent in G in rounds $[t, t + n^2)$. Since Algorithm Span-and-Prune terminates in $O(n^2 \log n)$ rounds, this implies the lemma.

Fix a round t and consider a snapshot of the network G in this round. Since components that were already dead in round t do not generate any messages in subsequent rounds, all messages sent in rounds $[t, t + n^2)$ are either due to dying components or to alive components.

Claim 1. *The number of messages sent in the network G in rounds $[t, t + n^2)$ due to dying components is in $O(n \log n)$.*

Proof of Claim 1. Let $D_1, D_2, \ldots, D_h$ be the dying components in round t. Since a dying component becomes dead within n rounds, all dying components were separated in round $t - n$. In n rounds, a component can conquer at most as many nodes as its size. Indeed, each tree inside the component can only conquer one node in n rounds and there cannot be more trees than nodes inside a component. Hence $|D_1| + |D_2| + \ldots + |D_h| \leq 2n$. Notice that the sum of sizes of dying components can grow above n, since nodes conquered by other trees are still counted in the dying component. Assume that each of the trees in the dying components conquers a new node before becoming dead. Hence the total number of nodes that the dying components can conquer before all of them die is bounded by $3n$. The number of messages sent in the network in rounds $[t, t+n^2)$ due to dying components is given by the sum of the number of messages sent for expanding these components plus the sum of the number of messages sent for killing them. Conquering one node for each tree costs as many messages as the number of trees. Hence at most n messages can be sent for conquering new nodes. Each node whose conquest failed would send either a kill or a prune message, for at most n additional messages.

Kill messages for a given component can travel at most twice along each edge of the component. Since the sum of numbers of dying components edges is bounded by $3n \log n$, at most $6n \log n$ kill messages can be sent due to dying components, which completes the proof of the claim.

Claim 2. *The number of messages sent in the network G in rounds $[t, t + n^2)$ due to alive components is in $O(n^{3/2} \log n)$.*

Proof of Claim 2. Let $S_1, S_2, \ldots, S_k$ be the alive components in round t. Let ϕ_i be the last phase whose Part 1 was completed by component S_i in round t. Hence each round in the segment $[t, t + n^2)$ is either in Part 2 of phase ϕ_i or in phase $\phi_i + 1$, for component S_i. If S_i and S_j are two distinct components whose nodes have the same signature, then S_i and S_j are separated by definition. If a tree in one component met a tree in another component whose nodes have a different signature, then one of the two components would be dying in round t. It follows that components $S_1, S_2, \ldots, S_k$ are pairwise separated. Hence $\sum_{i=1}^{k} |S_i| \leq n$. Moreover, for any $i \in [1, k]$, $S_i \geq 2^{\phi_i}$. Indeed, a single tree T_c in S_i would have grown to size 2^{ϕ_i} and any node that T_c failed to conquer must be in another tree $T_c' \in S_i$.

Let $\hat{S}_i$ be the set of nodes conquered by S_i by round $t + n^2$. Clearly $|\hat{S}_i| \le n$, since no component can grow outside of the graph G. Moreover, $|\hat{S}_i| \le 2^{\phi_i+1}|S_i| \le 2|S_i|^2$. Indeed, each tree T_c in S_i can grow at most to size 2^{ϕ_i+1} during phase $\phi_i + 1$ and there are at most $|S_i|$ trees in S_i.

Let S be the number of nodes conquered by an alive component by round $t + n^2$. Then at most $S + 4S \log n$ messages were sent in the network in rounds $[t, t + n^2)$ due to this component. Indeed each node that is conquered requires only 1 message, and each internal node can fail to conquer at most $\log n$ external nodes, totalling in less than $S + 2S \log n$ messages (at most S messages for conquering, at most $S \log n$ failed conquests each of which costs 1 message to attempt the conquest and one kill or prune message sent back by the unconquered node). Kill messages (sent in the case when some tree in the component met some other tree whose messages have a larger signature) are bounded by twice the number of component edges. This number is in turn bounded by $S \log n$.

If follows that the total number of messages sent due to alive components in rounds $[t, t + n^2)$ is upper bounded by $\sum_{i=1}^{k} |\hat{S}_i| + 4|\hat{S}_i| \log n < 8 \log n \sum_{i=1}^{k} |\hat{S}_i|$, while the following conditions hold.

(1) $\sum_{i=1}^{k} |S_i| \le n$. (2) $\forall i \in [1, k]$ $|\hat{S}_i| \le 2|S_i|^2$. (3) $\forall i \in [1, k]$ $|\hat{S}_i| \le n$.

Let $A = \{i \le k : |S_i| \le \sqrt{n/2}\}$ and let $B = \{i \le k : |S_i| > \sqrt{n/2}\}$.

We have that $\sum_{i=1}^{k} |\hat{S}_i| = \sum_{i \in A} |\hat{S}_i| + \sum_{i \in B} |\hat{S}_i| \le \sum_{i \in A} |\hat{S}_i| + |B|n$, where the second inequality follows from condition (3). Condition (1) implies $|B| \le \sqrt{2n}$. Moreover, by condition (2) we have that $\sum_{i \in A} |\hat{S}_i| \le \sum_{i \in A} 2|S_i|^2$, which, under condition (1), is maximized if $|A| = \sqrt{2n}$ and $|S_i| = \sqrt{n/2}$, for all $i \in A$. Hence $\sum_{i \in A} |\hat{S}_i| + \sum_{i \in B} |\hat{S}_i| \le 2n\sqrt{2n} < 4n^{3/2}$, which proves the claim.

Claims 1 and 2 show that $O(n^{3/2} \log n)$ messages are sent in any segment of n^2 rounds. Since the total number of rounds used by Algorithm Span-and-Prune is in $O(n^2 \log n)$, this concludes the proof.

3.2   Arbitrary Networks

Algorithm Extended Span-and-Prune, presented below, achieves consensus in arbitrary networks.

For each node u, assign a unique number in $[0, n - 1]$ to each node in G according to a breadth-first visit of the graph G starting from node u. Neighbors of a node v are visited according to the increasing order of port numbers, at node v, of the edges connecting them to v. Assign a label ℓ_u to each node u as follows. Perform a breadth-first visit of the graph G; the visit starts from u, and the first term of ℓ_u is 0. Let v be the current node in the visit and let $v_1, v_2, \ldots, v_\delta$ be the nodes connected to v by edges having port numbers $1, 2, \ldots, \delta$ at v, respectively. Let r_i, be the number assigned to v_i. For i going from 1 to δ, append r_i to ℓ_u. Clearly, two nodes in the same isomorphism class are assigned the same label by the above procedure. On the other hand, if two nodes u and v are assigned the same label $\ell_u = \ell_v$, then a port-preserving automorphism of G mapping u to v

can be constructed by mapping node with number r_i in the breadth-first visit of G starting from u to the node having the same number in the breadth-first visit of G starting from v. It follows that nodes in different isomorphism classes are assigned different labels. This in turn implies that it is possible to uniquely define the class $C = \{c_1, c_2, \ldots, c_s\}$ which is the isomorphism class corresponding to the lexicographically smallest label ℓ. Nodes in C will be called *chiefs*.

Each node u not in C selects as its chief the node in C closest to u, and among those with minimum distance, it selects the one with the shortest path of minimum code (see the previous subsection for the definition of a code). For each node having chief c_i (including c_i itself) construct a spanning tree of the whole network, rooted at c_i, according to Rule 1, assign ranks to nodes in this spanning tree according to Rule 2, and prune it of all nodes that have smaller ranks in the spanning tree rooted at another chief c_j. At the end of this process, each resulting tree T_i contains exactly one node from each isomorphism class (n/s nodes for each tree). The trees rooted at chiefs $c_1, \ldots, c_s$ are disjoint and constitute a spanning forest of G.

Construct a s-node graph S (in general not simple) as follows. Nodes of S are the trees T_i and there is an edge e' between T_i and T_j in S, if and only if, there are two nodes u and v, respectively in T_i and T_j, that are connected by an edge e in G. Edge e' in S is labeled with the set of pairs $\{(r_{c_i}(u), p_u(e)), (r_{c_j}(v), p_v(e))\}$, where $r_{c_i}(u)$ and $r_{c_j}(v)$ are the ranks of the endpoints of the edge e in the trees of their respective chiefs and $p_u(e)$ and $p_v(e)$ are the port numbers, respectively, at endpoints u and v of edge e. The graph S is fully symmetric. Hence, by Lemma 2, a fully symmetric spanning subgraph S' of S can be constructed, having at most $s\lfloor \log s \rfloor$ edges. (Notice that Lemma 2 holds for non simple graphs as well.)

Consensus on S can be achieved by applying Algorithm Span-and-Prune. Algorithm Extended Span-and-Prune simulates an execution of Algorithm Span-and-Prune on network S in order to achieve consensus on network G. Below is a detailed description of the algorithm.

Algorithm Extended Span-and-Prune

Each node, when woken up, computes the isomorphism classes, identifies its chief (or selects itself as a chief) and constructs the rooted tree T_i to which it belongs. No message exchange is needed to perform these tasks, as each node can perform this computation locally using its map of the network.

Nodes in a tree T_i woken up by the adversary send their value to their parent in T_i. These nodes never relay any value from other nodes in their tree. Nodes in T_i woken up by a message containing the value from another node in T_i only relay to their parent one value (i.e., the largest one they received in the round they got woken up). The root c_i decides on a value which is among those first obtained (i.e., its own value in the case when it is woken up by the adversary, or the largest value received in the first non silent round). Once the chief c_i of a tree is woken up and has decided on a value, c_i informs all nodes in its tree T_i (exactly n/s rounds are allotted to this task). Then all nodes in T_i set their age counters to zero and T_i starts the simulation of Algorithm Span-and-Prune. Let τ_d and τ_c be respectively the designed and the conquered trees (these are subtrees of the graph S', hence their nodes are the trees T_i) built during the simulation of Algorithm Span-and-Prune. Each round of the simulated algorithm takes n/s rounds. Hence, each phase takes $2s^2 \cdot (n/s) = 2ns$ rounds. A message sent by node T_i in round t of phase i over an edge e' having label $\{(r_{c_i}(u), p_u(e)), (r_{c_j}(v), p_v(e))\}$, is simulated by sending the same message from node u to node v over edge e in round $(n/s) \cdot t$ of simulated phase i. Messages are relayed by the receiving node to all neighbors in its tree T_i and relayed (inside T_i) in consecutive rounds, thus flooding the whole tree in at most n/s rounds. Hence all nodes in T_i are informed of the received message by the time when one of them has to send another message in the next simulated round.

A node terminates executing the algorithm and decides on the value from the signature of messages coming from the root of the tree τ_c to which it belongs, when the age in the signature of these messages reaches $2ns\lceil \log s \rceil$. ∎

Theorem 3. *Let G be an arbitrary n-node network. Algorithm Extended Span-and-Prune achieves consensus in G using $O(n^{3/2} \log^2 n)$ messages.*

3.3   Lower Bound

We now establish a lower bound on the message complexity of consensus in known networks. The result also holds for the classic, weaker version of the consensus problem.

Theorem 4. *For every positive integer k and for $n = 2^k$, there exists a n-node network for which every consensus algorithm requires $\Omega(n \log n)$ messages.*

4   Do Labels Help?

In this section we answer the question whether the availability of distinct labels of nodes, with each node knowing its label, permits to decrease the complexity of consensus with respect to the anonymous setting. It turns out that the answer is negative for unknown networks and positive for known networks. First observe that if the network is unknown, then a slight modification of the argument for anonymous networks from the proof of Proposition 1 gives a lower bound $\Omega(n^2)$ on message complexity of consensus for the following class of n-node networks, even if all nodes have distinct labels. A network in this class is defined as follows. Take two cliques on disjoint sets A and B, each of size $\Theta(n)$. Replace a pair of edges $\{a, a'\}$ in A and $\{b, b'\}$ in B by the pair of "bridges" $\{a, b\}$ and $\{a', b'\}$. For any consensus algorithm, the adversary can label ports, so as to delay message transmissions through both bridges until $\Omega(n^2)$ messages have been sent.

On the other hand, if the (labeled) network is known to the nodes, i.e., if a labeled map of the network (an isomorphic copy of it with all node labels marked) is available to all nodes, then consensus can be done more efficiently than in the anonymous setting with known network. Recall that in the anonymous setting with known network, some networks required $\Omega(n \log n)$ messages for consensus. By contrast, in the labeled setting we have the following proposition.

Proposition 2. *If all nodes have distinct labels and are provided with a labeled map of the network, then consensus can be done with $O(n)$ messages for n-node networks. This complexity is optimal.*

Proof. It is straightforward that at least $n-1$ messages have to be used (even for the classic, weaker version of consensus, and also for establishing global time). In order to give a consensus algorithm using $O(n)$ messages, first observe that all nodes can find a common rooted spanning tree T without any message exchange, by choosing the node with the largest label as the root and applying a fixed spanning tree construction procedure. Once the tree T is fixed, consensus can be achieved as follows. Any node woken up by the adversary sends its input value to its parent in T. Any node other than the root relays only one value to its parent (the one it got first and if it got many values simultaneously first, then it relays the largest of them). The root adopts the value received first (if it was woken up by the adversary, this is its own input value), and if it got many values simultaneously first, then it adopts the largest of them). Then the root sends the adopted value down the tree and consensus is made on this value. Together with the value, the root sends a message "consensus will be achieved in x rounds" with the counter x initialized to n. At each transmission the counter is decreased by 1, and nodes use their local clocks to make the agreement in the same round. At most two messages travel on each edge of the tree: one up and one down. Hence the number of messages is at most $2n - 2$.

5 Conclusion

We gave bounds on the message complexity of consensus in anonymous message passing systems. For unknown networks our bounds are tight and give $\Theta(n^2)$ complexity. For known networks we showed that the complexity of consensus is significantly smaller. In this scenario our bounds differ by a factor $\sqrt{n}\log n$: the upper bound is $O(n^{3/2}\log^2 n)$ while the lower bound is $\Omega(n\log n)$. Closing this gap is a natural open problem.

References

1. Angluin, D.: Local and global properties in networks of processors. In: Proc. 12th Annual ACM Symposium on Theory of Computing (STOC), pp. 82–93 (1980)
2. Attiya, H., Snir, M., Warmuth, M.: Computing on an Anonymous Ring. Journal of the ACM 35, 845–875 (1988)
3. Attiya, H., Snir, M.: Better Computing on the Anonymous Ring. Journal of Algorithms 12, 204–238 (1991)
4. Attiya, H., Welch, J.: Distributed Computing. John Wiley and Sons, Inc. (2004)
5. Boldi, P., Vigna, S.: Computing anonymously with arbitrary knowledge. In: Proc. 18th ACM Symp. on Principles of Distributed Computing (PODC), pp. 181–188 (1999)
6. Burns, J.E.: A formal model for message passing systems, Tech. Report TR-91, Computer Science Department. Indiana University, Bloomington (September 1980)

7. Chlebus, B.S., Gasieniec, L., Gibbons, A., Pelc, A., Rytter, W.: Deterministic broadcasting in ad hoc radio networks. Distributed Computing 15, 27–38 (2002)
8. Chlebus, B.S., Gąsieniec, L., Kowalski, D.R., Radzik, T.: On the Wake-Up Problem in Radio Networks. In: Caires, L., Italiano, G.F., Monteiro, L., Palamidessi, C., Yung, M. (eds.) ICALP 2005. LNCS, vol. 3580, pp. 347–359. Springer, Heidelberg (2005)
9. Chlebus, B.S., Kowalski, D.R.: A better wake-up in radio networks. In: Proc. 23rd ACM Symposium on Principles of Distributed Computing (PODC), pp. 266–274 (2004)
10. Chlebus, B.S., Kowalski, D.R., Strojnowski, M.: Scalable Quantum Consensus for Crash Failures. In: Lynch, N.A., Shvartsman, A.A. (eds.) DISC 2010. LNCS, vol. 6343, pp. 236–250. Springer, Heidelberg (2010)
11. Chockler, G., Demirbas, M., Gilbert, S., Lynch, N.A., Newport, C.C., Nolte, T.: Consensus and collision detectors in radio networks. Distributed Computing 21, 55–84 (2008)
12. Chrobak, M., Gasieniec, L., Kowalski, D.R.: The wake-up problem in multi-hop radio networks. In: Proc. 15th ACM-SIAM Symposium on Discrete Algorithms (SODA), pp. 985–993 (2004)
13. Codenotti, B., Gemmell, P., Simon, J.: Symmetry breaking in anonymous networks: characterizations. In: Proc. 4th Israel Symposium on Theory of Computing and Systems (ISTCS 1996), pp. 16–26 (1996)
14. Czumaj, A., Rytter, W.: Broadcasting algorithms in radio networks with un- known topology. In: Proc. 44th IEEE Symposium on Foundations of Computer Science (FOCS 2003), pp. 492–501 (2003)
15. Czyzowicz, J., Gąsieniec, L., Kowalski, D.R., Pelc, A.: Consensus and Mutual Exclusion in a Multiple Access Channel. In: Keidar, I. (ed.) DISC 2009. LNCS, vol. 5805, pp. 512–526. Springer, Heidelberg (2009)
16. Diks, K., Kranakis, E., Malinowski, A., Pelc, A.: Anonymous wireless rings. Theoretical Computer Science 145, 95–109 (1995)
17. Gasieniec, L., Pelc, A., Peleg, D.: The wakeup problem in synchronous broadcast systems. SIAM Journal on Discrete Mathematics 14, 207–222 (2001)
18. Gilbert, S., Kowalski, D.R.: Distributed agreement with optimal communication complexity. In: Proc. 21st ACM-SIAM Symposium on Discrete Algorithms (SODA 2010), pp. 965–977 (2010)
19. Kranakis, E.: Symmetry and Computability in Anonymous Networks: A Brief Survey. In: Proc. 3rd Int. Conf. on Structural Information and Communication Complexity, pp. 1–16 (1997)
20. Kranakis, E., Krizanc, D., van der Berg, J.: Computing Boolean Functions on Anonymous Networks. Information and Computation 114, 214–236 (1994)
21. Lynch, N.A.: Distributed Algorithms. Morgan Kaufmann Publ., Inc. (1996)
22. Pease, M.C., Shostak, R.E., Lamport, L.: Reaching agreement in the presence of faults. J. ACM 27, 228–234 (1980)
23. Raynal, M.: Fault-Tolerant Agreement in Synchronous Distributed Systems. Morgan & Claypool Publishers (2010)
24. Sakamoto, N.: Comparison of Initial Conditions for Distributed Algorithms on Anonymous Networks. In: Proc. 18th ACM Symp. on Principles of Distributed Computing (PODC 1999), pp. 173–179 (1999)
25. Yamashita, M., Kameda, T.: Computing on anonymous networks. In: Proc. 7th ACM Symp. on Principles of Distributed Computing (PODC 1988), pp. 117–130 (1988)
26. Yamashita, M., Kameda, T.: Computing on anonymous networks: Part I - characterizing the solvable cases. IEEE Trans. Parallel and Distributed Systems 7, 69–89 (1996)

Non-blocking k-ary Search Trees

Trevor Brown[1] and Joanna Helga[2]

[1] Theory Group, Dept. of Computer Science, University of Toronto
tabrown@cs.toronto.edu
[2] DisCoVeri Group, Dept. of Computer Science and Engineering, York University
helga@yorku.ca

Abstract. This paper presents the first concurrent non-blocking k-ary search tree. Our data structure generalizes the recent non-blocking binary search tree of Ellen et al. [5] to trees in which each internal node has k children. Larger values of k decrease the depth of the tree, but lead to higher contention among processes performing updates to the tree. Our Java implementation uses single-word compare-and-set operations to coordinate updates to the tree. We present experimental results from a 16-core Sun machine with 128 hardware contexts, which show that our implementation achieves higher throughput than the non-blocking skip list of the Java class library and the leading lock-based concurrent search tree of Bronson et al. [3].

Keywords: data structures, non-blocking, concurrency, binary search tree, set.

1 Introduction

With the arrival of machines with many cores, there is a need for efficient, scalable linearizable concurrent implementations of often-used abstract data types (ADTs) such as the set. Most existing concurrent implementations of the set ADT are lock-based (e.g., [3,9]). However, locks have some disadvantages (see [7]). Other implementations use operations not directly supported by most hardware, such as load-link/store-conditional [2] and multi-word compare-and-swap (CAS) [8]. Software transactional memory (STM) has been used to implement the set ADT (e.g., [10]), but this approach is currently inefficient [3].

Most multicore machines support (single-word) CAS operations. Non-blocking implementations of dictionaries have been given based on skip lists and binary search tree structures. Sundell and Tsigas [12], Fomitchev and Ruppert [6], and Fraser [8] have implemented a skip list using CAS operations. A binary search tree implementation using only CAS operations was sketched by Valois [13], but the first complete algorithm is due to Ellen et al. [5]. The non-blocking property ensures by definition that, while a single operation may be delayed, the system as a whole will always make progress. (Some refer to this property as lock-freedom.)

In this paper, we generalize the binary search tree of Ellen et al. (BST) to a k-ary search tree (k-ST) in which nodes have up to $k-1$ keys and k children. This

A. Fernández Anta, G. Lipari, and M. Roy (Eds.): OPODIS 2011, LNCS 7109, pp. 207–221, 2011.
© Springer-Verlag Berlin Heidelberg 2011

requires generalizing the existing BST update operations to k-ary trees, creating new kinds of updates to handle insertion and deletion of keys from nodes, and verifying that the coordination scheme works with the new updates. Using larger values of k decreases the average depth of nodes, but increases the local work done at each internal node in routing searches and performing updates to the tree. However, the increased work at each node is offset by the improved spatial locality offered by larger nodes. By varying k, we can balance these factors to suit a particular system architecture, expected level of contention, or ratio of updates to searches. Searches are extremely simple and fast. Oblivious to concurrent updates, they behave exactly as they would in the sequential case.

We have implemented both the BST and our k-ST in Java, and have compared these implementations with ConcurrentSkipListMap (SL) of the Java class library, and the lock-based AVL tree of Bronson et al. (AVL) [3]. The AVL tree is the leading concurrent search tree implementation. It has been compared in [3] with SL, a lock-based red-black tree, and a red-black tree implemented using STM. Since SL and AVL drastically outperform the red-black tree implementations, we have not included the latter in our comparison. In our experiments, the BST and 4-ST (k-ST with $k = 4$) algorithms are top performers in both high and low contention cases. We did not observe significant benefits when using values of $k > 4$, but we expect this would change with algorithmic improvements to the management of keys within nodes. This paper also provides the first performance data for the BST of Ellen et al. [5].

The BST and k-ST are both unbalanced trees. All performance tests in this paper use uniformly distributed random keys. If keys are not random then, in certain cases, SL (which uses randomization to maintain balance) and AVL (a balanced tree) will take the lead. Extending the techniques in this paper to provide balanced trees is the subject of current work.

2 k-ary Search Trees

2.1   The Structure

We use a leaf-oriented, non-blocking k-ST to implement the set ADT. A set stores a set of keys from an ordered universe. It does not admit duplicate keys. Here, we define the operations on the ADT to be FIND(key), INSERT(key), and DELETE(key). The FIND operation returns TRUE if key is in the set, and FALSE otherwise. An INSERT(key) operation returns FALSE if key was already present in the set. Otherwise, it adds key to the set and returns TRUE. A DELETE(key) returns FALSE if key was not present. Otherwise, it removes key and returns TRUE. The other implementations we compare to the k-ST and BST can additionally associate a value with each key, and it is a simple task to modify our structure to do so (as discussed in [4]).

The k-ST is leaf-oriented, meaning that at all times, the keys in the set ADT are the keys in the leaves of the tree. Keys in internal nodes of the k-ST serve only to direct searches down the tree.

Each leaf in a BST contains one key. Each internal node has exactly two children and one key. In our k-ST, each leaf has at most $k-1$ keys. It is permitted for a leaf to have zero keys, in which case it is said to be an empty leaf. Each internal node has exactly k children and $k-1$ keys. Inside each node, keys are maintained in increasing order.

The search tree property for k-STs is a natural generalization of the familiar BST property. For any internal node with keys $a_1, a_2, ..., a_{k-1}$, sub-tree 1 (left-most) contains keys $a < a_1$, sub-tree k (rightmost) contains keys $a > a_{k-1}$, and sub-tree $1 < i < k$ contains keys a with $a_i \leq a < a_{i+1}$.

2.2   Modifications to the Tree

We first describe a sequential implementation of the set operations, and subsequently transform it into a concurrent and non-blocking implementation. Since the k-ST is leaf-oriented, the INSERT and DELETE procedures always operate on leaves. Inserting a key into the set replaces a leaf by a larger leaf (with one more key), or by a new sub-tree if the leaf is full (has $k-1$ keys). Deleting a key replaces a leaf by a smaller leaf (without the deleted key), or prunes the leaf and its parent out of the tree.

More precisely, the operation INSERT(key) first searches for key. If it is found, the INSERT returns FALSE. Otherwise, it proceeds according to two cases as follows (see Fig. 1). Let l be the leaf into which key should be inserted. If l is full (has $k-1$ keys) then INSERT replaces l by a *newly created* sub-tree of $k+1$ nodes. This sub-tree consists of an internal node n whose keys are the $k-1$ greatest out of the $k-1$ keys in l and the new key key. The children of n are k new nodes, each containing one of the k aforementioned keys. We call this first type of insertion a *sprouting insertion*. Otherwise, if l has fewer than $k-1$ keys, INSERT simply replaces l by a *new* leaf that includes key in addition to all of the keys that were in l. We call this second type of insertion a *simple insertion*.

The operation DELETE(key) first searches for key. If it is not found, then FALSE is returned. Otherwise, it proceeds according to two cases (see Fig. 1). Let l be the leaf from which key should be deleted. If l has only one key and the parent of l has exactly two non-empty children, then the entire leaf l can

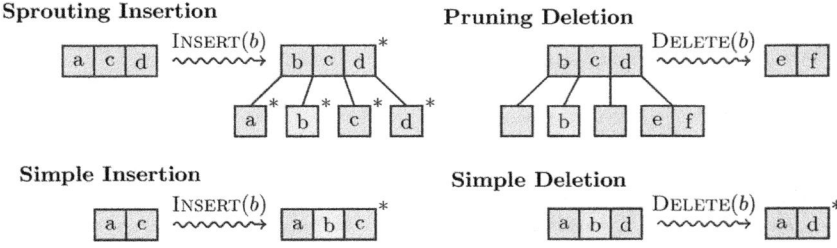

Fig. 1. The four types of modifications performed on the tree by an insertion or deletion. Asterisks indicate that nodes are newly created in freshly allocated memory.

be deleted (since it will be empty after the deletion) and, because it has only one non-empty sibling s, the parent node is no longer useful (since its keys just direct searches). Thus, the DELETE procedure simply replaces the parent with s. We call this first type of deletion a *pruning deletion*. Otherwise, if l has more than one key or the parent of l has more than two non-empty children, DELETE replaces l by a *new* leaf with all of the keys of l except for key. We call this second type of deletion a *simple deletion*. Simple deletion can yield empty leaves. However, with this insertion and deletion scheme an internal node always has at least two non-empty children. Note that if NULL were used instead of empty children, then the ABA problem would occur on child pointers.

Note that a pruning deletion changes a child pointer of the grandparent of l to point to l's only non-empty sibling. To avoid dealing with degenerate cases when there is no parent or grandparent of l, we initialize the tree with two dummy internal nodes and $2k - 1$ empty leaves at the top, as shown in Fig. 2(a). These internal nodes will not be deleted or replaced by an insertion. When $k = 2$, our algorithm is simply the BST of Ellen et al. [5], with some slight modifications, where all insertions and deletions are sprouting insertions and pruning deletions, except for an INSERT into an empty tree and a DELETE on the last key in a tree.

2.3   Coordination between Updates

Without some form of coordination, interactions between concurrent updates would produce incorrect results. Suppose that a pruning deletion and a simple insertion are performed concurrently in the 2-ary tree on the left in Fig. 2(b). If the steps of the INSERT(d) and DELETE(b) are interleaved in a particular order, key d may be inserted as a grandchild of p, and erroneously deleted along with b.

To avoid situations such as this, each internal node is augmented to contain an UpdateStep object that indicates an operation has exclusive access to the

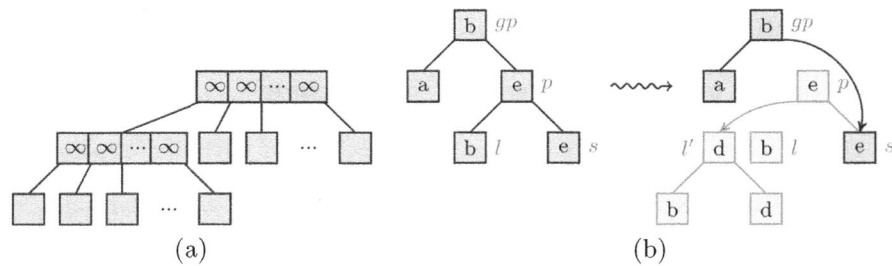

(a)   (b)

Fig. 2. (a) The initial state of the k-ary search tree. The root and its leftmost child have $k - 1$ keys valued ∞ (a special key, larger than any key in the set). All other children of these nodes are empty leaves. Keys in the set are stored in the sub-tree rooted at the leftmost grandchild of the root. (b) Example of the danger of uncoordinated concurrent updates. Faintly shaded nodes are no longer in the tree. If gp's right child is changed to s by a DELETE(b), and p's left child is changed to l' by a concurrent INSERT(d), then the new key d is lost.

child pointers of a node. This coordination scheme extends the work of Ellen et al. [5]. UpdateStep objects serve as something similar to locks, because all processes operate under the following agreement. When an operation intends to modify a child pointer of an internal node n, it first stores an UpdateStep object at n (using CAS). An operation cannot store an UpdateStep at node n if another operation x has already stored an UpdateStep at n, until x has relinquished control of n. Thus, UpdateStep objects behave like locks that are owned by an operation, rather than by a process, and this allows us to guarantee the non-blocking property by using the helping mechanism described in Sec. 2.4.

UpdateStep objects are divided into Flags and Marks. A Flag is placed on a node to reserve its child pointers for exclusive access, indicating that one will be changed by an operation. A Mark is similar to a Flag except, where a Flag is temporary (removed once a modification is completed), a Mark is permanent, and is placed on a node that is to be removed from the tree. The Mark permanently prevents the child pointers of the node from ever changing after it is removed. The final type of UpdateStep object is the Clean object which, if stored at a node x, indicates that no operation has exclusive access to x, and any operation is allowed to store a Flag or Mark there.

The details of the INSERT and DELETE operations, including flagging and marking steps, are as follows. In the following, l is the target leaf for insertion or deletion of a key, p is its parent, and gp is its grandparent. A simple insertion or simple deletion (see Fig. 1) creates the new leaf, flags p with a ReplaceFlag object (with a *Flag CAS*), changes the child pointer of p (with a *Child CAS*), and unflags p (with an *Unflag CAS*) by writing a new Clean object. Similarly, a sprouting insertion creates the new sub-tree, flags p with a new ReplaceFlag object, changes the child pointer of p, and unflags p. A pruning deletion flags gp with a PruneFlag object, then attempts to mark p with a *Mark CAS*. If the *Mark CAS* is successful, then the child pointer of gp is changed, finishing the deletion, and gp is unflagged. Otherwise, if the marking step fails, then the DELETE must unflag gp (with a *Backtrack CAS*) and try again from scratch.

We now return to Fig. 2(b) to illustrate how flagging and marking resolves the issue. After DELETE(b) has successfully stored a PruneFlag at gp, it must store a Mark at p. Say the Mark is successfully stored at p. Then it is safe to prune l and p out of the tree, since no child pointer of p will ever change, and l is a leaf (which has no mutable fields). Once l and p are pruned out of the tree, gp is unflagged by an *Unflag CAS* that replaces the PruneFlag stored by DELETE(b) by a newly created Clean object. If INSERT(a) subsequently tries to change a child reference belonging to p, it will first have to store a ReplaceFlag at p, which is impossible, since p is already marked. Otherwise, if the Mark cannot be successfully be stored at p (because p is already flagged or marked by another operation), then DELETE(b) will execute a *Backtrack CAS*, storing a new Clean object at gp (relinquishing control of gp to allow other operations to work with it), and retry from scratch.

2.4　Helping

To overcome the threat of deadlock that is created by the exclusive access that flags and marks grant to a single operation, we follow the approach taken by Ellen et al. [5], which has some similarities to Barnes' cooperative technique [1]. Suppose that a process P flags or marks a node hoping to complete some tree modification C. The Flag or Mark object is augmented to contain sufficient information so that any process can read the Flag or Mark and complete C on P's behalf. This allows the entire system to make progress even if individual processes are stalled indefinitely.

Unfortunately, while helping guarantees progress, it can mean duplication of effort. Several processes may come across the same UpdateStep object and perform the work necessary to advance the operation by performing some local work, followed by a *Mark CAS*, *Child CAS*, or *Unflag CAS*, but only one process can successfully perform each CAS, so the work performed by all other processes is wasted. For this reason it is advantageous to limit helping as much as possible. To this end, a search ignores flags and marks in our implementation, and proceeds down the tree without helping any operation. An INSERT or DELETE helps only those operations that interfere with its own completion. Thus, an INSERT will only help an operation that has flagged or marked p, and a DELETE will only help an operation that has flagged or marked p or gp (although they may help other operations recursively). After an INSERT or DELETE helps another operation, it restarts, performing another search from the top of the tree. An INSERT or DELETE operation is repeatedly attempted until it successfully modifies the tree or finds that it can return FALSE.

```
1    ▷ Type definitions:
2    type Node {
3        final Key ∪ {∞}  a₁, ..., a_{k-1}
4    }
5    subtype Leaf of Node {
6        final int keyCount
7    }
8    subtype Internal of Node {
9        Node c₁, ..., c_k
10       UpdateStep pending
            ▷ (initially a new Clean() object)
11   }

12   type UpdateStep { }
13   subtype ReplaceFlag of UpdateStep {
14       final Node l, p, newChild
15       final int pindex
16   }

17   subtype PruneFlag of UpdateStep {
18       final Node l, p, gp
19       final UpdateStep ppending
20       final int gpindex
21   }
22   subtype Mark of UpdateStep {
23       final PruneFlag pending
24   }
25   subtype Clean of UpdateStep { }

26   ▷ Initialization:
27   shared Internal root := the structure
         described in Fig. 2(a), with the pending
         fields of root and root.c₁ set to refer to
         new Clean objects.
```

Fig. 3. Type definitions and initialization

2.5　Pseudocode

Java-like pseudocode for all operations is found in Fig. 3 through Fig. 5. We borrow the concept of a *reference* type from Java. Any variable x of type C, where C is a type defined in Fig. 3, is a reference to an instance (or object) of type C. Such a variable x behaves like a C pointer, but does not require explicit dereferencing. References can point to an object or take on the value NULL, and management of their memory is automatic: memory is garbage-collected once it is unreachable from any executing thread. We use $a.b$ to refer to field b of the object referred to by a. We also adopt a Java-like definition of CAS: it atomically compares a field R with an expected value exp and either writes a new value and returns true (if R contains exp), or returns false (otherwise).

The SEARCH(key) operation is straightforward. Beginning at the leftmost child of $root$ (line 29) and continuing until it reaches a leaf (line 32), it compares its argument key with the key stored at each node and follows the appropriate child reference (line 36), saving some information along the way. (The keys of a node can be naively inspected in sequence because they never change.) The FIND(key) operation returns TRUE if SEARCH finds a leaf containing key; otherwise it returns FALSE. FIND can actually call a highly optimized version of SEARCH, given in Appendix A (see [4]).

To perform an INSERT(key), a process P locates the leaf l and its parent p, and stores the parent's $pending$ field in $ppending$ and the index of the child reference of gp that contained p in $pindex$ (line 49). If key is already in l, then the operation simply returns FALSE (line 50). Otherwise, P checks whether the parent's $pending$ field was of type Clean when it was read (line 51). If not, then $p.pending$ was occupied by a Flag or Mark belonging to some other operation x in progress at p. P helps x complete, and then re-attempts its own operation from scratch. Otherwise, if $p.pending$ was Clean, P tries to flag p by creating $newChild$, a new leaf or sub-tree depending on which insertion case applies (lines 54 to 58), creating the ReplaceFlag object op (line 59), and executing an $Rflag\ CAS$ to store it in the $pending$ field of p (line 60). If the $Rflag\ CAS$ succeeds, P calls HELPREPLACE(op) to finish the insertion (line 62) and the operation returns TRUE. Otherwise, if the $Rflag\ CAS$ failed, another process must have changed p's $pending$ field to a ReplaceFlag object, a PruneFlag object, a Mark object, or a new Clean object (different from the one read at line 49). Process P helps this other operation (if not a Clean object) complete, and then re-attempts its own operation. A call to HELPREPLACE executes a $Child\ CAS$ to change the appropriate child pointer of p from l to $newChild$ (line 116), and executes an $Runflag\ CAS$ to unflag p (line 117).

When process P performs a DELETE(key) operation, it first locates the leaf l, its parent p and grandparent gp, and stores the parent's and grandparent's $pending$ fields in $ppending$ and $gppending$, and the indices of the child references of gp and p that contained p and l, respectively, in $gpindex$ and $pindex$ (line 78). If l does not contain key, then the operation simply returns FALSE (line 79). Otherwise, P checks $gppending$ and $ppending$ to determine whether gp and p were Clean when their $pending$ fields were read (lines 80 and 82). If either has been flagged or marked by another operation, P helps complete this operation and re-attempts its own operation from scratch. Otherwise, it counts the number of non-empty children of p to determine the deletion case to apply. We shall

28   SEARCH(Key key) : ⟨Internal, Internal, Leaf, UpdateStep, UpdateStep⟩ {
       ▷ Used by INSERT, DELETE and FIND to traverse the k-ST
       ▷ SEARCH satisfies following *postconditions:*
       ▷ (1) $leaf$ points to a Leaf node, and $parent$ and $gparent$ point to Internal nodes
       ▷ (2) $parent.c_{pindex}$ has contained $leaf$, and $gparent.c_{gpindex}$ has contained $parent$
       ▷ (3) $parent.pending$ has contained $ppending$,
              and $gparent.pending$ has contained $gppending$
29   Node $gparent, parent := root, leaf := parent.c1$
30   UpdateStep $gppending, ppending := parent.pending$
31   int $gpindex, pindex := 1$
32   while $type(leaf)$ = Internal {   ▷ Save details for parent and grandparent of leaf
33       $gparent := parent; gppending := ppending$
34       $parent := leaf; ppending := parent.pending$
35       $gpindex := pindex$
36   ⟨$leaf, pindex$⟩ := ⟨appropriate child of $parent$ by the search tree property,
                        index such that $parent.c_{pindex}$ is read and stored in $leaf$⟩
37   }
38   return ⟨$gparent, parent, leaf, ppending, gppending, pindex, gpindex$⟩
39   }

40   FIND(Key key) : boolean {
41   if Leaf returned by SEARCH(key) contains key, then return TRUE, else return FALSE
42   }

43   INSERT(Key key) : boolean {
44   Node $p, newChild$
45   Leaf l
46   UpdateStep $ppending$
47   int $pindex$
48   while TRUE {
49   ⟨$-, p, l, ppending, -, pindex, -$⟩ := SEARCH($key$)
50   if l already contains key then return FALSE
51   if $type(ppending) \neq$ Clean then {
52   HELP($ppending$)   ▷ Help the operation pending on p
53   } else {
54   if l contains $k - 1$ keys {   ▷ **Sprouting insertion**
55           $newChild$:= new Internal node with $pending := new\ Clean()$,
                  and with the $k - 1$ largest keys in $S = \{key\} \cup$ keys of l,
                  and k new children, sorted by keys, each having one key from S
56   } else {   ▷ **Simple insertion**
57           $newChild$:= new Leaf node with keys: $\{key\} \cup$ keys of l
58   }
59   ReplaceFlag op := new ReplaceFlag($l, p, newChild, pindex$)
60   boolean $result$:= CAS($p.pending, ppending, op$)   ▷ **Rflag CAS**
61   if $result$ then {   ▷ *Rflag CAS* succeeded
62   HELPREPLACE(op)   ▷ Finish the insertion
63   return TRUE
64   } else {   ▷ *Rflag CAS* failed
65   HELP($p.pending$)   ▷ Help the operation pending on p
66   } } } }

67   HELP(UpdateStep op) {
       ▷ *Precondition: $op \neq$ NULL has appeared in $x.pending$ for some internal node x*
68   if $type(op)$ = ReplaceFlag then HELPREPLACE(op)
69   else if $type(op)$ = PruneFlag then HELPPRUNE(op)
70   else if $type(op)$ = Mark then HELPMARKED($op.pending$)
71   }

Fig. 4. Pseudocode for SEARCH, FIND, INSERT and HELP

```
72   DELETE(Key key) : boolean {
73     Node gp, p
74     UpdateStep gppending, ppending
75     Leaf l
76     int pindex, gpindex
77     while TRUE {
78       ⟨gp, p, l, ppending, gppending, pindex, gpindex⟩ := SEARCH(key)
79       if l does not contain key, then return FALSE
80       if type(gppending) ≠ Clean then {
81         HELP(gppending)                              ▷ Help the operation pending on gp
82       } else if type(ppending) ≠ Clean then {
83         HELP(ppending)                               ▷ Help the operation pending on p
84       } else {                                       ▷ Try to flag gp
85         int ccount := number of non-empty children of p (by checking them in sequence)
86         if ccount = 2 and l has one key then    ▷ Pruning deletion
87           PruneFlag op := new PruneFlag(l, p, gp, ppending, gpindex)
88           boolean result = CAS(gp.pending, gppending, op)              ▷ Pflag CAS
89           if result then {                      ▷ Pflag CAS successful–now delete or unflag
90             if HELPPRUNE(op) then return TRUE;
91           } else {                              ▷ Pflag CAS failed
92             HELP(gp.pending)                    ▷ Help the operation pending on gp
93           }
94         } else {                                ▷ Simple deletion
95           Node newChild := new copy of l with key removed
96           ReplaceFlag op := new ReplaceFlag(l, p, newChild, pindex)
97           boolean result := CAS(p.pending, ppending, op)              ▷ Rflag CAS
98           if result then {                      ▷ Rflag CAS succeeded
99             HELPREPLACE(op)                     ▷ Finish inserting the replacement leaf
100            return TRUE
101          } else {                              ▷ Rflag CAS failed
102            HELP(p.pending)                     ▷ Help the operation pending on p
103 } } } } }
104  HELPPRUNE(PruneFlag op) : boolean {          ▷ Precondition: op is not NULL
105    boolean result := CAS(op.p.pending, op.ppending, new Mark(op))   ▷ Mark CAS
106    UpdateStep newValue := op.p.pending
107    if result or newValue is a Mark with newValue.pending = op then {
108      HELPMARKED(op)                            ▷ Marking successful–complete the deletion
109      return TRUE
110    } else {                                    ▷ Marking failed
111      HELP(newValue)                            ▷ Help the operation pending on p
112      CAS(op.gp.pending, op, new Clean())       ▷ Unflag op.gp       ▷ Backtrack CAS
113      return FALSE
114  } }
115  HELPREPLACE(ReplaceFlag op) {                ▷ Precondition: op is not NULL
116    CAS(op.p.c_op.pindex, op.l, op.newChild)      ▷ Replace l by newChild  ▷ Rchild CAS
117    CAS(op.p.pending, op, new Clean())          ▷ Unflag p            ▷ Runflag CAS
118  }
119  HELPMARKED(PruneFlag op) {                   ▷ Precondition: op is not NULL
120    Node other := any non-empty child of op.p, or op.p.c_1 if none
              (found by visiting each child of op.p)
121    CAS(op.gp.c_op.gpindex, op.p, other)          ▷ Replace l by other  ▷ Pchild CAS
122    CAS(op.gp.pending, op, new Clean())         ▷ Unflag gp           ▷ Punflag CAS
123  }
```

Fig. 5. Pseudocode for DELETE, HELPPRUNE, HELPREPLACE and HELPMARKED

explain why counting the children in sequence is not problematic when we discuss correctness. We consider the two types of deletion separately.

If the operation is a simple deletion (line 94), it creates *newChild*, a new copy of leaf *l* with *key* removed, and a new ReplaceFlag object *op* to facilitate helping (line 96). Next, *P* attempts an *Rflag CAS* to store *op* in *p.pending* (line 97) and, if it succeeds, it calls HELPREPLACE to finish the deletion (line 99). Otherwise, if the *Rflag CAS* fails, *P* helps any operation that may be pending on *p*. After helping, *P* retries its own operation from scratch. Note that, apart from the creation of the new leaf, this is identical to simple insertion.

If the operation is a pruning deletion (line 86), *P* creates a PruneFlag object (line 87), then attempts a *Pflag CAS* to store it in the *pending* field of *gp* (line 88). If the *Pflag CAS* succeeds, *P* calls HELPPRUNE(*op*) to finish the deletion (line 90) and the operation returns TRUE (more on HELPPRUNE later). Otherwise, if the *Pflag CAS* fails, another process must have changed *gp*'s *pending* field to a ReplaceFlag object, a PruneFlag object, a Mark object, or a new Clean object (different from the one read at line 78). To help any other operation pending on *gp* to make progress, *P* calls HELP(*gp.pending*) (line 92) before retrying its own operation from scratch.

The HELPPRUNE procedure, invoked by the DELETE operation (and by HELP), attempts the second (marking) CAS step of a pruning deletion. Recall that *op*, created in the DELETE routine, contains pointers to *l*, the leaf containing the key to be deleted, its parent *p*, and its grandparent *gp*. The HELPPRUNE procedure begins by attempting to mark the parent *op.p* (line 105). If the CAS successfully marks *op.p*, or another helping process already stored a Mark for this operation, then the mark is considered to be successful. In this case, HELPMARKED is called to finish the pruning deletion (line 108), and TRUE is returned. Otherwise, if the CAS failed and the Mark was not already stored by a helping process, then another operation involving *op.p* has interfered with the DELETE. If the other operation is still in progress, it is helped (line 111), and then the operation backtracks, unflagging the grandparent *op.gp* (line 112), and HELPPRUNE returns FALSE. The process that invoked the DELETE procedure will ultimately retry the operation from scratch.

The HELPMARKED procedure performs the final step of a pruning deletion, pruning out some dead wood by changing the appropriate child pointer of *op.gp* from *op.p* to point to the only non-empty sibling of *op.l*. This sibling of *op.l* is found at line 120. (It is explained in Sec. 2.6 why this can be found simply by visiting each child of *op.p*.) The CAS-CHILD routine is invoked to change the child pointer of *op.gp* (line 121), and an *Unflag CAS* is executed to unflag *op.gp* (line 122).

2.6   Correctness

It can be demonstrated that our algorithm exhibits linearizability (defined in [11]), and the argument is very similar to the one made in the proof in [5]. We simply give the linearization points of operations here. See [4] for the complete proof of correctness. Consider some invocation of SEARCH(*key*). It can be proved that each node visited by SEARCH was on the search path for key in the tree at some time during the execution of SEARCH, so we linearize SEARCH at a point when the leaf it returns was on the search path. An invocation of FIND(key) is linearized at the point its corresponding SEARCH was linearized. It can be proved that each INSERT or DELETE invocation that returns TRUE has executed a successful *Child CAS*. An invocation of INSERT(key) or DELETE(key) that returns TRUE is linearized at this *Child CAS*; an invocation that returns FALSE is linearized at the same point as the corresponding SEARCH that discovered key was already in the tree, or was not in the tree, respectively.

The k-ST algorithm differs significantly from the BST algorithm at lines 85 and 120, which both involve accessing several children in sequence. Let P be a process executing line 85. We note that no flagging or marking has yet been attempted by P, and the expected values to be used by the CASs at lines 88 and 97 were verified to be Clean a few lines prior. Further, if any process Q wants to add or remove a key from a child x of p that P will read at line 85, it must replace x, changing a child pointer of p. However, it must flag p to change its child pointers, overwriting the Clean object that was read earlier by P to be used as the expected value for its *Flag CAS*. It is easy to prove that there is no ABA problem on pending fields, which implies that the expected value used by P for the CAS can never appear in $p.pending$ again, so P's CAS must fail, and the operation will be retried. It can then be shown that if an operation op successfully flags or marks p, $ccount$ contains the number of non-empty children of p until a *Child CAS* is executed for op, and that only the first *Child CAS* will be successful (occurring immediately after line 120). Thus, it can be shown that when line 120 is executed, the children of $op.p$ are precisely $op.l$ and one other leaf, or else the *Child CAS* will fail, so the value of *other* is irrelevant. This is rigorously demonstrated in the detailed proof of correctness presented in Appendix A (see [4]).

3   Experiments

In this section we present results from experiments comparing the performance of the BST of Ellen et al. [5], our k-ST algorithm, ConcurrentSkipListMap (SL) of the Java class library and the lock-based AVL tree (AVL) of Bronson et al. [3]. Experiments on each structure used put-if-absent and delete-if-present (set functions), returning TRUE if the operation could be completed, and FALSE otherwise. Preliminary experiments were run to tune the parameters of the final experimental set to maximize trial length while keeping standard deviations reasonable. The final experiments each consisted of selecting a particular algorithm and executing a sequence of 17 three-second trials, in which a fixed number of threads randomly perform INSERTs, DELETEs and FINDs according to a desired probability distribution (e.g., 5% INSERT, 5% DELETE, 90% FIND), on uniformly distributed random keys, drawn from a particular key range (e.g., the integers from 0 to 10^6). The average throughput (operations per second) was recorded for each trial, and the first few trials were discarded to account for the few seconds of "warm-up" time that the Java Virtual Machine (VM) needs to perform just-in-time compilation and optimization. We observed that throughput stabilized after the first three to five seconds of execution, so the first two trials (six seconds) of each experiment were discarded. Garbage collection was also triggered in between trials to minimize its haphazard impact on measurements.

Our experiments were run on a Sun machine at the University of Rochester, with two UltraSPARC-III CPUs, each having eight 1.2GHz cores capable of running 8 hardware threads apiece (totalling 128 hardware contexts), and 32GB of RAM, running Sun's Solaris 10 and the Java 64-bit VM version 1.6.0_21 (with 15GB initial and maximum heap sizes).

We call the probability distribution of INSERTs and DELETEs a *ratio*, and denote an experiment with $x\%$ INSERTs, $y\%$ DELETEs and $(100 - x - y)\%$ FINDs as, simply $xi\text{-}yd$. We denote the key range of integers from 0 to $10^x - 1$ by $[0, 10^x)$. The experimental results we present herein used algorithms BST, 4-ST, SL and AVL, key ranges $[0, 10^2)$ and $[0, 10^6)$ and ratios $0i\text{-}0d$, $5i\text{-}5d$, $8i\text{-}2d$ and $50i\text{-}50d$. The key ranges induce high and low levels of contention, respectively, with small trees increasing the probability that operations on random keys will coincide. The four ratios represent situations in which operations consist (1) entirely of searching, (2) mostly of searching, (3) mostly of searching, but with far more INSERTs than DELETEs, and (4) entirely of updates. Initially, each data structure was empty for each trial, except when the ratio was $0i\text{-}0d$, since that would mean performing all operations on an empty tree. In this case, each structure was pre-filled at the beginning of each trial by performing random operations in the ratio $50i\text{-}50d$ until the structure's size stabilized (to within 5% of the expected half-full). Additional results, including more operation mixes and key ranges, and results from a 32-core system at Intel's Multicore Testing Lab can be found in [4]. For implementations of the BST and k-ST, see [4].

We now discuss the graphs presented in Fig. 6. The $[0, 10^2)$ key range represents very high contention. There were at most 10^2 keys in the set, and as many as 128 threads accessing the tree. Under this load, BST was the top performer in all experiments. The low degree of BST's nodes permits many simultaneous updates to different parts of the tree, and its simplicity offers strong performance. 4-ST matched BST's performance in the $0i\text{-}0d$ and $8i\text{-}2d$ cases, indicating that, in the absence of many deletions, it can perform just as well under extremely high contention. For the other two ratios, 4-ST's performance was similar to the lock-free SL, surpassing AVL by a fair margin. BST scaled very well in all cases; 4-ST scaled equally well when deletions were few.

The $[0, 10^6)$ key range represents low contention: with as many as one million keys and only 128 threads, the chance of collisions in random keys is quite small. With this level of contention, 4-ST exhibits strong performance, surpassing BST, and the other algorithms. This is in line with expectations; as the size of the tree increases, the higher degree of the 4-ST affords it a shallower depth, allowing all operations to complete more quickly. Unlike the $[0, 10^2)$ case, all algorithms scale reasonably well in the $[0, 10^6)$ case, approaching linear improvement in throughput with an increase in the number of hardware threads.

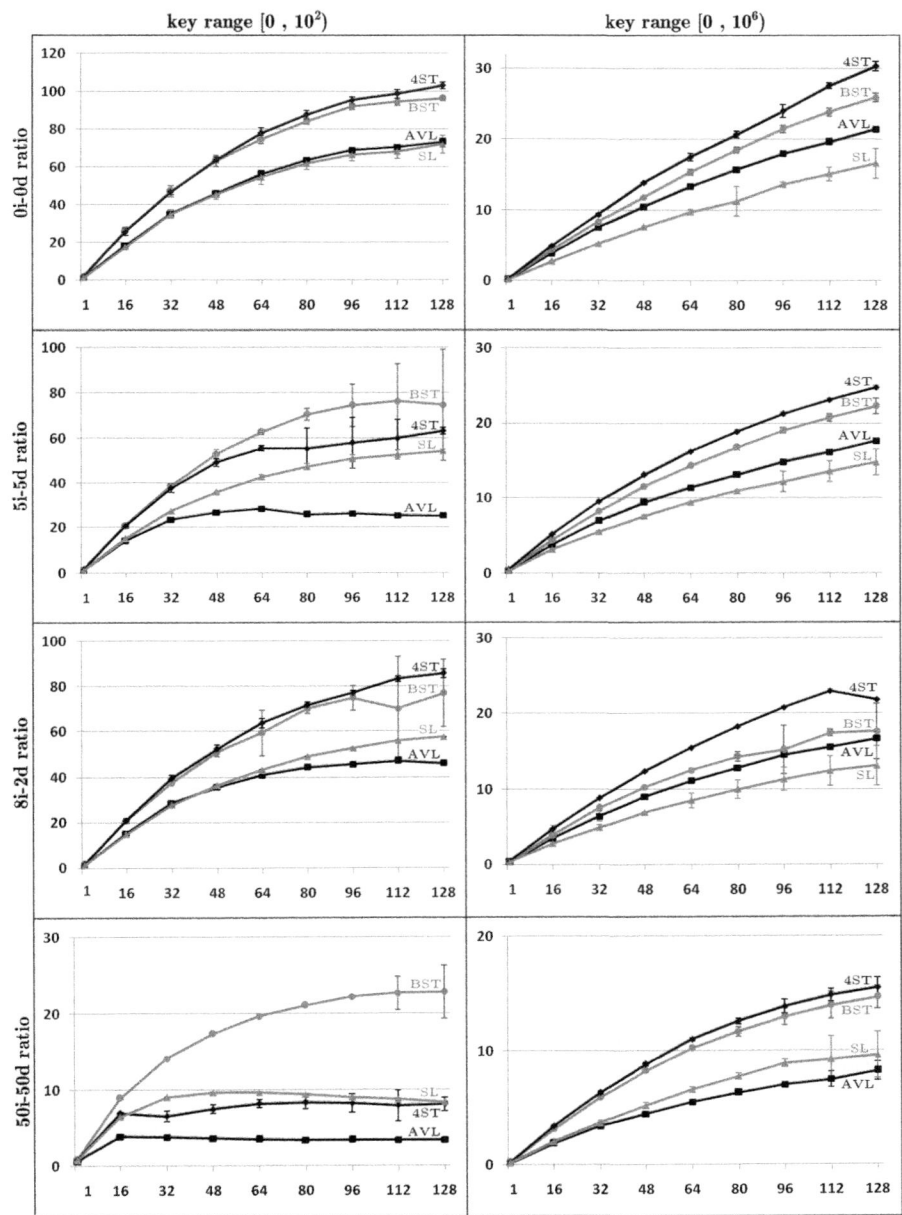

Fig. 6. Experimental results. Error bars are drawn to represent one standard deviation from the mean. Columns display ranges from which random keys are drawn. Rows display ratios of INSERTs to DELETEs to FINDs. The y-axis displays average throughput (millions of operations/sec.), and the x-axis displays the number of hardware threads.

4   Conclusion and Future Work

BST has the greatest advantage in high contention settings. Its simplicity pushes its performance beyond the other algorithms. As trees get larger and contention decreases, 4-ST surpasses BST to become the top performer. Similar to 4-ST, AVL also performs well as the size of the data structure increases. SL tends to performs well when its set of keys is small.

AVL is a balanced tree, so it does some extra work in maintaining this property. However, since our experiments insert random keys, 4-ST and BST also are nearly balanced. In this experimental setting, the balancing work of AVL does not pay off. In a situation where the keys inserted are not random, AVL would have a significant advantage over 4ST and BST. Since in many cases BST and 4ST outperform AVL and SL by a fair margin, we believe that it may be possible to add balancing and remain competitive, while offering a non-blocking progress guarantee.

Acknowledgments. We thank Michael L. Scott for providing access to the multi-core machine at the University of Rochester. Financial support for this research was provided by NSERC. We also thank Eric Ruppert and Franck van Breugel for their supervision and assistance in the preparation of this paper. Finally, we thank the anonymous OPODIS reviewers for their comments.

References

1. Barnes, G.: A method for implementing lock-free data structures. In: Proc. 5th ACM Symposium on Parallel Algorithms and Architectures, pp. 261–270 (1993)
2. Bender, M.A., Fineman, J.T., Gilbert, S., Kuszmaul, B.C.: Concurrent cache-oblivious B-trees. In: Proc. 17th ACM Symposium on Parallel Algorithms and Architectures, pp. 228–237 (2005)
3. Bronson, N.G., Casper, J., Chafi, H., Olukotun, K.: A practical concurrent binary search tree. In: Proc. 15th ACM Symposium on Principles and Practice of Parallel Programming, pp. 257–268 (2010)
4. Brown, T., Helga, J.: Non-blocking k-ary search trees. Technical Report CSE-2011-04, York University (2011), Appendix (with complete proof) and code available at http://www.cs.toronto.edu/~tabrown/ksts/
5. Ellen, F., Fatourou, P., Ruppert, E., van Breugel, F.: Non-blocking binary search trees. In: Proc. 29th ACM Symposium on Principles of Distributed Computing, pp. 131–140 (2010); Full version in Tech. Report CSE-2010-04, York University
6. Fomitchev, M., Ruppert, E.: Lock-free linked lists and skip lists. In: Proc. 23rd ACM Symposium on Principles of Distributed Computing, pp. 50–59 (2004)
7. Fraser, K., Harris, T.: Concurrent programming without locks. ACM Transactions on Computer Systems 25(2), 5 (2007)
8. Fraser, K.A.: Practical lock-freedom. PhD thesis, University of Cambridge (2003)
9. Guibas, L.J., Sedgewick, R.: A dichromatic framework for balanced trees. In: Proc. 19th IEEE Symp. on Foundations of Computer Science, pp. 8–21 (1978)

10. Herlihy, M., Luchangco, V., Moir, M., Scherer III, W.N.: Software transactional memory for dynamic-sized data structures. In: Proc. 22nd ACM Symposium on Principles of Distributed Computing, pp. 92–101 (2003)
11. Herlihy, M.P., Wing, J.M.: Linearizability: a correctness condition for concurrent objects. ACM Transactions on Programming Languages and Systems 12(3), 463–492 (1990)
12. Sundell, H., Tsigas, P.: Scalable and lock-free concurrent dictionaries. In: Proc. 19th ACM Symposium on Applied Computing, pp. 1438–1445 (2004)
13. Valois, J.D.: Lock-free linked lists using compare-and-swap. In: Proc. 14th ACM Symposium on Principles of Distributed Computing, pp. 214–222 (1995)

Probabilistic Compositional Reasoning
for Guaranteeing Fault Tolerance Properties

Jan Olaf Blech

fortiss GmbH

Abstract. We present a framework to formally describe system behavior and symbolically reason about possible failures. We regard systems which are composed of different units: sensors, computational parts and actuators. Considering worst-case failure behavior of system components, our framework is used to derive reliability guarantees for composed systems. The behavior of system components is modeled using monad like constructs that serve as an abstract representation for system behavior. We introduce rules to reason about these representations and derive results like, e.g., guaranteed upper bounds for system failure. Our approach is characterized by the fact that we do not just map a certain component to a failure probability, but regard distributions of error behavior. These serve as basis for deriving failure probabilities.

1 Introduction

The need for analysis of failure probabilities arises in many domains connected to safety critical embedded systems. Guaranteeing worst-case failure probabilities is an important prerequisite for certification of safety critical systems.

In this paper we present a new framework to model systems and their failure behavior. Our framework represents distinct parts of system behavior in an abstract monadic [16] way. We allow the modeling of behavioral entities with probabilistic distributions representing possible failures or uncertainties. When composing a system from different components, our approach allows modeling the propagation of failures through components by monadic composition of behavior associated with the components.

The second ingredient of our framework comprises rules to reason about systems. Our rules allow determining the semantic equivalence of systems and the reduction of systems to other systems such that certain properties are guaranteed to be preserved. The reduction of systems into simpler systems may be used to analyze and optimize systems.

Our approach comprises the following characteristics that all together distinguish it from existing approaches:

- Modeling of system behavior and possible faults using monad like constructs.
- Representation of uncertain/faulty behavior as distributions of possible behavior.
- Rules to reason about system behavior and distributions of values that appear in this system.

The work presented in this paper presents a complete framework and a case studies. The main intended purpose is the usage in safety critical industrial automation systems.

A. Fernández Anta, G. Lipari, and M. Roy (Eds.): OPODIS 2011, LNCS 7109, pp. 222–234, 2011.

1.1 Related Work

Early work establishing fault tolerance guarantees using theorem proving techniques is presented in [10]. Based on a formalism using extended petri nets, properties of (digital) hardware systems are shown. Furthermore, [13] describes work on guaranteeing fault-tolerance related properties using the PVS theorem prover. Here, systems and constraints are ported and proved in PVS. The presented examples come from the microprocessor and avionics domains.

Abstractions for reasoning about fault-tolerant systems in a higher-order theorem prover are presented and discussed in [17]. The abstraction aims at facilitating and standardizing the use of formal methods, especially higher-order theorem provers. Abstractions for individual message passing, faults, fault-masking, and further communication aspects are regarded.

Other early work comprises [2] which presents a Specification and Design Language (SDL) based framework. Other related work for guaranteeing fault tolerant properties using formal methods comprise the use of model-checking techniques [14] and concentrate on formal specification techniques [9]. The analysis of probabilistic system behavior is the goal of probabilistic model-checkers, like PRISM [11].

Modeling and reasoning aspects about probabilistic programs have been extensively studied in [12]. Here a language is introduced to describe probabilistic programs and reasoning about programs has a strong connection to this language. Further work on formal analysis of probabilistic systems has been done in the event-B context [7].

Like the work focused on theorem proving techniques, but unlike the (probabilistic) model checking approaches, we have a strong focus on symbolic reasoning. Unlike the existing theorem prover based work we have a strong focus on symbolically representing distributions of values and combining them. Handling errors and varying values as distributions as we do in this work allows a much richer failure analysis than assigning failure probabilities to distinct system components. For example, it allows the specification and handling of ranges in which a deviation from an optimal value is acceptable.

The framework presented in this paper builds upon work for a monadic representation of probabilities in programs [1] and an application for the analysis of cryptographic protocols [3]. Like in the analysis of cryptographic protocols, we regard possible computations that are associated with certain distributions of values. Furthermore, we regard rules that allow reasoning about sequences of such computations. In [3] and in this work the use of a monadic representation was chosen because:

- It gives a syntactic representation of the semantics of non-deterministic systems. This non-determinism can be "quantified" in the sense that different possibilities in the system execution can be assigned to different probabilities. This is achieved in combination with the use of distributions. Finally, it enables us to even specify an infinite amount of possibilities and reason about probabilities by using continuous probability distributions.
- The syntactic representation is well suited to match rules against and reason about it in a symbolic way.
- As used in [3] and as a possible future extension this gives us the possibility to reason about our system in an automatic or interactive way using, e.g., a higher-order theorem prover.

A future goal of our work is giving rise to certifying properties of systems in a scenario similar to [5].

As an amendment to our work on guaranteeing distinct probabilities, patterns for achieving fault tolerance have been extensively studied (e.g., [8]).

1.2 Overview

We present prerequisites like our monad and basic facts about probabilities in Section 2. The modeling framework is presented in Section 3. Section 4 features the rules to reason about system descriptions modeled in our framework. In Section 5 we present a case study from the industrial control domain. Finally, Section 6 gives a conclusion.

2 Prerequisites

In this section we describe a monad like construct to formalize computations [1,3]. This is needed to represent system behavior in a compositional way. The idea is to divide system behavior into different computation steps which correspond to distinct system components. These steps realize state transitions. Traditionally, a state comprises a kind of memory, e.g., variables which are associated with values. Unlike this, in our work, we consider states in which a variable is associated with a probabilistic distribution of possible values rather than a single value. Furthermore, we present some probabilistic background knowledge.

Distributions. Distributions may be either discrete or continuous. For a finite type T the (discrete) uniformly random distribution is denoted $\$_T$. For a given value val associated with a type T the distribution that contains just this value (probability 100 percent) is denoted $\mathcal{U}_T(val)$. We omit the T if the type is obvious from the context.

In the case of discrete distributions, the distribution can be regarded as a function that maps an element to its probability. For a given discrete distribution, the probability of an element x from D is denoted $D(x)$.

For the non-discrete case, P_D denotes the density function of the distribution D. Assuming a total order on the elements of D, the probability of all elements in D that are less or equal than x is denoted $P_D(x)$, In the case of normal distributions we use $\mathcal{N}(\mu, \sigma^2)$ for a normal distribution with mean μ and variance σ^2.

Monads and their composition. In our formalization, behavior is formalized using abstract computations. These are based on the *abstract computation monad* (cf. [1,3]) used for representing the changes of variable distributions. Here, we use a slightly adapted definition:

Definition 1 (Abstract Computation). *For a set of variables V an abstract computation M_V is defined by two functors* unit *and* bind.

$$\text{unit} : (V \rightarrow C_V) \rightarrow M_V \qquad \text{bind} : M_V \rightarrow V \rightarrow U \rightarrow M_V$$

unit *comprises a set of initial variables distributions: $(V \rightarrow C_V)$ is a mapping from*

variable names of type V to their distributions of type C_V. bind *comprises an abstract computation, an update set of variables of type V: V and a set of updates U. The update set contains all variables that have to be updated. An update is a tuple:*
$$(V \times ((V \rightarrow Val_V) \rightarrow D_V))$$

It comprises a variable to be updated of type V and a function that takes a mapping from variables of type V to values of type Val_V and returns a distribution of type D_V.

The unit constructor formalizes initial variable distributions. A bind is a single distinct computation step: The variable to distributions mapping resulting from evaluating the abstract computation (first argument of bind) is taken and for all variables in the update set (second argument of bind) the appropriate update function (in the third argument of bind) is performed resulting in new distributions for these variables.

Example. The semantics of the term

bind (bind (unit $\{(v \mapsto \$_{T_v})\}$) $\{v\}$ $\{(v, v \mapsto \mathcal{U}_{T_v}(f(v)))\}$) $\{v\}$ $\{(v, v \mapsto \mathcal{U}_{T_v}(g(v)))\}$

is that a value is drawn from a uniformly distributed distribution $\$_{T_v}$. A function is applied to this value, thus obtaining $f(v)$. and a distribution is made of this inner bind statement. In the next step a value is drawn from this distribution (associated with variable v) and a function g is applied to it. The entire term again denotes a distribution. $v \mapsto D$ is used to denote a function that maps a variable v to a distribution D.

If f and g are permutations the resulting distribution will be again uniformly distributed. In the remainder of this paper the "," is used to denote monadic composition of terms composed of bind and unit.

As a second example, we present an equation: a value is drawn from a distribution and one builds a new distribution that contains just this value. This denotes the original distribution:

$$(v, D), (v, \mathcal{U}_{T_v}(v)) = (v, D)$$

Events and Probabilities. We define probabilistic *events* on abstract computations. Events take values associated with variables $val_1, ..., val_n$ and return a truth value:

Definition 2 (Event). *An event E is a function:*
$$E(val_{v_1}, ..., val_{v_n}) \rightarrow \{true, false\}$$
with $v_1, ..., v_n \in V$ and $val_{v_1} \in T_{v_1}, ..., val_{v_n} \in T_{v_n}$.

An Event E can be applied to an abstract computation C thereby specifying a value between 0 and 1 stating the probability that E does hold after the computation of C. We denote this: $Pr([C]E) \in [0, 1]$.

Example. An event that states that a value drawn from a boolean distribution associated with a variable b is true is formalized as $b = true$. The probability of this event for a uniform boolean distribution:

$$Pr([(b, \$_{\{true, false\}})](b = true)) = 0.5$$

3   Our Modeling Framework

In this section we describe the framework for specification of systems using abstract computations from Section 2.

3.1   Monadic Representation of Probabilistic Systems

Complex systems composed of different components can be described with abstract computations. The components are modeled as functions that take some values bound to distinct variables and map them to their distributions.

Different components realizing a certain behavior may be modeled independently. They may be composed so that a system description can be realized. In the industrial automation domain – which we focus on – typical components for modeling systems comprise sensors, actuators, and plain computations. Sensors and actuators comprise a distinct failure behavior which can be modeled by using distributions. Computations may also be formalized in a way that they are associated with a failure behavior. This can be used, e.g., to model potential failures in the underlying hardware.

We give a small overview on modeling system behavior using our formalism.

Sequential Composition. Given two system components A and B performing a certain task and formalized using our abstract computation. To activate A and B sequentially we may combine them using standard monadic composition:

$$A , B$$

A gets executed before B and all values computed in A are accessible in B.

Parallel Composition. Two components A, B can be combined in parallel composition. This is denoted:

$$\begin{pmatrix} A \\ B \end{pmatrix}$$

A and B should not depend on each other, i.e., no write access to a value which is used by the other component. Thus, it is possible to linearize parallel composition to:

$$A , B \quad \text{or} \quad B , A$$

Conditional Structures. Conditional structures depending on an expression e can be realized in a straightforward way:

$$if \ e \ then \ A_1, ..., A_n \ else \ B_1, ...B_n$$

Distributions of variables' values can be effected in different ways by different branches of conditional expressions. For this reason, rather complex definitions of distributions may occur due to conditional structures.

Looping Structures. Given a loop body A, looping structures may be formalized in the following way, resembling process algebras:

$$L := A, L$$

Our framework bears some similarity to the way one describes Programmable Logic Controllers (PLCs) by, e.g., using the IEC-61131–3 standard [15]. It is intended to be an extension of the property certification proposed in [4] for PLCs.

3.2 An Example

Here we present a generic example that realizes a composed component made from sensors and a computation.

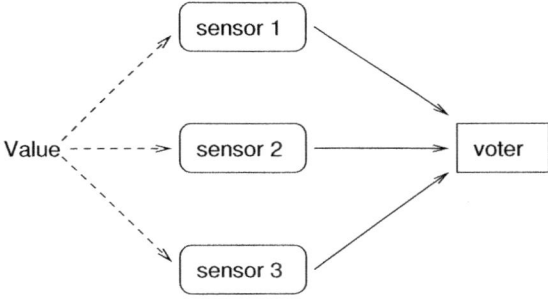

Fig. 1. Voting Example

The example system in Figure 1 realizes a voting. This is a fault tolerance mechanism that aims at eliminating errors occurring in individual components by replicating them. In our case a sensor is replicated. One physical value x is read by three different sensors. Each of these sensors may read a wrong value: some noise is added to x which corresponds to the distribution $\mathcal{N}_E$ and is associated with the variable names e_1, e_2 and e_3. Thus, this noise is independent for each sensor. The voting reads the three sensor values v_1, v_2, v_3 and can, e.g., be realized by computing the arithmetic mean r.

$voter\_mean(x) \equiv$

$$\begin{pmatrix} e_1, \mathcal{N}_E \\ e_2, \mathcal{N}_E \\ e_3, \mathcal{N}_E \end{pmatrix}, \begin{pmatrix} v_1, \mathcal{U}_T(x + e_1) \\ v_2, \mathcal{U}_T(x + e_2) \\ v_3, \mathcal{U}_T(x + e_3) \end{pmatrix}, (r, \mathcal{U}_{T_r}\left(\frac{v_1 + v_2 + v_3}{3}\right))$$

An alternative realization – in the case of discrete noise given by the distribution E – is given below:

$voter\_2(x) \equiv$

$$
\begin{pmatrix} e_1, E \\ e_2, E \\ e_3, E \end{pmatrix} , \begin{pmatrix} v_1, \mathcal{U}_T(x + e_1) \\ v_2, \mathcal{U}_T(x + e_2) \\ v_3, \mathcal{U}_T(x + e_3) \end{pmatrix} , (r, \mathcal{U}_{T_r} \begin{pmatrix} if & v_1 = v_2 & then & v_1 \\ if & v_1 = v_3 & then & v_1 \\ if & v_2 = v_3 & then & v_2 \\ & & else & v_3 \end{pmatrix})
$$

Here we compare the values from the three different sensors. If two are equal we return one of these equal values. If all three are different, we return the value of the third sensor.

4   Deductive Rules

We have established rules to reason about system descriptions which are based on abstract computations. Two goals can be distinguished:

– Rules that transform system descriptions into other system descriptions. A special case of transformation rules perform semantically equivalent transformations.
– Rules that allow reasoning about probabilities of certain events. Some of these rules allow the transformation of systems with respect to certain events.

4.1   Soundness and Semantics of Rules

Our abstract computations represent both: system components and a semantical representation of them. Further elements that carry semantical meaning are the application of an event to an abstract computation and the probability function Pr. For this reason, proving soundness of our rules to reason about abstract computations does not need to take a transformation between syntax and semantics into account. In order to prove soundness one proves that certain semantical aspects are preserved during a rule application. In particular our rules are proven sound with respect to the following notions of correctness:

– Rules that transform system descriptions into other system descriptions: These rules have the following form:

$$
\frac{\begin{array}{c} AbstractComputation(val_1, ..., val_n) \\ \text{additional assumptions } (val_1, ..., val_n, val_1', ..., val_n') \end{array}}{AbstractComputation'(val_1', ..., val_n')} \lesssim
$$

Soundness of such a rule is established by proving the following lemma:

$$
\forall E \; val_1 \; ... \; val_n \; val_1' \; ... \; val_n' \; \epsilon.
$$
$$
\text{additional assumptions } (val_1, ..., val_n, val_1', ..., val_n') \wedge
$$
$$
Pr([AbstractComputation(val_1, ..., val_n)]E) = \epsilon
$$
$$
\longrightarrow
$$
$$
Pr([AbstractComputation'(val_1', ..., val_n')]E) = \epsilon
$$

Thus, our notion of correctness states that the probability of all possible events is preserved while transforming an abstract computation. We use $\lesssim$ in this paper to denote probability preservation for all possible events.
– Rules that allow reasoning about probabilities of certain events directly correspond to a lemma stating a fact on abstract computations.

Rules are applied by matching the bottom part of the rule to a system description and event. The application reduces the expression to the upper part of the rule.

4.2   Basic Rules

Here we present basic rules to handle system descriptions within our framework. Their soundness is proved in our scheme by using the properties of abstract computations.

Function Propagation Rule.   We have established rules to reason and simplify our monadic system descriptions. Simple rules comprise, e.g., function propagation.

$$\frac{\forall i \leq m, x_i' \neq x' \qquad (x'', \mathcal{U}_T(g(f(x_1, ..., x_n), x_1', ...x_m')))}{(x', \mathcal{U}_T(f(x_1, ..., x_n))) \,, \, (x'', \mathcal{U}_T(g(x', x_1', ..., x_m')))} \lesssim$$

Omitting Unused Parts Rule.   Unused parts of an expression may be omitted.

$$\frac{\begin{array}{c} \forall i \leq n, x_i' \neq x \\ (x, X'(x_1', ..., x_n')) \end{array}}{(x, X(x_1, ..., x_n)) \,, \, (x, X'(x_1', ..., x_n'))} \lesssim$$

This rule performs a kind of dead-code elimination

Congruence Exchange Rule.   Semantical equivalent parts may be replaced by each other.

$$\frac{A, B, C \qquad B \lesssim B'}{A, B', C} \lesssim$$

Permutation Rule.   Parts may be permuted if they do not depend on each other.

$$\frac{\begin{array}{c} \forall i \leq n, x_i \neq x \\ \forall i' \leq m, x_i' \neq x' \\ (x, X(x_1, ..., x_n)) \,, \, (x', X'(x_1', ..., x_m')) \end{array}}{(x', X'(x_1', ..., x_m')) \,, \, (x, X(x_1, ..., x_n))} \lesssim$$

4.3   General Rules Relating Events and Probabilities

The following rules bridge the gaps between abstract computations, events and probabilities. Soundness is established by using the definitions from Section 2.

Event Approximation for Continuous Distributions Rule.   The following rule allows the numerical approximation of a probability. Given the continuous distribution D, its probability density function P_D and an order on the elements with this distribution $\leq$. We can use an approximation of $P_D : P_A$ in order to guarantee a certain maximal probability of an event that checks whether a certain value drawn from D is below some upper bound a.

$$\frac{\forall v . P_D(v) \leq P_A(v) \qquad P_A(a) < \epsilon}{Pr([(x, D)](x \leq a)) < \epsilon}$$

We can use this rule to simplify expressions and leave the subgoal $P_A(a) < \epsilon$ for numerical approximation. Similarly the following rule holds:

$$\frac{\forall v . P_A(v) \leq P_D(v) \qquad 1 - P_A(a) < \epsilon}{Pr([(x, D)](x \geq a)) < \epsilon}$$

Note that for the first rule the approximation $\int_{-\infty}^{\infty} P_A(v) \, dv$ will be greater or equal than 1 while in the second rule it while be smaller or equal than one. This is ensured by the first condition in both rules.

Range Event Splitting Rule. The following rule may be used to split an event stating that a variable is outside a certain range into two independent subevents.

$$\frac{Pr([(x, D)](x \geq a)) < \epsilon_1 \qquad Pr([(x, D)](x \leq b)) < \epsilon_2}{Pr([(x, D)](x \geq a \vee x \leq b)) < \epsilon_1 + \epsilon_2}$$

4.4   Rules for Normal Distributions

Here we present a few rules valid for normal distributions. Their soundness can be easily proven since they correspond to well known facts about normal distributions.

Normal Distribution Rule. We have established a rule for combining values originating from different normal distributions $\mathcal{N}(\mu_i, \sigma_i^2)$.

$$\frac{(x', \mathcal{N}(\mu_1 + ... + \mu_n, \sigma_1^2 + ... + \sigma_n^2))}{\begin{pmatrix} x_1, \mathcal{N}(\mu_1, \sigma_1^2) \\ ... \\ x_n, \mathcal{N}(\mu_n, \sigma_n^2) \end{pmatrix}, (x', \mathcal{U}_T(x_1 + ... + x_n))} \gtrsim$$

Normal Distribution Probability Event Rule. Another rule relates normal distributions, events, and probabilities.

$$\frac{Pr([(x, \mathcal{N}(\mu, \sigma^2))](x \leq \mu - a)) < \epsilon \qquad a \leq \sigma \qquad \sigma \leq \sigma'}{Pr([(x, \mathcal{N}(\mu, \sigma'^2))](x \leq \mu - a)) < \epsilon}$$

It corresponds to standard facts on normal distributions. Likewise the following rule holds.

$$\frac{Pr([(x, \mathcal{N}(\mu, \sigma^2))](x \geq \mu + a)) < \epsilon \qquad a \leq \sigma \qquad \sigma \leq \sigma'}{Pr([(x, \mathcal{N}(\mu, \sigma'^2))](x \leq \mu + a)) < \epsilon}$$

These rules relate abstract computations with typical events that check whether a normally distributed variable has a value in a certain range. Their correctness is established by looking at the probability density function for the normal distribution,

$$P_{\mathcal{N}}(x) = \frac{1}{\sigma\sqrt{2\pi}} exp\left(-\frac{1}{2}\left(\frac{x-\mu}{\sigma}\right)^2\right)$$

its derivation and especially the points $\mu - \sigma$ and $\mu + \sigma$.

Voting Abstraction Rule. A specialized rule for simplifying the semantics of voting can be established. Setups containing a voting computing the mean of several values which may be influenced by $\mathcal{N}$ distributed errors can be simplified by using the following rule:

$$\frac{((e, \mathcal{N}(\frac{\mu_1+...+\mu_n}{n}, \frac{\sigma_1^2+...+\sigma_n^2}{n^2})) , (r, \mathcal{U}_{T_r}(x+e)))}{\begin{pmatrix} e_1, \mathcal{N}(\mu_1, \sigma_1^2) \\ ... \\ e_n, \mathcal{N}(\mu_n, \sigma_n^2) \end{pmatrix}, \begin{pmatrix} v_1, \mathcal{U}_T(x+e_1) \\ ... \\ v_n, \mathcal{U}_T(x+e_n) \end{pmatrix}, (r, \mathcal{U}_{T_r}\left(\frac{v_1+...+v_n}{n}\right))} \lesssim$$

The soundness of this rule is derived from the Normal Distribution Rule and the Function Propagation Rule.

Additional Rules and Approximations. Our presented rules may be used to perform simplifications in order to discover a certain correctness result. Additionally, after these simplifications we may use numerical methods, e.g., to approximate probabilities of expressions that are not easily handled in an algebraic way.

5   Case Study

In this case study we regard a work piece on a conveyor belt. Actuators and sensors are used to bring it close to a desired position. A version of the voting element (cf. Section 3.2) with two sensors is used in our case study:

$$vote2(x) \equiv$$
$$\begin{pmatrix} e_1, \mathcal{N}(\mu_E, \sigma_E^2) \\ e_2, \mathcal{N}(\mu_E, \sigma_E^2) \end{pmatrix} , \begin{pmatrix} v_1, x+e_1 \\ v_2, x+e_2 \end{pmatrix} , (r, \mathcal{U}_{T_r}\left(\frac{v_1+v_2}{2}\right))$$

In the actual case study a work piece is put on a conveyor belt. Our goal is to bring this work piece close to a position p. We can measure the position x on the conveyor belt by using two sensors. The sensors can be influenced by normal distributed errors. We can activate the conveyor belt to adjust the position. This repositioning will also be influenced by some possible error.

Our conveyor belt case study is shown in Figure 2. In our formalism it is described as shown in Figure 3. The following sequence is performed two times in order to achieve a good positioning of the work piece:

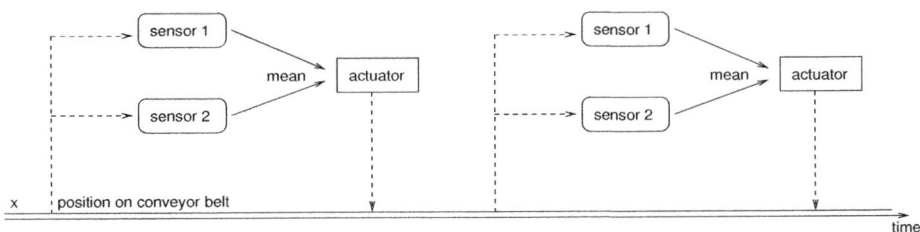

Fig. 2. Conveyor belt control

$conv\_belt(x, p) \equiv$

$$\begin{pmatrix} e_1, \mathcal{N}(\mu_E, \sigma_E^2) \\ e_2, \mathcal{N}(\mu_E, \sigma_E^2) \end{pmatrix}, \begin{pmatrix} v_1, x + e_1 \\ v_2, x + e_2 \end{pmatrix}, \left(r, \mathcal{U}_{T_r}\left(\frac{v_1+v_2}{2}\right)\right), (p', \mathcal{U}_T(p-r)),$$

$$(e, \mathcal{N}(\mu_E', \sigma_E'^2)), (x, \mathcal{U}_T(p' \cdot (1 + e))), \begin{pmatrix} e_1, \mathcal{N}(\mu_E, \sigma_E^2) \\ e_2, \mathcal{N}(\mu_E, \sigma_E^2) \end{pmatrix}, \begin{pmatrix} v_1, x + e_1 \\ v_2, x + e_2 \end{pmatrix},$$

$$\left(r, \mathcal{U}_{T_r}\left(\frac{v_1+v_2}{2}\right)\right), (p', \mathcal{U}_T(p-r)), (e, \mathcal{N}(\mu_E', \sigma_E'^2)), (x, \mathcal{U}_T(p' \cdot (1 + e)))$$

Fig. 3. Conveyor belt control (formal description)

- We read the value of the work piece via two different sensors. The value of the position of the work piece x may be effected by $\mathcal{N}(\sigma_E, \sigma_E)$ distributed sensor errors e_1, e_2 while reading them with the sensors. The sensors write the values which they have read to variables v_1, v_2. We perform a mean voting on their results and store it in a variable r.
- We activate the conveyor belt in order to move the work piece close to p by giving it $p - r$ as repositioning information.
- The conveyor belt is not necessarily moved exactly by the requested distance, but again, the actuator introduces an error which is $\mathcal{N}(\sigma_E', \sigma_E')$ distributed.

Our goal is to ensure that the conveyor belt will be in a close range to a distinct position p with a certain probability. We are interested in questions like: what is the probability that the difference between optimal and actual position is smaller than a given constant l: $p - x < l$:

$$Pr([conv\_belt(x, p)](l \leq p - x)) < \epsilon$$

The maximal probability that this is not the case shall be ϵ.

In order to do so, we apply our rules defined in Section 4 to the system definition. The goal is to derive the distribution of p once the system has reached its terminal state. In a first step, we simplify the voting of the sensors (Congruence Exchange, Voting Abstraction, Permutation, associativity of monads). Thus, we derive the following simplified system description:

$$\left(r, \mathcal{N}\left(x + \mu_E, \frac{\sigma_E^2}{2}\right)\right), (p', \mathcal{U}_T(p-r)), (e, \mathcal{N}(\mu_E', \sigma_E'^2)), (x, \mathcal{U}_T(p' \cdot (1 + e))),$$

$$\left(r, \mathcal{N}\left(x + \mu_E, \frac{\sigma_E^2}{2}\right)\right), (p', \mathcal{U}_T(p-r)), (e, \mathcal{N}(\mu_E', \sigma_E'^2)), (x, \mathcal{U}_T(p' \cdot (1 + e)))$$

This can be further simplified (Congruence Exchange, Function Propagation):

$$\left(r, \mathcal{N}\left(x + \mu_E, \frac{\sigma_E^2}{2}\right)\right), (e, \mathcal{N}(\mu_E', \sigma_E'^2)), (x, \mathcal{U}_T((p - r) \cdot (1 + e))),$$

$$\left(r, \mathcal{N}\left(x + \mu_E, \frac{\sigma_E^2}{2}\right)\right), (e, \mathcal{N}(\mu_E', \sigma_E'^2)), (x, \mathcal{U}_T((p - r) \cdot (1 + e)))$$

We do not touch the expression containing the product of normal distributed variables.

Since the initial position x and p are known, at this stage one can numerically handle the expression in order to convince oneself that numerical constraints on these distributions are met. Computer algebra systems allow for an over-approximation of the distribution above. Thus, one can apply the Event Approximation for Continuous Distributions Rule and ensure that the given failure probability ϵ is met.

Furthermore, given the expression above one can consider possible optimization alternatives, like, e.g., updating the sensors or actuators so that they feature a better error distribution. Another optimization possibility would be to replicate the sensor- actuator part another time. One can numerically recalculate the the results and convince oneself that they meet a certain ϵ.

6 Conclusion

We propose a framework to specify systems and their behavior and possible divergences that might occur in these systems. Our framework is build upon a monadic representation of execution steps which allows the representation of possible errors and their probabilities. A rule based logic is presented to reason about our system description and perform algebraic simplifications. Ultimately we derive guarantees for the system description. Our rules encapsulate common tasks, reasoning is not limited to these rules. In particular we allow numerical approximations. We present two case studies to demonstrate possible usage scenarios of our framework.

Future Work. As a long term goal, our framework is intended to be used with the Coq [6] theorem prover. Depending on the usage scenario, an implementation in another verification / analysis tool might also be an option. Currently we are also looking at additional case studies.

Acknowledgment. This work has been supported by the European research project ACROSS under the Grant Agreement ARTEMIS-2009-1-100208.

References

1. Audebaud, P., Paulin-Mohring, C.: Proofs of randomized algorithms in Coq. Science of Computer Programming (2008)
2. Ayache, S., Conquet, E., Humbert, P., Rodriguez, C., Sifakis, J., Gerlich, R.: Formal methods for the validation of fault tolerance in autonomous spacecraft. In: International Symposium on Fault-Tolerant Computing, FTCS 1996 (1996)
3. Blech, J.O.: Proving the Security of ElGamal Encryption Via Indistinguishability Logic. In: ACM Symposium On Applied Computing (2011)
4. Blech, J.O., Hattendorf, A., Huang, J.: An Invariant Preserving Transformation for PLC Models. In: IEEE International Workshop on Model-Based Engineering for Real-Time Embedded Systems Design (2011)
5. Blech, J.O., Périn, M.: Generating Invariant-based Certificates for Embedded Systems. ACM Transactions on Embedded Computing Systems (TECS) (to appear)
6. The Coq development team: The Coq Proof Assistant Reference Manual v8.3 (2010), http://coq.inria.fr
7. Hallerstede, S., Hoang, T.S.: Qualitative Probabilistic Modelling in Event-B*. In: Davies, J., Gibbons, J. (eds.) IFM 2007. LNCS, vol. 4591, pp. 293–312. Springer, Heidelberg (2007)
8. Hanmer, R.: Patterns for Fault Tolerant Software. Wiley (October 2007) ISBN: 978-0-470-31979-6

9. Jeffords, R., Heitmeyer, C., Archer, M., Leonard, E.: A Formal Method for Developing Provably Correct Fault-Tolerant Systems Using Partial Refinement and Composition. In: Cavalcanti, A., Dams, D.R. (eds.) FM 2009. LNCS, vol. 5850, pp. 173–189. Springer, Heidelberg (2009)
10. Kljaich, J., Smith, B.T., Wojcik, A.S.: Formal Verification of Fault Tolerance Using Theorem-Proving Techniques. IEEE Transactions on Computers 38(3) (March 1989)
11. Kwiatkowska, M., Norman, G., Parker, D.: PRISM: Probabilistic Symbolic Model Checker. In: Field, T., Harrison, P.G., Bradley, J., Harder, U. (eds.) TOOLS 2002. LNCS, vol. 2324, pp. 200–204. Springer, Heidelberg (2002)
12. McIver, A., Morgan, C.: Abstraction, Refinement and Proof for Probabilistic Systems. Springer, Heidelberg (2005)
13. Owre, S., Rushby, J., Shankar, N., von Henke, F.: Formal verification for fault-tolerant architectures: prolegomena to the design of PVS. IEEE Transactions on Software Engineering (February 1995)
14. Steiner, W., Rushby, J., Sorea, M., Pfeifer, H.: Model Checking a Fault-Tolerant Startup Algorithm: From Design Exploration To Exhaustive Fault Simulation. In: The International Conference on Dependable Systems and Networks. IEEE Computer Society (2004)
15. Programmable controllers - Part 3: Programming languages, IEC 61131-3: 1993, International Electrotechnical Commission (1993)
16. Wadler, P.: The essence of functional programming. In: 19'th Symposium on Principles of Programming Languages. ACM Press (January 1992)
17. Pike, L., Maddalon, J., Miner, P., Geser, A.: Abstractions for Fault-Tolerant Distributed System Verification. In: Slind, K., Bunker, A., Gopalakrishnan, G.C. (eds.) TPHOLs 2004. LNCS, vol. 3223, pp. 257–270. Springer, Heidelberg (2004)

Self-stabilizing Mutual Exclusion and Group Mutual Exclusion for Population Protocols with Covering

Joffroy Beauquier[1,*] and Janna Burman[2,**]

[1] LRI, University Paris-Sud 11, France
joffroy.beauquier@lri.fr
[2] MASCOTTE, INRIA, I3S (CNRS/University of Nice Sophia-Antipolis), France
janna.burman@inria.fr.

Abstract. This paper presents and proves correct two *self-stabilizing* deterministic algorithms solving the *mutual exclusion* and the *group mutual exclusion* problems in the model of *population protocols* with *covering*. In this variant of the population protocol model, a *local fairness* is used and bounded state anonymous mobile agents interact in pairs according to constraints expressed in terms of their *cover times*. The cover time is an indicator of the "time" for an agent to communicate with all the other agents. This indicator is expressed in the number of the pairwise communications (*events*) and is unknown to agents. In the model, we also assume the existence of a particular agent, the *base station*. In contrast with the other agents, it has a memory size proportional to the number of agents. We prove that without this kind of assumption, the mutual exclusion problem has no solution.

The algorithms in the paper use a *phase clock* tool. This is a synchronization tool that was recently proposed in the model we use. For our needs, we extend the functionality of this tool to support also phases with unbounded (but finite) duration. This extension seems to be useful also in the future works.

Keywords: distributed algorithms, mobile agent networks, population protocols, cover times, self-stabilization, synchronization, (group) mutual exclusion.

1 Introduction

Population protocols is an elegant communication model [2] specially designed for large cheap sensor networks with resource-limited mobile agents. In the original model, each agent is represented by a finite state machine. Agents are anonymous and move in an asynchronous way. When two agents come into range of each other (meet), they can exchange information (communicate). It is important to note that in this model, a type of *global fairness* condition (in the sense of, e.g., [3, 17]) is imposed on the scheduler. According to this condition, a configuration that can be reached infinitely often during the execution is reached infinitely often. This is in contrast to a weaker type of condition called *local fairness* which is generally assumed in the theoretical literature on

---

* The work of this author was partially supported by grants from Grand Large project, INRIA Saclay.
** The work of this author was supported by Chateaubriand grant from the French Government.

A. Fernández Anta, G. Lipari, and M. Roy (Eds.): OPODIS 2011, LNCS 7109, pp. 235–250, 2011.
© Springer-Verlag Berlin Heidelberg 2011

distributed computing. Informally, with local fairness, in an infinite fair execution, each agent satisfying certain conditions is given a turn infinitely often. In contrast to local fairness, global fairness brings an effect of randomization of the scheduling and hence has the ability to circumvent many impossibility results. For example, self-stabilizing algorithms which are proved correct in [3] will fail if local fairness is assumed (refer to [3, 17] for examples and a more detailed discussion on fairness).

In the considered model of population protocols with covering [8], a type of local fairness is used. According to this fairness, there is an indicator called *cover time* associated to each agent. A cover time of an agent x, cv_x, is the minimum (unknown to agents) number of global pairwise interactions (or events) that should happen in the system for being certain that x has met each other agent. A scheduler schedules the next event according to the cover times of agents. Even though this type of the local fairness condition seems relatively strong, many problems stay impossible to solve in such a model (see, e.g., [6, 9]). In this work, we prove a similar impossibility result stating that in the model with only bounded state agents, mutual exclusion has no solution (Sec. 6). This result can be easily extended to the group mutual exclusion problem as well. For circumventing the impossibility result, as in [6, 9, 10], we introduce a special agent, the base station (BS). We assume that BS has a memory size proportional to the number of the other system agents.

Like in the original model of population protocols [2], we assume that the number of agents in a system is unknown, that the agents, except BS, are anonymous (no identifiers, uniform code) and have only a few bits of memory (independent of the number of agents). As in [2], we assume a complete communication topology, where each agent has the ability to communicate sometime with any other agent.

We note that the fairness condition expressed in terms of the bounded cover times is supported by recent experimental and analytical studies. The assumption that an agent communicates with all other agents periodically, within a bounded period of real time, has been experimentally justified for some types of mobility such as the human or animal mobility within a bounded area or mobility with "home coming tendency" (the tendency to return periodically to some specific places, e.g., agents' homes). In these cases, a statistical analysis of experimental data sets confirms this assumption. These data sets concern students in a campus [1, 24], participants in a network conference [25], visitors at Disney World and more. All exhibit the fact that the *Inter Contact Time* (ICT - the time period between two successive contacts of the same two mobile agents) follows a so called truncated Pareto distribution [11, 20, 23]. In particular, this involves that ICT is practically bounded. Thus, ICT is also bounded when measured in the number of events. In our model, the cover time of an agent can be expressed in terms of ICTs. Hence, a cover time can be indeed bounded in practice.

The algorithms presented in this paper are self-stabilizing [14]. Such algorithms have the important property of operating correctly (except for some finite period), regardless of their initial configuration. In practice, self-stabilizing algorithms adjust themselves automatically to any change or corruption of the network components (excluding the algorithm's code). Those changes are assumed to cease for some sufficiently long time. Self-stabilization is considered here for two reasons. First, mobile agents are generally fragile, subject to failures and hard to initialize. Second, systems of mobile agents are by essence dynamic, some agents may leave the system while new ones can be introduced. Self-stabilization is a well adapted framework for dealing with such situations.

In this paper, for the model of population protocols with covering, we present self-stabilizing deterministic solutions for two basic and classical problems in distributed computing, mutual exclusion [21, 26] (Sec. 4) and group mutual exclusion - GME [18, 22] (Sec. 5). Self-stabilizing mutual exclusion in population protocols has been addressed in [3] and in [12]. In [3], a solution is given for Dijkstra-style token circulation in oriented ring communication topology and assumes global fairness (it does not work in the considered case of local fairness). In [12], a deterministic solution assuming local fairness is given. However, it uses strong schemes of "mobile agents" and of oracles. In [6], a simple self-stabilizing algorithm for GME is proposed for the same model as here. However, in this solution, agents can spend only some bounded predefined period in the *critical session* (CS). This is in contrast with the classical definition of GME, according to which they are allowed to spend an unbounded but finite amount of time in CS. Refer to Sec. 2.3 for the specifications we consider for mutual exclusion and for GME.

We design our algorithms in combination with the phase clock tool proposed in [6]. This synchronization tool provides an easy way to organize protocol execution into phases. However, the tool in [6] allows to define and run only phases with bounded duration. In order to be able to grant to an agent an unbounded (but finite) amount of time in CS, we extend the phase clock with the mechanism allowing it to manage also the unbounded duration phases (Sec. 3).[1]

The technique we use for stabilizing the algorithms is similar to the one used in [5, 15]. The variables of the algorithm are regularly reset to predetermined values and an instance of a protocol is repeatedly executed. Note that the automatic self-stabilizing transformer of [9], which uses the same technique, cannot be useful here, because it applies only to static problems.

The problems are not as simple as they appear at first sight, because of the self-stabilization requirement, unknown population size, asynchrony and the anonymity of agents, in particular, when fairness of access to a resource is concerned.

2 Model and Problem Specifications

Basically, the model is as in [2], with the addition of the cover times (Sec. 2.2) and BS.

2.1 Transition System

Let A be the set of all the agents in system $\mathcal{S}$, where the (population) size of $\mathcal{S}$ is $|A| = n$, and n is unknown to agents. Among the agents, there may be a distinguishable one, the *base station* (BS), which is said *non-mobile*[2] (and which can have an unbounded memory in contrast with the other agents). All the other agents are bounded state, anonymous (have uniform codes and no identifiers) and are referred as *mobile*.

---

[1] Note however that this extension does not imply an automatic or easy extension of the GME solution in [6], mainly due to the requirements of CS access fairness in GME.

[2] If BS is actually mobile, it will not change the analysis in this paper.

Population protocols can be modeled as transition systems. We adopt the common definitions of the following: *state* of an agent (a vector of the values of its variables), *configuration* (a vector of states of all the agents), *transitions* (possible atomic steps of agents and their associated state changes), *execution* (a possibly infinite sequence of configurations in which each element follows its predecessor by a transition). For the formal definitions, refer, e.g., to [26]. We refine and add some terms below.

A *local transition system* of an agent x is defined by a set of states and a set of transitions between states.

An event (x, y) is a pairwise communication (meeting) of two agents x and y. During an event, a transition of the form $(\pi_x, \pi_y) \to (\pi'_x, \pi'_y)$ is executed, where π_x and π_y are the states of x and y before the event, and π'_x and π'_y are the states of x and y after the event (the states after the event may not change). We assume symmetrical communication during (x, y), and (x, y) is an unordered pair. However, the results in the paper can be adapted to asymmetrical communications, in which the state of only x (or y) is modified during an event (x, y).

We extend the transitions between states to configurations as follows. Without loss of generality and as in [2], we assume that no two events happen "simultaneously". Then, there is a transition between two configurations C and C', iff there is a transition $(\pi_x, \pi_y) \to (\pi'_x, \pi'_y)$, from C and resulting in C', for some two agents x and y, and such that the states of all the other (than x and y) agents are identical in C and C'. Note that each execution corresponds to a unique sequence of events.

Intuitively, it is convenient to view executions as if a *scheduler* (an adversary) "chooses" which two agents participate in the next event. Formally, a scheduler $\mathcal{D}$ is a predicate on the sequences of events. A *schedule* of $\mathcal{D}$ is a sequence of events that satisfies predicate $\mathcal{D}$. To each schedule s corresponds a unique execution of the system (if it is deterministic, and possibly several executions, if it is not). We say that this (or these) execution(s) is (are) *induced* by the schedule s. Let a *period* or a *segment* of a schedule (or of an execution) s be any consecutive sequence of events.

For some l ($\in \mathbb{N}_0$) and agent x, let l *local events at* x, denoted $[l]^x$, be l consecutive (from x's "point of view") events in which agent x participates. This stands in contrast to l *global events* (or just events) which are l consecutive events in an execution. Note that if $[l]^x$ events occurred, then at least l global events occurred.

As in [26], a specification $\mathcal{P}$ of a problem is a predicate on the executions. We say that a system (algorithm) $\mathcal{S}$ solves a specification $\mathcal{P}$, iff any execution of $\mathcal{S}$ satisfies $\mathcal{P}$.

A transition system (or an algorithm) is said to be *self-stabilizing* for a specification $\mathcal{P}$ iff there exists a subset of the set of configurations, called *legitimate configurations*, such that starting from an arbitrary configuration, any execution reaches a legitimate configuration and any execution starting from a legitimate configuration satisfies $\mathcal{P}$. When an execution reaches a legitimate configuration, we say that a system (algorithm) *stabilizes* for $\mathcal{P}$ or just that *stabilization* has occurred. The maximum number of events until stabilization is the *stabilization time* of the protocol. More formal definitions can be found in [26].

2.2 The Cover Time Property (Covering)

Definition 1. *Given a system $\mathcal{S}$ with n agents, a vector $\overline{cv} = (cv_1, cv_2, \ldots, cv_n)$ of positive integers (the* cover times*) and a scheduler $\mathcal{D}$, $\mathcal{D}$ (and any of its schedules) is said to satisfy the* cover time property *for $\overline{cv}$ (in $\mathcal{S}$), if and only if any segment of cv_i ($\forall i : i \in \{1 \ldots n\}$) consecutive events of each schedule of $\mathcal{D}$ contains at least one event of an agent x_i with every other agent.*
- *Any* execution *of $\mathcal{S}$ under such a scheduler is said to* satisfy the cover time property for $\overline{cv}$.
- *The minimum cover time value in $\overline{cv}$ is denoted by cv_{min} and the maximum one by cv_{max}. A* fastest / slowest *agent x has $cv_x = cv_{min}$ / $cv_x = cv_{max}$.*

Note that agents are not assumed to know cover times. They are (usually) unable to store them as their memory is bounded (note that each cover time depends on n).

During the analysis, we consider all the executions that satisfy the cover time property for a given vector of cover times. However, for some vectors of cover times, there is no schedule satisfying the cover time property (take, for instance, the vector $(4, 6, 11, 11)$). A *vector of cover times $\overline{cv}$ is* acceptable *if and only if there exists at least one schedule satisfying the cover time property for $\overline{cv}$. In the sequel, we will only consider acceptable vectors of cover times.

Remark 1. Let $\overline{cv}$ be a vector of cover times in system $\mathcal{S}$. Assume that in $\mathcal{S}$, there exists a schedule s of length cv_{min} containing at least once every possible event. Then, the infinite schedule s_∞ resulting from the infinite iteration of s satisfies the cover time property of $\mathcal{S}$.

In our algorithms, for the purpose of event counting and for being able to use the phase clock tool from [6], BS should be able to estimate the upper bounds on the values of cv_{min} and cv_{max} in the system. The self-stabilizing algorithms that estimate cv_{min} and cv_{max} are presented in [9]. They are executed at BS and stabilize in $O([cv_{min}]^{BS})$ events. We assume that those algorithms provide the upper bounds of cv_{min} and cv_{max} in variables cv_{min}^* and cv_{max}^* respectively.

2.3 Specifications

2.3.1 The Mutual Exclusion Problem
We adopt here the presentation of the mutual exclusion problem in [21]. Thus, it is assumed that each mobile agent has a section of code programmed to use some shared resource. This section of code is called a *critical section* (CS). For the mutual exclusion, it is required that at most one mobile agent executes CS at any given time. An execution of CS is preceded by an *entry section* and followed by an *exit section*. An agent executing its entry section is said to ask for entering CS. After entering CS, an agent is assumed to leave it after some finite but unbounded period. Agents have to deliver the requests one at a time. Once a request is delivered, an agent has to access its CS before making another request.

Definition 2 (Specification of Mutual Exclusion).
- **Safety:** *In any configuration, no two agents are executing CS.*

- **Bounded waiting:** *There exists some bound B such that when an agent asks for entering CS, it can be preceded by at most B other agents, but eventually enters CS.*

The problem of mutual exclusion can be also presented and solved by introducing a token that circulates between the agents and grants the access to CS (see, e.g., [14]). Thus, the mutual exclusion problem is sometimes named a problem of *token circulation*. Our solution to the problem use the circulating token as well.

2.3.2   The Group Mutual Exclusion Problem (GME)

GME was introduced by Joung [22]. The problem deals with sharing r mutually exclusive resources between n processors (agents, in our case). There has been some discussions about a precise specification of GME. We adopt here the one given in [18], which is the most precise. As in the mutual exclusion problem, each mobile agent has a section of code called a critical section (CS), which is preceded by an entry section and followed by an exit section. After entering CS, an agent is assumed to leave it after a finite but unbounded period. In [18], an agent wishing to enter CS is said to request a *session*. Sessions represent resources. One resource can be used simultaneously by an arbitrary number of mobile agents, but two or more resources cannot.[3] Thus, agents that have requested different sessions cannot be in CS simultaneously, but agents that have requested the same session can. Each agent requests only one session at a time and it cannot request another session while a request it made is still pending. In [18], the remaining part of the code of an agent, which is outside the critical section, the entry section and the exit section, is called a *noncritical section* (NCS).

Definition 3 (Specification of Group Mutual Exclusion).

- **Mutual exclusion:** *If two agents are in CS at the same time, then they request the same session.*
- **Lockout freedom:** *If an agent enters its entry section, then it eventually enters CS.*
- **Bounded exit:** *If an agent enters the exit section then it enters NCS within a bounded number of its own steps.*
- **Concurrent entering:** *If an agent x requests a session and no other agent requests a different session, then x enters CS within a bounded number of its own steps.*

3   Self-stabilizing Phase Clock Tool

The algorithms we propose use the self-stabilizing bounded phase clock designed in [6]. The specification it satisfies is given below and follows a conventional definition of a phase clock (see, e.g., [4, 13]) with some adaptation (as the *frequency of progress* condition) to better suit the model of population protocols with covering.

Definition 4 (Specification of a Phase Clock [6]).
A bounded phase clock (the clock size $\mathbf{K}$ *is definable) provides each agent* x *with a clock/phase value in the variable* $\mathtt{clock}_x$ *subject to the following conditions.*

---

[3] In the following, session is also referred as the *period* during which agents can share the same resource.

- **Progress:** *In any execution, every variable* clock *is updated infinitely often and each time, according to the assignment statement* clock := (clock + 1) mod **K** *only.*
- **Frequency of progress:** *In any execution, after every update of* clock$_x$, *the next update cannot happen before* β(clock$_x$) *events, where* β *is a predefined function.*
- **Asynchronous unison:** *In any configuration reached by an execution, the clock values of any two agents differ by no more than* 1 (mod **K**). *That is, for any two agents x and y, the following predicate is true:* (clock$_x$ = clock$_y$)$\lor$ (clock$_x$ = (clock$_y$ + 1) mod **K**) $\lor$ (clock$_y$ = (clock$_x$ + 1) mod **K**)

In the phase clock algorithm of [6], agents synchronize their clocks with BS. According to the algorithm, BS is the only agent that increments the clock value, which is then propagated to the other agents. A maximal period (a segment of an execution) during which clock$_{BS}$ = p (for some p) is called *phase p*. A period between two increments of the clock at BS is called a *complete phase p* if after the first increment clock$_{BS}$ = p. *Incomplete phases* arise from a bad (faulty) initialization.

Remark 2. A useful property of the phase clock after stabilization is that during a complete phase p, there is a period of at least $\beta(p)$ events, where all the agents have the same clock value p (Lem. 7 [6]).
In addition, by Lem. 1 [6], every phase p is bounded, if $\beta(p)$ is bounded. Every complete phase p lasts $[\beta(p) + \min(2 \cdot \mathbf{cv}^*_{\min}, \mathbf{cv}^*_{\max})]^{BS}$ events.

Extending the Phase Clock of [6]. To use the phase clock of [6], one should define the size of the clock **K** (the number of the required phases) and the duration β for every phase. However, the ability to define only a bounded phase duration does not seem to be good enough for our purposes here. For being able to grant an agent a finite but unbounded period in CS, we should be able to define an unbounded phase that could be preempted externally, after some finite period. Then, the *progress* condition (in Def. 4) could be still satisfied. Thus, we slightly extend the phase clock of [6] to provide this functionality. We change appropriately the *frequency of progress* condition in the specification.

Definition 5 (New *frequency of progress* condition).
Frequency of progress: *In any execution, after every update of* clock$_x$, *the next update cannot happen before* β(clock$_x$) *events, if* β(clock$_x$) $\neq \infty$, *where* β *is a predefined function. Otherwise* (β(clock$_x$) = ∞), *the next update happens in a finite period.*

To satisfy the new specification, we adopt the same implementation of the phase clock as in [6] and we add only two things. First, we allow to define an unbounded duration for a phase p, by defining $\beta(p) = \infty$. Second, we design a user interface procedure called switch() which, when invoked, causes the phase clock to switch to the next phase in a finite period. The switch() procedure is implemented as follows: switch() = {event_ctr $\leftarrow$ min(2 · $\mathbf{cv}^*_{\min}$, $\mathbf{cv}^*_{\max}$)}, where event_ctr is the counter variable of the phase clock in [6] which is managed by BS and holds the number of the events that are still to count before the next switch (increment) of clock$_{BS}$. This counter is decremented on each event of BS with some agent. Thus, whenever a user calls switch(), in $[\min(2 \cdot \mathbf{cv}^*_{\min}, \mathbf{cv}^*_{\max})]^{BS}$ events, the phase clock switches to the next phase. Note that in switch(), event_ctr is not set to 0. This is to ensure that each phase lasts at least $\min(2 \cdot \mathbf{cv}^*_{\min}, \mathbf{cv}^*_{\max})$ events, as required in [6].

Remark 3. It is easy to verify that the correctness proofs of the new phase clock are similar to those in [6], with the only difference that now a phase may be defined to have some unbounded but finite duration. To guarantee the stabilization of the phase clock, the user should ensure that if for some p, $\beta(p) = \infty$, then phase p is indeed finite (e.g., ensure that switch() is called after a finite period).

The Extended Phase Clock in a Composition. Our algorithms for mutual exclusion (Sec. 4) and for group mutual exclusion (Sec. 5) use the extended phase clock tool as a module. The (group) mutual exclusion module reads only the clock variables (clock) of the phase clock module and invokes the switch() procedure when appropriate. The modules are composed in a *strict interleaving*. That means that during each event, the codes of both modules are executed one after the other. This composition is not *fair* in the sense of [19] or [26], so that general results about *fair composition* cannot be applied. However, the main result remains true and comes from Remark 3. That is, for proving that the composition of the two modules is self-stabilizing, it suffices to prove the self-stabilization of the (group) mutual exclusion module, assuming that the phase clock is already stabilized. In particular, this implies that one should first prove that every phase is finite.

For the phase clock module used in our algorithms, we define $\mathbf{K} = 3$ (that is, clock $\in \{0, 1, 2\}$), $\beta(0) = \mathbf{cv}^*_{\max}$, $\beta(1) = \infty$ and $\beta(2) = \mathbf{cv}^*_{\min}$. Thus, to prove stabilization, the first step is to prove that phase 1 is finite (see Lem. 1 and Lem. 7 [7]).

4   A Self-stabilizing Solution to Mutual Exclusion

In this section we present a self-stabilizing algorithm, Alg. 1, solving the mutual exclusion problem in the model of population protocols with covering (for any given acceptable vector of cover times) and with BS. As there is no such algorithm if all agents have a bounded state (Corollary 1), we assume that BS has a memory size proportional to the number of agents. The codes of the mobile agents are identical, but BS has a special code. Our solution uses a phase clock tool as explained in Sec. 3.

First, we describe the algorithm once stabilization has occurred. It operates by infinite iteration of a succession of three phases. A first phase (phase 0; lines 6-8), called a *request phase*, has a phase duration (see Sec. 3) of $\mathbf{cv}^*_{\max}$ events. An agent requesting CS (state = request), delivers a request to BS (and becomes **registered**) when it meets BS during phase 0 and its clock is equal to $0(\mod 3)$. During this phase, BS counts the number of requests (in a variable req_ctr). Each request phase is followed by a finite but unbounded phase, which is controlled by BS and called an *access phase*. During an access phase, BS gives the token to the first registered agent visiting it (lines 9-14), waits for the token to be returned (remember that as a basic assumption, an agent uses its CS for a finite time), and then decrements req_ctr (lines 15-20). Then, BS waits for another registered agent and so on, until req_ctr goes down to 0. Then, a *sweeping phase* (having a phase duration of $\mathbf{cv}^*_{\min}$ events) resets all the counters and the states of all the agents (to a **neutral** state) and a new request phase can begin (lines 34-43).

For the protocol to be self-stabilizing, we use the technique of re-initializing periodically the variables of the algorithm. After the re-initialization in the sweeping phase, the algorithm executes as described above. However, due to a bad initialization of the variables, the execution could stay forever in phase 1 and never reach a sweeping phase, and hence, never reach stabilization. As $\texttt{req\_ctr} \leq 0$ is the natural condition for ending phase 1 (line 32), we examine the cases in which $\texttt{req\_ctr}$ could stay strictly positive forever. We identify two such cases. The first one is when the number of the registered agents is lesser than the value of $\texttt{req\_ctr}$ (lines 20 and 32-33). The second case is when there is no token, neither at BS, nor at any mobile agent (consequently, BS will never be able to decrease $\texttt{req\_ctr}$ and switch to the sweeping phase; see lines 10,15, 20 and 32-33). We treat separately these two cases. We introduce a variable $\texttt{no\_req\_evntctr}$ for checking that there are still registered agents (or agents in CS) when $\texttt{req\_ctr} > 0$ (lines 21-25), and a variable $\texttt{no\_token\_evntctr}$ for checking that the system is not in the second case (lines 26-30). These variables are event counters of the local events at BS. Each counter counts till $\texttt{cv}^*_{\max}$ events to ensure that BS has met all the mobile agents during the last period, but neither the registered ones, nor the ones in CS or the ones in the CS exit section (that is, one of the cases above is satisfied). Then, BS calls $\texttt{switch}()$ (line 32-33) to switch to the next (sweeping) phase to reset the algorithm and to stabilize.

4.1 Proving Correctness

First, according to Remark 3, we prove that phase 1 is finite. Recall that phases 0 and 2 are finite (and bounded) by the correctness of the phase clock (see Remark 2).

Lemma 1. *In Alg. 1, every phase 1 is finite.*

Proof: At the beginning of a phase 1 (complete or incomplete), there are two cases:

(A) there is at least one token either at BS or at some agent which is in state **in** or **out**;
(B) there is no such agent holding a token (however, there may be an agent in a different state, holding a token).

In case (A) there are two possibilities:

(a) the number of the registered agents is greater or equal to the value of $\texttt{req\_ctr}$;
(b) the number of the registered agents is strictly lesser than the value of $\texttt{req\_ctr}$.

In case (a), if BS does not hold a token, then during the next event of BS with an agent in state **out** holding a token, $\texttt{token}_{BS}$ becomes true (line 16) and $\texttt{req\_ctr}$ is decremented (line 20). From this point, BS can dispatch the token to different registered agents until $\texttt{req\_ctr} \leq 0$ (lines 9-20). Then, the end of the phase condition, in line 32, triggers.

In case (b), the same happens, but once all the registered agents have visited BS and received the requested CS, $\texttt{req\_ctr}$ stays strictly positive. However, there are no more registered agents. Then, BS counts $\texttt{cv}^*_{\max}$ events without seeing a registered agent (lines 21-25) and the condition $\texttt{no\_req\_evntctr} \geq \texttt{cv}^*_{\max}$, in line 32, causes the end of the phase.

Algorithm 1. Self-stabilizing Mutual Exclusion

Memory in a mobile agent $x \neq$ **BS**

  $\text{token}_x : boolean$

  $\text{state}_x \in \{\textbf{neutral}, \textbf{request}, \textbf{registered}, \textbf{in}, \textbf{out}\}$

Memory in BS

  $\text{token}_{BS} : boolean$

  $\text{req\_ctr} : integer$

  $\text{no\_req\_evntctr} : integer$

  $\text{no\_token\_evntctr} : integer$

1: **when** <u>agent x enters its entry section</u> **do**
2:     $\text{state}_x \leftarrow \textbf{request}$
3: **when** <u>agent x enters its exit section</u> **do**
4:     $\text{state}_x \leftarrow \textbf{out}$
5: **when** agent x communicates with BS - event (x, BS) **do**
6:     **if** $(\text{clock}_{BS} = \text{clock}_x = 0 \mod 3 \wedge \text{state}_x = \textbf{request})$ **then** // *Request Phase*
7:        $\text{state}_x \leftarrow \textbf{registered}$
8:        $\text{req\_ctr} \leftarrow \text{req\_ctr} + 1$
9:     **if** $(\text{clock}_{BS} = \text{clock}_x = 1 \mod 3 \wedge \text{req\_ctr} > 0)$ **then** // *Access Phase*
10:        **if** $(\text{token}_{BS} \wedge \text{state}_x = \textbf{registered})$ **then** // *entering CS*
11:           $\text{token}_{BS} \leftarrow \textbf{false}$
12:           $\text{token}_x \leftarrow \textbf{true}$
13:           $\text{state}_x \leftarrow \textbf{in}$
14:           $\langle$ x enters CS $\rangle$
15:     **if** $(\text{clock}_{BS} = \text{clock}_x = 1 \mod 3 \wedge \text{token}_x \wedge \text{state}_x = \textbf{out})$ **then** // *exiting CS*
16:        $\text{token}_{BS} \leftarrow \textbf{true}$
17:        $\text{token}_x \leftarrow \textbf{false}$
18:        $\text{state}_x \leftarrow \textbf{neutral}$
19:        **if** $(\text{req\_ctr} > 0)$ **then**
20:           $\text{req\_ctr} \leftarrow \text{req\_ctr} - 1$
21:     **if** $(\text{clock}_{BS} = 1 \mod 3 \wedge \text{req\_ctr} > 0)$ **then** // *control of requests*
22:        **if** $(\text{state}_x = \textbf{registered} \vee \text{state}_x = \textbf{in})$ **then**
23:           $\text{no\_req\_evntctr} \leftarrow 0$
24:        **else**
25:           $\text{no\_req\_evntctr} \leftarrow \text{no\_req\_evntctr} + 1$
26:     **if** $(\text{clock}_{BS} = 1 \mod 3)$ **then** // *control of tokens*
27:        **if** $(\text{token}_{BS} \vee (\text{token}_x \wedge (\text{state}_x = \textbf{in} \vee \text{state}_x = \textbf{out})))$ **then**
28:           $\text{no\_token\_evntctr} \leftarrow 0$
29:        **else**
30:           $\text{no\_token\_evntctr} \leftarrow \text{no\_token\_evntctr} + 1$
31:     **if** $(\text{clock}_{BS} = 1 \mod 3)$ **then** // *end of access phase*
32:        **if** $(\text{req\_ctr} \leq 0) \vee (\text{no\_req\_evntctr} \geq \text{cv}^*_{\max}) \vee (\text{no\_token\_evntctr} \geq \text{cv}^*_{\max})$ **then**
33:           $\text{switch}()$
34:     **if** $(\text{clock}_{BS} = \text{clock}_x = 2 \mod 3)$ **then** // *Sweeping Phase*
35:        $\text{token}_{BS} \leftarrow \textbf{true}$
36:        $\text{req\_ctr} \leftarrow 0$
37:        $\text{no\_req\_evntctr} \leftarrow 0$
38:        $\text{no\_token\_evntctr} \leftarrow 0$
39: **when** <u>two mobile agents x and y communicate - event (x, y)</u> **do**
40:     **if** $(\text{clock}_x = \text{clock}_y = 2 \mod 3)$ **then** // *Sweeping Phase*
41:        **if** $\text{state}_x \neq \textbf{request}$ **then**
42:           $\text{state}_x \leftarrow \textbf{neutral}$
43:        $\text{token}_x \leftarrow \textbf{false}$

In case (B), BS has no token and cannot receive one, since the only possibility to receive a token is by executing lines 15-20. In this case, these lines cannot be executed, since for any agent x, the condition $\text{token}_x \wedge \text{state}_x = \textbf{out}$ (line 15) is false. Hence, during each meeting with BS, line 30 is executed and no_token_evntctr is incremented until the end of the phase condition (in line 32) triggers.   ∎

Lem. 1 is the key lemma of the proof of correctness, since it ensures that whatever the initial configuration is, a sweeping phase is eventually reached and causes all the variables to reset. Then, from Remark 3, there is no loss of generality in assuming that at the beginning of an execution, the phase clock is stabilized.

Lemma 2. *In any execution of Alg. 1, at the end of a complete phase 2 (the sweeping phase),* $\text{token}_{BS} = \textbf{true}$ *and, for every mobile agent* x, $\text{token}_x = \textbf{false}$.

Proof: By Remark 2, during phase 2, there is a period of at least $\textbf{cv}_{\min}^*$ events where all the agents have the clock value 2. During $\textbf{cv}_{\min}^*$ events happening in the system, at least $\textbf{cv}_{\min}$ events occur. During $\textbf{cv}_{\min}$ events, at least one agent (a fastest one) meets every other agent, including BS. Hence, during phase 2, for every mobile agent, the condition in line 40 and, for BS, the condition in line 34 become **true** at least once. Thus, by the end of phase 2, for every mobile agent x, $\text{token}_x = \textbf{false}$ due to the execution of line 43, and $\text{token}_{BS} = \textbf{true}$ due to the execution of line 35.   ∎

Lemma 3. *In Alg. 1, after the end of the first complete phase 2, in each configuration during a (complete) phase 1, there is exactly one agent* x *with* $\text{token}_x = \textbf{true}$, *and at the end of the phase 1,* $\text{token}_{BS} = \textbf{true}$.

Proof: By Lem. 2 and the fact that during a request phase the token variables are not updated (lines 6-8), at the end of a request phase (0) the only token is in BS. Right after, in phase 1, the update of the token variables can be done only by executing lines 9-14. There, BS and some mobile agent x in an event exchange the token. Then, the token can be exchanged again, only between the same agents (x returns the token to BS) in lines 15-20. Later in phase 1, the token can move according to the same unique scenario only (with a mobile agent $y \neq x$), because the token variables are updated in lines 9-14 and 15-20 only. Hence, during phase 1 (that comes after the first complete phase 2), there is only one token in the system.
After the sequence of the complete phases 2 and then, 0 (at the beginning of the phase 1), req_ctr equals the number of the registered agents. Hence, the scenario described in phase 1 will repeat to the very end of this phase (till req_ctr becomes 0). Thus, at the end of phase 1, $\text{token}_{BS} = \textbf{true}$.   ∎

Lemma 4 (safety). *In Alg. 1, in each configuration after the end of the first complete phase 2, there is exactly one agent* x *with* $\text{token}_x = \textbf{true}$.

Proof: By Lem. 2 and the fact that during a request phase the token variables are not updated (lines 6-8), during the whole phase 0 (following a complete phase 2) the only token is in BS. Then, in phase 1, the lemma is correct by Lem. 3. By the same lemma, at the end of this phase 1, the only token is in BS. Then, during the whole phase 2, the token stays in BS (lines 35 and 43). Then, the whole scenario of the 3 complete phases repeats and hence, the lemma follows.   ∎

Lemma 5 (bounded waiting). *Assume that in Alg. 1, following a complete phase 2, a mobile agent x asks to enter CS (by line 2). Then, x enters CS (in line 14) during the next (or after the next) phase 1.*

Proof: Starting from the event when x asks to enter CS, x meets BS during phase 0. If the closest phase 0 is incomplete and x does not meet BS, then it meets BS during the next complete phase 0. This is because the duration of the complete phase 0 is cv_{max}^{*} events (during which every mobile agent meets BS). During the event of x with BS in phase 0, x becomes registered and req_ctr is incremented (lines 6-8). Moreover, there are exactly req_ctr registered agents at the end of this phase. During the following phase 1 (this is case (A)-(a) of Lem. 1 proof), there are req_ctr entries in CS (in line 14) by the req_ctr different agents, including x. In the worst case, x has asked to enter CS during phase 0 in which it was not registered. Even in this case, it has not waited more than $2 \cdot (n - 1)$ entries of the other agents before entering CS. ∎

Lemmas 1-5 yield the following theorem.

Theorem 1. *Alg. 1 (composed with the extended phase clock from Sec. 3) is a self-stabilizing solution to the mutual exclusion problem.*

Proof: Define the legitimate configurations as those reached in an execution after a complete phase 2. Lem. 1 involves that starting from an arbitrary configuration, any execution reaches a legitimate configuration. Lem. 2-5 involve that any execution, starting from a legitimate configuration, satisfies the mutual exclusion specification - Def. 2. ∎

5   From Mutual Exclusion to Group Mutual Exclusion

Our solution to group mutual exclusion is given in Alg. 2 and follows a similar scheme as the solution for mutual exclusion (Alg. 1). It also iterates three phases: sweeping phase (phase 2 of the phase clock), request (phase 0) and access phase (phase 1). The difference here is that each such iteration represents a session and has a session number $j \in \{1, 2, \ldots, r\}$ during the execution. BS executes sessions successively for resources $\{1, 2, \ldots, r\}$ in a repetitive way. Session j starts with the first event of phase 2, where the session number is incremented to value j (line 33) and it ends with the last event of the next phase 1.

During session j, in the access phase, agents that have requested an access to session j (by going into the state **request**$_j$), receive a token from BS and access CS as in Alg. 1. The difference here is that several tokens can be given at the same time. This allows several agents requesting the same resource to enter CS, what (partially) satisfies the *concurrent entering* condition of GME. For each session, the request and the sweeping phases operate similarly as in Alg. 1 with the following differences. In the request phase, BS registers only agents having requested the current session j (those which are in state request$_j$; see line 6). In the sweeping phase, BS also advances to the next session by incrementing a session number j by 1 (mod r), in line 33.

Again, as in Alg. 1, to deal with bad initialization and to achieve self-stabilization, every execution should reach a complete sweeping phase. This is ensured by exactly the same mechanism as in Alg. 1 (lines 19-31). The *mutual exclusion* condition of GME is satisfied because two successive access phases are totally disjoint in time (by the correctness of the phase clock). All the other conditions of GME are mainly or partially satisfied by ensuring that each phase in the algorithm is finite. The proof of the algorithm follows almost the same lines as the proof of Alg. 1. Due to the lack of space, it is provided in [7].

6   Impossibility Result

We prove that if local transition systems of all agents are bounded in memory, the mutual exclusion problem has no self-stabilizing solution (which applies to an infinite family of systems) in the model of population protocols with covering. This impossibility result can be easily extended to GME. It justifies the strong assumption of the existence of BS in the paper.

Similar impossibility results are already known in the classical distributed models and also in the population protocol models without cover times (see [3, 12]). However, since the cover time property of the scheduler has similarities with *partial synchrony* [16] (cover times impose restrictions on scheduling of agents), previous results may appear to be different. Moreover, "bad execution" demonstrating the impossibility should be proved to satisfy the cover time property. Hence, for completeness, we provide the impossibility proof for the mutual exclusion problem in the considered model. We begin by giving some definitions required for proving and for stating the result.

Since the mobile agents are anonymous and their codes (the set of transitions) are uniform, their local transition systems are identical. Thus, in the model we use, a (global transition) system is entirely characterized by a vector of cover times and the local transition systems of a mobile agent and of BS, if BS exists. Two systems S_i and S_j, with possibly different number of agents and/or cover time vectors, are said to be *similar*, if and only if they have identical local transition systems.

A *local transition system* of an agent x is said *bounded* if and only if the number of the states (and the transitions) of x is upper bounded by some (predefined) integer constant. A *(global transition) system is bounded*, if and only if the local transition systems of a mobile agent and of BS (if BS exist) are bounded.

A *generic solution* to a problem P is a relation that associates to any positive integer n and to any acceptable vector of cover times $\overline{cv}$ (of size n) a system $S_{n,\overline{cv}}^P$, with n agents and a vector of cover times $\overline{cv}$ (the system scheduler satisfies the cover time property according to $\overline{cv}$), such that $S_{n,\overline{cv}}^P$ solves P. The set of all such systems is called an *image of the generic solution*. A *generic solution* is said *bounded* if and only if every element of its image is bounded. A *self-stabilizing generic solution to the mutual exclusion problem* is a generic solution, such that every system in the image is self-stabilizing for the specification of mutual exclusion in Def. 2.

The first lemma is a simple application of a well known combinatorial property - the pigeonhole principle.

Lemma 6. *Let A be a bounded generic solution to a problem P and let I_A be its image. Then, there exists an infinite subset I_A^{sim} of I_A of similar systems such that for any integer k and for any vector of cover times $\overline{cv}$, there exists in I_A^{sim} a system of size strictly greater than k with a vector of cover times strictly greater (component by component) than $\overline{cv}$.*

Algorithm 2. Self-stabilizing Group Mutual Exclusion

Memory in a mobile agent $x \neq \mathbf{BS}$
  $\mathrm{token}_x : boolean$
  $\mathrm{state}_x \in \{\mathbf{neutral}, \{\mathbf{request}_{j'} | j' = 1, 2, \ldots, \mathbf{r}\}, \mathbf{registered}, \mathbf{in}, \mathbf{out}\}$
Memory in BS
  $\mathrm{req\_ctr} : integer$
  $\mathrm{no\_req\_evntctr} : integer$
  $\mathrm{no\_token\_evntctr} : integer$
  $j \in \{1, 2, \ldots, \mathbf{r}\}$

```
 1: when agent x enters its entry section for resource j' do
 2:     state_x ← request_j'
 3: when agent x enters its exit section do
 4:     state_x ← out
 5: when agent x communicates with BS - event (x, BS) do
 6:     if (clock_BS = clock_x = 0  mod 3 ∧ state_x = request_j) then  // Request Phase
 7:         state_x ← registered
 8:         req_ctr ← req_ctr + 1
 9:     if (clock_BS = clock_x = 1  mod 3 ∧ req_ctr > 0) then  // Access Phase
10:         if (state_x = registered) then  // entering CS
11:             token_x ← true
12:             state_x ← in
13:             ⟨ x accesses resource j ⟩
14:     if (clock_BS = clock_x = 1  mod 3 ∧ token_x ∧ state_x = out) then  // exiting CS
15:         token_x ← false
16:         state_x ← neutral
17:         if (req_ctr > 0) then
18:             req_ctr ← req_ctr − 1
19:     if (clock_BS = 1  mod 3 ∧ req_ctr > 0) then  // control of requests
20:         if (state_x = registered ∨ state_x = in) then
21:             no_req_evntctr ← 0
22:         else
23:             no_req_evntctr ← no_req_evntctr + 1
24:     if (clock_BS = 1  mod 3) then  // control of tokens
25:         if (token_x ∧ (state_x = in ∨ state_x = out)) then
26:             no_token_evntctr ← 0
27:         else
28:             no_token_evntctr ← no_token_evntctr + 1
29:     if (clock_BS = 1  mod 3) then  // end of access phase
30:         if (req_ctr ≤ 0) ∨ (no_req_evntctr ≥ cv*_max) ∨ (no_token_evntctr ≥ cv*_max) then
31:             switch()
32:     if (clock_BS = clock_x = 2  mod 3) then  // Sweeping Phase
33:         j ← (j + 1)  mod r  // advance to the next session
34:         req_ctr ← 0
35:         no_req_evntctr ← 0
36:         no_token_evntctr ← 0
37: when two mobile agents x and y communicate - event (x, y) do
38:     if (clock_x = clock_y = 2  mod 3) then  // Sweeping Phase
39:         if state_x ∉ {request_j'} then  // for any j'
40:             state_x ← neutral
41:         token_x ← false
```

The intuitive justification for the lemma is the following. Since every system in the image of A is bounded, there are only a finite number of possible choices for the local transition systems. Thus, these choices have necessarily to be made infinitely many times, both for $\mathbf{n}$ and for $\overline{\mathbf{cv}}$.

Theorem 2. *There exists no self-stabilizing bounded generic solution to the mutual exclusion problem (in the model of population protocols with covering).*

Proof: Assume by contradiction that there exists a self-stabilizing bounded generic solution to mutual exclusion. Lem. 6 shows that there exists an infinite subset I^{sim} of similar systems self-stabilizing to the mutual exclusion specification (in Def. 2). Without loss of generality, let us assume that the circulating token is used in these systems. Consider some system $\mathcal{S}_1 \in I^{sim}$ that applies to a population of size $\mathbf{n}_1$ (> 1) with a vector of cover times $\overline{\mathbf{cv}}_1$. Let T be the stabilization time of $\mathcal{S}_1$.

By Lem. 6, in I^{sim}, there is another system $\mathcal{S}_2$ self-stabilizing to the mutual exclusion and similar to $\mathcal{S}_1$, with $\mathbf{n}_2$ agents and a cover time vector $\overline{\mathbf{cv}}_2$ such that $\mathbf{n}_2 > \mathbf{n}_1$ and $\mathbf{cv}_{\min}(\mathcal{S}_2) > T + \frac{\mathbf{n}_2 \cdot (\mathbf{n}_2 - 1)}{2}$ (where $\mathbf{cv}_{\min}(\mathcal{S}_2)$ is $\mathbf{cv}_{\min}$ in $\mathcal{S}_2$).

Consider a legitimate configuration C' of $\mathcal{S}_2$ in which there is (exactly) one token in some mobile agent x, and let C be the projection of C' on $\mathbf{n}_1$ arbitrary agents not holding a token. We consider C as the initial configuration of an execution e in $\mathcal{S}_1$. Note that C is effectively a configuration of $\mathcal{S}_1$, since $\mathcal{S}_1$ and $\mathcal{S}_2$ are similar. Moreover, execution e does exist, because $\overline{\mathbf{cv}}_1$ is acceptable. Since $\mathcal{S}_1$ is assumed to be a self-stabilizing solution, in the execution e from C, after at most T events, e reaches a configuration in which one token has been created. Let us denote by $<token\_creation>$ the step during which this token is created. Then $e = e_1 <token\_creation> e_2$ and the length of e_1 is at most T. We claim that $e_1 <token\_creation>$ is the prefix of a possible (infinite) execution of $\mathcal{S}_2$, from configuration C'. First, all the transitions used in this prefix are transitions of $\mathcal{S}_2$, since $\mathcal{S}_2$ and $\mathcal{S}_1$ are similar. Second, since $\mathbf{cv}_{\min}(\mathcal{S}_2) > |e1 <token\_creation>| + \frac{\mathbf{n}_2 \cdot (\mathbf{n}_2 - 1)}{2}$, the fact that the $n_2 - n_1$ agents in $\mathcal{S}_2$, but not in $\mathcal{S}_1$, that have no events in $e_1 <token\_creation>$, does not violate the cover time property. Third, $e_1 <token\_creation>$ can be completed to the segment e' of length $\mathbf{cv}_{\min}(\mathcal{S}_2)$ events by adding all the missing (according to the cover time property) meetings in at least $\frac{\mathbf{n}_2 \cdot (\mathbf{n}_2 - 1)}{2}$ next events. Now, if we repeat indefinitely many times the schedule of e', we can get an infinite execution e'_∞ of $\mathcal{S}_2$ that satisfies the cover time property (see Remark 1). However, in the configuration reached by this execution after the prefix $e_1 <token\_creation>$, there are two tokens - the token that is just created and the token at agent x. That contradicts the assumption that C' is legitimate and that $\mathcal{S}_2$ is a self-stabilizing solution. That proves the theorem. ∎

Corollary 1. *If all agents have a bounded state (bounded by some predefined integer constant, independent of the population size $\mathbf{n}$), there is no generic self-stabilizing solution (and, in particular, no self-stabilizing algorithm) to the mutual exclusion problem.*

References

1. The Dartmouth wireless trace archive - Dartmouth College (2007),
   http://crawdad.cs.dartmouth.edu/

2. Angluin, D., Aspnes, J., Diamadi, Z., Fischer, M.J., Peralta, R.: Computation in networks of passively mobile finite-state sensors. DC 18(4), 235–253 (2006)
3. Angluin, D., Aspnes, J., Fischer, M.J., Jiang, H.: Self-stabilizing population protocols. TAAS 3(4) (2008)
4. Awerbuch, B., Kutten, S., Mansour, Y., Patt-Shamir, B., Varghese, G.: A time-optimal self-stabilizing synchronizer using a phase clock. IEEE TDSC 4(3), 180–190 (2007)
5. Awerbuch, B., Varghese, G.: Distributed program checking: a paradigm for building self-stabilizing distributed protocols (extended abstract). In: FOCS, pp. 258–267 (1991)
6. Beauquier, J., Burman, J.: Self-Stabilizing Synchronization in Mobile Sensor Networks with Covering. In: Rajaraman, R., Moscibroda, T., Dunkels, A., Scaglione, A. (eds.) DCOSS 2010. LNCS, vol. 6131, pp. 362–378. Springer, Heidelberg (2010)
7. Beauquier, J., Burman, J.: Self-stabilizing mutual exclusion and group mutual exclusion for population protocols with covering (extended version). Technical Report Inria-00625838, INRIA (2011), http://hal.inria.fr/inria-00625838/en/
8. Beauquier, J., Burman, J., Clement, J., Kutten, S.: On utilizing speed in networks of mobile agents. In: PODC, pp. 305–314 (2010)
9. Beauquier, J., Burman, J., Kutten, S.: A self-stabilizing transformer for population protocols with covering. Theor. Comput. Sci. 412(33), 4247–4259 (2011)
10. Beauquier, J., Clement, J., Messika, S., Rosaz, L., Rozoy, B.: Self-Stabilizing Counting in Mobile Sensor Networks with a Base Station. In: Pelc, A. (ed.) DISC 2007. LNCS, vol. 4731, pp. 63–76. Springer, Heidelberg (2007)
11. Cai, H., Eun, D.Y.: Crossing over the bounded domain: from exponential to power-law inter-meeting time in MANET. In: MOBICOM, pp. 159–170 (2007)
12. Canepa, D., Gradinariu Potop-Butucaru, M.: Self-stabilizing tiny interaction protocols. In: WRAS, pp. 10:1–10:6 (2010)
13. Couvreur, J.-M., Francez, N., Gouda, M.G.: Asynchronous unison (extended abstract). In: ICDCS, pp. 486–493 (1992)
14. Dijkstra, E.W.: Self-stabilizing systems in spite of distributed control. Commun. of the ACM 17(11), 643–644 (1974)
15. Dolev, S.: Self-Stabilization. The MIT Press (2000)
16. Dwork, C., Lynch, N.A., Stockmeyer, L.J.: Consensus in the presence of partial synchrony. J. ACM 35(2), 288–323 (1988)
17. Fischer, M., Jiang, H.: Self-Stabilizing Leader Election in Networks of Finite-State Anonymous Agents. In: Shvartsman, M.M.A.A. (ed.) OPODIS 2006. LNCS, vol. 4305, pp. 395–409. Springer, Heidelberg (2006)
18. Hadzilacos, V.: A note on group mutual exclusion. In: PODC, pp. 100–106 (2001)
19. Herman, T.: Adaptivity through Distributed Convergence (Ph.D. Thesis). University of Texas at Austin (1991)
20. Hong, S., Rhee, I., Joon Kim, S., Lee, K., Chong, S.: Routing performance analysis of human-driven delay tolerant networks using the truncated levy walk model. In: Mobility-Models, pp. 25–32 (2008)
21. Peterson, J.L., Silberschatz, A.: Operating system concepts. Addison-Wesley (1985)
22. Joung, Y.-J.: Asynchronous group mutual exclusion. Distributed Computing 13(4), 189–206 (2000)
23. Karagiannis, T., Le Boudec, J., Vojnovic, M.: Power law and exponential decay of inter contact times between mobile devices. In: MOBICOM, pp. 183–194 (2007)
24. McNett, M., Voelker, G.M.: Access and mobility of wireless PDA users, vol. 9, pp. 40–55 (2005)
25. Rhee, I., Shin, M., Hong, S., Lee, K., Chong, S.: On the levy-walk nature of human mobility. In: INFOCOM, pp. 924–932 (2008)
26. Tel, G.: Introduction to Distributed Algorithms, 2nd edn. Cambridge University Press (2000)

Asynchronous Exclusive Perpetual Grid Exploration without Sense of Direction

François Bonnet[1,*], Alessia Milani[2], Maria Potop-Butucaru[3,**], and Sébastien Tixeuil[3]

[1] School of Information Science, JAIST
[2] Université de Bordeaux 1, LaBRI
[3] UPMC Sorbone Universités, LIP6

Abstract. In this paper, we investigate the exclusive perpetual exploration of grid shaped networks using anonymous, oblivious and fully asynchronous robots. Our results hold for robots without sense of direction (*i.e.* they do not agree on a common North, nor do they agree on a common left and right ; furthermore, the "North" and "left" of each robot is decided by an adversary that schedules robots for execution, and may change between invocations of particular robots). We focus on the minimal number of robots that are necessary and sufficient to solve the problem in general grids.

In more details, we prove that three deterministic robots are necessary and sufficient, provided that the size of the grid is $n \times m$ with $3 \leq n \leq m$ or $n = 2$ and $m \geq 4$. Perhaps surprisingly, and unlike results for the exploration with stop problem (where grids are "easier" to explore and stop than rings with respect to the number of robots), exclusive perpetual exploration requires as many robots in the ring as in the grid.

Furthermore, we propose a classification of configurations such that the space of configurations to be checked is drastically reduced. This preprocessing lays the bases for the automated verification of our algorithm for general grids as it permits to avoid combinatorial explosion.

1 Introduction

We consider a set of autonomous robots that have to collaborate to perpetually explore an area modeled as a grid graph. Each robot has to visit each vertex of the grid infinitely many times, with the additional constraint that no two robots may be present at the same vertex at the same time or may switch their positions by crossing the same edge. Introduced in [1], this problem is called the *exclusive perpetual exploration.*

Robots are endowed with visibility sensors and motion actuators and operate in *cycles* that comprise *look, compute,* and *move* phases. The look phase consists

---

* Supported by the JSPS Postdoctoral Fellowship for Foreign Researchers and by MEXT KAKENHI (No. 22-00720).
** Supported by the French ANR Project R-DISCOVER.

A. Fernández Anta, G. Lipari, and M. Roy (Eds.): OPODIS 2011, LNCS 7109, pp. 251–265, 2011.

in taking a snapshot of the other robots positions using its visibility sensors. In the compute phase a robot computes a target destination based on the previous observation. The move phase consists in moving toward the computed destination using motion actuators. We consider an asynchronous scheduling model, *i.e.* a finite yet unbounded amount of time may elapse between any two phases of a robot's cycle.

The difficulty to solve the *exclusive perpetual exploration* on the grid, depends on the restricted robot capabilities and on their asynchronous behavior. In particular, robots are *anonymous* (they execute the same protocol and have no mean to distinguish themselves from other robots), *oblivious* (their memory is not persistent between cycles), and they have no sense of direction (*i.e.* they may not agree on a common direction or orientation in the grid as each local direction and orientation are chosen by an adversary and may change between cycles). Asynchrony makes the problem harder because robots have to coordinate their movements despite the fact that a robot can decide to move according to an old snapshot of the system and that different robots may be in different phases of their cycles at the same time.

The robots' positions in the grid is the only information that can be use to decide moving. In this context, the number of robots on the grid directly impacts on the ability to break initial symmetry and to provide a grid reference "milestone" to help robots in their exploration process. Observe that due to the mutual exclusion constraints on the vertices occupancy, we cannot exploit the usual approach [7] of creating "towers" (having more than one robot on the same node) to construct a reference milestone in the graph to be explored.

Related work. While the vast majority of literature on coordinated distributed robots considers that those robots are evolving in a *continuous* two-dimensional Euclidean space and use visual sensors with perfect accuracy that permit to locate other robots with infinite precision, a recent trend was to shift from the classical continuous model to the *discrete* model. In the discrete model, space is partitioned into a *finite* number of locations. This setting is conveniently represented by a graph, where nodes represent locations that can be sensed, and where edges represent the possibility for a robot to move from one location to the other. For each location, a robot is able to sense if the location is empty or if robots are positioned on it (instead of sensing the exact position of a robot). Also, a robot is not able to move from a position to another unless there is explicit indication to do so (*i.e.*, the two locations are connected by an edge in the representing graph). The discrete model permits to simplify many robot protocols by reasoning on finite structures (*i.e.*, graphs) rather than on infinite ones. However, as noted in most related papers [6,7,8,10,11,12,13], this simplicity comes with the cost of extra symmetry possibilities, especially when the authorized paths are also symmetric.

Assuming visibility capabilities, anonymous and oblivious robots, the three main problems that have been studied in the discrete robot model are *gathering* [10,11,12] (all robots are requested to reach a single node, not known beforehand), exploration with stop [5,6,7,8,13] (all nodes must be visited by at least one robot, and eventually all robots must stop moving forever), and exclusive perpetual exploration [1,2] (all nodes must be visited by all robots infinitely often, and no node or edge should be occupied by more than one robot at any time).

For the exploration with stop problem, the fact that robots need to stop after exploring all locations requires robots to "remember" how much of the graph was explored, *i.e.*, be able to distinguish between various stages of the exploration process since robots have no persistent memory. As configurations can be distinguished only by robot positions, the main complexity measure is then the number of robots that are needed to explore a given graph. The vast number of symmetric situations induces a large number of required robots. For tree networks, [8] shows that $\Omega(n)$ robots are necessary for most n-sized tree, and that sublinear robot complexity is possible only if the maximum degree of the tree is 3. In uniform rings, [7] proves that the necessary and sufficient number of robots is $\Theta(\log n)$, although it is required that the number k of robots and the size n of the ring are coprime. Note that both approaches are *deterministic*, *i.e.*, if a robot is presented twice the same situation, its behavior is the same in both cases. In [4,6], the authors propose to adopt a *probabilistic* approach to lift constraints and to obtain tighter bounds. They show that *four* identical probabilistic robots are necessary and sufficient to solve the exploration problem in any anonymous unoriented ring, also removing the coprime constraint between the number of robots and the ring size. The impossibility presented in [6] for three robots to explore a ring is for the semi-synchronous model and naturally extends to the asynchronous model. This impossibility result extends to four robots in the deterministic setting, while five robots are sufficient to explore and stop a ring deterministically [13]. By contrast, it was recently pointed out that in the general case three robots are necessary and sufficient to explore an $n \times m$ grid-shaped network with $m > 3$ [5]. So, with respect to the required number of robots to explore and stop a graph, grid exploration is easier that ring exploration.

Most related to our task is the exclusive perpetual exploration of anonymous graphs [1,2]. Baldoni *et al.* [1] prove that the mutual exclusion constraint and the underlying graph structure drive an upper bound on the number of robots that can perpetually and exclusively explore a graph. They also provide a method to compute this upper bound for a given graph. In the same paper, they provide an algorithm to perpetually and exclusively explore any partial grid provided that the number of robots enables a solution. However, the considered scheduling model is synchronous and contrary to the vast majority of existing solutions in the literature [5,6,7,8,10,11,12,13], robots have a strong sense of direction as they agree on the four basic directions: North, South, East, and West. This strong settings permits to break all cases of initial symmetry since a global total order can be inferred on the vertices of the graph. Finally, Blin *et al.* [2] investigate exclusive perpetual exploration of an anonymous ring. Considering asynchronous scheduling and no sense of direction, they prove that three robots are necessary and sufficient to perpetually and exclusively explore a ring.

Our contribution. First, we prove that three deterministic robots are necessary and sufficient, provided that the size of the grid is $n \times m$ with $3 \leq n \leq m$ or $n = 2$ and $m \geq 4$. Perhaps surprisingly, and unlike results for the exploration with stop problem (where grids are "easier" to explore and stop than rings with respect to the number of robots), exclusive perpetual exploration requires as many robots in the ring and in the grid.

Second, we propose a general classification of all possible system configurations. Observing that in each symmetric configuration, deterministic robots act in the same manner, we consider a single representative configuration for each set of such symmetric configurations. Moreover, we classify representative configurations with respect to the positions of robots and their mobility constraints. Some configurations are more complex to manage than others, as several robots may be scheduled to move concurrently. Then, due to asynchrony one robot can move before the other, leading to a new unstable configuration where the second robot may move according to an old snapshot. We expect that our classification benefits protocol designers to ease the proving of exploration process and lays the basis of their automatic verification.

2   Model and Problem Specification

We consider a distributed system of k mobile robots that are scattered on an $n \times m$ grid graph where $2 \leq n \leq m$. The grid is *anonymous*, *i.e.*, there exists no labeling to distinguish nodes or edges. The robots are *identical*, *i.e.*, they cannot be distinguished using their appearance and they execute the same protocol. Moreover, the robots are *oblivious*, *i.e.*, they have no memory of their past actions, and they neither have a common North direction, nor a common chirality (handedness). Robots cannot explicitly communicate, but have the ability to sense their environment and see the position of the other robots.

Robots operate in three phase cycles: Look, Compute and Move. During the Look phase robots take a snapshot of the grid together with the robots. The collected information (position of the other robots in the egocentered view) are used in the compute phase in which robots decide to move or to stay idle. In the move phase, robots may move to one of their adjacent nodes computed in the previous phase.

The computational model we consider is the asynchronous CORDA model [9,14] in a discrete setting. Thus, when a robot takes a snapshot of the grid, it sees the other robots on nodes only. On the other hand, the time between Look, Compute, and Move operations is finite but unbounded, and it is decided by the adversary for each action of each robot. Thus, because of the asynchrony, different robots can execute concurrently different phases (*e.g.*, a robot can perform a look operation while another robot is moving), and a robot can use an outdated snapshot of the grid to compute where to move and whether to move. A configuration at a given time is defined by the positions of all robots at that time. We assume that initially every vertex of the grid contains at most one robot.

Exclusive Perpetual Exploration Problem

We study the *Exclusive Perpetual Exploration* problem introduced by Baldoni *et al.* [1] for the synchronous setting.

Definition 1 (Perpetual Exclusive Grid Exploration). *For any grid G of size n × m and for any initial configuration where all robots are located on different vertices, an algorithm solves the perpetual exclusive grid exploration problem if it guarantees the following two properties:*

- *liveness Each robot visits each vertex in G infinitely often.*
- *no collision No two robots visit the same vertex or traverse the same edge at the same time.*

3   Classification of Configurations

The lack of a global orientation implies that many configurations are alike from a robot's point of view. Two configurations are *indistinguishable* if they are symmetric in the robots egocentered view (see Figure 1). However, when the grid is not square (that is, when $n \neq m$) robots can distinguish landscape and portrait orientation. In this paper, we suppose *w.l.o.g.* that all configurations are in a landscape orientation.

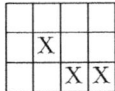 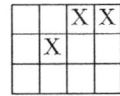 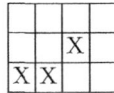

Fig. 1. Three indistinguishable configurations on a 3 × 4 grid

Representative Configuration. For notation purpose, we label each cell of a grid $n \times m$ with an integer in the set $\{0, \ldots, nm - 1\}$ according to the left-to-right-top-to-bottom reading direction. This labeling is not available to the robots.

Given an $n \times m$ grid, any configuration involving three robots on different vertices is conveniently represented by a unique sorted sequence $s = (s_1, s_2, s_3)$ where each $s_i \in \{0, \ldots, nm - 1\}$ is different and represents one of the cells occupied by the robots. The three configurations on Figure 1, for example, are respectively denoted by the sorted sequences $(5, 10, 11)$, $(2, 3, 5)$, and $(6, 8, 9)$.

Since robots are identical, deterministic, and execute the same code, in every indistinguishable configuration, robots perform the same actions. In each set of indistinguishable configurations, we choose a *representative*.

Definition 2 (representative configuration). *Let G be a grid with left-to-right-top-to-bottom labeling, and any non-empty set $\mathcal{I}$ of indistinguishable configurations of G, we say that a configuration $s \in \mathcal{I}$ is the representative configuration of every configuration in $\mathcal{I}$ iff s is the smallest in $\mathcal{I}$ w.r.t. the lexicographic order, $\prec$. That is, there is no $s' \in \mathcal{I}$ such that $s' \prec s$.*

Considering Figure 1, the natural labeling is

0	1	2	3
4	5	6	7
8	9	10	11

and

X	X	
		X

is the representative configuration, whose corresponding sorted sequence is $(0, 1, 6)$, which is the smallest amongst all indistinguishable configurations. In the sequel, we consider only representative configurations that are thus simply called configurations.

Classes of Configurations (for grids with 3 robots)

Given a particular configuration, we say that two robots are *indistinguishable* iff there is no deterministic way of distinguishing them. For example, in the second grid on Figure 2, the robots respectively located at cell 5 and cell 9 are indistinguishable.

Formally, in a configuration C there are two *indistinguishable* robots iff C is axially symmetric and two robots are symmetric in this axial symmetry[1]. For the second grid on Figure 2, the axis of symmetry is the vertical line that goes through the middle of the third column. For the third grid on Figure 2, the axis of symmetry is the diagonal line that goes from top-left to bottom-right.

Breaking and avoiding symmetric configurations is crucial to guarantee both liveness and the absence of collisions. Thus, we classify the set of configurations into four classes, depending on the amount of symmetry (with respect to the three robots) and the move possibilities of robots:

1. *Asymmetrical configuration*: every robot is uniquely distinguishable. In this case, even an adversarial scheduling cannot lead to ambiguous situations.
2. *Semi-symmetrical configuration*: two robots have symmetrical views but the third. This third robot can break the symmetry, possibly in more than one step.
3. *Semi-symmetrical blocked configuration*: two robots are indistinguishable, the other robot cannot move, but the two indistinguishable robots can move, without violating the vertex mutual exclusion (the no collision property). These configurations are difficult to deal with, since there is no immediate way of breaking the symmetry. One can nevertheless observe that any algorithm solving the perpetual exclusive exploration problem has to enable the movement of the two indistinguishable robots (the third robot is blocked).
4. *Symmetrical blocked configuration*: no possible move. For any possible move, an adversarial scheduling can decide an activation order that would lead to a collision.

From an algorithmic point of view, configurations of type 1 and 2 are ideal, while the configuration of type 4 is a dead-end. Thus, when designing an algorithm one should focus on configurations of type 3 which are the most complex

---

[1] If such an axis exists, it is necessarily the perpendicular bisector of the segment whose extremities are the two indistinguishable robots. The third robot lies on that axis.

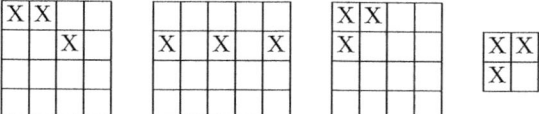

Fig. 2. The four types of configurations

to deal with. The complexity arises from the fact that the two indistinguishable robots may be scheduled to execute concurrently, but due to the asynchrony one robot may move first, bringing the system to a new configuration, while the second robot will move later and according to an old snapshot. This means that the system may transiently be in configurations of type 1 or 2 but having a robot move later because of a snapshot it took when in a configuration of type 3. The algorithm should be resilient to this unexpected movement.

4   Algorithms for the Perpetual Grid Exploration

4.1   Impossibility Results

Theorem 1. *There is no algorithm that solves the perpetual exploration problem*

- *with one robot for any grid but the 1×1 and 1×2 grid,*
- *with two robots for any grid.*

Proof. With one robot, since there are less configurations than the number of vertices, it is not possible to design a deterministic algorithm[2].

With two robots, there exist symmetric configurations, *e.g.* robots on two corners. If the scheduler always activate simultaneously both robots, it is not possible to break the symmetry and thus robots cannot explore the grid; each robot will stay in its half-side of the grid.

Theorem 2. *There is no algorithm that solves the perpetual exploration problem with three robots in the 2×2 grid or 2×3 grid.*

Due to page limitations, the proof appears in [3]. The proof simply consists in an exhaustive exploration of all possible moves.

4.2   Decomposition in Two Sub-problems

The exclusive perpetual exploration problem as defined in Section 2 can be decomposed into two independent sub-problems. To complete the exclusive perpetual exploration, one needs to solve both problems. The first sub-problem is directly related to the perpetual exploration, while the second sub-problem deals with the transient period starting from the (arbitrary) initial configuration.

---

[2] One probabilistic robot can explore any grid using a simple random walk.

1. Find a sequence of asymmetrical configurations (*i.e.* configurations of type 1) such that when robots "execute" this sequence, each robot visits each cell. This sequence must start and end with the same configuration, and in each step of the sequence exactly one robot moves.
2. From any initial configuration, find an algorithm that allows robots to reach, in a finite number of steps, a configuration that belongs to the previous sequence.

4.3   Grid 3 × 3

Theorem 3. *There exists an algorithm that solves the exclusive perpetual exploration with three robots starting from any initial configuration in the 3 × 3 grid.*

Proof. Sub-problem 1 The Figure 3 describes how the perpetual exploration is executed. For each configuration, the blue robot represents the robot that is allowed to move in that configuration. After 4 moves, the initial configuration is reached again with a rotation of the grid by 90° and a circular permutation of the positions of robots. In this process, one robot (the one initially on position 5) has visited the four colored cells. The union of this set of cells and the same sets of cells after rotations of 90°, 180°, and 270° covers the entire grid. Thus after 12 iterations[3] of this sequence (*i.e.* 48 moves) every robot has visited every cell of the grid.

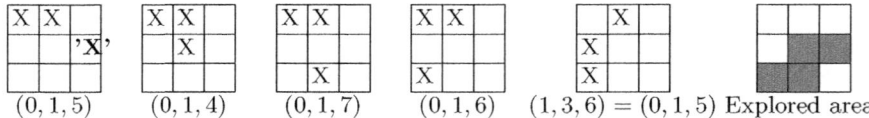

Fig. 3. Sequence that achieves the perpetual exploration for the 3 × 3 grid

Sub-problem 2 There is no configuration of type 4 and exactly one of type 3. For all other configurations (*i.e.* configurations of type 2, there is no problem since an algorithm can deterministically select one robot and make it move to reach a configuration of type 1; for example from configuration $(0, 1, 8)$, the system evolves to configuration $(0, 1, 7)$. Figure 4 explains how the algorithm deals with the single configuration of type 3. The main problem lies in the fact that only the two indistinguishable robots can move but due to asynchrony, it is possible that *(a)* only one robot sees this configuration and moves (the other one never sees this configuration); the system arrives in $(0, 1, 6)$, or *(b)* both robots see the configuration, compute their moves and then one robot moves; the system arrives in $(0, 1, 6)$. The color red for the robot in position 1 means that the system is in the configuration where this robot has computed its move

---

[3] $12 = 3$positions×4orientations: It guarantees that each robot starts in position 5 for each of the four orientations of the grid.

(phase look/compute) according to a previous configuration but has not yet accomplished its move (phase move).

This is problematic because the system arrives in a configuration $(0, 1, 6)$ where the robot located at position 0 is supposed to be the (only) robot to move (from our sequence of moves described in Figure 3). Thus from configuration $(0, 1, 6)$, three cases are possible depending on the first moving robot:

- The robot in 1 moves before the robot in 0 sees the current configuration. The system reaches $(0, 2, 6)$.
- The robot in 1 moves after the robot in 0 sees the current configuration. The system reaches $(0, 2, 6)$ and then $(0, 1, 8)$.
- The robot in 0 sees the current configuration and moves before the robot in 1 moves. The system arrives in $(0, 1, 5)$ and then $(0, 1, 8)$.

In all cases, the system leaves the "bad" configuration $(0, 1, 3)$ of type 3 and arrives, in at most three steps, in a configuration of type 1 or 2 without any "red" robot. The complete list of moves appears in [3].

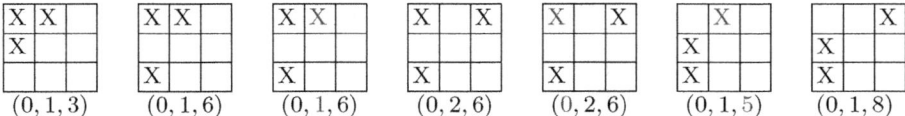

Fig. 4. From the configuration $(0, 1, 3)$ of type 3 to a configuration of type 1 or 2

4.4   Grids $2 \times m$ with $m \geq 4$

Theorem 4. *There exists an algorithm that solves the exclusive perpetual exploration with three robots starting from any initial configuration in $2 \times m$ grids, with $m \geq 4$.*

Proof. Sub-problem 1 One valid sequence is described informally on Figure 5 and the formal definition appears in [3]. After $2m$ moves, the system returns to the initial configuration with a circular permutation of the robot positions. In this process, one robot (the one initially in position 1) has visited all positions but 0 (Position 0 will be visited by this robot within the next $2m$ moves). Thus, after 3 iterations of this sequence (*i.e.* $3 \times 2m$ moves), every robot has visited every cell of the grid.

Sub-problem 2 There is no configuration of type 3 or 4. There exist configurations of type 2 only when m is odd; from these configurations it is easy to break the symmetry by moving the robot which is uniquely distinguishable. From any configuration of type 1, it is sufficient to move successively one robot to position 0, then one robot to position 1, then the last one in a position between $m - 1$ and $2m - 1$. The total number of moves in this process is bounded by $m + m + 1$. The complete list of moves appears in [3].

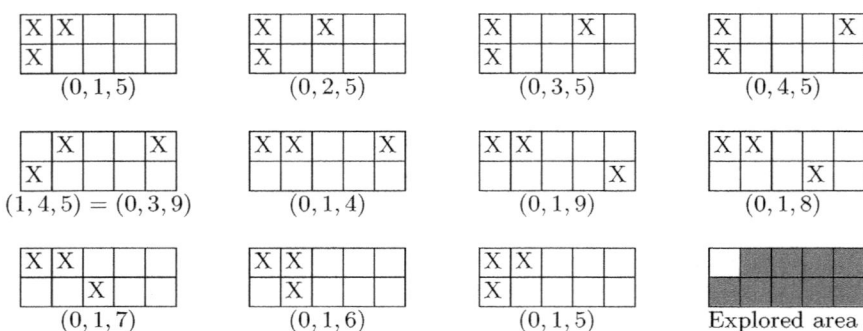

Fig. 5. Sequence for the perpetual exploration for $2 \times m$ grids ($m \geq 4$). Here $m = 5$

4.5   Grids $n \times m$ with $3 \leq n < m$

Theorem 5. *There exists an algorithm that solves the exclusive perpetual exploration with three robots starting from any initial configuration in $n \times m$ grids, with $3 \leq n < m$.*

Proof. Sub-problem 1 One valid sequence is described informally on Figure 6 and formally in [3]. After $(2n - 2) \times (m - 3) + n + 1 + (n - 3) + (m - 2) + (n - 1) + 1$ moves, the configuration of the system returns to the initial configuration with a circular permutation of the three robots and a rotation of the grid by $180°$. Doing zigzag moves, one robot (the one initially located on position $2m - 1$) explores all the grid but the first line and the two first columns. Thus, after 6 iterations[4] every robot has visited every vertex of the grid.

Sub-problem 2 There exists no configuration of type 3 or 4. There exist configurations of type 2 only when m or n is odd; from these configurations it is easy to break the symmetry by moving the robot which is uniquely distinguishable. As for the previous case, we show that, for any configuration of type 1, it is possible to reach one configuration used during the perpetual exploration. It is sufficient to move successively one robot to the position 0, then one robot to the position 1 and finally one robot to the position $(n - 1)m$ to reach configuration $C = (0, 1, (n - 1)m)$. However, depending on the initial configuration, it may happen that, while executing these moves, the system reaches a configuration C' used in the perpetual exploration phase; in that case, the sub-problem 2 is solved and robots start the perpetual exploration from this configuration C', without reaching C. Indeed, a move is already defined from C' and it is not possible to define a different one for this phase since robots are oblivious. The total number of moves is grossly bounded by $nm + nm + nm$. The complete list of moves appears in [3].

---

[4] $6 = 3$ positions $\times$ 2 orientations: It guarantees that each robot starts in position $2m - 1$ for each of the two orientations of the grid.

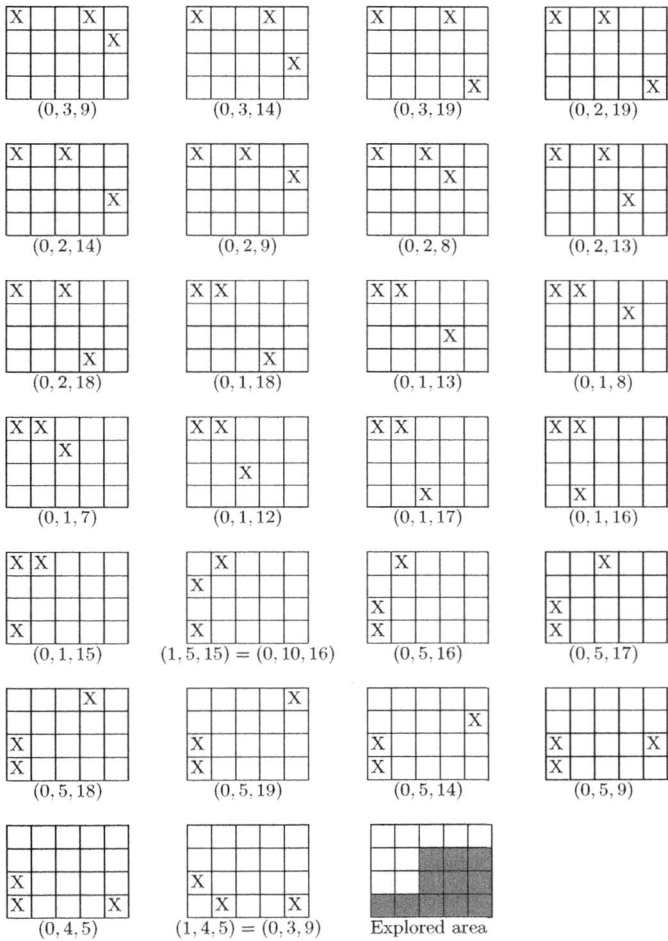

Fig. 6. Sequence for the perpetual exploration of $n \times m$ grids. Here $(n, m) = (4, 5)$

4.6 Grids $n \times n$ with $n \geq 4$

Theorem 6. *There exists an algorithm that solves the exclusive perpetual exploration with three robots starting from any initial configuration in $n \times n$ grids, with $n \geq 4$.*

Proof. Sub-problem 1 One valid sequence is described informally on Figure 7 and formally in [3]. After $(2n - 2) \times (n - 3) + n + 1$ moves, the configuration of the system returns to the initial configuration with a rotation of the grid by $90°$ and a circular permutation of the positions of robots. In this process, one robot (the one initially in $2n - 1$) has explored all the grid but the first line and the

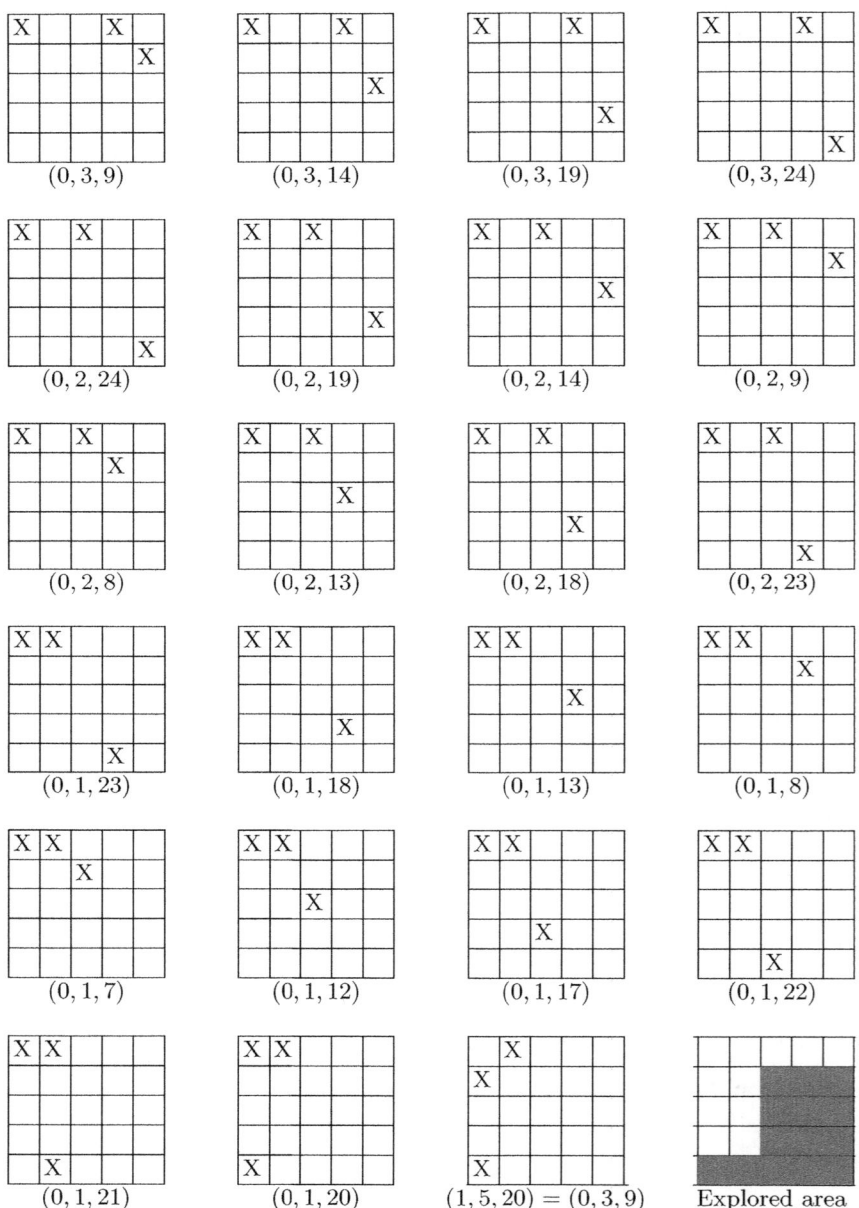

Fig. 7. Sequence for the perpetual exploration for $n \times n$ grids ($n \geq 4$). Here $n = 5$.

two first columns. Since $n \geq 4$, after 12 iterations every robot has visited every vertex of the grid[5].

Sub-problem 2 In the transient phase, robots try to reach configuration $C = (0, 1, (n-1)n)$ by successively moving one robot to position 0, then one to position 1, and finally the last one to position $(n-1)n$. There is however one main difference with the previous case: there exist symmetric configurations where the three robots are not on the same line/column, namely configurations $(0, x, xn)$, with $1 \leq x \leq n-1$ and in particular configuration $(0, 1, n)$, which is of type 3. To deal with these "bad" initial configurations we need a special behavior that is described on Figure 8. Informally, we first move the robot located in 0 to break the symmetry[6] and then we move another robot away from the side of the grid, so that a bad configuration cannot be reached again when the system evolves towards the configuration $(0, 1, (n-1)n)$. The total number of moves is grossly bounded by $4 + n^2 + n^2 + n^2$. The complete list of moves appears in [3].

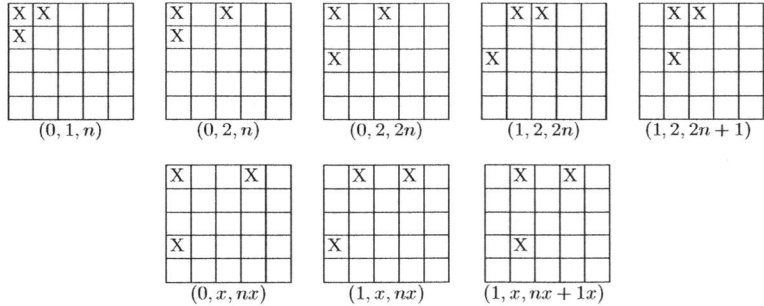

Fig. 8. From the configurations $(0, 1, n)$ and $(0, x, xn)$ with $2 \leq x \leq n-1$

5 Conclusions and Open Problems

We proved that three robots are necessary and sufficient to solve the problem of the exclusive perpetual exploration of grids. Similarly to [1], it would be interesting to generalize our result to any partial grid. Of course, only partial grids that preserve symmetry patterns are to be considered, as asymmetric partial grids give a global sense of direction for free.

---

[5] $n \geq 4$ is necessary since for $n = 3$, the central vertex is never visited with this algorithm as "avoiding" the first line and the two first columns prevents robots to visit the central vertex.

[6] Contrary to the 3×3 grid, configuration $(0, 1, n)$ does not raise issues since configuration $(0, 1, 2n)$ (reached after one move) does not belong to the set of configurations used during the perpetual exploration. From this configuration, we can decide to move the robot in position 2 to position 3, thus this robot executes the same move if it sees the previous configuration or the current configuration.

Studying the coordination of robots to solve basic tasks, such as exploration, in a distributed manner requires considering all possible reachable configurations. Our proposal for defining and classifying configurations considerably simplifies the design and verification of our algorithms. We believe it can be extended to address an arbitrary number of robots and be a first step in providing a complete framework to study coordination problems in mobile robots networks.

References

1. Baldoni, R., Bonnet, F., Milani, A., Raynal, M.: On the Solvability of Anonymous Partial Grids Exploration by Mobile Robots. In: Baker, T.P., Bui, A., Tixeuil, S. (eds.) OPODIS 2008. LNCS, vol. 5401, pp. 428–445. Springer, Heidelberg (2008)
2. Blin, L., Milani, A., Potop-Butucaru, M., Tixeuil, S.: Exclusive Perpetual Ring Exploration Without Chirality. In: Lynch, N.A., Shvartsman, A.A. (eds.) DISC 2010. LNCS, vol. 6343, pp. 312–327. Springer, Heidelberg (2010)
3. Bonnet, F., Milani, A., Potop-Butucaru, M., Tixeuil, S.: Asynchronous exclusive perpetual grid exploration without sense of direction. Tech. rep. (2011), http://hal.archives-ouvertes.fr/hal-00626155/fr/
4. Devismes, S.: Optimal exploration of small rings. In: Proceedings of the Third International Workshop on Reliability, Availability, and Security, WRAS 2010, pp. 9:1–9:6. ACM, New York (2010), http://doi.acm.org/10.1145/1953563.1953571
5. Devismes, S., Lamani, A., Petit, F., Raymond, P., Tixeuil, S.: Optimal grid exploration by asynchronous oblivious robots. Tech. rep., CoRR (2011), http://arxiv.org/abs/1105.2461
6. Devismes, S., Petit, F., Tixeuil, S.: Optimal Probabilistic Ring Exploration by Semi-Synchronous Oblivious Robots. In: Kutten, S., Žerovnik, J. (eds.) SIROCCO 2009. LNCS, vol. 5869, pp. 195–208. Springer, Heidelberg (2010), http://hal.inria.fr/inria-00360305/fr/
7. Flocchini, P., Ilcinkas, D., Pelc, A., Santoro, N.: Computing Without Communicating: Ring Exploration by Asynchronous Oblivious Robots. In: Tovar, E., Tsigas, P., Fouchal, H. (eds.) OPODIS 2007. LNCS, vol. 4878, pp. 105–118. Springer, Heidelberg (2007)
8. Flocchini, P., Ilcinkas, D., Pelc, A., Santoro, N.: Remembering Without Memory: Tree Exploration by Asynchronous Oblivious Robots. In: Shvartsman, A.A., Felber, P. (eds.) SIROCCO 2008. LNCS, vol. 5058, pp. 33–47. Springer, Heidelberg (2008)
9. Flocchini, P., Prencipe, G., Santoro, N., Widmayer, P.: Arbitrary pattern formation by asynchronous, anonymous, oblivious robots. Theor. Comput. Sci. 407(1-3), 412–447 (2008)
10. Kamei, S., Lamani, A., Ooshita, F., Tixeuil, S.: Asynchronous Mobile Robot Gathering from Symmetric Configurations Without Global Multiplicity Detection. In: Kosowski, A., Yamashita, M. (eds.) SIROCCO 2011. LNCS, vol. 6796, pp. 150–161. Springer, Heidelberg (2011), http://arxiv.org/abs/1104.5660
11. Klasing, R., Kosowski, A., Navarra, A.: Taking Advantage of Symmetries: Gathering of Asynchronous Oblivious Robots on a Ring. In: Baker, T.P., Bui, A., Tixeuil, S. (eds.) OPODIS 2008. LNCS, vol. 5401, pp. 446–462. Springer, Heidelberg (2008)

12. Klasing, R., Markou, E., Pelc, A.: Gathering asynchronous oblivious mobile robots in a ring. Theor. Comput. Sci. 390(1), 27–39 (2008)
13. Lamani, A., Potop-Butucaru, M.G., Tixeuil, S.: Optimal Deterministic Ring Exploration with Oblivious Asynchronous Robots. In: Patt-Shamir, B., Ekim, T. (eds.) SIROCCO 2010. LNCS, vol. 6058, pp. 183–196. Springer, Heidelberg (2010)
14. Prencipe, G.: *InstantaneousActions* vs. *FullAsynchronicity*: Controlling and Coordinating a Set of Autonomous Mobile Robots. In: Restivo, A., Ronchi Della Rocca, S., Roversi, L. (eds.) ICTCS 2001. LNCS, vol. 2202, pp. 154–171. Springer, Heidelberg (2001)

Fused State Machines for Fault Tolerance
in Distributed Systems

Bharath Balasubramanian and Vijay K. Garg [*]

Parallel and Distributed Systems Laboratory,
Dept. of Electrical and Computer Engineering,
The University of Texas at Austin,
Austin, TX 78712
bbharath@mail.utexas.edu, garg@ece.utexas.edu

Abstract. Replication is a standard technique for fault-tolerance in distributed systems modeled as deterministic finite state machines (DFSMs or machines). To correct f crash faults among n machines, replication requires nf additional backup machines. We present a fusion-based solution that requires just f additional backup machines (called fusions or fused backups). In this paper, we first propose a fundamental problem regarding DFSMs, independent of fault tolerance, that has not been explored in the literature so far: Given a machine M, with a set of states and a set of events, can we *replace* it with machines each containing fewer events than M? To formalize this we define a (k,e)-event decomposition of a given machine M, that is a set of k machines each with at least e events fewer than the event set of M, that acting in parallel, are equivalent to M. We present an algorithm to generate such machines with time complexity $O(|X_M|^3 |\Sigma_M|^e)$, where X_M is the set of states and Σ_M the set of events of M. Second, we use our event decomposition algorithm to generate fused backups that can correct faults among a given set of machines. We show that these backups are *minimal* w.r.t the number of states they contain and the number of events in their event set. Third, we use the notion of *locality sensitive hashing* to present algorithms for the detection and correction of faults for the fusion-based solution. The algorithm for the detection of Byzantine faults has time complexity $O(nf)$ on average, which is the same as that for replication. The algorithm for the correction of both crash and Byzantine faults has time complexity $O(n\rho f)$ with high probability (w.h.p), where ρ is the average state reduction achieved by fusion. We show that for small values of n (for most practical systems, $n < 10$) and ρ (average value of $\rho < 2$ in our experiments), this results in almost no overhead as compared to replication. Finally, we evaluate fusion on the widely used MCNC'91 benchmarks for DFSMs and results show that the average state space savings in fusion (over replication) is 38% (range 0-99%), while the average event-reduction is 4% (range 0-45%).

Keywords: Distributed Systems, Fault Tolerance, Finite State Machines.

1   Introduction

Distributed applications often use deterministic finite state machines (or just *machines*) to model computations such as regular expressions for pattern detection, syntactical

---

[*] Supported by NSF CNS-0718990, NSF CNS-1115808 and the Cullen Trust for Higher Education Endowed Professorship.

A. Fernández Anta, G. Lipari, and M. Roy (Eds.): OPODIS 2011, LNCS 7109, pp. 266–282, 2011.

analysis of documents or mining algorithms for large data sets. These machines executing on distinct distributed processes are often prone to faults. Traditional solutions to this problem involve some form of replication, in which to correct f crash faults [21] among n given machines (referred to as *primaries*), f copies of each primary are maintained [14,23,22]. If the backups start from the same initial state as the corresponding primaries and act on the same events, then in the case of faults, the state of the failed machines can be recovered from one of the remaining copies. These backups can also correct $\lfloor f/2 \rfloor$ Byzantine faults [15], where the processes lie about the state of the machine, since a majority of truthful machines is always available. This approach is expensive both in terms of the total number of backup machines, nf and the total backup state space.

Consider a distributed application that is searching for three different string patterns among a huge file, as modeled by the state machines A, B and C shown in Fig. 1. A state machine in our system consists of a finite set of states (including the initial execution state) and a finite set of events. On application of an event, the state machine transitions to the next state based on the state-transition function. For example, machine A in Fig. 1 contains the states $\{a^0, a^1\}$, events $\{0, 2\}$ and the initial state, shown by the dark ended arrow, is a^0. The state transitions are shown by the arrows from one state to another. Hence, if A is in state a^0 and event 0 is applied to it, then it transitions to state a^1. In this example, A checks the parity of $\{0, 2\}$ and so, if it is in state a^0, then an even number of $0s$ or $2s$ have been applied to the machine and if it is in state a^1, then an odd number of the inputs have been applied. Machines B and C check for the parity of $\{1, 2\}$ and $\{0\}$ respectively.

To correct one crash fault among these machines, replication requires a copy of each of them, resulting in three backup machines, consuming total state space of eight (2^3). Rather than replicate the machines, we can correct one fault by maintaining just one additional machine F_1 shown in Fig. 1. The relevant events from the client (or environment) are applied to all the machines. So if the event sequence 0, 0, 1, 2 is applied on all the machines, A, B, C and F_1 will be in states a^1, b^0, c^0 and f_1^1 respectively. Assume a crash fault in C. Given the parity of 1s (state of F_1) and the parity of 1s or 2s (state of B), we can first determine the parity of 2s. Using this, and the parity of 0s or 2s (state of A), we can determine the parity of 0s (state of C). Hence, we can correct the crash fault in C using A, B and F_1. This argument can be extended to correcting one fault among any of the machines in $\{A, B, C, F_1\}$. This approach consumes fewer backups than replication (one vs. three) and less backup state space (two vs. eight).

However, it is not always possible to design these backups merely by inspection. In Fig. 1, it may not be obvious that F_1 and F_2 can correct two crash faults among the primaries. In [18], we present the theory and algorithm to automatically generate f backup machines (called *fusions*) for any given set of primaries that can correct f crash faults (or $\lfloor f/2 \rfloor$ Byzantine faults). In this paper, we focus on the three main challenges faced by fusion which are the large event-sets of the fusions, the high time complexity for the generation of fusions and the high cost for detecting and correcting faults. To summarize our contributions in this paper:

Event-based Decomposition. We start with a question that is fundamental to the understanding of DFSMs, independent of fault-tolerance: Given a machine M, can it be

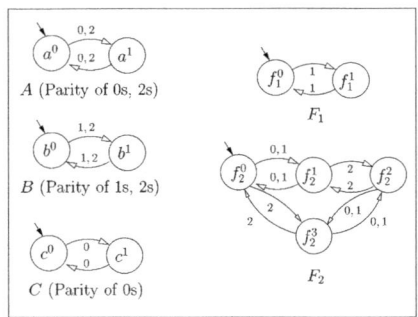

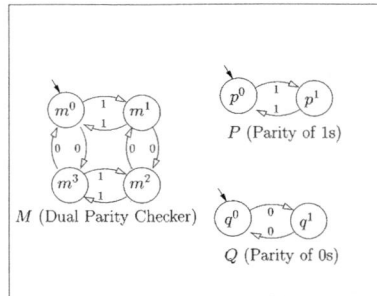

Fig. 1. Fused-Backups for Fault Tolerance   **Fig. 2.** Event-based Decomposition

replaced by two or more machines executing in parallel, each containing fewer events than M? In other words, given the state of these fewer-event machines, can we uniquely determine the state of M? In Fig. 2, the 2-event machine M (it contains events 0 and 1 in its event set), checks for the parity of 0s *and* 1s. M can be replaced by two 1-event machines P and Q, that check for the parity of just 1s or 0s respectively. Given the state of P and Q, we can determine the state of M. How can we generate these event-reduced machines (if they exist) for any given machine? While there has been work on both the state-based decomposition [11,16] and the minimization of completely specified machines [13,12], this is the first paper that presents the problem of event-reduction.

In this paper, we define the concept of a (k,e)-event decomposition of a machine M that is a set of k machines, each with at least e events fewer than the event set of M, such that given the state of these machines, we can determine the state of M. We present an algorithm to generate such machines with time complexity $O(|X_M|^3|\Sigma_M|^e)$, where X_M is the set of states and Σ_M the set of events of M. The load on a process running a machine is directly proportional to the number of events in the event-set of the machine. Hence, this decomposition is crucial for applications such as sensor networks in which there are strict limits on the number of events that each process can service.

Space-Event Optimized Fusion Algorithm. We apply our event-decomposition algorithm to generate backups for fault tolerance that are optimized for both events and states. In Fig. 1, it is better to choose the 1-event F_1 over the 3-event F_2 as a backup machine to correct one fault. We show that if our solution achieves no event-reduction, then no solution with the same number of backups achieves it. Further, we present an incremental approach for generating the fusions that improves the time complexity by a factor of ρ^n, where ρ is the average state savings achieved by fusion.

Efficient Algorithms for Detection/Correction of Faults. In [18], the algorithm for the correction of crash and Byzantine faults, has time complexity $O(n^2\rho + n\rho f + s^n)$, where n is the number of primaries, f is the number of crash faults, s is the maximum number of states among primaries and ρ is the average state savings achieved by fusion. In this paper, we present a Byzantine detection algorithm with time complexity $O(nf)$ on average, which is the same as the time complexity of detection for replication. Hence, for a

Table 1. Symbols/Notation used in the paper

$\mathcal{P}$	Set of primaries	n	Number of primaries
RCP	Reachable Cross Product of $\mathcal{P}$	N	Number of states in the RCP
f	No. of crash faults	s	Maximum number of states among primaries
$\mathcal{F}$	Set of fusions/backups	ρ	Average State Reduction in fusion
Σ	Union of primary event-sets	β	Event-Reduction parameter

Table 2. Fusion vs. Replication (n primaries, $O(s)$ states each, f faults, $|\Sigma|$ total events, average state reduction ρ)

	Replication	Fusion
Number of Backups	nf	f
Backup Space	$O(s^{nf})$	$O((s/\rho)^{nf})$
Backup Generation Time Complexity	$O(nsf)$	$O(s^n|\Sigma|f/\rho^n)$
Maximum Events/Backup	Maximum Events/primary	Minimal for f backups
Byzantine Detection Time Complexity	$O(nf)$	$O(nf)$ on average
Crash Correction Time Complexity	$\theta(f)$	$O(n\rho f)$ w.h.p
Byzantine Correction Time Complexity	$O(nf)$	$O(n\rho f)$ w.h.p

system that needs to periodically detect liars, fusion causes no additional overhead. We reduce the problem of fault correction to one of finding points within a certain Hamming distance of a given query point in n-dimensional space and present algorithms to correct crash and Byzantine faults with time complexity $O(n\rho f)$ with high probability. The time complexity for crash and Byzantine correction in replication is $\theta(f)$ and $O(nf)$ respectively. Hence, for small values of n and ρ, fusion causes almost no overhead for recovery. In Table 1 we summarize the notation used in this paper and in Table 2 we compare replication and the current version of fusion.

Evaluation of Fusion. In [18], we evaluated fusion on simple examples such as counters and dividers. In this paper, we evaluate our fusion algorithm on the MCNC'91 [24] benchmarks for DFSMs, that are widely used in the fields of logic synthesis and circuit design. Our results show that the average state space savings in fusion (over replication) is 38% (range 0-99%), while the average event-reduction is 4% (range 0-45%). Further, the average savings in time by the incremental approach for generating the fusions (over the non-incremental approach) is 8%. To illustrate the practical use of fusion, we apply its design to the *grep* application of the MapReduce framework [6]. Using a simple example, we show that the currently used checkpointing approach for fault tolerance needs 600,000 map tasks causing high latency, while replication requires 1200,000 tasks with minimum latency. Fusion offers a compromise with just 800,000 tasks but smaller latency than the checkpointing approach.

2   Model

The DFSMs in our system execute on separate processes with no shared state or communication. Clients of the state machines issue the events (or commands) to the concerned primaries and backups, all of which act on them in the same relative order. We assume loss-less FIFO communication links with a strict upper bound on the time taken for message delivery. Faults in our system are of two types: crash faults, resulting in a loss of the execution state of the machines and Byzantine faults resulting in an arbitrary execution state. Henceforth in the paper, when we simply say faults, we refer to crash faults. When faults are detected by a trusted recovery agent using timeouts (crash faults) or a detection algorithm (Byzantine faults) no further events are sent by any client to these machines. After the machines act on all events sent to them thus far, the recovery agent obtains their states, and recovers the correct execution states of all faulty machines. Since we assume a trusted recovery agent, the work on consensus in the presence of Byzantine faults [7,20], does not apply to our paper. In the following section, we summarize the relevant concepts and results introduced in our previous work.

3 Background [18]

State-based Decomposition. A DFSM, denoted by R, consists of a set of states X_R, set of events Σ_R, transition function $\alpha_R : X_R \times \Sigma_R \to X_R$ and initial state x_R^0. The size of R, denoted by $|R|$ is the number of states in R. We can partition the state space of R such that the transition function α_R, maps each block of the partition to another block for all events in Σ_R [11,16]. In other words, we combine the states of R to generate machines that are consistent to the transition function. The set of all machines generated by combining the states of R is called the *closed partition set* of R (example in Fig. 3).

Consider machine M_2 in Fig. 3, generated by combining the states r^0 and r^2 of R. On event 0, $\{r^0, r^2\}$ self-transitions to $\{r^0, r^2\}$ (self transitions not shown). However, since r^0 and r^2 transition to r^1 and r^3 respectively on event 1, we need to combine the states r^1 and r^3. Continuing this procedure, we obtain the combined states in M_2. We can define an order ($\leq$) among any two machines P and Q in this set as follows: $P \leq Q$, if each block of Q is contained in a block of P (shown by an arrow from P to Q). P and Q are incomparable, i.e., $P \| Q$, if $P \not\leq Q$ and $Q \not\leq P$. In Fig. 3, $F_1 < M_2$, while $M_1 \| M_2$.

Minimum Hamming distance for DFSMs (d_{min}). Consider a set of machines $\mathcal{R}$ each less than R, i.e., machines belonging to the closed partition set of R. We define the Hamming distance [10] between each $r^i, r^j \in X_R$, denoted $d(r^i, r^j)$, as the number of machines in $\mathcal{R}$ that contain r^i and r^j in different blocks (*separate r^i and r^j*). The minimum Hamming distance across all such pairs is denoted $d_{min}(\mathcal{R})$ or just d_{min}. In Fig. 3, if $\mathcal{R} = \{A, B\}$, $d(r^0, r^1) = 1$ (B separates them), while $d(r^0, r^2) = 0$ and hence $d_{min} = 0$.

Given the state of the machines in $\mathcal{R}$ we can determine the state of R if there is at least one machine in $\mathcal{R}$ to distinguish between each pair of states in X_R, or in other words, $d_{min} > 0$. In Fig. 3 if $\mathcal{R} = \{A, B\}$ and A and B are in states $a^0 = \{r^0, r^1, r^7, r^6\}$ and $b^0 = \{r^0, r^2, r^7, r^5\}$, we cannot determine if R is in state r^0 or r^7 (intersection of a^0 and b^0). However, if $\mathcal{R} = \{A, B, C\}$ ($d_{min} = 1$), then given that A, B and C are in a^0, b^0 and c^0, we can determine that R is in state r^0 (only state common to a^0, b^0 and c^0).

Fault Tolerance in DFSMs. To generate the backups (or fusions) for a set of machines, we first construct their *reachable cross product*. Given any two machines

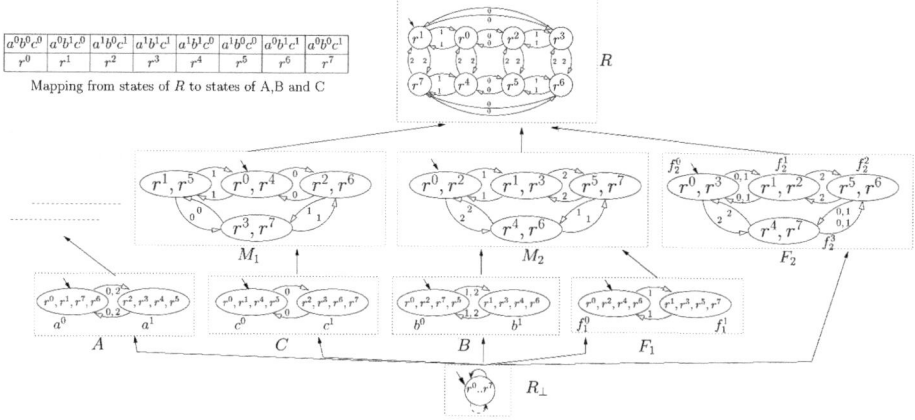

Fig. 3. Set of Machines less than R (all machines not shown due to space constraint)

$A = (X_A, \Sigma_A, \alpha_A, x_A^0)$ and $B = (X_B, \Sigma_B, \alpha_B, x_B^0)$, their reachable cross product, denoted RCP($\{A, B\}$) is the machine which consists of all the states in the product set of X_A and X_B reachable from the initial state $\{x_A^0, x_B^0\}$, with the transition function $\alpha_{RCP}(\{a, b\}, \sigma) = \{\alpha_A(a, \sigma), \alpha_B(b, \sigma)\}$ for all reachable states $\{a, b\} \in X_A \times X_B$ and $\sigma \in \Sigma_A \cup \Sigma_B$. Given a set of n primaries $\mathcal{P}$, their reachable cross product is denoted RCP $(X_{RCP}, \Sigma, \alpha_{RCP}, r^0)$, where Σ is the union of the event sets of all primary machines. The machine R in Fig. 3, is in fact the RCP of $\mathcal{P} = \{A, B, C\}$ shown in Fig. 1. For convenience, we label the states of the RCP, $r^0 \dots r^7$, where each $r^i \in X_{RCP}$ is a tuple consisting of the primary states (mapping shown in Fig. 3). The closed partition set of the RCP always includes the primary machines and its states correspond to the RCP states that contains it. In Fig. 3, $a^0 = \{a^0b^0c^0, a^0b^1c^0, a^0b^1c^1, a^0b^0c^1\}$.

Given the state of the RCP, the state of the primaries can be determined. The basic goal of fault tolerance is to generate a set of machines $\mathcal{F}$, each less than the RCP, so that despite f crash faults, there are sufficient machines in $\mathcal{P} \cup \mathcal{F}$, i.e., among the primaries and backups, whose $d_{min} > 0$. In other words, a set of machines in $\mathcal{P} \cup \mathcal{F}$ can correct f crash faults iff $d_{min}(\mathcal{P} \cup \mathcal{F}) > f$. In Fig. 3, for $\mathcal{P} = \{A, B, C\}$ and $\mathcal{F} = \{F_1, F_2\}$, it can be seen that $d_{min}(\mathcal{P} \cup \mathcal{F}) > 2$. Consider the state of the machines after the application of the event sequence 0, 1, 1 on the machines in $\mathcal{P} \cup \mathcal{F}$. Assume that B and C crash and we need to recover their state. Given the state of A, F_1 and F_2 as $a^1 = \{r^2, r^3, r^4, r^5\}$, $f_1^0 = \{r^0, r^2, r^4, r^6\}$ and $f_2^1 = \{r^1, r^2\}$, we can determine the state of the RCP as r^2 (only state common to a^1, f_1^0 and f_2^1). Since $r^2 = a^1b^0c^1$, we can recover the states of B and C as b^0 and c^1 respectively.

When $|\mathcal{F}| = f$, we call it the f-fusion of $\mathcal{P}$ and call the machines in $\mathcal{F}$, fused-backups or just *fusions*. An f-fusion is *minimal* if there exists no other f-fusion $\mathcal{G}$ in which every machine is less than or equal to some machine in $\mathcal{F}$ and at least one machine is strictly less than some machine in $\mathcal{F}$. In section 6, we describe how an f-fusion can also detect f Byzantine faults or correct $\lfloor f/2 \rfloor$ Byzantine faults.

Coding theory is often used in data fault tolerance for reducing redundancy [19,5]. In our previous work, we present coding-theoretic solutions to fault tolerance in data structures [2] and infinite state machines [8]. However, a direct coding-theoretic approach to DFSMs, in which we maintain the parity of the states of each machine would be too expensive in terms of communication and computation, since after every event transition, the machine needs to sends its state and the parity needs to be recalculated. Instead, we use our Hamming distance metric to construct backups that independently act on events.

4   Event-Based Decomposition of Machines

In this section, we explore the problem of replacing a given machine M with two or more machines, each containing fewer events than M. We present an algorithm to generate such event-reduced machines with time complexity polynomial in the size of M. This is important for applications with limits on the number of events each individual process running a DFSM can service. Note that, the contributions in this section are independent of fault tolerance. We first define the notion of event-based decomposition.

Definition 1. *A (k,e)-event decomposition of a machine* $M(X_M, \alpha_M, \Sigma_M, m^0)$ *is a set of* k *machines* $\mathcal{E}$, *each less than* M, *such that* $d_{min}(\mathcal{E}) > 0$ *and* $\forall P(X_P, \alpha_P, \Sigma_P, p^0) \in \mathcal{E}$, $|\Sigma_P| \le |\Sigma_M| - e$.

As $d_{min}(\mathcal{E}) > 0$, given the state of the machines in $\mathcal{E}$, the state of M can be determined (section 3). So, the machines in $\mathcal{E}$, each containing at most $|\Sigma_M| - e$ events, can effectively replace M. In Fig. 4, we present the *eventDecompose* algorithm that takes as input, machine M, parameter e, and returns a (k,e)-event decomposition of M (if it exists) for some $k \le |X_M|^2$.

In each iteration, Loop 1 generates machines that contain at least one event less than the machines of the previous iteration. So, starting with M in the first iteration, at the end of e iterations, $\mathcal{M}$ contains the set of largest machines (according to the order $\le$ defined in 3) less than M, each containing at most $|\Sigma_M| - e$ events. Loop 2, iterates through each machine P generated in the previous iteration, and uses the *reduceEvent* algorithm to generate the set of largest machines less than P containing at least one event less than Σ_P. To generate a machine less than P, that does not contain an event σ in its event set, the *reduceEvent* algorithm combines the states such that they loop onto themselves on σ. The algorithm then constructs the largest machine that contains these states in the combined form. This machine, in effect, ignores σ. This procedure is repeated for all events in Σ_P and the incomparable machines among them are returned. Loop 3 constructs an event-decomposition $\mathcal{E}$ of M, by iteratively adding at least one machine from $\mathcal{M}$ to separate each pair of states in M, thereby ensuring that $d_{min}(\mathcal{E}) > 0$[1].

---

[1] Since each machine added to $\mathcal{E}$ can separate more than one pair of states, an efficient way to implement Loop 3 is to check for the pairs that still need to be separated in each iteration and add machines till no pair remains.

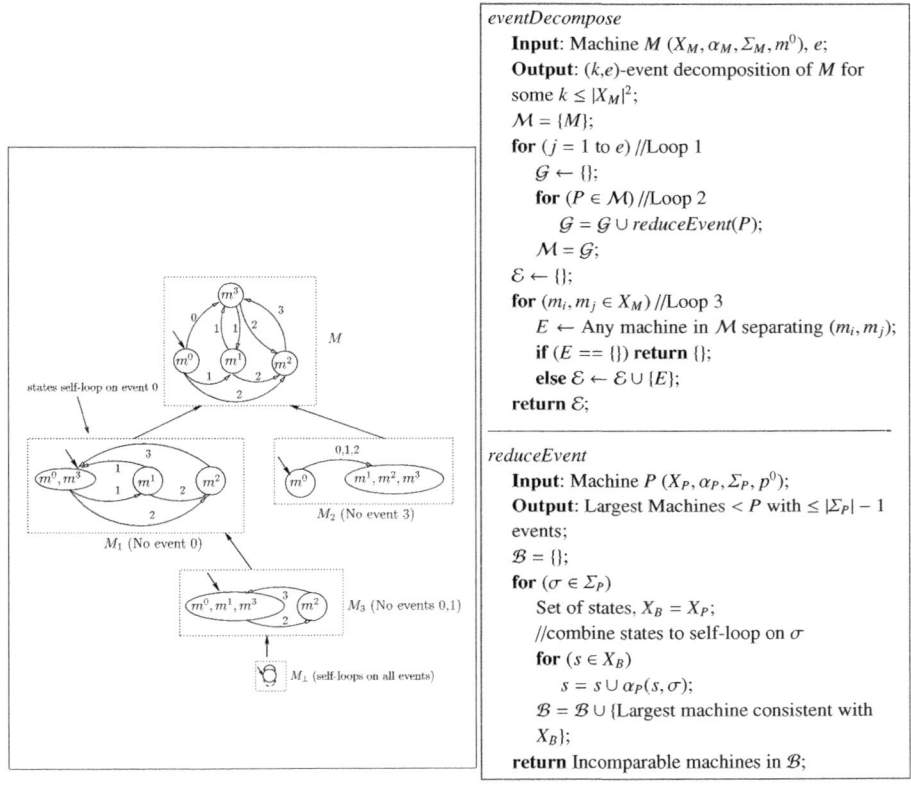

Fig. 4. Event-based Decomposition

Let the 4-event machine M shown in Fig. 4 be the input to the *eventDecompose* algorithm with $e = 1$. In the first and only iteration of Loop 1, $P = M$ and the *reduceEvent* algorithm generates the set of largest 3-event machines less than M, by successively eliminating each event. To eliminate event 0, since m^0 transitions to m^3 on event 0, these two states are combined. This is repeated for all states and the largest machine containing all the combined states self looping on event 0 is M_1. Similarly, the largest machines not acting on events 3,1 and 2 are M_2, M_3 and $M_\perp$ respectively. The *reduceEvent* algorithm returns M_1 and M_2 as the only incomparable machines in this set. The *eventDecompose* algorithm returns $\mathcal{E} = \{M_1, M_2\}$, since each pair of states in M are separated by M_1 or M_2. Hence, the 4-event M can be replaced by the 3-event M_1 and M_2, i.e., $\mathcal{E} = \{M_1, M_2\}$ is a (2,1)-event decomposition of M. We show in the technical report [4], that the *eventDecompose* algorithm has time complexity $O(|X_M|^3|\Sigma_M|^e)$ and also present the proof for the following theorem.

Theorem 1. *Given machine M $(X_M, \alpha_M, \Sigma_M, m^0)$, the* eventDecompose *algorithm generates a (k,e)-event decomposition of M (if it exists) for some $k \leq |X_M|^2$.*

5   State-Event Optimized Fusions

Given a set of n primaries $\mathcal{P}$, we present an algorithm in [18] to generate a minimal f-fusion of $\mathcal{P}$. In this paper, we present an algorithm to generate fusions that are optimized for both states and events. We show that if each fusion in our solution contains more than $\Sigma - \beta$ events, then no f-fusion of $\mathcal{P}$ contains a machine with less than or equal to $\Sigma - \beta$ events, where β is a user defined parameter. Further, we present an incremental approach to this problem that improves the time complexity by a factor of ρ^n, where ρ is the average state reduction achieved by fusion, i.e., (|RCP|/Average size of a fusion).

The $genFusion$ algorithm that generates the fusion machines is shown in Fig. 5. Starting with the RCP of the primaries, $RCP(\mathcal{P})$, the algorithm generates one machine for each iteration of Loop 1 that increases d_{min} by 1 and at the end of f iterations we have f machines in $\mathcal{F}$ such that $d_{min}(\mathcal{P} \cup \mathcal{F}) > f$. Loops 2 and 3 reduce the events and states of the fusion machines.

Loop 2, Event Reduction: Starting with the RCP, which always increases d_{min} by one, Loop 2 uses the *reduceEvent* algorithm in Fig. 4 to iteratively generate reduced event machines that increase d_{min} by one. In each iteration of Loop 2, we generate the set of

```
genFusion
    Input: Primaries P, faults f, event depth β;
    Output: f-fusion of P;
    F ← {};
    for (i = 1 to f) //Loop 1
        M ← {RCP(P)};
        for (j = 1 to β) //Loop 2
            G ← {};
            for (M ∈ M)
                G = G ∪ reduceEvent(M);
            M = Machines in G that increment d_min;
        M ← Any machine in M;
        while (M ≠ RCP(P)⊥) //Loop 3
            C ← reduceState(M);
            M = Machine in C that increments d_min;
        F ← {M} ∪ F;
    return F;
```

```
reduceState
    Input: Machine P (X_P, α_P, Σ_P, p^0);
    Output: Largest Machines with ≤ |X_P| − 1
    states;
    B = {};
    for (s_i, s_j ∈ X_P)
        //combine states s_i and s_j
        Set of states, X_B = X_P with (s_i, s_j)
        combined;
        B = B ∪ {Largest machine consistent
        with X_B};
    return Incomparable machines in B;
```

```
incFusion
    Input: Primaries P, faults f, event depth β;
    Output: f-fusion of P;
    F ← {};
    for each (P_i ∈ P)
        F ← genFusion({P_i} ∪ RCP(F), f, β);
    return F;
```

Fig. 5. Optimized Fusion Algorithm

machines that contain one event less than the machines in the previous iteration and increase d_{min} by one. At the end of β iterations, we generate machine M that increases d_{min} by one and contains at most $\Sigma - \beta$ events, if such a machine exists. At any stage, if no valid machine was found, we exit the loop and select a machine from the previous iteration.

Loop 3, State Reduction [18]: In Loop 3, we try to find a minimal machine less than the event-reduced M that increases d_{min} by one. Starting with M, the *reduceState* algorithm in Fig. 5 generates the set of largest machines less than M in which at least two states of M are combined. We choose a machine in that set that increases d_{min} and reduce it until no further state reduction is possible (hit the bottom machine $RCP(\mathcal{P})_{\perp}$).

In Fig. 3, let $\mathcal{P} = \{A, B, C\}, f = 1, \beta = 2$. Since, $d_{min}(\mathcal{P}) = 1$, we need to add a machine that increases d_{min} to two. The set of machines containing one event less than the RCP are M_1 and M_2 among which only M_2 increases d_{min}. Reducing the event-set of M_2, at the end of $\beta = 2$ iterations, $M = F_1$. Since there is no machine less than F_1 that increases d_{min}, no state reduction is possible and the *genFusion* algorithm returns F_1. Note that, for $\beta = 0$ (no event-reduction), the *genFusion* algorithm is identical to the one in [18]. However, without event-reduction, the state reduction algorithm can combine r^0 and r^3 into a single block and generate F_2 as the largest machine containing this block. Since this is a minimal machine, the *genFusion* algorithm can return this 3-event machine. The event-reduction in the current version forces the algorithm to pick the 1-event machine F_1. In the technical report, we show that the time complexity of *genFusion* is $O(N^2|\Sigma|^{\beta}f + N^3|\Sigma|f)$, where $N = |RCP|$ and present a proof for the following theorem.

Theorem 2. *Given a set of n machines $\mathcal{P}$, the* genFusion *algorithm generates a minimal f-fusion (state minimality) of $\mathcal{P}$ such that if each machine in $\mathcal{F}$ contains more than $|\Sigma|-\beta$ events, then no f-fusion of $\mathcal{P}$ contains a machine with less than or equal to $|\Sigma|-\beta$ events (event minimality).*

Incremental Approach. Given n primaries each of size s, the *genFusion* algorithm generates their RCP, that has size $O(s^n)$, and hence the algorithm can have very high execution times. In Fig. 5, we present an incremental approach to generate the fusions, referred to as the *incFusion* algorithm in which we may never have to reduce the RCP of all the primaries. In each iteration, we generate the fusion corresponding to a new primary and the RCP of the (possibly small) fusions generated for the set of primaries in the previous iteration.

In Fig. 6, rather than generate a fusion by reducing the 8-state RCP of $\{A, B, C\}$, we can reduce the 4-state RCP of $\{A, B\}$ to generate fusion F' and then reduce the 4-state RCP of $\{C, F'\}$ to generate fusion F. In the technical report, we present the proof of correctness for the incremental approach and show that it has time complexity ρ^n times better than that of the *genFusion* algorithm, where ρ is the average state reduction achieved by fusion.

6   Detection and Correction of Faults

In [18], the time complexity to detect and correct faults is $O(n^2\rho + n\rho f + N)$, where n is the number of primaries, f is the number of crash faults, s is the size of each machine,

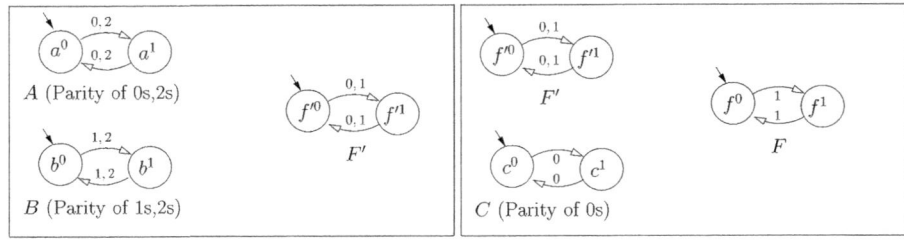

Fig. 6. Incremental Approach: First generate F' and then F

N is the size of the RCP and ρ is the average state reduction achieved by fusion. In this section, we provide algorithms to detect Byzantine faults with time complexity $O(nf)$, on average, and correct crash/Byzantine faults with time complexity $O(n\rho f)$, with high probability. Throughout this section, we refer to Fig. 3, with primaries, $\mathcal{P} = \{A, B, C\}$ and backups $\mathcal{F} = \{F_1, F_2\}$, that can correct two crash faults. The execution state of the primaries is represented collectively as a n-tuple (*primary tuple*) while the state of each backup is represented as the set of primary tuples it corresponds to (*tuple-set*). In Fig. 3, if A, B, C and F_1 are in their initial states, then the primary tuple is $a^0 b^0 c^0$ and the state of F_1 is $f_1^0 = \{a^0 b^0 c^0, a^1 b^0 c^1, a^1 b^1 c^0, a^0 b^1 c^1\}$ (which corresponds to $\{r^0, r^2, r^4, r^6\}$).

6.1   Detection of Byzantine Faults

Given the primary tuple and the tuple-sets corresponding to the backup states, the *detectByz* algorithm in Fig. 7 detects up to f Byzantine faults (liars). Assuming that the tuple-set of each backup state is stored in a permanent hash table at the recovery agent, the *detectByz* algorithm simply checks if the primary tuple r is present in each backup tuple-set b. In Fig. 3, if the states of machines A, B, C, F_1 and F_2 are a^1, b^1, c^0, f_1^1 and f_2^1 respectively, then the algorithm flags a Byzantine fault, since $a^1 b^1 c^0$ is not present in either $f_1^1 = \{a^0 b^1 c^0, a^1 b^1 c^1, a^1 b^0 c^0, a^0 b^0 c^1\}$ or $f_2^1 = \{a^0 b^1 c^0, a^1 b^0 c^1\}$. In the following theorem we show that if there are liars in the system, then the primary tuple will not be present in at least one of the backup tuple-sets.

Theorem 3. *Given a set of n machines $\mathcal{P}$ and an f-fusion $\mathcal{F}$ corresponding to it, the* detectByz *algorithm detects up to f Byzantine faults among them.*

In the technical report we present the proof for this theorem and also show that the space complexity for the *detectByz* algorithm is $O(Nfn \log s)$ while its time complexity is $O(nf)$ (on average). Even for replication, the recovery agent needs to compare the state of n primaries with the state of each of its f replicas, giving time complexity $O(nf)$.

6.2   Correction of Faults

Given the primary tuple and the tuple-sets of the backup states, to correct f crash faults (or $\lfloor f/2 \rfloor$ Byzantine faults), we first need to find the tuples among the backup tuple-sets that are within Hamming distance of f ($\lfloor f/2 \rfloor$ for Byzantine faults) from

detectByz	correctByz
Input: set of of fusion states B, primary tuple r;	**Input**: set of of fusion states B, primary tuple r;
Output: *true* or *false*	**Output**: corrected primary n-tuple;
for $(b \in B)$	$D \leftarrow \{\}$ //list of tuple-sets
if $\neg(hash\_table(b) \cdot contains(r))$	**for** $(b \in B)$
return *false*;	//tuples in b within Hamming distance
return *true*;	$\lfloor f/2 \rfloor$ of r
	$S \leftarrow lsh\_tables(b) \cdot search(r, \lfloor f/2 \rfloor)$;
correctCrash	$D \cdot add(S)$;
Input: set of of fusion states B, primary tuple r,	$G \leftarrow$ Set of tuples that appear in D;
crash faults among the primaries c $(\leq f)$;	$V \leftarrow$ Vote array of size $|G|$;
Output: corrected primary n-tuple;	**for** $(g \in G)$
$D \leftarrow \{\}$ //list of tuple-sets	// get votes from fusions
for $(b \in B)$	$V[g] \leftarrow$ Number of times g appears
//tuples in b within Hamming distance c	in D;
of r	// get votes from primaries
$S \leftarrow lsh\_tables(b) \cdot search(r, c)$;	**for** $(i = 1$ to $n)$
$D \cdot add(S)$;	**if**$(r[i] \in g)$
return Intersection of sets in D;	$V[g] + +$;
	return Tuple $g : V[g] \geq n + \lfloor f/2 \rfloor$;

Fig. 7. Detection and Correction of Faults

the primary tuple (explained in sections 6.2 and 6.2). In Fig. 3, the tuples in $f_1^0 = \{a^0 b^0 c^0, a^1 b^0 c^1, a^1 b^1 c^0, a^0 b^1 c^1\}$ that are within Hamming distance one of a primary tuple $a^0 b^0 c^1$ are $a^0 b^0 c^0$, $a^1 b^0 c^1$ and $a^0 b^1 c^1$. An efficient solution to finding the points among a large set within a certain Hamming distance of a query point is *locality sensitive hashing* (LSH) [1,9]. Based on this, we maintain L hash tables, $\{g_1 \ldots g_L\}$, for each fusion state at the recovery agent. The hash function for g_j, takes as input an n-tuple, selects k coordinates uniformly at random from them and returns the concatenated bit representation of these coordinates. In the example shown in Fig. 8(i), the tuple $a^1 b^0 c^1$ of f_1^0, is hashed into the 2nd bucket of g_1 and the 3rd bucket of g_2.

Given a point q and distance f, we obtain the points found in the buckets $g_j(q)$ for $j = 1 \ldots L$, and return those that are within distance of f from q. For example, in Fig. 8(i), given $q = a^0 b^0 c^0$, $f = 2$, this point hashes into the 1st bucket of g_1 and the 0th bucket of g_2 and hence the points returned are $a^0 b^1 c^1$ and $a^0 b^0 c^0$ respectively. If we set $L = \log_{1-\gamma^k} \delta$, where $\gamma = 1 - f/n$, such that $(1 - \gamma^k)^L < \delta$, then any f-neighbor of a point q is returned with probability at least $1 - \delta$ [1,9]. In the following sections, we present algorithms for the correction of crash and Byzantine faults based on these LSH functions.

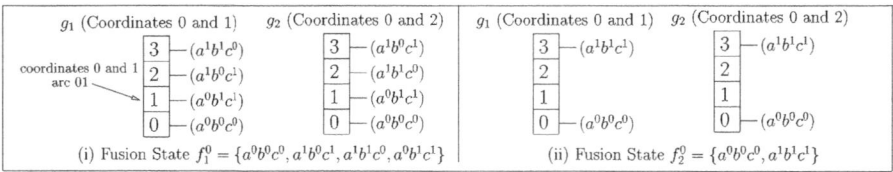

Fig. 8. LSH Example for fusion states in Fig. 3 with $k = 2, L = 2$

Crash Correction. Given the primary tuple (with possible gaps because of faults) and the tuple-sets of the available backup states, the *correctCrash* algorithm in Fig. 7 corrects up to f crash faults. The algorithm finds the tuples in the tuple-sets of each fusion state b that are within a Hamming distance c (actual number of faults) of the primary tuple r using the LSH tables for each fusion state. If the intersection of these sets is singleton, then we return that as the correct primary tuple. When the intersection is not singleton, we need to exhaustively search each fusion state for points within distance c of r (LSH has not returned all of them), but this happens with a very low probability [1,9]. In Fig. 3, assume crash faults in primaries B and C among $\{A, B, C\}$. Given the states of A, F_1 and F_2 as a^0, f_1^0 and f_2^0 respectively, the tuples within Hamming distance two of $r = a^0\{\}\{\}$ among $f_1^0 = \{a^0b^0c^0, a^1b^0c^1, a^1b^1c^0, a^0b^1c^1\}$ and $f_2^0 = \{a^0b^0c^0, a^1b^1c^1\}$ are $\{a^0b^0c^0, a^0b^1c^1\}$ and $\{a^0b^0c^0\}$ respectively. The algorithm returns their intersection, $a^0b^0c^0$ as the corrected primary tuple. In the following theorem, we prove that the *correctCrash* algorithm returns a unique primary tuple.

Theorem 4. *Given a set of n machines $\mathcal{P}$ and an f-fusion $\mathcal{F}$ corresponding to it, the* correctCrash *algorithm corrects up to f crash faults among them.*

In the technical report, we present the proof for this theorem and show that the space complexity of the *correctCrash* algorithm is $O(Nfn \log s)$ and its time complexity is $O(n\rho f)$ w.h.p. Crash correction in replication simply involves copying the state of the replicas of f failed primaries which has time complexity $O(f)$.

Byzantine Correction. Given the primary tuple and the tuple-sets of the backup states, the *correctByz* algorithm in Fig. 7 corrects up to $\lfloor f/2 \rfloor$ Byzantine faults. The algorithm finds the set of tuples among the tuple-sets of each fusion state that are within Hamming distance $\lfloor f/2 \rfloor$ of the primary tuple r using the LSH tables and stores them in list D. It then constructs a vote vector V for each unique tuple in this list. The votes for each tuple $g \in V$ is the number of times it appears in D plus the number of primary states of r that appear in g. The tuple with greater than or equal to $n + \lfloor f/2 \rfloor$ votes is the correct primary tuple. When there is no such tuple, we need to exhaustively search each fusion state for points within distance $\lfloor f/2 \rfloor$ of r (LSH has not returned all of them). In Fig. 3, let the states of machines A, B, C F_1 and F_2 are a^0, b^1, c^0, f_1^0 and f_2^0 respectively, with one liar among them ($\lfloor f/2 \rfloor = 1$). The tuples within Hamming distance one of $r = a^0b^1c^0$ among $f_1^0 = \{a^0b^0c^0, a^1b^0c^1, a^1b^1c^0, a^0b^1c^1\}$ and $f_2^0 = \{a^0b^0c^0, a^1b^1c^1\}$ are $\{a^0b^0c^0, a^1b^1c^0, a^0b^1c^1\}$ and $\{a^0b^0c^0\}$ respectively. The algorithm returns $a^0b^0c^0$, with four votes in total (one each from A, C, F_1 and F_2), since $n + \lfloor f/2 \rfloor = 3 + 1 = 4$. We show in the following theorem that there are enough machines separating each pair of tuples and even with liars the true primary tuple will get sufficient votes.

Theorem 5. *Given a set of n machines $\mathcal{P}$ and a f-fusion $\mathcal{F}$ corresponding to it, the* correctByz *algorithm corrects up to $\lfloor f/2 \rfloor$ Byzantine faults among them.*

In the technical report, we present a proof for the following theorem and show that the space complexity of the *correctByz* algorithm is $O(N f n \log s)$ and its time complexity of is $O(n \rho f)$ w.h.p. In the case of replication, we just need to obtain the majority across f copies of each primary with time complexity $O(nf)$.

7   Evaluation

7.1   Experimental Results

In [18], we evaluate fusion for simple examples such as counters and dividers. In this section, we evaluate fusion using the MCNC'91 benchmarks [24] for DFSMs, widely used for research in the fields of logic synthesis and finite state machine synthesis [17,25]. We implemented the *incFusion* algorithm of Fig. 5 in Java 1.6 and compared the performance of fusion with replication for 100 different combinations of the benchmark machines, with $n = 3$, $f = 2$, $\beta = 3$ and present some of the results in Table 3. The machine descriptions, implementation and detailed results are available in [3].

Let the primaries be denoted P_1, P_2 and P_3 and the fused-backups F_1 and F_2. Column 1 of Table 3 specifies the names of three primary DFSMs. Column 2 specifies the backup space required for replication ($\prod_{i=1}^{1=3} |P_i|^f$), column 3 specifies the backup space for fusion ($\prod_{i=1}^{i=2} |F_i|$) and column 4 specifies the percentage state space savings ((column 2-column 3)* 100/column 2). Column 5 specifies the total number of primary events, column 6 specifies the average number of events across F_1 and F_2 and the last column specifies the percentage reduction in events ((column 5-column 6)*100/column 5).

The average state space savings in fusion (over replication) is 38% (range 0-99%) over the 100 combination of benchmark machines, while the average event-reduction is 4% (range 0-45%). We also present results in [3] that show that the average savings in

Table 3. Evaluation of Fusion on the MCNC'91 Benchmarks

Machines	Replication State Space	Fusion State Space	% Savings State Space	Primary Events	Fusion Events	% Reduction Events
dk15, bbara, mc	25600	19600	23.44	16	10	37.5
lion, bbtas, mc	9216	8464	8.16	8	7	12.5
lion, tav, modulo12	36864	9216	75	16	16	0
lion, bbara, mc	25600	25600	0	16	9	43.75
tav, beecount, lion	12544	10816	13.78	16	16	0
mc, bbtas, shiftreg	36864	26896	27.04	8	7	12.5
tav, bbara, mc	25600	25600	0	16	16	0
dk15, modulo12, mc	36864	28224	23.44	8	8	0
modulo12, lion, mc	36864	36864	0	8	7	12.5

time by the incremental approach for generating the fusions (over the non-incremental approach) is 8%. Hence, fusion achieves significant savings in space for standard benchmarks, while the event-reduction indicates that for many cases, the backups will not contain a large number of events.

7.2 Practical Example: MapReduce

To motivate the practical use of fusion, we discuss its application to the MapReduce framework which is used to model large scale distributed computations. Typically, the Map-Reduce framework is built using the master-worker configuration where the master assigns the map and reduce tasks to various workers. Due to high cost of resources in replication, handling faults among the map workers is primarily based on checkpointing in which the processes periodically write to permanent storage. In the case of faults, the tasks are restarted from the last available state. This approach increases latency and may be inadequate for some applications.

Consider a distributed grep application over large files, where the master assigns three map tasks, each searching for one of the string patterns modeled by $\{A, B, C\}$ in Fig. 1. When the input files are partitioned into 200,000 chunks of data (the usual number in [6]), the current checkpointing-based approach requires 200,00*3 = 600,000 tasks in total, while causing high latency. A replication-based solution for correcting just one fault will involve creating a replica of each of the tasks A, B and C for each chunk of data, requiring 1200,000 tasks in total. A fusion-based approach needs to run only one additional backup task for each chunk of data, running F_1 shown in Fig. 1. Though recovery is costlier than replication, this approach requires only 800,000 tasks with much better latency than checkpointing.

8 Conclusion

We challenge the traditional approach of replication that requires nf backups to correct f crash faults among n machines and present a fusion-based solution that requires only f backups consuming considerably lesser state space. We present a problem that is fundamental to DFSMs: Can we replace a given DFSM with DFSMs containing fewer events? To formalize this, we introduce the concept of a (k,e)-event decomposition of a given machine and present efficient algorithms to generate such a decomposition. Based on this, we describe an algorithm to generate fused backups for a given set of machines that is optimized for both states and events.

Further, we present efficient algorithms to detect and correct faults in a system with fused backups. The algorithm for the detection of Byzantine faults has time complexity $O(nf)$ (on average), which is the same as that for replication. We apply the concept of locality sensitive hashing to the correction of faults and the time complexity for the correction of crash and Byzantine faults is $O(n\rho f)$ w.h.p. For relatively small values of n and ρ, fusion causes almost no overhead for recovery. Finally, we evaluate fusion on standard benchmarks for DFSMs and the results confirm that fusion achieves significant savings in space over replication. The event-reduction algorithm ensures that for many examples, the fused backups contain small event sets. Hence, in addition to our results on the theoretical optimality of the fused backups, we have illustrated the practical usefulness of fusion.

References

[1] Andoni, A., Indyk, P.: Near-optimal hashing algorithms for approximate nearest neighbor in high dimensions. Commun. ACM 51(1), 117–122 (2008)

[2] Balasubramanian, B., Garg, V.K.: A fusion-based approach for handling multiple faults in data structures. Technical Report ECE-PDS-2009-001, Parallel and Distributed Systems Laboratory, ECE Dept. University of Texas at Austin (2009)

[3] Balasubramanian, B., Garg, V.K.: Fsm backup library (implemented in java 1.6). In: Parallel and Distributed Systems Laboratory (2011), http://maple.ece.utexas.edu

[4] Balasubramanian, B., Garg, V.K.: A report on fused state machines for fault tolerance in distributed systems. Technical Report TR-PDS-2011-002 Parallel and Distributed Systems Laboratory, The University of Texas at Austin (2011), http://pdsl.ece.utexas.edu/TechReports/2011/TR-PDS-2011-002.pdf

[5] Chen, P.M., Lee, E.K., Gibson, G.A., Katz, R.H., Patterson, D.A.: Raid: high-performance, reliable secondary storage. ACM Comput. Surv. 26(2), 145–185 (1994)

[6] Dean, J., Ghemawat, S.: Mapreduce: simplified data processing on large clusters. Commun. ACM 51, 107–113 (2008)

[7] Fischer, M.J., Lynch, N., Paterson, M.: Impossibility of distributed consensus with one faulty process. Journal of the ACM 32(2) (April 1985)

[8] Garg, V.K.: Implementing Fault-Tolerant Services Using State Machines: Beyond Replication. In: Lynch, N.A., Shvartsman, A.A. (eds.) DISC 2010. LNCS, vol. 6343, pp. 450–464. Springer, Heidelberg (2010)

[9] Gionis, A., Indyk, P., Motwani, R.: Similarity search in high dimensions via hashing. In: VLDB 1999: Proceedings of the 25th International Conference on Very Large Data Bases, pp. 518–529. Morgan Kaufmann Publishers Inc., San Francisco (1999)

[10] Hamming, R.: Error-detecting and error-correcting codes. Bell System Technical Journal 29(2), 147–160 (1950)

[11] Hartmanis, J., Stearns, R.E.: Algebraic structure theory of sequential machines. Prentice-Hall international series in applied mathematics. Prentice-Hall, Inc., Upper Saddle River (1966)

[12] Hopcroft, J.E.: An n log n algorithm for minimizing states in a finite automaton. Technical report, Stanford, CA, USA (1971)

[13] Huffman, D.A.: The synthesis of sequential switching circuits. Technical report, Massachusetts, USA (1954)

[14] Lamport, L.: The implementation of reliable distributed multiprocess systems. Computer Networks 22, 95–114 (1978)

[15] Lamport, L., Shostak, R., Pease, M.: The byzantine generals problem. ACM Transactions on Programming Languages and Systems 4, 382–401 (1982)

[16] Lee, D., Yannakakis, M.: Closed partition lattice and machine decomposition. IEEE Trans. Comput. 51(2), 216–228 (2002)

[17] Mishchenko, A., Chatterjee, S., Brayton, R.: Dag-aware aig rewriting: A fresh look at combinational logic synthesis. In: DAC 2006: Proceedings of the 43rd Annual Conference on Design Automation, pp. 532–536. ACM Press (2006)

[18] Ogale, V., Balasubramanian, B., Garg, V.K.: A fusion-based approach for tolerating faults in finite state machines. In: International Parallel and Distributed Processing Symposium, pp. 1–11 (2009)

[19] Patterson, D.A., Gibson, G., Katz, R.H.: A case for redundant arrays of inexpensive disks (raid). In: SIGMOD 1988: Proceedings of the 1988 ACM SIGMOD International Conference on Management of Data, pp. 109–116. ACM Press, New York (1988)

[20] Pease, M., Lamport, L.: Reaching agreement in the presence of faults. Journal of the ACM 27, 228–234 (1980)

[21] Schneider, F.B.: Byzantine generals in action: implementing fail-stop processors. ACM Trans. Comput. Syst. 2(2), 145–154 (1984)

[22] Schneider, F.B.: Implementing fault-tolerant services using the state machine approach: A tutorial. ACM Computing Surveys 22(4), 299–319 (1990)

[23] Tenzakhti, F., Day, K., Ould-Khaoua, M.: Replication algorithms for the world-wide web. J. Syst. Archit. 50(10), 591–605 (2004)

[24] Yang, S.: Logic synthesis and optimization benchmarks user guide version 3.0 (1991)

[25] Youra, H., Inoue, T., Masuzawa, T., Fujiwara, H.: On the synthesis of synchronizable finite state machines with partial scan. Systems and Computers in Japan 29(1), 53–62 (1998)

Fork-Consistent Constructions from Registers*

Matthias Majuntke[1], Dan Dobre[2], Christian Cachin[3], and Neeraj Suri[1]

[1] Technische Universität Darmstadt, Hochschulstraße 10, 64289 Darmstadt, Germany
{majuntke,suri}@cs.tu-darmstadt.de
[2] NEC Laboratories Europe, Kurfürsten-Anlage 36, 69115 Heidelberg, Germany
dan.dobre@neclab.eu
[3] IBM Research – Zurich, Säumerstrasse 4, 8803 Rüschlikon, Switzerland
cca@zurich.ibm.com

Abstract. Users increasingly execute services online at remote providers, but they may have security concerns and not always trust the providers. Fork-consistent emulations offer one way to protect the clients of a remote service, which is usually correct but may suffer from Byzantine faults. They feature linearizability as long as the service behaves correctly, and gracefully degrade to fork-consistent semantics in case the service becomes faulty. This guarantees data integrity and service consistency to the clients.

All currently known fork-consistent emulations require the execution of non-trivial computation steps by the service. From a theoretical viewpoint, such a service constitutes a *read-modify-write* object, representing the strongest object in Herlihy's wait-free hierarchy [1]. A read-modify-write object is much more powerful than a shared memory made of so-called *registers*, which lie in the weakest class of all shared objects in this hierarchy. In practical terms, it is important to reduce the complexity and cost of a remote service implementation as computation resources are typically more expensive than storage resources.

In this paper, we address the fundamental structure of a fork-consistent emulation and ask the question: Can one provide a fork-consistent emulation in which the service does not execute computation steps, but can be realized only by a shared memory? Surprisingly, the answer is yes. Specifically, we provide two such algorithms that can be built only from registers: A fork-linearizable construction of a universal type, in which operations are allowed to abort under concurrency, and a weakly fork-linearizable emulation of a shared memory that ensures wait-freedom when the registers are correct.

Keywords: distributed system, shared memory, fork-consistency, universal object, atomic register, Byzantine faults.

1 Introduction

The increasing trend of executing services online "in the cloud" [2] offers many economic advantages, but also raises the challenge of guaranteeing security and strong consistency to its users. As the service is provided by a remote entity that wants to retain its customers, the service usually acts as specified. But online services may fail for

---

* Research funded in part by DFG GRK 1362 (TUD GKmM).

A. Fernández Anta, G. Lipari, and M. Roy (Eds.): OPODIS 2011, LNCS 7109, pp. 283–298, 2011.

various reasons, ranging from simply closing down (corresponding to a crash fault) to deliberate and sometimes malicious behavior (corresponding to a Byzantine fault).

For some kinds of services, cryptographic techniques can prevent a malicious provider from forging responses or snooping on customer data. But other violations are still possible in the asynchronous model considered here: for instance, when multiple isolated clients interact only through a remote provider, the latter may send diverging and inconsistent replies to the clients. In this context, "forking" consistency conditions [3,4] offer a gracefully degrading solution because they make it much easier for the clients to detect such violations. More precisely, they ensure that if a Byzantine provider even *once* sent a wrong response to some client, then this client becomes *forever isolated* or *forked* from those other clients to which the provider responded differently. With this notion, clients may easily detect service misbehavior from a single inconsistent operation, e.g. by out-of-band communication.

Forking consistency conditions are often encapsulated in the notion of a *Byzantine emulation* [4], which ensures graceful degradation of the service's semantics: If the service is correct, then operations execute atomically. In any other case, the clients still observe operations according to the forking consistency notion. Fork-consistency represents a safety property — after all, a faulty service may simply stop. The liveness property in a Byzantine emulation refers to the good case when the service behaves correctly.

Fork-linearizability [3,4] ensures that clients always observe linearizable [5] service behavior and that two clients, once forked, will never again see each other's updates to the system (i.e. they share the same history prefix up to the forking point). However, it has been found that fork-linearizable Byzantine emulations of a shared memory *cannot* always provide *wait-free* operations [4], i.e., some clients may be blocked because of other clients that execute operations concurrently. An escape is offered by the weaker liveness property of abortable emulations, which allow client operations to *abort* under contention [6]. As another alternative, the notion of *weak fork-linearizability* relaxes fork-linearizability in order to allow wait-free client operations in Byzantine emulations [7]. *Weak fork-linearizability* [7] allows two clients, after being forked, to observe a single operation of the other one (at-most-one-join), and that the real-time order induced by linearizability may be violated by the last operation of each client (weak real-time order).

In this paper, we explore the fundamental assumptions required for building a Byzantine service emulation. Up to now, all fork-consistent emulation protocols have required the service to execute non-trivial computation steps, i.e., the service must be implemented by an object of *universal* type [1], capable of *read-modify-write* operations [8]. We show the surprising result that this requirement can be dropped, and implement fork-consistent emulation protocols only from memory objects, so-called *registers*. They provide simple read and write operations and represent one of the weakest forms of computational objects. A long tradition of research has already addressed how to realize powerful abstractions from weaker base objects (e.g., [1,9]).

Specifically, we propose the first *fork-linearizable Byzantine emulation* of a universal object only from *registers*. Our algorithm necessarily offers abortable operations because a wait-free construction of a universal object from registers is not possible in an asynchronous system using only registers [1]. Moreover, we give an algorithm for a *weakly fork-linearizable Byzantine emulation* of a shared memory only from registers. It allows wait-free client operations when the underlying registers are correct.

Our two algorithms may directly replace the computation-based constructions in the existing respective emulations of shared memory on Byzantine servers [6,7,10]. For instance, our second construction, which yields a weakly fork-linearizable Byzantine emulation, allows to eliminate the server code from Venus [10]. Currently, Venus runs server code implemented by a *cloud computing* service, but our construction may realize it from a *cloud storage* service. For practical systems this can make a big difference in cost because full-fledged servers or virtual machines (e.g., Amazon EC2) are typically more expensive than simple disks or cloud-based key-value stores (e.g., Amazon S3).

Note that although our approach uses a collection of registers, we refrain from making more specific failure assumptions on them. Our remote service is comprised of registers, and as soon as one register is faulty, we consider the service to be faulty. It is conceivable to use fault-prone registers in our algorithms. Standard methods implementing robust shared registers from fault-prone base registers show how to *tolerate* up to a fraction of Byzantine base registers [11]. This extension, which is orthogonal to our work, would further refine our notion of graceful service degradation with faulty base objects.

Related Work. The notion of fork-linearizability was introduced by Mazières and Shasha [3]. They implemented a fork-linearizable multi-user storage system called SUNDR. An improved fork-linearizable storage protocol is described by Cachin *et al.* [4]; it reduces the communication complexity compared to SUNDR from $\mathcal{O}(n^2)$ to $\mathcal{O}(n)$. More recently, fork-linearizable Byzantine emulations have been extended to *universal services* [12]. All fork-linearizable emulations are blocking and sometimes require one client to wait for another client to complete [4].

In order to circumvent blocking the clients, Majuntke *et al.* [6] propose the first *abortable* fork-linearizable storage implementations. Their work takes up the notion of an abortable object introduced by Aguilera *et al.* [13]. They demonstrated, for the first time, how an abortable (and, hence, obstruction-free [14]) universal object can be constructed from abortable registers, which are base objects weaker than registers. In more recent work, it has been shown that abortable objects can be boosted to wait-free objects in a partially synchronous system [15]. This makes our Byzantine emulations of abortable objects very attractive in practical systems.

Actually implemented systems offering data storage integrity through forking consistency semantics include SUNDR (LKMS) [16], which realizes the protocol of Mazières and Shasha [3]. Furthermore, Cachin *et al.* [17] add fork-linearizable semantics to the Subversion revision control system, such that integrity and consistency of the server can be verified. The "blind stone tablet" of Williams *et al.* [18] provides fork-linearizable semantics for an untrusted database server; it may abort conflicting operations. Using a relaxation of fork-linearizability, called *fork-* consistency*, Feldman *et al.* [19] introduce a lock-free implementation for online collaboration that protects consistency and integrity of the service against a malicious provider.

Cachin *et al.* [7] present the storage service FAUST, which emulates a shared memory in a wait-free manner by exploiting the notion of *weak fork-linearizability*. It relaxes fork-linearizability in two fundamental ways: (1) after being forked, two clients may observe each others' operations once more and (2) the real-time order of the last operation of each client is not preserved. FAUST incorporates client-to-client communication in a higher layer, which ensures that all operations become eventually consistent over time (or the server is detected to misbehave). The Venus system [10] implements the mechanisms behind FAUST and describes a practical solution for ensuring integrity and consistency to the users of cloud storage.

Li and Mazières [20] study storage systems, built from $3f + 1$ server replicas, where more than f replicas are Byzantine faulty. Their storage protocol ensures *fork-* consistency*. Similar to weak fork-linearizability, fork-* consistency allows that two forked clients observe again at most one common operation.

Contributions We present, for the first time, Byzantine emulations with forking consistency conditions only from *registers*, instead of more powerful computation objects. Any number of registers may be affected by Byzantine failures. Our constructions are linearizable provided that the base registers are correct. The constructions are:

- A register-based abortable Byzantine emulation of a fork-linearizable universal type.
- A register-based wait-free Byzantine emulation of weak fork-linearizable shared memory.

In Section 1, we discuss related work; Section 2 introduces the underlying system model. The two main constructions are given in Sections 3 and 4. The paper concludes in Section 5. The correctness proofs of the protocols can be found in our Technical Report [21].

2   System Model

We consider a distributed system consisting of $n > 1$ *clients* $C_1, \ldots, C_n$ that communicate through shared *objects*. Each such base object has a *type* which is given by a set of *invocations*, a set of *responses*, and by its *sequential specification*. The sequential specification defines the allowed sequences of invocations and responses. An *invocation* and the corresponding *response* constitute an *operation* of an object. A collection of base objects is used to implement high-level objects, where clients execute algorithm A, consisting of n state machines $A_1, \ldots, A_n$ (where C_i implements A_i). When client C_i receives an *invocation* of an operation to the high-level object, it takes steps of A_i, where it (1) either invokes an operation on some base object, (2) or receives the response to its previous invocation to a base object, (3) or it performs some local computation. At the end of a step, C_i changes its local state and possibly returns a response to the pending high-level operation.

An *execution* of algorithm A is defined as the (interleaved) sequence of invocation and response events. Every execution induces a *history* which is the sequence of invocations and responses of the high-level operations. If σ is a history of an execution of algorithm A, then $\sigma|_{C_i}$ denotes the subsequence of σ containing all events of client C_i. For sequence σ and operation o, $\sigma|^o$ denotes the prefix of σ that ends with the last event of o. We say that a response *matches* an invocation, if both are events of the same operation. An operation is called *complete*, if there exists a matching response to its invocation, else *incomplete*. We assume that each client invokes a new operation only after the previous operation has completed. A history consisting only of matching invocation/response pairs is called *well-formed*. Operation o *precedes* operation o' in a sequence of events σ ($o <_\sigma o'$) iff o is complete and the response of o happens before the invocation of o'. If o precedes o' we denote o and o' as *sequential*, if neither one precedes the other, then o and o' are said to be *concurrent*.

For the proposed *abortable* construction (Sec. 3), we introduce the special response ABORT. A complete operation o is called *unsuccessful* ("o is aborted"), if it returns ABORT, else it is called *successful* ("o successfully completes"). The formal definition of an *abortable* object comprises a non-triviality property which allows aborts only under concurrency [13].

Clients may fail by *crashing*, i.e. they stop taking steps and hence, the last operation of each client might be *incomplete*. Base objects may deviate arbitrarily from their specification exhibiting *non-responsive-arbitrary faults* [22] (called *Byzantine*). Clients have access to a digital signature scheme used by each client to *sign* its data such that any other client can determine the authenticity of a datum by *verifying* the corresponding signature. We assume that signatures cannot be forged.

All constructions appearing in this paper are based on *atomic registers*. An atomic register provides two operations, *read* and *write*[1]. Operation $write(v)$ stores value v from domain *Values* into the register. A call of *read*() returns the latest written value from the register or the special value $\perp$ if no value has been written. As the register is atomic, its history satisfies linearizability [1], i.e. operations seem to appear as sequential, atomic events[2]. Further, the atomic registers used allow single-writer-multiple-reader access (SWMR), i.e. to each register we assign a dedicated client that may call *write* and *read*, while all other clients may only call *read* to that register.

A sequence of operations π satisfies *weak real-time* order of σ if π, excluding the last operation of each client in π, satisfies real-time order of σ. *Causality* between two operations depends on the type of the implemented object[3]. For two operations of a shared memory o and o' in σ, o *causally precedes* o' ($o \rightarrow_\sigma o'$), if o, o' are called by the same client and o happens before o', or if o' is a READ operation that returns the value written by WRITE operation o. The next definition formalizes the notion of *fork-linearizability* [4] and *weak fork-linearizability* [7]; for a formal definition of the term *possible view* as well as the above-mentioned notions we refer to the Technical Report [21].

Definition 1. Let σ be a history of an object of type T and for each client C_i there exists a sequence of events π_i such that π_i is a possible view of σ at C_i with respect to T.

History σ is *fork-linearizable* with respect to object type T if for each client C_i:

---

[1] We type operation calls to base registers in *italic* font and calls to constructed objects in CAPITALS.

[2] Hence, the "latest written value" is well-defined.

[3] As causality is needed to define *weak fork-linearizability*, here, we give causality for a *shared memory*, which is the type we implement with weak fork-linearizability.

1. π_i preserves the real-time order of σ, and
2. for every client C_j and for every $o \in \pi_i \cap \pi_j$, it holds $\pi_i|^o = \pi_j|^o$.

History σ is *weak fork-linearizable* with respect to object type T if for each client C_i:

1. π_i preserves the weak real-time order of σ, and
2. for every operation $o \in \pi_i$ and every operation $o' \in \sigma$ such that $o' \to_\sigma o$, it holds that $o' \in \pi_i$ and that $o' <_{\pi_i} o$, and
3. (At-most-one-join) for every client C_j and every two operations $o, o' \in \pi_i \cap \pi_j$ by the same client such that $o <_\sigma o'$, it holds $\pi_i|^o = \pi_j|^o$.

The notion of a *Byzantine emulation* [4] allows us to formally define the safety and liveness properties of our protocols. Note that the liveness condition of abortable operations is weaker than *wait-freedom* but still not weaker than *obstruction-freedom* [13].

Definition 2. An algorithm A *emulates* an object of type T on a set of Byzantine base objects B with {fork|weak fork}-linearizability whenever following conditions hold:

1. If all objects in set B are correct, the history of every fair[4] and well-formed execution of A is linearizable with respect to type T, and
2. the history of every fair and well-formed execution of A is {fork|weak fork}-linearizable with respect to type T.

Such an emulation is *wait-free* (*abortable* resp.), iff every fair and well-formed execution of the protocol with correct base objects is wait-free [1] (abortable [13] resp.).

3   A Fork-Linearizable Universal Type

In this section we present as our first main contribution an abortable fork-linearizable Byzantine emulation of a universal type implemented from atomic registers. The shared object ensures fork-linearizability in the presence of any number of faulty base registers. High-level operations are *abortable* [13], i.e. under concurrency, the special response ABORT may be returned. The functionality of a universal type T is encoded in the procedure APPLY$_T$. For client C_i, state s and operation o, APPLY$_T(s, o, i)$ returns (s', res), where s' is the new state of the universal object, res the computation result, and where the sequence of invoking APPLY$_T(s, o, i)$ and returning (s', res) is defined by the sequential specification of type T.

Our algorithm uses timestamp vectors called *versions* whose order reflects the real-time order in which operations are applied to the shared object. Each operation carries a version and the linearization of operations is achieved through the use of an INC&READ counter object C with two atomic operations INC&READ and READ. An invocation to INC&READ(C) advances the counter object C and returns a value which is higher than any value returned before, and READ(C) returns the current value of the counter object. An implementation of the INC&READ counter is given in the Technical Report [21] together with its formal properties. Our implementation uses wait-free atomic registers as base objects which makes it a wait-free variant of the abortable INC&READ counter described by Aguilera *et al.* [13].

---

[4] For a formal definition we refer to standard literature [23].

3.1 Algorithm Ideas

Universal Type. To implement universal type T, we use n SWMR registers $R_1, \ldots, R_n$ such that client C_i can read from all registers but may write only to R_i. The registers store states of the universal object. To implement high-level operations, client C_i reads from the register which holds the most current state, applies the relevant state transformation, and writes the new state to R_i. Note, that all information are digitally signed by the clients as base objects are untrusted. Thereby, operations "affect" each other which leads to the following relation on operations: Operation o of C_i *affects* operation o' of C_j, if during o', C_j is able to verify the signature of C_i on state s that has been written during o and if C_j executes APPLY$_T$ on s during o'; further, an operation of C_i *affects* each later operation of C_i.

Concurrency detection. We allow operations to abort under concurrency for two reasons: there is no wait-free construction of a universal type from registers, as shown by Herlihy [1], and no fork-linearizable protocol can be wait-free in all executions, as shown in a more recent work of Cachin *et al.* [4]. Cachin's impossibility is based on two runs, indistinguishable for the reader: In the first run a READ operation does not return value v as it is concurrently written, while in the second run v has been previously written and is hidden by malicious registers. To avoid such a situation, our protocol implements a concurrency detection mechanism [13] using INC&READ counter object C. If concurrency is detected, a pending operation is aborted. At the invocation of a high-level operation o, our protocol calls INC&READ(C) and remembers the timestamp returned. At the end of o, READ(C) is executed to check whether counter C still returns the same timestamp. If not, another operation o' was invoked during o — thus, o is aborted. Else, if at the end of o C has not been changed, all successful operations either terminated before o or will be invoked after o has terminated. This is because the timestamps, returned from INC&READ, are used to linearize operations: The current state is written together with the timestamp, and the timestamp is used to determine the most recent state. Hence, all other operations invoked so far write a state with a lower timestamp than o. Consequently, such operations are linearized before o and only the state written by o can be read by later operations.

Fork-Linearizability. In addition to the timestamp from INC&READ counter C, each operation is assigned a vector of timestamps of length n, called *version*. The order relation $\leq$ defined on versions respects real-time order and the "affected by" relation on operations. The idea is that each operation reads the most recent version from the storage, increments its own entry and writes the new version back to the storage. Thereby, each operation checks, if the version it reads, has been affected by the version of its own last successful operation, i.e. one which was not aborted. If the last successful operation of client C_i is hidden from C_j, then C_i does not accept operations of C_j as they have *not* been affected by the last successful operation of C_i. This ensures that the views of the clients after a forking attack are not rejoined. This principle is based on ideas of Mazières and Shasha [3], and Cachin *et al.* [4]. To apply it to this work, we have to add a specific handling for aborted operations: If operation o of client C_i is aborted, C_i cannot expect that o will affect later operations. However, it is still possible that some operation of C_j is affected by aborted o. In this case we call o *relevant* for C_j (refer to the Technical Report [21] for a formal definition).

3.2 Description of Algorithm 1

We now describe the steps preformed by client C_i when executing high-level operation o. The algorithm is given as Algorithm 1, the variables used are collected in Figure 1.

The protocol is framed by INC&READ(C) and READ(C) calls to the counter object C implementing the concurrency detection mechanism (lines 1.2 and 1.14). If the returned timestamps are not equal, the operation is aborted in line 1.16. In lines 1.3–1.5, the client reads from all atomic registers $R_1, \ldots, R_n$ and determines by means of the assigned timestamps the index l of the register holding the latest written data $\langle ts_l, V_l, s_l, sig_l \rangle$, where ts_l is a timestamp, V_l is the version, s_l is the state and sig_l is a signature. If some data have been written to R_l, the signature of the content of R_l is verified (line 1.6). Then, client C_i checks whether the read version V_l is not smaller than V_{suc} the version of its own last successful operation (line 1.7). When the check is passed the new state of the universal object and the computation result is computed by calling APPLY$_T(s_l, o, i)$ (line 1.8). Finally the new version for operation o has to be computed. This is done by taking the per-entry maximum of version V, which is the local version of C_i, and V_l, and by incrementing the ith entry (lines 1.9–1.11). After signing the current timestamp, the new version V, and new state s in line 1.12, client C_i writes ts, V, s and the signature into register R_i (line 1.13). If operation o is successful, version V is stored as last successful version V_{suc} and the computation result is returned (lines 1.17–1.19).

C INC&READ counter object, initially 0
$R_1, \ldots R_n$ SWMR atomic register, initially $\langle 0, (0, \ldots, 0), \bot, \bot \rangle$ /*
`ts+version+state+sig */`
ts, ts', ts_l, cn integer, initially 0   `/* timestamp & counter */`
$V[1..n], V_l[1..n], V_{suc}[1..n]$ array of integers, intially $(0, \ldots, 0)$ `/* version */`
s, s_l state, initially $\bot$ `/* state */`
res operation result, initially $\bot$ `/* return value */`
sig, sig_l signature, initially $\bot$ `/* signature */`

Fig. 1. Variables used in Algorithm 1

3.3 Correctness Arguments

In this section we argue why Algorithm 1 satisfies fork-linearizability. The goal is to construct for each client C_i a view π_i of σ that satisfies the properties of fork-linearizability. To construct π_i, we simplify our argumentation by ignoring operations that are not relevant for C_i. Recall, any operation is *relevant* for client C_i that affects C_i's last successful operation. Hence, operations that are not relevant for client C_i do not change the object's state from C_i's point of view. Thus, we can order them arbitrarily among the operations in π_i and the resulting sequences still satisfy fork-linearizability.

Algorithm 1. Universal Object Implementation, Code of Client i

---

1.1 EXECUTE(o) **do**

1.2 $\quad ts \leftarrow$ INC&READ(C) /* increment and read from counter */

1.3 $\quad$ **for** $j = 1, \ldots, n$ **do**

1.4 $\quad\quad \langle ts_j, V_j, s_j, sig_j \rangle \leftarrow read(R_j)$ /* low-level atomic read */

1.5 $\quad$ let l be such that $ts_l = \max_{1 \leq j \leq n}(ts_j)$ /* find register with most recent data */

1.6 $\quad$ **if** $V_l \neq [0 \ldots 0] \wedge \neg \text{verify}_l(sig_l, \langle ts_l, V_l, s_l \rangle)$ **then halt** /* signature verified? */

1.7 $\quad$ **if** $\exists k : V_{suc}[k] > V_l[k]$ **then halt** /* fork-linearizability check passed? */

1.8 $\quad \langle s, res \rangle \leftarrow$ APPLY$_T(s_l, o, i)$ /* compute new state + result */

1.9 $\quad$ **for** $j = 1, \ldots, n, j \neq i$ **do**

1.10 $\quad\quad V[j] \leftarrow \max(V[j], V_l[j])$ /* determine

1.11 $\quad V[i] \leftarrow V[i] + 1$ new version */

1.12 $\quad sig \leftarrow \text{sign}_i(ts\|V\|s)$ /* signature on ts, version, state */

1.13 $\quad write(R_i, \langle ts, V, s, sig \rangle)$ /* low-level atomic write */

1.14 $\quad ts' \leftarrow$ READ(C) /* read from counter */

1.15 $\quad$ **if** $ts \neq ts'$ **then**

1.16 $\quad\quad$ **return** ABORT /* concurrency detected */

1.17 $\quad$ **else**

1.18 $\quad\quad V_{suc} \leftarrow V$ /* reset last successful version */

1.19 $\quad\quad$ **return** res /* return result */

The idea behind the construction of the π_i in the proof is that operations are ordered according to their assigned versions. The proof shows that this order respects the "affected by" relation, the sequential specification of a universal type, and the real-time order. As during an operation the new version is computed using the client's last version and the read version, proving "affected by" and real-time order is straightforward. The core of the proof is to show that the order of version also respects the sequential specification. We sketch the intuition behind this with the following argument leading to a contradiction:

Assume that some operation o_c is not affected by the most recent state of the universal object, which has been written by relevant operation o_b, but is affected by an older state written by operation o_a. In this case, the clients of o_b and o_c are forked, and neither o_b nor o_c affect each other. We argue, that in such a situation, there is no relevant operation that has been affected by both o_b and o_c, as such an operation would join the two clients violating fork-consistency. We assume for contradiction, that a relevant operation o_{join} of client C_{join}, affected by o_b and o_c exists which is also the first among such operations (see Figure 2). Operation o_{join} is affected by o_{join_suc}, the last successful operation of C_{join} previous to o_{join}, *and* by o_r that wrote the state which is read during o_{join}. Hence, without loss of generality o_{join_suc} is affected by o_b while o_r is affected by o_c. During operation o_{join_suc}, client C_{join} raises its value in the version to $V[join]_{join\_suc}$. This implies that o_{join} only accepts versions where the $join$th entry is at least $V[join]_{join\_suc}$ (line 1.7). As o_{join_suc} is not on the path of "affected

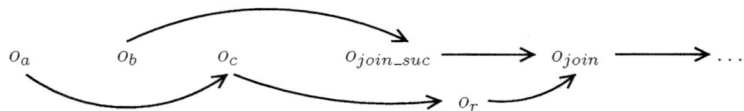

Fig. 2. Correctness Idea of Algorithm 1. Arrows denote the "affected by" relation.

by" relations from o_c to o_r, o_{join} would block while reading the state of o_r which is a contradiction. Thus, o_{join} does not exist.

Finally, it follows directly from the described construction, that sequences π_i satisfy the no-join property. To complete the correctness proof of the Byzantine emulation, we show that when all base objects are correct, no operation blocks and that no operation trivially aborts.

4   A Weak Fork-Linearizable Shared Memory

In this section we describe as our second contribution a wait-free, weak fork-linearizable Byzantine emulation of a shared memory implemented from atomic registers. The presented construction satisfies weak fork-linearizability in the presence of any number of faulty base objects. The implemented shared memory provides n atomic registers, such that each client can write to one dedicated register exclusively and may read from all registers. Operation WRITE(v), called by client C_i, writes value v to C_i's register. Operation READ(i) returns the last written value from C_i's register, and may be called by any client. Our algorithm makes use of an atomic single-writer snapshot object S with n components [24,25]. Snapshot object S provides two atomic operations: UPDATE(d, S, i), that changes the state of component i of S to d, and SCAN(S) that returns vector $(d_1, \ldots, d_n)$ such that d_i is the state of component i of S, $i = 1 \ldots, n$. Formally, d_i is the state written by the last UPDATE to component i prior to SCAN. It has been shown, that such a shared snapshot object can be wait-free implemented only from registers [24,25].

4.1   Algorithm Ideas

Each client locally maintains a timestamp that respects causality and real-time order of its *own* operations. As the basic principle, during each operation this timestamp is written to the shared memory and timestamps left by other operations are read. For each client C_i our implementation uses two registers only C_i may write to, but which can be read by all clients. The first one is needed to store value and timestamp written by C_i's WRITE operations and is implemented by a SWMR atomic register W_i (i.e. registers $W_1, \ldots, W_n$ in total). The second "register" is required to store the latest timestamp of C_i's READ operations. It is implemented as the ith component within the single-writer snapshot object with n components, S.

During READ(j) operation of C_i, C_i's current timestamp is written to S using UPDATE, thereafter, C_i reads a timestamp-value pair from register W_j (using low-level *read*). High-level WRITE(v) of C_i proceeds analogously: C_i writes its current timestamp plus value v to register W_i using low-level *write*, thereafter, it reads *all*

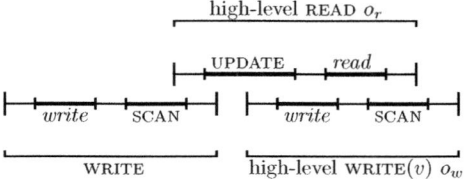

Fig. 3. Basic principle implemented by Algorithm 2

components from S using SCAN. By this, operations are able to observe each other, as expressed in the relation "seen": We say that a WRITE operation o_w of C_j *sees* a READ operation o_r of C_i with timestamp ts if C_i digitally signed ts and updated the ith component of S by signed ts during o_r and, if during o_w, C_j scanned S and was able to verify the signature of C_i on ts; READ operation o_r *sees* WRITE operation o_w if o_r returns the value written by o_w.

This construction guarantees the following property on interleaved high-level operations: Whenever high-level READ(j) o_r of C_i and WRITE(v) o_w of C_j appear in an execution such that o_r does not return v but a value written before v, then, by regularity of the atomic base registers, $o_w.write$[5] does not precede $o_r.read$, i.e., $o_r.read$ has been invoked *before* $o_w.write$ finishes. Consequently, o_r.UPDATE precedes o_w.SCAN (see Figure 3). Thus, if o_r does not "see" o_w, then o_w "sees" o_r. A similar property on interleaving operations has also been leveraged in our previous work [26] as well as by Aguilera *et al.* [9].

We can expect that client C_j writes information during its next WRITE operation such that future operations of C_i may verify whether operation o_w actually has seen operation o_r. More concrete, if READ o_r has seen WRITE o_w then the client checks during o_r whether the next WRITE operation after o_w (of the same client as o_w), has seen READ operation o_r or a newer one. Else, the base objects are faulty, as shown in the following example: Let o_w and o_w' be two sequential WRITE operations of C_i, o_w' precedes READ operation o_r of C_j but it is hidden by the malicious base objects such that o_r sees only o_w. As o_w' precedes o_r, o_w' cannot see o_r. However, as o_r sees o_w, it expects that o_w' will see o_r. The next WRITE operation o_w'' of C_i will write this information. If client C_j sees o_w'', which would violate weak fork-linearizablility, the check, explained above, is not passed.

4.2 Description of Algorithm 2

This section explains the steps taken by client C_i to implement high-level READ and WRITE operations. The algorithm is given as Algorithm 2, its variables in Figure 2.

At invocation of high-level READ(j), client C_i increments its local timestamp and generates a digital signature of it. The signed timestamp is stored to snapshot object S using operation UPDATE($(ots, sig), S, i$) (lines 2.2–2.4). Then, client C_i reads register W_j and verifies the signature (line 2.5–2.6). The content of register W_j contains the written value wv, the corresponding timestamp wts, as well as two matrices

---

[5] The notation $x.y$ denotes the call of low-level operation y during high-level operation x.

S, atomic snapshot object with n componenets, initially $((0, \perp), ..., (0, \perp))$ /*
`timestamp+sig */`
$W_1, \ldots, W_n$, SWMR atomic registers, initially $(\perp, 0, \emptyset, \emptyset, \perp)$ /*
`val+ts+rs+ws+sig */`
v, wv value, initially $\perp$ /* value written to storage */
$wts, ots, i, k, r, r', w, w', tmp_1, \ldots, tmp_n$ integer, initially 0   /* timestamps +
`temp. variables */`
$read\_seen[1..n][1..n]$, $write\_seen[1..n][1..n]$,   /* matrices of seen
     $r\_write\_seen[1..n][1..n]$, matrix of sets of pairs (integer, integer), initially $\emptyset$
`operations */`
$sig, sig_1, \ldots, sig_n$ signature, initially $\perp$ /* signatures */

Fig. 4. Variables used in Algorithm 2

r_read_seen and r_write_seen. Both matrices are of size $n \times n$ where each entry holds a set of integer pairs (r, w). Client C_i maintains a variable $read\_seen$ of the same type, where a pair $(r, w) \in read\_seen[i][j]$ denotes that READ of client C_i with timestamp r has seen WRITE of client C_j with timestamp w. Analogously, client C_i maintains a second matrix $write\_seen$, where $(r, w) \in write\_seen[i][j]$ denotes that WRITE of client C_i with timestamp w has seen READ of client C_j with timestamp r. In the next step (line 2.7), client C_i "merges" variables r_read_seen and $read\_seen$. The merge procedure returns for each entry of two $n \times n$ set matrices A, B set $A[i][j] \cup B[i][j]$, $i, j = 1, \ldots, n$. Then, C_i adds a pair consisting of its current timestamp and timestamp wts from W_j to $read\_seen[i][j]$. To ensure weak fork-linearizability, client C_i calls procedure "check" (line 2.9). If all checks are passed, C_i merges r_write_seen and $write\_seen$ and returns value wv (lines 2.10–2.11).

At invocation of WRITE(v), client C_i increments its timestamp (line 2.13). It digitally signs value v, its timestamp, and variables $read\_seen$ and $write\_seen$ to write to register W_i (lines 2.14–2.15). Next, it reads all timestamps of READs by calling SCAN to snapshot object S (line 2.16). All entries in S are digitally signed and thus client C_i verifies the signatures (line 2.18). Then, it adds to all sets $write\_seen[i][k]$ ($k = 1, \ldots, n$) a pair consisting of the timestamp of the kth component of S and C_i's current timestamp (line 2.19). Finally, client C_i successfully returns (line 2.20).

Procedure "check" implements the principle sketched in section 4.1 for n clients. It ensures that *weak fork-linearizability* is never violated. The procedure, called by C_i during READ(j) (line 2.21), moves through a loop performing two checks: The first check (line 2.24–2.25) considers the information left by clients during READ(i) operations (this information is stored in the ith column of $read\_seen$). If READ(i) with timestamp r of client C_k has seen WRITE of C_i with timestamp w, then it is tested whether the next WRITE of C_i has read (using SCAN) timestamp r or higher of client C_k. The check uses the local $write\_seen$ variable of C_i. The second check (line 2.27–2.28) reviews the information left by client C_i during any READ(k) (which is kept in the ith row of $read\_seen$). If READ(k) with timestamp r of client C_i has seen WRITE of C_k with timestamp w, then we check whether the next WRITE of C_k has read (using SCAN) timestamp r or higher of client C_i. This check requires matrix r_write_seen, which has been fetched from W_j in line 2.5 before procedure "check" is called.

Algorithm 2. Weak Fork-Linearizable Memory for n Clients, Code of Client C_i

---

2.1 READ(j) **do**
2.2 $ots \leftarrow ots + 1$ /* increment timestamp */
2.3 $sig \leftarrow \text{sign}_i(ots)$ /* signature on timestamp */
2.4   UPDATE($(ots, sig), S, i$)   /* update call to snapshot object */
2.5 $(wv, wts, r\_read\_seen, r\_write\_seen, sig) \leftarrow read(W_j)$ /* low-level atomic read */
2.6   **if not** $\text{verify}_j(sig)$ **then halt**   /* signature verified? */
2.7 $read\_seen \leftarrow \text{merge}(read\_seen, r\_read\_seen)$ /* update read_seen */
2.8 $read\_seen[i][j] \leftarrow read\_seen[i][j].\text{add}((ots, wts))$ /* add seen write */
2.9   check()   /* check passed? */
2.10 $write\_seen \leftarrow \text{merge}(write\_seen, r\_write\_seen)$ /* update write_seen */
2.11   return wv /* return read value */

2.12 WRITE(v) **do**
2.13 $ots \leftarrow ots + 1$ /* increment timestamp */
2.14 $sig \leftarrow \text{sign}_i(v, ots, read\_seen, write\_seen)$ /* signature on timestamp */
2.15 $write((v, ots, read\_seen, write\_seen, sig), W_i)$ /* low-level atomic write */
2.16 $\langle (tmp_1, sig_1), \ldots, (tmp_n, sig_n) \rangle \leftarrow \text{SCAN}(S)$ /* scan call to snapshot object */
2.17   **for** $k = 1, \ldots, n$ **do**
2.18   **if not** $\text{verify}_k(sig_k)$ **then halt**   /* signature verified? */
2.19 $write\_seen[i][k] \leftarrow write\_seen[i][k].\text{add}((tmp_k, ots))$ /* add all seen reads */
2.20   return OK   /* successfully return */

2.21 check() **do**
2.22   **for** $k = 1, \ldots, n$ **do**
2.23   **forall** $(r, w) \in read\_seen[k][i]$ **do**
                /* check if own writes have seen read operations
                   reading my values   */
2.24   **if** $\exists (r', w') \in write\_seen[i][k]$ s.t. $w' > w$ and w' minimal **then**
2.25   **if** $r' < r$ **then halt**
2.26   **forall** $(r, w) \in read\_seen[i][k]$ **do**
                /* check if own reads have been seen by other's
                   write operations   */
2.27   **if** $\exists (r', w') \in r\_write\_seen[k][i]$ s.t. $w' > w$ and w' minimal **then**
2.28   **if** $r' < r$ **then halt**

4.3   Correctness Arguments

In this section we give the intuition why Algorithm 2 satisfies the properties of a wait-free Byzantine emulation of a shared memory with weak fork-linearizability. Intuitively, the definition of weak fork-linearizability requires for each client C_i to construct a sequence π_i such that causality among operations, the sequential specification a shared memory, and weak real-time order is satisfied, and that two sequences π_i and π_j share

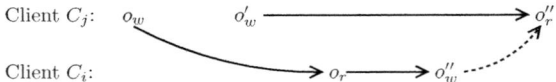

Fig. 5. Correctness Ideas of Algorithm 2. Arrows denote the "seen" relation.

the same prefix up to the second last common operation (at-most-one-join). The proof proceeds in steps, where in the first step all operations that have to be included in sequence π_i are causally ordered. Next, this order is extended such that it additionally respects the sequential specification. Intuitively, as all written values are digitally signed, the sequential specification never interferes with causality. The hardest step is to prove, that this order can be further refined such that it does not violate the weak real-time order. The intuition for this is given below as a proof by contradiction:

We assume that READ(j) operation o_r of client C_i does not return the latest value, written by WRITE operation o'_w, but an older value written by operation o_w (see Figure 5). Further, let o_r be not the last operation of C_i. During operation o_r, the pair $(r, w)^6$ is added to set $read\_seen[i][j]$. The data written by the next WRITE operation o''_w of C_i contains this information. Now, the algorithm prevents client C_j from reading the value written by o'_w which would violate weak real-time order (as o_r is ordered before o'_w according to the sequential specification). When during o''_r C_j sees operation o''_w, it finds the pair (r, w) in r_read_seen. As o'_w precedes o_r, it could not have seen o_r, thus $write\_seen[j][i]$ contains a pair (r', w') such that $r' < r$ and the check in line 2.25 is not passed. Hence, operation o''_r of client C_j would block — a contradiction. This implies that such a situation does not appear and the constructed order of operations also satisfies weak real-time order.

As the last step, showing that the sequences π_i satisfy the at-most-one-join property follows directly from a simple construction argument. To prove liveness, as required in the definition of a Byzantine emulation (Definition 2), we show that no operation blocks when all base objects are correct, which follows from the principle sketched in section 4.1 as in this case all checks are passed.

5   Analysis and Conclusions

The abortable construction in Algorithm 1 requires n atomic registers plus n additional ones to implement the INC&READ counter. The presented construction has an overall communication complexity of $O(n^2)$, as the size of the version vectors used in Algorithm 1 is linear in the number of clients n and as a linear number of such version vectors are exchanged per operation. In contrast, the *lock-step* protocol of Cachin *et al.* [4], also based on linear size version vectors, has an overall communication complexity of $O(n)$. This difference results from the fact that the server objects used by Cachin *et al.* are computationally strong enough to select the latest written version vector while in Algorithm 1 the client is required to read from *all* register objects to find the latest one by itself. For the implementation of Algorithm 2, we need n atomic registers plus $2n$

---

^{6} We assume that operation o_x is assigned timestamp x.

additional ones for the atomic snapshot object. Algorithm 2, uses matrices of size $n \times n$ where the size of each entry depends on the total number of operations N, resulting in a communication complexity of $O(N \cdot n^2)$. We leave for future research whether this complexity can be reduced by implementing a "garbage collection". However, both of our algorithms require only a linear number of base registers.

We have shown by ways of two protocols as a first known result that fork-consistent semantics can be implemented only from registers. Our first protocol satisfies fork-linearizability and implements a shared object of universal type. Similar to non-fork-consistent universal constructions from registers, our protocol may abort operations under concurrency. Hence, fork-linearizability may be "added" to such protocols without making additional assumptions. Our second protocol implements a shared memory object that ensures *weak* fork-linearizability and where operations are wait-free as long as the base registers behave correctly. Weak fork-linearizability is the strongest known fork-consistency property that may be implemented in a wait-free manner. Although it weakens fork-linearizability, it has shown to be of practical relevance [7]. Moreover, our second algorithm shows for the first time that registers are sufficient to implement a fork-consistent shared memory. So far, all existing implementations are based on computationally stronger objects (featuring read-modify-write operations [8]). We leave as an open question whether there is a weak fork-linearizable construction of a universal type providing a stronger liveness condition than *abortable* in the fault-free case.

References

1. Herlihy, M.: Wait-Free Synchronization. ACM Trans. Program. Lang. Syst. 13(1), 124–149 (1991)
2. Mell, P., Grance, T.: The NIST Definition of Cloud Computing. Report, National Institute of Standards and Technology, NIST (January 2011),
   `http://csrc.nist.gov/publications/drafts/800-145/`
   `Draft-SP-800-145_cloud-definition.pdf`
3. Mazières, D., Shasha, D.: Building Secure File Systems out of Byzantine Storage. In: PODC, pp. 108–117. ACM, New York (2002)
4. Cachin, C., Shelat, A., Shraer, A.: Efficient Fork-Linearizable Access to Untrusted Shared Memory. In: PODC, pp. 129–138. ACM, New York (2007)
5. Herlihy, M.P., Wing, J.M.: Linearizability: A Correctness Condition for Concurrent Objects. ACM Trans. Program. Lang. Syst. 12(3), 463–492 (1990)
6. Majuntke, M., Dobre, D., Serafini, M., Suri, N.: Abortable Fork-Linearizable Storage. In: Abdelzaher, T., Raynal, M., Santoro, N. (eds.) Proceedings of the 13th International Conference on Principles of Distributed Systems, OPODIS 2009. LNCS, vol. 5923, pp. 255–269. Springer, Heidelberg (2009)
7. Cachin, C., Keidar, I., Shraer, A.: Fail-Aware Untrusted Storage. SIAM Journal on Computing 40, 493–533 (2011)
8. Kruskal, C.P., Rudolph, L., Snir, M.: Efficient Synchronization of Multiprocessors with Shared Memory. ACM Trans. Program. Lang. Syst. 10, 579–601 (1988)
9. Aguilera, M.K., Keidar, I., Malkhi, D., Shraer, A.: Dynamic Atomic Storage Without Consensus. J. ACM 58, 7:1–7:32 (2011)
10. Shraer, A., Cachin, C., Cidon, A., Keidar, I., Michalevsky, Y., Shaket, D.: Venus: Verification for Untrusted Cloud Storage. In: Proceedings of the 2010 ACM Workshop on Cloud Computing Security, CCSW 2010, pp. 19–30. ACM, New York (2010)

298   M. Majuntke et al.

11. Malkhi, D., Reiter, M.K.: Byzantine Quorum Systems. Distributed Computing 11(4), 203–213 (1998)
12. Cachin, C.: Integrity and Consistency for Untrusted Services. In: Černá, I., Gyimóthy, T., Hromkovič, J., Jefferey, K., Královič, R., Vukolić, M., Wolf, S. (eds.) SOFSEM 2011. LNCS, vol. 6543, pp. 1–14. Springer, Heidelberg (2011)
13. Aguilera, M.K., Frolund, S., Hadzilacos, V., Horn, S.L., Toueg, S.: Abortable and Query-Abortable Objects and Their Efficient Implementation. In: PODC: Principles of Distributed Computing, pp. 23–32. ACM, New York (2007)
14. Herlihy, M., Luchangco, V., Moir, M.: Obstruction-Free Synchronization: Double-Ended Queues as an Example. In: ICDCS, p. 522. IEEE Computer Society, Washington, DC (2003)
15. Aguilera, M.K., Toueg, S.: Timeliness-Based Wait-Freedom: A Gracefully Degrading Progress Condition. In: PODC 2008: Proceedings of the Twenty-Seventh ACM Symposium on Principles of Distributed Computing, pp. 305–314. ACM, New York (2008)
16. Li, J., Krohn, M., Mazières, D., Shasha, D.: Secure Untrusted Data Repository (SUNDR). In: Proc. 6th Symp. Operating Systems Design and Implementation (OSDI 2004), pp. 121–136 (2004)
17. Cachin, C., Geisler, M.: Integrity Protection for Revision Control. In: Abdalla, M., Pointcheval, D., Fouque, P.-A., Vergnaud, D. (eds.) ACNS 2009. LNCS, vol. 5536, pp. 382–399. Springer, Heidelberg (2009)
18. Williams, P., Sion, R., Shasha, D.: The Blind Stone Tablet: Outsourcing Durability to Untrusted Parties. In: Proc. NDSS (2009)
19. Feldman, A.J., Zeller, W.P., Freedman, M.J., Felten, E.W.: SPORC: Group Collaboration on Untrusted Resources. In: Proc. 9th Symposium on Operating Systems Design and Implementation (OSDI 2010), Vancouver, BC (2010)
20. Li, J., Mazières, D.: Beyond One-Third Faulty Replicas in Byzantine Fault Tolerant Systems. In: Proc. NSDI (2007)
21. Majuntke, M., Dobre, D., Cachin, C., Suri, N.: Fork-Consistent Constructions from Registers. In: Technical Report TR-TUD-DEEDS-09-01-2011 (September 2011), http://www.deeds.informatik.tu-darmstadt.de/matze/fc_wo_sc_2011.pdf
22. Jayanti, P., Chandra, T.D., Toueg, S.: Fault-tolerant Wait-free Shared Objects. J. ACM 45(3), 451–500 (1998)
23. Lynch, N.A.: Distributed Algorithms. Morgan Kaufmann (1998)
24. Attiya, H., Guerraoui, R., Ruppert, E.: Partial Snapshot Objects. In: Proc. SPAA, pp. 336–343 (2008)
25. Fich, F.E.: How Hard Is It to Take a Snapshot? In: Vojtáš, P., Bieliková, M., Charron-Bost, B., Sýkora, O. (eds.) SOFSEM 2005. LNCS, vol. 3381, pp. 28–37. Springer, Heidelberg (2005)
26. Dobre, D., Majuntke, M., Suri, N.: On the Time-Complexity of Robust and Amnesic Storage. In: Baker, T.P., Bui, A., Tixeuil, S. (eds.) OPODIS 2008. LNCS, vol. 5401, pp. 197–216. Springer, Heidelberg (2008)

Easy Impossibility Proofs for k-Set Agreement in Message Passing Systems*

Martin Biely[1], Peter Robinson[2], and Ulrich Schmid[3]

[1] EPFL, Switzerland
martin.biely@epfl.ch
[2] Nanyang Technological University, Division of Mathematical Sciences, Singapore
peter.robinson@ntu.edu.sg
[3] Technische Universität Wien, Embedded Computing Systems Group, Austria
s@ecs.tuwien.ac.at

Abstract. Despite of being quite similar (agreement) problems, 1-set agreement (consensus) and general k-set agreement require surprisingly different techniques for proving the impossibility in asynchronous systems with crash failures: Rather than the relatively simple bivalence arguments as in the impossibility proof for consensus in the presence of a single crash failure, known proofs for the impossibility of k-set agreement in shared memory systems with $f \geqslant k > 1$ crash failures use algebraic topology or a variant of Sperner's Lemma. In this paper, we present a generic theorem for proving the impossibility of k-set agreement in various message passing settings, which is based on a reduction to the consensus impossibility in a certain subsystem resulting from a partitioning argument.

We demonstrate the broad applicability of our result by exploring the possibility/impossibility border of k-set agreement in several message-passing system models: (i) asynchronous systems with crash failures, (ii) partially synchronous processes with (initial) crash failures, and, most importantly, (iii) asynchronous systems augmented with failure detectors. In (i), (ii), and (iii), the impossibility part is an instantiation of our main theorem, whereas the possibility of achieving k-set agreement in (ii) follows by generalizing the consensus algorithm for initial crashes by Fisher, Lynch and Patterson. In (iii), applying our technique reveals the exact border for the parameter k where k-set agreement is solvable with the failure detector class $(\Sigma_k, \Omega_k)_{1 \leqslant k \leqslant n-1}$ of Bonnet and Raynal. As Σ_k was shown to be necessary for solving k-set agreement, this result yields new insights on the quest for the weakest failure detector.

Keywords: k-set agreement, failure detectors, consensus, impossibility proofs.

* This work has been supported by the Austrian Science Foundation (FWF) project P20529. Peter Robinson has also been supported by the Nanyang Technological University grant M58110000.

A. Fernández Anta, G. Lipari, and M. Roy (Eds.): OPODIS 2011, LNCS 7109, pp. 299–312, 2011.

1 Introduction

Agreement problems like consensus and set agreement are undoubtly the most prominent target for exploring the solvability/impossibility border in fault-tolerant distributed computing. In such problems, every process p_i, $1 \leqslant i \leqslant n$, in a distributed system owns a local proposal value x_i, and the problem is to irrevocably compute local output values (also called decision values) y_i that satisfy certain properties. For consensus, no two processes may decide on different values, for set agreement, the number of different decision values must be at most $n - 1$ system-wide. An obvious generalization is k-set agreement, which requires that the number of different decision values is at most k; clearly, consensus is just 1-set agreement, whereas set agreement is equivalent to $(n-1)$-set agreement.

Due to the landmark FLP impossibility result [1], which employs (now classic) combinatorial arguments (bivalence proofs), it is well-known that consensus is impossible to solve in asynchronous systems if a single process may crash. The corresponding result for general k-set agreement is the impossibility of solving this problem in asynchronous systems if $f \geqslant k$ processes may crash. Surprisingly, establishing this result requires quite involved techniques based on algebraic topology or a variant of Sperner's lemma [2, 3, 4].

A well-known general technique for establishing impossibility results are partitioning arguments, which have been used successfully for many distributed computing problems [5]. Essentially, a partitioning argument exploits the fact that one cannot guarantee agreement among those processes of a distributed system that never, neither directly nor indirectly, communicate with each other. In this paper, we use partitioning arguments in a—to the best of our knowledge—new way, namely, as a means for reduction.

More specifically, we present a surprisingly generic theorem that reduces the impossibility of k-set agreement to the impossibility of achieving consensus in a certain subsystem: In a nutshell, if failures and asynchrony in a model allow for runs where the system partitions into k parts, the processes must decide on their own in every partition. By choosing distinct proposal values, solving k-set agreement in such runs requires solving consensus in every partition. Consequently, the impossibility of k-set agreement can be proved by showing that it is impossible to reach consensus in at least one of these partitions.

Detailed Contributions: We present a generic impossibility result (Theorem 1) for k-set agreement that can be applied to a wide variety of message-passing system models and failure assumptions: It neither restricts the asynchrony of the model nor the types of failures that can occur. In Section 4, we revisit the impossibility of k-set agreement in asynchronous systems with crash failures (some of which are not initial crashes), with (and without) partially synchronous processes. Applying our generic theorem reveals the border that separates impossibility and possibility in this setting. Furthermore, by extending the algorithm for initial crashes of [1] to general k-set agreement, we show that the impossibility border is tightly matched. In Section 6, we present a technically involved application of our theorem to derive new results for asynchronous systems with failure detectors: We use our theorem to show that (Σ_k, Ω_k) is too weak for solving k-set agreement for $1 < k < n - 1$. Considering that Σ_k was shown to be necessary for solving k-set agreement with any failure detector, this provides new insights on the quest for the weakest failure detector for k-set agreement.

Related work: We are not aware of much research that uses similar ideas: We have employed reduction already in [6] to show that consensus is impossible in certain partially synchronous models, and to prove the tightness of our generalized loneliness failure detector $\mathcal{L}(k)$ for k-set agreement. Similar reduction arguments are employed in [7] and, in particular, in [8], where certain k-set agreement runs with disjoint participants are pasted together in order to prove the necessity of the generalized quorum failure detector Σ_k for solving k-set agreement. In [9], reduction to asynchronous set agreement is used to derive a lower bound on the minimum size of a "synchronous window" that is necessary for k-set agreement.

Unlike existing approaches based on algebraic topology and related techniques [2, 3, 4], our work provides an easy way of determining the impossibility in previously unexplored settings with varying degrees of synchrony and different system assumptions (e.g., models augmented with failure detectors and initial crash failures); it is not clear whether and how the modeling and analysis of [2, 3, 4] could be extended to such settings. On the other hand, our results do not subsume the impossibility results developed via such involved techniques for specific models, i.e., the impossibility of k-set agreement in the presence of up to $f = k$ crash failures. Our corresponding result, established in Section 4, holds only for relatively large values of f as compared to k: Consequently, according to Condition (1) on page 306, Theorem 2 does not cover the impossibility of k-set agreement in the presence of up to $f = k$ crash failures in case of $2 \leqslant k \leqslant n - 1$. It is unclear whether this is just due to a technical limitation of the applicability of our generic theorem or rather some different cause of impossibility that cannot be captured by a partitioning argument; this topic is a subject of further research.

2 System Models and Failure Assumptions

We use the computing model of [10], extended with the possibility of querying failure detectors. For the sake of brevity, we will not re-state the whole formal model of [10] here. Instead, we just introduce the necessary notations and explain the changes required for dealing with k-set agreement.

We consider a system $\Pi = \{p_1, \ldots, p_n\}$ of n processes with unique id's $\{1, \ldots, n\}$ that communicate via message-passing, using messages taken from some (possibly infinite) universe M. The communication subsystem is modeled by one buffer per process, which contains messages that have been sent to that process but not yet received. Every process $p \in \Pi$ is modeled as a deterministic state machine, which has a local state (program counter, local variables) that incorporates an input value x_p initialized to some value from a finite set of values

V, and a write-once output value $y_p \in V \cup \{\bot\}$ initialized to $\bot \notin V$. All other components of the local state are initialized to some fixed value.

State transitions are guided by a transition relation, which atomically takes the current local state of p, a (possibly empty) subset of messages L from p's current message buffer, and, in case of failure detectors, a value from the failure detector's domain, and provides a new local state. Sending of messages is guided by a deterministic message sending function, which determines a possibly empty set of messages that are to be sent to the processes in the system, i.e., maps the current state and the subset of messages L to a subset of $\Pi \times M$. Every message (q, m) in this subset is sent by just putting m into q's buffer.

A configuration of the system consists of the vector of local states and the message buffers of all the processes; in the initial configuration, all processes are in an initial state and the message buffers are empty. A *run* $\rho = (C_0, C_1, \dots)$ is an infinite sequence of configurations that starts from an initial configuration C_0, and C_{i+1} results from a legitimate (according to the transition relation and message sending function) step of a single process p in configuration C_i.

The above basic model is strengthened by restricting the set of runs by some admissibility conditions that depend on the particular system model $\mathcal{M}$ used. For example, the FLP model [1], denoted as $\mathcal{M}^{\text{ASYNC}}$, requires that (1) every correct process takes an infinite number of steps, (2) faulty processes execute only finitely many steps and may omit sending messages to a subset of receivers in the very last step, and (3) every message sent by a process to a correct receiver process is eventually received. With the exception of Section 3, we will assume systems adhering to the asynchronous model $\mathcal{M}^{\text{ASYNC}}$, sometimes augmented with a failure detector (Section 2.3) or with the assumption of partially synchronous processes (Section 4). The notation $\mathcal{M}_A$ will be used to denote the set of runs of algorithm A in model $\mathcal{M}$.

2.1 k-Set Agreement

We study distributed algorithms that solve agreement problems, namely, k-set agreement. Their purpose is to compute and irrevocably set the output y_i of process p_i to some decision value, based on the proposal values $x_i \in V$, for $|V| \geqslant n$,[1] which must satisfy the following properties:

k-Agreement: Processes must decide on at most k different values.
Validity: If a process decides on v, then v was proposed by some process.
Termination: Every correct process must eventually decide.
Note that the agreement property binds together the decision values of all (correct or faulty) processes. For $k = 1$, k-set agreement is hence equivalent to uniform consensus [11]. It follows from [1] that non-uniform and hence also uniform consensus cannot be solved in asynchronous systems if just one process may crash.

---

[1] The assumption $|V| \geqslant n$ allows runs where all processes start with different propose values.

2.2   Restrictions of Algorithms and Indistinguishability of Runs

We will occasionally use a subsystem $\mathcal{M}'$ that is a *restriction* of $\mathcal{M}$, in the sense that it consists of a subset of processes in Π, while using the same mode of computation (atomicity of computing steps, etc.) as $\mathcal{M}$. We make this explicit by using the notation $\mathcal{M} = \langle \Pi \rangle$ and $\mathcal{M}' = \langle D \rangle$, for some set of processes $D \subseteq \Pi$. Note that this definition does not imply anything about the synchrony assumptions which hold in $\mathcal{M}'$. All that is required is that $\mathcal{M}'$ is computationally compatible with $\mathcal{M}$: Any algorithm designed for $\mathcal{M}$ can also be run in $\mathcal{M}'$, albeit on a smaller set of processes.

Definition 1 (Restriction of an Algorithm). *Let A be an algorithm that works in system $\mathcal{M} = \langle \Pi \rangle$ and let $D \subseteq \Pi$ be a nonempty set of processes. Consider a restricted system $\mathcal{M}' = \langle D \rangle$. The restricted algorithm $A_{|D}$ for system $\mathcal{M}'$ is constructed by dropping all messages sent to processes outside D in the message sending function of A.*

Note that we do not change the actual code of algorithm A in any way. This means that—for example—the restricted algorithm still uses the value of $|\Pi|$ for the size of the system, even though the real size of D might be much smaller.

Whereas this is sufficient for running an algorithm designed for $\mathcal{M}$ in the restricted system $\mathcal{M}'$, in practice, one would also remove any dead code (resulting from state transitions triggered by message arrivals from processes in $\Pi \setminus D$) from the transition relation of A to obtain the actual transition relation of $A_{|D}$.

We will use a concept of indistinguishability of runs that is slightly weaker than the usual notion [12, Page 21], as we require the same states only *until* a decision state is reached. This makes a difference for algorithms where p can help others in reaching their decision after p has decided, e.g., by forwarding messages.

Definition 2 (Indistinguishability of Runs). *Two runs α and β are indistinguishable (until decision) for a process p, if p has the same sequence of states in α and β until p decides. By $\alpha \overset{p}{\sim} \beta$ we denote the fact that α and β are indistinguishable (until decision) for every $p \in D$.*

Definition 3 (Compatibility of Runs). *Let $\mathcal{R}$ and $\mathcal{R}'$ be sets of runs. We say that runs $\mathcal{R}'$ are compatible with runs $\mathcal{R}$ for processes in D, denoted by $\mathcal{R}' \preccurlyeq_D \mathcal{R}$, if $\forall \alpha \in \mathcal{R}' \; \exists \beta \in \mathcal{R} \colon \alpha \overset{D}{\sim} \beta$.*

2.3   Failure Detectors

A failure detector $\mathcal{D}$ is an oracle that can be queried by processes in any step, before making a state transition [13]. We assume familiarity with the notions of *failure environment*, the *failure pattern* $F(t)$ of a run, and the "weaker than" relation on FDs.[2]

---

[2] Due to lacking space, we had to relegate the detailed definitions, including the ones of the well-known generalized quorum FD Σ_k [8] and the generalized leader oracle Ω_k, to the full paper [14].

We denote the augmented asynchronous model, where runs are admissible in $\mathcal{M}^{\text{ASYNC}}$ and processes can query failure detector $\mathcal{D}$ in any step, as $\langle \mathcal{M}^{\text{ASYNC}}, \mathcal{D} \rangle$. While the weakest failure detector for message passing k-set agreement is still unknown (except for $k \in \{1, n-1\}$), the *quorum family* Σ_k was shown in [8] to be necessary for solving k-set agreement with any failure detector $\mathcal{X}$, in the sense that there is a transformation that implements Σ_k in the system $\langle \mathcal{M}^{\text{ASYNC}}, \mathcal{X} \rangle$; see [15] for a recent overview of failure detectors for k-set agreement.

3   The Impossibility Theorem

In this section, we present our general k-set agreement impossibility theorem. Due to its very broad applicability, it is stated in a highly generic and somewhat abstract way. It captures a reasonably simple idea, however, which boils down to extracting a consensus algorithm for a certain subsystem where consensus is unsolvable: Suppose that a given k-set agreement algorithm A for some system model $\mathcal{M}$ has runs, where processes start with distinct values, and k partitions $D_1, \ldots, D_{k-1}$ and $\overline{D}$ can be formed: Processes in the $k-1$ partitions D_i decide on (at least) $k-1$ different values, and no process in partition $\overline{D}$ ever hears from any process in D_i before it decides. Note carefully that processes in $\overline{D}$ can communicate arbitrarily within $\overline{D}$. Then, the ability of A to solve k-set agreement would imply that the restricted algorithm $A_{|\overline{D}}$ can solve consensus in the restricted model $\mathcal{M}' = \langle \overline{D} \rangle$. However, if the synchrony and failure assumptions are such that consensus cannot be solved in $\mathcal{M}'$, this is a contradiction. This intuition will become completely clear when we apply Theorem 1 in Sections 4 and 6.

Theorem 1 (k-Set Agreement Impossibility). *Let $\mathcal{M} = \langle \Pi \rangle$ be a system model and consider the runs $\mathcal{M}_A$ that are generated by some fixed algorithm A in $\mathcal{M}$, where every process starts with a distinct input value. Fix some nonempty disjoint sets of processes $D_1, \ldots, D_{k-1}$, and a set of distinct decision values $\{v_1, \ldots, v_{k-1}\}$. Moreover, let $D = \bigcup_{1 \leqslant i < k} D_i$ and $\overline{D} = \Pi \setminus D$. Consider the following two properties:*

(dec-D). *For every set D_i, value v_i was proposed by some $p \in D$, and there is some $q \in D_i$ that decides v_i.*

(dec-$\overline{D}$). *If $p_j \in \overline{D}$ then p_j receives no messages from any process in D until every process in $\overline{D}$ has decided.*

Let $\mathcal{R}_{(\overline{D})} \subseteq \mathcal{M}_A$ and $\mathcal{R}_{(D, \overline{D})} \subseteq \mathcal{M}_A$ be the sets of runs of A where (dec-$\overline{D}$) respectively both, (dec-D) and (dec-$\overline{D}$), hold.[3] Suppose that the following conditions are satisfied:

(A) $\mathcal{R}_{(\overline{D})}$ *is nonempty.*
(B) $\mathcal{R}_{(\overline{D})} \preccurlyeq_{\overline{D}} \mathcal{R}_{(D, \overline{D})}$.

---

[3] Note that $\mathcal{R}_{(\overline{D})}$ is by definition compatible with the runs of the restricted algorithm $A_{|\overline{D}}$.

In addition, consider $\mathcal{M}' = \langle \overline{D} \rangle$ *such that the following holds:*

(C) *There is no algorithm that solves consensus in* $\mathcal{M}'$.
(D) $\mathcal{M}'_{A_{|\overline{D}}} \preceq_{\overline{D}} \mathcal{M}_A$.

Then, A does not solve k-set agreement in $\mathcal{M}$.

Proof. For the sake of a contradiction, assume that there is a k-set agreement algorithm A for model $\mathcal{M}$, sets of runs $\mathcal{R}_{(\overline{D})}$ and $\mathcal{R}_{(D,\overline{D})}$ and some sets of processes $D_1, \ldots, D_{k-1}$ such that conditions (A)–(D) hold. Due to (A) we have $\mathcal{R}_{(\overline{D})} \neq \emptyset$; then, (B) implies that $\mathcal{R}_{(D,\overline{D})}$ is nonempty too. Observe that (dec-D) ensures that there are at least $k-1$ distinct decision values among the processes in D, in every run in $\mathcal{R}_{(D,\overline{D})}$. Since algorithm A satisfies k-agreement, the compatibility requirement (B) between runs $\mathcal{R}_{(\overline{D})}$ and $\mathcal{R}_{(D,\overline{D})}$ for processes in $\overline{D}$ implies the following constraint:

(Fact 1). *In each run in* $\mathcal{R}_{(\overline{D})}$, *all processes in* $\overline{D}$ *must decide on a common value.*

We will now show that this fact yields a contradiction. Starting from $\mathcal{M}'_{A_{|\overline{D}}}$, i.e., the set of runs of the restricted algorithm in model $\mathcal{M}'$, we know by (D) that for each $\rho' \in \mathcal{M}'_{A_{|\overline{D}}}$, there exists a run $\rho \in \mathcal{M}_A$ such that $\rho' \overset{\overline{D}}{\sim} \rho$. Obviously, no process $p \in \overline{D}$ receives messages from a process $q \in D$ in ρ' before p's decision, as such a process q does not exist in the restricted model $\mathcal{M}'$. Clearly, the same is true for the indistinguishable run ρ (even though such a process q does exist in model $\mathcal{M}$). Therefore, we have that, in fact, $\rho \in \mathcal{R}_{(\overline{D})}$, and due to (Fact 1), it follows that in each run $\rho' \in \mathcal{M}'_{A_{|\overline{D}}}$ all processes decide on the same value. This, however, means that we could employ $A_{|\overline{D}}$ to solve consensus in $\mathcal{M}'$, which is a contradiction to (C). □

There are several noteworthy points about Theorem 1:
- The proof neither restricts the types of failures that can occur in $\mathcal{M}$ nor the underlying synchrony assumptions of $\mathcal{M}$ in any way.
- Our impossibility result uses a 2-partitioning argument but does *not* require the system to (temporarily or permanently) decompose into $k+1$ partitions. In particular, there is no further restriction on the communication among processes within D and within $\overline{D}$.
- At a first glance, requirement (B) might appear to be redundant. After all, it should always be possible to find a run in $\mathcal{R}_{(D,\overline{D})}$ that is indistinguishable for the processes in $\overline{D}$, given some run in $\mathcal{R}_{(\overline{D})}$. To see why (B) is necessary, first consider some run γ (of some algorithm in some model $\mathcal{M}$) that satisfies property (dec-D). This stipulates $k - 1$ distinct decision values among the processes in D, which essentially means that γ was a quite "asynchronous" run for the processes in D. It could therefore be the case that the synchrony assumptions of $\mathcal{M}$ require γ to be "synchronous" for the processes in $\overline{D}$.

Now suppose that we are given a run $\alpha \in \mathcal{R}_{(\overline{D})}$ and we need to find a run $\beta \in \mathcal{R}_{(D,\overline{D})}$ that is indistinguishable for processes in $\overline{D}$, in order to make (B) hold. If α is an "asynchronous" run for the processes in $\overline{D}$, we might not be able to find a matching run $\beta \in \mathcal{R}_{(D,\overline{D})}$, as the above setting requires such runs to be "synchronous" for the processes in $\overline{D}$. Consider, for example, the model where computing speed and communication among processes in $\overline{D}$ is synchronous in a run if and only if the processes in D decide on at least $k - 1$ distinct values. Clearly $\mathcal{R}_{(\overline{D})} \preccurlyeq_{\overline{D}} \mathcal{R}_{(D,\overline{D})}$ does not hold here. (See Theorem 4 for a less artificial example.)

4   Impossibility in Partially Synchronous and Asynchronous Systems

It is easy to show that k-set agreement is impossible in the purely asynchronous model, if we assume a wait-free environment: It suffices to simply delay all communication until every process has decided on its own value. When the number of failures is somewhat restricted and/or the model is partially synchronous, however, a more involved argument is necessary.[4]

Theorem 2. *There is no algorithm that solves k-set agreement in a system $\mathcal{M}$ of n processes where processes are synchronous, communication is asynchronous, a process can broadcast a message in an atomic step, and receiving and sending are part of the same atomic step, for any*

$$k \leqslant \frac{n-1}{n-f}, \tag{1}$$

even if, of the f possibly faulty processes, $f - 1$ can fail by crashing initially and only one process can crash during the execution.

Proof. Assume in contradiction that some f-resilient algorithm A solves k-set agreement. We will show that conditions (A)–(D) of Theorem 1 are satisfied, thus yielding a contradiction.

As a first step, we will identify suitable sets D_i such that (A)–(B) hold for the runs in $\mathcal{R}_{(\overline{D})}$ and $\mathcal{R}_{(D,\overline{D})}$, respectively. Let $\ell = n - f$; for $1 \leqslant i < k$, define $D_i = \{p_{(i-1)\ell+1}, \ldots, p_{i\ell}\}$ and let $D = \bigcup_{1 \leqslant i \leqslant k-1} D_i$. Since $|D| = (k-1)\ell$ and (1) can be rewritten as $k\ell \leqslant n - 1$, we easily obtain the following lemma about the process set sizes.

Lemma 1. *The set $\overline{D}$ contains at least $n - f + 1$ processes, and every D_i, $1 \leqslant i < k$, contains exactly $\ell = n - f$ processes.*

We can now establish the conditions of Theorem 1:

---

[4] Due to lacking space, we only present the major issues here; all the details and proofs can be found in the full paper [14].

(A), (B) These two conditions follow from asynchrony of communication in conjunction with Lemma 1.

(C) Consider a system $\mathcal{M}' = \langle \overline{D} \rangle$ that has the same system assumptions as $\mathcal{M}$, with the restriction that at most one process can crash in $\mathcal{M}'$ (at any time). Condition (C) follows immediately from the result of [10, Table I], since $|\overline{D}| \geqslant n - f + 1 \geqslant 2$ (see Lemma 1) and one process can crash in the runs of $\mathcal{M}'$.

(D) We will show that for every run $\rho' \in \mathcal{M}'_{A_{|\overline{D}}}$, there is a corresponding run $\rho \in \mathcal{M}_A$ such that $\rho' \overset{\overline{D}}{\sim} \rho$. Fix any $\rho' \in \mathcal{M}'_{A_{|\overline{D}}}$ and consider the run $\rho \in \mathcal{M}_A$ where every correct process in $\overline{D}$ has the same sequence of states in ρ as in ρ', and all remaining processes—of which there are $\leqslant f - 1$—are initially dead in ρ. Such a run ρ exists, since $A_{|\overline{D}}$ is the restriction of A (see Definition 1). We can therefore apply Theorem 1 and conclude that A does not solve k-set agreement. □

Corollary 1. *The impossibility of k-set agreement from Theorem 2 continues to hold under weaker assumptions, in particular, if processes are asynchronous, broadcasts are not possible in one step, sending and receiving within one atomic step is not possible, and all f processes may fail by crashing.*

5 Possibility of k-Set Agreement with Initially Dead Processes

In this section, we will show that Theorem 2 tightly captures the impossibility of k-set agreement, by presenting a matching bound for the solvability of k-set agreement in asynchronous systems with f initial crashes.

For the consensus case $k = 1$, we know from [1] that it is sufficient for a majority of processes to be correct. The protocol of [1] operates in two stages: In the first stage, each process broadcasts a message (containing its process id). Every process then waits until it has received $L - 1$ (where L is $\lceil (n + 1)/2 \rceil$) messages. In the second stage, every process broadcasts a message containing its initial value and the list of $L - 1$ processes it has received messages from in the first stage. Then it waits for messages from those $L - 1$ processes it has received messages from in the first stage, and for a message from every remote process mentioned in one of the lists it receives.

Now consider a directed simple graph, in which each node corresponds to a process and there is an edge from u to w iff the process corresponding to w has received a message from the process corresponding to u in the first stage. Let us call this graph G. Clearly, every node in G has in-degree $L - 1$. Another (less obvious) feature of G is the existence of a *source component*. We call a strongly connected component C of a directed graph source component if, in the directed acyclic graph (DAG) generated by contracting all vertices of the strongly connected components of G into single vertices, the vertex corresponding to C is a source, i.e., has in-degree 0. As we will see below, $n > 2f$ implies that there is only one source component.

While processes only know the in-arcs of their own node in G after the first stage, after the second stage, they have got complete and consistent knowledge of the source component. Therefore, a deterministic rule for choosing one of the proposal values of the processes in C (e.g., the value proposed by the process whose identifier is minimal in the source component) can be used as the decision value of every process.

For the general case $k \geqslant 1$, we can use the same algorithm if we can make sure that each process can determine at least one of at most k source components. We will now determine a value for L, which guarantees this for some given k. Note that the ability to select a value for L is also restricted by f, to avoid blocking. Thus, by combining the relations between L and k and f, respectively, we will be able to determine the range of f for which k-set agreement is solvable.

Lemma 2. *Consider a finite directed simple graph G, where each vertex has at least in-degree $\delta > 0$. In each weakly connected component (WCC) of G, there exists at least one source component of size at least $\delta + 1$.*

From this lemma, it follows that every process has (at least) one directed incoming path from all the processes in (at least) one source component. Moreover, it is easy to see that there can be at most $\lfloor n/(\delta + 1) \rfloor$ source components. Returning to the algorithm from [1], waiting for $L - 1$ messages in the first stage clearly induces a graph G with $\delta = L - 1$, and thus at most $\lfloor n/L \rfloor$ source components. From this it follows that processes will decide on at most $\lfloor n/L \rfloor$ values, so k-set agreement with $k \geqslant \lfloor n/L \rfloor$ is indeed solvable.

As our last step, we have to relate L to the bound on the number of initially crashed processes f. On one hand, we want L to be as large as possible in order to decrease the number of source components. On the other hand, since processes wait until they have received a message from $L - 1$ remote processes in the first stage, it is clearly not advisable to choose $L - 1 \geqslant n - f$. Therefore, we now fix $L = n - f$, which leads to k-set agreement being solvable when $k \geqslant \lfloor n/(n - f) \rfloor$. Since n, f, and k are all integers, we get that $k + 1 > n/(n - f)$ and hence $kn > (k + 1)f$. Note that, for $k = 1$, this matches the requirement of a majority of correct processes.

Considering the border case $kn = (k + 1)f$, we get $n - f = n/(k + 1)$. A standard partitioning argument reveals that k-set agreement is impossible in this case: Assume that there is an algorithm A that solves k-set agreement in such a system. The above condition on n and f implies that we can partition the system into $k + 1$ disjoint groups of processes $\Pi_0, \ldots, \Pi_k$. From the set of possible input values V, choose any $v_0, \ldots, v_k$, s.t., $v_i = v_j \Leftrightarrow i = j$. Clearly, for each i, there is an execution ε_i of A where all processes in Π_i have initial value v_i and all processes in $\Pi \setminus \Pi_i$ are initially dead. Since A solves k-set agreement, all processes in Π_i have to eventually decide on v_i in ε_i. Therefore, by delaying messages between the partitions Π_i sufficiently long, it is easy to construct an execution ε without any initial crashes, which is indistinguishable (until decision) for all $p \in \Pi_i$ from ε_i, $0 \leqslant i \leqslant k$. But now we have $k + 1$ different decision values (i.e., $v_0, \ldots, v_k$) in ε, which contradicts the assumption that A solves k-set agreement. Therefore, we have get the following result:

Theorem 3. *In an asynchronous system with n processes up to f of which may be initially dead, k-set agreement is solvable if and only if $kn > (k+1)f$ or, equivalently, $k > f/(n-f)$.*

Comparison with this theorem reveals that the bound (1) in Theorem 2 is indeed tight.

6 Impossibility with Failure Detector (Σ_k, Ω_k)

In this section, we will demonstrate the full power of Theorem 1 by deriving a new result: We prove the impossibility of achieving k-set agreement with failure detector (Σ_k, Ω_k), for all $1 < k < n-1$. In [7, Theorem 2], it was shown that k-set agreement is impossible with (Σ_k, Ω_k) if $1 < 2k^2 \leqslant n$, which is a much more restrictive bound than the one given by Theorem 4 below. For our impossibility proof, we will make use of a certain stronger failure detector that nevertheless allows up to k partitions.

Definition 4. *Let $\{D_1, \dots, D_{k-1}, D_k\}$ be a partitioning of the processes in Π, and let $\overline{D} = D_k$. The partition failure detector (Σ_k', Ω_k') provides failure detector histories with the following properties; note that we call a history of (Σ_k', Ω_k') a partitioning history:*

1. *For $1 \leqslant i \leqslant k$, the output of Σ_k' at every process in D_i is a valid history for Σ $(= \Sigma_1)$ in the restricted model $\mathcal{M}_i = \langle D_i \rangle$ (where only processes from D_i are ever output by Σ), with an additional condition: Let t_j be the earliest point in time when p_j failed, i.e., $p_j \in F(t_j)$, for any $p_j \in D_i$. If t_j is finite, then $\forall t \geqslant t_j$ it holds that the output of Σ_k' at p_j is defined to be the whole set Π (rather than D_i).*
2. *Ω_k' is the same as Ω_k.*

Lemma 3. *Failure detector (Σ_k, Ω_k) is weaker than (Σ_k', Ω_k').*

Theorem 4. *There is no $(n-1)$-resilient algorithm that solves k-set agreement in an asynchronous system with failure detector (Σ_k, Ω_k), for all $2 \leqslant k \leqslant n-2$.*

Proof. The restriction of the range of k implies that there are exactly $n = k-1+j$ processes in the system, for some $j \geqslant 3$. Consider the following partitioning of Π: Let $\overline{D} = \{p_1, \dots, p_j\}$ and choose $D_1, \dots, D_{k-1}$ such that they partition the set $\Pi \setminus \overline{D}$; since we have $D = \bigcup_{1 \leqslant i < k} D_i$, and therefore $|D| = n - j = k - 1$, such a partitioning exists.

We assume by contradiction that there is an algorithm A that solves k-set agreement using (Σ_k', Ω_k'). Applying Lemma 3 will then complete the proof of Theorem 4.

We start with two technical lemmas, which justify why we call histories of (Σ_k', Ω_k') partitioning histories: Intuitively speaking, it is straightforward to combine histories at different processes. The first lemma proves that we can "paste together" different executions at partition boundaries. Let $\mathcal{R} \subseteq \mathcal{R}_{(D, \overline{D})}$ be the set of runs where all communication between the sets of processes $D_1, \ldots, D_{k-1}, \overline{D}$ is delayed until every correct process has decided, and assume that $\mathcal{R} \neq \emptyset$ (which will be proved in Lemma 5).

Lemma 4. *Let* $\beta \in \mathcal{R}$ *(and hence* $\beta \in \mathcal{R}_{(D, \overline{D})}$ *and* $\beta \in \mathcal{R}_{(\overline{D})}$*) and* $\alpha \in \mathcal{R}_{(\overline{D})}$ *be given, where* t^{α}_{dec} *resp.* t^{β}_{dec} *denotes the time when the last process in* $\overline{D}$ *has crashed or decided in* α *resp.* β*. Then, a run* β' *which fulfills* $\beta' \in \mathcal{R}$ *and* $\beta' \overset{\overline{D}}{\sim} \alpha$ *is obtained from* β *by*

1. *replacing* $H_\beta(p, t)$ *by* $H_\alpha(p, t)$ *at all processes* $p \in \overline{D}$ *at all times* $t \geqslant 0$,
2. *setting* $F_{\beta'}(t) = (F_\beta(t) \cap (\Pi \setminus \overline{D})) \cup (F_\alpha(t) \cap \overline{D})$ *at all times* $t \geqslant 0$, *and hence* $F_{\beta'} = (F_\beta \cap (\Pi \setminus \overline{D})) \cup (F_\alpha \cap \overline{D})$,
3. *letting the processes in* $\overline{D}$ *receive messages and perform their steps exactly as in* α,
4. *delivering messages between* $D_1, \ldots, D_k$ *only after all correct processes have decided in* β',
5. *choosing some (arbitrarily large)* $t_{GST} \geqslant \max\{t^{\alpha}_{dec}, t^{\beta}_{dec}\}$ *and some set* LD *that satisfies* $LD \cap (\Pi \setminus F_{\beta'}) \neq \emptyset$, *and setting* $LD^t_j = LD$ *in* $H_{\beta'}$ *for all processes* $p_j \in \Pi \setminus F_{\beta'}$ *and all* $t \geqslant t_{GST}$.

Lemma 5. $\mathcal{R}_{(D, \overline{D})} \neq \emptyset$, *in particular,* $\mathcal{R} \subset \mathcal{R}_{(D, \overline{D})}$ *is nonempty.*

Equipped with these results, we can establish the conditions required for applying Theorem 2:

(A): Consider the run α_k where all processes outside $\overline{D}$ are initially dead, then clearly processes in $\overline{D}$ decide before receiving a message from processes outside $\overline{D}$. Since $\alpha_k \in \mathcal{R}_{(\overline{D})}$, we obviously have $\mathcal{R}_{(\overline{D})} \neq \emptyset$.

(B): Consider any run $\alpha \in \mathcal{R}_{(\overline{D})}$, then we can use Lemma 5 to obtain some $\beta \in \mathcal{R}$ from which we can construct $\beta' \in \mathcal{R}$, s.t., $\alpha \overset{\overline{D}}{\sim} \beta'$ using Lemma 4. As $\beta' \in \mathcal{R} \subseteq \mathcal{R}_{(D, \overline{D})}$, we have $\mathcal{R}_{(\overline{D})} \preccurlyeq_{\overline{D}} \mathcal{R}_{(D, \overline{D})}$.

(C): We will first choose an appropriately restricted model $\mathcal{M}'$: Since $|\overline{D}| = j \geqslant 3$, let $\mathcal{M}' = \langle \overline{D} \rangle$ be an asynchronous system where up to $j - 1$ processes may fail by crashing. Moreover, $\mathcal{M}'$ is augmented with a failure detector that is compatible to (Σ'_k, Ω'_k), in the sense that its failure detector histories can be extended to match an admissible history of (Σ'_k, Ω'_k) in $\mathcal{M}$, without changing the output at processes in $\overline{D}$: Considering Definition 4, we just assume that processes in $\mathcal{M}'$ effectively access a failure detector (Σ, Γ), where Γ satisfies the part of Definition 4 that concerns Ω_k in the following constrained way, for all processes in $\overline{D}$: Γ outputs a possibly changing set of k process ids in the range of Π, which eventually stabilizes on some set LD that intersects $\overline{D}$ in exactly two processes p_s and p_t. Obviously, this restriction is compatible with Ω'_k. Note that one of p_s and p_t (but not necessarily both) may be faulty. In $\mathcal{M}'$ we can easily implement Ω_2 using Γ (the transformations uses Γ to eventually choose two fixed processes from $\overline{D}$) and vice versa (by just extending the output of Γ with processes from D), thus (Σ, Γ) is and (Σ, Ω_2) are equally strong. Moreover, (Σ, Ω_2) is strictly weaker than

(Σ, Ω), as there is no transformation T providing the properties of (Σ, Ω) from those of (Σ, Ω_2): If T existed, we could use it to obtain a wait-free transformation T' for shared memory to obtain Ω from Ω_2 (by simulating an asynchronous message passing system equipped with Σ, cf. [16]) which contradicts the results of [17]. Since (Σ, Ω) is the weakest failure detector for solving consensus, we can therefore conclude that (Σ, Γ) is too weak for solving consensus in $\mathcal{M}'$.

(D): Finally, for any run in $\mathcal{M}'_{A_{|\overline{D}}}$, there is obviously a run in $\mathcal{R}_{(\overline{D})}$ where all processes in D are initially dead, the processes in $\overline{D}$ take identical steps, fail at the same time, and receive the same failure detector output and the same messages. Hence, $\mathcal{M}'_{A_{|\overline{D}}} \preccurlyeq_{\overline{D}} \mathcal{R}_{(\overline{D})}$ and, by transitivity, $\mathcal{M}'_{A_{|\overline{D}}} \preccurlyeq_{\overline{D}} \mathcal{M}_A$. Applying Theorem 1 thus yields the required contradiction. $\quad\square$

From [8], we know that Σ_{n-1} (and thus also $(\Sigma_{n-1}, \Omega_{n-1})$) is sufficient for solving $(n-1)$-set agreement, Together with the fact that (Σ_1, Ω_1) is sufficient for solving consensus [16], we have the following result:

Corollary 2. *There is an $(n-1)$-resilient algorithm that solves k-set agreement with failure detector class $(\Sigma_k, \Omega_k)_{1 \leqslant k \leqslant n-1}$ in an asynchronous system, if and only if $k = 1$ or $k = n - 1$.*

7 Discussion

In this paper, we introduced a reduction to consensus for generically characterizing the impossibility of k-set agreement in message passing systems. The main advantage of our approach is that we are independent of a specific system model, since Theorem 1 neither makes assumptions on the available synchrony, nor on the power of computing steps and communication primitives available.A particularly promising application of our theorem is as both a guidance and quick verification tool for finding new models and algorithms for k-set agreement. This is particularly true for the quest for the (still unknown) weakest failure detector for solving message-passing k-set agreement: As we have shown, Σ_k, which is known to be necessary for k-set agreement in [8], is not powerful enough for overcoming the fatal partitioning into k subsystems. So what can be learned from our result is that, whatever one adds to Σ_k, it has to allow solving consensus in each partition.

References

1. Fischer, M.J., Lynch, N.A., Paterson, M.S.: Impossibility of distributed consensus with one faulty process. Journal of the ACM 32(2), 374–382 (1985)
2. Borowsky, E., Gafni, E.: Generalized FLP impossibility result for t-resilient asynchronous computations. In: STOC 1993: Proceedings of the ACM Symposium on Theory of Computing, pp. 91–100. ACM, New York (1993)
3. Herlihy, M., Shavit, N.: The asynchronous computability theorem for t-resilient tasks. In: STOC 1993: Proceedings of the Twenty-Fifth Annual ACM Symposium on Theory of Computing, pp. 111–120. ACM, New York (1993)

 4. Saks, M., Zaharoglou, F.: Wait-free k-set agreement is impossible: The topology of public knowledge. SIAM J. Comput. 29(5), 1449–1483 (2000)
 5. Fich, F., Ruppert, E.: Hundreds of impossibility results for distributed computing. Distributed Computing 16, 121–163 (2003)
 6. Biely, M., Robinson, P., Schmid, U.: Weak synchrony models and failure detectors for message passing k-set agreement. In: Abdelzaher, T., Raynal, M., Santoro, N. (eds.) OPODIS 2010. LNCS, vol. 5923, pp. 285–299. Springer, Heidelberg (2009)
 7. Bouzid, Z., Travers, C.: (anti−Ω^x × Σ_z)-Based k-Set Agreement Algorithms. In: Lu, C., Masuzawa, T., Mosbah, M. (eds.) OPODIS 2010. LNCS, vol. 6490, pp. 189–204. Springer, Heidelberg (2010)
 8. Bonnet, F., Raynal, M.: On the road to the weakest failure detector for k-set agreement in message-passing systems. Theoretical Computer Science 412(33), 4273–4284 (2011)
 9. Alistarh, D., Gilbert, S., Guerraoui, R., Travers, C.: Brief Announcement: New Bounds for Partially Synchronous Set Agreement. In: Lynch, N.A., Shvartsman, A.A. (eds.) DISC 2010. LNCS, vol. 6343, pp. 404–405. Springer, Heidelberg (2010)
10. Dolev, D., Dwork, C., Stockmeyer, L.: On the minimal synchronism needed for distributed consensus. Journal of the ACM 34(1), 77–97 (1987)
11. Charron-Bost, B., Schiper, A.: Uniform consensus is harder than consensus. J. Algorithms 51(1), 15–37 (2004); Also published as Tech. Rep. DSC/2000/028, EPFL
12. Lynch, N.: Distributed Algorithms. Morgan Kaufmann (1996)
13. Chandra, T.D., Toueg, S.: Unreliable failure detectors for reliable distributed systems. Journal of the ACM 43(2), 225–267 (1996)
14. Biely, M., Robinson, P., Schmid, U.: Easy impossibility proofs for k-set agreement in message passing systems. Research Report 2/2011, Technische Universität Wien, Institut für Technische Informatik (2011)
15. Raynal, M.: Failure detectors to solve asynchronous k-set agreement: a glimpse of recent results. Bulletin of the EACTS (103), 74–95 (2011)
16. Delporte-Gallet, C., Fauconnier, H., Guerraoui, R.: Tight failure detection bounds on atomic object implementations. J. ACM 57, 22:1–22:32 (2010)
17. Neiger, G.: Failure detectors and the wait-free hierarchy (extended abstract). In: Proceedings of the Fourteenth Annual ACM Symposium on Principles of Distributed Computing, PODC 1995, pp. 100–109. ACM, New York (1995)

On the Nature of Progress

Maurice Herlihy[1,*] and Nir Shavit[2]

[1] Brown University Computer Science
herlihy@cs.brown.edu
[2] MIT CSAIL
shanir@csail.mit.edu

Abstract. We identify a simple relationship that unifies seemingly unrelated progress conditions ranging from the deadlock-free and starvation-free properties common to lock-based systems, to non-blocking conditions such as obstruction-freedom, lock-freedom, and wait-freedom.

Properties can be classified along two dimensions based on the demands they make on the operating system scheduler. A gap in the classification reveals a new non-blocking progress condition, weaker than obstruction-freedom, which we call clash-freedom.

The classification provides an intuitively-appealing explanation why programmers continue to devise data structures that mix both blocking and non-blocking progress conditions. It also explains why the wait-free property is a natural basis for the consensus hierarchy: a theory of shared-memory computation requires an independent progress condition, not one that makes demands of the operating system scheduler.

1 Introduction

The advent of multicore architectures has brought about a renewed interest in concurrent data structures and algorithms, whose behavior is captured, in addition to safety properties, by their progress conditions. The literature encompasses a bewildering array of progress conditions. Some ("non-blocking") conditions guarantee progress even if one or more threads halt, while others do not. Some blocking conditions guarantee that threads will not deadlock, and some go further and rule out starvation.

On modern multiprocessor machines, programmers often use a variety of lock-based and non-blocking algorithms, sometimes mixing and matching progress conditions within a single system. (For example, consider lock-free, obstruction-free, and lock-based software transactional memory systems [13]). How can these data structures and algorithms work well together when they make incomparable and incompatible progress guarantees?

This paper proposes a novel *unified explanation* that ties together these seemingly unrelated progress conditions, ranging from the deadlock-free and starvation-free properties common to lock-based data structures, to the obstruction-free, lock-free, and wait-free properties that have been the focus

---
* Supported by NSF 0811289.

A. Fernández Anta, G. Lipari, and M. Roy (Eds.): OPODIS 2011, LNCS 7109, pp. 313–328, 2011.

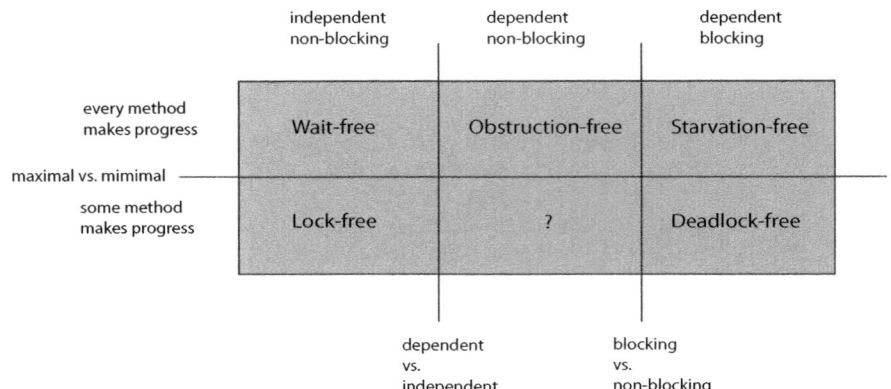

Fig. 1. A "Periodic Table" Style Chart of Progress Conditions

of so much recent research. We are deliberately not presenting a "unified *theory*", (even though our explanation is not difficult to formalize), because our primary goal is to provide a clear, simple, and intuitively-appealing explanation how these dissimilar properties actually fit together. These ideas may seem straightforward, perhaps even obvious, but we have never seen this formulation in any published work.

We show that progress conditions can be classified as shown in Figure 1. The horizontal line separates properties that ensure *maximal* progress, that is, progress for all threads, from properties that ensure *minimal* progress, progress for only some threads. The vertical lines separate properties that depend on different kinds of guarantees provided by the operating system (OS) scheduler.

It is important to distinguish between *dependent* and *independent* progress conditions. At one extreme, the wait-free and lock-free properties are *independent* of the OS scheduler: they guarantee progress as long as threads are scheduled, but no matter how they are scheduled. The other properties are *dependent*: they rely on the OS scheduler to satisfy certain properties. The deadlock-free and starvation-free properties guarantee progress only if each thread eventually leaves each critical section, and the obstruction-free property [8] requires the scheduler to allow each thread to run in isolation for a sufficient duration.

If we further restrict our attention to schedulers that satisfy a *benevolent* property defined below, then the distinction between minimal and maximal progress along the horizontal axis vanishes: any algorithm that provides minimal progress provides maximal progress as long as the scheduler is benevolent. This is why algorithms that (in principle) permit starvation are so widely used in practice: programmers implicitly (and reasonably) assume that OS schedulers are benevolent in practice.

Here is how to unify the disparate progress conditions in the literature. Instead of analyzing each algorithm and its progress properties in isolation, focus on the interaction between the algorithm and the guarantees provided by the OS scheduler. Implicitly, programmers, whether they design starvation-free, deadlock-free, obstruction-free, lock-free, or wait-free data structures, all want the same thing: maximal progress[1] They differ only in the assumptions they make about the OS scheduler.

One way to test an ambitious hypothesis is by its predictive power. Figure 1 contains a hole: the obstruction-free property has no minimal counterpart. We define a new *clash-free* property to fill this gap, and show it is strictly weaker than the obstruction-free property (addressing an open question due to Herlihy, Luchangco, and Moir [8]).

Finally, we observe that our classification explains why the wait-free property is a natural basis for the consensus hierarchy [7]: a theory of shared-memory computation requires an independent progress condition, not one that makes demands of the OS scheduler.

The remainder of this paper expands these observations. It builds on many papers, and a comprehensive survey of relevant literature would take up too much space. Instead, we refer the reader to books by Attiya and Welch [4], Lynch [14], Taubenfeld [17], and to references cited later.

2   Conventional Explanations

We start with a review of the conventional view of progress conditions taken from the literature. We then reformulate these notions in our unified model.

An *object* is a container for data. Each object provides a set of *methods* which are the only way to manipulate that object. Each object has a *class*, which defines the object's methods and how they behave. An object has a well-defined *state* (for example, a FIFO queue's current sequence of items).

The simplest way to synchronize concurrent access to an object is to associate a mutual exclusion lock with the object. Each method acquires the lock when it is called, and releases the lock when it returns. (We postpone consideration of methods that need to block, waiting until a condition is satisfied.)

Perhaps the weakest progress condition one could demand of a method that employs locks is that the method be *deadlock-free*, meaning that some thread trying to acquire the lock eventually succeeds. This condition guarantees that the system as a whole makes progress, but does not guarantee progress to individual threads. For example, a test-and-set spin lock is deadlock-free, because some thread will acquire a free lock. Here is an important point that we will explore later on: a deadlock-free lock guarantees progress only if every thread that acquires the lock eventually releases it. This requirement constrains both the scheduler, which cannot halt a thread in a critical section, and the software, which must use the lock correctly.

Sometimes we would like locks to have an even stronger property. A lock is *starvation-free* if every attempt to acquire the lock eventually succeeds. For example, a test-and-set spin lock is not starvation-free, because it is possible

---

[1] "Purity of heart is to will one thing" – Sören Kierkegaard.

(though unlikely) that some thread's attempts to acquire the lock repeatedly fail. By contrast, queue locks [15] are typically starvation-free because threads acquire locks in the order they are requested. Like deadlock-free locks, starvation-free locks make sense only if every thread that acquires a lock eventually releases it.

We described the deadlock and starvation-free properties directly in terms of classical mechanisms such as locks and critical sections because that is usually how these properties are used in the literature [3]. Later on, when we make these notions precise, we will see that this approach is unsatisfactory for several reasons. First, it is relatively easy to devise obfuscated object implementations where it is difficult to identify a particular field as a lock and particular statements as critical sections. Second, it is unclear how to compare such a property to non-blocking properties that, by definition, do not use locks and critical sections. Finally, progress should not be defined in terms of locks, which are low-level mechanisms, but in a more general way in terms of completed method calls.

While operating system schedulers rarely, if ever, halt threads holding locks, it is possible that preemption might well delay a thread holding a lock, effectively blocking progress by other threads. To address such issues, a number of *non-blocking* progress conditions have emerged. A non-blocking condition ensures that an arbitrary and unexpected delay by any thread (say, one holding a lock) does not prevent other threads from making progress.

A method is *lock-free* if some thread that calls that method eventually returns. A method is *wait-free* if every thread that calls that method eventually returns.

There is another non-blocking progress condition. We say that a method call executes *in isolation* for a duration if no other threads take steps during that time. A method is *obstruction-free* if every thread that calls that method returns if that thread executes in isolation for long enough. This condition is non-blocking, and is strictly weaker than the lock-free condition. It rules out the use of locks and mutual exclusion, but does not guarantee progress when multiple threads execute concurrently. Obstruction-free algorithms typically rely on a *contention manager* [9] module to delay threads so that a given thread can make progress. For example, a contention manager might employ a *back-off* delay policy: a thread that is about to conflict with another pauses to give the earlier thread time to finish.

3   Modeling Progress

Our model is adapted from Herlihy and Wing, assuming *linearizability* [11] as our basic correctness condition. We are interested in progress conditions for methods of abstract objects. A given object has a set of different methods, each of which can be invoked many times during an execution.

An execution of a concurrent object is modeled by a *history*, a sequence of method *invocation* and *response events*. A *subhistory* of a history H is a subsequence of the events of H. An *interval* is a finite subhistory consisting of contiguous events.

We focus on two-level implementations that include an abstract object (the one being implemented) and concrete ones (the ones used in the implementation). Informally, each abstract method call is implemented by the sequence of concrete method calls it encompasses. We use this two-level approach because we care about the number of concrete steps needed to implement an abstract method call.

An abstract method call that never returns could happen in two ways: if it encompasses an infinite number of concrete steps, then the thread starved, but if it encompasses only a finite number of concrete steps, then the thread halted in the middle of the call. These situations are different, and must be distinguished.

A thread is *active* if it takes an infinite number of concrete steps (and is *suspended* if not), and an invocation is *active* if it is made by an active thread. To avoid clutter, we focus on implementation histories of a single abstract object with a single method, which is repeatedly called by all threads. It is easy to generalize these definitions to encompass multiple objects and methods, and to allow threads to shut down gracefully.

3.1   Minimal and Maximal Progress

In some sense, the weakest interesting notion of progress requires that the system as a whole continues to advance. Consider a fixed history H. An abstract method provides *minimal progress* in H if, in every suffix of H, some pending active invocation has a matching response. In other words, there is no point in the history where all threads that called the abstract method take an infinite number of concrete steps without returning. This condition might, for example, be useful for a thread pool, where we care about advancing the overall computation, but do not care whether individual threads are underutilized.

The strongest notion of progress, and arguably the one most programmers actually want, requires that each individual thread continues to advance. An abstract method provides *maximal progress* in a history H if in every suffix of H, every pending active invocation has a matching response. In other words, there is no point in the history where a thread that calls the abstract method takes an infinite number of concrete steps without returning. This condition might be useful for a web server, where each thread represents a customer request, and we care about advancing each individual computation.

3.2   The Scheduler's Role

A history is *fair* if each thread takes an infinite number of concrete steps. A history is *uniformly isolating* if, for every $k > 0$, any thread that takes an infinite number of steps has an interval where it takes at least k concrete contiguous steps (that is, not interleaved with any other thread). Exponential back-off [1] is one possible mechanism to make schedules uniformly isolating (with high probability). Threads back off until all but one are inactive. Back-off durations can be controlled by the programmer.

We are now ready to reformulate the definitions of the progress properties surveyed in Section 2.

Definition 1. *A method implementation is* deadlock-free *if it guarantees minimal progress in every fair history, and maximal progress in some fair history.*

The restriction to fair histories captures the informal requirement that each thread eventually leaves its critical section. The definition does not mention locks or critical sections because progress should be defined in terms of completed method calls, not low-level mechanisms. Moreover, as noted, not all deadlock-free object implementations will have easily recognizable locks and critical sections.

The requirement that the implementation will provide maximal progress in some fair history is intended to rule out certain pathological cases. For example, the first thread to access an object might lock it and never release the lock. Such an implementation guarantees minimal progress (for the thread holding the lock) in every fair execution, but does not provide maximal progress in any execution. Clearly, such an implementation would not be considered acceptable in practice and is of no interest to us.

The starvation-free property is now straightforward:

Definition 2. *A method implementation is* starvation-free *if it guarantees maximal progress in every fair history.*

These properties are *dependent*: they are restricted to the subset of fair histories. Informally, these properties depend on a well-behaved operating system scheduler. We can capture the notion of dependency as follows:

Definition 3. *A progress condition is* dependent *if it does not guarantee minimal progress in every history, and is* independent *if it does.*

Here are the non-blocking properties.

Definition 4. *A method implementation is* lock-free *if it guarantees minimal progress in every history, and maximal progress in some history.*

Definition 5. *A method implementation is* wait-free *if it guarantees maximal progress in every history.*

The two properties above are independent: they apply to all histories. There is however a dependent non-blocking property:

Definition 6. *A method implementation is* obstruction-free *if it guarantees maximal progress in every uniformly isolating history.*

4   The Structure of Progress

Although these progress conditions may have seemed quite different, each provides either minimal or maximal progress with respect to some set of histories. The result is a simple and regular structure illustrated in the "periodic table" style chart shown in Figure 1 (and its more complete counterpart in Figure 2). These observations may appear so simple as to be obvious in retrospect, but we have never seen them described in this way.

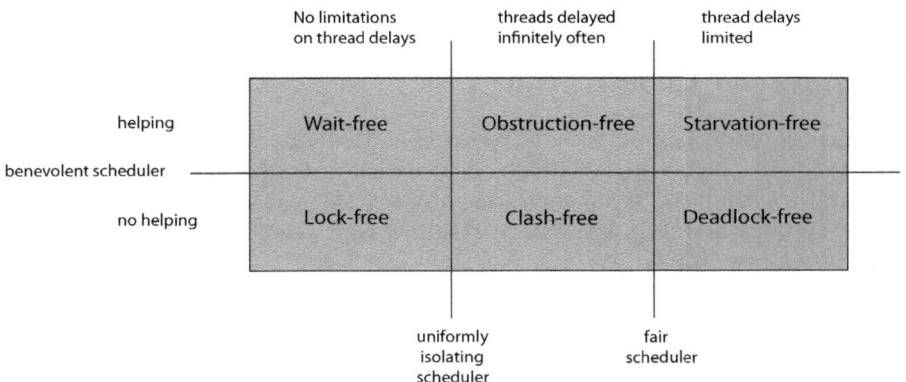

Fig. 2. Clash-freedom: the missing element

There are three dividing lines, two vertical and one horizontal, that split the five conditions. The leftmost vertical line separates dependent conditions from the rest. The lock-free and wait-free properties apply to any histories, while obstruction-freedom, starvation-freedom, and deadlock-freedom require some kind of external scheduler support to guarantee progress.

The rightmost vertical line separates the blocking and non-blocking conditions. The lock-free, wait-free, and obstruction-free conditions are non-blocking: if a suspended thread stops at an arbitrary point in a method call, at least some active threads can make progress. The deadlock-free and starvation-free conditions do not have this property.

Finally, the horizontal line separates the minimal and maximal progress conditions. The minimal conditions guarantee the system as a whole makes progress while the maximal conditions guarantee that each thread makes progress. For brevity, *minimal* progress properties encompass the lock-free and deadlock-free properties, while *maximal* properties encompass the wait-free, starvation-free, and obstruction-free properties. Later we will see several ways to cross this line: "helping" (Section 5) and benevolent schedulers (Section 8). Helping [7] is an algorithmic mechanism which has threads avoid being delayed by others that are slow by completing the slow threads' work in their place. Benevolence is an assumption on the scheduler behavior that allows one to avoid the high communication costs associated with helping.

There is a hole in Figure 1: a conspicuous empty slot occupied by a dependent, non-blocking progress property that guarantees minimal progress in uniformly-isolating histories.

Definition 7. *A method implementation is* clash-free *if it guarantees minimal progress in every uniformly isolating history, and maximal progress in some such history.*

In the next two sections we show that being clash-free is strictly weaker than being obstruction-free[2].

5   Universal Constructions

In this section we define two universal constructions that transform any sequential object into a linearizable concurrent object satisfying the same minimal progress condition as its consensus objects. We will use them, among other things, to demonstrate a separation result: a clash-free object implementation that is not obstruction-free.

The construction relies on a supply of one-time consensus objects (see Figure 3), ones in which each thread can call the decide() method at most once in any execution history. Interestingly enough, the minimal progress universal construction provides the minimal form of whatever progress guarantee is provided by the consensus objects: it is lock-free if the consensus objects are lock-free or wait-free, clash-free if they are clash-free or obstruction-free, and deadlock-free if they are deadlock-free or starvation-free.

The maximal-progress universal construction does the same, except that it provides the maximal form of the consensus objects' progress guarantee: it is wait-free if the consensus objects are lock-free or wait-free, obstruction-free if they are clash-free or obstruction-free, and starvation-free if they are deadlock-free or starvation-free (notice that for one-time objects the minimal progress conditions are by definition equal to the maximal progress conditions).

Our two constructions are adapted from Herlihy and Shavit [10], which contains their proofs of correctness.

```
1   public interface Consensus<T> {
2     T decide(T value);
3   }
```

Fig. 3. Consensus Object Interface

```
1   public interface SeqObject {
2     public abstract Response apply(Invocation invoc);
3   }
```

Fig. 4. A Generic Sequential Object: the apply() method applies the invocation and returns a response

Figure 4 shows a *generic* definition for a sequential object. Each object is created in a fixed initial state. The apply() method takes as argument an *invocation* which describes the method being called and its arguments, and returns

---

[2] Clash-freedom is arguably the Einsteinium of progress conditions. Like Einsteinium, symbol *Es*, atomic number 99, it fills a vacant table slot, yet does not occur naturally in any measurable quantities and has no commercial value.

a *response*, containing the call's termination condition (normal or exceptional) and the return value, if any. For example, a stack invocation might be push() with an argument, and the corresponding response would be normal and *void*.

Figures 5 and 6 show a universal construction that transforms any sequential object into a linearizable concurrent object satisfying the same minimal progress condition as its consensus objects.

For simplicity, this construction assumes that sequential objects are *deterministic*: if we apply a method to an object in a particular state, then there is only one possible response and one possible new object state. We can represent any object as a combination of a sequential object in its initial state and a *log*: a linked list of nodes representing the sequence of method calls applied to the object (and hence the object's sequence of state transitions). A thread executes a method call by scanning the log, starting at the oldest node (the tail), until it finds the newest node (the head). It then uses the node's consensus object to append its own node to the list. It then retraverses the log, applying the method calls to a private copy of the object. The thread finally returns the result of applying its own operation. It is important to understand that only the head of the log is mutable: the initial state and nodes following the head never change.

This algorithm works even when apply() calls are concurrent because the prefix of the log up to the thread's own node never changes. The losing threads, who failed to append their own nodes, must start over.

The maximal-progress universal construction appears in Figure 7. We must guarantee that every thread completes an apply() call within a finite number of steps, that is, no thread starves. To guarantee this property, threads making

```
1   public class Node {
2      public Invoc invoc;                        // method name and args
3      public Consensus<Node> decideNext;        // decide next Node in list
4      public Node next;                          // the next node
5      public int seq;                            // sequence number
6      public Node(Invoc invoc) {
7         invoc = invoc;
8         decideNext = new Consensus<Node>()
9         seq = 0;
10     }
11     public static Node max(Node[] array) {
12        Node max = array[0];
13        for (int i = 1; i < array.length; i++)
14           if (max.seq < array[i].seq)
15              max = array[i];
16        return max;
17     }
18  }
```

Fig. 5. The Node class

```
1   public class MinUniversal {
2     private Node tail;
3     public MinUniversal() {
4       tail = new Node();
5       tail.seq = 1;
6     }
7     public Response apply(Invoc invoc) {
8       int i = ThreadID.get();
9       Node prefer = new Node(invoc);
10      Node head = tail;
11      while ( prefer.seq == 0) {
12        while (head.next != null) {
13          head = head.next;
14        }
15        Node after = head.decideNext.decide( prefer );
16        head.next = after;
17        after.seq = head.seq + 1;
18      }
19      SeqObject myObject = new SeqObject();
20      current = tail.next;
21      while ( current != prefer ){
22        myObject.apply(current.invoc );
23        current = current.next;
24      }
25      return myObject.apply(current.invoc );
26    }
27  }
```

Fig. 6. The minimal-progress universal construction

progress must help less fortunate threads to complete their calls.

To allow helping, each thread shares with other threads the apply() call that it is trying to complete. We add an n-element announce[] array, where announce[i] is the node thread i is currently trying to append to the list. Initially all entries refer to the sentinel node, which has a sequence number 1. A thread i *announces* a node when it stores the node in announce[i].

To execute apply(), a thread first announces its new node. This step ensures that if the thread itself does not succeed in appending its node onto the list, some other thread will append that node on it's behalf. It then proceeds as before, attempting to append the node into the log.

6   Separation Results

We can now use our universal constructions to prove the following theorem:

Theorem 1. *There exists a clash-free object implementation that is not obstruction-free.*

```
1    public class MaxUniversal {
2      private Node[] announce; // array added to coordinate helping
3      private Node[] head;
4      private Node tail = new node(); tail.seq = 1;
5      for (int j=0; j < n; j++){head[j] = tail; announce[j] = tail };
6      public Response apply(Invoc invoc) {
7        int i = ThreadID.get();
8        announce[i] = new Node(invoc);
9        head[i] = Node.max(head);
10       while (announce[i].seq == 0) {
11         Node before = head[i];
12         Node help = announce[(before.seq + 1 % n)];
13         if (help.seq == 0)
14           prefer = help;
15         else
16           prefer = announce[i];
17         after = before.decideNext.decide(prefer);
18         before.next = after;
19         after.seq = before.seq + 1;
20         head[i] = after;
21       }
22       SeqObject MyObject = new SeqObject();
23       current = tail.next;
24       while (current != announce[i]){
25         MyObject.apply(current.invoc);
26         current = current.next;
27       }
28       head[i] = announce[i];
29       return MyObject.apply(current.invoc);
30     }
31   }
```

Fig. 7. The maximal-progress universal construction

Proof. Herlihy, Luchangco, and Moir [8] observe that one can implement an obstruction-free (and hence clash-free) one-time consensus object by derandomizing the randomized consensus protocol of Aspnes and Herlihy [2] (replacing the random coin by a deterministic one). [3] The minimal-progress universal construction using such an obstruction-free consensus object is easily shown to be clash-free.

We now construct a history in which the minimal-progress universal construction using an obstruction-free consensus object is not obstruction-free.

The line numbers in the next paragraph refer to Figure 6. Pick one favored thread A. Run each thread until it reaches Line 12. When all n threads have

---

[3] Similarly, it is easy to implement a deadlock-free consensus object using a mutual exclusion lock.

arrived, run each one through the entire loop between Lines 12 and 14. After all the threads have executed the loop, allow A to call and return from the consensus object, completing its own call. The others call the consensus object after A's call has returned, so A succeeds while the others fail.

If we repeat this interleaving, the result is a uniformly-isolating history, because each thread scans the log in isolation, and each time the log is longer. However, all threads but A will never succeed and so the implementation is not obstruction-free.

It follows that being clash-free is a weaker condition than being obstruction-free.

7   Partial Methods

So far we have considered only *total methods*, methods that are always capable of returning a response. Much of concurrent programming, however, makes use of *partial methods* that block when called in certain states. For example, Figure 8 shows how one might implement a partial FIFO queue in the Java^{TM} programming language. The deq() method is *synchronized*: it acquires an implicit lock

```
1    public class Queue<T> {
2      T items [];
3      int head, size ;
4      int capacity ;
5      public Queue(int capacity) {
6        items = (T[]) new Object[capacity];
7        head = size = 0;
8      }
9      public synchronized T deq() {
10       while ( size  == 0) {
11         wait ();
12       }
13       notifyAll ();
14       size --;
15       return items [head++];
16     }
17     public synchronized void enq(T x) {
18       while ( size  == capacity) {
19         wait ();
20       }
21       notifyAll ();
22       items [( head + size) % capacity]  = x;
23       size ++;
24     }
25   }
```

Fig. 8. A FIFO queue with partial methods

when it is called and releases it when it returns. If it encounters an empty queue (Line 10) the method temporarily releases the lock and suspends itself. Later, if an enq() call adds an item to the queue, it calls notifyAll () to wake up any suspended dequeuers. These threads reacquire the lock and retest whether the queue is empty. We say that an invocation is *enabled* if there is a response it could return.

It is not obvious how to define minimal and maximal progress for partial methods. For example, one might be tempted to say that a method provides maximal progress in a history if no pending invocation is infinitely often enabled. (In other words, any invocation enabled often enough will return.) The following example illustrates why this definition is problematic. Consider an empty FIFO queue, where thread A calls a blocking deq(). Because the queue is empty, the method cannot return, so A's invocation is disabled, and A blocks. Thread B then enqueues an item, enabling A's invocation, but then immediately dequeues that item, again disabling A's invocation. If B repeats this sequence forever, then A's invocation is infinitely often enabled, yet A never returns. Should we deem this history as not providing maximal progress?

The problem with rejecting such a history is that it is permitted by all threads packages of which we are aware. For example, in the Queue implementation of Figure 8, A's deq() call releases the lock and waits. B's enq() call notifies A asynchronously, but before the operating system reschedules A, B's deq() removes the item. (Similar behavior can occur also with the *Pthreads* and *.Net* threads libraries.) We should avoid any definition of maximal progress that cannot be implemented.

A pending invocation is *continually enabled* in H if it is enabled at every step in some suffix of H. Once an invocation becomes continually enabled, then when its thread is awakened and resumed, however asynchronously, it is certain to discover a response.

We are ready to propose another definition. A method provides *minimal progress* in H if, in every suffix of H where some active invocation is continually enabled, some pending active invocation has a matching response. In other words, at no point in H does the method have continually-enabled active invocations, none of which ever returns. Similarly, a method provides *maximal progress* in H if it has no continually-enabled active invocations ever.

8   Benevolent Schedulers

In practice, programmers often use implementations that guarantee only minimal progress, not because they do not care about lack of progress by individual threads, but because such lack of progress almost never happens under normal circumstances. For example, while programs that use spin locks are deadlock free, they are not starvation-free because the scheduler might schedule one particular thread only when the lock is held by another thread. In practice, few programmers worry about this prospect because they do not expect schedulers to persecute individual threads.

Let us make this notion more precise. Consider an algorithm that guarantees a minimal progress condition. A scheduler is *benevolent* for that algorithm if it guarantees maximal progress for that algorithm in every history it permits. Such a guarantee can also be probabilistic in nature.

For example, an *oblivious scheduler* is a fair scheduler that chooses the next thread to take a step uniformly at random. Now consider a deadlock-free spin lock algorithm where each thread repeatedly acquires the lock (by spinning), executes an operation, and releases the lock.

Theorem 2. *An oblivious scheduler is benevolent (with probability one) for any deadlock-free spin-lock algorithm.*

Proof. Because the scheduler is fair, the lock must become free an infinite number of times. Each time the lock becomes free, that thread is chosen with probability at least $1/n$, implying that the thread starves with probability measure zero.

Along the same lines, we can use exponential back-off [1] to make lock-free algorithms wait-free.

We have barely scratched the surface with these theorems, and we leave it as an open question to derive more theorems of this nature. In particular, this approach provides a new way to think about *contention managers* [9], application-specific modules that modify the behavior of schedulers.

9   Foundations of Shared-Memory Computability

Our classification of dependent progress conditions has implications for the foundations of shared-memory computability. Lamport's register-based approach [12] to read-write memory computability is based on wait-free implementations of one register type from another. Similarly, Herlihy's consensus hierarchy [7] applies to wait-free or lock-free object implementations. Combined, these structures form the basis of a *theory of concurrent shared-memory computability* [10] that explains what objects can be used to implement other objects in an asynchronous shared memory multiprocessor environment.

One might ask, however, why such a theory should rest on non-blocking progress conditions (that is, wait-free or lock-free) and not on locks. After all, locking implementations are common in practice. Moreover, the obstruction-free condition is a non-blocking progress condition where read-write registers are universal [8], effectively leveling the consensus hierarchy.

We are now in a position to address this question. Perhaps surprisingly, Figure 2 suggests that the lock-free and wait-free conditions provide a sound basis for a concurrent computability theory because they are *independent* progress conditions that do not rely on the good behavior of the operating system scheduler. A theory based on a dependent condition would require strong, perhaps arbitrary assumptions about the environment in which programs were executed.

When studying the computational power of synchronization primitives, it is unsatisfactory to rely on the operating system to ensure progress, both because it obscures the inherent synchronization power of the primitives, and because we might want to use such primitives in the construction of the operating system itself. For these reasons, a satisfactory theory of shared-memory computability should rely on independent progress conditions such as the wait-free or lock-free properties, and not on the other, dependent properties.

We have discussed progress properties of individual methods, not of entire objects, because it is often useful for objects to provide different methods that satisfy different progress properties. For example, Heller et al. [6]) describe a linked-list that supports starvation-free lock-based insertion and removal, but with a wait-free search.

Our definitions are easily generalized to *collections* of methods ranging from a single method to all of an object's methods. For example, the Harris-Michael lock-free list [5,16] provides add(), remove(), and contains() methods. Each method on its own is obstruction-free, but the collection of all methods taken together is lock-free.

10   Conclusions

This paper proposes a novel way to impose order on the previously unstructured world of progress conditions for algorithms on multicore machines. Much, however, remains to be done.

For example, it would be of great interest to identify new classes of benevolent schedulers. It could be of practical importance to understand how *contention managers* [9], application-specific modules that modify the behavior of schedulers, serve in making them benevolent. It would be interesting to better understand the role of "helping" in overcoming scheduler limitations, possibly finding lower bounds on the cost of universal helping, a cost that perhaps captures the value of the benevolence scheduling property.

Finally, our approach implies that real-world operating system designers should be aware of the progress guarantees that their systems and services provide. Perhaps it is time that these criteria be formally stated and made available to the user in a manner similar to how memory models are defined with respect to correctness.

Acknowledgments. We are grateful to Ori Shalev and Victor Luchangco for their help in formulating and crystallizing some of the concepts presented here.

References

1. Agarwal, A., Cherian, M.: Adaptive backoff synchronization techniques. In: Proceedings of the 16th International Symposium on Computer Architecture, pp. 396–406 (May 1989)
2. Aspnes, J., Herlihy, M.: Fast randomized consensus using shared memory. J. Algorithms 11(3), 441–461 (1990)
3. Attiya, H., Welch, J.: Distributed Computing: Fundamentals, Simulations, and Advanced Topics. John Wiley and Sons (2004)

4. Attiya, H., Welch, J.: Distributed Computing: Fundamentals, Simulations, and Advanced Topics, 2nd edn. John Wiley & Sons, Inc., New York (2004)
5. Harris, T.L.: A Pragmatic Implementation of Non-Blocking Linked-Lists. In: Welch, J.L. (ed.) DISC 2001. LNCS, vol. 2180, pp. 300–314. Springer, Heidelberg (2001)
6. Heller, S., Herlihy, M., Luchangco, V., Moir, M., Scherer III, W.N., Shavit, N.: A Lazy Concurrent List-Based Set Algorithm. In: Anderson, J.H., Prencipe, G., Wattenhofer, R. (eds.) OPODIS 2005. LNCS, vol. 3974, pp. 3–16. Springer, Heidelberg (2006)
7. Herlihy, M.: Wait-free synchronization. ACM Transactions on Programming Languages and Systems 13(1), 124–149 (1991)
8. Herlihy, M., Luchangco, V., Moir, M.: Obstruction-free synchronization: Double-ended queues as an example. In: Proceedings of the 23rd International Conference on Distributed Computing Systems, pp. 522–529. IEEE (2003)
9. Herlihy, M., Luchangco, V., Moir, M., Scherer, W.: Software transactional memory for dynamic-sized data structures. In: PODC 2003: Proceedings of the Twenty-Second Annual Symposium on Principles of Distributed Computing, pp. 92–101. ACM, New York (2003)
10. Herlihy, M., Shavit, N.: The Art of Multiprocessor Programming. Morgan Kaufmann Publishers, San Mateo (2008)
11. Herlihy, M.P., Wing, J.M.: Linearizability: a correctness condition for concurrent objects. ACM Transactions on Programming Languages and Systems (TOPLAS) 12(3), 463–492 (1990)
12. Lamport, L.: On interprocess communication. part ii: Algorithms. Distributed Computing 1(2), 86–101 (1986)
13. Larus, J.R., Rajwar, R.: Transactional Memory. Morgan and Claypool (2006)
14. Lynch, N.: Distributed Algorithms. Morgan Kaufmann Publishers, San Mateo (2008)
15. Mellor-Crummey, J., Scott, M.: Algorithms for scalable synchronization on shared-memory multiprocessors. ACM Transactions on Computer Systems 9(1), 21–65 (1991)
16. Michael, M.M.: High performance dynamic lock-free hash tables and list-based sets. In: Proceedings of the Fourteenth Annual ACM Symposium on Parallel Algorithms and Architectures, pp. 73–82. ACM Press (2002)
17. Taubenfeld, G.: Synchronization Algorithms and Concurrent Programming. Prentice-Hall, Inc., Upper Saddle River (2006)

Byzantine Fault-Tolerance with Commutative Commands

Pavel Raykov[1], Nicolas Schiper[2], and Fernando Pedone[2]

[1] Swiss Federal Institute of Technology (ETH)
Zurich, Switzerland
[2] University of Lugano (USI)
Lugano, Switzerland

Abstract. State machine replication is a popular approach to increasing the availability of computer services. While it has been largely studied in the presence of crash-stop failures and malicious failures, all existing state machine replication protocols that provide byzantine fault-tolerance implement some variant of atomic broadcast. In this context, this paper makes two contributions. First, it presents the first byzantine fault-tolerant generic broadcast protocol. Generic broadcast is more general than atomic broadcast, in that it allows applications to deliver commutative commands out of order—delivering a command out of order can be done in fewer communication steps than delivering a command in the same order. Second, the paper presents an efficient state machine replication protocol that tolerates byzantine failures. Our protocol requires fewer message delays than the best existing solutions under similar conditions. Moreover, processing of commutative commands on replicas requires only two MAC operations. The protocol is speculative in that it may rollback non-commutative commands.

1 Introduction

State machine replication is a popular approach to increasing the availability of computer services [1,2]. By replicating a service on multiple machines, hardware and software failures can be tolerated. Although state machine replication has been largely studied in the presence of crash-stop failures and malicious failures, all existing protocols that provide byzantine fault-tolerance (BFT) (e.g., [3,4,5,6,7]) implement some variant of atomic broadcast, a group communication primitive that guarantees agreement on the set of commands delivered and on their order. In this context, this paper makes two contributions.

The first contribution of this paper is a byzantine fault-tolerant generic broadcast protocol. Generic broadcast defines a conflict relation on messages, or commands, and only orders messages that conflict. Two messages conflict if their associated commands do not commute. For instance, two increment operations of some variable x commute since the final value of x is independent of the execution order of these operations. Generic broadcast generalizes atomic broadcast—the two problems are equivalent when every two messages conflict. Previous generic

A. Fernández Anta, G. Lipari, and M. Roy (Eds.): OPODIS 2011, LNCS 7109, pp. 329–342, 2011.
© Springer-Verlag Berlin Heidelberg 2011

broadcast protocols appeared in the crash-stop model [8,9,10]; ours is the first to tolerate malicious failures. The difficulty with generic broadcast stems from the need to deliver commutative commands in two communication delays and ensure that their delivery order, with respect to non-commutative commands, is the same at all correct processes. To address this challenge under byzantine failures we define Recovery Consensus, an abstraction that ensures proper ordering between conflicting and non-conflicting messages. The proposed protocol requires $n \geq 5f + 1$ replicas to tolerate f byzantine failures. We use Recovery Consensus at the core of our generic broadcast protocol.

The second contribution of this paper is a state machine replication protocol that generalizes and improves current byzantine fault-tolerant state machine replication protocols. Our protocol builds on our generic broadcast algorithm. A naive implementation of state machine replication based on generic broadcast to propagate commands to servers would lead to a best latency of three communication delays. We rely on speculative execution to provide an efficient algorithm that executes commutative commands in two communication delays. The algorithm is speculative in that it may rollback commands in some cases (i.e., when non-commutative commands are issued). To summarize, the principal advantage of the proposed state machine replication protocol is to allow fast execution of commutative commands in two message delays. Moreover, when commands commute servers only need to execute two MAC operations per command.

The remainder of the paper is structured as follows. Section 2 defines the system model. Sections 3 and 4, respectively, present the Recovery Consensus and generic broadcast protocols. We extend our generic broadcast protocol to provide state machine replication in Section 5. Section 6 discusses related work and Section 7 concludes the paper. Correctness proofs of the protocols can be found in the appendix of the full version of this paper [11].

2   System Model and Definitions

We consider an asynchronous message passing system composed of n processes $\Pi = \{p_1, \ldots, p_n\}$, out of which f are *byzantine* (i.e., they can behave arbitrarily). A process that is not byzantine is *correct*. The adversary that controls byzantine processes is computationally bounded (i.e., it cannot break cryptographic primitives) and cannot change the content of messages sent by one correct process to another correct process. The network is fully connected and *quasi-reliable*: if a correct process p sends a message m to a correct process q, then q receives m.[1] We make use of public-key signatures to allow a process to sign a message m [12]. We denote message m signed by process p_i as $\langle m \rangle_{\sigma_i}$. We also use HMACs [13] to establish a bidirectional authenticated channel between any two processes p_x and p_y, with the notation $\langle m \rangle_{\sigma_{xy}}$ indicating a message m signed with a secret key shared between processes p_x and p_y.

---

[1] The presented algorithms can trivially be modified to tolerate *fair-lossy* links, links that may drop messages but guarantee delivery of a message m if m is repeatedly sent. We assume quasi-reliable links to simplify the presentation of the algorithms.

Due to the impossibility to solve consensus in asynchronous systems prone to crash failures [14], it is also impossible to solve atomic broadcast and generic broadcast [15]. This impossibility is typically overcome by strengthening the model with further assumptions (e.g., [16,17,18]). In this paper we assume the existence of an atomic broadcast oracle [19]. Atomic broadcast is defined by the primitives A-Bcast(m) and A-Deliver(m), where m is a message. It guarantees the following properties:

- (Validity) If a correct process p A-Bcasts a message m, then p eventually A-Delivers m.
- (Agreement) If a correct process p A-Delivers a message m, then every correct process q eventually A-Delivers m.
- (Integrity) For any message m, every correct process p A-Delivers m at most once.
- (Order) If correct processes p and q both A-Deliver messages m and m', then p and q A-Deliver them in the same order.

3 Recovery Consensus

In this section we introduce Recovery Consensus, an abstraction used by generic broadcast to order messages whose associated commands do not commute, also denoted as conflicting messages. Below, we provide an implementation of Recovery Consensus that employs digital signatures for message authentication.

3.1 Problem Definition

Recovery Consensus allows each process p_i to propose a set of non-conflicting messages $NCSet_i$ and a set of conflicting messages $CSet_i$. The set $NCSet_i$ is called non-conflicting since every pair of messages in it does not conflict, and the set $CSet_i$ is called conflicting since for every message $m \in CSet_i$ there is a message $m' \in NCSet_i$ such that m and m' conflict. Recovery Consensus ensures agreement on a set of non-conflicting messages $NCSet$ and on a set of conflicting messages $CSet$. Additionally, it guarantees that if $n_{chk} - f$ correct processes p_i propose a message m, i.e., m belongs to either $NCSet_i$ or $CSet_i$, where n_{chk} is a parameter of the problem, m will be part of either $NCSet$ or $CSet$.

More formally, Recovery Consensus is defined by primitives proposeRC($NCSet_i, CSet_i$) and decideRC($NCSet, CSet$). Provided that every correct process p_i invokes proposeRC($NCSet_i$, $CSet_i$) and there are no conflicting messages in $NCSet_i$, the following properties are guaranteed:

- (Termination) Every correct process eventually decides on some pair of message sets.
- (Agreement) If two correct processes decide on pairs of message sets ($NCSet_1$, $CSet_1$) and ($NCSet_2, CSet_2$), then $NCSet_1 = NCSet_2$ and $CSet_1 = CSet_2$.
- (Validity) If a correct process invokes decideRC($NCSet, CSet$), then:
    1. $NCSet \cap CSet = \varnothing$.

2. If a message m belongs to $n_{chk} - f$ $NCSet_i$ sets of correct processes, then $m \in NCSet$.
3. No two messages in $NCSet$ conflict.
4. If a message m is in $n_{chk} - f$ sets $NCSet_i \cup CSet_i$ of correct processes, then $m \in NCSet \cup CSet$.

3.2 Solving Recovery Consensus

Algorithm $\mathcal{C}_{absign}$ requires at least n_{chk} correct processes, where $n_{chk} \leq n - f$. It consists of a single task and works as follows. Each process p_i starts by atomically broadcasting the pair $(NCSet_i, CSet_i)$ signed with p_i's signature—in the algorithm, the signed message is denoted as $\langle NCSet_i, CSet_i \rangle_{\sigma_i}$ (line 2). Process p_i then waits until it A-Delivered n_{chk} unique and valid messages, that is, messages from distinct sources that do not contain conflicting messages in $NCSet_j$ (line 3). Detecting unique messages is done with signatures: if p_i A-Delivers two messages from the same source p_j, then p_i discards both messages since p_j is byzantine. By considering the first n_{chk} unique and valid messages, this ensures that at least $n_{chk} - f$ messages A-Delivered by p_i were broadcast by correct processes.

Algorithm $\mathcal{C}_{absign}$
Process p_i Recovery Consensus algorithm with atomic broadcast and signatures

---

1: Procedure proposeRC($NCSet_i, CSet_i$)
2:     A-Bcast($\langle NCSet_i, CSet_i \rangle_{\sigma_i}$)
3:     **wait until** [$|GS| = n_{chk} : GS \overset{def}{=} \{(NCSet_j, CSet_j) \mid$ A-Delivered unique and
4:                                 valid $\langle NCSet_j, CSet_j \rangle_{\sigma_j}$ from p_j $\}$]
5:     $NCSet \leftarrow \{m \mid \exists \; \lceil \frac{n_{chk}+1}{2} \rceil \; NCSet_j : (NCSet_j, \cdot) \in GS$ **and** $m \in NCSet_j \}$
6:     $CSet \leftarrow (\bigcup_{(NCSet_j, CSet_j) \in GS} NCSet_j \cup CSet_j) \setminus NCSet$
7:     decideRC($NCSet, CSet$)

Any message that appears in a majority of the n_{chk} $NCSet_j$ sets will appear in $NCSet$ (line 5). This guarantees the third validity property, namely that no two messages in $NCSet$ conflict, since (i) the considered $NCSet_j$ sets at line 3 do not contain conflicting messages and (ii) any message in $NCSet$ belongs to a majority of $NCSet_j$ sets.

Let $Q_{n_{chk}-f}$ be a quorum of $n_{chk} - f$ correct processes that propose a message m as part of $NCSet_i$, and let $Q_{n_{chk}}$ be a quorum of n_{chk} processes, the number of unique and valid A-Delivered messages processes consider at line 3. To ensure that we include m in $NCSet$ if m belongs to $n_{chk} - f$ $NCSet_i$ sets of correct processes, the minimum size of the intersection between $Q_{n_{chk}-f}$ and $Q_{n_{chk}}$ must be $\lceil \frac{n_{chk}+1}{2} \rceil$. Hence, n_{chk} must satisfy inequality $(n_{chk}-f)+n_{chk} \geq n+\lceil \frac{n_{chk}+1}{2} \rceil$. Since $n_{chk} \leq n - f$, we conclude that $\mathcal{C}_{absign}$ requires $n > 5f$.

Finally, the set of conflicting messages $CSet$ consists of messages gathered using atomic broadcast that are not part of $NCSet$ (line 6). From the total order property of atomic broadcast, correct processes gather the same set of pairs $(NCSet_j, CSet_j)$. Since $NCSet$ and $CSet$ are constructed from the gathered pairs using a deterministic procedure, all correct processes agree on these two sets.

4   BFT Generic Broadcast

We present a byzantine fault-tolerant generic broadcast protocol that we denote as $\mathcal{PGB}$. This protocol relies on Recovery Consensus to handle conflicting messages. We first define generic broadcast in our model and then present the algorithm.

4.1   Generic Broadcast

Generic broadcast is defined by the primitives g-Broadcast(m) and g-Deliver(m), where m is a message from the predefined set $\mathcal{M}$, to which all messages belong. We assume that each message broadcast has a unique identifier. Generic broadcast is parameterized by a symmetric relation $\sim$ on $\mathcal{M} \times \mathcal{M}$. If $(m, m') \in \sim$, or $m \sim m'$ for short, we say that m and m' *conflict* or are *conflicting messages*. If m and m' conflict, then generic broadcast will order m and m'. If m and m' do not conflict, they can be delivered in any order.

Generic broadcast guarantees the following properties, adapted from [10] to the byzantine failure model:

- (Validity) If a correct process p g-Broadcasts a message m, then p eventually g-Delivers m.
- (Agreement) If a correct process p g-Delivers a message m, then every correct process q eventually g-Delivers m.
- (Integrity) For any message m, every correct process p g-Delivers m at most once.
- (Order) If correct processes p and q both g-Deliver conflicting messages m and m' $(m \sim m')$, then p and q g-Deliver them in the same order.

As noted in [10], atomic broadcast is a special case of generic broadcast when all messages conflict with all messages, that is, $\sim = \mathcal{M} \times \mathcal{M}$. Thus, one could question the difficulty of implementing generic broadcast since we assume the existence of an atomic broadcast primitive that could be used to implement generic broadcast. The main idea of our generic broadcast protocol is that it allows *fast delivery* (i.e., in two communication delays) of non-conflicting messages, a bound that no atomic broadcast protocol can achieve in the general case [20].

4.2   Solving Generic Broadcast

The protocol is composed of two phases: an acknowledgment phase (ACK) and a check phase (CHK). Consecutive ACK and CHK phases form a "round". During the ACK phase processes g-Deliver non-conflicting messages in two message delays.

Algorithm $\mathcal{PGB}$

Process p_i generic broadcast algorithm

1: Initialization:
2:     $Received \leftarrow \varnothing,\ G\_del \leftarrow \varnothing,\ pending^1 \leftarrow \varnothing,\ gAck\_del^1 \leftarrow \varnothing,\ k \leftarrow 1$
3: To execute $g\text{-}Broadcast(m)$:   {Task 1}
4:     $send(m)$ to all
5: $g\text{-}Deliver(m)$ occurs as follows:
6: **when** $receive(m)$ **do**   {Task 2a}
7:     $Received \leftarrow Received \cup \{m\}$
8: **when** $receive(\langle k,\ pending_j^k,\ \textsc{ack} \rangle_{\sigma_{ij}})$ **do**   {Task 2b}
9:     $Received \leftarrow Received \cup pending_j^k$
10: **when** $receive(k, S_j, \textsc{chk})$ **do**   {Task 2c}
11:     $Received \leftarrow Received \cup S_j$
12: **when** $\big(Received \smallsetminus (G\_del \cup pending^k) \neq \varnothing\big)$ **do**   {Task 3}
13:     **if** $(\forall\ m, m' \in (Received \smallsetminus G\_del)\ :\ m \not\sim m')$ **then**
14:         $pending^k \leftarrow Received \smallsetminus G\_del$
15:         $send(\langle k,\ pending^k,\ \textsc{ack} \rangle_{\sigma_{ij}})$ to all processes p_j
16:     **else**
17:         $send(k, (Received \smallsetminus G\_del), \textsc{chk})$ to all   ▷ start of $\textsc{chk}$ phase
18:         $proposeRC(k, pending^k, (Received \smallsetminus (G\_del \cup pending^k)))$
19:         **wait until** $decideRC(k, NCSet^k, CSet^k)$
20:         **for each** $m \in NCSet^k \smallsetminus (G\_del \cup gAck\_del^k)$ **do** $g\text{-}Deliver(m)$ ⎫
21:         **for each** $m \in CSet^k \smallsetminus (G\_del \cup gAck\_del^k)$ **in ID** order **do** ⎪
22:             $g\text{-}Deliver(m)$ ⎬ atomic
23:         $G\_del \leftarrow G\_del \cup NCSet^k \cup CSet^k$ ⎪
24:         $k \leftarrow k + 1,\ pending^k \leftarrow \varnothing,\ gAck\_del^k \leftarrow \varnothing$ ▷ end of ⎭ $\textsc{chk}$ phase
25:     **end if**
26: **when** $\exists\ m : [$ for n_{ack} processes $p_j :$ received $\langle k,\ pending_j^k,\ \textsc{ack} \rangle_{\sigma_{ij}}$ ⎫ {Task 4}
27:     from p_j **and** $m \in (pending_j^k \smallsetminus gAck\_del^k) \cap pending^k\ ]$ **do** ⎬ atomic
28:     $gAck\_del^k \leftarrow gAck\_del^k \cup \{m\}$ ⎪
29:     $g\text{-}Deliver(m)$ ⎭

In the $\textsc{chk}$ phase, the protocol orders conflicting messages. Notice that $\mathcal{PGB}$ does not require signatures to deliver non-conflicting messages.

Algorithm $\mathcal{PGB}$ consists of six concurrent tasks. Each line of the algorithm, lines 20–24, and lines 26–29 are executed atomically. The following variables are used by the algorithm: k defines the current round number, $Received$ contains all the g-Broadcast messages that the process has received so far, G_del contains all the messages that have been g-Delivered in the previous rounds, $pending^k$ defines the set of non-conflicting messages acknowledged by the process in the current round, and $gAck\_del^k$ is the set of messages g-Delivered in the $\textsc{ack}$ phase of the current round.

When a process p wishes to g-Broadcast a message m, p sends m to all (line 4). When receiving m, a process q adds m to its $Received$ set (line 7) and eventually checks whether m conflicts with any message that was received but not delivered yet (line 13). If it is not the case, then q adds m to its $pending^k$ set and acknowledges all messages in this set by sending $pending^k$ to all (lines 14–15).[2] A process q g-Delivers m in the ACK phase when q receives n_{ack} acknowledgments for m (lines 26–29). To prevent conflicting messages from being g-Delivered in the ACK phase despite f byzantine processes, n_{ack} must be greater than $(n + f)/2$.

It is possible that q receives a message m' that conflicts with m before receiving n_{ack} acknowledgments for m. In that case, q proceeds to the CHK phase. At this point, processes start by exchanging all messages that they received but did not deliver in previous rounds (line 17). In doing so, all correct processes eventually receive m and m' despite potentially faulty senders, and enter the CHK phase. $\mathcal{PGB}$ then relies on Recovery Consensus to ensure agreement on the set of non-conflicting messages $NCSet^k$ that were potentially g-Delivered in the ACK phase, and the set $CSet^k$ of conflicting messages to deliver at the end of the current round. Correct processes invoke $proposeRC$ with round number k, the set of non-conflicting messages $pending^k$, and all the other messages that were received but not delivered so far, denoted as $Received \setminus (G\_del \cup pending^k)$ (line 18), and decide on sets $NCSet^k$ and $CSet^k$ (line 19). Processes deliver non-conflicting messages in $NCSet^k$ that they had not delivered so far (line 20), and then deliver conflicting messages $CSet^k$ (line 22).

To ensure that if a message m was delivered in the ACK phase m will appear in set $NCSet^k$ decided by Recovery Consensus, m must be proposed by $n_{chk} - f$ correct processes in $pending^k$ at line 18. If m was delivered in the ACK phase, at least $n_{ack} - f$ correct processes propose m to Recovery Consensus. Hence, to maximize resilience we set n_{ack} equal to n_{chk}.

Note that Recovery Consensus (Algorithm $\mathcal{C}_{absign}$) runs n atomic broadcasts in parallel. Hence, when conflicting messages are issued, $\mathcal{PGB}$ has message complexity n times bigger than a usual atomic broadcast protocol. $\mathcal{PGB}$ is optimized to perform well when non-conflicting messages are broadcast and Recovery Consensus is invoked rarely.

5   State Machine Replication

5.1   A Trivial Algorithm

Implementing state machine replication [1,2] using the generic broadcast algorithm of Section 4 is straightforward: each command of the state machine corresponds to a message in the set $\mathcal{M}$ and the conflict relation on messages is defined such that two messages conflict if and only if their associated commands do not commute. For instance, if replicas store bank accounts, two deposit commands on the same account commute since their execution order does not have an effect on the final state of the state machine nor on the respective outputs of these commands, which in this case only contain an acknowledgment that the

---

[2] To ensure that messages are not acknowledged twice and improve the efficiency of the algorithm, processes can remember the set of messages that were acknowledged and only acknowledge them once.

Algorithm $\mathcal{SMR}_{client}$
Client c algorithm

1: To execute command m:
2:     send($\langle m \rangle_{\sigma_c}$) to all replicas
3:     **wait until** [$\exists\, k$, s.t. received from different replicas
4:         n_{ack} $\langle k, m, res(m), \text{ACK} \rangle_{\sigma_{r_i c}}$ **or** $f + 1$ $\langle k, m, res(m), \text{CHK} \rangle_{\sigma_{r_i c}}$]
5:     **return** res(m)

---

operations were successfully executed. Clients can then directly broadcast commands to replicas using Algorithm $\mathcal{PGB}$. Once replicas deliver a command, they execute it and send back the result to the client. When $f + 1$ identical replies are received by the client, the result of the command is known. This technique guarantees a form of linearizability [21,22].

5.2   An Optimal Algorithm

The above algorithm allows clients to learn the outcome of a command cmd in three communication delays if cmd commutes with concurrent commands. As we show next, a lower latency can be achieved by modifying Algorithm $\mathcal{PGB}$ and speculatively executing commands.

Before presenting the algorithm, we extend the system model of Section 2. We assume a population of n *replicas*, aforenamed processes, and a set of *clients*. Any number of clients may be byzantine and f bounds the number of faulty replicas. The latter execute a command cmd of the state machine by invoking $execute(cmd)$. This invocation modifies the state of the replica and returns a result. Commands are deterministic, that is, they produce a new state and a result only based on the current state. Our protocol speculatively executes commands and may require *rolling back* some commands if their speculative order does not correspond to their definitive order. The effect of operation $rollback(cmd)$ is such that if a sequence of commands Seq is executed between $execute(cmd)$ and $rollback(cmd)$ then the replica's state is as if only commands in sequence Seq were executed. Notice that although replicas may rollback some commands, clients always see the definitive result of a command (i.e., clients do not perform rollbacks).

The protocols for clients and replicas are presented in Algorithms $\mathcal{SMR}_{client}$ and $\mathcal{SMR}_{replica}$ respectively. The replica's algorithm is similar to $\mathcal{PGB}$, except for the handling of acknowledgment messages, which is moved to the client. We highlight in gray the differences between $\mathcal{SMR}_{replica}$ and $\mathcal{PGB}$. Similarly to $\mathcal{PGB}$, replicas do not need to sign any messages when clients issue commutative commands.

When a client c invokes a command m, c sends $\langle m \rangle_{\sigma_c}$ to all replicas ($\mathcal{SMR}_{client}$, line 2). A replica includes message $\langle m \rangle_{\sigma_c}$ in the *Received* set at lines 4,6,8 if m's signature is valid. When m arrives at a replica r, one of two things can happen: either (a) m does not conflict with any other command that

---

Algorithm $\mathcal{SMR}_{replica}$
Replica r algorithm (the differences with Algorithm $\mathcal{PGB}$ are highlighted in gray)

1: Initialization:
2:     $Received \leftarrow \varnothing,\ G\_del \leftarrow \varnothing,\ pending^1 \leftarrow \varnothing,\ k \leftarrow 1,\ Res \leftarrow \varnothing$
3: **when** receive($\langle m \rangle_{\sigma_c}$) **do**   {Task 1a}
4:     $Received \leftarrow Received \cup \{\langle m \rangle_{\sigma_c}\}$
5: **when** receive($k,\ pending_j^k,\ \text{ACK}$) **do**   {Task 1b}
6:     $Received \leftarrow Received \cup pending_j^k$
7: **when** receive(k, S_j, CHK) **do**   {Task 1c}
8:     $Received \leftarrow Received \cup S_j$
9: **when**$\big(Received \smallsetminus (G\_del \cup pending^k) \neq \varnothing\big)$ **do**   {Task 2}
10:     **if** $(\forall\, m, m' \in (Received \smallsetminus G\_del) : m \not\sim m')$ **then**
11:         **for each** $m \in (Received \smallsetminus (G\_del \cup pending^k))$ **do**
12:             $res(m) \leftarrow$ execute(m), $Res \leftarrow Res \cup (m, res(m))$
13:             send($\langle k, m, res(m), \text{ACK} \rangle_{\sigma_{rc}}$) to client($m$)
14:         $pending^k \leftarrow Received \smallsetminus G\_del$
15:         send($k, pending^k, \text{ACK}$) to all replicas
16:     **else**
17:         send($k, (Received \smallsetminus G\_del), \text{CHK}$) to all replicas   ▷ start of CHK phase
18:         proposeRC($k, pending^k, (Received \smallsetminus (G\_del \cup pending^k))$)
19:         **wait until** decideRC($k, NCSet^k, CSet^k$)
20:         **for each** $m \in pending^k \smallsetminus NCSet^k$ **do**
21:             $rollback(m)$, remove($m, res(m)$) from Res
22:         **for each** $m \in NCSet^k \smallsetminus G\_del$ **do**
23:             **if** $m \notin pending^k$ **then** $res(m) \leftarrow$ execute(m)
24:             send($\langle k, m, res(m), \text{CHK} \rangle_{\sigma_{rc}}$) to client($m$)   ▷ $res(m)$ is retrieved from Res if needed
25:         **in ID order: for each** $m \in CSet^k \smallsetminus G\_del$ **do**
26:             $res(m) \leftarrow$ execute(m), send($\langle k, m, res(m), \text{CHK} \rangle_{\sigma_{rc}}$) to client($m$)
27:         $G\_del \leftarrow G\_del \cup NCSet^k \cup CSet^k$
28:         $k \leftarrow k + 1$, $pending^k \leftarrow \varnothing$, $Res \leftarrow \varnothing$ ▷ end of CHK phase
29:     **end if**

---

r received in the current round or (b) m conflicts with a command received in the same round.

In case (a), r speculatively executes m, stores the result in set Res, and sends the result back to the client as an acknowledgment message ($\mathcal{SMR}_{replica}$, lines 10–12). We use a function client(m) defining for a given message m the client that issued m. If client c receives n_{ack} identical acknowledgment messages for m, c learns the result of command m ($\mathcal{SMR}_{client}$, lines 3–5)—this is a similar condition under which a process can g-Deliver a message in the ACK phase of Algorithm $\mathcal{PGB}$.

In case (b), command m conflicts with a command received in the current round. Similarly to $\mathcal{PGB}$, each replica r uses Recovery Consensus to order these commands. For each command m' that was received by r in the ACK phase but that does not appear in the decided $NCSet$, r rollbacks m' and deletes the corresponding entry from Res—the speculative execution order of m' differs from its final execution order ($\mathcal{SMR}_{replica}$, line 21).

Then, commands in $NCSet$ are executed if they were not acknowledged in the ACK phase, and the results of these commands are sent to the corresponding clients (lines 22–24). Similar actions are done for the conflicting commands of $CSet$ (lines 25–26). A client learns the result of a command m that was executed in the CHK phase after receiving $f+1$ identical replies for m ($\mathcal{SMR}_{client}$, line 4).

5.3   Optimizations

We briefly discuss two optimizations allowing Algorithms $\mathcal{SMR}_{client}$ and $\mathcal{SMR}_{replica}$ (a) to achieve the optimal latency of two communication delays in executions without *contention*, defined next, and (b) to avoid message signing by clients.

No contention. Assume that pending sets contain the order in which commands were received and executed by the replicas; essentially, a pending set becomes a command sequence. We say that two pending sets conflict if they contain two conflicting messages executed in a different order. When no pending sets conflict, we say that there is no *contention*.

The main idea behind this optimization is that now we consider the conflicts between pending sets instead of the conflicts between individual messages. In the optimized Algorithm $\mathcal{SMR}_{client}$, a client c learns the result $res(m)$ of the execution of command m if: (1) c received n_{ack} *non-conflicting* pending sets with $res(m)$ or (2) c received $f+1$ CHK messages with $res(m)$. A replica enters the CHK phase if: (1) it has received two conflicting *pending* sets or (2) it has received a CHK message indicating that some other replica entered the CHK phase in the current round.

Since conflicting commands can be executed in the ACK phase, provided that they are executed in the same order, replicas include the execution order of commands in sets $NCSet_i$ proposed to Recovery Consensus. Hence, the Recovery Consensus algorithm must be modified and $NCSet$ essentially contains commands proposed by $\lceil \frac{n_{chk}+1}{2} \rceil$ replicas r_i as part of $NCSet_i$, such that no two pending sets containing m include two non-commutative commands m_1 and m_2 that were executed before m and in different orders.

Avoiding message signing by clients. Digital signatures based on asymmetric cryptography can be expensive to generate or verify, let alone the problem of distributing and refreshing key pairs. Instead of signing a message m, clients can use an authenticator (a list of HMACs) to authenticate m [5].

We modify Algorithm $\mathcal{SMR}_{replica}$ as follows: (1) during the ACK phase replica r_j puts message m at lines 4,6,8 in the $Received$ set only if m's authenticator contains a valid HMAC entry for r_j; (2) during the CHK phase, we change the way $CSet$ is built in the underlying protocol $\mathcal{C}_{absign}$: message m is included in $CSet$ only if it belongs to $f+1$ different $NCSet_j \cup CSet_j$. This guarantees

Table 1. Byzantine fault-tolerant replication protocols ("bf":"byzantine failures")

Protocol	PBFT [5]	Zyzzyva [3]	HQ [6]	Q/U [4]	Aliph [7]	this paper
Resilience	$f < n/3$	$f < n/3$	$f < n/3$	$f < n/5$	$f < n/3$	$f < n/5$
Best-case latency	4	3	4	2	2	2
Best-case latency in the absence of...	bf	bf slow links	bf contention	bf contention	bf contention slow links	bf contention
MAC operations at bottleneck server	2+8f	2+3f	2+4f	2+4f	2^3	2
Command classification	read-only/ mutative	read-only/ mutative	read-only/ mutative	read-only/ mutative	none	by conflict relation $\sim$
Client-based recovery	no	yes	yes	yes	yes	no

that only client c can issue commands with c's identifier, i.e., it is impossible to impersonate client c.

Unfortunately these modifications are more difficult to apply in Recovery Consensus. To avoid expensive signing during Recovery Consensus one could use matrix signatures [23] or employ the approach described in [6] for signing certificates, both of which essentially trade off signatures for additional network delays.

Digital signatures scale better than authenticators, whose size grows linearly with the number of replicas, so deciding which technique to apply depends on the specific system settings. In any case, we note that by design, Algorithms $\mathcal{SMR}_{replica}$ and $\mathcal{SMR}_{client}$ optimize the ACK phase, since this is the case we expect to happen more often.

6   Related Work

In the following we compare our BFT state machine replication protocol to the related work (see Table 1). To the best of our knowledge, this paper is the first to present an implementation of byzantine generic broadcast. All BFT state machine replication protocols we are aware of have a "fast mode"—analogous to the ACK phase, where messages are delivered fast under certain assumptions (also called "best-case"), and a recovery mechanism to switch to a "slow mode"—analogous to the CHK phase that resolves possible problems, usually contention or failures. Despite these similarities, existing protocols differ from each other in a number of aspects:

- PBFT was the first practical work on BFT state-machine replication. The best-case latency of four message delays is achieved when there are no byzantine failures. For read-only operations, the protocol can be optimized to achieve a latency of two message delays.

---

[3] In Table 2 of paper [7], Aliph's latency and throughput represent two different sub-protocols: *Chain* and *Quorum*. We here show the number of MAC operations that *Quorum* uses since only *Quorum* achieves the best-case latency of two network delays.

- Zyzzyva employs tentative execution to improve the best-case delay of PBFT. It executes commands in three network delays when there are no byzantine failures and links are timely. Otherwise, the protocol requires five network delays. Like PBFT, it can be optimized to execute read-only operations in two message delays.
- HQ, a descendant of PBFT, is optimized to execute read-only commands in two message delays and update commands in four message delays when the execution is contention-free.
- Q/U was the first protocol to achieve the best-case latency of two network delays for all commands when replicas are failure-free and updates do not access the same object concurrently.
- In [7], the authors propose a modular approach to build BFT services based on the concept of abstract instances. An abstract instance is a BFT replication protocol optimized for specific system conditions that can abort commands. In this context, the authors propose Aliph, a composition of three abstract instances: *Quorum*, *Chain*, and *PBFT*. Quorum is optimized for latency and allows command execution in two network delays when links are timely and the execution is contention- and failure-free. Chain, on the other hand, is optimized for throughput and achieves a latency of $f + 2$ network delays when there are no failures.

The protocol presented in this paper is the first to achieve a latency of two network delays when the execution is failure-free but concurrent commutative commands are submitted. Under the same conditions, PBFT and Zyzzyva achieve latency of four and three message delays respectively, while HQ, Q/U and Aliph run an additional protocol to resolve contention.

Q/U [4] and HQ [6] proposed a simplified version of the conflict relation: all commands are either *reads* or *writes* [6] (respectively, *queries* and *updates* in [4]); reads do not conflict with reads, and writes conflict with reads and writes. This is more restrictive than a conflict relation, as mutative operations on the same object do not necessarily conflict (e.g., incrementing a variable).

Zyzzyva, Q/U, and Aliph, more specifically the *Quorum* instance, do not use inter-replica communication to agree on the order of commands; instead they assume that it is the client's responsibility to resolve contention by collecting authenticated responses from replicas and distributing a valid certificate to the replicas. PBFT and the state machine replication protocol presented in this paper rely on inter-replica communication to serialize commands, which allows a lightweight protocol for clients. HQ uses a hybrid approach: it uses inter-replica communication only when clients demand to resolve contention explicitly, while in the "fast case" clients coordinate the execution.

Finally, we note that although [24] executes commutative commands in parallel, all commands are totally ordered using PBFT, resulting in a higher latency than our protocol in the aforementioned scenario.

7   Final Remarks

This paper introduces the first generic broadcast algorithm that tolerates byzantine failures. Generic broadcast is based on message conflicts, a notion that is more general than read and write operations [4,6]. The proposed algorithm is modular and relies on an abstraction called Recovery Consensus, used to ensure that correct processes (i) deliver the same set of non-conflicting messages in each round, and (ii) agree on the delivery order of conflicting messages. A modular approach facilitates the understanding of the ACK phase and CHK phase of Algorithm $\mathcal{PGB}$, and allows to explore various implementations of Recovery Consensus. We provided an implementation of Recovery Consensus that is based on atomic broadcast and digital signatures and requires at least $5f + 1$ processes. Finally, we extended the proposed generic broadcast algorithm to provide state machine replication. The resulting protocol, with its optimizations, can execute commands in two message delays under weaker assumptions than state-of-the-art algorithms.

The Aliph protocol [7] opened new directions in the development of state machine replication protocols: it is now possible to combine different protocols in one that switches through given implementations of state machine replication under certain policies to speed up the execution. Hence, it could be an interesting task to implement the protocol proposed in this paper as an **Abstract** instance (*Generic*) and see the behavior of the resulting algorithm (e.g., *Generic-Chain-Quorum-Backup*).

References

1. Lamport, L.: Time, clocks, and the ordering of events in a distributed system. Communications of the ACM 21, 558–565 (1978)
2. Schneider, F.B.: Implementing fault-tolerant services using the state machine approach: A tutorial 22, 299–319 (1990)
3. Kotla, R., Alvisi, L., Dahlin, M., Clement, A., Wong, E.: Zyzzyva: Speculative byzantine fault tolerance. ACM Transactions on Computer Systems 27, 1–39 (2009)
4. Abd-El-Malek, M., Ganger, G.R., Goodson, G.R., Reiter, M.K., Wylie, J.J.: Fault-scalable byzantine fault-tolerant services. In: SOSP 2005: Proceedings of the Twentieth ACM Symposium on Operating Systems Principles, pp. 59–74. ACM, New York (2005)
5. Castro, M., Liskov, B.: Practical byzantine fault tolerance and proactive recovery. ACM Transactions on Computer Systems 20, 398–461 (2002)
6. Cowling, J., Myers, D., Liskov, B., Rodrigues, R., Shrira, L.: HQ replication: a hybrid quorum protocol for byzantine fault tolerance. In: OSDI 2006: Proceedings of the 7th Symposium on Operating Systems Design and Implementation, pp. 177–190. USENIX Association, Berkeley (2006)
7. Guerraoui, R., Knežević, N., Quéma, V., Vukolić, M.: The next 700 bft protocols. In: EuroSys 2010: Proceedings of the 5th European Conference on Computer Systems, pp. 363–376. ACM, New York (2010)
8. Aguilera, M.K., Delporte-Gallet, C., Fauconnier, H., Toueg, S.: Thrifty Generic Broadcast. In: Herlihy, M.P. (ed.) DISC 2000. LNCS, vol. 1914, pp. 268–283. Springer, Heidelberg (2000)

9. Lamport, L.: Generalized consensus and paxos. Technical report, Microsoft Research MSR-TR-2005-33 (2005)
10. Pedone, F., Schiper, A.: Handling message semantics with generic broadcast protocols. Distributed Computing 15, 97–107 (2002)
11. Raykov, P., Schiper, N., Pedone, F.: Byzantine fault-tolerance with commutative commands. Technical report, University of Lugano (2011),
    http://www.inf.usi.ch/faculty/pedone/Paper/2011/2011OPODIS-full.pdf
12. Rivest, R.L., Shamir, A., Adleman, L.: A method for obtaining digital signatures and public-key cryptosystems. Communications of the ACM 26, 96–99 (1983)
13. Bellare, M., Canetti, R., Krawczyk, H.: Keying Hash Functions for Message Authentication. In: Koblitz, N. (ed.) CRYPTO 1996. LNCS, vol. 1109, pp. 1–15. Springer, Heidelberg (1996)
14. Fischer, M., Lynch, N., Paterson, M.: Impossibility of distributed consensus with one faulty process. Journal of the ACM 32, 374–382 (1985)
15. Chandra, T.D., Toueg, S.: Unreliable failure detectors for reliable distributed systems. Journal of the ACM 43, 225–267 (1996)
16. Ben-Or, M.: Another advantage of free choice (extended abstract): Completely asynchronous agreement protocols. In: PODC 1983: Proceedings of the Second Annual ACM Symposium on Principles of Distributed Computing, pp. 27–30. ACM, New York (1983)
17. Dwork, C., Lynch, N., Stockmeyer, L.: Consensus in the presence of partial synchrony. Journal of the ACM 35, 288–323 (1988)
18. Toueg, S.: Randomized byzantine agreements. In: PODC 1984: Proceedings of the Third Annual ACM Symposium on Principles of Distributed Computing, pp. 163–178. ACM, New York (1984)
19. Cachin, C., Kursawe, K., Petzold, F., Shoup, V.: Secure and Efficient Asynchronous Broadcast Protocols. In: Kilian, J. (ed.) CRYPTO 2001. LNCS, vol. 2139, pp. 524–541. Springer, Heidelberg (2001)
20. Lamport, L.: Lower bounds for asynchronous consensus. Distributed Computing 19, 104–125 (2006)
21. Herlihy, M.P., Wing, J.M.: Linearizability: a correctness condition for concurrent objects. ACM Trans. Program. Lang. Syst. 12, 463–492 (1990)
22. Malkhi, D., Reiter, M., Lynch, N.: A correctness condition for memory shared by byzantine processes (1998) (unpublished manuscript)
23. Aiyer, A.S., Alvisi, L., Bazzi, R.A., Clement, A.: Matrix Signatures: From MACs to Digital Signatures in Distributed Systems. In: Taubenfeld, G. (ed.) DISC 2008. LNCS, vol. 5218, pp. 16–31. Springer, Heidelberg (2008)
24. Kotla, R., Dahlin, M.: High-throughput byzantine fault tolerance. In: International Conference on Dependable Systems and Networks, DSN (2004)

Partially Non-Preemptive Dual Priority Multiprocessor Scheduling

Chiahsun Ho and Shelby H. Funk

University of Georgia
Department of Computer Science
Athens, GA 30602
{ho,shelby}@cs.uga.edu

Abstract. We propose a partially non-preemptive dual priority scheduling algorithm (PNPDP) for multiprocessors. In dual priority scheduling, each task has two fixed priorities. When a job is released, it executes at its task's lower priority. After some fixed amount of time, its priority is promoted. Our approach is to prevent lower priority jobs from preempting one another. We use the tasks' Worst Case Response Times to determine when a promotion must occur in order to guarantee all deadlines will be met. During execution, this promotion time is adjusted to extend non-preemptive execution of lower priority tasks whenever possible. Tasks executing at their promoted priorities are scheduled using preemptive fixed priority (FP) scheduling algorithm. Experimental results demonstrate that this approach reduces the preemption and migration overheads by as much as 90%. Moreover we found that many FP-unschedulable task sets are PNPDP-schedulable.

Keywords: Dual Priority Scheduling, Multiprocessor Scheduling, Procrastination, Delayed preemption, Cooperative Scheduling.

1 Introduction

Consider the scheduling of hard real-time systems on multiprocessor platforms. Such systems may be scheduled with or without allowing preemptions. Completely prohibiting preemptions can make schedulability impossible — particularly if some jobs have very long execution times. On the other hand, using a global scheduling strategy can dramatically degrade schedulibility [6]. Hence, it is important to develop algorithms that avoid preemptions as much as possible without causing deadline misses.

Often real-time jobs execute repeatedly and at regular intervals. We call such sequences periodic tasks. In general, there are two distinct strategies for scheduling task sets: fixed-priority and dynamic-priority. In a fixed-priority scheduling scheme, each task is assigned a priority level and all jobs generated by a given task are executed at that task's priority. In a dynamic-priority scheme, the tasks' priorities may change over time. The flexibility of using dynamic priorities may increase the likelihood that all deadlines will be met. However, maintaining the

A. Fernández Anta, G. Lipari, and M. Roy (Eds.): OPODIS 2011, LNCS 7109, pp. 343–356, 2011.
© Springer-Verlag Berlin Heidelberg 2011

priorities of the tasks adds to the complexity of the scheduling algorithm. Hence, fixed-priority scheduling is often preferred in practice.

In dual priority scheduling, each task has two priorities – a high priority and a low priority. When a task releases a new job, it is initially assigned the lower priority. If the job does not complete execution within a specified amount of time, the job's priority is reassigned to the higher priority level. Because each task has exactly two priority levels, we observe that dual priority scheduling is in some sense the least dynamic of the dynamic priority scheduling approaches[1].

This paper considers a variation of dual priority scheduling for multiprocessor platforms. With the advent of multicore processors, the study of multiprocessor systems has become increasingly important. Multiprocessor scheduling is particularly difficult in the presence of hard real-time constraints. Real-time scheduling algorithms that are known to perform very well on uniprocessor systems, such as Earliest Deadline First (EDF) and rate monotonic (RM), do not perform as well on multiprocessor platforms. As a result, the real-time scheduling community has devoted quite a bit of attention to multiprocessor scheduling recently.

We consider a partially non-preemptive variation of dual priority scheduling algorithm for multiprocessor platforms. Our variation applies two significant changes to the standard dual priority algorithm. First, we only allow tasks to preempt one another when executing at their higher priority level. In standard dual priority scheduling, the highest-priority jobs are selected to execute at all times. Thus, preemptions may occur either when a task releases a new job (if its lower priority level is higher than that of some executing job) or when a task's priority is promoted to its higher level. Our partially non-preemptive dual priority (PNPDP) approach does not allow tasks to preempt one another when they are executing at their lower priority level. Thus, preemptions are deferred and can occur *only* when a tasks priority gets promoted. Second, we delay promotion times of tasks that execute at their lower priority level. Because tasks cannot preempt unless they are executing at their higher priority level, this delay of priority promotion also delays preemptions.

The contributions of this paper are as follows:

- We introduce a variation of dual priority scheduling for multiprocessor platforms that prevents tasks from initiating preemptions before they are promoted.
- We present an online method of delaying priority promotion times.
- We analytically demonstrate that any system that meets all deadlines using standard preemptive fixed priority scheduling will also meet all deadlines using the PNPDP approach.
- We experimentally demonstrate that this approach can significantly reduce the context switch overheads (i.e., preemptions and migrations) as compared to both fixed priority and dual priority schedules.
- We demonstrate that the PNPDP approach can sometimes prevent deadline misses that might occur in a standard fixed priority system.

---

[1] When considered in terms of *jobs*, EDF might be considered less dynamic. However, there are advantages to managing priorities at the task level.

The remainder of this paper is organized as follows. Section 2 presents other research on dual priority scheduling and delayed preemption strategies. Section 3 introduces our model and notation in more detail. Section 4 analytically proves that the validity of the PNPDP scheduling strategy. Section 5 experimentally demonstrates that this strategy can significantly reduce scheduling overheads and improve schedulability. Finally, Section 6 provides some concluding remarks and ideas for further research.

2   Related Work

Because we do not allow tasks to initiate preemptions when executing at their low priority level, the PNPDP algorithm has elements of both the dual priority and deferred preemption scheduling strategies. Below, we discuss some important results pertaining to each of these approaches.

2.1   Uniprocessor Scheduling Results

In 1995, Davis et al. [10] presented dual priority scheduling for uniprocessors in order to improve aperiodic task response times. In [10], there are three distinct priority ranges – upper, middle and lower. Periodic tasks are assigned priorities in the upper and lower ranges, and aperiodic tasks are assigned priorities in the middle range. Periodic tasks initially execute at lower priority level. If a job does not complete execution within a specified amount of time, its priority is promoted to the upper level.

Jejurikar et al. [16] extended the uniprocessor dual priority scheduling approach to reduce power consumption. In their model, task execution is procrastinated when the processor is in the sleep state. The algorithm extends the sleep state as long as possible while still ensuring every task still meets its deadline. Their goal is to reduce the number of transitions from sleep state to busy state as these transitions consume energy.

Gopalakrishnan et al. [13] determined utilization bounds for EDF and RM scheduling with deferred preemption. They assume tasks are scheduled in discrete quanta and can only be preempted at quantum boundaries. They developed tests to check if a task system can be scheduled with a given quantum length using either EDF or RM.

Bril et al. [8] explore worst-case response time (WCRT) analysis of uniprocessor fixed priority systems using a deferred preemption strategy. Their work explores various types of overhead resulting from preemption (such as cache misses and reloads). They analyze the WCRT of a higher priority job which may be blocked by a lower priority job.

Baruah [5] and Yao [21] both analyze systems that alternate between preemptive and non-preemptive scheduling, providing a balance between feasibility and overheads on uniprocessor platforms. In both cases, the authors present analysis finding the longest duration of a non-preemptive interval that does not cause deadlines to be missed. In [5], Baruah considers the EDF scheduling of sporadic tasks. In [21], Yao et al. consider fixed priority scheduling of periodic tasks.

2.2   Multiprocessor Scheduling Results

All of the above approaches apply only to uniprocessor platforms. Recent research has also considered dual priority scheduling and delayed preemption approaches on multiprocessor platforms.

Davis, et al. [12] introduce the fixed priority with zero laxity algorithm (FPZL). We say a job has zero laxity if its remaining execution time is equal to the amount of time that remains before its deadline. Any such job must execute immediately in order to avoid missing its deadline. FPZL schedules tasks in a standard fixed priority manner unless a job's laxity is zero. All zero laxity jobs are given the highest possible priority. Thus, FPZL may be viewed as a dual priority algorithm where promotion times occur only when not executing a job is certain to result in a missed deadline. FPZL is a fully preemptive algorithm that was developed solely to correct fixed priority schedules in which jobs may miss their deadlines.

Tumeo et al. [20] consider deferred preemption on multiprocessor platforms using a global fixed priority scheduling algorithm. They consider systems that execute both periodic and aperiodic tasks on FPGA multiprocessor platforms. Similar to [10], Tumeo et al. use three priority ranges. In addition, each task is assigned to a particular processor. While tasks can execute on any processor prior to promotion time, they must execute on their assigned processor once they get promoted. As a result, priority promotion times can be determined using uniprocessor WCRT analysis.

Banus et al. [4] also introduce a hybrid architecture combining global and partitioned scheduling for multiprocessor platforms. Like [20], their system partitions the higher priority tasks onto the processors and uses uniprocessor WCRT analysis to determine priority promotion times. Middle and lower priority tasks are scheduled using the earliest promotion time first (EPF) strategy, a dynamic scheduling approach. They introduce an acceptance test for aperiodic tasks with deadlines. Their aim is to schedule as many hard aperiodic tasks as possible while still trying to minimize the response times of the soft aperiodic tasks (which have no deadlines).

Both of the above approaches focus on improving the response time of aperiodic jobs. By contrast, we aim to reduce overhead due to preemptions. Therefore, we employ a non-preemptive approach until a task's priority is promoted. Furthermore, we allow tasks to migrate at any point in time (even after a job is promoted). With a partitioned approach, WCRT values are determined using uniprocessor analysis. Instead, we determine WCRT values using multiprocessor analysis developed by Guan et al. [14]. Partitioning tasks onto processors is similar to the bin-packing problem, which is known to be NP-complete. Therefore, both of the above approaches may be unable to schedule task sets that could be scheduled using a global approach. Finally, none of the above approaches employ the online adjustment of priority promotion times.

3   Model and Definition

Real-time processes often recur at regular intervals. Hence, the periodic [18] and sporadic task models have proven very useful for the modeling and analysis of real-time systems. We consider the scheduling tasks[2] on multiprocessor platforms comprised of m identical processors.

Let $\tau \equiv \{T_1, T_2, \ldots, T_n\}$ be a collection of n tasks. Each task T_i is characterized by six parameters: worst case execution requirement (e_i), period (p_i), two distinct priorities ($\pi_{i,high}$ and $\pi_{i,low}$), relative deadline (D_i), and promotion offset (Λ_i). Each task T_i generates an infinite sequence of jobs $T_{i,0}, T_{i,1}, \ldots, T_{i,k}, \ldots$. A periodic task T_i generates each job $T_{i,k}$ at time $a_{i,k} = k \cdot p_i$, for all non-negative integers k. The sporadic task model allows tasks to diverge from the strict arrival pattern. Hence, consecutive arrivals of a sporadic task T_i occur *at least* p_i time units apart – i.e., $a_{i,0} \geq 0$ and $a_{i,k+1} \geq a_{i,k} + p_i$ for $k > 0$. Each job $T_{i,k}$ needs to execute for e_i units of time by its deadline of $d_{i,k} = a_{i,k} + D_i$. If $p_i = D_i$, we say T_i has an implicit deadline and use the notation $T_i = (p_i, e_i)$.

Each task T_i is assigned two priorities, $\pi_{i,high}$ and $\pi_{i,low}$, and a promotion offset Λ_i. Priorities are assigned in two bands — i.e., for every pair of tasks T_i and T_j, $\pi_{i,high}$ is higher than $\pi_{j,low}$. We assume tasks are indexed according to their higher priority levels with $\pi_{1,high}$ being the highest and $\pi_{n,high}$ being the lowest. When T_i releases a new job, it is initially assigned priority level $\pi_{i,low}$. If it does not complete execution within Λ_i time units, its priority is promoted to $\pi_{i,high}$. When task T_i generates a job $T_{i,k}$, we let $\lambda_{i,k}$ denote this job's *latest promotion time*. Thus, $\lambda_{i,k} = a_{i,k} + \Lambda_i$. At any time t, we let $\pi_i(t)$ denote T_i's current priority.

We assume a preemptive schedule which permits migration. Thus, jobs generated by higher-priority tasks can preempt (interrupt) a currently executing lower-priority job and the preempted job can restart on any processor. However, our scheduling strategy will not allow a higher-priority task to preempt a lower-priority task immediately. Whenever the scheduler executes a lower priority job while a higher priority job is waiting, we say a *priority inversion* occurs. Priority inversions must be limited because they could lead to deadline misses. However, as long as no deadline will be missed, we can use priority inversions to our advantage.

While our goal is to reduce preemptions, we recognize that some preemptions must occur in order to ensure all deadlines are met. Our aim is to delay preemptions until further delay may cause a deadline to be missed. We show that permitting priority inversions for limited periods of time may remove the need for at least some preemptions. First, though, we describe our strategy and demonstrate that it is correct.

---

[2] The term "task" with no modifier indicates that the task may be either periodic or sporadic.

4 Partially Non-Preemptive Dual Priority Scheduling Strategy

As described above, while both promoted and non-promoted tasks can be pre-empted, only promoted tasks can *initiate* preemptions. If task T_i is promoted at time t, it will initiate a preemption if and only if $\pi_{i,high}$ is higher than $\pi_j(t)$ for some executing task T_j.

We use WCRT analysis to determine the promotion offset Λ_i for each task T_i. Lehoczky et al. [17] developed the Time Demand Analysis (TDA) for fixed priority schedules on uniprocessor platforms. This method was extended for mul-tiprocessor platforms [1,7,9,14,19]. We use the approach presented by Guan et al. [14], which examines the maximum interference from higher priority jobs more precisely than the earlier extensions and applies to sporadic task sets with unconstrained deadlines.

Below we demonstrate that our PNPDP approach will not introduce dead-line misses if τ is FP-schedulable. We first consider PNPDP schedule without adjusting priority promotion times. In Section 4.1, we demonstrate that priority promotion times may be delayed.

Lemma 1. *Assume τ is FP-schedulable when using all tasks' high priority lev-els. For each task T_i, let R_i be T_i's WCRT in the FP schedule. If $\Lambda_i = D_i - R_i$ for all $T_i \in \tau$ then τ is PNPDP schedulable regardless of how lower priority levels are assigned.*

Proof. Consider the sporadic task system τ' where $D'_i = R_i$, $p'_i = p_i$ and $e'_i = e_i$ for all tasks T_i. Let S be any PNPDP schedule of τ. Construct a corresponding FP schedule S' of τ' where each job $T_{i,k}$ arrives Λ_i time units later in S' (i.e., $a'_{i,k} = a_{i,k} + \Lambda_i$) as illustrated in Figure 1. By assumption, the WCRT of each job T_i is R_i. Therefore, all jobs meet their deadlines in the schedule S' (because $D'_i = R_i$). Note that

$$d_{i,k} = a_{i,k} + D_i = (a_{i,k} + \Lambda_i) + (D_i - \Lambda_i) = a'_{i,k} + R_i = d'_{i,k}.$$

Hence, each job in S' has the same deadline as the corresponding job in S, but has Λ_i fewer time units to execute.

Now let $x_{i,k}$ and $y_{i,k}$ be the amount of time job $T_{i,k}$ executes in schedule S before and after being promoted, respectively. Create a new schedule S'' of τ' where each job $T_{i,k}$ executes for $y_{i,k}$ time units. Clearly, $y_{i,k} \leq e_i$, so no execution times are longer in S'' than in S'. By the predictability of FP scheduling [15], all jobs meet their deadlines in the schedule S''.

Let $S|_{high}$ be the high-priority portion of the schedule S — i.e., $S|_{high}$ replaces any non-promoted execution with idle time. Note that $S|_{high}$ executes each job $T_{i,k}$ for $y_{i,k}$ time units at priority $\pi_{i,high}$. Moreover, because all tasks' low priority levels are lower than any task's high priority level, none of the non-promoted execution in S can interfere with the promoted execution. Hence, $S|_{high}$ and S'' are precisely the same schedule, so all jobs meet their deadlines in $S|_{high}$ (and also S). □

4.1 Adjusting Task Promotion Times

Intuitively, we expect WCRT to decrease as execution time decreases. We use this intuition to incorporate an online adjustment to tasks' priority promotion times.

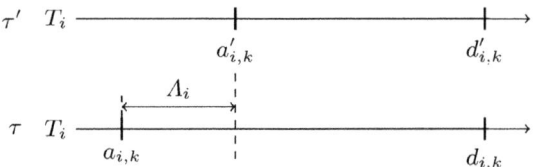

Fig. 1. FP and PNPDP schedule comparison

All the TDA methods determine the maximum amount of time that a task T_i will be forced to wait while higher priority tasks execute. We call this T_i's *interference time*, denoted I_i. Then T_i's WCRT is simply the amount of time T_i executes plus its worst-case interference time. I.e., $R_i = I_i + e_i$. Note that if a job $T_{i,k}$ executes for ξ_i time units before its promotion time, its interference time will not increase and its remaining execution time is decreased by ξ_i time units. Hence, we can delay a job's promotion time if it executes before being promoted.

We first show that delaying the promotion time of a job will not cause that job to miss its deadline. We then show that the delayed promotion will not cause any other jobs to miss their deadlines.

Lemma 2. *Assume $T_{i,k}$ will be able meet its deadline if its priority is promoted at time $\lambda_{i,k}$ and it does not execute during $[a_{i,k}, \lambda_{i,k})$. Then it will still be able to meet its deadline if its promotion time is increased at a rate of 1 whenever $T_{i,k}$ executes.*

Proof. Let $I_{i,k}$ be the amount of time that $T_{i,k}$ will have to wait for higher priority jobs if it executes for e_i time units after being promoted at time $\lambda_{i,k}$. Because $T_{i,k}$ will meet its deadline we know that $\lambda_{i,k} + I_{i,k} + e_i \le d_{i,k}$.

Now consider what happens if $T_{i,k}$'s remaining work decreases. Because changing T_i's execution has no impact on the behavior of higher priority tasks, its worst case interference cannot be larger than it would have been had T_i executed for the full e_i time units. By definition, I_i is the maximum amount of time the $(i-1)$ highest priority tasks can keep all m processors busy during an interval $[\lambda_{i,k}, d_{i,k})$. Therefore, if T_i's execution decreases by ξ_i time units, the remaining $(e_i - \xi_i)$ time units cannot have a response time larger than $I_i + (e_i - \xi_i)$. Let $C_{i,k}$ denote $T_{i,k}$'s completion time. Then

$$C_{i,k} \le (\lambda_{i,k} + \xi_i) + (e_i - \xi_i) + I_i$$
$$= \lambda_{i,k} + I_{i,k} + e_i$$
$$\le d_{i,k}$$

so T_i meets its deadline.   □

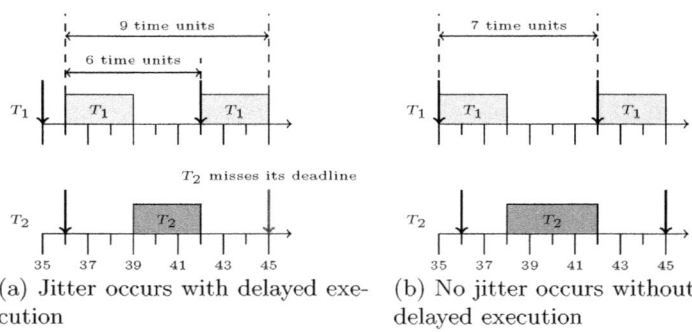

(a) Jitter occurs with delayed exe-   (b) No jitter occurs without
cution   delayed execution

Fig. 2. Jitter may increase the response time of lower priority tasks

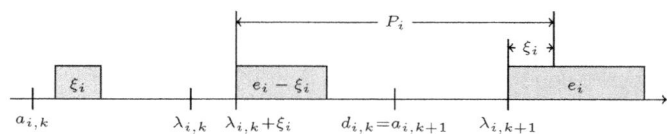

Fig. 3. Extending promotion time by ξ_i time units

One issue that we need to consider when changing promotion times is a jitter in the release times. Such release time jitter can potentially cause the system to fail due to a task imposing extra interference on lower priority tasks as the following example illustrates.

Example 1. Consider two tasks $T_1 = (7,3)$ and $T_2 = (9,4)$. Because T_1 should execute for at most 3 time units in any interval of length 7, we see that $R_2 = 7$. If, however, T_1 delays execution for 1 time unit as illustrated in Fig. 2, it may execute for 6 out of 7 consecutive time units, thereby causing T_2 to miss a deadline.   □

The jitter issue illustrated in the above example occurs as a result of T_1's delaying its start time and executing for the full 3 time units. If T_1 had executed for only 2 time units (i.e., if execution time had been decreased by the duration of the delay), T_2 would have met its deadline.

If priority promotions are not adjusted then each task T_i will trigger priority promotions exactly p_i time units apart. However, our adjustment strategy may cause a task T_i to have less time between consecutive promotion times. We need to ensure that this adjustment cannot increase the response time of lower priority jobs as illustrated in the example above. The following lemma addresses this issue.

Lemma 3. *Assume $T_{i,k}$'s priority promotion time is delayed as described in Lemma 2. Then the maximum demand that $T_{i,k}$ and $T_{i,k+1}$ can impose during the p_i time units after $T_{i,k}$'s promotion time is e_i.*

Proof. Assume $T_{i,k}$ executes for ξ_i time units at priority $\pi_{i,low}$, and let $\lambda'_{i,k}$ be $T_{i,k}$'s delayed promotion time. Then $\lambda'_{i,k} = \lambda_{i,k} + \xi_i = a_{i,k} + \Lambda_i + \xi_i$.

Clearly, $T_{i,k}$ executes for at most $e_i - \xi_i$ time units during the interval $[\lambda'_{i,k}, d_{i,k})$. Let $\lambda'_{i,k+1}$ be $T_{i,k+1}$'s priority promotion time (which may or may not be delayed) and let $\xi_{i,k+1}$ be the amount of time $T_{i,k+1}$ executes during the interval $[\lambda'_{i,k}, \lambda'_{i,k} + p_i)$. Then,

$$
\begin{aligned}
\xi_{i,k+1} &\leq (\lambda'_{i,k} + p_i) - \lambda'_{i,k+1} \\
&\leq (\lambda'_{i,k} + p_i) - (a_{i,k+1} + \Lambda_i) \\
&= (a_{i,k} + \Lambda_i + \xi_i + p_i) - (a_{i,k+1} + \Lambda_i) \\
&= \xi_i + (p_i - (a_{i,k+1} + a_{i,k})) \\
&\leq \xi_i.
\end{aligned}
$$

Therefore, $T_{i,k}$ and $T_{i,k+1}$ have a combined demand of at most $(e_i - \xi_i) + \xi_i = e_i$ time units during the interval $[\lambda'_{i,k}, \lambda'_{i,k} + p_i)$. Figure 3 illustrates this result.

We now prove that if Λ_i is determined using the TDA method introduced by Guan et al. [14], then delaying priority promotion times will not cause any deadline misses.

Theorem 1. *Assume τ is any task set and TDA analysis determines each task T_i has WCRT $R_i \leq D_i$. For each task T_i, let Λ_i be the promotion offset such that $\Lambda_i \leq D_i - R_i$. If τ is scheduled in a PNPDP manner and promotion times are delayed as described in Lemma 2, then τ will meet all deadlines.*

Proof. Let I_i be T_i's worst case interference from higher priority tasks in the FP schedule as determined by Guan's TDA method. Let I'_i and I''_i be T_i's worst case interference from higher priority tasks while executing at priority $\pi_{i,high}$ in the PNPDP schedule without and with adjusting priority promotion times, respectively. By Lemma 2, we know that $I'_i \leq I_i$. We now show that $I''_i \leq I'_i$ using induction.

Clearly, $I''_1 = I'_1 = 0$. Assume, $I''_i \leq I'_i$ for all $i \leq k$. We need to show that $I''_{k+1} \leq I'_{k+1}$.

All of the TDA techniques including Guan's find the WCRT by determining the maximum possible interference that can be caused by each higher priority task T_i during an interval of length x using the formula $\lfloor \frac{x}{p_i} \rfloor \cdot e_i + \delta \cdot e_i$, where δ measures the fraction of the worst case demand of some job of T_i that is not completely contained within an interval of length x. If τ_i arrives before the beginning of the interval we say $\delta_i \cdot e_i$ is T_i's "carry-in". If T_i has a deadline after the end of the interval we say $\delta_i \cdot e_i$ is T_i's "carry-out".

As discussed above, task T_i will execute at promoted proirity for at most e_i of the p_i time units after promotion even if promotion times are adjusted. By our induction hypothesis, for each $i \leq k$ task T_i has a worst case completion time in the PNPDP schedule with adjusted promotion times of at most $e_i + I''_i \leq e_i + I'_i$. I.e., adjusting promotion times cannot increase T_i's WCRT. Therefore, T_i's carry-in in the PNPDP schedule cannot increase either. On the other hand, T_i's carry-out *can* increase if its promotion time was delayed. However, the increased carry-out cannot be more than the decreased demand during the delayed priority promotion[3] Therefore, the maximum amount of time each task T_i can interfere with task T_{k+1} when promotion times can be delayed, cannot be more than the interference when using the original promotion times — i.e., $I''_{k+1} \leq I'_{k+1} \leq I_{k+1}$, as desired.

We now show that every task completes execution before its deadline. Let R'_i be T_i's WCRT in the PNPDP schedule with online priority promotion adjustment. Then

$$
\begin{aligned}
R'_i &= \Lambda_i + e_i + I''_i \\
&\leq \Lambda_i + e_i + I_i \\
&\leq D_i - R_i + e_i + I_i \\
&= D_i,
\end{aligned}
$$

where the last step follows because $R_i = e_i + I_i$. Thus every task meets its deadline.   □

In summary, we present a variation of the dual priority scheduling algorithm for multiprocessor platforms. This algorithm differs from standard dual priority scheduling in two ways.

- In an effort to reduce preemptions. No job executing at its lower (non-promoted) priority will preempt another job. Hence, we do not always execute m highest priority jobs.
- Whenever a non-promoted job executes, we delay its promotion time accordingly.

In this section, we have proved that neither of these changes will cause deadlines to be missed. In particular, if τ will meet all deadlines in a fixed priority schedule then the PNPDP schedule with online adjustment of priority promotions will also meet all deadlines. Below, we experimentally demonstrate the our strategy can significantly reduce preemption overhead. Additionally, we demonstrate that sometimes the PNPDP scheduling algorithm can successfully schedule task sets even when the fixed priority schedule misses some deadlines.

5   Experimental Results

We use Baker's method for randomly generating task sets [3] for platforms containing 2, 4, 8, 16 and 32 identical processors. While the above analysis applies to task sets with unconstrained deadlines, in this experiment we assume deadlines are equal to periods. We use the maximum utilization to characterize a task set is light, medium or heavy (e.g., a task set with maximum utilization less

---

[3] This follows from reasoning similar to that used to prove Lemma 3.

than 30% is considered to be light; a task set with maximum utilization greater than 60% is considered to be heavy). Heavy task sets potentially have higher preemption overheads than light task sets. For every light, medium, and heavy scenario with a given number of processors, we randomly generate 1,000 task sets with total utilization uniformly distributed from $0.025 \times m$ to $0.975 \times m$.

After generating task sets, we assigned the priority of each task using four different strategies: Rate-Monotonic [18], TkC [11], Slack-Monotonic [2], and OPA [11]. For standard dual priority (SDP) and PNPDP, the lower priorities and the higher priorities both use the same order. We compute WCRT by using Guan's TDA analysis for multiprocessor platforms [14] for all four priority assignment methods. Because this is a pessimistic calculation, we may have $R_i > D_i$ even though the task will never miss a deadline. In this case, we simply set R_i to be equal to D_i. Hence, these tasks only execute at their promoted priority level.

Because task periods can greatly impact the frequency of preemptions, we consider both harmonic task sets and randomly generated periods. Harmonic task sets have periods uniformly selected from the set $\{4, 8, 16, 32, 64, 128, 256, 512, 1024, 2048\}$. Randomly generated periods are selected within the interval $[10, 200]$.

For each priority assignment of each task set, we simulate the schedule using SDP, PNPDP, fixed priority (FP) and fixed priority until zero laxity (FPZL) and divided the task sets into two categories:

- Type I contains the prioritized task sets that are FP-schedulable — i.e., FP simulation of these task sets met all deadlines at the given priority assignment.
- Type II contains the prioritized task sets that are not FP-schedulable.

In each simulation we kept track of the number of preemptions and migrations. As migration costs have been shown to be similar to preemption costs [6], we use a simple count to determine the overhead of each algorithm.

For the Type I task sets, we compare the overheads of both SDP and PNPDP to FP[4]. An average savings of 100% means all of the preemptions and migrations in the fixed-priority schedule were avoided. By contrast, average savings of 0% means that the SDP or PNPDP algorithm have the same number of preemptions and migrations as the FP schedule.

Due to space limitations, we only present the result for 4 processors. Results are similar for other platforms as well. Figure 4 illustrates the percentage of task sets able to achieve a given level of overhead reduction. For instance, 20% of the task sets with harmonic periods reduced overhead by at least 10% when scheduled by SDP under OPA priority assignment. By contrast, more than 90% of task sets achieve at least 10% savings when scheduled using PNPDP. We see that the PNPDP strategy dramatically reduces overheads in all cases, with higher reductions for randomly generated periods.

---

[4] When FP schedules a system successfully, both FP and FPZL generate the same schedule. Hence, we do not consider algorithm FPZL with the Type I task sets.

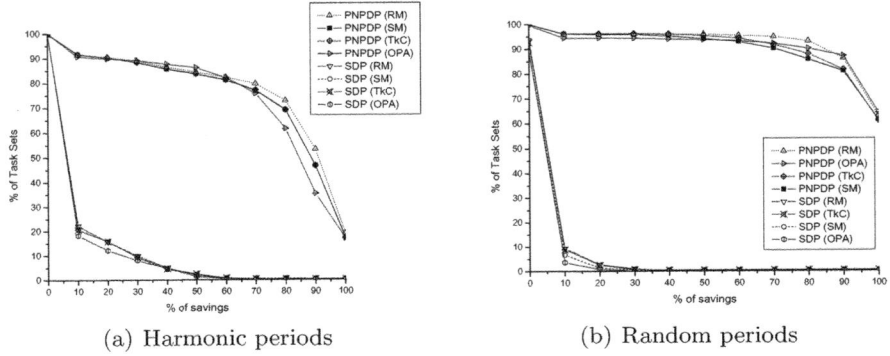

(a) Harmonic periods　　　　　　　　　　(b) Random periods

Fig. 4. Overhead savings of PNPDP and SDP schedules as compared to FP schedule using all four priority settings

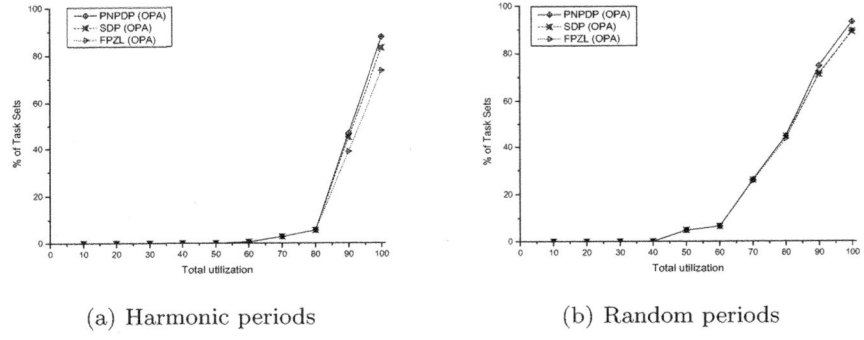

(a) Harmonic periods　　　　　　　　　　(b) Random periods

Fig. 5. Feasibility improvement of PNPDP, SDP and FPZL as compared to FP on 8 processors under OPA priority assignment

For the Type II task sets, we also examined the degree to which priority adjustments can help meet deadlines that are missed in the FP schedule. We executed these task sets using SDP, PNPDP and FPZL. We found that there is a huge feasibility improvement in all cases, with PNPDP having slightly better improvements than the other two algorithms. Figure 5, shows the total percentage of infeasible task sets that become feasible using these algorithms under the OPA priority assignment. These improvements are typical of all Type II scenarios. For instance, the feasibility improvement for task sets with harmonic periods become significant when the total utilization is over 80%. By contrast, the feasibility improvement for task sets with random periods become significant when the total utilization is over 60%.

6   Conclusion

We consider a multiprocessor scheduling algorithm using a partially non-preemptable dual priority strategy. Experimental results indicate that this approach incurs significantly less overhead than both FP scheduling and standard dual priority scheduling under a variety of priority assignment schemes. We analytically prove that PNPDP scheduling will never cause a deadline to be missed that would not be missed under FP scheduling. In addition, we demonstrate that systems that miss deadlines in an FP schedule might be schedulable using a PNPDP approach.

In future we plan to consider methods of further delaying the priority promotion times. Specifically, we would like to analyze how much work each task performs at its low priority level and calculate R_i using the worst-case *promoted* execution time. We would also like to explore priority assignment strategies that can reduce overhead even more. In particular we plan to examine different methods of assigning tasks' lower priorities.

Acknowledgments. The authors wish to thank the reviewers for their helpful comments.

References

1. Andersson, B., Jonsson, J.: Fixed-priority preemptive multiprocessor scheduling: to partition or not to partition. In: IEEE Embedded and Real-Time Computing Systems and Applications (RTCSA), Cheju Island, South Korea, pp. 337–346 (2000)
2. Andersson, B.: Global Static-Priority Preemptive Multiprocessor Scheduling with Utilization Bound 38%. In: Baker, T.P., Bui, A., Tixeuil, S. (eds.) OPODIS 2008. LNCS, vol. 5401, pp. 73–88. Springer, Heidelberg (2008)
3. Baker, T.: An analysis of fixed-priority schedulability on a multiprocessor. Real-Time Systems 32, 49–71 (2006)
4. Banus, J.M., Arenas, A., Labarta, J.: Extended global dual priority algorithm for multiprocessor scheduling in hard real-time systems. In: Euromicro Conference on Real-Time Systems WIP (ECRTS), Palma de Mallorca, Spain, pp. 13–16 (2005)
5. Baruah, S.: The limited-preemption uniprocessor scheduling of sporadic task systems. In: IEEE Real-Time Systems Symposium (RTSS), Miami, Florida, USA, pp. 137–144 (2005)
6. Bastoni, A., Brandenburg, B.B., Anderson, J.H.: An empirical comparison of global, partitioned, and clustered multiprocessor EDF schedulers. In: IEEE Real-Time Systems Symposium (RTSS), San Diego, CA, USA, pp. 14–24 (2010)
7. Bertogna, M., Cirinei, M.: Response-time analysis for globally scheduled symmetric multiprocessor platforms. In: IEEE Real-Time Systems Symposium (RTSS), Tucson, AZ, USA, pp. 149–160 (2007)
8. Bril, R.J., Lukkien, J.J., Verhaegh, W.F.J.: Worst-case response time analysis of real-time tasks under fixed-priority scheduling with deferred preemption revisited. In: Euromicro Conference on Real-Time Systems (ECRTS), Pisa, Italy, pp. 269–279 (2007)
9. Burns, A., Wellings, A.: Real-Time Systems and Programming Languages, 3rd edn. Addison-Wesley (2001)

10. Davis, R., Wellings, A.: Dual priority scheduling. In: IEEE Real-Time Systems Symposium (RTSS), Pisa, Italy, pp. 100–109 (1995)
11. Davis, R.I., Burns, A.: Priority assignment for global fixed priority pre-emptive scheduling in multiprocessor real-time systems. In: IEEE Real-Time Systems Symposium (RTSS), Washington, D.C., USA, pp. 398–409 (2009)
12. Davis, R.I., Burns, A.: FPZL schedulability analysis. In: IEEE Real-Time and Embedded Technology and Applications Symposium (RTAS), Chicago, IL, USA, pp. 245–256 (2011)
13. Gopalakrishnan, R., Parulkar, G.M.: Bringing real-time scheduling theory and practice closer for multimedia computing. In: ACM SIGMETRICS International Conference on Measurement and Modeling of Computer Systems (SIGMETRICS), Philadelphia, PA, USA, pp. 1–12 (1996)
14. Guan, N., Stigge, M., Yi, W., Yu, G.: New response time bounds for fixed priority multiprocessor scheduling. In: IEEE Real-Time Systems Symposium (RTSS), Washington, DC, USA, pp. 387–397 (2009)
15. Ha, R., Liu, J.W.S.: Validating timing constraints in multiprocessor and distributed real-time systems. In: Proceedings of the 14th International Conference on Distributed Computing Systems (ICDCS), Poznan, Poland, pp. 162–171 (1994)
16. Jejurikar, R., Gupta, R.: Procrastination scheduling in fixed priority real-time systems. In: Proceedings of the 2004 ACM SIGPLAN/SIGBED Conference on Languages, Compilers, and Tools for Embedded Systems (LCTES), Washington, DC, USA, pp. 57–66 (2004)
17. Lehoczky, J., Sha, L., Ding, Y.: The rate monotonic scheduling algorithm: Exact characterization and average case behavior. In: IEEE Real-Time Systems Symposium (RTSS), Santa Monica, California, USA, pp. 166–171 (1989)
18. Liu, C.L., Layland, J.W.: Scheduling algorithms for multiprogramming in a hard real-time environment. Journal of the ACM 20(1), 46–61 (1973)
19. Lundberg, L.: Multiprocessor scheduling of age constraint processes. In: IEEE Embedded and Real-Time Computing Systems and Applications (RTCSA), Hiroshima, Japan, pp. 42–47 (1998)
20. Tumeo, A., Branca, M., Camerini, L., Ceriani, M., Palermo, G., Ferrandi, F., Sciuto, D., Monchiero, M.: A dual-priority real-time multiprocessor system on FPGA for automotive applications. In: Proceedings of the Conference on Design, Automation and Test in Europe (DATE), Munich, Germany, pp. 1039–1044 (2008)
21. Yao, G., Buttazzo, G., Bertogna, M.: Bounding the maximum length of nonpreemptive regions under fixed priority scheduling. In: IEEE Embedded and Real-Time Computing Systems and Applications (RTCSA), Beijing, China, pp. 351–360 (2009)

Private Similarity Computation in Distributed Systems: From Cryptography to Differential Privacy*

Mohammad Alaggan[1], Sébastien Gambs[2], and Anne-Marie Kermarrec[3]

[1] Université Rennes 1 – IRISA, Rennes, France
[2] Université de Rennes 1 – INRIA/IRISA, Rennes, France
[3] INRIA Rennes Bretagne-Atlantique, Rennes, France

Abstract. In this paper, we address the problem of computing the similarity between two users (according to their profiles) while preserving their privacy in a fully decentralized system and for the passive adversary model. First, we introduce a two-party protocol for privately computing a threshold version of the similarity and apply it to well-known similarity measures such as the scalar product and the cosine similarity. The output of this protocol is only one bit of information telling whether or not two users are similar beyond a predetermined threshold. Afterwards, we explore the computation of the exact and threshold similarity within the context of differential privacy. Differential privacy is a recent notion developed within the field of private data analysis guaranteeing that an adversary that observes the output of the differentially private mechanism, will only gain a negligible advantage (up to a privacy parameter) from the presence (or absence) of a particular item in the profile of a user. This provides a strong privacy guarantee that holds independently of the auxiliary knowledge that the adversary might have. More specifically, we design several differentially private variants of the exact and threshold protocols that rely on the addition of random noise tailored to the sensitivity of the considered similarity measure. We also analyze their complexity as well as their impact on the utility of the resulting similarity measure. Finally, we provide experimental results validating the effectiveness of the proposed approach on real datasets.

Keywords: Privacy, similarity measure, homomorphic encryption, differential privacy.

1   Introduction

In the Web 2.0, more and more personal data are released by users (queries, social network, geolocated data...), which creates a huge pool of useful information to leverage in the context of search or recommendation for instance. In fully decentralized systems, tapping on the power of this information usually involves some kind of clustering process that relies on an exchange of personal data

---

* Supported by the ERC GOSSPLE project.

A. Fernández Anta, G. Lipari, and M. Roy (Eds.): OPODIS 2011, LNCS 7109, pp. 357–377, 2011.

(such as profiles) to compute similarity between users [2]. In this paper, we address the problem of *computing similarity between users while preserving their privacy* and *without relying on a central entity*. Dissociating the identifiers of users from their data, through the use of pseudonyms for instance, is clearly not sufficient to protect their privacy. In fact, just looking at these *Personal Identifiable Information* (PII) may sometimes be enough to infer the identity of the associated users thus causing a privacy breach [3,22,21]. Moreover, to preserve the fully distributed nature of such systems, no trusted third party (e.g. central server) should be required.

In this paper, we propose a protocol based on cryptographic primitives that computes the similarity between two user profiles (represented as vectors) in such a way that each user only learns the output of the similarity computation but not the profiles themselves. The novelty of our approach is twofold. First, considering well-known similarity metrics, namely scalar product and cosine similarity, we propose a two-party *threshold similarity* protocol for these metrics and prove its security against a passive adversary. Instead of revealing the exact value of the similarity, this protocol outputs only one bit of information stating whether or not two users are similar beyond a predetermined threshold. Compared to the exact similarity computation from which more information can be extracted, this protocol is more privacy-preserving in the sense that it reveals less information. While, we focus on the scalar product and the cosine similarity for illustration purpose, our method is generic enough to be applied to other similarity metrics.

Second, we go beyond the traditional cryptographic framework by analyzing the similarity computation within the context of *differential privacy* [9]. In a nutshell, *differential privacy* is an orthogonal and complementary notion to cryptography that, by adding random noise to the output of a function, provides strong privacy guarantees with respect to how well an adversary observing the output of the function can deduce the presence (or absence) of a specific item in a profile. We design a differentially private protocol for the exact and threshold similarity and analyze their impact with respect to utility. To the best of our knowledge, this is the first attempt to address differential privacy in the context of distributed similarity computation. More specifically, we first analyze the sensitivity of these similarity metrics in the context of a protocol computing exactly the similarity between two user profiles. Finally, we also study the impact of the differential privacy (which requires the addition of random noise) on the resulting *utility* of the similarity measure, both through a theoretical analysis and experimental validation.

The paper is organized as follows. Section 2 describes the system model and provides the required background. In Section 3, we introduce the threshold similarity protocol and prove its security with respect to a passive adversary. In Section 4, we describe differentially-private protocols for the exact and threshold similarity, while in Section 5, we provide a theoretical analysis of the impact on utility of the differentially-private protocol as well as experimental results. Finally, we briefly review related work in Section 6 before concluding.

2   Preliminaries

In this section, we describe the system model, the definitions of the similarity metrics considered, and the background in cryptography required in the context of our contributions.

System model. We consider a distributed system of n nodes, connected via an unstructured network [17]. (Each node is typically connected to $O(\log n)$ other nodes picked uniformly at random [4].) The nodes need to periodically run a clustering protocol that requires computing similarity between pairs of nodes. This semantic clustering can later be used to improve the search, provide content recommendation or personalized query expansion. Nodes are characterized by their profile representing their interests. For example, a node's profile can be a vector of items the associated user has tagged using a collaborative system [1] such as delicious[1]. We assume that for two different nodes A and B, their profiles S_A and S_B can be represented as binary vectors of size l, where l is the size of the domain. More precisely, $S_A = \{a_1, \ldots, a_l\}$ and $S_B = \{b_1, \ldots, b_l\}$, such that $a_i = 1$ if item i is in A's profile and 0 otherwise (b_i is defined similarly for the second node). For illustration purpose, we shall call the first node Alice and the second node Bob in the rest of the paper.

The profile is a personal and private information that should be protected, and therefore our main concern is *how to compute the similarity measure while preserving its privacy*. In this context, this means not revealing the content of the profile and restricting the possibility for an adversary to infer the presence or absence of a particular item in this profile. Moreover, besides the private computation of the similarity, we also assume the existence of a bidirectional anonymous lossless channel to break the link between a node's identity and its profile. Although, it is not the focus of this paper to detail how such a channel could be implemented in practice, we describe in Appendix A a simple implementation of this channel called *gossip-on-behalf*[2] that relies on the use of a third node acting as an anonymizer to break the link between the two nodes computing their similarity. Obviously, other implementations of the bidirectional anonymous channel are possible but they require non-trivial modifications of current anonymous communication networks [5,26,8] and are beyond the scope of this paper[3]. In order to guarantee a high level of anonymity, as measured for instance by the size of the anonymity set, it is also necessary to assume that the size of the network is sufficiently large ($n \gg 3$). Moreover, in order to avoid the possibility for an adversary to query several times the similarity computation with different forged profiles, it is also necessary to restrict to limit the use of a particular bidirectional anonymous channel (for instance to use it only once).

---

[1] http://delicious.com/

[2] The protocol described here is a modification of a protocol published earlier [4].

[3] However, see http://www.torproject.org/docs/hidden-services.html for a description of how to build an anonymous server that can be accessed by anonymous users within the network of the Tor project.

Similarity measures. Nodes aim at detecting the most similar other nodes (i.e. those which share similar interests). Thus, we assume the existence of similarity measures that can be used by the two nodes to quantify how similar they are. A *similarity measure* sim is a function that takes as input two sets S_A and S_B representing the profiles of users Alice and Bob and outputs a value in the range between 0 and 1 (i.e. $\mathsf{sim}(S_A, S_B) \in [0, 1]$), where 0 indicates that the sets are entirely different (the profiles have no items in common) while 1 means that the sets are identical (and therefore the users can be considered as sharing exactly the same interests).

The *cosine similarity* is commonly used to assess the similarity between two sets [4] and can be seen as a normalized overlap between the sets. Formally, it is defined as

$$\frac{|S_A \cap S_B|}{\sqrt{|S_A| \times |S_B|}} \ , \tag{1}$$

where S_A and S_B are the private sets of the first and second node respectively and $|S_A|$ and $|S_B|$ their corresponding sizes (i.e. the number of 1s in their profiles for binary vectors). The *size of the set intersection* between S_A and S_B (i.e. $|S_A \cap S_B|$) is equivalent to the *scalar product* in the case where the sets are represented as binary vectors. For instance, the scalar product of two vectors of length l, $a = (a_1, \cdots, a_\ell)$ and $b = (b_1, \cdots, b_\ell)$, is defined as $\sum_{i=1}^{\ell} a_i b_i$. Other similarity metrics can be considered such as the Jaccard index [16], but for the sake of clarity, we focus on the cosine similarity metric and the scalar product in the sequel.

Cryptographic background. In this paper, we only consider privacy against a computationally-bounded *passive adversary* (also sometimes called *semi-honest* or *honest-but-curious*) that can control a fraction of the nodes (see [14] for a formal cryptographic definition). Note that in this model (contrary to the active one), nodes do not misbehave and follow the recipe of the protocol. However, they may try to infer as much information as possible regarding the private inputs of other participants from the interactions and messages they have seen and recorded.

Definition 1 (Privacy – passive adversary [14]). *A protocol is said to be private with respect to passive adversary controlling a node (or a collusion of nodes), if this adversary cannot learn (except with negligible probability) more information from the execution of the protocol that it could from its own input (i.e. the inputs of the nodes he controls) and the output of the protocol.*

In our work, we rely on a cryptographic primitive known as *homomorphic encryption*, which allows to perform arithmetic operations (such as addition and/or multiplication) on encrypted values.

Definition 2 (Homomorphic cryptosystem). *Consider a public-key (asymmetric) cryptosystem where (1) $\mathsf{Enc}_{pk}(a)$ denotes the encryption of the message a under the public key pk and (2) $\mathsf{Dec}_{sk}(a) = a$ is the decryption of* *this message with the secret key[4] sk. A cryptosystem is* additively *homomorphic if there is an efficient operation* $\oplus$ *on two encrypted messages such that* $\mathsf{Dec}(\mathsf{Enc}(a) \oplus \mathsf{Enc}(b)) = a + b$. *Moreover, such an encryption scheme is called* affine *if there is also an efficient scalaring operation* $\odot$ *taking as input a ciphertext and a plaintext, such that* $\mathsf{Dec}(\mathsf{Enc}(c) \odot a) = c \times a$.

Besides, the elementary operations of addition and multiplication, more complex arithmetic operations can also be performed on the ciphertexts, such as for instance protocols for the comparison of integers [12,23]. These protocols take as input two encrypted integers and output whether or not they correspond to the same integer or which one is greater than the other, but without revealing the corresponding plaintexts (i.e. values of the integers).

Paillier's cryptosystem [25] is an instance of a homomorphic encryption scheme that is both additive and affine. Moreover, Paillier's cryptosystem is also *semantically secure* [14], which means that a computationally-bounded adversary cannot derive non-trivial information about the plain text m encrypted from the cipher text $\mathsf{Enc}(m)$ and the public key pk. For instance, a computationally-bounded adversary who is given two different cipher texts encrypted with the same key of a semantic cryptosystem, cannot even decide with non-negligible probability if the two cipher texts correspond to the encryption of the same plain text or not. This is because a semantically secure cryptosystem is by essence *probabilistic*, meaning that even if the same message is encrypted twice, the two resulting ciphertexts will be different except with negligible probability. In this paper, we also use a *threshold version* of the Paillier's cryptosystem [7].

Definition 3 (Threshold cryptosystem). *A* (t, n) *threshold cryptosystem is a public cryptosystem where at least* $t > 1$ *nodes out of* n *need to actively cooperate in order to decrypt an encrypted message. In particular, no collusion of even* $(t - 1)$ *nodes can decrypt a cipher text. However, any node may encrypt a value on its own using the public-key* pk. *After the threshold cryptosystem has been set up, each node* i *gets as a result his own secret key* sk_i *(for* $1 \leq i \leq n$).

The cooperation between nodes for the decryption usually involves an interactive cryptographic protocol during which several nodes need to combine their own secret keys with an encrypted value to be able to perform the corresponding decryption.

3   Threshold Similarity Protocol

The threshold similarity protocol preserves privacy by outputting only one bit of information stating whether (or not) the similarity between two profiles is above some well-chosen threshold τ. To this end, we define thereafter the notion of threshold similarity.

---

[4] In order to simplify the notation, we drop the indices and write $\mathsf{Enc}(a)$ instead of $\mathsf{Enc}_{pk}(a)$ and $\mathsf{Dec}(a)$ instead of $\mathsf{Dec}_{sk}(a)$ for the rest of the paper.

Definition 4 (Threshold similarity). *Two nodes are τ-similar if the output of applying a similarity measure* sim *on their respective profiles is above a certain threshold $0 \le \tau \le 1$ (i.e.* sim$(S_A, S_B) > \tau$).

A threshold similarity protocol takes as input two profiles S_A and S_B (one profile per node) and outputs one bit of information, which is 1 if S_A and S_B are τ-similar (i.e. sim$(S_A, S_B) > \tau$ for sim a predefined similarity measure and τ the value of the threshold) and 0 otherwise. In practice, the value of the threshold τ is application dependent and is set empirically so as to be significantly above the average similarity between nodes in the population. The threshold similarity is very appealing with respect to privacy as it guarantees that the output of the similarity computation only reveals one bit of information, which is potentially much less than disclosing the exact value of the similarity measure. As a practical illustration, we show how to compute privately the cosine similarity between two profiles (Equation 1) represented as binary vectors using an algorithm that we called ThresholdCosine. Note that our approach is generic enough to accommodate other similarity metrics such as for example Jaccard index or Hamming distance. The value of the threshold is set once and for all in advance and therefore the adversary cannot perform a kind of binary with different values for τ.

As a preprocessing step to this protocol, the two nodes engage in the setup phase of a distributed key generation protocol of a threshold affine homomorphic cryptosystem [7] (see for instance [24] for a detailed description of a distributed key generation protocol without a trusted third party for the Paillier cryptosystem). At the end of this key generation phase, both nodes receive the same public key pk and each one of them gets as private input a different secret key, respectively sk_A for the first node and sk_B for the second node. The threshold cryptosystem[5] is such that any node can encrypt a value using the public key pk but that the decryption of a homomorphically encrypted value requires the active cooperation of the two nodes.

At the beginning of the protocol, the two nodes compute the (encrypted) size of the set intersection of their two profiles S_A and S_B by using one of the several algorithms that can be found in the literature. Once this is done, the nodes only receive as output a ciphertext that is an encrypted version of the size (and not the size itself in plaintext). Let k denote the number of items in a profile and l is the size of the domain (e.g. in the dataset delicious k is around 200 items and l is approximately 1 million items). Some of the state-of-the-art algorithms work directly with profiles represented as sets while others are specifically designed to compute the scalar product when profiles are represented as binary vectors. For instance, the two-party scalar product protocol proposed by Goethals [13] provides semantic security for one node and information-theoretic security for the other one, for a communication cost of $O(l)$ bits and a computational complexity in terms of cryptographic operations of $O(l)$ for each node. Other recent protocols for scalar product can be found in the litterature [28,29], but they have roughly the same complexities as Goethals' protocol. Regarding the cardinality of the

---

[5] The threshold cryptosystem should not be confused with the threshold similarity.

set intersection, a protocol presented in [19] also provides semantic security for a communication cost of $O(k \log l)$ and a computational complexity of $O(k^2)$. Apart from those specific algorithms, generic techniques from secure multiparty computation could also be used but in general they are less efficient (see for instance an analysis in [19]). In the rest of the paper, we denote by ScalarProduct the subroutine corresponding to the use of the protocol of Goethals [13].

Afterwards, instead of computing directly the cosine similarity as denoted in Equation (1), we avoid the need for performing a square root on encrypted values (an operation which is non-trivial and often costly) by squaring the whole equation. The squaring operation renders the next cryptographic operations easier while preserving at the same time the order relation. Formally, the similarity metric effectively used in ThresholdCosine is

$$
\frac{|S_A \cap S_B|^2}{|S_A| \times |S_B|} . \tag{2}
$$

On one hand for obtaining the numerator, we square the output of the scalar product by applying the multiplication gate from [6] to multiply it by itself. On the other hand, the denominator can be computed by the first node sending its homomorphically-encrypted set cardinality to the second node (i.e. $\mathsf{Enc}(|S_A|)$), who scalarizes it by its own set cardinality by doing $\mathsf{Enc}(|S_A|) \odot |S_B|$ to obtain $\mathsf{Enc}(|S_A| \times |S_B|)$. Recall, that the objective of the ThresholdCosine protocol is only to learn if the similarity between S_A and S_B is above a certain (publicly known) threshold τ. We assume that the threshold can be represented as a fraction $\tau = a/b$ and therefore our goal is to verify whether or not the following condition holds

$$
\frac{|S_A \cap S_B|^2}{|S_A| \times |S_B|} > \frac{a}{b} \Leftrightarrow b|S_A \cap S_B|^2 > a|S_A| \times |S_B|. \tag{3}
$$

The left side and right side of the inequality can be compared by using secure protocols for integer comparison [12,23]. We choose to apply specifically the comparison technique from [23] as it does not require knowledge of the input as well as a full bit decomposition of the input. Although this protocol was developed initially for secret-sharing, it can be implemented with homomorphic encryption as well. The output of this comparison step is one bit stating whether or not the (squared) cosine similarity is above the threshold τ.

Theorem 1 (Threshold cosine similarity). *The protocol ThresholdCosine is private with respect to a passive adversary and returns 1 if two nodes are τ-similar and 0 otherwise. The protocol has a communication complexity of $O(l)$ bits and a computational cost of $O(l)$, for l being the size of the binary vectors representing the profiles.*

Proof. *All the communication exchanged between Alice and Bob is done using a homomorphic encryption scheme with semantic security, therefore the encrypted*

Algorithm 1. ThresholdCosine(S_A,S_B)

1: Alice and Bob generate the keys of the threshold homomorphic encryption
2: Alice receives sk_a, Bob receives sk_b and they both get the public key pk
3: Alice and Bob compute $\mathsf{Enc}(|S_A \cap S_B|) = \mathsf{ScalarProduct}(S_A, S_B)$
4: Alice applies the multiplication gate from [6] to obtain $\mathsf{Enc}(|S_A \cap S_B|^2)$
5: Alice computes $\mathsf{Enc}(|S_A|)$ and sends it to Bob
6: Bob computes $\mathsf{Enc}(|S_A|) \odot |S_B| = \mathsf{Enc}(|S_A| \times |S_B|)$
7: Alice computes $\mathsf{Enc}(|S_A \cap S_B|^2) \odot b = \mathsf{Enc}(b|S_A \cap S_B|^2)$
8: Bob computes $\mathsf{Enc}(|S_A| \times |S_B|) \odot a = \mathsf{Enc}(a|S_A| \times |S_B|)$
9: Alice and Bob use the integer comparison protocol of [23] on $\mathsf{Enc}(b|S_A \cap S_B|^2)$
    and $\mathsf{Enc}(a|S_A| \times |S_B|)$
10: **if** $\mathsf{Enc}(b|S_A \cap S_B|^2) > \mathsf{Enc}(a|S_A| \times |S_B|)$ **then**
11:     output 1 to state that Alice and Bob are τ-similar
12: **else**
13:     output 0
14: **end if**

---

*messages exchanged do not leak any information about their content. Moreover
as the encryption scheme is a threshold version, neither Alice nor Bob alone can
decrypt the messages and learn their content. The multiplication gate [6] as well
as the integer comparison protocol [23] are also semantically secure, which therefore guarantees that the protocol is secure against a passive adversary. Regarding
the correctness, it is easy to see from the execution of the protocol that if Alice and
Bob are τ-similar then this will result in $\mathsf{Enc}(b|S_A \cap S_B|^2) > \mathsf{Enc}(a|S_A| \times |S_B|)$
when the integer comparison protocol is executed (and therefore an output of 1)
and in 0 otherwise. The multiplication gate and the integer comparison protocols
are independent of l and can be considered as having constant complexity (both in
terms of communication and computation) for the analysis. On the other hand,
the protocol ScalarProduct requires the exchange of $O(l)$ bits between Alice and
Bob as well as $O(l)$ computations [13]. This results in a similar complexity for
the global protocol ThresholdCosine.*

4   Differentially Private Similarity Computation

Cryptography gives us the tools to compute any distributed function without
revealing any other information than the output of the function itself and while
removing the need for a trusted third party. This is a strong privacy guarantee
but at the same time, this does not preclude the possibility that the output itself
might leak information about the private inputs of participants. For instance,
suppose that a deterministic computation of the similarity is performed and
that it outputs 1 as similarity value. In this situation, both nodes know that
they exactly have the same profile. Differential privacy [9] precisely aims at
addressing the problem of what can be inferred about the inputs from the output
of a computation by adding some randomization to it. In that respect, differential
privacy can be seen as an orthogonal but complementary notion to cryptography

as it addresses a different issue. Therefore, in order to get the best of both worlds, the main idea is to combine them by using cryptographic techniques to compute securely a differentially private algorithm.

4.1 Differential Privacy

Apart from the traditional cryptographic definition of privacy, we are also interested in a recent notion called *differential privacy* [9]. Two inputs X_A and X_B are said to *differ in at most one element* if they are both equal except for possibly one entry of the inputs. For instance, if X_A and X_B would be databases, it would mean that they are identical except for one row.

Definition 5 (Differential privacy [9]). *A randomized function K gives ϵ−differential privacy if for all possible inputs X_A and X_B differing in at most one element, and all $S \subseteq Range(K)$,*

$$\Pr[K(X_A) \in S)] \leq \exp(\epsilon) \times \Pr[K(X_B) \in S)]. \tag{4}$$

This probability is taken over all the coin tosses of K. (Range(K) is the range of the function K and exp *refers to the exponential function.)*

Originally, differential privacy was developed within the context of private data analysis and the main guarantee is that if a differentially private mechanism is applied on a dataset composed of the personal data of individuals, no output would become significantly more (or less) probable whether or not a participant removes his data from the dataset. This means that for an adversary observing the output of the mechanism, he only gains a negligible advantage from the presence (or absence) of a particular individual in the database. This statement is a statistical property about the behavior of the mechanism (function) and holds independently of the auxiliary knowledge that the adversary might have gathered. More specifically, even if the adversary knows the whole database but one individual row, a mechanism satisfying differential privacy still protects the privacy of this individual. The parameter ϵ is public and may take different values depending on the application (for instance it could be 0.01, 0.1 or even 0.25). Dwork, McSherry, Nissim and Smith have designed a general technique, called *Laplacian mechanism* [11], that achieves ϵ-differential privacy for a function f by adding random noise to the true answer. The amount of noise that has to be added is directly proportional to the *sensitivity* of the function, which measures how much the output of a function can change with respect to a small change in the input [11].

Definition 6 ((Global) sensitivity [11]). *For $f : D \rightarrow \mathbb{R}$, the sensitivity of f is*

$$GS(f) = \max_{X_A, X_B \in D} \|f(X_A) - f(X_B)\|_1 \tag{5}$$

for all X_A, X_B differing in at most one element, where D is the domain of the function (for instance for binary vectors of l bits, $D = \{0, 1\}^l$).

The Laplacian mechanism achieves ϵ-differential privacy by adding noise directly proportional to $\mathsf{GS}(f)$ and ϵ.

Theorem 2 (Laplacian mechanism [11]). *For $f : D \to R$, a randomized function K achieves ϵ-differential privacy if it releases on input x*

$$K(x) = f(x) + \mathsf{Lap}(\frac{\mathsf{GS}(f)}{\epsilon}) \tag{6}$$

for $\mathsf{GS}(f)$ the sensitivity of the function f and Lap is a randomly generated noise according to the Laplacian distribution parametrized by $\frac{\mathsf{GS}(f)}{\epsilon}$.

The smaller the value of ϵ, the higher the privacy but also, as a result, the higher the impact might be on the utility of the resulting output. The following lemma also shows that differential privacy is a "natural" notion that composes well.

Lemma 1 (Composition and post-processing [18]). *If a randomized algorithm A runs k algorithms $A_1, \ldots, A_k$ where each A_i is ϵ-differentially private, and outputs a function of the results (i.e $A(x) = g(A_1(x), \ldots, A_k(x)$ for some probabilistic algorithm g) then A is $k\epsilon$-differentially private.*

4.2 Differentially Private Similarity

We define two profiles S_A and S_B as *neighbors* if they are the same except for one particular item. Note that for simplicity and without loss of generality, we consider only neighboring profiles of the same size. For instance, S_A is a neighbor of S_B (and vice versa) if it is identical except for one item that may have been replaced to obtain the profile S_B. If the two profiles are represented as binary vectors, they are neighbors if their Hamming distance is 0 or 2 (i.e. $\|S_A \oplus S_B\| \in \{0, 2\}$). The following lemma states the sensitivity of the squared cosine similarity. (Treatment for the differentially private computation of the scalar product can be found in Appendix B.)

Lemma 2 (Sensitivity – squared cosine similarity).
The sensitivity of the function
ExactSquaredCosine *is at most* $\frac{2\min(|S_A|,|S_B|)+1}{|S_A| \times |S_B|}$.

Proof. Consider three different profiles S_A, S_B and S_C, represented as binary vectors of same size, such that S_B and S_C are neighbors. The computation of the cosine similarity between S_A and S_B requires to compute two quantities: (1) the squared size of the set intersection $|S_A \cap S_B|^2$ and (2) the multiplication of the lengths of S_A and S_B (i.e. $|S_A| \times |S_B|$). Replacing an object from S_B to obtain S_C will only increase (or decrease) the value of the set intersection by 1 at most. Moreover, replacing an object from S_B will not change size of the profile $|S_C|$. Therefore

$$\mathsf{GS}(Cosine^2) = \max_{\substack{S_A, S_B, S_C \\ S_B, S_C \text{ neighbors}}} \|\mathsf{sim}(S_A, S_B) - \mathsf{sim}(S_A, S_C)\| = \max_{\substack{S_A, S_B, S_C \\ S_B, S_C \text{ neighbors}}} \left\| \frac{|S_A \cap S_B|^2 - |S_A \cap S_C|^2}{|S_A| \times |S_B|} \right\|$$

$$= \max_{S_A, S_B} \left\| \frac{|S_A \cap S_B|^2 - (|S_A \cap S_B| \pm 1)^2}{|S_A| \times |S_B|} \right\| = \max_{S_A, S_B} \left\| \frac{\pm 2|S_A \cap S_B| - 1}{|S_A| \times |S_B|} \right\|$$

And then substituting the max *quantifier yields:*

$$= \frac{2|S_A \cap S_B| + 1}{|S_A| \times |S_B|} \leq \frac{2\min(|S_A|, |S_B|) + 1}{|S_A| \times |S_B|} .$$

There are several ways to achieve ϵ-differential privacy in a distributed context. For instance, if we assume that Alice and Bob have access to a *semi-trusted party* that does not collude with any of the two nodes that computes their similarity, it can be used to help Alice and Bob during the similarity computation. This is the case for instance in the gossip-on-behalf protocol (Appendix A) in which another node acts as an anonymizer. The anonymizer is semi-trusted because although it is used to connect anonymously two nodes, it is not trusted to the point of having access to the content of the messages exchanged between them due to the semantic encryption scheme used. Another possible way to achieve differential privacy would be for the two nodes to add the noise themselves directly when executing the protocol for similarity computation.

Differential privacy via two-party computation. For instance, suppose that Alice and Bob want to release the result of the scalar product between their two profiles. At the end of the protocol, Alice and Bob could both simply add independently generated random noise with distribution $\mathsf{Lap}(\frac{1}{\epsilon})$ using the homomorphic property of the encryption scheme. Afterwards, they could cooperate to perform the threshold decryption (which remember is not the same as the threshold similarity computation) and they would both get to learn the perturbed scalar product. Finally, Alice may subtract her own noise from the released output to recover only a version of the similarity that has been randomized with Bob's noise (which she cannot remove).

Differential privacy via semi-trusted third party. In the context of gossip-on-behalf (Appendix A), the node that acts as an anonymizer to set up the bidirectional anonymous channel could also generate some random noise and add it to the similarity value that has been computed by using the homomorphic property of the cryptosystem. Afterwards, the two nodes that have been involved in the similarity computation would recover the result using the threshold decryption. The following algorithm describes this procedure.

Theorem 3 (Protocol for differential squared cosine). *The protocol DifferentialSquaredCosine is private with respect to a passive adversary and ϵ-differentially private. The protocol has a communication complexity of $O(l)$ bits and a computational cost of $O(l)$, for l the size of the binary vectors representing the profiles.*

Proof. All the communication exchanged between Alice and Bob are done using a homomorphic encryption scheme with semantic security, therefore the encrypted messages exchanged do not leak any information about their content. Moreover as the encryption scheme is a threshold version, it means that neither Alice nor Bob alone can decrypt the messages and learn their content. At the end of the protocol providing that the anonymizer does not collude either with Alice or Bob, Alice

---

Algorithm 2. DifferentialSquaredCosine(S_A,S_B,ϵ)

1: Alice and Bob generate the keys of the threshold homomorphic encryption
2: Alice receives sk_A, Bob receives sk_B and they both get the public key pk
3: Alice and Bob compute $\mathsf{Enc}(|S_A \cap S_B|^2) = \mathsf{ScalarProduct}(S_A, S_B)^2$
4: Alice and Bob gives to the node acting as the anonymizer $\mathsf{Enc}(|S_A \cap S_B|^2)$ as well as the sizes of their profiles $|S_A|$ and $|S_B|$
5: The anonymizer computes the squared cosine similarity $\mathsf{Enc}(\frac{|S_A \cap S_B|^2}{|S_A| \times |S_B|})$ and adds Laplacian noise parametrized by $\frac{\mathsf{GS}(Cosine^2)}{\epsilon} = \frac{2\min(|S_A|,|S_B|)+1}{\epsilon \times |S_1| \times |S_2|}$ using the homomorphic property
6: The anonymizer sends the perturbed squared cosine similarity (which is homomorphically encrypted) to Alice and Bob
7: Alice and Bob cooperate to decrypt the homomorphically encrypted value and get as output $(\mathsf{ExactSquaredCosine}(S_A, S_B) + \mathsf{Lap}(\frac{2\min(|S_A|,|S_B|)+1}{\epsilon \times |S_A| \times |S_B|}))$

---

and Bob only get to learn $(\mathsf{ExactSquaredCosine}(S_A, S_B) + \mathsf{Lap}(\frac{2\min(|S_A|,|S_B|)+1}{\epsilon \times |S_A| \times |S_B|}))$, which ensures the ϵ-differential property of the protocol. Moreover, because of the use of the protocol *ScalarProduct* as a subroutine, the protocol *DifferentialSquaredCosine* has a communication cost of $O(l)$ bits as well as a computational cost of $O(l)$ (we consider here that the threshold decryption has constant complexity and is negligible with respect to the cost of the scalar product).

Note that in this protocol, where the noise needed to reach differential privacy is added by the semi-trusted third party, it needs to know the value of $|S_A|$ and $|S_B|$ (or at least an upper bound on these values) to be able to add noise tailored to the sensitivity of the function.

Differentially private threshold similarity. Regarding the threshold similarity, it is important to notice that it is meaningless to add some random noise to a binary value (for instance the output of the threshold similarity), because it amounts to flipping this value with some non-negligible probability. Instead, the most direct way to achieve differential privacy is to add the noise before the application of the threshold function. The following observation states that this does not hurt the privacy guarantee obtained.

Observation 1 (Impact of threshold on privacy). *Applying the Laplacian mechanism before the threshold function does not hurt the differential privacy guarantee.*

Proof. Suppose that we have some output of a function f to which we have added some Laplacian noise calibrated to the sensitivity $\mathsf{GS}(f)$ of the function as well as ϵ. As stated by Lemma 1 as this output is ϵ-differentially private, performing some pre-determined post-processing on it such as applying a threshold function before releasing it has no impact on the privacy guarantees. Therefore, the threshold function is by itself ϵ-differentially private if it is fed with some similarity measure that has been computed with a ϵ-differentially private algorithm.

In the previous observation as well as in the context of Lemma 1, note that $k = 1$ as only one differential privacy mechanism (namely $A_1 = f$) is applied. The threshold function itself corresponds to g as it only counts as a post-processing step whose input is not the original profiles of nodes but rather the output of a differentially private mechanism.

5   Utility Analysis

In this section, we are interested in evaluating the impact of differential privacy on the utility of the application.

Theoretical analysis. In particular, we are interested in measuring the amount of false negatives induced by applying differential privacy on a specific similarity metric. A false negative arises when the protocol outputs that two nodes are not τ-similar while in fact they are. In particular, we have derived an equation that takes the threshold value (τ) and the privacy parameter (ϵ) as parameters and computes the probability of having false negatives when we use the differentially private similarity metric. This equation may be use to guide and set up the different parameters of the algorithms and also to measure the achievable trade-off between utility and privacy. We focus primarily on false negatives for the analysis because we believe that a high rate of false negatives will have a big impact on the utility while a high rate of false positives will mainly hurt privacy. However, the rate of false positives can be also derived straightforwardly by following the same approach we used to compute the rate of false negatives.

In our model, the parameter l is the total number of items in the domain of items (which we assume to be a finite domain) and $l_A = |S_A|$ and $l_B = |S_B|$ are the sizes of the profiles of Alice and Bob. The random variable S represents the number of items in common between any two profiles picked at random with the given sizes l_A and l_B (i.e the size of the set intersection $|S_A \cap S_B|$).

Lemma 3 (Hypergeometric distribution [15]). $S \sim Hypergeometric(\max (l_A, l_B), \min(l_A, l_B), l)$, where l is the total number of items in the domain (which is the size of the binary vectors).

Proof. Let min be the set which has the smallest size among the two sets (we assumed it is S_A without loss of generality). Fix min, and let Bob (owner of set S_B), pick l_B items from the domain of size l without replacements. A pick is successful if the item picked is also contained within the set min, hence the number of possible success is at most l_B. This corresponds exactly to the definition of the Hypergeometric distribution.*

Remember that the utility is measured as the percentage of similarity measures that does *not* count as a false negative after the noise has been added and that the similarity value is $S^2/(l_A l_B)$.

Definition 7 (Utility function). *The utility function is:*

$$u(l_A, l_B, l, \tau, \epsilon) = 1 - P(N \leq \tau - \frac{S^2}{l_A l_B} | \frac{S^2}{l_A l_B} > \tau) = \quad = 1 - \sum_{s=\lceil \sqrt{l_A l_B \tau} \rceil}^{\min(l_A, l_B)} \frac{f_S(s) F_N(\tau - \frac{s^2}{l_A l_B})}{1 - F_S\left(\sqrt{l_A l_B \tau}\right)} ,$$

for $S \sim Hypergeometric(\max(l_A, l_B), \min(l_A, l_B), l)$. The τ parameter can be chosen by substituting the desired acceptance rate of the threshold similarity (without taking into account the error caused by the addition of noise) into the inverse cumulative density function (CDF) of the Hypergeometric distribution. This is a function of l_A and l_B assuming that l is fixed a priori and it corresponds to an integer, which when divided by the minimum size among both sets, gives the threshold τ. To summarize, we have $\tau = CDF^{-1}(\max(l_A, l_B), \min(l_A, l_B), l, r) / \min(l_A, l_B)$, where r is the desired acceptance rate. The utility function can be used by nodes to set the privacy parameter ϵ dynamically depending on the size of the sets of the two nodes (see Appendix C for more details).

Experimental evaluation. We have also studied experimentally the proposed mechanisms in the context of a fully decentralized clustering algorithm [4] and evaluate the achievable trade-off between utility (as measured by the quality of the global clustering) and privacy. The clustering algorithm groups nodes according to their interests. In the baseline implementation of the clustering algorithm (which we refer simply as "baseline" in the sequel), each node samples the network and exchanges a digest of its profile that is a Bloom filter representation of its vector profile, which it uses to compute its squared cosine similarity with other nodes. Based on that value, each node iteratively sorts its clustering view and retains the c closest nodes according to the computed similarity metric (c is set to 10 in our experiments). After a predetermined number of cycles when the protocol converges, each node should end up with the c most similar (closest) nodes in its view. (More details about this algorithm are available in [4].)

We used a dataset from delicious in which users tags items (i.e. URLs). The user profile is represented as a vector of tagged items such that there is a 1 in each vector entry corresponding to an item a user has tagged and 0 otherwise. In our experiments, we compare two models against the baseline model. The first one is a *threshold* similarity protocol, where nodes exchange their Bloom filters only if the threshold protocol presented in Algorithm 2, outputs 1. This protocol computes privately the similarity measure and outputs 1 if the similarity between the two nodes exceeds the predetermined threshold τ. If a node has in its view less than c nodes whose similarity is above τ, the rest of the view is chosen at random and the Bloom filters are not transmitted. The second model, which is the *threshold differentially private protocol (TDP)*, is a variant of the threshold version in which we added the property of differential privacy to the cryptographic protocol. Computing similarity between two nodes requires $O(k)$ bits, where k is the size of the Bloom filter in the baseline model, while using homomorphic encryption to encrypt each bitresult in an expansion factor of ~ 2048 due to the size of the generated ciphertexts.

Experimental setup. Evaluations are conducted through the simulation of a network of 500 nodes. Each node represents a user, selected randomly from a dataset of 20,000 users from a delicious trace crawled in 2009. The resulting domain of items is a set of 1,144,000 URLs. In this datatset, the average number of items tagged by a user is 323 and the average similarity between pair of users is 0.00004 (which explained why the chosen values for τ may seem relatively low). Specifically in the experiments, we have set $\tau \in \{10^{-3}, 10^{-4}, 7 \times 10^{-5}, 10^{-5}, 1 \times 10^{-5}, 5 \times 10^{-6}\}$ and $\epsilon \in \{0.001, 0.01, 0.1, 1, 10, 100\}$. We evaluate our two models (threshold and TDP) according to the following metrics: the *quality of the clustering* and the *level of privacy*. The quality of the clustering is measured by *(i)* the difference between the cluster view obtained by Threshold and TDP protocols compared to the baseline and, *(ii)* the recall when looking for items (previously removed from the profiles) in the profiles of the c closest nodes. More specifically, clustering is done based on 90% of the tagged items of each user and the remaining 10% is used to measure the recall by comparing how many of them are in the set of tagged items of the view the node ends up with. The level of privacy is measured as the number of Bloom filters exchanged as well as the chosen value for ϵ.

Results. In Figure 1, the x-axis represents the privacy parameter ϵ. The larger its value, the less privacy (i.e. noise) is provided. We plot the experiments as a constant function with respect to ϵ. The threshold line for a given τ should be interpreted as the upper bound of the performance of the "private experiments" with the same value for τ. Therefore, the less the number of the Bloom filters exchanged, the better for the users' privacy. The results obtained demonstrate that the number of Bloom filters exchanged are up to half that of the baseline. Yet, for most of our choices of τ, the recall and view quality in the threshold experiment are close to the one obtained with the baseline. Note that, as observed on Figure 1b, the recall of the baseline is 0.26, which is mainly due to the sparsity of the dataset. The exception being for the value of $\tau = 0.001$ which turns out to be much higher than the average similarity (0.00004), resulting in almost no exchange of Bloom filters (which has the same effect has letting the nodes chosen their view at random). We observe that for some choices of τ, adding privacy (in terms of noise) can even enhance the utility. For instance, when we add a large amount of noise and that the threshold is extremely low, this will increase the number of false positives, thus resulting in more exchanges of Bloom filters than with the use of the threshold alone without the addition of noise. Finally, Figure 1d displays the convergence time obtained with $\tau = 0.00007$. The convergence plots for the view and the Bloom exchanges are similar to the one presented before. Moreover, we observe that in all runs, the private protocols converges almost as fast as the baseline (in less than 25 cycles) to their optimal value, with respect to the quality of the view and the recall. To summarize, applying the threshold (respectively the TDP) protocol impacts only slightly the recall by 4% (respectively 12%) but reduces up to 80% the number of Bloom filters exchanged, thus providing a higher privacy. Therefore, we can conclude that it is possible to achieve reliable clustering and high recall even if instead of exchanging Bloom filters, we use a differentially private threshold mechanism for computing the similarity between nodes.

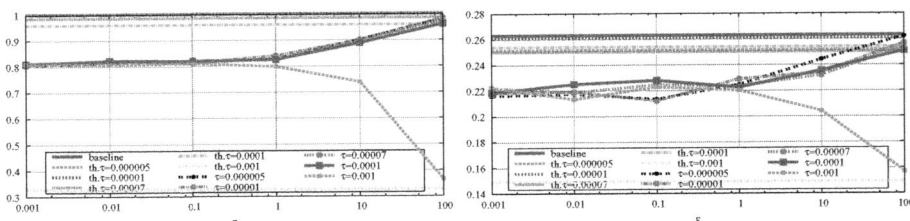

(a) Difference of view with the baseline.   (b) Quality of clustering as measured by the recall.

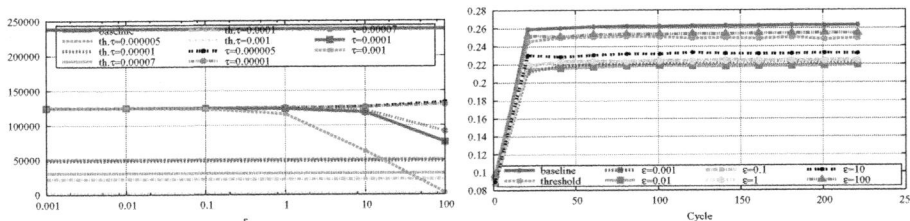

(c) Number of Bloom filters exchanged.   (d) Recall obtained for a fixed $\tau = 0.00007$.

Fig. 1. Experimental results obtained with 500 nodes from Delicious for different values of ϵ and τ

6   Conclusion

Main results. The Web 2.0 has recently witnessed a proliferation of user generated content including a large proportion of personal data. Preserving privacy is a major issue to be able to leverage this information to provide personalized services. Fully decentralized systems somehow protect users privacy to be exposed to large companies, avoiding the *"Big brother is watching you"* syndrome. However, in some sense this is an illusion as they might expose personal data to other users in the network. In this paper, we have addressed this challenge by providing users with a way to compute their similarity with respect to other users while preserving the privacy of their profiles. More precisely, we have introduced a two-party threshold similarity protocol enabling a user to quantify her similarity with another user, without revealing her profile and without requiring a trusted third party. We proved that the proposed protocol is secure in the presence of a passive adversary. We have also proposed differentially private protocols for the exact and threshold similarity and studied the impact of the noise generation on the utility of the resulting similarity. To summarize, our work highlights the fact that *cryptography and differential privacy are two different but complementary notions*. On one hand, differential privacy gives strong privacy guarantees with respect to how much information can be learned about the inputs of the participants from the (perturbed) output of a function. Thus, differential privacy helps us to reason on which type of information can be safely released with respect to privacy. On the other hand, cryptography, and more specifically secure multi-party computation, gives us the tools to compute a distributed function in a secure and robust way and removes the need for a trusted third party, which is of paramount importance in a decentralized setting. When possible, it seems therefore natural to combine differential privacy and cryptography into an integrated approach as we have done for private similarity computation in distributed systems.

Related work. Distributed noise generation has been addressed in the context of the secure multi-party and differential privacy [10]. The resulting protocol has a greater complexity than our approach, but on the other hand is secure against active (Byzantine) adversary. A protocol for nearest neighbor search in distributed settings has been proposed in [27] but it was designed only within the cryptography framework and not the differential privacy context. Therefore, if it is possible that some privacy breach related to specific item in the profile may arise if the adversary has some background knowledge. Differential private protocols have also been considered in centralized systems such as [20] for analyzing the recommender system of Netflix with respect to differential privacy.

Future work. Currently, we have mainly focused on providing security with respect to a passive adversary, which can be seen as a privacy analysis of how much knowledge can be inferred by an adversary following the rules of the protocol but trying to extract as much information as possible from the transcript of the communications seen, the output of the protocol and its own input. While this is a first step, we plan as future work to address malicious participants (modeled by active adversaries) that can cheat during the execution of the protocol. It might also be possible to add the unlinkability property to the bidirectional anonymous channel to prevent an adversary from linking two queries to the same honest node.

Acknowledgements. We are very grateful to the anonymous reviewers for their constructive comments that have help us to improve the quality of this paper.

References

1. Amer-Yahia, S., Benedikt, M., Lakshmanan, L.V.S., Stoyanovich, J.: Efficient Network Aware Search in Collaborative Tagging Sites. PVLDB 2008 1(1) (August 2008)
2. Bai, X., Bertier, M., Guerraoui, R., Kermarrec, A.-M., Leroy, V.: Gossiping Personalized Queries. In: EDBT 2010, Lausanne, Switzerland, March 22-26 (2010)
3. Barbaro, M., Zeller, T.: A face is exposed for AOL searcher No. 4417749. New York Times (2006)
4. Bertier, M., Frey, D., Guerraoui, R., Kermarrec, A.-M., Leroy, V.: The Gossple Anonymous Social Network. In: Gupta, I., Mascolo, C. (eds.) Middleware 2010. LNCS, vol. 6452, pp. 191–211. Springer, Heidelberg (2010)
5. Chaum, D.: Untraceable Electronic Mail, Return Addresses, and Digital Pseudonyms. CACM 24(2) (February 1982)

6. Cramer, R., Damgård, I., Nielsen, J.B.: Multiparty Computation from Threshold Homomorphic Encryption. In: Pfitzmann, B. (ed.) EUROCRYPT 2001. LNCS, vol. 2045, pp. 280–300. Springer, Heidelberg (2001)

7. Damgård, I., Jurik, M.: A Generalisation, a Simplification and Some Applications of Paillier's Probabilistic Public-Key System. In: Kim, K. (ed.) PKC 2001. LNCS, vol. 1992, pp. 119–136. Springer, Heidelberg (2001)

8. Dingledine, R., Mathewson, N., Syverson, P.F.: Tor: The Second-Generation Onion Router. In: Proceedings of the 13th USENIX Security Symposium, San Diego, California, USA, August 9-13 (2004)

9. Dwork, C.: Differential Privacy: A Survey of Results. In: Agrawal, M., Du, D.-Z., Duan, Z., Li, A. (eds.) TAMC 2008. LNCS, vol. 4978, pp. 1–19. Springer, Heidelberg (2008)

10. Dwork, C., Kenthapadi, K., McSherry, F., Mironov, I., Naor, M.: Our Data, Ourselves: Privacy Via Distributed Noise Generation. In: Vaudenay, S. (ed.) EUROCRYPT 2006. LNCS, vol. 4004, pp. 486–503. Springer, Heidelberg (2006)

11. Dwork, C., McSherry, F., Nissim, K., Smith, A.: Calibrating Noise to Sensitivity in Private Data Analysis. In: Halevi, S., Rabin, T. (eds.) TCC 2006. LNCS, vol. 3876, pp. 265–284. Springer, Heidelberg (2006)

12. Garay, J.A., Schoenmakers, B., Villegas, J.: Practical and Secure Solutions for Integer Comparison. In: Okamoto, T., Wang, X. (eds.) PKC 2007. LNCS, vol. 4450, pp. 330–342. Springer, Heidelberg (2007)

13. Goethals, B., Laur, S., Lipmaa, H., Mielikäinen, T.: On Private Scalar Product Computation for Privacy-Preserving Data Mining. In: Park, C., Chee, S. (eds.) ICISC 2004. LNCS, vol. 3506, pp. 104–120. Springer, Heidelberg (2005)

14. Goldreich, O.: Foundations of Cryptography. Cambridge University Press (2001)

15. Harkness, W.L.: Properties of the Extended Hypergeometric Distribution. The Annals of Mathematical Statistics 36(3) (June 1965)

16. Jaccard, P.: Étude Comparative de la Distribution Florale dans une Portion des Alpes et des Jura. Bulletin de la Société Vaudoise des Sciences Naturelles 37(142) (1901)

17. Jelasity, M., Voulgaris, S., Guerraoui, R., Kermarrec, A.-M., van Steen, M.: Gossip-based Peer Sampling. In: TOCS 2007, vol. 25(3) (August 2007)

18. Kasiviswanathan, S.P., Lee, H.K., Nissim, K., Raskhodnikova, S., Smith, A.: What Can We Learn Privately?. In: FOCS 2008, Philadelphia, Pennsylvania, USA, October 25-28 (2008)

19. Kissner, L., Song, D.X.: Privacy-Preserving Set Operations. In: Shoup, V. (ed.) CRYPTO 2005. LNCS, vol. 3621, pp. 241–257. Springer, Heidelberg (2005)

20. McSherry, F., Mironov, I.: Differentially Private Recommender Systems: Building Privacy into the Netflix Prize Contenders. In: SIGKDD 2009, June 28-July 1. ACM, Paris (2009)

21. Narayanan, A., Shmatikov, V.: Robust De-anonymization of Large Sparse Datasets. In: Proceedings of the 29th IEEE Symposium on Security and Privacy, Oakland, California, USA, May 18-21 (2008)

22. Narayanan, A., Shmatikov, V.: De-anonymizing Social Networks. In: Proceedings of the 30th IEEE Symposium on Security and Privacy, Oakland, California, USA, May 17-20 (2009)

23. Nishide, T., Ohta, K.: Multiparty Computation for Interval, Equality, and Comparison Without Bit-Decomposition Protocol. In: Okamoto, T., Wang, X. (eds.) PKC 2007. LNCS, vol. 4450, pp. 343–360. Springer, Heidelberg (2007)

24. Nishide, T., Sakurai, K.: Distributed Paillier Cryptosystem without Trusted Dealer. In: WISA 2010, Jeju Island, Korea, August 24-26 (2010)

25. Paillier, P.: Public-Key Cryptosystems Based on Composite Degree Residuosity Classes. In: Stern, J. (ed.) EUROCRYPT 1999. LNCS, vol. 1592, pp. 223–238. Springer, Heidelberg (1999)
26. Reiter, M.K., Rubin, A.D.: Crowds: Anonymity for Web Transactions. TISSEC 1(1) (November 1998)
27. Shaneck, M., Kim, Y., Kumar, V.: Privacy Preserving Nearest Neighbor Search. In: ICDM 2006, December 18-22. IEEE, Hong Kong (2006)
28. Wright, R.N., Yang, Z.: Privacy-Preserving Bayesian Network Structure Computation on Distributed Heterogeneous Data. In: SIGKDD 2004, August 22-25. ACM, Seattle (2004)
29. Yao, D., Tamassia, R., Proctor, S.: Private Distributed Scalar Product Protocol With Application To Privacy-Preserving Computation of Trust. In: IFIPTM 2007, Moncton, New Brunswick, Canada, July 30-August 2 (2007)

A   Gossip-on-Behalf

In *gossip-on-behalf*, the first node, Alice, starts by choosing at random another node C (that we refer thereafter as Charlie) from her random sample (typically provided by a random peer sampling service [17]). She then generates a pair of public key/secret key for this session and asks Charlie to select a node at random (that we called Bob) as the second node that will be involved in the similarity computation. Charlie does not disclose the identity of Bob to Alice (and vice versa), therefore acting as an anonymizer. Afterwards, Charlie transmits the public key of Alice to Bob that will use it to encrypt any data exchanged with Alice. Finally, Bob either generates also a pair of public key/secret key for this session or a secret that will be used as the key of a symmetric cryptosystem (such as AES) and transmits this to Bob encrypted with his public key via Charlie as a relay (this is similar in spirit to the SSL authentication protocol). Alice and Bob now share a secure anonymous channel. The communication between Alice and Bob goes through Charlie but as it is encrypted, this forbids Charlie from learning any information exchanged during their interactions. One important security assumption is that Charlie does not collude neither with Alice nor with Bob, or otherwise this would break the anonymity property of the channel.

B   Scalar Product

Lemma 4 (Sensitivity – scalar product). *The sensitivity of the function* ScalarProduct *is 1.*

Proof. Consider three different profiles S_A, S_B *and* S_C, *represented as binary vectors of the same size, such that* S_B *and* S_C *are neighbors. Replacing an object from* S_B *by another object to obtain* S_C *increases (or decrease) the value of the scalar product by 1 at most, and therefore the sensitivity of the scalar product is 1.*

Algorithm 3. DifferentialScalarProduct(S_A,S_B,ϵ)

1: Alice and Bob generate the keys of the threshold homomorphic encryption
2: Alice receives sk_A, Bob receives sk_B and they both get the public key pk
3: Alice and Bob compute $\mathsf{Enc}(|S_A \cap S_B|) = \mathsf{ScalarProduct}(S_A, S_B)$
4: Alice generates Laplacian noise parametrized by $\mathsf{Lap_A}(\frac{1}{\epsilon})$ and computes
   $\mathsf{Enc}(|S_A \cap S_B|) \oplus \mathsf{Enc}(\mathsf{Lap_A}(\frac{1}{\epsilon})) = \mathsf{Enc}(|S_A \cap S_B| + \mathsf{Lap_A}(\frac{1}{\epsilon}))$ and sends the result
   to Bob
5: Bob generates Laplacian noise parametrized by $\mathsf{Lap_B}(\frac{1}{\epsilon})$ and computes
   $\mathsf{Enc}(|S_A \cap S_B| + \mathsf{Lap_A}(\frac{1}{\epsilon})) \oplus \mathsf{Enc}(\mathsf{Lap_B}(\frac{1}{\epsilon})) = \mathsf{Enc}(|S_A \cap S_B| + \mathsf{Lap_A}(\frac{1}{\epsilon}) + \mathsf{Lap_B}(\frac{1}{\epsilon}))$
6: Alice and Bob cooperate to decrypt the homomorphically encrypted value and
   get as output $(|S_A \cap S_B| + \mathsf{Lap_A}(\frac{1}{\epsilon}) + \mathsf{Lap_B}(\frac{1}{\epsilon}))$

Theorem 4 (Protocol for differential scalar product). *The protocol DifferentialScalarProduct is private with respect to a passive adversary and ϵ-differentially private. The protocol has a communication complexity of $O(l)$ bits and a computational cost of $O(l)$, for l the size of the binary vectors representing the profiles.*

Proof. All the communication exchanged between Alice and Bob are done using an homomorphic encryption scheme with semantic security, therefore the encrypted messages exchanged do not leak any information about their content. Moreover as the encryption scheme is a threshold version, it means that neither Alice nor Bob alone can decrypt the messages and learn their content. At the end of the protocol, Alice and Bob only get to learn $(|S_A \cap S_B| + \mathsf{Lap_A}(\frac{1}{\epsilon}) + \mathsf{Lap_B}(\frac{1}{\epsilon}))$ which ensures the ϵ-differential property of the protocol[6]. Moreover, because of the use of the protocol ScalarProduct as a subroutine, the protocol DifferentialScalarProduct has a communication cost of $O(l)$ bits and as well as a computational cost of $O(l)$ (we consider here that the threshold decryption and the generation of Laplacian noise have constant complexity and are negligible with respect to the cost of the scalar product).

C   Utility Analysis

The utility function can be used by nodes to set the privacy parameter ϵ dynamically depending on the size of the sets of the two nodes. The probability that a node gets accepted is $P(S^2/(l_A l_B) > \tau)$, where τ is the public threshold value while the probability of getting rejected (false negative rate) after adding the Laplacian noise is $P(S^2/(l_A l_B) + N \leq \tau) = P(N \leq \tau - S^2/(l_A l_B))$. This result in the following utility function:

---

[6] More precisely, Alice can learn $(|S_A \cap S_B| + \mathsf{Lap_B}(\frac{1}{\epsilon}))$ if she subtracts her own noise but this still preserves ϵ-differential privacy (the same reasoning can be made for Bob).

$$1 - P(N \le \gamma | \frac{S^2}{l_A l_B} > \tau) = 1 - P(N \le \gamma | S > \theta) = 1 - \frac{P(N \le \gamma \wedge S > \theta)}{1 - F_s(\theta)} = 1 - \sum_{s > \theta} \frac{\int_{-\infty}^{\gamma} f_{N,S}(n,s)\, dn}{1 - F_s(\theta)}$$

$$= 1 - \sum_{s > \theta} \frac{\int_{-\infty}^{\gamma} f_N(n) f_S(s)\, dn}{1 - F_S(\theta)} = 1 - \sum_{s > \theta} \frac{f_S(s) \int_{-\infty}^{\gamma} f_N(n)\, dn}{1 - F_S(\theta)} = 1 - \sum_{s > \theta} \frac{f_S(s) F_N(\gamma)}{1 - F_S(\theta)} \quad ,$$

where $\theta = \sqrt{l_A l_B \tau}$ and $\gamma = \tau - \frac{S^2}{l_A l_B}$. The upper limit of this sum is $\min(l_A, l_B)$ whereas the lower limit is $\lceil \theta \rceil$. The above equation (which is a function of l_A, l_B, τ, ϵ, and l), when plotted with different values of l_A and l_B, shows the effect of the privacy parameter ϵ for a given τ and domain cardinality l. Alternatively, we can also derive the probability of not having false positives or the probability of not having false decisions as described below.

Definition 8 (Utility as the probability of not having false positives.).

$$\mathcal{U}_+(l_A, l_B, l, \tau, \epsilon) = 1 - \sum_{s=0}^{\lfloor \sqrt{l_A l_B \tau} \rfloor} \frac{f_S(s) F_N(\frac{s^2}{l_A l_B} - \tau)}{F_S(\lfloor \sqrt{l_A l_B \tau} \rfloor)} \quad .$$

Definition 9 (Utility as the probability of not having false decisions.).

$$\mathcal{U}_\dagger = \sum_{s=0}^{\min(l_A, l_B)} f_S(s)\, F_N\left(d(s)\left(\tau - \frac{s^2}{l_A l_B} \right) \right) \quad ,$$

where

$$d(s) = \begin{cases} 1 & if\, s \le \lfloor \sqrt{l_A l_B \tau} \rfloor \\ -1 & if\, s > \lfloor \sqrt{l_A l_B \tau} \rfloor \end{cases} \quad .$$

The Impact of Edge Deletions on the Number of Errors in Networks*

Christian Glacet, Nicolas Hanusse, and David Ilcinkas

LaBRI, University of Bordeaux, CNRS, INRIA

Abstract. In this paper, we deal with an error model in distributed networks. For a target t, every node is assumed to give an *advice*, *ie.*to point to a neighbour that take closer to the destination. Any node giving a bad advice is called *a liar*. Starting from a situation without any liar, we study the impact of topology changes on the number of liars.

More precisely, we establish a relationship between the number of liars and the number of distance changes after one edge deletion. Whenever ℓ deleted edges are chosen uniformly at random, for any graph with n nodes, m edges and diameter D, we prove that the expected number of liars and distance changes is $O(\frac{\ell^2 Dn}{m})$ in the resulting graph. The result is tight for $\ell = 1$. For some specific topologies, we give more precise bounds.

Keywords: dynamic graph, errors and faults, shortest path and routing.

1 Introduction

1.1 The Search Problem

Everyone has already faced the problem of reaching a destination in an uncertain network. This is typically the case whenever you are in an unknown city, without a map, and you aim at reaching, let us say, the closest cash machine. The only thing you can do is ask for some information from people in the street. Unfortunately, there is no evidence that all the information you get is reliable.

Nowadays, in a communication network, a corresponding situation can occur. Let us consider the routing task. Due to its dynamicity (change of topology, time required to update local information) and its large-scale size, current networks are not immune to faults and crashes. It is no more realistic to blindly trust the data stored locally at each node. For instance, the Border Gate Protocol (BGP) used in Internet to route messages between autonomous systems implicitly assumes that some paths are known to reach any target. Ideally, these paths are as short as possible. Unfortunately, many messages do not reach their destination because no paths are temporally known although some paths could exist. Is there a way to find such paths ?

In the following, for a given target t, we informally refer to a *liar* as a node containing bad information about the location of t. The word liar is used even if nodes have not necessarily malicious intentions, but are simply ignorant.

---

* This work is granted by the european project EULER.

A. Fernández Anta, G. Lipari, and M. Roy (Eds.): OPODIS 2011, LNCS 7109, pp. 378–391, 2011.
© Springer-Verlag Berlin Heidelberg 2011

A series of papers [HKK04,HKKK08,HIKN10] tackle the problem of locating a target (node, resource, data, ...) in presence of liars.

A first model was introduced by Kranakis and Krizanc [KK99]. They designed algorithms for searching in distributed networks having the ring or the torus topology, when a node has a constant probability of being a liar. A more realistic model was proposed by Hanusse *et al.* [HKK04]: the number of liars is a parameter k and during a routing query, the information stored at every node is unchanged. The main performance measure is the number of edge traversals during a request. Several algorithms, either generic or dedicated to some topologies, and bounds are presented in [HKK04,HKKK08,HIKN10] and are typically of the form $O(d + k^{O(1)})$ (for path,grids, expanders,...) or $\Theta(d + 2^{O(k)})$ for bounded degree graphs, d being the distance between the source and the target.

In these papers, there is an implicit assumption: the number of liars is small. Our goal is to evaluate whether this is realistic or not. Starting from a network without any liar, we aim at estimating bounds on the number of liars obtained after few changes of topology. It turns out that this problem is related to the problem of estimating the number of distance changes after few edge/node deletions or insertions. In this paper, we focus on edge deletions for the following reasons: it is a more atomic event than node deletion (any node deletion can be represented as a sequence of edge deletions) and a deletion is much more dramatic than an insertion in our context. On the one hand, after one deletion, there is potentially no *known or existing* path toward the target and on the other hand, after one insertion, we could only miss a shortcut.

1.2   Related Works

The influence of topology changes on graph parameters is studied in several works. In [CG84,SBvL87], it is proved that for any sequence of ℓ edge deletions that do not disconnect the graph, the diameter D of any unweighted graph turns to be less than $D(\ell+1)$. Our work is also related to the computation of the most vital node of a shortest path [NPW03], that is the node whose removal results in the largest increase of the distance for a given pair of source/target, and the Vickrey pricing of edges [HS01].

Recently, some work on *dynamic* data structures for shortest paths/distance computation problems has been proposed. By dynamic, we mean that the data structures can tolerate some topology changes in a given network. A dynamic network model defines how the underlying graph changes/evolves over time. More precisely, the following type of models are usually considered:

- *Evolving models without constraint*: it consists in an "online" insertion and/or suppression of links and/or nodes. Roughly speaking, if $G(t)$ is the network at time t then $G(0)$ and $G(t)$ can be quite different.
- *Failure model*: $G(t)$ is a subgraph of $G(0)$. In practice, we consider that few nodes/links are removed from $G(0)$.

The most standard model of dynamic network is the following: starting from an initial graph, a sequence of ℓ insertions/deletions of edges/nodes is done.

Each query has to be answered taking into account the ℓ updates. The most naive solution consists in recomputing all shortest paths after any update but it is generally quite costly. For instance, the update time of the fastest dynamic algorithms for the all-pairs shortest path takes $O(n^2 \text{polylog}(n))$ [DI04,Tho04]. It turns out that in the failure model, it is not always necessary to recompute all shortest paths. Some solutions provide efficient data structures dedicated to the problem of reporting shortest path or distance queries for $\ell = 1$. More precisely, we can distinguish data structures dedicated to *exact* solution [DTCR08,BK09] or *constant approximation of the solution* [KB10,CLPR10], that is a constant factor of shortest path/distance after one edge/node deletion. The challenge is to handle efficiently more than $\ell > 1$ updates. To our knowledge, the more general result is the ℓ-sensitivity distance [CLPR10] oracle for which a data structure of size $O(\ell s n^{1+1/s} \log n)$ is able to approximate the distance between any node pairs within a factor $O(s \cdot \ell)$ for undirected graphs in $O(\ell \log^{O(1)} n)$ time. Note that the data structures report distances / routing paths, given the knowledge of the ℓ nodes/edges to avoid. They provide a similar result for weighted graphs and, only if $\ell \leq 2$, for compact routing.

In these works, the implicit model is the one of a *strong adversary model*: the worst sequence of updates. This is sometimes too pessimistic to explain and to model macroscopic observations done on real dynamic networks. In the following, we will also consider the *random fault adversary model*: any sequence of ℓ updates has the same probability to occur. Estimating the number of distance changes in a dynamic network can be used to get a tight analysis of the update time. In King's algorithm analysis ([Kin99] - section 2.1 or [Ber09]), the update time to maintain a shortest path tree turns to be $O(D \cdot \#\text{number of distance changes from the root})$ for connected bounded degree graphs whenever $\ell = 1$. Our results allow to analyse the random fault case.

1.3   Contribution

Models. The network is modelled by a graph $G = (V, E)$ of $|V| = n$ nodes and $|E| = m$ edges. G is assumed to be unweighted and can be disconnected. Note that D correspond to the maximum diameter of all the connected components. The neighbourhood of vertex u is noted $\Gamma(u)$ and includes u itself. Given a target located at a node t, each node $u \in V \setminus \{t\}$ has an *advice* $\mathsf{Adv}(u) \in \Gamma(u) \setminus \{u\}$. Node u is a *truthteller* if $\mathsf{Adv}(u)$ belongs to a shortest path from u to t and otherwise u is a liar. The set of advice A can also define a directed subgraph of G, noted G_A. There is an arc (u, v) in G_A if and only if $v = \mathsf{Adv}(u)$. Whenever there exists no *liar*, G_A is a shortest path spanning tree rooted at t.

We shall investigate two main parameters:

- The number of liars $k = k_G(A)$ for a set of advice A in graph G
- And the size of the set $\mathcal{S}$ of nodes whose distance to t has changed after one edge deletion.

For instance, in *Figure 2*, we have $n - D$ *lying* nodes pointing toward a dead-end in the rightmost drawing and $D - 1$ nodes whose distance to t has changed after one edge deletion.

Given a graph G *without any liar* and a target t, we aim at analysing the combined effect of the choice of set of advice A and the set of ℓ edges. Note that A is not arbitrary since we assume that G has no liar. After a deletion, it may happen that the resulting graph turns to be disconnected. Nodes that do not belong to the connected component of node t become liars. The set of advice is unchanged with a potential exception: if a deleted edge was used as an advice, one extremity needs to draw another advice among its current neighbours. We focus on two models:

- The **adversary model**: this model represents a worst-case analysis. An adversary has the capacity of choosing A, the set of edges to remove and the potential new advice to draw. Thus, k is maximal in this model.
- The **random fault model**: A is assumed to be chosen uniformly at random in the universe of set of advice without liars for the given graph. The set of edges to delete and the potential new advice are chosen uniformly at random.

$\widetilde{G}$ is the resulting subgraph of G after ℓ deletions.

Results. The majority of our results focus on the random fault model since most of the results in the adversary model are simpler. However, it is interesting to take the two models in order to see a potential gap between them.

More precisely, our main result deals with the random fault model : after ℓ deleted edges are chosen uniformly at random, for any graph of n nodes, m edges and diameter D, we prove that the expected number of liars, $\mathbb{E}(k)$, and the expected number of distance changes $\mathbb{E}(|\mathcal{S}|)$ is in $O(\frac{\ell^2 Dn}{m})$ in the resulting graph.

Table 1 shows our results after one deletion in both models. Note that the notation $\Theta(\cdot)$ simultaneously stands for a lower bound *and* an upper bound. The lower bound means that *there exists* a graph of the family for which the number of liars is in $\Omega(\cdot)$.

Note that an edge deletion does not necessarily imply the creation of a liar even if some nodes have changed their distance to t, for instance the complete graph [1]). Conversely, some liars can appear without any change of distance within the graph.

For the family of graphs of diameter D, it is easy to reach the bound for the adversary model : just take a path of D nodes and add a star of $n - D$ leaves to one extremity. If t is located to the other extremity, one edge deletion can disconnect the graph implying k and $|\mathcal{S}|$ to be of linear size. Even if somebody would restrict edge deletion to connected graphs, we can easily claim a lower bound of $\Omega(n - D - 1)$ (see *Figure 2*).

The structure of the paper is the following: we start by exhibiting a relationship between the number of distance changes and the number of liars induced

---

[1] In the complete graph, if an edge is removed,
  - Either this edge was used as an advice by node $u \in V$, in this case $d(u, t) = 2$ and any new advice takes closer to t ;
  - Or not and therefore no liar is created.

Topology	Adversary	Random fault
Graphs of diameter D	$\Theta(n)$	$\Theta(\frac{Dn}{m})$
Square Grid	$\Theta(\sqrt{n})$	$\Theta(1)$
ErdsRnyi model	$\frac{n-1}{4}+1$	$\Theta(\frac{1}{n})$
Hypercube	$\log n - 1$	$\Theta(\frac{1}{\log n})$

Fig. 1. Number of liars induced by a single edge deletion

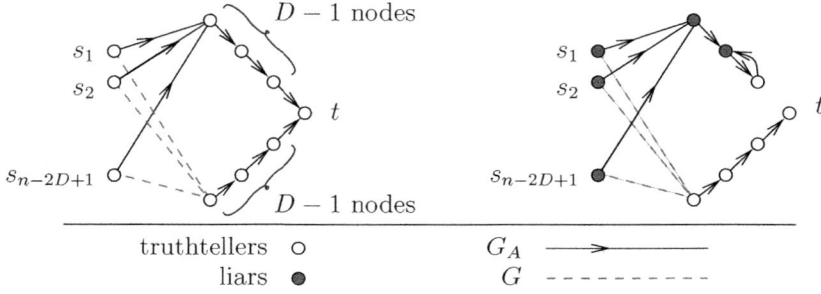

Fig. 2. An example of an edge deletion that creates $n - D - 1$ liars

by an arbitrary edge deletion (*Lemma 2*). Then, we prove that, in the random fault model, $\mathbb{E}(|S|) \leq \frac{Dn}{m}$. Combining with *Lemma 2*, we show that $\mathbb{E}(k) < 2D$ (*Theorem 1*). This result is then improved (*Theorem 2*) and generalized to ℓ edge deletions (*Theorem 4*). More precisely, we prove that the deletion of ℓ random edges creates at most $O(\frac{\ell^2 nD}{m})$ liars. In the last section, we give more precise bounds for specific topologies (see *Table 1*).

2   General Results

2.1   Preliminaries

We start by presenting some notations and some easy facts used in our paper.

$\widetilde{G}_e$ G after deletion of edge e, $\widetilde{G}_e = (V, E \setminus e)$, or simply $\widetilde{G}$.

$d(u, v)$ distance in G from u to v.

$d_{\widetilde{G}}(u, v)$ distance in $\widetilde{G}$ from u to v.

$\Gamma(X)$ X's neighbourhood in G, $\Gamma(X) = \bigcup_{x \in X} \Gamma(x)$

$\mathsf{Adv}^{-1}(X)$ set of nodes advising another node that belongs to X,

     $ie.\mathsf{Adv}^{-1}(X) = \{u \in V \mid \mathsf{Adv}(u) \in X\}$

$\mathcal{F}(e)$ indicates if edge $e = \{x, y\}$ belongs to the set of advised edges G_A.

     More precisely, $\mathcal{F}(e) = 1$ if $\mathsf{Adv}(x) = y \lor \mathsf{Adv}(y) = x$

     and $\mathcal{F}(e) = 0$ otherwise.

Many of our proofs are based on the notion of (s, t)-arterial edges:

Definition 1. *An edge $\{x, y\}$ is (s, t)-arterial if it belongs to* all *shortest paths from s to t*

The deletion of a (s, t)-arterial edge implies

$$\text{the event } \mathcal{E}_{s,t} : d_{\widetilde{G}}(s, t) > d(s, t) \tag{1}$$

Otherwise, there exists a shortest path from s to t which does not contain $\{x, y\}$. The set of arterial edges from s to t is denoted $\mathcal{C}_{s,t}$. It follows that

Lemma 1. *The distance from s to t is modified by a single edge deletion if and only if this edge belongs to $\mathcal{C}_{s,t}$.*

2.2 Relationships between the Number of Liars and the Number of Distance Changes

Let us denote $\mathcal{S} = \mathcal{S}_t^e = \{s \in V \mid \text{the deletion of } e \text{ implies } \mathcal{E}_{s,t}\}$ the set of nodes that have changed their distance to t after the deletion of some edge e.

Lemma 2. *In any graph containing k_0 liars, the number of liars k after deletion of an edge e always satisfies*

$$\left|Adv^{-1}(\mathcal{S}) \setminus \mathcal{S}\right| \leq k \leq \left|Adv^{-1}(\mathcal{S})\right| + \mathcal{F}(e) + k_0 \tag{2}$$

Proof. In any graph with k_0 liars, after one edge deletion, we study the impact for every node (*ie.*advice) on the resulting number of liars k. For every node u with $v \in V$ and $Adv(u) = v$, we have :

$$d_G(u, t) - d_G(v, t) \in \begin{cases} \{1\} & \text{if } u \text{ is a truthteller} \\ \{0, -1\} & \text{if } u \text{ is a liar} \end{cases}$$

If $u \notin \mathcal{S}$ and $v \in \mathcal{S}$ then

$$d_{\widetilde{G}}(u, t) - d_{\widetilde{G}}(v, t) \in \begin{cases} \{0, -1\} & \text{if } u \text{ was a truthteller} \\ \{-1\} & \text{if } u \text{ was a liar} \end{cases}$$

hence u becomes (or remains) a liar. The minimum number of liars after one deletion is then

$$k \geq \left|Adv^{-1}(\mathcal{S}) \setminus \mathcal{S}\right|$$

Let us now consider the upper bound. First assume that the removed edge $e \neq \{u, v\}$. If $v \notin \mathcal{S}$ then u remains a liar:

- $u \in \mathcal{S}$ and $v \notin \mathcal{S}$ then :

$$d_{\widetilde{G}}(u, t) - d_{\widetilde{G}}(v, t) \in \begin{cases} [2, \infty] \text{ (impossible}^2) & \text{if } u \text{ was a truthteller} \\ \{0, 1\} \text{ (could be a liar)} & \text{if } u \text{ was a liar} \end{cases}$$

---

2 impossible because u and v are neighbours.

- if $u \notin S$ and $v \notin S$ then $d_{\widetilde{G}}(u,t) - d_{\widetilde{G}}(v,t) = d_G(u,t) - d_G(v,t)$.

If $u \in S$ and $v \in S$, then $d_{\widetilde{G}}(u,t) - d_{\widetilde{G}}(v,t) \in \{1, 0, -1\}$, so u could be a liar or not independently of its previous state. So, the maximum number of liars added by one edge deletion is at most $\left| \mathsf{Adv}^{-1}(S) \right|$. Then

$$k \leq \left| \mathsf{Adv}^{-1}(S) \right| + k_0$$

Finally, if the removed edge $e = \{u, v\}$, $ie.\mathcal{F}(e) = 1$, then u has to change its advice and becomes a liar. In the worst situation, the number of liars is then increased by one. $\qquad \square$

2.3   Upper Bounds for $\ell = 1$ Deleted Edge in the Random Fault Model

According to our model, and as we have already seen in *Lemma 2*, liars apparition is due to distance changes and advice deletion.

Number of Distance Changes

Lemma 3. *In any m-edge graph $G = (V, E)$, if an edge, chosen uniformly at random, is removed from E then the number $|S|$ of distance changes satisfies*

$$\forall t \in V : \mathbb{E}(|S|) = \frac{1}{m} \sum_{s \in V \setminus \{t\}} |\mathcal{C}_{s,t}|. \tag{3}$$

Proof. From *Lemma 1*, if edge $\{x, y\}$ is chosen uniformly at random in E then $\forall s \in V$:

$$\mathbb{P}(\mathcal{E}_{s,t}) = \frac{|\mathcal{C}_{s,t}|}{m}$$

Let $X_{s,t}$ be a random variable defined by $X_{s,t} = 1$ if $\mathcal{E}_{s,t}$, and $X_{s,t} = 0$ otherwise. We get

$$\mathbb{E}(|S|) = \mathbb{E}\left(\sum_{s \in V \setminus \{t\}} X_{s,t} \right) = \sum_{s \in V \setminus \{t\}} \mathbb{E}(X_{s,t}) = \sum_{s \in V \setminus \{t\}} \mathbb{P}(\mathcal{E}_{s,t}) = \frac{1}{m} \sum_{s \in V \setminus \{t\}} |\mathcal{C}_{s,t}|$$

$\qquad \square$

Corollary 1. *For any n-node, m-edge graph of diameter D, after one random edge deletion, we have in the random fault model*

$$\mathbb{E}(|S|) \leq \frac{D(n-1)}{m}. \tag{4}$$

Proof. In a graph of diameter D, by definition, all shortest paths lengths are at most D. So, $\forall s \in V \setminus \{t\}$, there is at most D (s,t)-arterial edges in E. $\qquad \square$

Number of Liars. Applying *Lemma 2*, we get

Corollary 2. *For any n-node, m-edge graph of diameter D and maximal degree Δ without liar, after one random edge deletion, we have $\mathbb{E}(k) \leq \frac{(D\Delta+1)(n-1)}{m}$.*

This turns to be optimal up to a constant factor for bounded degree graphs (see *Theorem 3*). However, this is not the case whenever the graph has nodes of unbounded degree.

Theorem 1. *For graphs of diameter D without liar, after random one edge deletion , we have*

$$\mathbb{E}(k) \leq 2D \tag{5}$$

Proof. According to *Lemma 2*, for any edge e, if $|\mathcal{S}^e_{e,t}|$ nodes change their distance to t, then the number of added liars after deletion of edge e, is at most $|\mathsf{Adv}^{-1}(\mathcal{S})| + \mathcal{F}(e) \leq \sum_{s\in\mathcal{S}} |\mathsf{Adv}^{-1}(s)| + \mathcal{F}(e)$.

Take the possible m edge deletions trials and consider the m corresponding sets $\mathcal{S}_i$ for i going from 1 to m. In a given trial in which event $\mathcal{E}_{s,t}$ occurs, each node s adds at most $|\mathsf{Adv}^{-1}(s)| \leq \mathrm{degree}(s) - 1$ liars (excluding itself) since G contains initially no liar and at least one neighbour of s is closer to t than s. Since $\forall s \in V \setminus \{t\}$, event $\mathcal{E}_{s,t}$ can occur in at most $|\mathcal{C}_{s,t}| \leq D$ instances among the m ones. It follows that for given s, $\sum_{i:s\in\mathcal{S}_i} |\mathsf{Adv}^{-1}(s)| \leq D(\mathrm{degree}(s) - 1)$.

Thus, for any $i \in [1,m]$, we have $k_{\widetilde{G}_i} \leq |\mathsf{Adv}^{-1}(\mathcal{S}_i)| + \mathcal{F}(e) \leq \sum_{s\in\mathcal{S}_i} \mathrm{degree}(s)$. Summing over all values of i, we get

$$\sum_{i=1}^{m} k_{\widetilde{G}_i} \leq \sum_{i=1}^{m}\sum_{s\in\mathcal{S}_i} \mathrm{degree}(s) = \sum_{s\in V\setminus\{t\}}\sum_{i:s\in\mathcal{S}_i} \mathrm{degree}(s) \leq \sum_{s\in V\setminus\{t\}} D\cdot\mathrm{degree}(s) = 2m\cdot D$$

It turns out that $\mathbb{E}(k_{\widetilde{G}}) \leq \frac{2mD}{m} = 2D$. $\qquad\square$

A more precise bound can be found by reasoning on a hierarchical cutaway of G from distance 0 to D with respect to target t. The following part shows a detailed proof based on this principle to get a tighter upper bound ($\leq \frac{Dn}{m}$).

Nodes in danger Let $\mathcal{T}_{u,v}$ be the set of nodes that have at least one shortest path to $v \in V$ through $u \in V$. Let $L_i = \{x \in V \mid d(x,t) = i\}$ be the set of nodes at distance i from t. Every node $v \in \mathcal{T}_{x,t}$ with $x \in L_i$ is *in danger*[3] with respect to level i if and only if only one shortest path from x to t exists. In *Figure 3*, all nodes from sets $\mathcal{T}_{x_2,t}$ and $\mathcal{T}_{x_3,t}$ are in danger.

Distances and shortest paths Let $\mathcal{C}_i = \{\{x,y\} \mid x \in L_i, y \in L_{i-1} \wedge \Gamma(x)\cap L_{i-1} = \{y\}\}$ be the set of arterial edges between L_i and L_{i-1}. Let $B_t(i-1)$ be the set of nodes at distance at most $i-1$ from t. If G is not connected then the set of edges that does not belong to the connected component of t is $\mathcal{C}_\infty$.

---
[3] Can potentially turns into a liar.

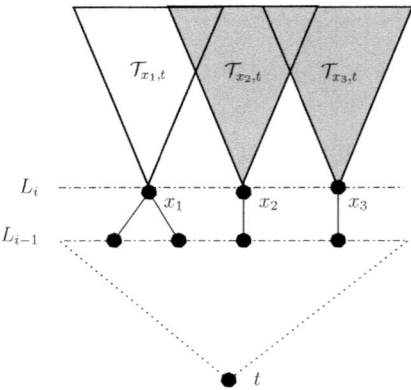

Fig. 3. G levels and nodes in *danger* (grey filled areas)

Lemma 4. *For any graph, containing an arbitrary number of liars k_0. If the edge $\{x, y\}$ is deleted uniformly at random between levels L_i and L_{i-1} and $i \leq D$ then the number of liars added k_{new}[4] is*

$$\mathbb{E}(k_{new} \mid \{x, y\} \in \mathcal{C}_i) \leq \frac{n - |B_t(i - 1)|}{|\mathcal{C}_i|} \qquad (6)$$

and $\mathbb{E}(k_{new} \mid \{x, y\} \in \mathcal{C}_\infty) = 0$.

Proof. The number of arterial edges between L_i and L_{i-1} is

$$|\mathcal{C}_i| = |\{x \in L_i, |\Gamma(x) \cap L_{i-1}| = 1\}|$$

Since all the nodes in danger belong to $\bigcup_{x \in L_i} \mathcal{T}_{x,t}$, the average number of liars added by a random deletion between levels L_i and L_{i-1} is at most

$$\mathbb{E}(k_{new} \mid \{x, y\} \in \mathcal{C}_i) \leq \frac{|\bigcup_{x \in L_i} \mathcal{T}_{x,t}|}{|\mathcal{C}_i|} \leq \frac{n - B_t(i - 1)}{|\mathcal{C}_i|}$$

Note that some of the k_0 liars could belong to $\bigcup_{x \in L_i} \mathcal{T}_{x,t}$. These liars will be counted twice.

$\square$

Theorem 2. *For $D \geq 2$, the numbers of liars added k_{new} by deleting an edge chosen uniformly at random in E is*

$$\mathbb{E}(k_{new}) \leq \frac{D(n - \frac{D-1}{2})}{m} \leq \frac{D(n - 1)}{m} \qquad (7)$$

For $D = 1$, $\mathbb{E}(k_{new}) = k_{new} = 0$. This result holds for arbitrary graphs, unnecessarily connected.

Proof. The average number of liars added is the sum of the expected number of liars induced by deletions between every levels $L_1, L_2, \ldots, L_D$

---

[4] Note that $k_{new} = k - k_0$.

$$\mathbb{E}(k_{new}) = \sum_{i=1}^{\infty}(\mathbb{E}(k_{new} \mid \{x,y\} \in \mathcal{C}_i) \times \mathbb{P}(\{x,y\} \in \mathcal{C}_i)) = \sum_{i=1}^{D}(\mathbb{E}(k_{new} \mid \{x,y\} \in \mathcal{C}_i) \times \mathbb{P}(\{x,y\} \in \mathcal{C}_i))$$

The probability of deleting an edge at level i is

$$\mathbb{P}(\{x,y\} \in \mathcal{C}_i) = \frac{|\mathcal{C}_i|}{m}$$

Thus, from *lemma 4*

$$\mathbb{E}(k_{new}) \le \sum_{i=1}^{D} \frac{n - |B_t(i-1)|}{|\mathcal{C}_i|} \times \frac{|\mathcal{C}_i|}{m} \le \frac{Dn}{m} - \frac{1}{m} \sum_{i=1}^{D} |B_t(i-1)|$$

$\forall i \in D, |B_t(i-1)| \ge i-1$, hence, the average number of liars added is

$$\mathbb{E}(k_{new}) \le \frac{Dn}{m} - \frac{D(D-1)}{2m} \le \frac{D(n - \frac{D-1}{2})}{m}$$

$\square$

2.4   Lower Bound for $\ell = 1$ in the Random Fault Model

Theorem 3. *For any integers n, m, D such that $m \ge n \ge 2D \ge 20$,*

- *There exists a graph of $n + O(1)$ nodes, $\Theta(m)$ edges and diameter D for which the expected number of liars after a random edge deletion is greater than $\frac{(D-8)n}{32m}$.*
- *There exists a graph of $\Theta(n)$ nodes, $\Theta(m)$ edges and diameter D for which the expected number of distance changes after a random edge deletion is $\Omega(\frac{Dn}{m})$.*

Proof. Let us consider a graph H (see H_1 in *Figure 4*) built in the following way: take a complete graph of size r and a stable of size r'. Add two extra nodes u, v and link them to the $r + r'$ nodes. This graph has diameter 2, $r + r' + 2$ nodes and $\frac{r(r-1)}{2} + 2(r + r')$ edges. Take now four copies of H named H_1, H_2, H_3 and H_4. For i going from 1 to 4, link u_i to $v_{(i \mod 4)+1}$ by a path of $D/2 - 4$ edges. The resulting graph G has diameter D. We set up $r = \left\lceil \sqrt{\frac{m-D}{2}} \right\rceil$ and $r' = \left\lceil \frac{n-D}{4} - r \right\rceil$. It follows that G has $n + O(1)$ nodes. The total number of edges is $\Theta(n)$. This graph is presented in *Figure 4*.

Without loss of generality, assume now that target t is either between u_1 and v_2 or belongs to H_1. In the first case, it follows that every node of H_3 (excluding v_3 and potentially u_3) has v_3 as advice toward t. The probability that the deleted random edge belongs to the path from u_2 to v_3 is $p = \frac{D-8}{2m}$. The expected number of liars/distance changes is at least $p(r + r') \ge \frac{(D-8)n}{16m}$.

For the second case, every node of H_3 excluding u_3 or v_3 can point arbitrarily to u_3 and v_3. Take the node given by the majority. If v_3 (resp. u_3) is chosen,

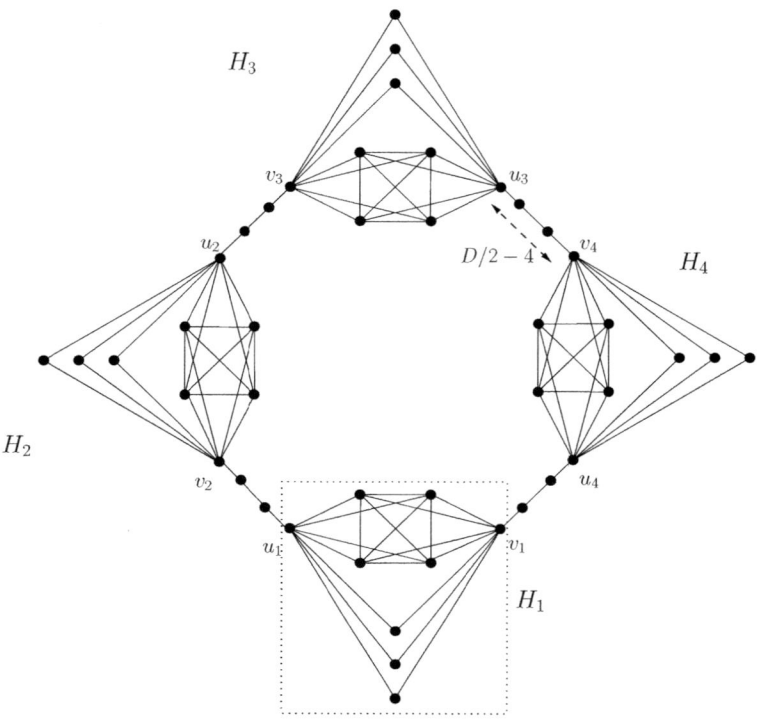

Fig. 4. Sample graph in which the lower bound is reached

then p corresponds to the probability that the deleted random edge belongs to the path from u_2 to v_3 (resp. v_3 to u_4). The expected number of liars turns to be greater than $p(\frac{r+r'}{2}) \geq \frac{(D-8)n}{32m}$.

In this last case, in order to get a similar lower bound for the expected number of distance changes, we just have to slightly modify each H_i copy. We just substitute each node of the stable set by an edge between two nodes. Each copy turns to have $r + 2r'$ nodes and $\frac{r(r-1)}{2} + 2r + 3r'$ edges. We only have to consider the distance change from t and r' nodes of this new set. To have $r' = \Theta(n)$, we might have to consider a graph G with $\Theta(n)$ nodes (at most $2n$ is enough).   □

3   Number of Liars after ℓ Deletions

Lemma 5. *After ℓ edge deletions in any graph G of diameter D, every connected component of the resulting graph have diameter at most $D(\ell + 1)$.*

Proof. As claimed in [SBvL87], given ℓ, the maximum diameter of the graph obtained by deleting ℓ edges from a graph G of diameter D is $D(\ell+1)$, assuming that the resulting graph is still connected. Now, if a single deletion disconnect in two parts a connected component of diameter D, both resulting components will have diameter at most D. So, after $\ell + 1$ deletions, any connected component has at most diameter $D(\ell + 1)$. $\qquad\square$

Theorem 4. *Let G be a n-nodes, m-edges graph of diameter D without any liars. For any $\ell \leq m$, after ℓ edges deletion uniformly at random in G the number of liars is $O(\frac{\ell^2 Dn}{m})$.*

Proof. As stated in *Theorem 2*, deleting one edge into a graph of diameter D creates an average of at most $D(n-1)/m$ liars. From *Lemma 5*, after the deletion of ℓ edges, the expected number of liars is

$$\mathbb{E}(k) \leq \sum_{i=1}^{\ell} \frac{Di(n-1)}{m-(i-1)} \leq \sum_{i=1}^{\ell} \frac{Di(n-1)}{m-(\ell-1)} \leq \frac{D(n-1)}{m-(\ell-1)} \sum_{i=1}^{\ell} i$$

or

$$\mathbb{E}(k) \leq \frac{D(n-1)}{m-(\ell-1)} \times \frac{\ell(\ell-1)}{2}$$

$\qquad\square$

4   Specific Topologies

In this section, we show how tight the bounds are for some specific topologies. We just briefly describe the sketch of proofs. The study gives a justification for the introduction of the adversary model. In order to get tight bounds in the random fault model, we exhibit the worst configurations of advice and evaluate their probabilities in the random fault model.

Theorem 5. *In the adversary model,*

- $k = \Theta(n)$ *for ErdsRnyi's random graphs with parameter $p = 1/2$;*
- $k = \Theta(\sqrt{n})$ *for square grids;*
- $k = \log_2 n - 1$ *for hypercube.*

In the random fault model,

- $k = \Theta(1/n)$ *for ErdsRnyi's random graphs with parameter $p = 1/2$;*
- $k = \Theta(1)$ *for square grids;*
- $k = \Theta(1/\log n)$ *for hypercube.*

Here is some clue about the behaviour of the different graph families in the adversary model :

- ErdsRnyi's random graphs: each pair of nodes is connected with probability p. For $p = 1/2$, almost all graphs have diameter 2. If the deleted edge is between L_1 and L_2 then only 1 node can turn into a liar. However, a deletion between $L_0 = \{t\}$ and L_1 can create $\Theta(n)$ liars since on average, there are $(n-1)/4$ neighbours in L_2 of any individual node of L_1.

- grids: only nodes that share a coordinate (same row or column) with t have (s,t)-arterial edges and thus can change their distance to t. The number of distance changes is then $|\mathcal{S}| = \Theta(\sqrt{n})$ for square grids. An adversary can force all neighbours of $\mathcal{S}$ to point to $\mathcal{S}$. From *Lemma 2*, we get that $k = \Theta(\sqrt{n})$.
- hypercube: only target's neighbours can increase their distance to t after one edge deletion, so $|\mathcal{S}| \leq 1$ and only $k \leq \log_2 n - 1$ nodes of level L_2 can become liars.

In order to get tight bounds for the random fault model, we simulate the m possible edge deletions and average k:

- ErdsRnyi's random graphs: only edges leading to advice deletion can create liars. Condition on this event, on average, only $\Theta(1)$ liars appear. However, this event occurs with probability $\Theta(1/n)$. In the other cases, no liar are obtained.
- grids: with probability $1 - \Theta(1/\sqrt{n})$, there is no (s,t)-arterial edge between a random node and t. It follows that, with probability $1 - \Theta(1/\sqrt{n})$, we have at most one new liar (if the deleted edge contains an advice). With probability $\Theta(1/\sqrt{n})$, we have $\Theta(\sqrt{n})$ liars.
- hypercube: only edges leading to an advice deletion or being neighbours of t can create liars. However neighbours of t can not become liars. For nodes of levels $L_{i \geq 2}$, there is no distance change after one edge deletion. Since $\mathbb{E}(\mathcal{F}(e)) = \frac{n-1}{n \log_2 n} = \Theta(1/\log n)$, we have $\mathbb{E}(k) = O(1/\log n)$. To get a lower bound of $\Omega(1/\log n)$, we just have to consider the $n/2$ closest nodes from t. The probability that the deleted edge is linked to one of these nodes is at least $1/2$ and condition on this event, with probability at least $\frac{1}{2 \log_2 n}$, a new advice is required and create a liar.

5   Conclusion

This work shows the importance of the diameter for the number of distance changes and liars appearances in a dynamic graph model. Of course, it would be interesting to consider edge/node addition. Contrary to edge deletion, an edge addition can drastically change the distance within the graph. Even for grids, the number of distance changes would be $\Omega(n)$ after a random edge addition.

References

Ber09. Bernstein, A.: Fully dynamic $(2 + \text{epsilon})$ approximate all-pairs shortest paths with fast query and close to linear update time. In: FOCS, pp. 693–702. IEEE Computer Society (2009)

BK09. Bernstein, A., Karger, D.R.: A nearly optimal oracle for avoiding failed vertices and edges. In: Mitzenmacher, M. (ed.) STOC, pp. 101–110. ACM (2009)

CG84. Chung, F.R.K., Garey, M.R.: Diameter bounds for altered graphs. Journal of Graph Theory 8(4), 511–534 (1984)

CLPR10. Chechik, S., Langberg, M., Peleg, D., Roditty, L.: f-Sensitivity Distance Oracles and Routing Schemes. In: de Berg, M., Meyer, U. (eds.) ESA 2010, Part I. LNCS, vol. 6346, pp. 84–96. Springer, Heidelberg (2010)

DI04. Demetrescu, C., Italiano, G.F.: A new approach to dynamic all pairs shortest paths. J. ACM 51(6), 968–992 (2004)

DTCR08. Demetrescu, C., Thorup, M., Chowdhury, R.A., Ramachandran, V.: Oracles for distances avoiding a failed node or link. SIAM J. Comput. 37, 1299–1318 (2008)

HIKN10. Hanusse, N., Ilcinkas, D., Kosowski, A., Nisse, N.: Locating a Target with an Agent Guided by Unreliable Local Advice. In: Proceedings of the 29th Annual ACM SIGACT-SIGOPS Symposium on Principles of Distributed Computing PODC 2010, Zurich, Suisse, pp. 355–364. ACM, New York (2010)

HKK04. Hanusse, N., Kranakis, E., Krizanc, D.: Searching with mobile agents in networks with liars. Discrete Applied Mathematics 137, 69–85 (2004)

HKKK08. Hanusse, N., Kavvadias, D.J., Kranakis, E., Krizanc, D.: Memoryless search algorithms in a network with faulty advice. Theor. Comput. Sci. 402(2-3), 190–198 (2008)

HS01. Hershberger, J., Suri, S.: Vickrey prices and shortest paths: What is an edge worth? In: FOCS, pp. 252–259 (2001)

KB10. Khanna, N., Baswana, S.: Approximate shortest paths avoiding a failed vertex: Optimal size data structures for unweighted graphs. In: Marion, J.-Y., Schwentick, T. (eds.) STACS. LIPIcs, pp. 513–524. Schloss Dagstuhl - Leibniz-Zentrum fuer Informatik (2010)

Kin99. King, V.: Fully dynamic algorithms for maintaining all-pairs shortest paths and transitive closure in digraphs. In: FOCS, pp. 81–91 (1999)

KK99. Kranakis, E., Krizanc, D.: Searching with uncertainty. In: Proc. SIROCCO 1999, pp. 194–203 (1999)

NPW03. Nardelli, E., Proietti, G., Widmayer, P.: Finding the most vital node of a shortest path. Theor. Comput. Sci. 296, 167–177 (2003)

SBvL87. Schoone, A., Bodlaender, H., van Leeuwen, J.: Improved Diameter Bounds for Altered Graphs. In: Tinhofer, G., Schmidt, G. (eds.) WG 1986. LNCS, vol. 246, pp. 227–236. Springer, Heidelberg (1987)

Tho04. Thorup, M.: Fully-Dynamic All-Pairs Shortest Paths: Faster and Allowing Negative Cycles. In: Hagerup, T., Katajainen, J. (eds.) SWAT 2004. LNCS, vol. 3111, pp. 384–396. Springer, Heidelberg (2004)

N-party BAR Transfer

Xavier Vilaça, João Leitão, Miguel Correia, and Luís Rodrigues

INESC-ID, Instituto Superior Técnico, Universidade Técnica de Lisboa

Abstract. We introduce the N-party BAR transfer problem that consists in reliably transferring arbitrarily large data from a set of N producers to a set of N consumers in the BAR model, i.e., in the presence of Byzantine, Altruistic, and Rational participants. The problem considers the existence of a trusted observer that gathers evidence to testify that the producers and consumers have participated in the transfer. We present an algorithm that solves the problem for $N \geq 2f + 1$, where f is the maximum number of Byzantine processes in each of the producer and consumer sets. We do not impose limits on the number of Rational participants, although they can deviate from the algorithm to improve their utility. We show that our algorithm provides a Nash equilibrium.

1 Introduction

Peer-to-peer systems may be used to provide temporary or long-term storage services. Such services are useful in a number of settings. For instance, peer-to-peer systems can be used to process large volumes of data using volunteer computation, as illustrated by projects such as SETI@home [4] and, more recently, by the Boinc infrastructure that supports several computationally intensive research projects [3]. If such computations are performed using MapReduce, information produced by mappers needs to be transferred to the reducers or to intermediate storage. Volunteer storage nodes may not be willing to store data indefinitely, so they have to transfer data to other nodes after serving the system for some time. In any case, volunteers expect to be recognized for their contribution, for instance by being awarded credits that make them appear in a chart with the top contributors of the project.

In scenarios such as the ones listed above, a reliable protocol to transfer data from a set of producers to a set of consumers is an important building block. Any realistic service for this environment has to consider the existence of both Byzantine and Rational nodes, i.e., of nodes that deviate from the protocol, respectively, in an arbitrary way (Byzantine) and with the purpose of gaining some measurable benefit like being listed as top contributors without really executing jobs (Rational). A system model that captures the existence of these different kinds of participants is the Byzantine-Altruistic-Rational (BAR) model [2].

This paper introduces the N-party BAR Transfer problem (BAR-Transfer). This problem can be informally defined as follows. There are N producers and N consumers, which we generically call processes. Up to f processes of each of these sets can be Byzantine; the remainder are either Altruistic or Rational.

A. Fernández Anta, G. Lipari, and M. Roy (Eds.): OPODIS 2011, LNCS 7109, pp. 392–408, 2011.

All non-Byzantine producers have the same piece of arbitrarily large data that they have to transfer to all non-Byzantine consumers. Altruistic processes follow the protocol, Byzantine processes deviate arbitrarily from the protocol (e.g., omitting or sending modified messages), and Rational processes deviate from the protocol following a strategy to increase their utility. There is an abstract trusted observer that is not involved in the transfer, but that collects evidence about it. BAR-Transfer is the problem of reliably transferring data from the producers to the consumers, while providing the trusted observer enough evidence to testify which processes participated in the transfer.

Systems not designed to cope with Rational behaviour may fall into the Tragedy of Commons [15]: the job is not done because all participants are Rational and aim for profit by not performing (part of) their role. To model Rational behaviour, we use an approach based on Game Theory [25]. The protocol executed by the processes is modelled as a game, in which each player (i.e., process) follows a strategy to increase its utility. To contradict this behaviour, an algorithm to solve BAR-Transfer should provide a Nash equilibrium, so that no Rational process has an incentive to deviate from the protocol. We model the BAR-Transfer problem as a strategic game, in which players choose a strategy simultaneously, once and for all [25], i.e., without knowledge of the others strategies and without the ability of changing it during the algorithm execution. This is not a restriction in the case of our algorithm as explained later. We do the usual assumption [9] that processes are risk-averse, i.e., that they do not follow a strategy that may put their profit at risk.

Besides introducing the BAR-Transfer problem, we present an algorithm that solves it in a synchronous message-passing distributed system. We prove its correctness and that it provides a Nash equilibrium which is a dominant strategy. Therefore, all Rational processes should follow our solution. The paper makes the following main contributions: i) defines the BAR-Transfer problem; ii) proposes an algorithm that solves BAR-Transfer; iii) proves the correctness of the algorithm and that it provides a Nash equilibrium.

The remaining of the paper is structured as follows. Section 2 compares this work with related work. Section 3 describes the system model and defines the BAR-Transfer problem. The algorithm to solve the problem is presented in Section 4. The correctness and cost of the algorithm are analysed in Section 5. Finally, Section 6 concludes the paper.

2   Related Work

The BAR-Transfer problem is related to classical distributed systems problems such as Byzantine Agreement (BA), Reliable Broadcast (RB), Terminating Reliable Broadcast (TRB), and Interactive Consistency (IC) [19,11,7]. A first and major difference is that these algorithms are executed among a single set of processes, while BAR-Transfer is about communication and agreement between two sets: producers and consumers. In that sense there is some resemblance with Paxos with its three process roles – proposers, acceptors, learners – but in

BAR-Transfer all producers are proposers of the same value, a notion that does not exist in Paxos [18]. An algorithm for solving BAR-Transfer might be implemented by running N instances of algorithms that solved these problems, or even a single one in the case of IC. However, these solutions would be very inefficient in terms of message, time, and bit complexity because they would not exploit the fact that all (non-Byzantine) producers send the same data. Furthermore, these problems do not consider the BAR model. If there were Rational processes, algorithms that solved these problems would not satisfy their properties. The same discussion applies to All-to-All Reliable Broadcast (ATA-RB) algorithms [20,14]. Although they reduce the number of messages sent when compared with parallel executions of BA, RB or TRB, to the best of our knowledge there is no work in ATA-RB algorithms in the BAR model. Besides, these algorithms only provide probabilistic guarantees.

Many Byzantine fault-tolerant algorithms have some relation to our work. Several papers presented implementations of registers based on Byzantine quorum systems [23,24]. Others presented algorithms to implement state machine replication, a generic solution to implement fault-tolerant distributed services [8,17]. In both cases the objective is to ensure that Byzantine nodes are unable to disrupt the consistency of the data stored in the servers or the service provided by the servers. In contrast, our work aims at ensuring the transference of a correct value from a set of nodes that produce the data independently, although following a deterministic function, to another set of nodes which have to determine which is the correct data. A third set of papers presented Byzantine fault-tolerant consensus algorithms for asynchronous systems, which might also be used as building blocks of less efficient solutions of BAR-Transfer [12,6,10]. Again, none of these works considers the BAR model.

Some works applied Game Theory to problems involving both Rational and Byzantine players. Eliaz introduced the notion of k-Fault Tolerant Nash equilibrium (k-FNTE), as an equilibrium in which no Rational participant has any incentive to unilaterally deviate from the expected behaviour, with up to k players whose strategy is arbitrary [13]. This concept was applied to auctions. Abraham et al. extended the work of [13] by introducing the notion of (k, t)-robustness, where k is the maximum number of colluding Rational participants and t is the upper limit to the number of Byzantine players [1]. The authors propose a solution for secret sharing that is (k, t)-robust. Contrary to our work, they assumed that the utility of each player depends only on the output of the algorithm, therefore ignoring communication costs. It has been proved that no non-trivial distributed protocol for which Rational nodes take into consideration communication costs can be (k, t)-robust [9].

The Byzantine-Altruistic-Rational (BAR) model was proposed as an abstraction for capturing these three distinct behaviours of processes [2]. The authors also proposed a general three-tied architecture for developing BAR-tolerant protocols, in cooperative distributed systems that span Multiple Administrative Domains (MAD). The first two levels of the proposed architecture implement a Replicated State Machine using a BAR-tolerant TRB protocol [9] and a mechanism than enforces periodic work and guarantees responses. Although this architecture might be used to solve the BAR-Transfer problem, the use of the TRB protocol for transferring arbitrarily large data is too costly and the guaranteed response mechanism requires the active participation of a witness, which must be either a centralized entity or implemented through message broadcast to all the remaining nodes. Furthermore, the proposed mechanisms are based on the long-term cooperation between participants modelled as a repeated game [25], which is not the case of the BAR-Transfer problem.

The same authors [9] have shown that the Dolev and Strong's TRB protocol [11] can be changed to provide a Nash equilibrium in the BAR model using ∞-tit-for-tat mechanisms [5]. The problem is modelled as a repeated game with an infinite number of rounds. Each round a different participant runs an instance of the protocol to broadcast its information to all the remaining non-Byzantine participants. They proved that Rational participants cannot expect any increase in their utility by omitting messages, even if a fraction of the participants is Byzantine. In this work we are interested in large peer-to-peer networks in which it is unlikely that the same participants interact more than once. For that reason we do not consider repeated executions, but model the algorithm instead in terms of a strategic game in which players interact only once. Therefore, in our case it is not possible to apply the incentive mechanisms of [2,22,21] based on tit-for-tat. Furthermore, none of these works addresses the problem of transferring an arbitrarily large value without using an active witness or direct reciprocity.

The BAR model has also been used with gossip data dissemination algorithms [22,21]. These algorithms are not directly applicable to solve BAR-Transfer as they assume that the source of the information is trusted and provide no guarantee that the disseminated information reaches its destination. Furthermore, data transfer between each pair of nodes is performed using direct reciprocity in a fair exchange process. This requires that each Rational participant has incentives to transfer data if it expects to receive an equivalent contribution from its peer. In addition, the pestering mechanism of BAR Gossip [22] only provides a Nash equilibrium if a certain fraction of the participants are Altruistic [26]. In BAR-Transfer, consumers do not possess any data that may serve as currency to pay the producers for the transfer, and no assumption is made about the presence of Altruistic participants. Equicast [16] also implements a dissemination protocol in an environment with selfish participants, which is proven to provide a Nash equilibrium. However, it assumes that Rational processes only deviate from the protocol by adjusting a cooperation factor.

3 System Model and Problem Statement

3.1 System Model

The BAR-Transfer problem involves a set of *producers* $\mathcal{P}$ of cardinality $N_{\mathcal{P}}$ and a set of *consumers* $\mathcal{C}$ of cardinality N_c. To simplify the description of our algorithm, in this paper, we consider that the cardinality of both sets is the same, i.e., $N_{\mathcal{P}} = N_{\mathcal{C}} = N$. We do not address the problem of forming these sets, in this paper. However, we assume that this mechanism ensures with high probability that the number of Byzantine processes is upper bounded and that processes cannot influence this mechanism. There is also a special process called *trusted observer* (TO). We use the words *processes* or *participants* to designate these entities. Sometimes we use the word *players* to designate producers and consumers, when we model their interaction as a game.

We assume that the system is synchronous (there are maximum communication and processing delays) and that all processes are fully connected by authenticated reliable channels. This is a reasonable assumption as we require that the transfer may terminate after a finite period of time such that Rational processes may have some guarantees that they will be eventually rewarded. However, it is not strictly necessary for the communication and processing delays to be upper bounded. Nevertheless, in order to simplify the description of our algorithm, we will make that assumption. We also assume that each process has a public-private key pair and that there is a public-key infrastructure in place, so every process has access to the public key of all others. Each process has access to a collision-resistant hash function (*hash*) and a signature function based on public-key cryptography (*sign, verifysig*).

Participants can be Byzantine, Altruistic, and Rational, in accordance with the BAR model. We assume that up to f elements of each of the $\mathcal{P}$ and $\mathcal{C}$ sets can be Byzantine. Any number of consumers and producers can be Altruistic or Rational. The trusted observer TO always follows its protocol.

An Altruistic process is one that follows the protocol. A Byzantine process can deviate arbitrarily from its behaviour, e.g., by sending or not sending certain messages, or by sending messages in a format or with content that is not according to the protocol. Byzantine processes however are not able to break the cryptographic mechanisms used in the algorithm (e.g., they are not able to generate signatures on behalf of Altruistic or Rational processes).

A Rational process is one that aims at maximizing a *utility function*, defined in terms of *benefits* and *costs*. A producer has a benefit by proving to the *TO* that it has contributed to the transfer; it incurs on the cost of sending the data. Consumers send to the *TO* acknowledgements of the reception of the data. A consumer benefits by obtaining the data and proving its reception to the *TO*; it incurs in the costs of receiving and processing messages and sending the acknowledgements to the *TO*. We assume that there is no collusion among Rational processes.

3.2   The BAR-Transfer Problem

The BAR-Transfer problem can be defined as follows. Each producer p has a value (or data) of arbitrary size v_p such that, for any two non-Byzantine producers p_i and p_j, $v_{p_i} = v_{p_j} = v$. Sometimes we refer to this value as the *correct value*, to denote that it is the value held by all non-Byzantine producers.

The algorithm terminates successfully when every non-Byzantine consumer consumes v. A consumer c is said to *consume* value v_c when the primitive *consume*(c, v_c) is called. All non-Byzantine producers start the algorithm by producing value v. A producer p is said to *produce* value v_p by calling the primitive $produce(p, v_p)$. The *TO* is said to *produce evidence* about the transfer by calling primitive $certify(TO, evidence)$. There are also two predicates *hasProduced(evidence, p_i)* and *hasAcknowledged(evidence, c_j)* that take as input the evidence produced by the *TO* to indicate, respectively, if producer p_i participated in the BAR-Transfer and if consumer c_j notified the reception of the correct value. The problem consists informally in i) transferring the value from the producers to the consumers; and ii) providing evidence about the transfer. More formally the problem is defined in terms of the following properties:

- **BAR-Transfer 1** *(Validity):* If a non-Byzantine consumer consumes v, then v was produced by some non-Byzantine producer.
- **BAR-Transfer 2** *(Integrity):* No non-Byzantine consumer consumes more than once.
- **BAR-Transfer 3** *(Agreement):* No two non-Byzantine consumers consume different values.
- **BAR-Transfer 4** *(Termination):* Every non-Byzantine consumer consumes a value.
- **BAR-Transfer 5** *(Evidence):* The trusted observer produces evidence about the transfer.
- **BAR-Transfer 6** *(Producer Certification):* if producer p is non-Byzantine, then *hasProduced(evidence, p)* is *true*.
- **BAR-Transfer 7** *(Consumer Certification):* if consumer c is non-Byzantine, then *hasAcknowledged(evidence, c)* is *true*.

With these definitions in mind, we can provide a more precise characterization of the *benefits* that Rational nodes aim to obtain. The benefit of a producer p is to have *hasProduced(evidence, p)* *true*. The benefit of a consumer c is twofold: i) to obtain the correct value and ii) to have *hasAcknowledged(evidence, c) true*.

4   BAR-Transfer Algorithm

We now present an algorithm that solves the BAR-Transfer problem (Alg. 1). The algorithm requires $N \geq 2f+1$ producers and consumers. The algorithm aims at ensuring that each consumer receives the value and can decide which is the correct value, in case it receives several different values (e.g., due to Byzantine producers). To satisfy this goal, each producer is not required to send a copy of the (possibly large) value to every consumer. In fact, it is enough that it sends the value to $f + 1$ consumers and a signed hash of the value to the remaining $N - f - 1$ consumers.

We define a deterministic function that returns the set of consumers that receive a copy of the value from producer p_i, denoted *consumerset$_i$*, as: $consumerset_i = \{c_j | j \in [i...(i + f) \mod N]\}$. The intuition behind this function is that the consumers are seen as a circular space where each producer is responsible for sending the value it has computed to a set of consecutive consumers of cardinality $f + 1$, which are shifted from one another by one position.

We model the operation of the algorithm in *rounds*. The round of a process is increased as result of a *nextRound* event. The system is synchronous, so non-Byzantine processes have their clocks synchronized and the *nextRound* event occurs simultaneously in all of them. The synchrony of the system and reliability of the channels ensure that if in response to event *nextRound(n)* a non-Byzantine process sends a message to another non-Byzantine process, that message is delivered to the destination before *nextRound(n+1)* is triggered. This implies that *nextround* events are triggered periodically with a period greater than the worst case latency of communication channels. The algorithm executes in three rounds. In round 0, the producers send values or hashes to consumers. In round 1, consumers send certificates of reception to the trusted observer. In round 2, the trusted observer produces the evidence.

In round 0, a producer computes the hash of the value and signs it (lines 106-108). When the first round starts, it sends the value, its hash, and signature to the consumers in *consumerset$_i$* (lines 111-113), but only the hash and signature to the remaining consumers (lines 114-116).

A consumer starts by waiting for signed values and hashes from producers in round 1 (lines 209 and 215). Each value, hash, and signature received is stored in an array named *values* (lines 214 and 219). If a node does not send the message it was supposed to in this round, or if the hash or signature are not valid, the entry in the values set for that producer remains with the special value $\perp$, which will serve to build a proof of misbehaviour for the TO (if $f+1$ consumers provide similar certificates).

When round 1 ends, the consumer picks the value v such that $hash(v)$ appears in more than f positions of the array (lines 222-223). There are at most f faulty producers in the system, thus there is at most one value that matches this condition. Then, the consumer prepares the *confirm* array to serve as a *certificate* that vouches for the correct or incorrect behaviour of all producers, and that simultaneously proves that it has received and picked the correct value as described below (lines 224-225). For each producer p_i, the consumer either stores in *confirm*: i) the received hash and corresponding signature (extracted from the values set) or ii) the special value $\perp$ when no data, or incorrect data, was received from that producer. The consumer then signs this data structure with its private key and sends it as a proof of reception to the trusted observer (lines 226-228). The consumer terminates by outputting the value (line 229).

The trusted observer waits for a certificate from each consumer in round 2 (line 306). The certificates are collected in an array called *evidence* (line 309). In the end, the trusted observer produces the array as evidence (line 311).

Considering the data structure that is created by the trusted observer as evidence, we can now define with more detail the predicates *hasProduced* and *hasAcknowledged*. Let $h(v)$ denote the hash of the value v and let $s_{p_k}(h(v))$ denote the hash of v signed by the producer p_k:

– *hasProduced(evidence, p_i)* is true if the following condition holds: there are at least $N-f$ consumers $c_k \in \mathcal{C}$: $evidence[c_k][p_i] = \langle h(v), s_{p_i}(h(v)) \rangle$. It is false otherwise.

Algorithm 1. BAR-Transfer Algorithm

producer p_i:

```
101    upon init do
102        myvalue := ⊥;
103        myhash :=⊥;
104        myhashsig := ⊥;
105        round := 0;
106    upon produce(p_i,myvalue) ∧ round = 0 do
107        myhash := hash(myvalue);
108        myhashsig := sign (p_i, myhash);
109    upon nextRound ∧ round = 0 do // start of round 1
110        round := 1;
111        msgsig := sign (p_i, VALUE || myvalue || myhash || myhashsig);
112        forall c_j ∈ consumerset_i do
113            send (p_i, c_j, [VALUE, myvalue, myhash, myhashsig, msgsig])
114        msgsig := sign (p_i, SUMMARY || myhash || myhashsig);
115        forall c_j ∈ C\consumerset_i do
116            send (p_i, c_j, [SUMMARY, myhash,myhashsig, msgsig])
```

consumer c_j:

```
201    upon init do
202        myvalue :=⊥;
203        myhash:=⊥;
204        confirm := [⊥]^P;
205        values := [⊥]^P;
206        round := 0;
207    upon nextRound ∧ round = 0 do // start of round 1
208        round := 1;
209    upon deliver (p_i, c_j, [VALUE, pvalue, phash, phashsig, msgsig]) ∧ round = 1 do
210        if (c_j ∈ consumerset_i)then
211            if verifysig(p_i, VALUE || pvalue || phash || phashsig, msgsig)then
212                if verifysig(p_i,phash, phashsig) then
213                    if verifyhash(pvalue, phash) then
214                        values[p_i] := ⟨pvalue, phash, phashsig⟩;
215    upon deliver (p_i, c_j, [SUMMARY, phash, phashsig, msgsig]) ∧ round = 1 do
216        if (c_j ∉ consumerset_i)then
217            if verifysig(p_i, SUMMARY ||phash || phashsig, msgsig) then
218                if verifysig(p_i, phash, phashsig) then
219                    values[p_i] := ⟨⊥, phash, phashsig⟩;
220    upon nextRound ∧ round = 1 do // start of round 2
221        round := 2;
222        myhash := h : #({p|value[p] = ⟨*, h, *⟩}) > f.
223        myvalue := v : {p|value[p] = ⟨v, myhash, *⟩}.
224        forall p_i: values[p_i] = ⟨*, myhash, *⟩ do
225            confirm[p_i] := ⟨values[p_i].hash, values[p_i].signature⟩;
226        confsig := sign (c_j, confirm);
227        msgsig := sign (c_j, CERTIFICATE||confirm||confsig);
228        send (c_j, TO, [CERTIFICATE, confirm, confsig, msgsig])
229        consume (c_j, myvalue);
```

trusted observer TO:

```
301    upon init do
302        evidence:= [⊥]^C;
303        round := 0;
304    upon nextRound ∧ round < 2 do
305        round := round+1;
306    upon deliver (c_j, TO, [CERTIFICATE, confirm, confsig, msgsig]) ∧ round = 2 do
307        if verifysig (c_j, CERTIFICATE||confirm||confsig, msgsig) then
308            if verifysig (c_j, confirm, confsig) then
309                evidence[c_j] := ⟨confirm, confsig⟩;
310    upon nextRound ∧ round = 2 do // start of round 3
311        certify (TO, evidence);
```

- *hasAcknowledged(evidence, c_j)* is true if exists a set of producers, named *correctset$_j$*, such that $|correctset_j| \geq N - f$ and for $\forall p_k \in correctset_j$ *hasProduced(evidence, p_k)* is true and $evidence[c_j][p_k] = \langle h(v), s_{p_k}(h(v)) \rangle$. It is false otherwise.

The algorithm does not require the observer to actively participate in the execution of the algorithm. Furthermore, the verification process performed by the trusted observer is independent for each transfer. Therefore many instances of BAR-Transfer can be executed in parallel under the jurisdiction of one or more trusted observers, without the trusted entity being a single point of failure or a bottleneck.

5   Analysis

The analysis of the algorithm has three parts. First, we prove its correctness. Then, we demonstrate that it is a Nash equilibrium. Finally, we perform a complexity analysis in terms of communication costs.

5.1   Correctness

This section provides a proof of the correctness of the algorithm, i.e., that it satisfies the properties BAR-Transfer 1-7. The proof assumes that at most f producers and f consumers are Byzantine and that the rest of the processes follow the algorithm, i.e., are Altruistic. The case of Rational processes is left for Section 5.2, in which we show that Rational processes also follow the algorithm.

We now show with the following Lemmas that the algorithm presented in Section 4, satisfies each of the BAR-Transfer properties.

Lemma 1. (Validity) *If a non-Byzantine consumer consumes v, then v was produced by some non-Byzantine producer.*
Proof. A non-Byzantine consumer c consumes v only if it receives a $hash(v)$ from at least $f + 1$ producers and v from at least one producer. There are at most f Byzantine producers, which implies that c receives $hash(v)$ from at least a non-Byzantine producer p_i. Thus, c consumes v only if v was input by p_i.

Lemma 2. (Integrity) *No non-Byzantine consumer consumes more than once.*
Proof. A consumer consumes a value when the *consume* primitive is called. A trivial inspection of the algorithm shows that this primitive can be called only once in a non-Byzantine consumer, thus it consumes the value no more than once.

Lemma 3. (Agreement) *No two non-Byzantine consumers consume different values.*
Proof. By Lemma 1, if a non-Byzantine consumer consumes v, then v was produced by some non-Byzantine producer. By assumption, every non-Byzantine producer produces the same value. Therefore, non-Byzantine consumers never deliver a value different from v.

Lemma 4. (Termination) *Every non-Byzantine consumer consumes a value.*

Proof. All the non-Byzantine producers produce and send v or its hash to all the consumers in the beginning of round 1. Given that channels are reliable and synchronous, all non-Byzantine consumers receive these values in that round. Therefore, when round 2 begins, every non-Byzantine consumer must possess both the correct value and $f + 1$ or more hashes of that value, so it executes the *consume* primitive which means consuming v.

These four properties ensure the reliable transfer of the correct value in the presence of Byzantine participants. In the following Lemmas, we prove that the properties BAR-Transfer 5-7 related to Rational behaviour are fulfilled, therefore ensuring that each node that obeys the protocol is rewarded after the completion of the transfer.

Lemma 5. (Evidence) *The trusted observer produces evidence about the transfer.*

Proof. A trivial inspection of the algorithm shows that *certify(TO, evidence)* is executed at the end of round 2, which is the same as saying the trusted observer produces evidence.

Lemma 6. (Producer Certification) *If producer p is non-Byzantine, then hasProduced(evidence, p) is true.*

Proof. By Lemmas 3 and 4, every non-Byzantine consumer delivers the same value v. Before delivering these consumers send their *confirm* vectors to the trusted observer. Therefore, there are at least $N - f$ non-Byzantine consumers $c_k \in \{c_1 \ldots c_{N-f}\}$ that send confirm vectors to the trusted observer at the start of round 2. If producer p_i followed the algorithm, each of these consumers c_k has received $hash(v)$ from p_i, and included $\langle h(v), s_{p_i}(h(v)) \rangle$ in the message sent to the trusted observer. Since all those messages are included in the evidence generated by the trusted observer, *hasProduced(evidence, p_i)* is true.

Lemma 7. (Consumer Certification) *If consumer c is non-Byzantine, then hasAcknowledged(evidence, c) is true.*

Proof. A non-Byzantine consumer sends its *confirm* vector to the trusted observer. Also, since there are at least $N - f$ non-Byzantine producers, consumer c_j includes $\langle h(v), s_{p_i}(h(v)) \rangle$ for each of these non-Byzantine producers p_i in the *confirm* vector sent to the trusted observer. According to Lemma 6, there exists a set *correctset* of at least $N - f$ producers $p_k \in \{p_1 \ldots p_{N-f}\}$ for which *hasProduced(evidence,p_k)* is true (the set of $N - f$ non-Byzantine producers). Therefore, *hasAcknowledged(evidence, c_j)* becomes true for any non-Byzantine consumer c_j.

Theorem 1. (Correctness) *If all non-Byzantine participants follow the protocol, then the provided algorithm solves the BAR-Transfer problem defined in terms of properties BAR-Transfer 1-7.*

Proof. The proof follows directly from Lemmas 1, 2, 3, 4, 5, 6, and 7.

5.2   Game Theoretic Analysis

To prove that the protocol provides a Nash equilibrium, we model the BAR-Transfer problem as a strategic game $\Gamma = (M, S_M, \boldsymbol{u})$, where $M = \mathcal{P} \bigcup \mathcal{C}$ is the set of players, S_M the set of all possible strategies, and $\boldsymbol{u}$ is a vector with the utility functions of all players.

Each player decides its *strategy* (or plan of action) once and it remains valid for all its actions during the execution of BAR-Transfer. These decisions about the strategy are made simultaneously and, as Rational players do not collude among themselves, without knowledge of the strategies selected by other players. The set of possible strategies for player i is denoted S_i. S_M consists on the set of all possible strategies, i.e., of S_i for all $i \in M$. The set of all possible strategies of *producers* is S_P. Altruistic producers send $hash(v)$ to all consumers and the value v to the consumers of $consumerset_i$. Rational producers send $hash(v)$ to any subset of $\mathcal{C}$ and the value to any subset $\mathcal{C}' \subseteq \mathcal{C}$. Similarly, S_C denotes the set of all possible strategies that can be followed by *consumers*. Altruistic consumers process all the information received from producers, send it to the TO, and consume one value. Rational consumers may or may not: consume a value, process all the values or hashes received from producers, and send the received information to the TO. Byzantine players follow an arbitrary strategy from $S_{\mathcal{F}}$, where $\mathcal{F} = \mathcal{F}_P \cup \mathcal{F}_C$ is the set of all Byzantine producers ($\mathcal{F}_P$) and Byzantine consumers ($\mathcal{F}_C$), such that $|\mathcal{F}_P| \leq f$ and $|\mathcal{F}_C| \leq f$. Notice that these are pure strategies, that is, the decisions about which strategy to follow is deterministic.

We now identify the reasons why modelling the BAR-Transfer problem as a strategic game is not a limitation of our analysis. Strategic games are appropriate for interactions between players where a player cannot form his expectation from the behaviour of the other players on the basis of information about the way that the game was played in the past. The information gathered by each process regarding the past behaviour of other processes is determined by the number of instances of BAR-Transfer in which those processes interacted, which depends on the mechanism used to form the sets of processes in each instance. This mechanism must ensure that with high probability the number of Byzantine processes of each set is upper bounded by f. Furthermore, in a large peer-to-peer network, it is true that N is much smaller than the total number of processes in the system, and the processes connected to the network during the periods when that mechanism is applied vary from instance to instance. Thus, it is reasonable to assume that processes interact with a very small frequency, which implies that the information of each process regarding the nature of other processes is limited and never certain. Since processes do not incur in risks, they cannot form their expectation from the behaviour of the other players on the basis of information about the way that the game was played in the past. Therefore, it is reasonable to model our solution as a strategic game.

In addition, it is only adequate to model a protocol as a strategic game if players do not change their strategy during the execution of the game, which is true in our protocol. Producers cannot increase their knowledge of the strategies of other players during the execution of the algorithm, as they do not obtain any information from any other participant. Thus, the initial chosen strategy remains adequate for the three rounds of the protocol. On the other hand, each consumer c_j learns about the behaviour of producers in round 1, so it can determine which producers adopted the strategy of sending it the value or its hash. However, according to the definition of *hasAcknowledged*, the *TO* rewards c_j based not only on the information included in the certificate sent by the consumer, but also based on the information the producers (certified by c_j) sent to the remaining consumers. Therefore, the information gathered by c_j is insufficient for the consumer to determine if an alternative strategy provides greater expected profits. For these reasons, it is reasonable to assume that Rational participants do not change their strategy during the execution of the protocol.

We define a *profile of strategies* as the correspondence between players and their respective strategy: $\sigma_M : M \mapsto S_M$. By definition, σ_i denotes the strategy followed by player $i \in M$. We define σ_M as the composition of different profile strategies for disjoint subsets of players $M_1, M_2, ..., M_N$: $\sigma_M = (\sigma_{M_1}^1, \sigma_{M_2}^2, ..., \sigma_{M_N}^N)$, where $\sigma_{M_i}^i$ is the strategy followed by all players of M_i and $M = M_1 \cup M_2 \cup \ldots \cup M_N$.

We also define an *utility function* $u_i(\sigma_M) = \beta_i(\sigma_M) - \nu_i(\sigma_M)$ as the profit that player i obtains when all the players follow the strategy specified by σ_M. β_i denotes the benefit obtained by player i. It is assumed that a producer p gets a benefit of ϕ_P only if *hasProduced(evidence, p)* holds true. Otherwise, the benefit is 0. A consumer c only gets a benefit ϕ_C if it consumes the correct value v (therefore, all non-Byzantine consumers consume the correct value) and if *hasAcknowledged(evidence, c)* holds true. The function $\nu_i(\sigma_M)$ maps the costs incurred by player i when every player follows the strategies specified by σ_M. We assume that $\phi_P > \nu_p(\sigma_M)$ and $\phi_C > \nu_c(\sigma_M)$ for any non-Byzantine producer p and consumer c, respectively. To distinguish the arbitrary behaviour of Byzantine players from the strategies of Altruistic and Rational players, we denote by $\pi_P \in \Pi_P$ the profile of strategies of Byzantine producers and by $\pi_C \in \Pi_C$ the profile of strategies of Byzantine consumers.

The remaining of this section provides a proof that the protocol provides a Nash equilibrium. In the BAR model, Rational players also take into consideration Altruistic and Byzantine behaviour [2]. A utility function for Rational player i that considers Byzantine, Altruistic and Rational behaviour, denoted by $\bar{u}_i$, is the expected utility for i if it obeys a given Rational strategy σ_i when all the remaining participants either obey a non-Byzantine strategy specified by the profile σ_M (that includes the Altruistic strategy of following the protocol) or follow a Byzantine strategy specified by the profile $\pi_{\mathcal{F}}$. Given that Byzantine participants may behave arbitrarily, in the definition of the expected utility function it is necessary to consider not only the expected number of Byzantine players but also the probability of each Byzantine player following each of the possible Byzantine strategies. In this work, we assume that players are risk-averse, therefore the expected utility considers the worst possible scenario of Rational and Byzantine behaviour, i.e., it assumes that all non-Byzantine players are Rational and all Byzantine players adopt a strategy that minimizes the utility of non-Byzantine players.

Hereupon, we provide a definition for the *expected utility* $\bar{u}_i(\boldsymbol{\sigma}_M)$ of player $i \in M$ when all Rational players follow the strategy specified by $\boldsymbol{\sigma}_M$. Let $\boldsymbol{\sigma}'_{M \setminus \mathcal{F}, \pi_{\mathcal{P}}, \pi_{\mathcal{C}}} = (\boldsymbol{\sigma}_{M \setminus \mathcal{F}}, \pi_{\mathcal{P}}, \pi_{\mathcal{C}})$ be a profile of strategies where no player in M is Altruistic, all Rational players follow the strategy specified by $\boldsymbol{\sigma}_{M \setminus \mathcal{F}}$, Byzantine producers follow the strategy specified by $\pi_{\mathcal{P}}$, and Byzantine consumers follow the strategy specified by $\pi_{\mathcal{C}}$. The expected utility of player i is given by the following equation:

$$\bar{u}_i(\boldsymbol{\sigma}_M) = \min_{\mathcal{F}_{\mathcal{P}}:|\mathcal{F}_{\mathcal{P}}| \le f, \mathcal{F}_{\mathcal{C}}:|\mathcal{F}_{\mathcal{C}}| \le f} \circ \min_{\pi_{\mathcal{P}} \in \Pi_{\mathcal{P}}, \pi_{\mathcal{C}} \in \Pi_{\mathcal{C}}} u_i(\boldsymbol{\sigma}'_{M \setminus \mathcal{F}, \pi_{\mathcal{P}}, \pi_{\mathcal{C}}}) \qquad (1)$$

Notice the distinction between the expected utility $\bar{u}_i(\boldsymbol{\sigma}_M)$, which denotes the minimum utility Rational player i expects to obtain when all Rational participants follow the strategy specified by $\boldsymbol{\sigma}_M$, and the effective utility $u_i(\boldsymbol{\sigma}'_{M \setminus \mathcal{F}, \pi_{\mathcal{P}}, \pi_{\mathcal{C}}})$, which is the difference between the benefits obtained and the costs incurred by i when Byzantine players follow the specific strategies specified by $\pi_{\mathcal{P}}$ and $\pi_{\mathcal{C}}$.

We can now define the functions $\bar{\beta}_i(\boldsymbol{\sigma}_M)$ and $\bar{\nu}_i(\boldsymbol{\sigma}_M)$ as the expected benefits and costs for the worst possible scenario of Rational and Byzantine behaviour. Thus, the expected utility of player $i \in M$ can also be defined as $\bar{u}_i(\boldsymbol{\sigma}_M) = \bar{\beta}_i(\boldsymbol{\sigma}_M) - \bar{\nu}_i(\boldsymbol{\sigma}_M)$.

We now introduce the notion of *Nash equilibrium*. Let $\boldsymbol{\sigma}^*_{M \setminus \{i\}, \sigma^*_i} = (\boldsymbol{\sigma}_{M \setminus \{i\}}, \sigma^*_i)$ denote the profile of strategies where all Rational players follow the strategy specified by $\boldsymbol{\sigma}_{M \setminus \{i\}}$ and player i follows a given strategy σ^*_i. A Nash equilibrium is a profile of strategies for which no player benefits from deviating from its strategy, which can be stated as follows:

Definition 1. $\boldsymbol{\sigma}_M$ *is a Nash equilibrium if* $\forall_{i \in M} \forall_{\sigma^*_i \in S_i} \bar{u}_i(\boldsymbol{\sigma}_M) \ge \bar{u}_i(\boldsymbol{\sigma}^*_{M \setminus \{i\}, \sigma^*_i})$.

The following Lemmas provide the complete proof that neither the producers nor the consumers benefit from deviating from the protocol. We use $\boldsymbol{\sigma}_{\mathcal{P}}$ and $\boldsymbol{\sigma}_{\mathcal{C}}$ to denote the profile of strategies of, respectively, producers and consumers that comply with the protocol. $\boldsymbol{\sigma}_M$ denotes the composition of the profiles of strategies $\boldsymbol{\sigma}_{\mathcal{P}}$ and $\boldsymbol{\sigma}_{\mathcal{C}}$, and $\boldsymbol{\sigma}^*_M$ denotes an alternative profile of strategies.

In the next Lemma and Corollary, we show that a producer does not benefit from sending the expected information to less than N consumers and from not sending the value to all consumers of *consumerset*. Then, in Theorem 2, we show that no producer can increase its utility by deviating from the protocol, when all consumers follow the expected strategy.

Lemma 8. *For each producer $p \in \mathcal{P}$, for each k such that $0 \le k < N$, let $\boldsymbol{\sigma}^*_M = (\boldsymbol{\sigma}_{\mathcal{P} \setminus \{p\}}, \boldsymbol{\sigma}_{\mathcal{C}}, \sigma^*_p)$ be a deviating profile of strategies, where σ^*_p is the strategy of sending a value or its signature to k consumers. Then, $\bar{\beta}_p(\boldsymbol{\sigma}^*) = 0$.*

Proof. According to the Equation 1, Rational players determine their utility considering the worst case scenario of Byzantine and Rational behaviour. Suppose the set of k consumers to which p sends the information includes all the

Byzantine players. According to the protocol, the trusted observer only receives vectors containing signed hashes in the second round. Hence, if p only sends the signature of the value or its hash to $k < N$ consumers at the beginning of the first round and if no Byzantine consumer sends their vectors to the trusted observer, the trusted observer only receives $max(k - f, 0) < N - f$ vectors that contain $\langle h(v), s_p(h(v)) \rangle$ in the second round. Therefore, *evidence* will not contain $N - f$ entries with $\langle h(v), s_p(h(v)) \rangle$, *hasProduced(evidence, p)* will hold false, and $\bar{\beta}_p(\boldsymbol{\sigma}^*) = 0$.

Corollary 1. *For each producer $p_i \in \mathcal{P}$, for each k such that $0 \leq k < f + 1$, let $\boldsymbol{\sigma}_M^* = (\boldsymbol{\sigma}_{\mathcal{P} \setminus \{p_i\}}, \boldsymbol{\sigma}_{\mathcal{C}}, \sigma_{p_i}^*)$ be a deviating profile of strategies, where $\sigma_{p_i}^*$ is the strategy of sending the value to k consumers. Then, $\bar{\beta}_{p_i}(\boldsymbol{\sigma}^*) = 0$.*

Proof. The proof comes trivially from the previous lemma.

Theorem 2. *No producer has any incentives to deviate from the protocol.*

Proof. It follows from Lemma 8 and Corollary 1 that if the producer p follows an alternative strategy specified by $\boldsymbol{\sigma}_M^*$, then $\bar{\beta}_p(\boldsymbol{\sigma}_M^*) = 0$, $\bar{u}_p(\boldsymbol{\sigma}_M^*) = -\bar{\nu}_p(\boldsymbol{\sigma}_M^*)$, and $\bar{u}_p(\boldsymbol{\sigma}_M^*) < 0$, for the worst possible scenario. According to the Theorem 1, $\bar{\beta}_p(\boldsymbol{\sigma}_M) = \phi_P$, $u_p(\boldsymbol{\sigma}_M) = \phi_P - \bar{\nu}_p(\boldsymbol{\sigma}_M)$, and $\bar{u}_p(\boldsymbol{\sigma}_M) > 0$, since $\phi_P > \bar{\nu}_p(\boldsymbol{\sigma}_M)$. Therefore, $\bar{u}_p(\boldsymbol{\sigma}_M) > \bar{u}_p(\boldsymbol{\sigma}_M^*)$. Since it is assumed that Rational participants are risk-averse, producers do not have incentives to deviate from the protocol.

We now show that no consumer benefits either by not sending the *confirm* vector to the *TO* or by not processing all the information it receives from the producers. Then, in Theorem 4, we prove that no consumer can increase its utility by deviating from the protocol, given that producers follow the expected behaviour.

Lemma 9. *For any consumer $c \in \mathcal{C}$, let $\boldsymbol{\sigma}_M^* = (\boldsymbol{\sigma}_{\mathcal{P}}, \boldsymbol{\sigma}_{\mathcal{C} \setminus \{c\}}, \sigma_c^*)$ be a deviating profile of strategies, where σ_c^* is the strategy of not sending its vector containing hashes sent by producers to the trusted observer in round 2. Then, $\bar{\beta}(\boldsymbol{\sigma}_M^*) = 0$.*

Proof. The proof derives directly from that fact that, if a consumer c does not send its vector, this information is not included in the evidence and *hasAcknowledged(evidence, c)* hods false. Hence, $\bar{\beta}_c(\boldsymbol{\sigma}_M^*) = 0$.

Lemma 10. *For any consumer $c \in \mathcal{C}$, let P_c be the set of producers that sent the correct value or hash to the consumer c, and let $\boldsymbol{\sigma}_M^* = (\boldsymbol{\sigma}_{\mathcal{P}}, \boldsymbol{\sigma}_{\mathcal{C} \setminus \{c\}}, \sigma_c^*)$ be a deviating profile of strategies, where σ_c^* is the strategy of sending an incomplete vector of hashes to the trusted observer with only $f + 1 \leq k < |P_c|$ entries different from the $\perp$ value. Then, $\bar{\beta}_c(\boldsymbol{\sigma}_M^*) = 0$.*

Proof. The worst possible scenario for a non-Byzantine consumer c_j occurs when $|F_{\mathcal{P}}| = f$ and for all these Byzantine producers *hasProduced* is *false*, while they still send valid information to c_j. In this case, there is only one set *correctset$_j$*, where, for all $p \in$ *correctset$_j$*, *hasProduced(evidence,p)* is true: the set of non-Byzantine producers. If c_j does not set $hashes[p_i] = \langle hash(v), s_{p_i}(hash(v)) \rangle$ and p_i is non-Byzantine, then, at the trusted observer, $evidence[c_j]$ will not contain the information of at least $N - f$ producers from *correctset$_j$*. Therefore, *hasAcknowledged(evidence,c)* holds *false*, and $\bar{\beta}_c(\boldsymbol{\sigma}_M^*) = 0$.

Theorem 3. *No consumer has any incentives to deviate from the protocol.*

Proof. It follows from Lemmas 9 and 10 that if the consumer c follows an alternative strategy specified by $\boldsymbol{\sigma}_M^*$, then $\bar{\beta}_c(\boldsymbol{\sigma}_M^*) = 0$, $\bar{u}_c(\boldsymbol{\sigma}_M^*) = -\bar{\nu}_c(\boldsymbol{\sigma}_M^*)$, and $\bar{u}_c(\boldsymbol{\sigma}_M^*) < 0$, for the worst possible scenario. According to the Theorem 1, $\bar{\beta}_c(\boldsymbol{\sigma}_M) = \phi_C$, $\bar{u}_c(\boldsymbol{\sigma}_M) = \phi_C - \bar{\nu}_c(\boldsymbol{\sigma}_M)$, and $\bar{u}_c(\boldsymbol{\sigma}_M) > 0$, since $\phi_C > \bar{\nu}_c(\boldsymbol{\sigma}_M)$. Therefore, $\bar{u}_c(\boldsymbol{\sigma}_M) > \bar{u}_c(\boldsymbol{\sigma}_M^*)$. Since it is assumed that Rational participants are risk-averse, consumers do not have any incentive to deviate from the protocol.

The following Theorem concludes that the protocol provides a Nash equilibrium.

Theorem 4. (Nash equilibrium) *The profile of strategies $\boldsymbol{\sigma}_M$ where every player follows the protocol is a Nash equilibrium.*

Proof. It follows from Theorems 2 and 4 that for every player $i \in M$ and, for all alternative profiles of strategies $\boldsymbol{\sigma}_M^*$ where i deviates from the protocol, $\bar{u}_i(\boldsymbol{\sigma}_M^*) < \bar{u}_i(\boldsymbol{\sigma}_M)$. Hence, $\boldsymbol{\sigma}_M$ is a Nash equilibrium.

From the previous proofs, it is possible to observe that our solution is a dominant strategy, that is, any other Nash equilibrium has a utility lower than the utility that each Rational process expects to obtain when following our solution. Hence, Rational processes should obey our algorithm.

5.3   Complexity Analysis

This section briefly evaluates the algorithm in terms of time, message, and bit complexity. The time complexity is the number of rounds for termination and in this case is constant: 3 rounds. For the other two we consider the case in which all processes follow the protocol. The message complexity, i.e., the number of messages sent by the algorithm, is $N^2 + N$, or $O(N^2)$. The bit complexity, i.e., the number of bits sent, is $O(Nfl_v + N^2 l_s)$, where l_v is the bit length of the value and l_s the bit length of a signature, assuming that $3l_s \gg 2l_h$, where l_h is the bit length of an hash.

6   Conclusions

In this paper we have introduced the BAR-Transfer problem that abstracts the problem of transferring data from a set of producers to a set of consumers under the BAR system model. We have presented an algorithm that solves the BAR-Transfer problem for $N \geq 2f + 1$, where N is the number of producers and consumers. We have shown that our algorithm is a Nash equilibrium, so Rational participants are unable to extract any benefit from deviating from the algorithm. BAR-Transfer is a powerful construct to build peer-to-peer systems that support distributed storage and parallel processing based on volunteer nodes. We are building such a system, based on a P2P architecture, which aims at supporting distributed computations using the MapReduce model.

Acknowledgment. This work was partially supported by the FCT (INESC-ID multi annual funding through the PIDDAC Program fund grant and by the project PTDC/EIA-EIA/102212/2008).

References

1. Abraham, I., Dolev, D., Gonen, R., Halpern, J.: Distributed computing meets game theory: robust mechanisms for rational secret sharing and multiparty computation. In: PODC 2006, Denver, USA, pp. 53–62 (July 2006)
2. Aiyer, S., Alvisi, L., Clement, A., Dahlin, M., Martin, J.-P., Porth, C.: BAR fault tolerance for cooperative services. In: SOSP 2005, Brighton, United Kingdom, pp. 45–58 (October 2005)
3. Anderson, D.: Boinc: A system for public-resource computing and storage. In: GRID 2004, Pittsburgh, USA, pp. 4–10 (November 2004)
4. Anderson, D., Cobb, J., Korpela, E., Lebofsky, M., Werthimer, D.: SETI@home: an experiment in public-resource computing. Communications of the ACM 45(11), 56–61 (2002)
5. Axelrod, R.: The Evolution of Cooperation. Basic Books, New York (1984)
6. Baldoni, R., Helary, J.-M., Raynal, M., Tanguy, L.: Consensus in Byzantine asynchronous systems. J. Discrete Algorithms 1(2), 185–210 (2003)
7. Canetti, R., Rabin, T.: Fast asynchronous Byzantine agreement with optimal resilience. In: STOC 1993, New York, USA, pp. 42–51 (1993)
8. Castro, M., Liskov, B.: Practical Byzantine fault tolerance and proactive recovery. ACM Transactions on Computer Systems 20(4), 398–461 (2002)
9. Clement, A., Napper, J., Li, H., Martin, J.-P., Alvisi, L., Dahlin, M.: Theory of bar games. In: PODC 2007, Portland, USA, pp. 358–359 (August 2007)
10. Correia, M., Neves, N.F., Lung, L.C., Verissimo, P.: Low complexity Byzantine-resilient consensus. Distributed Computing 17(3), 237–249 (2005)
11. Dolev, D., Strong, H.: Authenticated algorithms for Byzantine agreement. SIAM J. Comput. 12(4), 656–666 (1983)
12. Dwork, C., Lynch, N., Stockmeyer, L.: Consensus in the presence of partial synchrony. J. of ACM 35, 288–323 (1988)
13. Eliaz, K.: Fault tolerant implementation. Review of Economic Studies 69(3), 589–610 (2002)
14. Fraigniaud, P.: Asymptotically optimal broadcasting and gossiping in faulty hypercube multicomputers. IEEE Transactions on Computers 41(11), 1410–1419 (1992)
15. Hardin, G.: The tragedy of the commons. Science 162(3859), 1243–1247 (1968)
16. Keidar, I., Melamed, R., Orda, A.: Equicast: Scalable multicast with selfish users. In: PODC 2006, pp. 63–71 (July 2006)
17. Kotla, R., Alvisi, L., Dahlin, M., Clement, A., Wong, E.: Zyzzyva: speculative Byzantine fault tolerance. In: SOSP 2007, Stevenson, USA, pp. 45–58 (October 2007)
18. Lamport, L.: The part-time parliament. ACM Trans. on Computer Systems 16(2), 133–169 (1998)
19. Lamport, L., Shostak, R., Pease, M.: The Byzantine generals problem. ACM Trans. Program. Lang. Syst. 4, 382–401 (1982)
20. Lee, S., Shin, K.G.: Interleaved all-to-all reliable broadcast on meshes and hypercubes. IEEE Transactions on Parallel and Distributed Systems 5(5), 449–458 (1994)
21. Li, H., Clement, A., Marchetti, M., Kapritsos, M., Robison, L., Alvisi, L., Dahlin, M.: Flightpath: Obedience vs choice in cooperative services. In: OSDI 2008, San Diego, USA (December 2008)

22. Li, H., Clement, A., Wong, E., Napper, J., Roy, I., Alvisi, L., Dahlin, M.: BAR gossip. In: OSDI 2006, Seattle, USA, pp. 191–204 (November 2006)
23. Malkhi, D., Reiter, M.: Byzantine quorum systems. In: STOC 1997, El Paso, USA, pp. 569–578 (1997)
24. Martin, J.P., Alvisi, L., Dahlin, M.: Minimal Byzantine storage. In: Malkhi, D. (ed.) DISC 2002. LNCS, vol. 2508, pp. 311–325. Springer, Heidelberg (2002)
25. Martin, O., Ariel, R.: A Course in Game Theory. MIT Press (1994)
26. Wong, E.L., Leners, J.B., Alvisi, L.: It's on Me! The Benefit of Altruism in BAR Environments. In: Lynch, N.A., Shvartsman, A.A. (eds.) DISC 2010. LNCS, vol. 6343, pp. 406–420. Springer, Heidelberg (2010)

Computing with Pavlovian Populations[*]

Olivier Bournez[1], Jérémie Chalopin[2], Johanne Cohen[3],
Xavier Koegler[4], and Mikaël Rabie[5]

[1] Ecole Polytechnique & Laboratoire d'Informatique (LIX),
91128 Palaiseau Cedex, France
Olivier.Bournez@lix.polytechnique.fr
[2] Laboratoire d'Informatique Fondamentale de Marseille, CNRS & Aix-Marseille
Université, 39 rue Joliot Curie, 13453 Marseille Cedex 13, France
Jeremie.Chalopin@lif.univ-mrs.fr
[3] CNRS & PRiSM 45 Avenue des Etats Unis, 78000 Versailles, France
Johanne.Cohen@prism.uvsq.fr
[4] LIAFA & Université Paris Diderot - Paris 7,
75205 Paris Cedex 13, France
Xavier.Koegler@liafa.jussieu.fr
[5] ENS de Lyon & Laboratoire d'Informatique (LIX),
91128 Palaiseau Cedex, France
Mikael.Rabie@ens-lyon.fr

Abstract. Population protocols have been introduced by Angluin et al.
as a model of networks consisting of very limited mobile agents that
interact in pairs but with no control over their own movement. A collection of anonymous agents, modeled by finite automata, interact pairwise
according to some rules that update their states. Predicates on the initial configurations that can be computed by such protocols have been
characterized as semi-linear predicates.

In an orthogonal way, several distributed systems have been termed
in literature as being realizations of games in the sense of game theory.

We investigate under which conditions population protocols, or more
generally pairwise interaction rules, correspond to games.

We show that restricting to asymetric games is not really a restriction: all predicates computable by protocols can actually be computed
by protocols corresponding to games, i.e. any semi-linear predicate can
be computed by a Pavlovian population multi-protocol.

1 Introduction

The computational power of networks of anonymous resource-limited mobile
agents has been investigated recently. Angluin et al. proposed in [3] the model
of *population protocols* where finitely many finite-state agents interact in pairs
chosen by an adversary. Each interaction has the effect of updating the state
of the two agents according to a joint transition function. A protocol is said to

---

[*] This work and all authors were partly supported by ANR Project SHAMAN, Xavier
Koegler was supported by the ANR projects ALADDIN and PROSE.

A. Fernández Anta, G. Lipari, and M. Roy (Eds.): OPODIS 2011, LNCS 7109, pp. 409–420, 2011.
© Springer-Verlag Berlin Heidelberg 2011

(stably) compute a predicate on the initial states of the agents if, in any fair execution, after finitely many interactions, all agents reach a common output that corresponds to the value of the predicate.

The model has been originally proposed to model computations realized by sensor networks in which passive agents are carried along by other entities. Variants of the original model considered so far include restriction to one-way communications [1], restriction to particular interaction graphs [2], random interactions [3], with "speed" [7]. Various kinds of fault tolerance have been considered for population protocols [10], including the search for self-stabilizing solutions [5]. Solutions to classical problems of distributed algorithms have also been considered in this model (see [18]).

Most of the works so far on population protocols have concentrated on characterizing which predicates on the initial states can be computed in different variants of the model and under various assumptions [18]. In particular, the predicates computable by the unrestricted population protocols from [3] have been characterized as being precisely the semi-linear predicates, that is those predicates on counts of input agents definable in first-order Presburger arithmetic [3,4].

In an orthogonal way, pairwise interactions between finite-state agents are sometimes motivated by the study of the dynamics of particular two-player games from game theory. For example, the work in [11] considers the dynamics of the so-called *PAVLOV* behavior in the iterated Prisoners' Dilemma. Several results about the time of convergence of this particular dynamics towards the stable state can be found in [11], and [12], for rings, and complete graphs [16] with having various classes of adversarial schedulers [15].

Our purpose is to better understand whether and when pairwise interactions, and hence population protocols, can be considered as the result of a game. We prove the result that restricting to games is not really a restriction: all predicates computable by protocols can actually be computed by protocols corresponding to games, i.e. any semi-linear predicate can be computed by a Pavlovian population multi-protocol.

In Section 2, we recall population protocols. In Section 3, we give some basics from game theory. In Section 4, we discuss how a game can be turned into a dynamics, and introduce the notion of Pavlovian population. In Section 5 we state our main result: any semi-linear predicate can be computed by a Pavlovian population multi-protocol. Remaining sections correspond to its proof: we prove that threshold and modulo predicates can be computed respectively in Sections 6 and 7.

Related Works. As we already said, population protocols have been introduced in [3], and proved to compute all semi-linear predicates. They have been proved not to be able to compute more in [4]. Various restrictions on the initial model have been considered up to now. An survey can be found in [18].

More generally, population protocols arise as soon as populations of anonymous agents interact in pairs. Our original motivation was to consider rules corresponding to two-player games, and population protocols arose quite incidentally.

The main advantage of the [3] settings is that it provides a clear understanding of what is called a computation by the model. Many distributed systems have been described as the result of games, but as far as we know there has not been attempt to characterize what can be computed by games in the spirit of this computational model.

In this paper, we turn two players games into dynamics over agents, by considering *PAVLOV* behavior. This is inspired by [11,12,17] that consider the dynamics of a particular set of rules termed the *PAVLOV* behavior in the iterated Prisoners' Dilemma. The *PAVLOV* behavior is sometimes also termed *WIN-STAY, LOSE-SHIFT* [19,6]. Notice, that we extended it from two-strategies two-player games to n-strategies two-player games, whereas above references only talk about two-strategies two-player games, and mostly of the iterated Prisoners' Dilemma. This is clearly not the only way to associate a dynamic to a game. Alternatives to *PAVLOV* behavior could include *MYOPIC* dynamics (at each step each player chooses the best response to previously played strategy by its adversary), or the well-known and studied *FICTIOUS-PLAYER* dynamics (at each step each player chooses the best response to the statistics of the past history of strategies played by its adversary). We refer to [13,8] for a presentation of results known about the properties of the obtained dynamics according to the properties of the underlying game. This is clearly non-exhaustive, and we refer to [6] for a zoology of possible behaviors for the particular iterated Prisoners' Dilemma game, with discussions of their compared merits.

Recently Jaggard et al. [16] studied a distributed model similar to protocol populations where the interactions between pairs of agents correspond to a game. Unlike in our model, each agent has there its own pay-off matrix and has some knowledge of the history. This work gives several non-convergence results.

In this paper we consider possibly asymmetric games. In a recent paper [9] we discussed population protocols corresponding to Pavlovian strategies obtained from *symmetric* games and we gave some protocols to compute some basic predicates. Unlike what we obtain here, where we prove that any computable predicate is computable by a asymmetric Pavlovian population protocol, restricting to symmetric games seems a (too) strong restriction and most predicates (e.g. counting up to 5, to check where $x = 0 \mod 2$) seems not even computable.

2 Population Protocols

A protocol [3] is given by $(Q, \Sigma, \iota, \omega, \delta)$ with the following components. Q is a finite set of *states*. Σ is a finite set of *input symbols*. $\iota : \Sigma \to Q$ is the initial state mapping, and $\omega : Q \to \{0, 1\}$ is the individual output function. $\delta \subseteq Q^4$ is a joint transition relation that describes how pairs of agents can interact. Relation δ is sometimes described by listing all possible interactions using the notation $(q_1, q_2) \to (q_1', q_2')$, or even the notation $q_1 q_2 \to q_1' q_2'$, for $(q_1, q_2, q_1', q_2') \in \delta$ (with the convention that $(q_1, q_2) \to (q_1, q_2)$ when no rule is specified with (q_1, q_2) in the left-hand side). The protocol is termed *deterministic* if for all pairs (q_1, q_2) there is only one pair (q_1', q_2') with $(q_1, q_2) \to (q_1', q_2')$. In that case, we write $\delta_1(q_1, q_2)$ for the unique q_1' and $\delta_2(q_1, q_2)$ for the unique q_2'.

Computations of a protocol proceed in the following way. The computation takes place among n *agents*, where $n \geq 2$. A *configuration* of the system can be described by a vector of all the agents' states. The state of each agent is an element of Q. Because agents with the same states are indistinguishable, each configuration can be summarized as an unordered multiset of states, and hence of elements of Q. Each agent is given initially some input value from Σ: Each agent's initial state is determined by applying ι to its input value. This determines the initial configuration of the population.

An execution of a protocol proceeds from the initial configuration by interactions between pairs of agents. Suppose that two agents in state q_1 and q_2 meet and have an interaction. They can change into state q_1' and q_2' if (q_1, q_2, q_1', q_2') is in the transition relation δ. If C and C' are two configurations, we write $C \to C'$ if C' can be obtained from C by a single interaction of two agents: this means that C contains two states q_1 and q_2 and C' is obtained by replacing q_1 and q_2 by q_1' and q_2' in C, where $(q_1, q_2, q_1', q_2') \in \delta$. An *execution* of the protocol is an infinite sequence of configurations $C_0, C_1, C_2, \cdots$, where C_0 is an initial configuration and $C_i \to C_{i+1}$ for all $i \geq 0$. An execution is *fair* if for every configuration C that appears infinitely often in the execution, if $C \to C'$ for some configuration C', then C' appears infinitely often in the execution. As proved in [4], the fairness condition implies that any global configuration that is infinitely often reachable is eventually reached.

At any point during an execution, each agent's state determines its output at that time. If the agent is in state q, its output value is $\omega(q)$. The configuration output is 0 (resp. 1) if all the individual outputs are 0 (resp. 1). If the individual outputs are mixed 0s and 1s then the output of the configuration is undefined.

Let p be a predicate over multisets of elements of Σ. Predicate p can be considered as a function whose range is $\{0, 1\}$ and whose domain is the collection of these multisets. The predicate is said to be computed by the protocol if, for every multiset I, and every fair execution that starts from the initial configuration corresponding to I, the output value of every agent eventually stabilizes to $p(I)$. Predicates can also be considered as functions whose range is $\{0, 1\}$ and whose domain is $\mathbb{N}^{|\Sigma|}$. The following is then known.

Theorem 1 ([3,4]). *A predicate is computable in the population protocol model if and only if it is semilinear.*

Recall that semilinear sets are exactly the sets that are definable in first-order Presburger arithmetic [20].

3   Game Theory

We now recall the simplest concepts from Game Theory. We focus on non-cooperative games, with complete information, in normal form.

The simplest game is made up of two players, called I (or *initiator*) and R (or *responder*), with a finite set of actions, called *pure strategies*, $Strat(I)$ and $Strat(R)$. Denote by $A_{i,j}$ (resp. $B_{i,j}$) the score for player I (resp. R) when I uses strategy $i \in Strat(I)$ and R uses strategy $j \in Strat(R)$. The scores are given by $n \times m$ matrices A and B, where n and m are the cardinality of $Strat(I)$ and $Strat(R)$.

A strategy x in $Strat(I)$ is said to be a best response to strategy y in $Strat(R)$, denoted by $x \in BR_A(y)$ if $A_{z,y} \leq A_{x,y}$ for all strategies $z \in Strat(I)$. Conversely, a strategy $y \in Strat(R)$ satisfies $y \in BR_B(x)$ if $B_{x,z} \leq B_{x,y}$ for all strategies $z \in Strat(R)$. A pair (x,y) is a *(pure) Nash equilibrium* if $x \in BR_A(y)$ and $y \in BR_B(x)$. In other words, two strategies (x,y) form a Nash equilibrium if in that state neither of the players has a unilateral interest to deviate from it.

There are two main approaches to discuss dynamics of games. The first consists in repeating games [8]. The second in using models from evolutionary game theory. Refer to [14,21] for a presentation of this latter approach.

Repeating k times a game, is equivalent to extending the space of actions into $Strat(I)^k$ and $Strat(R)^k$: player I (respectively R) chooses his or her action $\mathbf{x}(t) \in Strat(I)$, (resp. $\mathbf{y}(t) \in Strat(R)$) at time t for $t = 1, 2, \cdots, k$. This is equivalent to a two-player game with respectively n^k and m^k choices for players.

In practice, player I (respectively R) has to solve the following problem at each time t: given the history of the game up to now, that is to say $X_{t-1} = \mathbf{x}(1), \cdots, \mathbf{x}(t-1)$ and $Y_{t-1} = \mathbf{y}(1), \cdots, \mathbf{y}(t-1)$ what should I (resp. R) play at time t? In other words, how to choose $\mathbf{x}(t) \in Strat(I)$? (resp. $\mathbf{y}(t) \in Strat(R)$?)

Is is natural to suppose that this is given by some behavior rules: $\mathbf{x}(t) = f(X_{t-1}, Y_{t-1})$ and $\mathbf{y}(t) = g(X_{t-1}, Y_{t-1})$ for some particular functions f and g.

The question of the best behavior rule to use in games, in particular for the Prisoners' Dilemma gave birth to an important literature. In particular, after the book [6], that describes the results of tournaments of behavior rules for the iterated Prisoners' Dilemma, and that argues that there exists a best behavior rule called $TIT - FOR - TAT$. This consists in cooperating at the first step, and then do the same thing as the adversary at subsequent times. A lot of other behaviors, most of them with very picturesque names have been proposed and studied: see for example [6].

Among possible behaviors there is $PAVLOV$ behavior: in the iterated Prisoners' Dilemma, a player cooperates if and only if both players opted for the same alternative in the previous move. This name [6,17,19] stems from the fact that this strategy embodies an almost reflex-like response to the payoff: it repeats its former move if it was rewarded above a threshold value, but switches behavior if it was punished by receiving under this value. Refer to [19] for some study of this strategy in the spirit of Axelrod's tournaments. The $PAVLOV$ behavior can also be termed $WIN\text{-}STAY, LOSE\text{-}SHIFT$ since if the play on the previous round results in a success, then the agent plays the same strategy on the next round. Alternatively, if the play resulted in a failure the agent switches to another action [6,19].

4 From Games to Population Protocols

In the spirit of the previous discussion, to any game, we can associate a population protocol as follows, corresponding to a $PAVLOV$(ian) behaviour:

Definition 1 (Associating a Protocol to a Game). *Assume a (possibly asymmetric) two-player game is given. Let A and B be the corresponding matrices. Let Δ be some threshold.*

The protocol associated to the game is a population protocol whose set of states is Q, where $Q = Strat(I) = Strat(R)$ is the set of strategies of the game, and whose transition rules δ are given as follows: $(q_1, q_2, q_1', q_2') \in \delta$ where
- $q_1' = q_1$ *when* $A_{q_1,q_2} \geq \Delta$, $-$ $q_2' = q_2$ *when* $B_{q_2,q_1} \geq \Delta$,
- $q_1' \in BR_A(q_2)$ *when* $A_{q_1,q_2} < \Delta$, $-$ $q_2' \in BR_B(q_1)$ *when* $B_{q_2,q_1} < \Delta$.

Definition 2 (Pavlovian Population Protocol). *A population protocol is Pavlovian if it can be obtained from a game as above.*

A population protocol obtained from a game as above will be termed *deterministic* if best responses are assumed to be unique; in this case, the rules are deterministic: for all q_1, q_2, there is a unique q_1' and a unique q_2' such that $(q_1, q_2, q_1', q_2') \in \delta$.

In order to avoid to talk about matrices, we start by stating some structural properties of Pavlovian population protocols.

Proposition 1. *Consider a set of rules. For all rules $ab \to a'b'$, we denote $\delta_a^I(b) = b'$ and $\delta_b^R(a) = a'$. Let $Stable^I(a) = \{x \in Q | \delta_a^I(x) = x\}$, and $Stable^R(a) = \{x \in Q | \delta_a^R(x) = x\}$.*

Then the set of rules is deterministic Pavlovian iff $\forall a \in Q \; \exists \; max^I(a) \in Stable^I(a)$ and $\exists \; max^R(a) \in Stable^R(a)$ such that for all states a,

1. $\forall b \notin Stable^I(a)$ implies $\delta_a^I(b) = max^I(a)$.
2. $\forall b \notin Stable^R(a)$ implies $\delta_a^R(b) = max^R(a)$.

Proof. First, we consider a Pavlovian population protocol P obtained from corresponding matrices A and B. Let Δ be the associated threshold. Let a be an arbitrary state in Q, and let q be the best response to strategy a for matrix B.

Focus on the rule $aq \to a'q'$ where $(a', q') \in Q^2$, i.e., focus on the case where player I plays a while player R plays q. As $q = BR_B(a)$, we have, by Definition 1, q' equals to q. Thus, $q \in Stable^I(a)$.

Now, let consider b such that $b \notin Stable^I(a)$. We focus on the rule $ab \to a''b'$ where $(a'', b') \in Q^2$. So by definition of set $Stable^I$, we have $b \neq b'$. Using Definition 1, we have $B_{b,a} < \Delta$ and $b' = BR_B(a)$. So $b' = BR_B(a) = q$. Thus, if we let $max^I(a) = q$, $max^I(a)$ satisfies the conditions of the proposition.

Using similar arguments, we can also prove that $\exists \; max^R(a) \in Stable^R(a)$ such that $\forall b \notin Stable^R(a)$ implies $\delta_a^I(b) = max^R(a)$. In fact, we can sum up the relationship between the game matrix and rules by the following: for any $a \in Q$, we have $Stable^I(a) = \{x \in Q | B_{x,a} \geq \Delta\} \cup \{BR_B(a)\}$ and $max^I(a) = BR_B(a)$ and $Stable^R(a) = \{x \in Q | A_{x,a} \geq \Delta\} \cup \{BR_A(a)\}$ and $max^R(a) = BR_A(a)$.

Conversely, consider a population protocol P satisfying the properties of the proposition. All rules $ab \to a'b'$ are such that $\delta_a^I(b) = b'$ and $\delta_b^R(a) = a'$. We focus on the construction on a two-player game having the corresponding matrices A, and B. We fix an arbitrary value Δ as the threshold of the corresponding game.

- If $Stable^I(a) \neq Q$, then $B_{max^I(a),a} = \Delta + 1$. If $x \in Stable^I(a)$ and if $x \neq max^I(a)$ then $B_{x,a} = \Delta$. If $x \notin Stable^I(a)$, then $B_{x,a} = \Delta - 1$.
- If $Stable^I(a) = Q$, then $\forall x \in Q$, $B_{x,a} = \Delta$.
- If $Stable^R(a) \neq Q$, then $A_{max^R(a),a} = \Delta + 1$. If $x \in Stable^R(a)$ and if $x \neq max^R(a)$ then $A_{x,a} = \Delta$. If $x \notin Stable^R(a)$, then $A_{x,a} = \Delta - 1$.
- If $Stable^R(a) = Q$, then $\forall x \in Q$, $A_{x,a} = \Delta$.

It is easy to see that this game describes all rules of P. So, P is a Pavlovian population Protocol.

5   Main Result

Inverting value of the individual output function, the class of predicates computable by a Pavlovian population protocol is clearly closed under negation. However, this is not clear that predicates computable by Pavlovian population protocols are closed under conjunction or disjunction.

This is true if one considers *multi-protocol*. The idea is to consider k (possibly asymmetric) two-player games. At each step, each player chooses a strategy for each of the k games. Now each of the k games is played independently when two agents meet. Formally:

Definition 3 (Multiprotocol). *Consider k (possibly asymmetric) two-player games. For game i, let Q^i be the corresponding states, A^i and B^i the corresponding matrices.*

The associated population protocol is the population protocol whose set of states is $Q = Q^1 \times Q^2 \times \ldots \times Q^k$, and whose transition rules are given as follows: $((q_1^1, \ldots, q_1^k), (q_2^1, \ldots, q_2^k), (q_1^{1'}, \ldots, q_1^{k'}), (q_2^{1'}, \ldots, q_1^{k'})) \in \delta$ where, for all $1 \leq i \leq k$, $(q_1^i, q_2^i, q_1^{i'}, q_2^{i'})$ is a transition of the Pavlovian population protocol associated to the i_{th} game.

Notice that, when considering population protocols, a multi-protocol is a particular population protocol. This is the key property used in [3] to prove that stably computable predicates are closed under boolean operations. When considering Pavlovian games, one can build multi-protocols that are not Pavlovian protocols, and it is not clear whether one can always transform any pavlovian multi-protocol into an equivalent pavlovian protocol.

As explained before, multisets of elements of $\Sigma = (\sigma_1, \ldots, \sigma_l)$ are in bijection with elements of $\mathbb{N}^l$, and can be represented by a vector $(x_1, \ldots, x_l)$ of non-negative integers where x_i is the number of occurrences of σ_i in the multiset. Thus, we consider predicates ψ over vectors of non-negative integers. We write $[\psi]$ for their characteristic functions. Recall that a predicate is semi-linear iff it is Presburger definable [20]. Semi-linear predicate correspond to boolean combinations of threshold predicates and modulo predicates defined as follows (variables x_i represent the number of agents initially in state σ_i): A *threshold* predicate is of the form $[\Sigma a_i x_i \geq k]$, where $\forall i, a_i \in \mathbb{Z}$, $k \in \mathbb{Z}$ and the x_is are variables. A *modulo* predicate is of the form $[\Sigma a_i x_i \equiv b \mod k]$, where $\forall i, a_i \in \mathbb{Z}$, $k \in \mathbb{N} \setminus \{0,1\}$, $b \in [1, k-1]$ and the x_is are variables.

We can then state our main result:

Theorem 2. *For any predicate ψ, the following conditions are equivalent:*

- ψ *is computable by a population protocol*
- ψ *is computable by a Pavlovian population multi-protocol*
- ψ *is semi-linear.*

Due to lack of space, the proof of the following proposition is omitted.

Proposition 2. *The class of predicates computable by multi-games are closed under boolean operations.*

As from Proposition 2, predicates computable by Pavlovian population multi-protocols are closed under boolean operations, and as a Pavlovian population protocol is a particular Pavlovian population multi-protocol, and as predicates computable by (general) population protocols are known to be exactly semi-linear predicates, to prove Theorem 2 we only need to prove that we can compute threshold predicates and modulo predicates by Pavlovian population protocols. This is the purpose of the following sections.

6   Threshold Predicates

In this section, we prove that we can compute threshold predicates using Pavlovian protocols.

Proposition 3. *For any integer k, and any integers $a_1, a_2, \cdots, a_m$ there exists a Pavlovian population protocol that computes $[\sum_{i=1}^{m} a_i x_i \geq k]$.*

First note, that we can assume without loss of generality that $k \geq 1$. Indeed, $[\Sigma a_i x_i \geq -k] = [\Sigma(-a_i)x_i \leq k] = [\Sigma(-a_i)x_i < k+1]$ which is the negation of $[\Sigma(-a_i)x_i \geq k+1]$. Thus from a population protocol computing $[\Sigma(-a_i)x_i \geq k+1]$ with $k \geq 0$, we just have to inverse the output function to obtain a population protocol that computes $[\Sigma a_i x_i \geq -k]$.

The purpose of the rest of this section is to prove Proposition 3. We first discuss some basic ideas: Our techniques are inspired by the work of Angluin et al. [1]. The set of states we use is the set of integers from $[-M, M]$ where $M = \max(|a_i|, 2k-1)$. Each agent with input σ_i is given an initial weight of a_i. During the execution, the sum of the weights over the whole population is preserved. In [4], the general idea is the following: two interacting agents with positive weights p and q such that $p + q \leq M$ are transformed into an agent with weight 0 and an agent with weight $p + q$, while two agents with weight p and q such that $p + q > M$ are transformed into two agents with weight $\lfloor (p+q)/2 \rfloor$ and $\lceil (p+q)/2 \rceil$ that are both greater or equal to k.

In our setting, we cannot use the same rules since all agents that change their states when they meet an agent in state p while being initiator (resp. responder) must take the same state that only depends of p. To avoid this problem, a trick is to use rules of the following form: $pq \to (p+1)(q-1)$. However, we also have to make sure that the protocol enables all agents to agree in the final configuration. Whereas this kind of consideration is easy in the classical population protocol model, this turns out to be tricky in our settings.

We describe a protocol that computes $[\sum_{i=1}^{m} a_i x_i \geq k]$. Our protocol is defined as follows: we consider $\Sigma = \{\sigma_1, \ldots, \sigma_l\}$, $Q = \{\top\} \cup [-M, M]$; for all i, $\iota(\sigma_i) = a_i$; and we take $\omega(\top) = 1$ and for any $p \in [-M, M]$, $\omega(p) = 1$ if and only if $p \geq 1$.

We distinguish two cases: either $k = 1$, or $k \geq 2$. We present two protocols here, because we need to have a mechanism in our protocols to enable to "broadcast" the result; this is not so difficult in the first case whereas it is more technical in the second one. Due to lack of space, we only give the rules for $k = 1$, but provide a full proof for the case $k \geq 2$.

Case k = 1. Our protocol computing $[\Sigma a_i x_i \geq 1]$ is defined as follows. The rules are the following.

$$\top\top \to \top\top \qquad\qquad \top x \to \quad\;\; \top x \qquad \forall x \in [-M, M]$$
$$1\top \to 1\top \qquad\qquad 1n \to (n+1)\top \qquad \forall n \in [-M, 0]$$
$$\qquad\qquad\qquad\qquad 1p \to \quad\;\; 1p \qquad \forall p \in [1, M]$$

$$\forall n \in [-M, 0], \forall p \in [2, M-1]$$
$$n\top \to n0 \qquad\qquad p\top \to \quad\;\; p\top$$
$$nx \to nx \;\; \forall x \in [-M, M], \qquad pn \to (n+1)(p-1)$$
$$\qquad\qquad\qquad\qquad\qquad\qquad pp' \to \quad\;\; pp' \qquad \forall p' \in [1, M]$$

Case k ≥ 2. Our protocol is deterministic and from Proposition 1 uniquely determined by the sets $Stable^I(q)$, $Stable^R(q)$, and by the values $max^I(q)$, $max^R(q)$ defined as follows.

$q \in Q$	$Stable^I(q)$	$max^I(q)$	$Stable^R(q)$	$max^R(q)$
$\top$	$\{\top\} \cup [-M, 0] \cup [k, M]$	-1	$\{\top\} \cup [-M, M]$	
$n \in [-M, -1]$	$[-M, M]$	0	$\{\top\} \cup [-M, 0]$	$(n+1)$
0	$[-M, M]$	0	$\{\top\} \cup [-M, k-1]$	1
1	$\{\top, 0, M\}$	$\top$	$[-M, 0]$	2
$p \in [2, k-1]$	$\{\top, 0, M\}$	$(p-1)$	$[-M, 0]$	$(p+1)$
$b \in [k, M-1]$	$\{\top\} \cup [k, M]$	$(b-1)$	$\{\top\} \cup [-M, 0] \cup [k, M]$	$(b+1)$
M	$\{\top\} \cup [k, M]$	$(M-1)$	$\{\top\} \cup [-M, M]$	

The transition rules we obtain from these sets and values are the following.

$$\top\top \to \quad\;\; \top\top \qquad\qquad\qquad \top x \to \top x \qquad\qquad \forall x \in [-M, 0] \cup [k, M]$$
$$\top p \to (p+1)(-1) \;\; \forall p \in [1, k-1]$$
$$1\top \to \quad\;\; 1\top \qquad\qquad\qquad 10 \to 10$$
$$1x \to (x+1)\top \quad \forall x \notin \{\top, 0, M\} \qquad 1M \to 1M$$

$$\forall p \in [2, k-1]$$
$$p\top \to \quad\;\; p\top \qquad\qquad\qquad px \to (x+1)(p-1) \; \forall x \notin \{\top, 0, M\}$$
$$p0 \to \quad\;\; p0 \qquad\qquad\qquad pM \to pM$$

$$\forall n \in [-M, 0]$$
$$n\top \to \quad\;\; n0 \qquad\qquad\qquad nx \to nx \qquad\qquad \forall x \in [-M, M]$$

$$\forall b \in [k, M]$$
$$b\top \to \quad\;\; b\top \qquad\qquad\qquad bb' \to bb' \qquad\qquad \forall b' \in [k, M]$$
$$bx \to (x+1)(b-1) \; \forall x \in [-M, k-1]$$

We say that an agent in state $x \in [-M, M]$ has weight x and that an agent in state $\top$ has weight 0. Note that in the initial configuration the sum of the weights of all agents is exactly $\Sigma a_i x_i$. Note that any of the rule of our protocol does not modify the total weight of the population, i.e., at any step of the execution, the sum of the weights of all agents is exactly $\Sigma a_i x_i$.

Note that the stable configurations, (i.e., the configurations where no rule can be applied to modify the state of any agent), are the following:

– every agent a is in some state $n(a) \in [-M, 0]$,
– a unique agent is in state $p \in [1, k - 1]$ and every other agent is in state 0.
– every agent a is either in some state $b(a) \in [k, M]$ or in state $\top$.

Note that no agent starts in state $\top$, and that no rule enables the two interacting agents to enter the state $\top$ except for the rule $\top\top \to \top\top$. Thus, we know that it is impossible that all agents are in state $\top$. Consequently, in the last case described, we know that there is at least one agent in a state $b \in [k, M]$.

Note that in any stable configuration, all agents have the same output; if $\Sigma a_i x_i \geq k$ then all agents output 1, while in all the other cases, the agents output 0. Thus, if the population reaches a stable configuration, we know that the computed output is correct and that it will not be modified any more. Now, we should prove that the fairness condition ensures that we always reach a stable configuration. In fact, it is sufficient to prove that from any reachable configuration, there exists an execution that reaches a stable configuration.

Consider any configuration reached during the execution. As long as there is an agent in state $p \in [1, M]$ and an agent in state $n \in [-M, -1]$, we apply $pn \to (n + 1)(p - 1)$. Thus we can always reach a configuration where the states of all agents are in $[-M, 0] \cup \{\top\}$ if $\Sigma a_i x_i \leq 0$, or in $[0, M] \cup \{\top\}$ otherwise.

If $\Sigma a_i x_i \leq 0$, then there is at least one agent in state $n \in [-M, 0]$, since all agents cannot be in state $\top$. In this case, applying iteratively the rule $n\top \to n0$, we reach a stable configuration where all agents have a state in $[-M, 0]$.

Suppose now that $\Sigma a_i x_i \in [1, k - 1]$. Since $\Sigma a_i x_i \in [1, k - 1]$, each agent with a positive weight is in a state in $[1, k - 1]$. Applying iteratively the rule $pp' \to (p-1)(p'+1)$ where $p, p' \in [1, k-1]$, we reach a configuration where there is exactly one agent in state $p \in [1, k - 1]$ while all the other agents are in state 0 or $\top$. Applying iteratively the rules $\top p \to (p + 1)(-1)$ and $(p + 1)(-1) \to 0p$, we reach a configuration where one agent is in state $p \in [1, k - 1]$ while all the other agents are in state 0.

Finally, assume that $\Sigma a_i x_i \geq k$. If there is an agent in state $p \in [1, k - 1]$, we know that there is at least another agent in state $q \in [1, M]$. If $p + q \leq M$, applying iteratively the rule $pq \to (p - 1)(q + 1)$ between these two agents, we reach a configuration where one of these two agents is in state 0 while the other is in state $p + q$. In this case, we have strictly reduced the number of agents in a state in $[1, k-1]$. If $p+q > M \geq 2k$, then $q \in [k, M]$, and applying iteratively the rule $qp \to (q - 1)(p + 1)$, we reach a configuration where one agent is in state k while the other agent is in state $p + q - k \in [k, 2M]$. Here again, we have strictly reduced the number of agents in a state in $[1, k-1]$. Applying these rules as long as there exists an agent in state $p \in [1, k - 1]$, we reach a configuration where all agents are either in a state in $[k, M]$, or in state 0 or $\top$. Since $\Sigma a_i x_i \in [k, M]$, we know there exists an agent in state $b \in [k, M]$. Applying iteratively the rules $b0 \to 1(b-1)$ and $1(b-1) \to b\top$, we reach a stable configuration where all agents are either in state $\top$ or in a state in $[k, M]$.

7 Modulo Counting

Proposition 4. *For any integers k, b, and any integers $a_1, a_2, \cdots, a_m$ there exists a Pavlovian population protocol that computes $[\sum_{i=1}^{m} a_i x_i \equiv b \mod k]$.*

Due to lack of space, we only give the rules of the protocol for the case when $b \in [1, k-1]$. In that case, our protocol is defined as follows: $\Sigma = \{\sigma_1, \ldots, \sigma_l\}$, $Q = \{\top\} \cup [0, k-1]$; for all i, let $\iota(\sigma_i) \equiv a_i \mod k$; let $\omega(\top) = 1$ and for any $p \in [0, k-1]$, let $\omega(p) = 1$ if and only if $p = b$.

The rules are the following:

$$
\begin{array}{llll}
\top\top & \to & \top\top & \top 0\to \quad 00 \\
b\top & \to & b\top & 0b\to \quad \top(k-1) \quad \text{if } b = k-1 \\
0\top & \to & 0\top & 0b\to (b+1)(k-1) \quad \text{if } b \neq k-1 \\
\end{array}
$$

$\forall p \in [1, k-1]$
$$
\begin{array}{llll}
\top p & \to & \top p & pp'\to \quad pp' \quad \forall p' \in [0, p-1] \\
p(k-1)\to\top(p-1) & & & pp'\to(p'+1)(p-1) \; \forall p' \in [p, k-2] \\
\end{array}
$$

$\forall p \in [0, k-1] \setminus \{b\}$
$$
\begin{array}{llll}
0p & \to & 0p & p\top\to \quad 1(p-1)
\end{array}
$$

8 Conclusion

In this work, we present some (original an non-trivial) Pavlovian population protocols that compute the general threshold and modulo predicates. From this, we deduced that a predicate is computable in the Pavlovian population multi-protocol model if and only if it is semilinear.

In other words, we proved that restricting to rules that correspond to asymmetric games in pairwise interactions is not a restriction.

We however needed to consider multi-protocols, that is to say multi-games. We conjecture that the Pavlovian population protocols (i.e. non-multi-protocol) can not compute all semilinear predicates. A point is that in such protocols the set of rules are very limited (see Proposition 1). In particular, it seems rather impossible to perform an "or" operation between two modulo predicates in the general case.

Notice that the hypothesis of asymmetric games seems also necessary. We studied symmetric Pavlovian population protocols in [9] where we demonstrated that some non-trivial predicates can be computed. However, even very basic predicates, like the threshold predicate counting up to 5, seems problematic to be computed by symmetric games. With asymmetric games, general threshold and modulo predicates can be computed.

References

1. Angluin, D., Aspnes, J., Eisenstat, D., Ruppert, E.: The computational power of population protocols. Distributed Computing 20(4), 279–304 (2007)
2. Angluin, D., Aspnes, J., Chan, M., Fischer, M.J., Jiang, H., Peralta, R.: Stably Computable Properties of Network Graphs. In: Prasanna, V.K., Iyengar, S.S., Spirakis, P.G., Welsh, M. (eds.) DCOSS 2005. LNCS, vol. 3560, pp. 63–74. Springer, Heidelberg (2005)
3. Angluin, D., Aspnes, J., Diamadi, Z., Fischer, M.J., Peralta, R.: Computation in networks of passively mobile finite-state sensors. In: PODC, pp. 290–299. ACM Press (2004)
4. Angluin, D., Aspnes, J., Eisenstat, D.: Stably computable predicates are semilinear. In: PODC 2006, pp. 292–299. ACM Press, New York (2006)
5. Angluin, D., Aspnes, J., Fischer, M.J., Jiang, H.: Self-Stabilizing Population Protocols. In: Anderson, J.H., Prencipe, G., Wattenhofer, R. (eds.) OPODIS 2005. LNCS, vol. 3974, pp. 103–117. Springer, Heidelberg (2006)
6. Axelrod, R.M.: The Evolution of Cooperation. Basic Books (1984)
7. Beauquier, J., Burman, J., Clément, J., Kutten, S.: On utilizing speed in networks of mobile agents. In: PODC, pp. 305–314 (2010)
8. Binmore, K.: Fun and Games. D.C. Heath and Company (1992)
9. Bournez, O., Chalopin, J., Cohen, J., Koegler, X.: Playing with population protocols. In: The Complexity of a Simple Program, Cork, Irland (2008)
10. Delporte-Gallet, C., Fauconnier, H., Guerraoui, R., Ruppert, E.: When Birds Die: Making Population Protocols Fault-Tolerant. In: Gibbons, P.B., Abdelzaher, T., Aspnes, J., Rao, R. (eds.) DCOSS 2006. LNCS, vol. 4026, pp. 51–66. Springer, Heidelberg (2006)
11. Dyer, M.E., Goldberg, L., Greenhill, C.S., Istrate, G., Jerrum, M.: Convergence of the iterated prisoner's dilemma game. Combinatorics, Probability & Computing 11(2) (2002)
12. Fribourg, L., Messika, S., Picaronny, C.: Coupling and self-stabilization. Distributed Computing 18(3), 221–232 (2006)
13. Fudenberg, D., Levine, D.K.: The Theory of Learning in Games, vol. 624 (1996), http://ideas.repec.org/p/cla/levarc/624.html
14. Hofbauer, J., Sigmund, K.: Evolutionary game dynamics. Bulletin of the American Mathematical Society 4, 479–519 (2003)
15. Istrate, G., Marathe, M.V., Ravi, S.S.: Adversarial scheduling in evolutionary game dynamics. CoRR, abs/0812.1194, informal publication (2008)
16. Jaggard, A.D., Schapira, M., Wright, R.N.: Distributed computing with adaptive heuristics. In: Proceedings of Innovations in Computer Science ICS (2011)
17. Kraines, D., Kraines, V.: Pavlov and the prisoner's dilemma. Theory and Decision 26, 47–79 (1988)
18. Michail, O., Chatzigiannakis, I., Spirakis, P.G.: New Models for Population Protocols. Morgan & Claypool Publishers (2011)
19. Nowak, M., Sigmund, K.: A strategy of win-stay, lose-shift that outperforms tit-for-tat in the Prisoner's Dilemma game. Nature 364(6432), 56–58 (1993)
20. Presburger, M.: Über die Vollständigkeit eines gewissen Systems der Arithmetik ganzer Zahlen, in welchem die Addition als einzige Operation hervortritt. In: Comptes-rendus du I Congres des Mathematicians des Pays Slaves, pp. 92–101 (1929)
21. Weibull, J.W.: Evolutionary Game Theory. The MIT Press (1995)

Asynchronous Rendezvous of Anonymous Agents in Arbitrary Graphs

Samuel Guilbault and Andrzej Pelc*

Département d'informatique, Université du Québec en Outaouais,
Gatineau, Québec J8X 3X7, Canada
samuel.guilbault@gmail.com, pelc@uqo.ca

Abstract. Two identical (anonymous) mobile agents have to meet in an arbitrary, possibly infinite, unknown connected graph. Agents are modeled as points, they start at nodes of the graph chosen by the adversary and the route of each of them only depends on the already traversed portion of the graph and, in the case of randomized rendezvous, on the result of coin tossing. The actual walk of each agent also depends on an asynchronous adversary that may arbitrarily vary the speed of the agent, stop it, or even move it back and forth, as long as the walk of the agent in each segment of its route is continuous, does not leave it and covers all of it. Meeting means that both agents must be at the same time in some node or in some point inside an edge of the graph.

In the deterministic scenario we characterize the initial positions of the agents for which rendezvous is feasible and we provide an algorithm guaranteeing asynchronous rendezvous from all such positions in an arbitrary connected graph. In the randomized scenario we show an algorithm that achieves asynchronous rendezvous with probability 1, for arbitrary initial positions in an arbitrary connected graph. In both cases the graph may be finite or (countably) infinite.

Keywords: rendezvous, anonymous agent, graph, asynchronous, deterministic, randomized.

1 Introduction

The problem and the model. Two mobile agents starting at different nodes of an unknown connected graph have to meet. This task is known in the literature as the rendezvous problem in networks. In this paper we study the *asynchronous* version of this problem, for identical (anonymous) agents. Each agent designs its route in the graph, which is a sequence of edges (consecutive edges being incident), and an adversary controls the speed of each agent, can vary this speed, stop the agent, or even move it back and forth, as long as the walk of the

---
* Research partly supported by NSERC discovery grant and by the Research Chair in Distributed Computing at the Université du Québec en Outaouais.

A. Fernández Anta, G. Lipari, and M. Roy (Eds.): OPODIS 2011, LNCS 7109, pp. 421–434, 2011.

agent in each edge is continuous, does not leave it and covers all of it.[1] In this asynchronous version of the rendezvous problem, meeting at a node may be impossible even in the two-node graph, as the adversary controlling the speed of the agents can make them visit nodes at different times. Thus it is necessary to relax the meeting requirement by allowing the agents to meet either in a node or inside an edge. Such a definition of meeting is natural, e.g., when agents are robots traveling in a labyrinth or people wandering along streets of an unknown town. Since agents can meet inside an edge, the graph has to be geometrically presented without edge crossings. (Such crossings would permit an "accidental" meeting of two agents situated in different edges, which should not be considered as a meeting in the graph.) Thus, we consider an embedding of the underlying graph in the three-dimensional Euclidean space, with nodes of the graph being points of the space and edges being pairwise disjoint line segments joining them. Agents are modeled as points moving inside this embedding. This model of asynchronous motion of agents that have to meet in a graph has been previously used in [7,11,12].

If nodes of the graph have unique labels then a simple rendezvous algorithm is to meet at the node with the smallest label, hence for finite graphs the rendezvous problem reduces to graph exploration. However, in many applications, when rendezvous is needed in a network of unknown topology, such labeling of nodes may be unavailable, agents may be unable to perceive such labels due to limited sensory capabilities, or nodes may be unwilling to reveal their labels, e.g., due to security reasons. Hence it is important to design rendezvous algorithms for agents operating in *anonymous* graphs, i.e., graphs without unique labeling of nodes. It is important to note that the agents have to be able to *locally* distinguish ports at a node: otherwise, the adversary could prevent an agent from choosing a particular edge, thus making rendezvous impossible even in the simple case of trees. This justifies a common assumption made in the literature: ports at a node of degree d are enumerated $1, \ldots, d$. Local labelings of ports at each node are fixed: every agent sees the same local labeling at each node of the graph. However, no coherence between those local labelings over all the graph is assumed. When an agent leaves a node, it is aware of the port number by which it leaves and when it enters a node, it learns the entry port number and the degree of the node. Agents know neither the topology of the graph nor the initial distance between them. They cannot mark the nodes or the edges in any way. Each agent stops at the time of meeting the other agent.

As opposed to [7,11,12], where agents were distinguishable either by their labels or by known coordinates of their initial positions, in this paper we assume that agents are anonymous, i.e., identical and that they execute the same algorithm. In order to accomplish rendezvous, such identical agents usually have to break symmetry to prevent executing identical moves that would keep them in

---

[1] Notice that this definition of the adversary is very strong. In fact, all our positive results (algorithms) are valid even with this powerful adversary, and our impossibility result holds even for the much weaker synchronous adversary that always moves the agent forward, keeping constant speed, which is equivalent to the synchronous model.

different locations at all times. In the deterministic setting this can be done by exploiting different views of both agents from their initial positions, and in the random setting different results of coin tosses can be used. It will turn out that in some cases even deterministic rendezvous of agents starting from symmetric positions is possible. We do not impose any restriction on the memory of the agents: from the computational point of view agents are Turing machines.

Two important notions used to describe movements of agents are the *route* of the agent and its *walk*. Roughly speaking, the agent chooses the route *where* it moves and the adversary describes the walk on this route, deciding *how* the agent moves. More precisely, these notions are defined as follows. The adversary initially places an agent at some node of the graph. The route is chosen by the agent and is defined as follows. The agent chooses one of the available ports at the current node. After getting to the other end of the corresponding edge, the agent learns the port number by which it enters and the degree of the entered node. Then it chooses one of the available ports at this node, and so on, indefinitely (until rendezvous). The resulting route of the agent is the corresponding sequence of edges $(e_1, e_2, \dots)$, such that e_i is incident to e_{i+1}. This sequence is a (not necessarily simple) path in the graph.

We now describe the walk f of an agent on its route. Let $R = (e_1, e_2, \dots)$ be the route of an agent. Let $e_i = \{v_{i-1}, v_i\}$. Let $(t_0, t_1, t_2, \dots)$, where $t_0 = 0$, be an increasing sequence of reals, chosen by the adversary, that represent points in time. Let $f_i : [t_i, t_{i+1}] \to [v_i, v_{i+1}]$ be any continuous function, chosen by the adversary, such that $f_i(t_i) = v_i$ and $f_i(t_{i+1}) = v_{i+1}$. For any $t \in [t_i, t_{i+1}]$, we define $f(t) = f_i(t)$. The interpretation of the walk f is as follows: at time t the agent is at the point $f(t)$ of its route. This general definition of the walk and the fact that (as opposed to the route) it is designed by the adversary, are a way to formalize the asynchronous characteristics of the process. The movement of the agent can be at arbitrary speed, the adversary may sometimes stop the agent or move it back and forth, as long as the walk in each edge of the route is continuous and covers all of it. This definition makes the adversary very powerful, and consequently rendezvous is hard to achieve.

Notice that the ability of the asynchronous adversary to produce any continuous walk inside edges of the routes determined by the agents implies the following significant difference with respect to the synchronous scenario. While in the latter scenario the relative movement of the agents depends only on their routes and hence is entirely controlled by the agents, in the asynchronous setting this relative movement is also controlled by the adversary.

Agents with routes R_1 and R_2 and with walks f_1 and f_2 meet at time t, if points $f_1(t)$ and $f_2(t)$ are identical. A rendezvous is guaranteed for routes R_1 and R_2, if the agents using these routes meet at some time t, regardless of the walks chosen by the adversary. A rendezvous algorithm executed by an agent in a graph produces the route of the agent, given its starting point (and results of coin tosses in the randomized scenario). We say that asynchronous rendezvous is *feasible* from given initial positions, if there exist routes R_1 and R_2 starting from these positions that guarantee rendezvous.

An important feature of our rendezvous algorithms is that in the choice of consecutive edges of its route an agent does not use the knowledge of the walk to date. Thus the route depends only on the graph and on the starting point chosen by the adversary (as well as on coin tosses for randomized algorithms), but not on other decisions of the adversary.

Finally, we need to mention the termination problem. Since agents are identical and they do not know any bound on the size of the graph (in fact, the graph can even be infinite), they are not able to recognize when rendezvous is impossible. For example, agents situated in an oriented ring, where rendezvous is impossible, cannot distinguish this situation from that of being in an oriented ring with one node distinguished by an addition of a single leaf adjacent to it, before visiting this special node. In the latter case rendezvous is possible. Hence agents are never able to tell that rendezvous is impossible in the first situation and in this case they walk indefinitely without meeting. We will show that when rendezvous is possible, our agents always eventually meet and hence they stop.

Our results. In the deterministic scenario we characterize the initial positions of the agents for which rendezvous is feasible. It turns out that this is the case when the views[2] from these initial positions are different or when these positions are connected by a path whose corresponding sequence of port numbers is a palindrome. We provide an algorithm guaranteeing deterministic asynchronous rendezvous from all such initial positions in an arbitrary connected graph. In the randomized scenario we show an algorithm that achieves asynchronous rendezvous with probability 1, for arbitrary initial positions in an arbitrary connected graph. In both cases the graph may be finite of arbitrary unknown size or (countably) infinite. [3]

Our result in the randomized scenario has the following, perhaps surprising, consequence. Fix an arbitrary positive constant ϵ. We show an algorithm guaranteeing that two identical asynchronous agents equipped with a compass, starting from arbitrary positions in the plane, will eventually get at distance at most ϵ with probability 1. For synchronous agents (in the case of the plane this restriction means that both agents move at constant identical speed), such a result follows from the fact that a random walk on an infinite 2-dimensional grid reaches any node of the grid with probability 1 [14]. Our algorithm permits to accomplish such an ϵ-approach with probability 1 in the much harder asynchronous setting, i.e., when each agent walks with arbitrary, possibly varying speed, decided by the adversary . Moreover, our algorithm also works in higher dimensions (e.g., in the 3-dimensional space), while random walks in an infinite grid of dimension > 2 cannot be used (it is well known [14] that reaching a given node of such grids by a random walk occurs with probability strictly smaller than 1). To the best of our knowledge there are no previously known methods of accomplishing an ϵ-approach of anonymous agents in the 3-dimensional space

---

[2] See Section 2 for a precise definition of a view.

[3] For simplicity, we assume that all node degrees are finite (although possibly unbounded). However, all our results can be easily generalized to graphs containing nodes of (countably) infinite degrees.

with probability 1, even in the synchronous scenario, while our algorithm permits to do it in the much harder asynchronous setting.

Related work. A detailed discussion of the large literature on rendezvous, especially in the randomized scenario, can be found in the excellent book [3]. The recent survey [22] covers a large part of the literature on deterministic rendezvous in graphs. Another line of research in this domain concerns the geometric scenario (rendezvous in the line, see, e.g., [8,16], or in the plane, see, e.g., [5,6,9,15,23]). The probabilistic scenario where inputs and/or rendezvous strategies are random was considered, e.g., in [1,2,4,8]. Randomized rendezvous strategies often use random walks in graphs, which were thoroughly investigated and applied also to other problems, such as on-line algorithms [10]. A generalization of the rendezvous problem is that of gathering [15,17,18,21,25], when more than 2 agents have to meet in one location.

If graphs are unlabeled, rendezvous requires breaking symmetry, which can be accomplished either by coin tossing or – in the deterministic scenario – by allowing marking nodes or by labeling the agents. Deterministic rendezvous with anonymous agents working in unlabeled graphs but equipped with tokens used to mark nodes was considered e.g., in [20]. In [27] the authors studied gathering many agents with unique labels. In [13,19,24] deterministic rendezvous in graphs with two labeled agents was considered. However, in all of the above papers, the synchronous setting was assumed. Asynchronous gathering of anonymous agents under geometric scenarios has been studied, e.g., in [9,15,23] in different models than ours: agents could not remember past events, but they were assumed to have at least partial visibility of the scene. Gathering many anonymous agents in a graph, under an asynchronous scenario similar to the above but contrasting with ours (no memory of past events but the whole graph can be seen by each agent) has been studied in [17,18].

The first paper to consider deterministic rendezvous in graphs under our model of asynchrony was [12]. The authors concentrated on complexity of rendezvous of labeled agents in simple graphs, such as the ring and the infinite line. They also showed feasibility of deterministic asynchronous rendezvous in arbitrary finite connected graphs with *known* upper bound on the size. Asynchronous rendezvous of labeled agents was studied in [11], both in arbitrary connected graphs and in connected terrains in the plane. The main result of [11] in the graph scenario was an algorithm guaranteeing asynchronous rendezvous of arbitrary agents *with distinct labels* in an arbitrary connected graph. In [7] the authors consider the asynchronous rendezvous problem in grids and in the plane under the additional assumption that each agent knows its initial position with respect to some common system of coordinates. In this stronger model they show that agents starting in arbitrary positions in the plane can get at distance 1 from each other at cost $O(d^2 polylog(d))$, where d is the initial distance between the agents. It should be stressed that both in [12,11] and in [7] agents were distinguishable: in the first case using labels, in the second case using coordinates. Here is where our present scenario differs sharply: our agents are identical.

2   Preliminary Notions and Results

We will use the following notion from [26]. Let G be a graph and v a node of G, of degree k. The *view* from v is an infinite rooted tree $\mathcal{V}(v)$ with labeled ports, defined recursively as follows. $\mathcal{V}(v)$ has the root x_0 corresponding to v. For every node v_i, $i = 1, \ldots, k$, adjacent to v, there is a neighbor x_i in $\mathcal{V}(v)$ such that the port number at v corresponding to edge $\{v, v_i\}$ is the same as the port number at x_0 corresponding to edge $\{x_0, x_i\}$, and the port number at v_i corresponding to edge $\{v, v_i\}$ is the same as the port number at x_i corresponding to edge $\{x_0, x_i\}$. Node x_i, for $i = 1, \ldots, k$, is now the root of the view from v_i.

Our algorithms are based on the notion of a *tunnel*, introduced in [11]. Consider any graph G and two routes R_1 and R_2 starting at nodes v and w, respectively. We say that these routes form a tunnel, if there exists a prefix $[e_1, e_2, \ldots, e_n]$ of route R_1 and a prefix $[e_n, e_{n-1}, \ldots, e_1]$ of route R_2, for some edges e_i in the graph, such that $e_i = \{v_i, v_{i+1}\}$, where $v_1 = v$ and $v_{n+1} = w$. Intuitively, the route R_1 has a prefix P ending at w and the route R_2 has a prefix which is the reverse of P, ending at v. For simplicity we will also say that prefixes $[e_1, e_2, \ldots, e_n]$ and $[e_n, e_{n-1}, \ldots, e_1]$ form a tunnel. The following proposition was proved in [11].

Proposition 1. *If routes R_1 and R_2 form a tunnel, then they guarantee rendezvous.*

We now briefly recall the idea of the rendezvous algorithm from [11], which, as opposed to our scenario, works for agents that have distinct positive integer labels. In [11] the authors assumed that each agent knows its own label but not that of the other agent. Their algorithm will be later used as a building block for our rendezvous of anonymous agents.

Let $G = (V, E)$ be the connected graph in which the rendezvous must be performed. Denote by $\mathbb{N}$ the set of positive integers. Let $\mathcal{S}$ be the set of all finite sequences of positive integers. Let $\mathcal{P} = \{(i, j, s', s'') \mid i, j \in \mathbb{N}, i < j$ and $s', s'' \in \mathcal{S}\}$. Observe that the set $\mathcal{P}$ is countable. Let $\varphi_1, \varphi_2, \ldots$ be a fixed enumeration of $\mathcal{P}$.

For a finite path r in G, denote by $\bar{r}$ the path with the same edges as in r, but in the reverse order. Remark that r and $\bar{r}$ form a tunnel. Consider a path $r = (e_1, e_2, \ldots, e_m)$, such that $e_i = \{v_i, v_{i+1}\}$. Let $s = (p_1, \ldots, p_m)$, where p_i is the port number at node v_i, corresponding to e_i. We say that the sequence s of port numbers induces path r.

The algorithm from [11] forces the routes of any two agents to form a tunnel for every possible combination of starting nodes and labels of the two agents. By Proposition 1, this suffices to guarantee rendezvous. Any starting configuration of agent i placed at node v and agent j placed at node w by the adversary corresponds to a quadruple (i, j, s', s'') where s' is a sequence of ports inducing a path from v to w and s'' is a sequence of ports inducing the reverse path from w to v.

Each agent constructs its route in phases. In the beginning and at the end of each phase the agent is in its starting node. In phase k the previously constructed initial part of the route is extended while the agent processes quadruple φ_k (some of the extensions are null). This extension guarantees that the routes of agents of the corresponding starting configuration will form a tunnel after both agents have processed the quadruple φ_k. When agent with label l processes quadruple $\varphi_k = (i, j, s', s'')$ nothing happens if $l \neq i$ and $l \neq j$. If $l = i$, agent i tries to extend its route to guarantee rendezvous with agent j under the hypothesis that a path q from v to w corresponds to the sequence s' of ports and the reverse path $\bar{q}$ corresponds to the sequence s''. For this to happen, the agent first tries to follow the path induced by the sequence s' of ports. This attempt is considered successful if the following conditions are satisfied:

- At consecutive nodes of the traversed path, ports with numbers from the sequence s' are available,
- The reverse path corresponds to the sequence s'' of ports.

When the attempt is successful, the agent is at node w and has already traversed the route $r_v \frown q$, where r_v is the entire route traversed by it in the $k - 1$ first phases. (The symbol $\frown$ stands for concatenation.) Now it simulates the first $k - 1$ phases of the execution of the algorithm by agent with label j starting from w. The effect of this simulation is the path r_w. Hence the agent traversed route $r_v \frown q \frown r_w$ and is again in w. Now the agent goes back to v using path $\bar{q}$, it traverses $\bar{r_v}$ getting back to v, traverses q again reaching w, uses $\bar{r_w}$ getting back to w and retracts to v using path $\bar{q}$.

If $l = j$, the above actions are performed with the roles of i and j reversed and the role of s' and s'' reversed. To summarize, after both agents with labels i and j have processed the quadruple $\varphi_k = (i, j, s', s'')$, where s' is the sequence of port numbers inducing q and s'' is the sequence of port numbers inducing $\bar{q}$, then agent i traversed the route $\rho = r_v \frown q \frown r_w \frown \bar{q} \frown \bar{r_v} \frown q \frown \bar{r_w} \frown \bar{q}$. and agent j traversed the route $\rho' = r_w \frown \bar{q} \frown r_v \frown q \frown \bar{r_w} \frown \bar{q} \frown \bar{r_v} \frown q$. By construction, the part $r_v \frown q \frown r_w \frown \bar{q} \frown \bar{r_v} \frown q \frown \bar{r_w}$ of ρ and the part $r_w \frown \bar{q} \frown r_v \frown q \frown \bar{r_w} \frown \bar{q} \frown \bar{r_v}$ of ρ' form a tunnel, which guarantees rendezvous.

In order to construct the reverse paths $\bar{q}$, $\bar{r_v}$, and $\bar{r_w}$, when the agent is traversing one of the paths q, r_v, or r_w, each time it reaches a new node, it stores the entry port number of the edge from which it arrives. The respective reverse path is obtained by taking the sequences of entry port numbers in the reverse order.

A finite path r in G is called a *palindrome*, if r and $\bar{r}$ are induced by the same sequence of port numbers. (An equivalent condition is that the sequence of all ports met when traversing this path is identical to its reverse sequence, i.e., to this sequence read from end to beginning.)

3   Deterministic Rendezvous

In this section we characterize the initial positions of the agents for which deterministic rendezvous is feasible. More precisely, we show that these are initial positions for which either views are different or which are connected by a path which is a palindrome. Moreover, we provide an algorithm guaranteeing deterministic asynchronous rendezvous from all such initial positions in an arbitrary connected graph.

The main difficulty in designing the algorithm is that, as opposed to [11], agents do not have labels allowing them to break symmetry. Hence symmetry can be broken only by inspecting the views of the agents, if these views are different. Even when they are different, the agents cannot know how deeply these views have to be explored to find the first difference. Thus the algorithm proceeds in epochs: in each consecutive epoch each agent explores its view more deeply, and creates a code of this truncated view, subsequently treating it as its temporary label and applying the procedure described in the previous section to a restricted list of quadruples. If views are different, a tunnel will be eventually created after an epoch with sufficiently high index. The algorithm has an additional feature permitting creation of a tunnel when views of the agents are the same but their initial positions are joined by a path which is a palindrome. Below is a detailed description of the algorithm.

Algorithm Deterministic-RV
We present the algorithm for an agent whose initial position is at node v. (The name of the node is for description only, as nodes do not have labels.) The algorithm proceeds in epochs numbered by consecutive integers $1, 2, \ldots$. In the beginning and end of each epoch, the agent is in node v. In epoch n the agent first performs a restricted depth-first search to depth n, leaving a visited node by all ports with numbers at most n, in increasing order. At the end the agent is back in v. Now the agent obtains the code $C(v, n)$ of this DFS traversal, defined as the sequence of port numbers it visited while performing it, with the provision that all port numbers larger than n (by which the agent may have entered a node) are replaced by 0. The code $C(v, n)$ is a sequence of integers from the set $\{0, 1, \ldots, n\}$ of length at most $4(n+1)^{n+1}$. Observe that if views $\mathcal{V}(v)$ and $\mathcal{V}(w)$ are different, then $C(v, n) \neq C(w, n)$ for some integer n.

Now consider the following set of quadruples. A quadruple (c_1, c_2, s_1, s_2) belongs to this set if c_1, c_2 are sequences of numbers $\{0, 1, \ldots, n\}$ of length at most $4(n+1)^{n+1}$, s_1 and s_2 are sequences of numbers $1, 2, \ldots, n$ of lengths at most n and either $c_1 \neq c_2$, or $c_1 = c_2$ and $s_1 = s_2$. The list of all such quadruples ordered lexicographically is denoted by Q_n. Let $\psi_{n,i}$ be the i-th quadruple in the list.

After obtaining the code $C = C(v, n)$, the rest of epoch n is devoted to processing quadruples from the list Q_n in order and proceeds in stages. Stage i is devoted to processing quadruple $\psi_{n,i} = (c_1, c_2, s_1, s_2)$. At the beginning and end of each stage, the agent is in node v. If C is different from c_1 and from c_2, nothing happens in stage i. If $C = c_1$, the agent processes the quadruple $\psi_{n,i}$ similarly as it was done for quadruple φ_k in the previous section. Let r_v denote the route traversed by the agent in all previous epochs and in the preceding part of the current epoch. The route r_v begins and ends at v. The agent tries to extend the route r_v to guarantee rendezvous with the other agent under the hypothesis that:

(a) this agent is situated at a node w,

(b) $c_2 = C(w, n)$,

(c) a path q from v to w is induced by the sequence s_1 of ports,

(d) the reverse path $\bar{q}$ is induced by the sequence s_2.

The agent first tries to follow the path induced by the sequence s_1 of ports. As before, this attempt is considered successful if:

(1) at consecutive nodes of the traversed path, ports with numbers from the sequence s_1 are available,

(2) the reverse path corresponds to the sequence s_2 of ports.

When the attempt is successful, the agent is at node w and has already traversed the route $r_v \frown q$. Now it simulates all previous epochs and the preceding part of the current epoch of the execution of the algorithm by an agent with code $c_2 = C(w, n)$ starting from w. Observe that, in order to perform this simulation, it is necessary to compute all codes $C(w, m)$ for $m < n$, where $c_2 = C(w, n)$, because an agent with code c_2 in epoch n would have code $C(w, m)$ in epoch $m < n$. However, all codes $C(w, m)$ for $m < n$ can be deduced from $C(w, n)$. This is done as follows. From the code $C(w, n)$ reconstruct the tree T_n corresponding to the restricted depth-first search traversal to depth n. Now construct a subtree T_m of T_n, by cutting off all subtrees rooted at children u of a node z such that the port at z corresponding to edge $\{z, u\}$ is larger than m. In this pruned tree replace all port numbers larger than m by 0. Now the code $C(w, m)$ is the sequence of port numbers encountered while visiting T_m in a depth-first manner, in increasing order of port numbers.

The effect of this simulation is the path r_w. Hence the agent traversed route $r_v \frown q \frown r_w$ and is again in w. Now the agent goes back to v using path $\bar{q}$, it traverses $\overline{r_v}$ getting back to v, traverses q again reaching w, uses $\overline{r_w}$ getting back to w and retracts to v using path $\bar{q}$.

If $C = c_2$, the above actions are performed with the roles of c_1 and c_2 reversed and the roles of s_1 and s_2 reversed. Notice that in our current algorithm the processed quadruple can be of the form (c, c, s_1, s_2) (which could not happen in the algorithm from [11], as labels where always assumed different). In this case there is no ambiguity concerning the actions of the agent, because then the equality $s_1 = s_2$ holds by the definition of the quadruples.

Upon completing all stages of epoch n, the agent starts epoch $n + 1$, until rendezvous or indefinitely, if rendezvous is impossible.

Lemma 1. *If views from the initial positions of the agents are different or if the initial positions are connected by a path which is a palindrome, then Algorithm Deterministic-RV guarantees asynchronous rendezvous.*

Proof. First suppose that the views $\mathcal{V}(v)$ and $\mathcal{V}(w)$ from the initial positions v and w of the agents are different. Consider the smallest n, such $C(v, n) \neq C(w, n)$ and there exists a path q of length at most n between v and w, such that all port numbers on q are at most n. Let $c_1 = C(v, n)$ and let $c_2 = C(w, n)$. Hence $c_1 \neq c_2$. Let s_1 be the sequence of port numbers inducing q and let s_2 be the sequence of port numbers inducing $\bar{q}$. Thus the quadruple (c_1, c_2, s_1, s_2) appears in the list

Q_n. Suppose that $\psi_{n,i} = (c_1, c_2, s_1, s_2)$. We show that upon completion of stage i of epoch n by both agents, a tunnel is formed and consequently asynchronous rendezvous is guaranteed.

Let r_v (resp. r_w) be the route of the agent starting at v (resp. of the agent starting at w) after completing all epochs $1, \ldots, n-1$ and all stages $1, \ldots, i-1$ of epoch n. Upon completing stage i of epoch n, the agent starting at v has traversed the route $\rho = r_v \frown q \frown r_w \frown \bar{q} \frown \overline{r_v} \frown q \frown \overline{r_w} \frown \bar{q}$ and the agent starting at w has traversed the route $\rho' = r_w \frown \bar{q} \frown r_v \frown q \frown \overline{r_w} \frown \bar{q} \frown \overline{r_v} \frown q$. By construction, the part $r_v \frown q \frown r_w \frown \bar{q} \frown \overline{r_v} \frown q \frown \overline{r_w}$ of ρ and the part $r_w \frown \bar{q} \frown r_v \frown q \frown \overline{r_w} \frown \bar{q} \frown \overline{r_v}$ of ρ' form a tunnel.

Next suppose that the views $\mathcal{V}(v)$ and $\mathcal{V}(w)$ from the initial positions v and w of the agents are identical but these initial positions are connected by a path q which is a palindrome. Let s be the sequence of port numbers inducing this path and inducing the reverse path $\bar{q}$. Let n be the larger of the two integers: the length of the path q and the largest term of the sequence s. Let $c = C(v, n) = C(w, n)$. Thus the quadruple (c, c, s, s) appears in the list Q_n. Suppose that $\psi_{n,j} = (c, c, s, s)$. We show that upon completion of stage j of epoch n by both agents, a tunnel is formed and consequently asynchronous rendezvous is guaranteed.

Let r_v (resp. r_w) be the route of the agent starting at v (resp. of the agent starting at w) after completing all epochs $1, \ldots, n-1$ and all stages $1, \ldots, j-1$ of epoch n.

Upon completing stage j of phase 1 of epoch n, one agent has traversed the route
$r_v \frown q \frown r_w \frown \bar{q} \frown \overline{r_v} \frown q \frown \overline{r_w} \frown \bar{q}$ and the other agent has traversed the route
$r_w \frown \bar{q} \frown r_v \frown q \frown \overline{r_w} \frown \bar{q} \frown \overline{r_v} \frown q$. As before, these routes form a tunnel.

The next lemma shows that unless the initial positions of the agents satisfy the condition of Lemma 1, asynchronous rendezvous cannot be guaranteed.

Lemma 2. *If views from the initial positions of the agents are identical and the initial positions are not connected by a path which is a palindrome, then asynchronous rendezvous cannot be guaranteed.*

Proof. Consider two agents starting at nodes v and w, such that $\mathcal{V}(v) = \mathcal{V}(w)$. Consider an arbitrary algorithm guaranteeing asynchronous rendezvous. It produces routes $R(v)$ and $R(w)$, respectively. Consider an adversary that moves these agents along their routes at constant identical speed. A meeting of the agents could either occur when they first get to some node u simultaneously, or when they traverse the same edge e in the opposite directions simultaneously (in which case rendezvous would occur in the middle of this edge). The first situation is impossible because this would imply that distinct edges incident to u have the same port number at u (in view of $\mathcal{V}(v) = \mathcal{V}(w)$). Hence we may assume that the second situation occurs. Let $\pi = (e_1, \ldots, e_k)$ and $\pi' = (e'_1, \ldots, e'_k)$ be the parts of the routes $R(v)$ and $R(w)$, respectively, before the agents enter edge e. In view of $\mathcal{V}(v) = \mathcal{V}(w)$, the sequence of ports encountered by the agents when they traverse these parts of their routes is the same and the ports at both extremities of edge e are identical. This implies that the path $\pi \frown \{e\} \frown \overline{\pi'}$ connecting v and w is a palindrome.

Lemmas 1 and 2 imply our main result for deterministic asynchronous rendezvous.

Theorem 1. *Deterministic asynchronous rendezvous of anonymous agents is feasible if and only if the views from the initial positions of the agents are different or the initial positions are connected by a path which is a palindrome. If this condition is satisfied, Algorithm Deterministic-RV guarantees asynchronous rendezvous of the agents.*

4   Randomized Rendezvous

In this section we show a randomized algorithm that achieves asynchronous rendezvous with probability 1, for arbitrary initial positions in an arbitrary connected graph.

Algorithm Randomized-RV
We present the algorithm for an agent whose initial position is at node v. (As before, the name of the node is for description only.) The algorithm proceeds in epochs numbered by consecutive integers $1, 2, \ldots$. In the beginning and end of each epoch, the agent is at node v. In the beginning of epoch n the agent has a code $C(v, n - 1)$ which is a binary sequence of length $n - 1$. The agent starts epoch n by choosing a random bit with probability $1/2$ and appending it to $C(v, n - 1)$, thus forming the code $C(v, n)$.

Now consider the following set of quadruples. A quadruple (c_1, c_2, s_1, s_2) belongs to this set if c_1 and c_2 are different binary sequences of length n and s_1 and s_2 are sequences of numbers $1, 2, \ldots, n$ of lengths at most n. The list of all such quadruples ordered lexicographically is denoted by P_n. Let $\lambda_{n,i}$ be the i-th quadruple in the list.

After obtaining the code $C = C(v, n)$, the rest of epoch n is devoted to processing quadruples from the list P_n in order, and proceeds in stages. Stage i is devoted to processing quadruple $\lambda_{n,i} = (c_1, c_2, s_1, s_2)$. In the beginning and end of each stage, the agent is in node v. If C is different from c_1 and from c_2, nothing happens in stage i. If $C = c_1$ or $C = c_2$, the agent processes the quadruple $\lambda_{n,i}$ in the same way as it was done for quadruple $\psi_{n,i}$ in Algorithm Deterministic-RV. Notice that in our current algorithm (as opposed to Algorithm Deterministic-RV) all processed quadruples satisfy $c_1 \neq c_2$.

Upon completing all stages of epoch n, the agent starts epoch $n + 1$, until rendezvous or indefinitely.

Theorem 2. *Algorithm Randomized-RV guarantees asynchronous rendezvous with probability 1, for arbitrary initial positions of the agents in an arbitrary connected graph.*

Proof. Consider agents initially situated in nodes v and w. First observe that the probability of the event that for some m the codes $C(v, m)$ and $C(w, m)$ are different, is 1. Hence it is enough to prove that at some point of the execution of the algorithm the routes of the agents form a tunnel, under the assumption that $C(v, m) \neq C(w, m)$ for some m.

Let p be the smallest integer for which $C(v, p) \neq C(w, p)$. Let $n \geq p$ be the smallest integer for which there exists a path q of length at most n joining v and w, such that all ports on path q have numbers at most n. Denote $c_1 = C(v, n)$ and $c_2 = C(w, n)$. Let s_1 be the sequence of port numbers inducing q and let s_2 be the sequence of port numbers inducing $\bar{q}$. Thus the quadruple (c_1, c_2, s_1, s_2) appears in the list P_n. Suppose that $\lambda_{n,i} = (c_1, c_2, s_1, s_2)$. The same argument as in the proof of Lemma 1 shows that upon completion of stage i of epoch n by both agents, a tunnel is formed.

We conclude this section by presenting an application of the above Algorithm Randomized-RV to the problem of approximate asynchronous rendezvous of anonymous agents in the plane. Two agents (modeled as moving points) equipped with compasses, start from arbitrary initial positions in the plane. For a fixed constant $\epsilon > 0$ known to the agents, ϵ-rendezvous consists in bringing the agents at distance at most ϵ.

This problem can be solved with probability 1, by applying the above Algorithm Randomized-RV as follows. Each agent executes this algorithm on a rectangular ϵ-grid one of whose nodes is the initial position of the agent. More precisely, neighbors of a given node in each grid are the four points North, East, South and West of the given point, at distance ϵ. Port numbers corresponding to these edges at each node of the grid are labeled $1, 2, 3, 4$. Since each agent has a compass, it can determine port numbers at each visited node of the grid.

Let v and w be the starting points of the agents and let G_v and G_w be the respective grids of the agents. Denote by w' the node of the grid G_v closest to w (take any such node, if there are more than 1). The distance between w and w' is less than ϵ. Simulate the moves of the agent starting at w by making a parallel shift by the vector $[w, w']$. The simulated moves are on grid G_v starting at w'. By Theorem 2, the agent starting at v and the (virtual) agent starting at w' will meet with probability 1 on grid G_v. At the time of their meeting, the (real) agent starting at w is at distance less than ϵ from the agent starting at v. Hence ϵ-rendezvous of the agents starting at v and at w is accomplished (with probability 1).

It is interesting to compare this application of Algorithm Randomized-RV to what can be done using other methods. For synchronous agents (i.e., agents moving in the plane at constant identical speed), ϵ-rendezvous in the plane with probability 1 follows from the fact that a random walk on an infinite 2-dimensional grid reaches any node of the grid with probability 1 [14]. Our algorithm permits to accomplish ϵ-rendezvous in the plane with probability 1 in the much harder asynchronous setting.

More importantly, our algorithm works also in higher dimensions (e.g., in the 3-dimensional space for agents with a "3-dimensional compass" that can establish directions North, East, South, West, Up and Down), while random walks in an infinite grid of dimension > 2 cannot be used (it is well known [14] that reaching a given node of such grids by a random walk occurs with probability strictly smaller than 1).

5   Conclusion

Our deterministic algorithm accomplishes asynchronous rendezvous of anonymous agents in arbitrary connected graphs for all initial positions for which such a rendezvous can be guaranteed. Our randomized algorithm accomplishes rendezvous of anonymous agents in arbitrary connected graphs for all initial positions, with probability 1. Hence, in terms of feasibility, the problem of asynchronous rendezvous of anonymous agents is completely solved.

However, both our algorithms are inefficient in terms of the number of edge traversals they use. The routes of the agents are obtained by appending simulated routes of potential agents one after another, resulting in cost at least exponential in the size of the graph, in case of finite graphs. Thus it is natural to ask if there exists a deterministic algorithm accomplishing asynchronous rendezvous of anonymous agents for feasible initial positions, or a randomized algorithm accomplishing rendezvous of anonymous agents for all initial positions, with probability 1, which uses a number of edge traversals polynomial in the size of the graph, for agents operating in finite graphs.

References

1. Alpern, S.: The rendezvous search problem. SIAM J. on Control and Optimization 33, 673–683 (1995)
2. Alpern, S.: Rendezvous search on labelled networks. Naval Reaserch Logistics 49, 256–274 (2002)
3. Alpern, S., Gal, S.: The theory of search games and rendezvous. Int. Series in Operations research and Management Science, vol. 55. Kluwer Academic Publishers (2002)
4. Anderson, E., Weber, R.: The rendezvous problem on discrete locations. Journal of Applied Probability 28, 839–851 (1990)
5. Anderson, E., Fekete, S.: Asymmetric rendezvous on the plane. In: Proc. 14th Annual ACM Symp. on Computational Geometry, pp. 365–373 (1998)
6. Anderson, E., Fekete, S.: Two-dimensional rendezvous search. Operations Research 49, 107–118 (2001)
7. Bampas, E., Czyzowicz, J., Gąsieniec, L., Ilcinkas, D., Labourel, A.: Almost Optimal Asynchronous Rendezvous in Infinite Multidimensional Grids. In: Lynch, N.A., Shvartsman, A.A. (eds.) DISC 2010. LNCS, vol. 6343, pp. 297–311. Springer, Heidelberg (2010)
8. Baston, V., Gal, S.: Rendezvous on the line when the players' initial distance is given by an unknown probability distribution. SIAM J. on Control and Optimization 36, 1880–1889 (1998)

9. Cieliebak, M., Flocchini, P., Prencipe, G., Santoro, N.: Solving the Robots Gathering Problem. In: Baeten, J.C.M., Lenstra, J.K., Parrow, J., Woeginger, G.J. (eds.) ICALP 2003. LNCS, vol. 2719, pp. 1181–1196. Springer, Heidelberg (2003)
10. Coppersmith, D., Doyle, P., Raghavan, P., Snir, M.: Random walks on weighted graphs, and applications to on-line algorithms. In: Proc. 22nd Annual ACM Symposium on Theory of Computing (STOC 1990), pp. 369–378 (1990)
11. Czyzowicz, J., Labourel, A., Pelc, A.: How to meet asynchronously (almost) everywhere. In: Proc. 21st Annual ACM-SIAM Symposium on Discrete Algorithms (SODA 2010), pp. 22–30 (2010)
12. De Marco, G., Gargano, L., Kranakis, E., Krizanc, D., Pelc, A., Vaccaro, U.: Asynchronous deterministic rendezvous in graphs. Theoretical Computer Science 355, 315–326 (2006)
13. Dessmark, A., Fraigniaud, P., Kowalski, D., Pelc, A.: Deterministic rendezvous in graphs. Algorithmica 46, 69–96 (2006)
14. Feller, W.: An introduction to probability theory and its applications. Wiley (1968)
15. Flocchini, P., Prencipe, G., Santoro, N., Widmayer, P.: Gathering of Asynchronous Oblivious Robots with Limited Visibility. In: Ferreira, A., Reichel, H. (eds.) STACS 2001. LNCS, vol. 2010, pp. 247–258. Springer, Heidelberg (2001)
16. Gal, S.: Rendezvous search on the line. Operations Research 47, 974–976 (1999)
17. Klasing, R., Kosowski, A., Navarra, A.: Taking Advantage of Symmetries: Gathering of Asynchronous Oblivious Robots on a Ring. In: Baker, T.P., Bui, A., Tixeuil, S. (eds.) OPODIS 2008. LNCS, vol. 5401, pp. 446–462. Springer, Heidelberg (2008)
18. Klasing, R., Markou, E., Pelc, A.: Gathering asynchronous oblivious mobile robots in a ring. Theoretical Computer Science 390, 27–39 (2008)
19. Kowalski, D., Malinowski, A.: How to meet in anonymous network. Theoretical Computer Science 399, 141–156 (2008)
20. Kranakis, E., Krizanc, D., Santoro, N., Sawchuk, C.: Mobile agent rendezvous in a ring. In: Proc. 23rd International Conference on Distributed Computing Systems (ICDCS 2003), pp. 592–599 (2003)
21. Lim, W., Alpern, S.: Minimax rendezvous on the line. SIAM J. on Control and Optimization 34, 1650–1665 (1996)
22. Pelc, A.: DISC 2011 Invited Lecture: Deterministic Rendezvous in Networks: Survey of Models and Results. In: Peleg, D. (ed.) DISC 2011. LNCS, vol. 6950, pp. 1–15. Springer, Heidelberg (2011)
23. Prencipe, G.: Impossibility of gathering by a set of autonomous mobile robots. Theoretical Computer Science 384, 222–231 (2007)
24. Ta-Shma, A., Zwick, U.: Deterministic rendezvous, treasure hunts and strongly universal exploration sequences. In: Proc. 18th Annual ACM-SIAM Symposium on Discrete Algorithms (SODA 2007), pp. 599–608 (2007)
25. Thomas, L.: Finding your kids when they are lost. Journal on Operational Res. Soc. 43, 637–639 (1992)
26. Yamashita, M., Kameda, T.: Computing on Anonymous Networks: Part I-Characterizing the Solvable Cases. IEEE Trans. Parallel Distrib. Syst. 7, 69–89 (1996)
27. Yu, X., Yung, M.: Agent Rendezvous: a Dynamic Symmetry-Breaking Problem. In: Meyer auf der Heide, F., Monien, B. (eds.) ICALP 1996. LNCS, vol. 1099, pp. 610–621. Springer, Heidelberg (1996)

Robust Network Supercomputing without Centralized Control*

Seda Davtyan[1], Kishori M. Konwar[2], and Alexander A. Shvartsman[1]

[1] Department of Computer Science & Engineering, University of Connecticut, Storrs
CT 06269, USA
{seda,aas}@engr.uconn.edu
[2] Department of Immunology and Microbiology, University of British Columbia,
Vancouver BC V6T 1Z3, Canada
kishori@interchange.ubc.ca

Abstract. Internet supercomputing provides means for harnessing the power of a vast number of interconnected computers. With this come the challenges of marshaling distributed resources and dealing with failures. Traditional centralized approaches employ a master processor and many worker processors that execute a collection of tasks on behalf of the master. Despite the simplicity and advantages of centralized schemes, the master processor is a performance bottleneck and a single point of failure. Additionally, a phenomenon of increasing concern is that workers may return incorrect results, e.g., due to unintended failures, over-clocked processors, or due to workers claiming to have performed work to obtain a high rank in the system. This paper develops an original approach that eliminates the master and instead uses a *decentralized* algorithm, where workers cooperate in performing tasks. The failure model assumes that the *average* probability of a worker returning a wrong result is inferior to $1/2$. We present a randomized synchronous algorithm for n processors and t tasks ($t \geq n$) achieving time complexity $\Theta(\frac{t}{n} \log n)$ and work $\Theta(t \log n)$. It is shown that upon termination the workers know the results of all tasks with high probability, and that these results are correct with high probability. The message complexity of the algorithm is $\Theta(n \log n)$, and the bit complexity is $O(tn \log^3 n)$. Simulations illustrate the behavior of the algorithm under realistic assumptions.

Keywords: Distributed Algorithms, Fault-Tolerance, Internet Supercomputing.

1 Introduction

Internet supercomputing is becoming a popular means for harnessing the computing power of an enormous number of processors around the world. A typical Internet supercomputer consists of a *master* computer and a large number of computers called *workers*. Applications submit the tasks to be performed to

---

* This work is supported in part by the NSF award 1017232.

A. Fernández Anta, G. Lipari, and M. Roy (Eds.): OPODIS 2011, LNCS 7109, pp. 435–450, 2011.
© Springer-Verlag Berlin Heidelberg 2011

the master that in turn directs the workers to perform the tasks and then collects the results. Several Internet Supercomputers are in existence today. For instance, Internet PrimeNet Server encompasses about 30,000 computers, achieving throughput of over 1 teraflop [1], and even higher throughput is reported by the SETI@home project [2].

A major concern in network supercomputing is the correctness of the results returned by the workers. While most workers may be reliable, workers have been known to return incorrect results. This may be due to unintended failures caused (e.g., by over-clocked processors), or the workers claiming to have performed assigned work so as to obtain incentives, such as getting higher rank on the SETI@home list of contributed units of work. Prior research developed models and algorithms for network supercomputing, e.g., [5,6,11]. In these models it is assumed that a reliable master and a collection of unreliable workers cooperatively perform a set of tasks. Using a variety of probabilistic failure models, the goal is to design algorithms that correctly perform all tasks with high probability. One drawback of this approach is the assumption of the existence of a reliable master processor. Despite the simplicity and advantages of this approach, the master is a single point of failure. The master is further assumed to be able to keep up with the large number of results returned by the workers, making such systems poorly scalable. In any message passing system, during some short time interval, a network node can maintain only a limited number of connections. Thus scalable distributed (i.e., not centralized) solutions are desirable.

In the current paper, we aim to remove the assumption of an infallible and bandwidth-unlimited master processor and consider a fully decentralized solution using just the cooperating workers.

Contributions. We consider the problem of performing t tasks in a distributed system of n workers. The tasks are independent, they admit at-least-once execution semantics, and each task can be performed by any worker in constant time. The workers either obtain the tasks from some repository or the tasks are initially known to all processors. The workers can return incorrect results and ultimately crash. The fully-connected message-passing system is synchronous, and the workers communicate using authenticated messages (to prevent malicious workers from impersonating other workers). Our system of autonomous processors is fully decentralized in the sense that it does not contain any distinguished participants (e.g., a master). We present an original randomized decentralized algorithm of logarithmic time complexity, where in each iteration of the algorithm each worker sends just one message. The algorithm works under several failure models differing in the assumptions about the fraction of possibly faulty workers and the failure probabilities. In more detail our contributions are as follows.

1. We define a general failure model $\mathcal{F}$, where each worker i ($i \in [n]$) independently returns an incorrect result, each time it performs a task, with probability p_i, such that $\frac{1}{n} \sum_{i \in [n]} p_i < \frac{1}{2} - \varepsilon$ for some $\varepsilon > 0$. (We show later how this model specializes to other intuitive models.)

2. We provide a n-processor, n-task decentralized randomized algorithm for model $\mathcal{F}$ that works in synchronous rounds. The number of rounds performed by the algorithm is an external (compile-time) parameter. Within each round each processor performs a random task (for some number of rounds), and communicates its cumulative knowledge to one randomly chosen processor. The algorithm naturally generalizes for t tasks, where $t \geq n$, by having processors work on groups of $\lceil t/n \rceil$ tasks instead of single tasks.

3. We analyze our algorithm under model $\mathcal{F}$ and show that it is sufficient for it to iterate for $\Theta(\log n)$ rounds in order to perform all tasks with high probability (whp). More specifically, we prove that after $\Theta(\log n)$ rounds every processor holds the array of computed results that are all correct whp, and that the arrays of results are consistent among all processors whp. With t tasks ($t \geq n$), the algorithm has time complexity $\Theta(\frac{t}{n} \log n)$, message complexity $\Theta(n \log n)$, bit communication complexity $O(t\, n \log^3 n)$, space complexity is $\Theta(t\, n \log^2 n)$, and work $\Theta(t \log n)$.

4. We show that failure model $\mathcal{F}$ can be extended to incorporate processor crashes in the way that does not require any changes to our algorithm. We also present three additional failure models that specialize model $\mathcal{F}$ and that are more intuitive. Since each of these models is a specialization of model $\mathcal{F}$, the same algorithm works under all these models and has the same (or better) complexity.

5. We present selected simulation results that illustrate and provide insights into the behavior of the algorithm.

Note that our problem is related to the Do-All problem [4,9] of using n processors to perform a collection of t independent tasks in the presence of adversity. However the two problems are not identical. In Do-All, the problem is solved when *some* correct processor knows that all tasks have been performed. In our problem, with the removal of the infallible master, a client application should be able to obtain the results from any worker. Thus the current problem is solved when *all* correct processors know that all tasks have been performed and are in the possession of the results of all tasks (whp in this work).

Consequently, an algorithm solving our problem is also an algorithm for Do-All, but not necessarily vice versa. Additionally a lower bound for Do-All is also a lower bound for the current problem. In [3] Chlebus and Kowalski give a lower bound $\Omega(t + n\frac{\log n}{\log\log n})$ on work of any algorithm solving Do-All, including randomized, against an adaptive linearly bounded adversary. This bound applies also to the current problem, and the work of our algorithm is close to this bound.

Prior work. Several approaches have been explored to improve the quality of the results obtained from untrusted workers. Fernandez, Georgiou, Lopez, and Santos [5,6] and Konwar, Rajasekaran, and Shvartsman [11] consider a distributed system consisting of a reliable master and a collection of workers that execute tasks on behalf of the master, where the workers may act maliciously by deliberately returning wrong results. Works [5,6,11] focus on designing algorithms that help the master determine the correct result with high probability, and at the least possible cost in terms of the total number of tasks executed.

The failure models assume that some fraction of processors can exhibit faulty behavior.

Gao and Malewicz [8] consider the problem of maximizing the expected number of correct results when the tasks have dependencies. Their distributed system is composed of a reliable server that coordinates unreliable workers that compute correctly with some probability, and where any incorrectly performed task corrupts all dependent tasks. The goal is to produce a schedule for task execution by the participants that maximizes the expected number of correct results under a constraint on the computation time.

Paquette and Pelc [13] consider a general model of a fault-prone system in which a decision has to be made on the basis of unreliable information. They assume that a Boolean value is conveyed to the deciding agent by several processors. An *a priori* probability distribution of this value is known to the agent and can be any arbitrary distribution. Relaying processors are assumed to fail independently with a known probability distribution. Fault-free processors relay the correct value, but faulty ones may behave arbitrarily. The deciding agent receives the vector of relayed values and must make a decision concerning the original value. The authors design a deterministic decision strategy with a high probability of correctness, and it is shown that a locally optimal decision strategy need not have the highest probability of correctness globally.

We have already mentioned the related problem of distributed cooperation called Do-All. Many algorithms, both deterministic and randomized, have been developed for Do-All in various models of computation, including message-passing and shared-memory models [10,9]. A related problem is the Omni-Do problem of performing a collection of tasks with the help of group communication services in partitionable networks [9].

Document structure. In Section 2 we give models of computation and failure, and measures of efficiency. Section 3 presents our algorithm. In Section 4 we carry out the analysis of the algorithm and derive complexity bounds. In Section 5 we deal with processor crashes. In Section 6 we present the simulation of the algorithm. We conclude in Section 7 with a discussion.

2   Model of Computation and Definitions

System model. We consider a set of n processors, or workers, each with a unique identifier (id) from set $\mathcal{P} = [n]$. We refer to the processor with id i as processor i. The system is synchronous and the processors communicate by exchanging reliable authenticated messages. Computation is structured in terms of synchronous *steps*, where in each step a processor can send or receive messages, and perform some local computation. The duration of each step depends on the algorithm and need not be constant (e.g., it may depend on n), but it is fixed at compile-time. Messages received by a processor in a given step include all messages sent to it in the previous step.

Tasks. There are t tasks to be performed, each with a unique id from set $\mathcal{T} = [t]$. We refer to the task with id i as $Task[i]$. Workers obtain tasks from some repository or workers initially know all tasks. The tasks are (a) similar, meaning that any task can be done in constant time by any processor, (b) independent, meaning that each task can be performed independently of other tasks, and (c) idempotent, meaning that the tasks admit at-least-once semantics and can be performed concurrently. For simplicity, we assume that the outcome of each task is a binary value. The problem is most interesting when there are at least as many tasks as there are processors, thus we only consider $t \geq n$.

Models of Failure. Some processors may exhibit faulty behavior by (maliciously) returning an incorrect result for a task. We assume that the result of each task is signed by the performing processor and that the signatures are unforgeable. The main failure model is defined as follows.

Model $\mathcal{F}$: *Each worker, independently of other workers, returns faulty results for a performed task with probability p_i, for $i \in [n]$, such that, $\frac{1}{n} \sum_i p_i < \frac{1}{2} - \varepsilon$ for some $\varepsilon > 0$.*

We use the constant ε to ensure that the average probability of worker misbehavior does not become arbitrarily close to $1/2$ as n tends to infinity.

In algorithm simulations we also use three related specialized models that were introduced in [5,11] in the context of the centralized master-worker setting.

Model $\mathcal{F}_a$: *Each worker, independently of other workers, returns faulty results for a task with probability $p < \frac{1}{2}$.*

Model $\mathcal{F}_b$: *A fixed fraction f of workers can return faulty results for any task with probability p, with $fp < \frac{1}{2}$.*

Model $\mathcal{F}_c$: *A fixed fraction f of workers can return faulty results for any task, with $f < \frac{1}{2}$.*

Observe that model $\mathcal{F}$ generalizes these specialized models since in all three cases the average probability of worker returning a wrong result is inferior to $1/2$. Thus any algorithm that solves our problem in model $\mathcal{F}$ also solves it in models $\mathcal{F}_a$, $\mathcal{F}_b$, and $\mathcal{F}_c$. Because the last three models are simpler to implement we use them in simulations.

Measures of efficiency. We use the conventional worst-case measures of *time complexity*, *work complexity*, and *space complexity*. *Message complexity* is the worst-case number of point-to-point messages sent in an execution, and *bit complexity* is the total number of bits sent in all messages.

Lastly, we use the common definition of *an event $\mathcal{E}$ occurring with high probability* (*whp*) to mean that $\mathbf{Pr}[\mathcal{E}] = 1 - O(n^{-\alpha})$ for some constant $\alpha > 0$.

3   Algorithm Description

In this section we present our decentralized algorithm A that employs no master and instead uses a gossip-based approach. We present the algorithm for n

procedure for processor i;
    external n /* the number of processors and tasks */
    external L /* 2L is the the number of rounds */
    $Task[1..n]$ /* set of tasks */
    $R_i[1..n]$ **init** $\emptyset^n$ /* set of collected results */
    $Results_i[1..n]$ /* array of results */

    Compute:
1:    Randomly select $j \in \mathcal{T}$ /* choose task id */
2:    Compute the result v_j for $Task[j]$
3:    $R_i[j] \leftarrow \{\langle v_j, i, 0 \rangle\}$ /* Record result for round 0 */

    for $r = 1$ **to** $2L$ **do**
        Send:
4:        Randomly select a processor $q \in \mathcal{P}$
5:        Send the array $R_i[\,]$ to processor q
        Receive:
6:        Let M be the set of received messages
7:        **for all** $j \in \mathcal{T}$
8:            $R_i[j] \leftarrow R_i[j] \cup \{R[j] : R[\,] \in M\}$
        Compute:
9:        **if** $r < L$ **then**
10:       Randomly select $j \in \mathcal{T}$ /* choose task id */
11:          Compute the result v_j for $Task[j]$
12:          $R_i[j] \leftarrow R_i[j] \cup \{\langle v_j, i, r \rangle\}$

13:    **for each** $j \in \mathcal{T}$
14:        $Results_i[j] \leftarrow u$ such that triples $\langle u, \_, \_ \rangle$ form a plurality in $R_i[j]$
    end

Fig. 1. Algorithm A at processor i for $i \in \mathcal{P}$, and $t = n$

processors and $t = n$ tasks. The algorithm naturally generalizes for t tasks, where $t \geq n$, by having processors perform work on fixed groups of $\lceil t/n \rceil$ tasks instead of single tasks (we discuss this in more detail at the conclusion of the analysis). Each processor (worker) maintains two arrays of size linear in n, one used to accumulate knowledge gathered from different processors, and another to store the results. The algorithm works in synchronous rounds. The number of rounds performed by the algorithm is an external (compile-time) parameter. Within each round a processor communicates its cumulative knowledge to one randomly chosen processor and performs a random task (for some determined number of rounds). The pseudocode for the algorithm is given in Figure 1, and we now detail it.

Local knowledge and state variables. Every processor i maintains the following:

- L, the external parameter that is used to control the number of iterations, i.e., $2L$, of the main loop; r is the current round (iteration) number.
- The array of results $R_i[1..n]$, where the element $R_i[j]$, for $j \in \mathcal{T}$, is the set of results for $Task[j]$. Each $R_i[j]$ is a set of triples $\langle v_j, i, r \rangle$ representing the result v_j computed for $Task[j]$ by processor i during round r. The use of such triples eliminates repeated inclusions of the results for the same task, in the same round, by the same processor.
- The array $Results_i[1..n]$ stores the final results.

Control flow. The algorithm contains the main for-loop, and we use the term *round* to refer to a single iteration of the loop. The loop contains three stages (or steps), viz., *Send*, *Receive*, and *Compute*. The algorithm starts by performing a single *Compute* stage, after which it enters the main loop. The algorithm uses an external parameter L (whose value is established in the analysis of the algorithm). The main loop iterates $2L$ times, where in the first L iterations all three stages are executed, and the final L iterations only the *Send* and *Receive* stages are executed. (We will prove that L needs to be $\Theta(\log n)$ to yield our high probability guarantee.)

We now describe the stages in more detail, starting with *Compute*. In *Compute* stage in round r processor i randomly selects a task j, computes the result v_j, and adds the triple $\langle v_j, i, r \rangle$ to the results set $R_i[j]$. This is done in the first L rounds.

In each *Send* stage, a processor choses a target processor q at random from the set of processors $\mathcal{P}$. The array of results $R[\,]$ is sent to processor q.

During the *Receive* stage processor i receives messages (if any) sent to it during the *Send* stage by other processors (including itself). Upon receiving the messages the processor updates its $R_i[j]$ (for each $j \in \mathcal{T}$) by taking a union with the triples for task j received in all messages.

When the main loop terminates after $2L$ rounds, each processor goes over the result set for every task and computes the result that corresponds to the plurality of the results (in the analysis we prove that in fact a majority exists). The results of the tasks are available locally in array $Results_i[1..n]$.

4   Algorithm Analysis

We now analyze algorithm A for $t = n$, then extend the analysis to $t \geq n$. We start by stating the Chernoff bound result that we use in several places.

Lemma 1 (Chernoff Bounds). *Let $X_1, X_2, \cdots, X_n$ be n independent Bernoulli random variables with $\mathbf{Pr}[X_i = 1] = p_i$ and $\mathbf{Pr}[X_i = 0] = 1 - p_i$, then it holds for $X = \sum_{i=1}^{n} X_i$ and $\mu = E[X] = \sum_{i=1}^{n} p_i$ that for all $\delta > 0$, (i) $\mathbf{Pr}[X \geq (1 + \delta)\mu] \leq e^{-\frac{\mu\delta^2}{3}}$, and (ii) $\mathbf{Pr}[X \leq (1 - \delta)\mu] \leq e^{-\frac{\mu\delta^2}{2}}$.*

The following lemma shows that within $\Theta(\log n)$ rounds of algorithm A every task τ is chosen for execution $\Theta(\log n)$ times *whp*. Weaker variations of Lemma 2 are known in the literature, e.g., see the *Occupancy Problem* [12]. We prove our

lemma for completeness, and more importantly, for acquiring a stronger bound
required for our complexity results.

Lemma 2. *In $\Theta(\log n)$ rounds of the algorithm every task is performed $\Theta(\log n)$ times* whp, *possibly by different processors.*

Proof. Let us assume that after $L = k \log n$ rounds of algorithm A, where k is a sufficiently large constant, there exists a task τ that is performed less than $(1 - \delta)L$ times among all workers, for some $\delta > 0$. We prove that *whp* such a task does not exist.

According to our assumption at the end of round L for some task τ, we have $|\cup_{j=1}^{n} R_j[\tau]| < (1-\delta)L$. Let X_i be a Bernoulli random variable such that $X_i = 1$ if the task was chosen to be performed in line 10 (only once the task is chosen in line 1) of the algorithm, and $X_i = 0$ otherwise.

Let us next define the random variable $X = X_1 + \cdots + X_{Ln}$ to count the total number of times task τ is performed by the end of L rounds of algorithm A.

Note that according to line 10 any worker picks a task uniformly at random. To be more specific let x be an index of one of Ln executions of line 10. Observe that for any x, $\mathbf{Pr}[X_x = 1] = \frac{1}{n}$ given that the workers choose task τ uniformly at random. Let $\mu = \mathbf{E}[X] = \sum_{x=1}^{Ln} \frac{1}{n} = L$, then by applying Chernoff bound, for the same $\delta > 0$ chosen as above, we have:

$$\mathbf{Pr}[X \leq (1 - \delta)L] \leq e^{-\frac{L\delta^2}{2}} \leq e^{-\frac{(k \log n)\delta^2}{2}} \leq \frac{1}{n^{\frac{c\delta^2}{2}}} \leq \frac{1}{n^\alpha}$$

where $\alpha > 1$ for some sufficiently large c. Now let us denote by $\mathcal{E}_\tau$ the fact that $|\cup_{i=1}^{n} R_i(\tau)| > (1-\delta)L$ by the round L of the algorithm and we denote by $\bar{\mathcal{E}}_\tau$ the complement of that event. Next by Boole's inequality we have $\mathbf{Pr}[\cup_\tau \bar{\mathcal{E}}_\tau] \leq \sum_\tau \mathbf{Pr}[\bar{\mathcal{E}}_\tau] \leq \frac{1}{n^\beta}$, where $\beta = \alpha - 1 > 0$. Hence each task is performed at least $\Theta(\log n)$ times *whp*, i.e., $\mathbf{Pr}[\cap_\tau \mathcal{E}_\tau] = \mathbf{Pr}[\overline{\cup_\tau \bar{\mathcal{E}}_\tau}] \geq 1 - \frac{1}{n^\beta}$.

The following lemma shows that *whp* after $\Theta(\log n)$ rounds of the algorithm every worker obtains every triple generated in the system by either generating it locally or by means of gossiping.

A somewhat similar result is shown by Fraigniaud and Glakkoupis [7] who study the communication complexity of rumor-spreading in the random phone-call model. They consider n players communicating in parallel rounds, where in each round every player u calls a randomly selected communication partner. Player u is allowed to exchange information with the partner, either by *pulling* or *pushing* information. In order to avoid repetition, we anchor part of our proof to their results related to the *push* part of their algorithm.

The following lemma, proved in [7], shows that every triple $\vartheta = \langle v_j, i, r \rangle$ (in their work a rumor ρ) is disseminated to at least $\frac{3}{4}n$ workers (in their work players) *whp*.

Lemma 3. *With probability $1 - n^{-3+o(1)}$, at least $\frac{3}{4}$ fraction of the players knows ρ at the end of round $\tau = \lg n + 3 \lg \lg n$.*

Our proof also makes use of the *Coupon Collector's* problem [12]:

Definition 1. *The Coupon Collector's Problem (CCP). There are n types of coupons and at each trial a coupon is chosen at random. Each random coupon is equally likely to be of any of the n types, and the random choices of the coupons are mutually independent. Let m be the number of trials. The goal is to study the relationship between m and the probability of having collected at least one copy of each of n types.*

In [12] it is shown that $E[X] = n \ln n + O(n)$ and that *whp* the number of trials for collecting all n coupon types lies in a small interval centered about its expected value. Now we state and prove the needed lemma.

Lemma 4. *If every task is performed $\Theta(\log n)$ times, then* whp *in $\Theta(\log n)$ rounds of the algorithm each worker acquires the results for every task.*

Proof. Let us assume that in some round r task j is performed by worker i; thus a triple $\vartheta \equiv \langle v_j, i, r \rangle$ is generated by worker i, where v_j is the calculated value of task j.

By applying Lemma 3 to our algorithm we infer that in $\Theta(\log n)$ rounds of algorithm A at least $\frac{3}{4}n$ of the workers become aware of triple ϑ *whp*. Next consider any round d such that at least $\frac{3}{4}n$ of the workers are aware of triple ϑ for the first time. Let us denote this subset of workers by S_d ($|S_d| \geq \frac{3}{4}n$.)

We denote by U_d the remaining fraction of the workers that are not aware of ϑ. We are interested in the number of rounds required for every worker in U_d to learn about ϑ *whp* by receiving a message from one of the workers in S_d in some round following d.

We show that, by the analysis very similar to CCP, in $\Theta(\log n)$ rounds triple ϑ is known to all workers *whp*. Every worker in $\mathcal{P}$ has a unique id, hence we can think of those workers as of different types of coupons and we assume that the workers in S_d collectively represent the coupon collector. In this case, however, we do not require that every worker in S_d contacts all workers in U_d *whp*. Instead, we require only that the workers in S_d *collectively* contact all workers in U_d *whp*. According to our algorithm in every round every worker in $\mathcal{P}$ ($S_d \subset \mathcal{P}$), selects a worker uniformly at random and sends all its data to it. Let us denote by m the collective number of trials by workers in S_d to contact workers in U_d. According to CCP if $m = O(n \ln n)$ then *whp* workers in S_d collectively contact every worker in $\mathcal{P}$, including those in U_d. Since there are at least $\frac{3}{4}n$ workers in S_d then in every round the number of trials is at least $\frac{3}{4}n$, hence in $O(\ln n)$ rounds *whp* all workers in U_d learn about ϑ. Therefore, in $\Theta(\log n)$ rounds *whp* all workers in U_d learn about ϑ.

Thus we showed that if a new triple is generated in the system then *whp* it will be known to all workers in $\Theta(\log n)$ rounds. Now by applying Boole's inequality we want to show that *whp* in $\Theta(\log n)$ rounds all generated triples are spread among all workers.

According to our algorithm every worker generates $L = \Theta(\log n)$ triples before it terminates. We have n workers which means that by the end of the algorithm the number of generated triples is $\Theta(n \log n)$. Let us denote the set of all generated triples by $\mathcal{V}$. Let $\overline{\mathcal{E}_\vartheta}$ be the event that some triple ϑ is not spread around among all workers when the algorithm terminates. In the preceding part of the proof we have shown that $\mathbf{Pr}[\overline{\mathcal{E}_\vartheta}] < \frac{1}{n^\beta}$, where $\beta > 1$. By Boole's inequality, the probability that there exists one triple that did not get spread to all workers, can be bounded as

$$\mathbf{Pr}[\cup_{\vartheta \in \mathcal{V}} \overline{\mathcal{E}_\vartheta}] \leq \Sigma_{\vartheta \in \mathcal{V}} \mathbf{Pr}[\overline{\mathcal{E}_\vartheta}] = \Theta(n \log n) \frac{1}{n^\beta} \leq \frac{1}{n^\gamma}$$

where $\gamma > 0$. This implies that upon termination every worker collects all $\Theta(n \log n)$ triples generated in the system *whp*.

Next theorem shows that at termination the correct result for each task is obtained from the collectively computed results, whether correct or incorrect.

Theorem 1. *In $\Theta(\log n)$ rounds algorithm A produces the results of all n tasks correctly at every processor* whp.

Proof. We first prove that at termination the algorithm computes correctly a majority of the results for any task τ *whp*. Then we argue that *whp* at termination the result computed for each task by any processor is correct.

In order to prove the first step we estimate (with a concentration bound) the number of times the results are computed correctly. Then we estimate the bound on total number of times task τ was computed (whether correctly or incorrectly), and we show that a majority of the results are computed correctly.

Let us consider random variables X_{ir} that denote the success or failure of correctly computing the result of some task τ in round r by worker i. Specifically, $X_{ir} = 1$ if in round r, worker i computes the result of task τ correctly, otherwise $X_{ir} = 0$. According to our algorithm we observe that $\mathbf{Pr}[X_{ir} = 1] = \frac{q_i}{n}$ and $\mathbf{Pr}[X_{ir} = 0] = 1 - \mathbf{Pr}[X_{ir} = 1]$, where $q_i \equiv 1 - p_i$.

Let $X_r \equiv \sum_{i=1}^{n} X_{ir}$ denote the number of correctly computed results for task t among all workers during round r. By linearity of expected values of a sum of random variables we have

$$\mathbf{E}[X_r] = \mathbf{E}[\sum_{i=1}^{n} X_{ir}] = \sum_{i=1}^{n} \mathbf{E}[X_{ir}] = \sum_{i=1}^{n} \frac{q_i}{n}$$

We denote by $X \equiv \sum_{r=1}^{L} X_r$ the number of correctly computed results for some task τ at termination. Again, using the linearity of expected values of a sum of random variables we have

$$\mathbf{E}[X] = \mathbf{E}[\sum_{i=1}^{n} \sum_{r=1}^{L} X_{ir}] = \frac{L}{n} \sum_{i=1}^{n} q_i$$

Note that since $\frac{1}{n} \sum_{i=1}^{n} q_i > \frac{1}{2} + \varepsilon$, for some fixed $\varepsilon > 0$, there exists some $\delta > 0$, such that, $(1 - \delta) \frac{L}{n} \sum_{i=1}^{n} q_i > (1 + \delta) \frac{L}{2}$. Also, observe that the random variables $X_1, X_2, \cdots, X_L$ are mutually independent. Therefore, by applying Chernoff bound on $X_1, X_2, \cdots, X_L$ we have

$$\mathbf{Pr}[X \leq (1-\delta)E[X]] \equiv \mathbf{Pr}[X \leq (1-\delta)\frac{L}{n}\sum_{i=1}^{n} q_i] \leq e^{-\frac{\delta^2 L(1+\delta)}{4(1-\delta)}} \leq \frac{1}{n^{\alpha_1}}$$

where $\alpha_1 > 1$ such that $L = k\log n$ for some sufficiently large constant $k > 0$.

Let us now count the total number of times task τ is chosen to be performed during the execution of the algorithm in the course of the first L rounds. We represent the choice of task τ by worker i during round r by the random variable Y_{ir}. We assume $Y_{ir} = 1$ if τ is chosen by worker i in round r, otherwise $Y_{ir} = 0$. Since Y_{ir}'s are mutually independent we have $\mathbf{E}[Y_{ir}] = \frac{1}{n}$. We denote by $Y \equiv \sum_{i=1}^{n}\sum_{r=1}^{L} Y_{ir}$ the number of times task t is computed at termination. By linearity of expected values we have $E[Y] = L$. Then by applying Chernoff bound for the same $\delta > 0$ chosen as above we have

$$\mathbf{Pr}[Y \geq (1+\delta)\mathbf{E}[Y]] \equiv \mathbf{Pr}[Y \geq (1+\delta)L] \leq e^{-\frac{\delta^2 L}{3}} \leq \frac{1}{n^{\alpha_2}}$$

for some $\alpha_2 > 1$. Hence, applying Boole's inequality to the bounds on the above two events

$$\mathbf{Pr}[\{X \leq (1-\delta)\frac{L}{n}\sum_{i=1}^{n} q_i\} \cup \{Y \geq (1+\delta)L\}] \leq \frac{2}{n^{\alpha}}$$

where $\alpha = \min\{\alpha_1, \alpha_2\} > 1$

Therefore, from above and by using $(1-\delta)\frac{L}{n}\sum_{i=1}^{n} q_i > (1+\delta)\frac{L}{2}$ we have

$$\mathbf{Pr}[Y/2 < X] \geq \mathbf{Pr}[\{Y < (1+\delta)L\} \cap \{X > (1-\delta)\frac{L}{n}\sum_{i=1}^{n} q_i\}]$$

$$= 1 - \mathbf{Pr}[\{Y \geq (1+\delta)L\} \cup \{X \leq (1-\delta)\frac{L}{n}\sum_{i=1}^{n} q_i\}]$$

$$\geq 1 - \frac{1}{n^{\beta}}$$

for some $\beta > 1$. Hence, at termination of the algorithm *whp* the majority of calculated results for task τ are correct. Let us denote this event by $\mathcal{E}_t$.

From above we have $\mathbf{Pr}[\overline{\mathcal{E}_\tau}] \leq \frac{1}{n^{\beta}}$. Now, by Boole's inequality we obtain

$$\mathbf{Pr}[\bigcup_{t \in \mathcal{T}} \overline{\mathcal{E}_\tau}] \leq \sum_{\tau \in \mathcal{T}} \mathbf{Pr}[\overline{\mathcal{E}_\tau}] \leq \frac{1}{n^{\beta-1}} \leq \frac{1}{n^{\gamma}}$$

where $\mathcal{T}$ is the set of all n tasks, and $\gamma > 0$.

By Lemma 4 *whp* all calculated results of every task are disseminated across all workers. Thus, the majority of the results computed for any task at any worker is the same among all workers, and moreover it is correct *whp*. Recall that according to our algorithm (line 14) every processor computes the result of every task by taking the plurality of calculated results, and hence the claim of the theorem.

Algorithm A terminates after $\Theta(\log n)$ rounds and thus every processor generates $\Theta(\log n)$ triples. This implies that at termination $\Theta(n \log n)$ triples are generated. To obtain consistent and correct results among all processors *whp* we want all processors to hold the same set of triples. Each triple consists of the calculated result of a task, the id of the processor that performed the task, and the round number. Thus $\Theta(\log n)$ bits are required to represent each triple. Next we assess work, message, bit, and space complexities.

Theorem 2. *Algorithm A has work complexity $\Theta(n \log n)$, message complexity $\Theta(n \log n)$, bit complexity $O(n^2 \log^3 n)$, and space complexity $\Theta(n^2 \log^2 n)$.*

Proof. Algorithm A terminates in $\Theta(\log n)$ rounds, thus its work is $\Theta(n \log n)$. In every round every worker sends one message to a randomly chosen worker (including itself). Hence, the message complexity is $\Theta(n \log n)$.

Now let us estimate bit complexity. For every performed task algorithm adds a triple to the result set, where $\Theta(\log n)$ bits are required to store a triple. According to our algorithm every processor sends $O(n \log^2 n)$ bits in every round, where the additional multiplicative $\log n$ factor represents the number of different triples per task. On the other hand, the algorithm terminates in $\Theta(\log n)$ rounds, hence every processor communicates $O(n \log^3 n)$ bits of information to other processors. Therefore, the bit complexity of the algorithm is $O(n^2 \log^3 n)$.

Finally, it is easy to see that space complexity of the algorithm is $\Theta(n^2 \log^2 n)$. Indeed, by termination of the algorithm every processor i holds an array of sets R_i and the result vector $Results_i$, for $i \in [n]$. The result vector consists of just n bits. On the other hand, according to Lemmas 2 and 4, after algorithm terminates each $R_i[j]$ contains $\Theta(\log n)$ triples *whp*, hence the number of bits required for each $R_i[j]$ is $\Theta(\log^2 n)$, where $i, j \in [n]$. Considering that the number of tasks and processors is n, the total bit complexity is $\Theta(n^2 \log^2 n)$.

Finally, we extend the algorithm to handle the number of tasks larger than the number of processors as follows. Let $\mathcal{T}' = [t]$ be the set of unique task identifiers, where $t \geq n$. We segment the t tasks into groups of $\lceil t/n \rceil$ tasks, and construct a new array of super-tasks with identifiers $\mathcal{T} = [n]$, where each super-task takes $\Theta(t/n)$ time to perform by any processor. For a super-task τ, the result v_τ is now a sequence of $\lceil t/n \rceil$ bits, instead of a single bit. We now use algorithm A, where the only difference is that each *Compute* stage takes $\Theta(t/n)$ time, and the data structures are larger to accommodate the results consisting of $\lceil t/n \rceil$ bits. We call the resulting algorithm A' and we show the following.

Theorem 3. *For $t \geq n$ algorithm A' has time complexity $\Theta(\frac{t}{n} \log n)$, work complexity $\Theta(t \log n)$, message complexity $\Theta(n \log n)$, bit complexity $O(t\, n \log^3 n)$, and space complexity $\Theta(t\, n \log^2 n)$.*

Proof sketch. As with algorithm A, algorithm A' takes $\Theta(\log n)$ iterations to produce the results *whp*, except that each iteration now takes $\Theta(t/n)$ time. This yields time complexity $\Theta(\frac{t}{n} \log n)$. Work complexity is then $n \cdot \Theta(\frac{t}{n} \log n) = \Theta(t \log n)$.

The message complexity remains the same at $\Theta(n \log n)$ as the number of messages does not change. The messages are larger, however, by a factor of t/n relative to the result of Theorem 2, thus the bit complexity is $O(t \; n \log^3 n)$. Lastly, the storage requirements are increased by the same factor, resulting in space complexity $\Theta(tn \log^2 n)$. $\Box$

In closing this section we note that the same results hold for models $\mathcal{F}_a$, $\mathcal{F}_b$, and $\mathcal{F}_c$, since they are direct specializations of model $\mathcal{F}$.

5   Tolerating Crash Failures

We now show that algorithm A correctly performs n tasks *whp* even if up to fn processors crash for a constant f, where $0 < f < 1$, under failure model $\mathcal{F}$. We prove that the asymptotics of the algorithm are unchanged if crashes do not invalidate the definition of model $\mathcal{F}$, meaning that the *average* probability of a non-crashed worker returning an incorrect result remains inferior to $1/2$. Specifically, we show that Lemmas 2 and 4, and Theorem 1 remain valid under this model.

In any execution of Algorithm A we denote the set of processors that do not crash by $\mathcal{P}'$, and we let $n' = |\mathcal{P}'|$. As before, we start with $t = n$.

Lemma 5. *In $\Theta(\log n)$ rounds of the algorithm every task is performed $\Theta(\log n)$ times* whp, *possibly by different processors when at most fn processors can crash.*

Proof sketch. In the worst case all failure prone processors will crash in the first round of the algorithm. Thus, it is sufficient to prove that *whp* every task is performed $\Theta(\log n)$ times among the processors in $\mathcal{P}'$. In order for every task to be performed $\Theta(\log n)$ times *whp* by processors in $\mathcal{P}'$ it is sufficient to increase the value of L by a factor $\lambda = \frac{1}{1-f}$ (compared to the case without crashes). Since all processors pick a new task to be performed from the set of n tasks uniformly at random (line 10 of algorithm A) we can prove the results by carrying out the computation using Chernoff bound as in the proof of Lemma 2. $\Box$

Now we prove that *whp* after $\Theta(\log n)$ rounds of the algorithm every worker in $\mathcal{P}'$ holds the same set of triples for every task.

Lemma 6. *If processors in $\mathcal{P}'$ collectively hold $\Theta(\log n)$ calculated results for every task, then* whp *in $\Theta(\log n)$ rounds of the algorithm each processor $i \in \mathcal{P}'$ obtains all $\Theta(\log n)$ triples for every task j, when at most fn processors crash.*

Proof sketch. Consider a triple ϑ that is generated (or obtained by gossiping) by some processor in $\mathcal{P}'$. The proof of Lemma 4 uses the results from Lemma 3 and CCP. Both of these results rely on the fact that there are $\Theta(n)$ participating processors, and since there are at most fn processors that crash we have $\Theta(n)$ processors left in $\mathcal{P}'$. Therefore, following a similar line of analysis we can claim the lemma with respect to the processors that do not crash until the end of algorithm A and the triples possessed by them. $\Box$

The final theorem shows that *whp* the correct results for each task are computed in $\Theta(\log n)$ rounds by the processors in $\mathcal{P}'$.

Theorem 4. *Algorithm A computes all n tasks correctly at every live processor in $\Theta(\log n)$ rounds whp and has work $\Theta(n \log n)$ in the presence of at most fn crashes.*

Proof sketch. To prove this we need to show that, at termination, for any task t the majority of the results are computed correctly *whp*. Note that if we consider only the results (triples) that are generated by the processors in $\mathcal{P}'$ then our high probability correctness results can be shown similarly to the proof of Theorem 1. Suppose that we also consider the triples that are generated by the processors that are not in $\mathcal{P}'$. Note that according to our assumption the *average* probability of a worker returning an incorrect result remains inferior to $1/2$ in spite of crashes. Hence, the probability of correctly choosing the result for a task is not affected. Since algorithm A terminates in $\Theta(\log n)$ rounds its work cannot exceed $\Theta(n \log n)$. □

Clearly in the presence of up to fn crashes the message and bit complexities, as well as the space complexity of the algorithm A remains unchanged. Although the complexity results do not change in the presence of crashes, it is important to note that the overall number of rounds may increase by a constant factor of $\lambda = \frac{1}{1-f}$.

Finally, the algorithm is extended as discussed in the previous section to deal with t tasks when $t \geq n$. Given Theorem 4, the complexity bounds established in Theorem 3 remain valid in the crash-extended failure model.

6   Simulation Results

To illustrate our analytical findings we present selected simulation results of algorithm A (for $t = n$) in model $\mathcal{F}$ and in model $\mathcal{F}_c$. We use model $\mathcal{F}$ as the most general model, and we use model $\mathcal{F}_c$ to show the behavior of the algorithm in one of the specialized settings. (We do not show simulations for all defined models for paucity of space.)

Theorems 1 and 4 show that algorithm A performs all n tasks correctly *whp* at every node in $\Theta(\log n)$ rounds. In simulations we let $L = k \log n$, where $k > 0$ is a constant. We carried out simulations for up to $n = 1000$ tasks and processors, and for modest values of $k \in \{2, 3, 4\}$. For every n paired with every k we ran the simulation for 100 times and graphed the average of the percentage of *incorrectly* calculated results as the function of n and k. In all simulations the calculated results are *always consistent* among all processors in every run of the algorithm as anticipated by Lemmas 4 and 6.

Figures 2 and 3 show results for model $\mathcal{F}_c$ and model $\mathcal{F}$ (without crashes) respectively. For model $\mathcal{F}_c$ we let $f = \frac{1}{4}$ of processors be faulty: these processors return incorrect results with probability $p = 1$. The rest of the processors are correct. For model $\mathcal{F}$ we assume that the average probability of returning incorrect results is inferior to 0.25. The results for models $\mathcal{F}_c$ and $\mathcal{F}$ are similar, showing the percentage of incorrect results is diminishing rapidly even for modest k. Analysis shows that this error can be made as small as necessary by

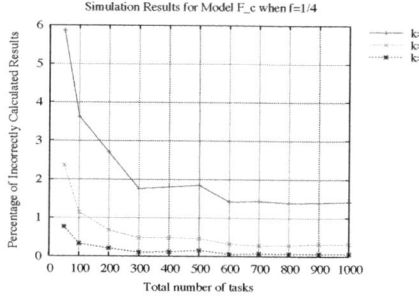

Fig. 2. Simulation results for model $\mathcal{F}_c$

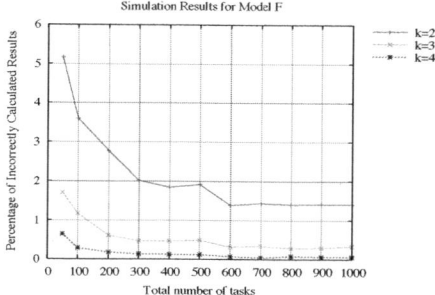

Fig. 3. Simulation results for model $\mathcal{F}$

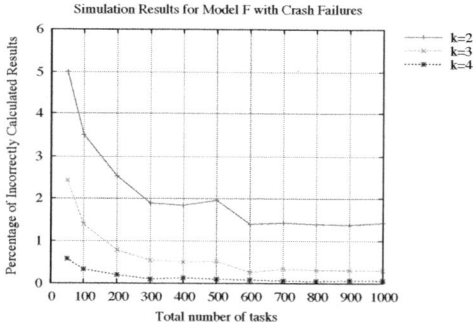

Fig. 4. Simulation for model $\mathcal{F}$ with crashes

increasing k (of course if the average probability of calculating results incorrectly tends to $\frac{1}{2}$, k may need to be substantial to guarantee the results).

Figure 4 shows the percentage of incorrectly calculated results in model $\mathcal{F}$ with crashes. Here we let $f = \frac{3}{5}$ fraction of processors be crash-prone, keeping similar probabilities of returning incorrect results as before. Hence, the average probability of returning an incorrect result is still inferior to 0.25 for all processors that do not crash. The results again show diminishing error as k grows.

7   Conclusion

Abstracting the setting of network supercomputing with untrusted workers, we defined a model of failures for workers that may return incorrect results, and we presented and analyzed a decentralized algorithm that allows correct workers to cooperatively perform a collection of tasks. The new algorithm breaks with tradition and removes the assumption of the central infallible master processor. The algorithm imposes only a logarithmic time overhead, while sharing information about the progress of computation by means of gossip. Noteworthy, each processor sends only one message for each iteration of the algorithm. We showed that the algorithm performs all tasks correctly *whp* and we developed a simulation of the algorithm to illustrate our analytical findings. Future work includes considering more virulent failure behaviors and task sets with inter-task dependencies.

References

1. Internet primenet server, http://mersenne.org/ips/stats.html
2. Seti@home, http://setiathome.ssl.berkeley.edu/
3. Chlebus, B., Kowalski, D.: Randomization helps to perform independent tasks reliably. Random Structures and Algorithms 24(1), 11–41 (2004)
4. Dwork, C., Halpern, J.Y., Waarts, O.: Performing work efficiently in the presence of faults. SIAM J. Comput. 27(5), 1457–1491 (1998)
5. Fernandez, A., Georgiou, C., Lopez, L., Santos, A.: Reliably executing tasks in the presence of untrusted entities. In: Proc. of the 25th IEEE Symposium on Reliable Distributed Systems, pp. 39–50 (2006)
6. Fernandez, A., Georgiou, C., Lopez, L., Santos, A.: Algorithmic mechanisms for internet-based master-worker computing with untrusted and selfish workers. Tech. rep., Proc. of the 24th IEEE Int'l Symposium on Parallel and Distributed Processing (2010)
7. Fraigniaud, P., Giakkoupis, G.: On the bit communication complexity of randomized rumor spreading. In: Proc. of the 22nd ACM Symposium on Parallelism in Algorithms and Architectures, SPAA 2010, pp. 134–143 (2010)
8. Gao, L., Malewicz, G.: Toward maximizing the quality of results of dependent tasks computed unreliably. Theory of Computing Systems 41(4), 731–752 (2007)
9. Georgiou, C., Shvartsman, A.A.: Do-All Computing in Distributed Systems: Cooperation in the Presence of Adversity. Springer, Heidelberg (2008)
10. Kanellakis, P.C., Shvartsman, A.A.: Fault-Tolerant Parallel Computation. Kluwer Academic Publishers (1997)
11. Konwar, K.M., Rajasekaran, S., Shvartsman, M.M.A.A.: Robust Network Supercomputing with Malicious Processes. In: Dolev, S. (ed.) DISC 2006. LNCS, vol. 4167, pp. 474–488. Springer, Heidelberg (2006)
12. Motwani, R., Raghavan, P.: Randomized Algorithms. Cambridge University Press (1995)
13. Paquette, M., Pelc, A.: Optimal decision strategies in byzantine environments. Parallel and Distributed Computing 66(3), 419–427 (2006)

On the Power of Waiting When Exploring Public Transportation Systems

David Ilcinkas* and Ahmed Mouhamadou Wade*

LaBRI, CNRS & Université de Bordeaux
{ilcinkas,wade}@labri.fr

Abstract. We study the problem of exploration by a mobile entity (agent) of a class of dynamic networks, namely the periodically-varying graphs (the PV-graphs, modeling public transportation systems, among others). These are defined by a set of carriers following infinitely their prescribed route along the stations of the network. Flocchini, Mans, and Santoro [FMS09] (ISAAC 2009) studied this problem in the case when the agent must always travel on the carriers and thus cannot wait on a station. They described the necessary and sufficient conditions for the problem to be solvable and proved that the optimal number of steps (and thus of moves) to explore a n-node PV-graph of k carriers and maximal period p is in $\Theta(k \cdot p^2)$ in the general case.

In this paper, we study the impact of the ability to wait at the stations. We exhibit the necessary and sufficient conditions for the problem to be solvable in this context, and we prove that waiting at the stations allows the agent to reduce the worst-case optimal number of moves by a multiplicative factor of at least $\Theta(p)$, while the time complexity is reduced to $\Theta(n \cdot p)$. (In any connected PV-graph, we have $n \leq k \cdot p$.) We also show some complementary optimal results in specific cases (same period for all carriers, highly connected PV-graphs). Finally this new ability allows the agent to completely map the PV-graph, in addition to just explore it.

Keywords: Exploration, Dynamic graphs, Mobile agent, PV-graph.

1 Introduction

1.1 The Problem

The problem of graph exploration consists, for a mobile entity, in exploring all nodes (or edges) of an a priori unknown graph. This problem being one of the most classical in the mobile agent computing framework, it has received a lot of attention so far. Time complexity, space complexity or impact of a priori knowledge have extensively been studied in the last 40 years (see, e.g., [PP99, Rei05, DP04]). However, the large majority of these works concern static

---

* Partially supported by the ANR project ALADDIN, the INRIA project CEPAGE, and the European project EULER.

A. Fernández Anta, G. Lipari, and M. Roy (Eds.): OPODIS 2011, LNCS 7109, pp. 451–464, 2011.

graphs. Considering nowadays networks, it is now common to deal with dynamic networks. In this paper, we study the graph exploration problem in one model of dynamic networks, namely the periodically-varying graph (PV-graph) model.

Roughly speaking, a PV-graph consists of a set of carriers, each following periodically its respective route among the sites of the system. This models in particular various types of public transportation systems like bus systems or subway systems for example. It also models low earth orbiting satellite systems, or security systems composed of security guards making tours in the place to be secured. Performing exploration in such systems may be useful for maintenance operations for example. Indeed, an agent can check that everything is in order during the exploration. This agent may be a piece of software, or a human being.

The exploration problem in the PV-graph model was already considered by Flocchini, Mans, and Santoro in [FMS09]. They considered that the agent cannot leave the carrier to stay on a site. Not being able to stay on a site is particularly legitimate in low earth orbiting satellite systems for example, where the sites do not correspond to any physical station. However, in most public transportation systems, it is possible for the agent (human or not) to stay on a site in order to wait for a (possibly different) carrier. In this paper, we consider the same problem but in the case when the agent can leave carriers to wait on a site. We study the impact of this new ability on the complexity (time and number of moves) of the PV-graph exploration problem.

1.2   Related Work

Motivated by the automatic exploration of the Web, Cooper and Frieze [CF03] studied the question of the minimum cover time of a graph that evolves over time. They considered a particular model of so-called web graphs and show that if after every constant number of steps of the walk a new node appears and is connected to the graph, a random walk does not visit a constant fraction of nodes. Kuhn, Lynch and Oshman [KLO10] introduced a stability property (intervals of connectivity). They assume that for any T consecutive rounds, there is a stable and connected common subgraph. In 2008, Avin, Koucky and Lotker [AKL08] showed that a random walk may have an exponential cover time in some dynamic graphs. They also show that a variant, the lazy random walk, has however a polynomial cover time in any dynamic graph.

In 2009, Flocchini, Mans and Santoro [FMS09] introduced a new model of dynamic networks, the PV-graph model. They first show that if the nodes of the PV-graph are labeled, the knowledge of an upper bound on the longer period or the exact knowledge of the number n of nodes is necessary and sufficient for an agent to explore the PV-graph. If the nodes of the PV-graph are anonymous, then the knowledge of an upper bound on the longer period is necessary and sufficient. In both settings, the time and move complexity of the agent is proved to be in $\Theta(k \cdot p^2)$, where k is the number of carriers and p the maximum period of the carriers. In the particular case of homogeneous PV-graphs (PV-graphs for which all carriers have the same period), the time and move complexity drops to $\Theta(k \cdot p)$.

Flocchini, Kellett, Mason, and Santoro [FKMS10] studied the mapping of a PV-graph containing black holes (sites destroying agents). They considered that several agents are operating in the PV-graph, and that they can leave messages on the sites. The goal of the agents is to construct the map of the PV-graph without losing too many agents. Casteigts, Flocchini, Santoro and Quattrociocchi [CFQS10] integrated a large collection of concepts, formalisms and results in the literature about dynamic graphs in an unified space.

1.3   Our Results

In this article, we extend the study of Flocchini, Mans and Santoro [FMS09] to the case when the agent can leave a carrier to stay at a site. This new ability allows the agent to explore PV-graphs that are less connected over time (formal definitions are given in Section 2). We prove that in the general case (so, even considering non highly-connected PV-graphs) the move complexity is reduced to $\Theta(\min\{k{\cdot}p, n{\cdot}p, n^2\})$, while the time complexity decreases to $\Theta(n{\cdot}p)$. (Note that in any connected PV-graph, we have $n \le k{\cdot}p$.) If the PV-graphs are restricted to be both homogeneous and highly-connected, then Flocchini, Mans and Santoro proved that the time complexity is in $O(k \cdot p)$. In this paper, we prove that if the PV-graphs satisfy only one of these restrictions, then the time complexity remains in $\Theta(n \cdot p)$. Besides, it turns out that our algorithm not only performs exploration but also performs mapping, i.e., it can output an isomorphic copy of the PV-graph. Finally, note that our algorithm does not use possible identifiers of the nodes, while all our lower bounds still hold when the agent has access to unique node identifiers.

2   Model and Definitions

We consider a system $S = \{s_1, \cdots, s_n\}$ of n *sites* among which k *carriers* are moving. Each carrier c has an identifier $\mathrm{Id}(c)$ and follows a finite sequence $R(c) = (s_{i_1}, \cdots, s_{i_{p(c)}})$ of sites, called its *route*, in a periodic manner. The positive integer $p(c)$ is called the *period* of the carrier c. More precisely, the carrier c starts at node s_{i_1} at time 0 and then proceeds along its route, moving to the next site at each time unit, in a cyclic manner (that is, when c is at node $s_{i_{p(c)}}$, it goes back to s_{i_1} and follows the route again and again).

A PV-graph (for periodically-varying graph) is a pair (S, C), where S is a set of sites, and C is a set of carriers operating among these sites. We will usually denote by n, k and p, respectively, the number of sites, the number of carriers and the maximum over the periods of the carriers. A PV-graph is said to be *homogeneous* if and only if all its carriers have the same period.

For any PV-graph G, we define two (classical) graphs $H_1(G)$ and $H_2(G)$ as follows. Both graphs have the set of carriers as the set of nodes. There is an edge in $H_1(G)$ between two carriers c and c' if and only if there exists a site appearing in the routes of c and c'. There is an edge in $H_2(G)$ between two carriers c and c' if and only if there exists a site s and a time $t \ge 0$ such that c and c' are both in s at time t. A PV-graph is said to be *connected* if and only if $H_1(G)$ is connected. A PV-graph is said to be *highly-connected* if and only if $H_2(G)$ is connected. In this paper, we will always consider PV-graphs that are at least connected. (Non-connected PV-graphs cannot be explored.) Furthermore note that, for any connected PV-graph, its parameters n (number of sites), k (number of carriers), and p (maximal period) satisfy the inequality $n - 1 \leq k(p - 1)$.

An entity, called *agent*, is operating on these PV-graphs. It can see the carriers and their identities. It can ride on a carrier to go from a site to another. Contrary to the model in [FMS09], the agent is allowed to leave a carrier, stay at the current site, and get back on a carrier (the same or another). We do not assume any restriction on the memory size of the agent or on its computational capabilities. We consider two models concerning the nodes' identities. In an *anonymous* PV-graph, the nodes do not have any identities, or the agent is not able to see them. In a *labeled* PV-graph, the nodes have distinct identities and the agent can see and memorize them.

We say that an agent *explores* a PV-graph if and only if, starting at time 0 on the starting site of the first carrier (this can be assumed without loss of generality), the agent eventually visits all sites of the PV-graph and switches afterwards to a terminal state. This terminal state expresses the fact that the agent knows that exploration has been completed.

3 Solvability

Similarly as in the case when the agent cannot wait, an agent without information on the PV-graphs it has to explore cannot explore all PV-graphs (even if restricted to the labeled homogeneous highly-connected ones).

Theorem 1. *There exists a family of labeled homogeneous highly-connected PV-graphs such that no agent can explore all the graphs of this family if it has no information on the PV-graphs it has to explore.*

Proof. Let $S = \{s_1, s_2, s_3\}$ be a set of three sites with distinct ids ($\mathrm{Id}(s_i) = i$). For $l > 0$, we define the PV-graph G_ℓ over the set S of sites composed of a single carrier. Its route is $(s_1, s_2, \cdots, s_1, s_2, s_1, s_3)$, where (s_1, s_2) is repeated exactly l times. Moreover, let G_0 be the PV-graph over the set of sites $\{s_1, s_2\}$ composed of a single carrier, whose route is (s_1, s_2). The family $\{G_0, G_1, \cdots\}$ is denoted $\mathcal{G}$.

Assume, for the purpose of contradiction, that there exists an algorithm solving the exploration problem in all the PV-graphs in $\mathcal{G}$, provided that the agent A running this algorithm does not receive any additional information. In particular, A explores G_0. Let m be the time at which A switches to the terminal state. Assume now that A is placed in G_m. For the first m time units, A cannot tell the difference between G_0 and G_m, because A has no information about the PV-graph it has to explore and in particular it does not know the number of sites or an upper bound on the system period. It will therefore act exactly the same in G_m than in G_0. In particular, it will switch to the terminal state at time m although the site s_3 has not yet been explored. This contradiction concludes the proof.   □

4   General Case

In this section, we make no assumption on the PV-graphs (except the connectedness assumption of course). We basically show that the ability to wait allows the agent to explore, and even map, all connected PV-graphs (not only the highly-connected ones), provided that the agent knows for each of them an upper bound on its maximal period. This can be done in only $\Theta(\min\{k \cdot p, n \cdot p, n^2\})$ moves, that is, at least p times less than when the agent cannot wait. Besides, the time complexity is reduced from $\Theta(k \cdot p^2)$ to $\Theta(n \cdot p)$.

4.1   Lower Bound on the Number of Moves

Flocchini, Mans and Santoro [FMS09] proved a lower bound $\Omega(k \cdot p)$ on the number of moves to explore the PV-graphs with k carriers and maximum period p (even if restricted to the labeled homogeneous highly-connected ones). This lower bound does not apply directly in our setting because the agent, having the possibility to wait, could potentially be able to explore in significantly less moves. We will prove later that this is actually the case: the move complexity of our algorithm is bounded by $O(\min\{k \cdot p, n \cdot p, n^2\})$. We prove here that this complexity is optimal.

Lemma 1. *For any n, k, and p sufficiently large, $p \geq \lfloor \frac{n-1}{k} \rfloor + 1$ (necessary for connectedness), there exists a labeled homogeneous highly-connected PV-graph $G_{n,k,p}$ with n sites, k carriers and period p such that any algorithm needs at least $\min\{k \cdot p - 1, \lfloor n/8 \rfloor \cdot p - 1, 7n/8 \cdot (\lfloor n/8 \rfloor - 1)\}$ moves to explore it.*

Proof. Fix any integers $n \geq 8$, $k \geq 8$, and $p \geq 1$ such that $p \geq \lfloor \frac{n-1}{k} \rfloor + 1$. First assume that $k \leq n/8$.

- Subcase 1: $p \leq \frac{n^2}{4k} - k$.
  Let $q = \lfloor n/2k \rfloor$. Note that $q \geq 4$ and $p \geq q$. We denote by r the non-negative integer $\lceil p/q \rceil q - p$. Let $S = \{s_1, s_2, \ldots, s_n\}$ be a set of n sites. We partition S into the sets S_0 and $S_{i,j}$, with $1 \leq i \leq k$ and $1 \leq j \leq q$, such that:
    - $S_0 = \{s_1, s_2, \ldots, s_{\lceil p/q \rceil - 1}\}$ and $S_{1,1} = \{s_{\lceil p/q \rceil}\}$;
    - for all $1 \leq i \leq k$ and $1 \leq j \leq q$, we have $S_{i,j} \neq \emptyset$;
    - for all $2 \leq i \leq k$, we have $|S_{i,1}| \leq \lceil p/q \rceil - 1$;
    - for all $1 \leq i \leq k$ and $2 \leq j \leq q - r$, we have $|S_{i,j}| \leq \lceil p/q \rceil$;
    - for all $1 \leq i \leq k$ and $q - r < j \leq q$, we have $|S_{i,j}| \leq \lfloor p/q \rfloor$;
  Note that such a partition is always possible when p satisfies our assumption $\lfloor \frac{n-1}{k} \rfloor + 1 \leq p \leq \frac{n^2}{4k} - k$.
  The PV-graph $G_{n,k,p}$ is now defined as follows. Let S be its set of sites and $C = \{c_1, c_2, \ldots, c_k\}$ be the set of its carriers. For every $1 \leq i \leq k$, the route $R(c_i)$ is defined as follows. The route starts at s_1 at time 0 and then visits $s_2, s_3, \cdots, s_l$, with $l = \lceil p/q \rceil - |S_{i,1}|$, followed by each site of the set $S_{i,1}$. The route continues by visiting, for successive values of j from 2 to q, the sites $s_1, s_2, \cdots, s_l$, with $l = \lceil p/q \rceil - |S_{i,j}|$ (or $l = \lfloor p/q \rfloor - |S_{i,j}|$ if $j > q - r$),

followed by each site of the set $S_{i,j}$. Note that $G_{n,k,p}$ is both homogeneous (of period p) and highly-connected.

The PV-graph $G_{n,k,p}$ is constructed in such a way that the agent basically has to follow each carrier's route entirely to visit all sites. More precisely, to visit the sites of any set $S_{i,j}$, the agent has to pay $\lceil p/q \rceil$ moves ($\lfloor p/q \rfloor$ if $j > q - r$). Hence the minimum number of moves an exploring agent has to perform in $G_{n,k,p}$ is $k \cdot p - 1$.

- Subcase 2: $p > \frac{n^2}{4k} - k$.

Let us first assume that $k = \lfloor n/8 \rfloor$. The PV-graph $G_{n,k,p}$ is defined in this case as follows. Let $S = \{s_1, s_2, \ldots, s_n\}$ be the set of its sites and let $C = \{c_1, c_2, \ldots, c_k\}$ be the set of its carriers. For every $1 \leq i \leq \lfloor n/8 \rfloor$, the route $R(c_i)$ is any route of period p going through (and only through) sites $s_1, s_2, \ldots, s_{n-\lfloor n/8 \rfloor}$ and s_{n-i+1}, such that c_i is only once per period in s_{n-i+1}, just after being in $s_{n-\lfloor n/8 \rfloor}$, and just before being in s_1. Moreover, if c_i is in some site s_j, $2 \leq j \leq n - \lfloor n/8 \rfloor - 1$, at some time t, then at time $t+1$ the carrier c_i can only be at s_{j-1}, s_j, or s_{j+1}. We further assume that all carriers are in s_1 at time 0. If k is smaller than $\lfloor n/8 \rfloor$, then each carrier has to deal with several sites of the form s_{n-i+1}, with $1 \leq i \leq \lfloor n/8 \rfloor$. This is always possible thanks to our assumption on p. Note that $G_{n,k,p}$ is both homogeneous and highly-connected.

By construction, all sites s_{n-i+1}, with $1 \leq i \leq \lfloor n/8 \rfloor$, are only accessible through $s_{n-\lfloor n/8 \rfloor}$ and the agent can only leave them by going to s_1 with some carrier. Again by construction, any agent willing to go from s_1 to $s_{n-\lfloor n/8 \rfloor}$ has to go through all the sites $s_1, s_2, \ldots, s_{n-\lfloor n/8 \rfloor}$. Therefore, for any i, j such that $1 \leq i \neq j \leq \lfloor n/8 \rfloor$, going from s_{n-i+1} to s_{n-j+1} requires any agent to perform at least $n - \lfloor n/8 \rfloor + 1$ moves. Since any agent performing exploration of the PV-graph must visit all its sites, any agent requires at least $(n - \lfloor n/8 \rfloor + 1)(\lfloor n/8 \rfloor - 1)$ moves to explore $G_{n,k,p}$.

Now assume that $k > n/8$. In this case, we simply use the above constructions for $\lfloor n/8 \rfloor$ carriers. All carriers c_i, with $i > \lfloor n/8 \rfloor$ are given the same route as c_1 for example. This gives us immediately a lower bound $\lfloor n/8 \rfloor \cdot p - 1$ for $p \leq \frac{n^2}{4\lfloor n/8 \rfloor} - \lfloor n/8 \rfloor$ and still the lower bound $7n/8 \cdot (\lfloor n/8 \rfloor - 1)$ for $p > \frac{n^2}{4\lfloor n/8 \rfloor} - \lfloor n/8 \rfloor$. $\qquad \square$

Summarizing the previous lemma by considering the asymptotic behavior, we directly obtain the following theorem.

Theorem 2. *The move complexity of the PV-graph exploration problem is in $\Omega(\min\{k \cdot p, n \cdot p, n^2\})$, where n, k, and p denote respectively the number of sites, the number of carriers, and the maximal period. This result holds even if the agent knows completely the PV-graph, has unlimited memory, and even in the labeled homogeneous highly-connected case.*

4.2   Lower Bound on Time

We can prove a larger lower bound for the time complexity than for the move complexity in the general case. More precisely, we have the following lemma.

Lemma 2. *Consider any n, k, and p, with $n \geq 6$, $2 \leq k \leq \frac{n-1}{3}$, and $p \geq \lfloor \frac{n-1}{k} \rfloor + 1$ (necessary for connectedness). There exists a family $\mathcal{G}_{n,p,k}$ of labeled homogeneous PV-graphs with n sites, k carriers and period p such that, for any algorithm, there exists a PV-graph in this family which cannot be explored by the algorithm using less than $(k-1)(p\lfloor \frac{n-1}{k} \rfloor - 1) + \lfloor \frac{n-1}{k} \rfloor - 1$ time steps.*

Proof. Fix any n, k, and p such that $n \geq 6$, $2 \leq k \leq \frac{n-1}{3}$, $p \geq \lfloor \frac{n-1}{k} \rfloor + 1$. Fix any $j_1, j_2, \ldots, j_{k-1}$ and $t_2, t_3, \ldots, t_k$ such that, for every $1 \leq i \leq k-1$, we have $(i-1)\lfloor \frac{n-1}{k} \rfloor + 2 \leq j_i \leq i\lfloor \frac{n-1}{k} \rfloor + 1$ and $1 \leq t_{i+1} \leq p$. The PV-graph $G((j_1, t_2), (j_2, t_3), \ldots, (j_{k-1}, t_k))$ is defined as follows.

Let $S = \{s_1, s_2, \ldots, s_n\}$ be the set of its sites and let $C = \{c_1, c_2, \ldots, c_k\}$ be the set of its carriers. Let us partition S into $k+1$ subsets $S_0, S_1, \ldots, S_{k-1}, S_k$ such that $S_0 = \{s_1\}$, $S_i = \{s_{(i-1)\lfloor \frac{n-1}{k} \rfloor + 2}, \ldots, s_{i\lfloor \frac{n-1}{k} \rfloor + 1}\}$, for $1 \leq i \leq k-1$, and S_k contains all the remaining sites.

Let $j_0 = 1$ and $t_1 = 0$. Consider any i such that $1 \leq i \leq k$. The route $R(c_i)$ is any route of period p going through (and only through) all the sites in $S_i \cup \{s_{j_{i-1}}\}$ satisfying the following two conditions. First, c_i visits $s_{j_{i-1}}$ only once per period, at all times equal to t_i modulo p. Second, the route $R(c_i)$ does not depend on the values j_l and t_{l+1}, for $l \geq i$.

The family $\mathcal{G}_{n,p,k}$ is defined as the set of all PV-graphs $G((j_1, t_2), (j_2, t_3), \ldots, (j_{k-1}, t_k))$ with, for every $1 \leq i \leq k-1$, $(i-1)\lfloor \frac{n-1}{k} \rfloor + 2 \leq j_i \leq i\lfloor \frac{n-1}{k} \rfloor + 1$ and $1 \leq t_{i+1} \leq p$. All these PV-graphs are labeled homogeneous PV-graphs with n sites, k carriers and period p.

Let A be any exploring agent (i.e. executing any exploration algorithm). Given $1 \leq i \leq k$ and G a PV-graph of $\mathcal{G}_{n,p,k}$, let $\mathcal{T}_i(G)$ be the first time at which the agent A, starting at s_1 at time 0 in G, sees the carrier c_i. Given q, $1 \leq q \leq k$, and $j_1, j_2, \ldots, j_{q-1}$ and $t_2, t_3, \ldots, t_q$ in the usual ranges, we define $\mathcal{G}_{n,p,k}((j_1, t_2), (j_2, t_3), \ldots, (j_{q-1}, t_q))$ as the set of all the PV-graphs $G((j_1, t_2), (j_2, t_3), \ldots, (j_{k-1}, t_k))$ with, for every $q \leq i \leq k-1$, $(i-1)\lfloor \frac{n-1}{k} \rfloor + 2 \leq j_i \leq i\lfloor \frac{n-1}{k} \rfloor + 1$ and $1 \leq t_{i+1} \leq p$.

Claim. For every q, $1 \leq q \leq k$, and every i, $1 \leq i \leq q-1$, there exist j_i and t_{i+1} satisfying $(i-1)\lfloor \frac{n-1}{k} \rfloor + 2 \leq j_i \leq i\lfloor \frac{n-1}{k} \rfloor + 1$ and $1 \leq t_{i+1} \leq p$ such that for every graph $G \in \mathcal{G}_{n,p,k}((j_1, t_2), (j_2, t_3), \ldots, (j_{q-1}, t_q))$ we have $\mathcal{T}_q(G) \geq (q-1)(p\lfloor \frac{n-1}{k} \rfloor - 1)$.

Proof of the Claim: We prove the claim by induction on q. The base case $q = 1$ is trivially true. Fix any q such that $1 \leq q \leq k-1$, and assume, by induction hypothesis, that the claim holds for the value q.

Let $\mathcal{G}_q$ be the family $\mathcal{G}_{n,p,k}((j_1, t_2), (j_2, t_3), \ldots, (j_{q-1}, t_q))$ whose existence is guaranteed by the induction hypothesis. Note that all PV-graphs in $\mathcal{G}_q$ have exactly the same routes $R(c_i)$, for $1 \leq i \leq q$. We can thus define H_q to be the PV-graph consisting only of the carriers c_1 to c_q of any PV-graph in $\mathcal{G}_q$. Let us consider now the agent A starting at s_1 at time 0 in H_q. By induction hypothesis and by construction of H_q, the agent A sees c_q for the first time at time t with $t \geq (q-1)(p\lfloor \frac{n-1}{k} \rfloor - 1)$ time steps. Thus there exists j_q and t_{q+1} satisfying $(q-1)\lfloor \frac{n-1}{k} \rfloor + 2 \leq j_q \leq q\lfloor \frac{n-1}{k} \rfloor + 1$ and $1 \leq t_{q+1} \leq p$ such that A is never at

s_{j_q} at a time equal to t_{q+1} modulo p before time $t + p\lfloor \frac{n-1}{k} \rfloor - 1$, and thus before time $q(p\lfloor \frac{n-1}{k} \rfloor - 1)$.

Consider now the agent A starting at s_1 at time 0 in any PV-graph G in $\mathcal{G}_{n,p,k}((j_1, t_2), (j_2, t_3), \ldots, (j_{q-1}, t_q), (j_q, t_{q+1}))$. Before time $q(p\lfloor \frac{n-1}{k} \rfloor - 1)$, the agent will behave exactly the same as in H_q and will not see the carrier c_{q+1}. This concludes the proof of the claim.   ◇

The theorem now follows by considering the claim for the last value $q = k - 1$, and by noting that the agent still has to visit all sites of S_k after reaching c_k, which requires additional $\lfloor \frac{n-1}{k} \rfloor - 1$ time steps.   □

Again, summarizing the previous lemma by considering the asymptotic behavior, we directly obtain the following theorem.

Theorem 3. *The time complexity of the PV-graph exploration problem is in $\Omega(n \cdot p)$ in the general case. This result holds even if the agent knows n, k, and p, has unlimited memory, and even in the labeled case.*

4.3   Our Algorithm

In the above part of the paper, we exhibited some necessary conditions on the existence of a solution. We then provided lower bounds on the move and time complexities. We now essentially prove that all these results are optimal by describing and proving a PV-graph exploration algorithm with matching upper bounds on the move and time complexities, provided that the agent knows a linear upper bound B on the maximum period p. As a consequence, we show that the ability to wait allows to decrease both the move and time complexities, the former by a multiplicative factor at least $\Theta(p)$.

Algorithm EXPLORE-WITH-WAIT

Our algorithm stores a matrix Mat where lines correspond to (known so far) carriers. The algorithm progressively fills in each line with the sequence of sites visited by the corresponding carrier. In order to do that, the agent stays $2B$ steps at each site, looking at each visit of the carriers at this site. From each partially filled in line, the algorithm computes a divisor of the period of the corresponding carrier, allowing the agent to predict the exact schedule of the carriers at the sites already known by the agent. The algorithm also maintains a tree of carriers, where a carrier c is a child of a carrier c' if c was discovered for the first time while visiting c'. The algorithm visits successively new sites until the whole matrix is filled in. Note that the completed matrix contains the complete schedule of all carriers. Hence one can easily extract a map of the PV-graph from the matrix.

Let a start with carrier c_1. Initially: $Home = c_1$; $parent(Home) := \emptyset$; $Visited := \emptyset$; $ToExplore := \{c_1\}$; $p := 1$.

Algorithm 1. EXPLORE-WITH-WAIT (c)

1: **if** $c = Home$ and $ToExplore = \emptyset$ **then**
2:　　Terminate
3: **else**
4:　　**if** $c \notin Visited$ **then**
5:　　　VISIT(c)
6:　　**end if**
7:　　$c' \leftarrow$ NEXT(c)
8:　　EXPLORE-WITH-WAIT (c')
9: **end if**

Algorithm 2. NEXT-EMPTY-CELL (Mat, j, c, v)

1: $p = \text{period}(Mat[idC])$
2: **while** $Mat[Id(c), j] \neq v$ **and** $j < B$ **do**
3:　　$u \leftarrow Mat[Id(c), j]$
4:　　$i \leftarrow j + 1$
5:　　**while** $Mat[Id(c), i] \neq u$ **and** $Mat[Id(c), i] \neq v$ **and** $i < B$ **do**
6:　　　$i \leftarrow i + 1$
7:　　**end while**
8:　　**if** Mat[Id(c), i]=u **then**
9:　　　Get on c at its next visit at the current site u
10:　　　$j \leftarrow i$
11:　　**else**
12:　　　**if** $Mat[Id(c), i] = v$ **then**
13:　　　　Get on c at the first time k such that $k \bmod p = j$
14:　　　　Do one move with c
15:　　　　$j \leftarrow j + 1$
16:　　　　**if** $Mat[Id(c), j] \neq v$ **then**
17:　　　　　Get off on the current site
18:　　　　**end if**
19:　　　**else**
20:　　　　$j \leftarrow i$
21:　　　**end if**
22:　　**end if**
23: **end while**
24: Return j

Algorithm 3. NEXT(c)

1: **if** $(N(c)) = \emptyset$ and $(c = c_0)$ **then**
2:     Return c_0
3: **else if** $c \neq c_0$ **then**
4:     $c' \leftarrow NEXT(parent(c))$
5: **else**
6:     $c' \leftarrow$ an element of $N(c) \cap ToExplore$
7: **end if**
8: $p = \text{period}(Mat[Id(c')])$
9: $i \leftarrow 0$
10: **while** $Mat[Id(c'), i] == \emptyset$ **do**
11:     $i \leftarrow i + 1$
12: **end while**
13: v=Mat[Id(c'), i]
14: $j \leftarrow$ NEXT-EMPTY-CELL (Mat, j, c, v)
15: Stay on site v
16: Get on c' at the first time k such that $k \bmod p = i$
17: $Parent($c'$) := c$
18: Return c'

Algorithm 4. VISIT (c, j)

1: $MyParent \leftarrow$ parent (c); $N(c) := \{MyParent\}$
2: $i \leftarrow j$
3: **while** $j < B + i$ **do**
4:     $u \leftarrow$ current site
5:     Get off on site u
6:     **if** $Mat[Id(c), j] == \emptyset$ **then**
7:         $Mat[Id(c), j] \leftarrow u$
8:         **while** $i < (2B + j)$ **do**
9:             Stay on u and at each step DO
10:             $i \leftarrow i + 1$
11:             **if** c visits u at this step **then**
12:                 $Mat[Id(c), i \bmod 2B] \leftarrow u$
13:             **else**
14:                 **if** the agent sees $c' \notin (ToExplore \cap Visit)$ **then**
15:                     $Mat[Id(c'), i \bmod 2B] \leftarrow u$
16:                     $ToExplore := ToExplore \cup \{c'\}$
17:                     $N(c) := N(c) \cup \{c'\}$
18:                 **end if**
19:             **end if**
20:             $p = \text{period}(Mat[Id(c)])$
21:         **end while**
22:     **else**
23:         $j \leftarrow$ NEXT $-$ EMPTY $-$ CELL(Mat, j, c, $\emptyset$)
24:     **end if**
25: **end while**
26: $Visit \leftarrow Visit \cup \{c\}$
27: $ToExplore \leftarrow ToExplore - \{c\}$

Correctness

Theorem 4. *Algorithm EXPLORE-WITH-WAIT correctly explores and maps in finite time any PV-graph, even anonymous, but provided that an upper bound on the maximum period is known.*

Proof. First observe that when an agent stays at a site for $2B$ steps, where B is the known upper bound on the maximum period, it sees all the carriers visiting that site. Moreover, after filling in the matrix with that information, it is able to predict at any point in the future which carrier will be at that site. Since the PV-graph is connected, the agent will miss no carriers and thus no sites either. At the end of the algorithm, the matrix will be completely filled in and it will be equivalent to a map of the PV-graph. $\qquad\square$

Move and Time Complexities

Theorem 5. *With the algorithm EXPLORE-WITH-WAIT, the agent makes at most $O(\min\{k \cdot p, n \cdot p, n^2\})$ moves to explore any n-site k-carrier PV-graph of maximum period p.*

Proof. Let us first prove that the move complexity is in $O(n^2)$. Obviously, the agent only moves when looking for the next empty cell. Since an empty cell always corresponds to a new unvisited site, looking for the next empty cell is done at most n times. The algorithm is done in such a way that, during the travel from the last visited site u to the following new site v, each site w is visited at most once. Indeed, it is always possible for the agent to wait on w for the appropriate carrier to come at w. Hence the number of moves is bounded by n^2.

We now prove that the move complexity is in $O(k \cdot p)$. During a single travel to go to the next empty cell, the agent may have to use several carriers. However, we visit the carriers following a DFS traversal of the tree of carriers. Hence in total the agent uses at most $2k$ carriers. When using a carrier, the agent does at most p moves. Hence the number of moves is bounded by $2k \cdot p$.

We finally prove that the move complexity is in $O(n \cdot p)$. This is done by refining the previous argument. A carrier is always added as a leaf to the tree of carriers. Moreover, a carrier is used only if the agent goes to an empty cell of this carrier. Since the agent goes to at most n empty cells, it means that at most n carriers of the tree are used. Hence the number of moves is bounded by $2n \cdot p$. $\qquad\square$

Theorem 6. *The algorithm EXPLORE-WITH-WAIT allows to explore any n-node PV-graph in $O(nB)$ time steps, where B is a known upper bound on p.*

Proof. A lot of time is spent by the agent by staying $O(B)$ steps on a site to note all passing carriers. Since there are n sites to visit, the agent spends at most $O(nB)$ time steps doing this. It turns out that this is the main cost of the algorithm in terms of time complexity. Indeed, as noticed in the previous proof, the agent uses at most $2\min\{k,n\}$ carriers when traveling. On each carrier, the agent uses not only at most p moves but also at most p time steps. Hence the completion time of the algorithm is at most $O(nB) + 2\min\{k,n\} \cdot p$. This proves the theorem. □

As noticed before, we have the following corollary.

Corollary 1. *Given the a priori knowledge of an upper bound $B = O(p)$ on the maximum period p, Algorithm $EXPLORE-WITH-WAIT$ is asymptotically optimal in the general case with respect to both the move and the time complexities. The optimal move complexity is in $\Theta(\min\{k\cdot p, n\cdot p, n^2\})$ while the optimal time complexity is in $\Theta(n\cdot p)$.*

5   Specific Cases

We showed in the previous section the optimal move and time complexities for the PV-graph exploration problem in the general case. This section is devoted to the specific cases of homogeneous or highly-connected PV-graphs. In both cases, we prove that the move and time complexities remain the same as in the general case. Note, however, that when considering PV-graphs being both homogeneous and highly-connected, we know from [FMS09] that the optimal time complexity is at most $O(k \cdot p)$, even when n is large.

5.1   The Homogeneous Case

If we consider the homogeneous PV-graphs (but possibly not highly-connected), the time and move complexities remain the same as in the general case.

Theorem 7. *Given the a priori knowledge of an upper bound $B = O(p)$ on the maximum period p, Algorithm $EXPLORE-WITH-WAIT$ is asymptotically optimal in the homogeneous case with respect to both the move and the time complexities. The optimal move complexity is in $\Theta(\min\{k\cdot p, n\cdot p, n^2\})$ while the optimal time complexity is in $\Theta(n \cdot p)$.*

Proof. The result directly follows from Lemma 2 and Corollary 1. □

5.2   The Highly-Connected Case

If we consider the highly-connected PV-graphs (but possibly not homogeneous), the time and move complexities remain the same as in the general case.

Lemma 3. *Consider any n, k, and p, with $n \geq 6$, $2 \leq k \leq \frac{n-1}{3}$, $p \geq \lfloor\frac{n-1}{k}\rfloor + 2$. There exists a family $\mathcal{G'}_{n,p,k}$ of labeled highly-connected PV-graphs with n sites, k carriers and maximum period p such that, for any algorithm, there exists a PV-graph in this family which cannot be explored by the algorithm using less than $\lceil\frac{k-1}{2}\rceil(p\lfloor\frac{n-1}{k}\rfloor - 1) + \lfloor\frac{k-1}{2}\rfloor((p-1)\lfloor\frac{n-1}{k}\rfloor - 1) + \lfloor\frac{n-1}{k}\rfloor - 1$ time steps.*

Proof. Fix any n, k, and p such that $n \geq 6$, $2 \leq k \leq \frac{n-1}{3}$, $p \geq \lfloor \frac{n-1}{k} \rfloor + 2$. Fix any $j_1, j_2, \ldots, j_{k-1}$ and $t_2, t_3, \ldots, t_k$ such that, for every $1 \leq i \leq k-1$, we have $(i-1)\lfloor \frac{n-1}{k} \rfloor + 2 \leq j_i \leq i\lfloor \frac{n-1}{k} \rfloor + 1$ and $1 \leq t_{i+1} \leq p$, if i is odd, $1 \leq t_{i+1} \leq p-1$, if i is even. The PV-graph $G((j_1, t_2), (j_2, t_3), \ldots, (j_{k-1}, t_k))$ is defined as follows.

Let $S = \{s_1, s_2, \ldots, s_n\}$ be the set of its sites and let $C = \{c_1, c_2, \ldots, c_k\}$ be the set of its carriers. Let us partition S into $k+1$ subsets $S_0, S_1, \ldots, S_{k-1}, S_k$ such that $S_0 = \{s_1\}$, $S_i = \{s_{(i-1)\lfloor \frac{n-1}{k} \rfloor + 2}, \ldots, s_{i\lfloor \frac{n-1}{k} \rfloor + 1}\}$, for $1 \leq i \leq k-1$, and S_k contains all the remaining sites.

Let $j_0 = 1$ and $t_1 = 0$. Consider any i such that $1 \leq i \leq k$. The route $R(c_i)$ is any route going through (and only through) all the sites in $S_i \cup \{s_{j_{i-1}}\}$ satisfying the following three conditions. First, c_i is of period $p-1$ if i is odd, and of period p if i is even. Second, c_i visits $s_{j_{i-1}}$ only once per period, at all times equal to t_i modulo its period. Third, the route $R(c_i)$ does not depend on the values j_l and t_{l+1}, for $l \geq i$.

The family $\mathcal{G}'_{n,p,k}$ is defined as the set of all PV-graphs $G((j_1, t_2), (j_2, t_3), \ldots, (j_{k-1}, t_k))$ with, for every $1 \leq i \leq k-1$, $(i-1)\lfloor \frac{n-1}{k} \rfloor + 2 \leq j_i \leq i\lfloor \frac{n-1}{k} \rfloor + 1$ and $1 \leq t_{i+1} \leq p$, if i is odd, $1 \leq t_{i+1} \leq p-1$, if i is even. All these PV-graphs are labeled highly-connected PV-graphs with n sites, k carriers and maximum period p. (Indeed, note that, for every $1 \leq i \leq k-1$, c_i and c_{i+1} meets at s_{j_i} at most every $p(p-1)$ steps.)

Let A be any exploring agent (i.e. executing any exploration algorithm). Given $1 \leq i \leq k$ and G a PV-graph of $\mathcal{G}'_{n,p,k}$, let $\mathcal{T}_i(G)$ be the first time at which the agent A, starting at s_1 at time 0 in G, sees the carrier c_i. Given q, $1 \leq q \leq k$, and $j_1, j_2, \ldots, j_{q-1}$ and $t_2, t_3, \ldots, t_q$ in the usual ranges, we define $\mathcal{G}'_{n,p,k}((j_1, t_2), (j_2, t_3), \ldots, (j_{q-1}, t_q))$ as the set of all the PV-graphs $G((j_1, t_2), (j_2, t_3), \ldots, (j_{k-1}, t_k))$ with, for every $q \leq i \leq k-1$, $(i-1)\lfloor \frac{n-1}{k} \rfloor + 2 \leq j_i \leq i\lfloor \frac{n-1}{k} \rfloor + 1$ and $1 \leq t_{i+1} \leq p$, if i is odd, $1 \leq t_{i+1} \leq p-1$, if i is even.

Claim. For every q, $1 \leq q \leq k$, and every i, $1 \leq i \leq q-1$, there exist j_i and t_{i+1} satisfying $(i-1)\lfloor \frac{n-1}{k} \rfloor + 2 \leq j_i \leq i\lfloor \frac{n-1}{k} \rfloor + 1$ and $1 \leq t_{i+1} \leq p$ ($t_{i+1} \leq p-1$ when i is even) such that for every graph $G \in \mathcal{G}'_{n,p,k}((j_1, t_2), (j_2, t_3), \ldots, (j_{q-1}, t_q))$ we have $\mathcal{T}_q(G) \geq \lceil \frac{q-1}{2} \rceil (p\lfloor \frac{n-1}{k} \rfloor - 1) + \lfloor \frac{q-1}{2} \rfloor ((p-1)\lfloor \frac{n-1}{k} \rfloor - 1)$.

Proof of the Claim: We prove the claim by induction on q. The base case $q = 1$ is trivially true. Fix any q such that $1 \leq q \leq k-1$, and assume, by induction hypothesis, that the claim holds for the value q.

Let $\mathcal{G}'_q$ be the family $\mathcal{G}'_{n,p,k}((j_1, t_2), (j_2, t_3), \ldots, (j_{q-1}, t_q))$ whose existence is guaranteed by the induction hypothesis. Note that all PV-graphs in $\mathcal{G}'_q$ have exactly the same routes $R(c_i)$, for $1 \leq i \leq q$. We can thus define H'_q to be the PV-graph consisting only of the carriers c_1 to c_q of any PV-graph in $\mathcal{G}'_q$. Let us consider now the agent A starting at s_1 at time 0 in H'_q. By induction hypothesis and by construction of H'_q, the agent A sees c_q for the first time at time t with $t \geq \lceil \frac{q-1}{2} \rceil (p\lfloor \frac{n-1}{k} \rfloor - 1) + \lfloor \frac{q-1}{2} \rfloor ((p-1)\lfloor \frac{n-1}{k} \rfloor - 1)$ time steps. Thus there exists j_q and t_{q+1} satisfying $(q-1)\lfloor \frac{n-1}{k} \rfloor + 2 \leq j_q \leq q\lfloor \frac{n-1}{k} \rfloor + 1$ and $1 \leq t_{q+1} \leq p$, if q is odd, $1 \leq t_{q+1} \leq p-1$, if q is even, such that A is never at s_{j_q} at a time equal to t_{q+1} modulo the period p' of c_{q+1} before time $t + p' \lfloor \frac{n-1}{k} \rfloor - 1$, and thus before time $\lceil \frac{q}{2} \rceil (p \lfloor \frac{n-1}{k} \rfloor - 1) + \lfloor \frac{q}{2} \rfloor ((p-1) \lfloor \frac{n-1}{k} \rfloor - 1)$.

Consider now the agent A starting at s_1 at time 0 in any PV-graph G in $\mathcal{G}'_{n,p,k}((j_1, t_2), (j_2, t_3), \ldots, (j_{q-1}, t_q), (j_q, t_{q+1}))$. Before time $\lceil \frac{q}{2} \rceil (p \lfloor \frac{n-1}{k} \rfloor - 1) + \lfloor \frac{q}{2} \rfloor ((p-1) \lfloor \frac{n-1}{k} \rfloor - 1)$, the agent will behave exactly the same as in H'_q and will not see the carrier c_{q+1}. This concludes the proof of the claim. $\Diamond$

The theorem now follows by considering the claim for the last value $q = k-1$, and by noting that the agent still has to visit all sites of S_k after reaching c_k, which requires additional $\lfloor \frac{n-1}{k} \rfloor - 1$ time steps. $\square$

Again, summarizing the previous lemma, using Corollary 1, and considering the asymptotic behavior, we obtain the following theorem.

Theorem 8. *Given the a priori knowledge of an upper bound $B = O(p)$ on the maximum period p, Algorithm $EXPLORE - WITH - WAIT$ is asymptotically optimal in the highly-connected case with respect to both the move and the time complexities. The optimal move complexity is in $\Theta(\min\{k \cdot p, n \cdot p, n^2\})$ while the optimal time complexity is in $\Theta(n \cdot p)$.*

References

[AKL08]   Avin, C., Koucký, M., Lotker, Z.: How to Explore a Fast-Changing World (Cover Time of a Simple Random Walk on Evolving Graphs). In: Aceto, L., Damgård, I., Goldberg, L.A., Halldórsson, M.M., Ingólfsdóttir, A., Walukiewicz, I. (eds.) ICALP 2008, Part I. LNCS, vol. 5125, pp. 121–132. Springer, Heidelberg (2008)

[CFQS10]   Casteigts, A., Flocchini, P., Quattrociocchi, W., Santoro, N.: Time-varying graphs and dynamic networks. CoRR, abs/1012.0009 (2010)

[CF03]   Cooper, C., Frieze, A.M.: Crawling on simple models of web graphs. Internet Mathematics 1(1) (2003)

[DP04]   Dessmark, A., Pelc, A.: Optimal graph exploration without good maps. Theor. Comput. Sci. 326(1-3), 343–362 (2004)

[FKMS10]   Flocchini, P., Kellett, M., Mason, P.C., Santoro, N.: Mapping an Unfriendly Subway System. In: Boldi, P. (ed.) FUN 2010. LNCS, vol. 6099, pp. 190–201. Springer, Heidelberg (2010)

[FMS09]   Flocchini, P., Mans, B., Santoro, N.: Exploration of Periodically Varying Graphs. In: Dong, Y., Du, D.-Z., Ibarra, O. (eds.) ISAAC 2009. LNCS, vol. 5878, pp. 534–543. Springer, Heidelberg (2009)

[KLO10]   Kuhn, F., Lynch, N.A., Oshman, R.: Distributed computation in dynamic networks. In: 42nd ACM Symposium on Theory of Computing (STOC), pp. 513–522 (2010)

[PP99]   Panaite, P., Pelc, A.: Exploring Unknown Undirected Graphs. J. Algorithms 33(2), 281–295 (1999)

[Rei05]   Reingold, O.: Undirected st-connectivity in log-space. In: 37th ACM Symposium on Theory of Computing (STOC), pp. 376–385 (2005)

Accurate Byzantine Agreement with Feedback

Vijay K. Garg⋆, John Bridgman, and Bharath Balasubramanian

Parallel and Distributed Systems Laboratory,
Dept. of Electrical and Computer Engineering,
The University of Texas at Austin,
Austin, TX 78712
garg@ece.utexas.edu, {johnfb,bbharath}@mail.utexas.edu

Abstract. The standard Byzantine Agreement (BA) problem requires non-faulty processes to agree on a common value. In many real-world applications, it is important that the processes agree on the *correct* value rather than any value. In this paper, we present a problem called Accurate Byzantine Agreement (ABA) in which all processes get a common feedback (or payoff) from the environment indicating if the value they agreed upon was correct or not. The solution to this problem, referred to as the ABA algorithm, requires the non-faulty processes to incorporate the feedback so that their chance of choosing the correct value improves over subsequent iterations of the algorithm. We present an algorithm that solves the ABA problem based on two key ingredients: a standard solution to the BA problem and a multiplicative method to maintain and update process weights indicative of how often they are correct. We give guarantees on the accuracy of the algorithm based on assumptions on the accuracy of the processes and the proportion of faulty and non-faulty processes in the system. For each iteration, if the weight of accurate processes is at least $3/4^{th}$ the weight of the non-faulty processes, the algorithm always decides on the correct value. When the non-faulty processes are accurate with probability greater than $1/2$, the algorithm decides on the correct value with very high probability after some initial number of mistakes. In fact, among n processes, if there exists even *one* process which is accurate for all iterations, the algorithm is wrong only $O(\log n)$ times for any large number of iterations of the algorithm.

Keywords: Byzantine Agreement, Weighted Majority, Multiplicative Update.

1 Introduction

In real-world applications, processes in a distributed system may be compromised, leading to malicious or arbitrary behavior. The Byzantine Agreement (BA) problem [20, 17, 10, 8, 12] requires all non-faulty processes to agree on a common binary value given that some of the processes may show arbitrary faulty or Byzantine behavior. In the standard version of the problem, the value that is agreed upon may be either of the binary values so long as it is proposed by at least one non-faulty process. In some scenarios, it is better for the system to agree on a specific value among the two binary

---

⋆ Supported by NSF CNS-0718990, NSF CNS-1115808 and the Cullen Trust for Higher Education Endowed Professorship.

A. Fernández Anta, G. Lipari, and M. Roy (Eds.): OPODIS 2011, LNCS 7109, pp. 465–480, 2011.

values. For example, suppose in a distributed control system a coordinated action needs to be taken (such as opening or closing a valve) depending upon the observations made by possibly faulty distributed processes. Depending upon the outcome of the action, the environment can provide a feedback if the action taken was correct or not. As another example, suppose that the system is making decision on whether to sell a stock based on recommendations made by multiple processes. The final closing price of the stock provides a feedback for the decision made. Thus, the system or the environment can usually provide feedback to the non-faulty processes about which of the values was preferred or correct for that iteration of the agreement algorithm. Can the non-faulty processes use this feedback in a way that the probability of choosing the correct value increases in subsequent iterations of the algorithm?

We refer to this version of the BA problem as Accurate Byzantine Agreement (ABA) and define it as follows. Assume a set of n processes among which at most f Byzantine faults can occur. All non-faulty processes are required to make decisions for multiple rounds or iterations. For each iteration, a process can propose a binary value 0 or 1. All nonfaulty processes must agree on each decision and must take finite time to agree. After each decision, the environment provides a common feedback to all processes indicating if their decision was correct or wrong. The goal is to design an algorithm that maximizes the (expected) number of correct decisions by non-faulty processes over iterations of the algorithm.

In this paper, we give an algorithm, referred to as the ABA algorithm for the ABA problem. Our method relies on maintaining a common weight vector at all processes and updating this vector based on the feedback for each iteration. Initially, the weight of each process is a non-negative value proportional to the trust of the system on that process. If there is no prior information available, then the weights can simply be initialized to $1/n$. We use a weighted majority rule to determine the agreed upon value for the ABA problem. Once the value is committed, the feedback determines whether the decided value was a mistake or not. An important aspect of the algorithm is how the weights are updated based on the feedback. One possibility is to penalize all processes that proposed a wrong value after each iteration. Another possibility is to penalize processes only if the value decided in that iteration was wrong. Somewhat surprisingly, the behavior of the ABA algorithm may crucially depend upon which rule is used. We provide guarantees on the accuracy of the algorithm based on different assumptions on the accuracy of the processes and different weight update rules.

Byzantine Agreement is a well-studied problem in the field of distributed computing with research in both the theoretical [16, 1, 14, 11] and practical aspects [5, 7, 6]. For the synchronous model of communication (as assumed in this paper), it is known that agreement can be achieved only when $n \geq 3f + 1$ [20]. In our work in [13], we present algorithms and bounds for weighted BA, where processes are assigned weights according to the application. In that paper, we give Byzantine agreement protocols that work even when $n < 3f + 1$, where f is the number of processes that have failed so long as the ratio of the weight of the failed processes to the weight of nonfaulty processes is at most $1/2$. We also present techniques to increase the weights of the non-faulty processes relative to that of the faulty processes based on detection of faulty behavior. Weighted BA problem does not have any notion of *accurate* value for agreement or environmental feeback as required for the ABA problem. It can be used as a subroutine in the ABA algorithm as shown in Section 5. Other approaches to BA include the use of artificial neural networks [22, 18], randomized algorithms [21, 4] or authentication based algorithms [9, 20]. None of these works explore the notion of accurate processes or the correct value for agreement. Our work can be applied to extend the results of these papers.

The concept of weighted majority and multiplicative weight update is used in many disciplines such as learning theory, game theory and linear programming [15, 19]. In the literature for this methodology, the experts are independent entities and there is no notion of liars that can collude and confuse other experts into suggesting the wrong value. In this paper, we assume the presence of malicious Byzantine experts and design algorithms to tolerate them. In summary, we make the following contributions:

- *The ABA Problem*: We introduce the problem of Accurate Byzantine Agreement, where the processes have to agree on a correct binary value as deemed by environmental feedback. The goal is to use this feedback to improve the accuracy of the algorithm in subsequent iterations.
- *The ABA Algorithm*: We present an algorithm to solve the ABA problem that uses a standard solution to the BA problem and a multiplicative method to maintain and update process weights. We make guarantees on the accuracy of the algorithm for the following models:
  - *Deterministic Accuracy*: We make assumptions on two ratios, the *accuracy ratio* (α) and the initial fault ratio (r_0). The accuracy ratio is the ratio of weight of the accurate processes to the weight of the non-faulty processes. The fault ratio r is the ratio of the weight of the faulty processes to that of the non-faulty processes. When $\alpha > 3/4$, the algorithm is always accurate if $r_0 < 1/2$. We relax this bound and show that when $\alpha > (1/2 + d)$, for any $0 \le d \le 1/2$, the algorithm is always accurate if $r_0 < 2d$.
  - *Probabilistic Accuracy*: We make assumptions on the probability with which non-faulty processes propose the correct value, β, and on the fault ratio r. When $\beta > 1/2 + d$ for any $0 < d < 1/2$, the probability of the algorithm being inaccurate is exponentially small if $r < 2d$.
  - *At-Least-One Accuracy*: If there exists at least one process such that it is inaccurate at most b times, then the ABA algorithm is inaccurate only $O(b + \log n)$ times. Hence, the algorithm tracks the most accurate process in the system.
- *Experimental Evaluation*: We present simulation results evaluating the performance of three distinct solutions: the ABA algorithm (with update on inaccuracy), the ABA algorithm with update on every iteration (always update) and the standard Byzantine Agreement (never update). While always-update and never-update perform very well for one of the models each, they perform poorly for the other one. The update on inaccuracy method performs well for both the models.

2   Model and Definitions

We consider a distributed system of n processes, $P_1 \ldots P_n$ with a completely connected topology. We assume that the underlying system is *synchronous* i.e., there is an upper bound on the message delay and on the duration of actions performed by processes. The communication system is assumed to be reliable and hence, no messages are dropped. The processes may undergo Byzantine failures, i.e., fail in an arbitrary fashion; in particular, they may lie and collude with other failed processes to foil any algorithm. However, they may not fake their identity.

We classify the processes in our system based on their behavior into non-faulty, accurate and faulty processes. While the notion of faulty and non-faulty processes is common to all BA problems, we introduce the concept of accurate processes that captures the idea of a correct proposal. A non-faulty process is considered *accurate* for an iteration if it proposes the correct value for that iteration.

In the standard BA problem, all non-faulty processes must agree on a common value. The only requirement on the decided value is that it must be proposed by a non-faulty process. In our proposal, the value decided by the algorithm is important as there is a reward function associated with the value decided, awarded by the environment or the system. The *correct* value is assigned 1 unit of reward and an incorrect value is assigned 0 units, i.e., no reward. Based on the reward, we replace the standard concept of validity with the notion of *accuracy*. Validity specifies that the value decided by the non-faulty processes must have been proposed by at least one of the non-faulty processes. This condition eliminates the trivial solution where all non-faulty processes agree on a fixed value all the time. In our system, the accuracy requirement eliminates the trivial solution. We define our problem below.

Definition 1. *(Accurate Byzantine Agreement with Feedback) Consider n processes consisting of non-faulty and faulty processes. There are multiple binary decisions that these n processes are required to make. For each possible decision (iteration of the ABA problem), each of the non-faulty processes proposes either 0 or 1. An algorithm that solves the Accurate Byzantine Agreement with Feedback (ABA) problem, must guarantee the following properties:*

- *Agreement: For each iteration, all non-faulty processes decide on the same value.*
- *Termination: The algorithm terminates in a finite number of rounds.*
- *Accuracy: The non-faulty processes agree on a value that is deemed correct by environmental feedback.*

To incorporate the feedback provided by the environment we assign a non-negative weight w_i to each process P_i that provides an estimate, possibly erroneous, of the trust placed on that process. We summarize our notation in table 1.

Table 1. Notation

n	Number of processes	f	Number of Byzantine faults
w_i	Weight of process P_i	a	Total weight of accurate processes
p	Total weight of non-faulty processes	q	Total weight of faulty processes
r	Fault Ratio (= q/p)	α	Accuracy ratio (= a/p)

3   The ABA Algorithm

In this section, we propose an algorithm (Fig. 1) for the ABA problem. The algorithm is identical at all processes and executes in synchronous iterations. At each process, we maintain two vectors W and V. Vector W stores the weight of each process while vector V stores the value proposed by each of them. Initially, the weight of each process is a non-negative value directly proportional to the initial trust on that process. In each iteration of the algorithm each non-faulty process proposes a value and executes Step 1 to Step 5 of the algorithm.

```
var
    W: array[1..n] of float initialized according to system trust (default value all 1/n);
    V: array[1..n] of {0, 1} initialized to 0;
    t: integer specifying the total number of iterations;

for iteration := 1 to t do
    V[i] = proposed value by P_i;

    // Step 1: Exchange values with all
    for j : 1 to n do
        send V[i] to P_j;
        receive V[j] from P_j; //if (no value received from P_j), V[j] = 0;

    // Step 2: Agree on V vector
        for j : 1 to n do: run standard Byzantine Agreement on V[j];

    // Step 3: Compute support for values 0 and 1 and choose the majority value
    float s_0 = Σ_j{W[j] | V[j] = 0};    float s_1 = Σ_j{W[j] | V[j] = 1};
    if (s_0 ≥ s_1) then decided := 0; else decided := 1;

    // Step 4: Wait for reward and determine the correct value based on the feedback
    if (reward = 1) then correctVal := decided;
    else correctVal := 1 − decided; //the process decided on the wrong value

    // Step 5: //multiplicative weight update on inaccuracy: ABA(UI)
    if (reward = 0) then
        for j : 1 to n do
            if (V[j] ≠ correctVal) then W[j] = (1 − ε) · W[j];

    // Alternative Step 5': //multiplicative weight update on all iterations ABA(UA)
    for j : 1 to n do
        if (V[j] ≠ correctVal) then W[j] = (1 − ε) · W[j];

endfor;
```

Fig. 1. The ABA Algorithm at P_i

In Step 1, all processes exchange their proposed values to populate V. If no value is received from some process, the corresponding entry is set to 0. Since faulty processes may send conflicting values to other processes, it is not guaranteed that the V vector is identical at all non-faulty processes after Step 1.

In Step 2, the algorithm requires all non-faulty processes to agree on the value proposed by every other process and thereby make the V vector identical at all non-faulty processes after Step 2 of any iteration. For this step, we can use any standard BA algorithm such as the King algorithm [2] that requires $n \geq 3f + 1$, or the Queen algorithm [3] that requires $n \geq 4f + 1$. The validity property satisfied by these algorithms ensures that the value of $V[i]$ for any non-faulty process P_i is exactly the value proposed by P_i.

In Step 3, processes determine the sum of weights of all processes that support value 0 or 1. The value with larger support, i.e., the weighted majority is chosen as the value in *decided*.

In Step 4, processes receive the common feedback from the environment to determine the correct value.

In Step 5, we carry out the update of weights. If the value decided was incorrect, then the weights of the processes that proposed an incorrect value is reduced by some constant proportion ϵ ($0 < \epsilon < 1$) of its previous weight (multiplicative update). As an alternative to step 5, in step 5', we carry out the weight update on all iterations irrespective of the reward value. If we update weights only on inaccuracy, we refer to the algorithm as ABA(UI) ("update on inaccuracy"). If we update weights on all iterations, we refer to the algorithm as ABA(UA) ("update always"). We now prove that both the versions of the algorithm guarantee the agreement and termination property specified in definition 1 independent of the assumptions on accuracy.

Theorem 1. *(Agreement & Termination) Assuming $n \geq 3f + 1$, all iterations of the ABA algorithm guarantee agreement and termination.*

Proof. Agreement: We show that after Step 2 of every iteration, all non-faulty processes have identical W and V vectors. The proof is by induction on the iteration number. At the first iteration, the vector W is identical at all non-faulty processes by the initialization. Now assume that the vector W is identical at the beginning of any iteration i. Because all processes agree on vector V using Byzantine agreement, all non-faulty processes will have identical V after Step 2. This implies that all non-faulty processes will have identical values of $s0$, $s1$, and *decided* after step 3 because these variables depend only on W and V. Since the reward function is assumed to be common, all non-faulty processes will have identical value of *correctVal* and therefore will update W in an identical manner. The value decided depends only on W and V vectors and hence all non-faulty processes agree on the same value.

Termination: This is a synchronous algorithm which executes in finite number of rounds and hence, termination is satisfied trivially.

The ABA algorithm guarantees another useful property: if a nonfaulty process proposes an accurate value, then it can never be penalized. This property exploits the validity condition satisfied by the BA algorithm used in Step 2. A non-faulty process P_i will send the same value to all non-faulty processes. Therefore, all non-faulty processes will have identical $V[i]$ when they invoke the BA algorithm. Therefore, by validity of the BA algorithm, $V[i]$ at all non-faulty processes will be identical to the one proposed by P_i.

4   Accuracy Guarantees of the ABA Algorithm

In the previous section, we have shown that ABA algorithm guarantees agreement and termination. This section focuses on the accuracy guarantees the algorithm can provide based on varying assumptions about the accuracy of the processes in the system. Since standard Byzantine agreement is used in Step 2, in this section we assume that $n \geq 3f + 1$, according to the lower bound for the BA problem [20]. In Section 5, we consider the case when $n \geq 3f + 1$ does not hold.

4.1   Deterministic Accuracy

For deterministic accuracy, we make guarantees based on the accuracy ratio α (ratio of the weight of accurate processes to the weight of non-faulty processes) and the fault ratio of the system r (ratio of the weight of faulty processes to the weight of non-faulty processes). We show that if $\alpha > 3/4$ for each iteration and if the initial fault ratio $r_0 < 1/2$, then the algorithm guarantees accuracy. Then we relax this requirement and show that it is sufficient that $\alpha > (1/2 + d)$ for each iteration such that $r_0 < 2d$, to guarantee accuracy.

We first show that as long as $\alpha > 1/2$ for each iteration, r never increases if we update weights only on error. This enables us to make guarantees just based on the initial fault ratio of the system. The proof crucially depends on the fact that we update the weights of inaccurate processes only when the algorithm chooses the incorrect value.

Lemma 1. *(Non-Increasing Fault Ratio) For any iteration, if the accuracy ratio $\alpha > 1/2$, then the fault ratio r cannot increase after that iteration of the ABA(UI) algorithm.*

Proof. In the ABA(UI) algorithm, the weights of the processes changes only when the algorithm makes a mistake. Consider the weight of the non-faulty processes, p. Since $\alpha > 1/2$, when the algorithm makes a mistake, greater than $p/2$ of the weight will be unaffected and less than $p/2$ of the weight will be reduced by a factor of $1 - \epsilon$. Hence, if p' is the weight of the non-faulty processes after a weight update,

$$p' > p/2 + (1 - \epsilon)p/2 = p(2 - \epsilon)/2 \tag{1}$$

Now consider the weight of the faulty processes q. The algorithm chooses the wrong value only when a majority weight, i.e. $> (p + q)/2$ of the weights are inaccurate. Since greater than $p/2$ of the weights are accurate, at least $q/2$ of the weights are inaccurate. Hence, if q' is the weight of the faulty processes after a weight update,

$$q' < q/2 + (1 - \epsilon)q/2 = q(2 - \epsilon)/2 \tag{2}$$

Dividing equation 2 by equation 1, we get, $q'/p' < q/p$.

Note that the proof for lemma 1 does not hold for the always update rule. If the faulty processes keep proposing the correct value, then the ABA(UA) algorithm will increase the relative weight of the faulty processes and consequently the fault ratio. If the fault ratio increases beyond 1 then Byzantine processes can force the ABA algorithm to choose incorrect value on crucial decisions.

In the following theorem we show that if $\alpha > 3/4$, then the ABA(UI) algorithm never makes a mistake as long as the initial fault ratio is less than $1/2$.

Lemma 2. *If the accuracy ratio $\alpha > 3/4$ for all iterations, and the initial fault ratio $r_0 < 1/2$, then the ABA(UI) algorithm always guarantees accuracy.*

Proof. If the accuracy ratio a/p is greater than $3/4$, then the weight of accurate proposals a is at least $3p/4$. This implies that the weight of inaccurate proposals is at most $p + q - 3p/4 = p/4 + q$. The algorithm selects the correct value if the accurate weight is more than the inaccurate weight. We need to show that, $p/4 + q < 3p/4$. Dividing both sides by p and rearranging, this is equivalent to showing that $r < 1/2$. Since $r_0 < 1/2$, from lemma 1, for all iterations, $r < 1/2$. Note that, for the ABA(UI) algorithm, we update the weights only when the algorithm makes a mistake. So for any iteration, if $\alpha > 3/4$ and $r < 1/2$, it will remain so for every subsequent iteration and hence the ABA(UI) algorithm never makes a mistake.

In the following theorem, we show that even if the accuracy ratio is just above $1/2$, the ABA algorithm never makes a mistake as long as the initial fault ratio is less than a certain threshold.

Theorem 2. *(Deterministic Accuracy) If the accuracy ratio $\alpha > 1/2 + d$ for all iterations and if the initial fault ratio $r_0 < 2d$, for any $0 \leq d \leq 1/4$, then the ABA(UI) algorithm always guarantees accuracy.*

Proof. If the weight of accurate proposals is at least $p(1/2 + d)$, then the weight of inaccurate proposals is at most $p(1/2 - d) + q$. The algorithm selects the correct value if the accurate weight is more than the inaccurate weight. Therefore, we need $p(1/2 - d) + q < p(1/2 + d)$. This condition is equivalent to $r < 2d$. Since $r_0 < 2d$, from lemma 1, for any iteration, $r < 2d$. Since the correct decision was made, the weights are not updated and the algorithm continues to chose the correct value in the subsequent iterations. makes a mistake.

Note that when d equals $1/4$, this theorem reduces to lemma 2. Thus, theorem 2 generalizes lemma 2, when $d < 1/4$. Accuracy of the ABA(UI) is guaranteed if either an overwhelming majority of non-faulty processes is accurate (d is large) or there is a large percentage of non-faulty processes (r_0 is small).

In the following theorem, we make guarantees based on the number of accurate processes and the number of faulty processes in the system.

Theorem 3. *If the number of accurate processes is greater than $1/2 + d$ times the number of nonfaulty processes for all iterations and if the initial number of faulty processes is less than $2d$ times the number of nonfaulty processes, for any $0 \leq d \leq 1/4$, then the ABA(UI) algorithm always guarantees accuracy.*

Proof. We initialize the weights of all processes to $1/n$. This proof follows directly from theorem 2. If the number of accurate processes is greater than $1/2 + d$ times the number of nonfaulty processes then the accuracy ratio $\alpha > 1/2 + d$, since the weights are equally initialized. Similarly, the initial fault ratio $r_0 < 2d$. Hence from theorem 2, the ABA(UI) algorithm guarantees always guarantees accuracy. As mentioned in the proof of theorem 2, since the algorithm decides on the correct value, the weights are not updated and hence the algorithm continues to chose the correct value in the subsequent iterations.

The following theorem handles the case when a majority of the nonfaulty processes are accurate but the fault ratio is not smaller than $2d$.

Theorem 4. *(Accuracy after some initial mistakes) If the accuracy ratio $\alpha > 1/2 + d$ for all iterations, for any $0 \leq d \leq 1/4$, then the ABA(UI) algorithm guarantees accuracy after some initial mistakes.*

Proof. (Sketch) Similar to the proof of lemma 1, it is easy to show that there exists a constant γ such that the fault ratio decreases by a factor of at least γ for any mistake. Therefore, eventually the fault ratio becomes less than $2d$. Subsequently, by theorem 2 the algorithm ABA(UI) does not make any mistake.

It is easy to show that ABA(UA) algorithm can be forced to make unbounded mistakes by the Byzantine processes for any accuracy ratio less than $3/4$. Byzantine processes may initially propose correct values to increase the fault ratio. Once the fault ratio is high, they can ensure that ABA makes mistakes. They can repeat this cycle forever.

4.2 Probabilistic Accuracy

For probabilistic accuracy, we make guarantees based on the probability of accuracy of each non-faulty process β, and the fault ratio r, of the system. We show that if $\beta > 1/2 + d$ $(0 < d < 1/2)$, and $r < 2d$, the ABA algorithm guarantees accuracy with high probability.

Theorem 5. *(Probabilistic Accuracy) Let all weights in the system be in $[0, 1]$. If the accuracy probability of non-faulty processes $\beta > 1/2 + d$ and the fault ratio $r < 2d$, for any $(0 < d < 1/2)$, for all iterations, then the ABA algorithm guarantees accuracy with probability greater than $1 - (\frac{e^{-\delta}}{(1-\delta)^{(1-\delta)}})^\mu$, where $\mu = p(1/2 + d)$ and $\delta = (2d - r)/(2d + 1)$.*

Proof. Let X_i be the random variable indicating the non-faulty process P_i making an accurate proposal. Let $X = \sum_i w_i X_i$, where w_i is the weight of process P_i. We have $E[X_i] = 1/2 + d$. Therefore, $E[X] = (1/2 + d) \sum_i w_i = p(1/2 + d)$.

Let $\mu = E[X]$. We now show that $(1 - \delta)\mu = (p + q)/2$.

$(1-\delta)\mu = (2d+1-(2d-r))/(2d+1) \cdot p \cdot (2d+1)/2 = (1+r)p/2 = (1+q/p)p/2 = (p+q)/2$.

When $r < 2d \leq 1$, we get that $0 < \delta < 1$. Hence, from Chernoff's bound, we have,

Pr[ABA algorithm makes wrong decision]

$= Pr$[sum of all weights supporting the correct value $< (p + q)/2$] { From the algorithm}

$\leq Pr[X < (p + q)/2)]$ { Considering only non-faulty processes }

$= Pr[X < (1 - \delta)\mu]$ { Shown above}
$< (\frac{e^{-\delta}}{(1-\delta)^{(1-\delta)}})^{\mu}$, { From Chernoff's bound and $0 < \delta < 1$ }.

In Theorem 5, the error probability depends upon $\delta = (2d - r)/(2d + 1)$. As r decreases, δ increases. We now show that for the ABA(UA) algorithm the ratio r is expected to decrease exponentially with increasing iterations.

Theorem 6. *(ABA(UA): Exponentially Decreasing Expected Fault Ratio) If the accuracy probability of non-faulty processes is at least $1/2 + d$, and the accuracy probability of faulty processes is at most $1/2 - d$, then there exists $k > 1$, such that after j iterations of the ABA(UA) algorithm, the expected ratio of the weight of the non-faulty processes to the weight of the faulty processes is at least k^j/r_0.*

Proof. We first show a bound on the expected weight of a non-faulty process after j iterations. Let the initial weight of a nonfaulty process be w_0. Let M_i be the random variable denoting the multiplicative factor at iteration i for a non-faulty process. Let W_j be the random variable denoting weight of a non-faulty process after j iterations. It is clear that for ABA(UA) algorithm, $W_j = w_0 \Pi_{i=1}^{i=j} M_i$. The multiplicative factor for any iteration depends on the environmental feedback and is independent of other iterations. Hence, $E[W_j] = w_0 \Pi_{i=1}^{i=j} E[M_i] \geq w_0 \Pi_{i=1}^{i=j}((1/2 + d) \cdot 1 + (1/2 - d) \cdot (1 - \epsilon)) = w_0(1 - \epsilon/2 + d\epsilon)^j$.

Similarly, since the probability that a faulty process makes a correct proposal is at most $1/2 - d$, the expected weight of a faulty process after j iterations of the ABA(UA) algorithm is at most $(1 - \epsilon/2 - d\epsilon)^j$ times its original weight.

We now show that the expected fault ratio decreases exponentially with the number of iterations. Let p_0 and q_0 be the intial weights of non-faulty and faulty processes such that $q_0/p_0 = r_0$. Let S_j and T_j be the random variables to denote weights of the non-faulty processes and faulty processes after j iterations of the ABA(UA) algorithm. Since the expected weight of each non-faulty process after j iterations is at least $(1 - \epsilon/2 + d\epsilon)^j$ times its original weight; by linearity of expectation, $E[S_j] \geq p_0 \cdot (1 - \epsilon/2 + d\epsilon)^j$. Similarly, $E[T_j] \leq q_0 \cdot (1 - \epsilon/2 - d\epsilon)^j$. We now bound $E[S_j/T_j]$. Using independence of S_j and T_j, we get that $E[S_j/T_j] = E[S_j] \cdot E[1/T_j]$. We now use the fact that for any nonnegative random variable X, $E[1/X] \geq 1/E[X]$ which can be shown using Jensen's inequality $(E[f(X)] \geq f(E[X])$ for convex f). Therefore,
$E[S_j] \cdot E[1/T_j] \geq E[S_j] \cdot 1/E[T_j]$
$\geq \frac{p_0 \cdot (1-\epsilon/2+d\epsilon)^j}{q_0 \cdot (1-\epsilon/2-d\epsilon)^j} = 1/r_0 \cdot (1 + \frac{2d\epsilon}{1-\epsilon/2-d\epsilon})^j$

By defining $k = (1 + \frac{2d\epsilon}{1-\epsilon/2-d\epsilon})$, we get the desired result. Because $0 < \epsilon < 1$ and $0 < d < 1/2$, $(1 - \epsilon/2 - d\epsilon)$ is guaranteed to be positive which ensures $k > 1$.

Remark: The above theorem can be generalized to the case when non-faulty processes are accurate with probability at least $1/2 + d_1$ and faulty processes are accurate with probability at most $1/2 - d_2$. In this case, $k = (1 + \frac{\epsilon(d_1+d_2)}{1-\epsilon/2-d_2\epsilon})$. When $d_1 = d_2$ we get the original value of k. Also when $d_2 = -d_1$ (faulty processes are as accurate as non-faulty processes), we get k equals 1.

4.3 At-Least-One Accuracy

For this section, we assume that there is at least one process in the system that is inaccurate *only* for a small number of iterations of the ABA algorithm. This assumption is sufficient to guarantee *cumulative accuracy*, i.e., a bound on the total number of mistakes made by the algorithm. Our results are based on the method of weighted majority with multiplicative updates [15]. We first consider ABA(UI) algorithm. In the following theorem, we show that ABA(UI) guarantees accuracy for a large number of iterations.

Theorem 7. *(At-Least-One Accuracy, ABA(UI)) Assume $n \geq 3f + 1$. If there exists at least one process such that is inaccurate at most b out of j iterations of the ABA(UI) algorithm, then the algorithm is inaccurate at most $2(1 + \epsilon)b + (2/\epsilon)\log n$ times.*

Proof. The proof follows from standard arguments in multiplicative update method [15]. We initialize the weights of all the processes to $1/n$. Let $\phi(i)$ be the sum of all the weights of the processes at the end of iteration i. Suppose that for any iteration i, the ABA(UI) algorithm is wrong. This means that the weighted majority of the values in the proposed vector were wrong and hence a majority of the weights will decrease by $(1-\epsilon)$ of their previous value. Therefore, $\phi(i) \leq \phi(i-1)/2 + \phi(i-1)/2 \cdot (1-\epsilon) = \phi(i-1)(1-\epsilon/2)$. The total weight, at the beginning of the algorithm, $\phi(0)$ is equal to one. Suppose that the ABA(UI) algorithm makes $m(j)$ mistakes in the first j iterations. After j iterations of ABA(UI), we get $\phi(j) \leq \phi(0)(1 - \epsilon/2)^{m(j)} = (1 - \epsilon/2)^{m(j)}$.

Now consider a nonfaulty process that is inaccurate at most b out of j iterations. In spite of the presence of Byzantine processes, ABA(UI) algorithm guarantees that this process is never penalized when it is accurate. After j iterations, the weight of this process is at least $(1 - \epsilon)^b \cdot$(its initial weight) $= (1 - \epsilon)^b/n$. This weight is less than the total weight. Therefore, $(1 - \epsilon)^b/n < (1 - \epsilon/2)^{m(j)}$.

Taking log on both sides and shifting n to the right hand side, we get $b \log(1 - \epsilon) < \log n + m(j)\log(1 - \epsilon/2)$.

Dividing both sides by $\log(1 - \epsilon/2)$ which is a negative quantity and rearranging gives us

$$m(j) < b \log(1 - \epsilon)/\log(1 - \epsilon/2) - \log n/\log(1 - \epsilon/2).$$

In the following part of the proof, we use two inequalities: $-\log(1 - \epsilon) \leq \epsilon + \epsilon^2$ and $-\log(1 - \epsilon/2) \geq \epsilon/2$ that require $\epsilon < 0.684$. Applying these inequalities we get, $m(j) < b \cdot 2 \cdot (\epsilon + \epsilon^2)/\epsilon + 2 \log n/\epsilon$.

Therefore, $m(j) < 2(1 + \epsilon)b + 2/\epsilon \log n$. $\qquad\qquad\qquad\qquad\qquad\qquad\qquad\square$

Interestingly, the result holds even when we use ABA(UA).

Theorem 8. *(At-Least-One Accuracy, ABA(UA)) Assume $n \geq 3f + 1$. If there exists at least one process such that is inaccurate at most b out of j iterations of the ABA(UA) algorithm, then the algorithm is inaccurate at most $2(1 + \epsilon)b + (2\epsilon)\log n$ times.*

Proof. Note that even when we update weights on all iterations, the following inequalities hold. The total weight in the system, $\phi(j) \leq \phi(0)(1 - \epsilon/2)^{m(j)} = (1 - \epsilon/2)^{m(j)}$. The weight of the process that is wrong b out of j iterations is $(1 - \epsilon)^b \cdot$(its initial weight) $= (1 - \epsilon)^b/n$. Hence, the previous proof applies. $\qquad\qquad\qquad\qquad\square$

Substituting $b = 0$ in the above theorem, i.e., when at least one process is accurate for all j iterations of the algorithm, the ABA algorithm makes a mistake only $O(\log n)$ times. Note that this is independent of the number of iterations and hence, if the ABA algorithm is run for a large number of iterations ($j \gg \log n$), then it guarantees accuracy in most of them. Or in other words, the ABA algorithm is approximately as accurate as the most accurate process in the system.

5   ABA Algorithm with Weighted Byzantine Agreement

In the ABA algorithm proposed in Fig. 1, we have used standard Byzantine Agreement in Step 2. Since standard Byzantine Agreement assumes $n \geq 3f + 1$, the ABA algorithm also made the same assumption. This assumption is crucial for correctness of the ABA algorithm because agreement requires that processes have identical V vector after step 2. We now consider the case when $n < 3f + 1$, but the initial fault ratio is less than $1/2$. Thus, more than a third of the processes may be faulty but the total weight of the faulty processes is still less than $1/2$ of the weight of the nonfaulty processes. Under this scenario, we propose an alternative to ABA algorithm by replacing Step 2 of the ABA algorithm by

> // Step 2'(Alternative to Step 2) : Agree on V vector
> **for** j : 1 to n do: run *Weighted* Byzantine Agreement [13] on $V[j]$

Thus, we use the weight vector even to agree on the value of $V[j]$ (as used by the algorithms in [13]). We refer to this algorithm as ABAW algorithm. Since the fault ratio decreases under various accuracy assumptions, the ABAW algorithm works correctly even when the set of processes that act Byzantine increases with time so long as the fault ratio stays less than $1/2$. The following theorem can be shown for the ABAW algorithm analogous to that for ABA algorithm.

Theorem 9. *Assuming $r_0 < 1/2$, all iterations of the ABAW algorithm guarantee agreement and termination if the weight update method ensures $r < 1/2$.*

For the deterministic accuracy property, we have the following theorem.

Theorem 10. *If the accuracy ratio $\alpha > 1/2 + d$ and if the initial fault ratio $r_0 < 2d$, for any $0 \leq d \leq 1/4$, for all iterations, then the ABAW(UI) algorithm guarantees accuracy.*

Note that Theorem 4 does not hold for ABAW(UI) because we require $r_0 < 1/2$. Theorem 5 holds for ABAW(UA) without the assumption of $n \geq 3f + 1$ (assuming $r < 1/2$). Theorem 7 does not hold for ABAW(UI) or ABAW(UA) because the fault ratio may increase beyond $1/2$ even if one process is accurate most of the times.

6   Experimental Evaluation of ABA Algorithm

The experimental evaluation compares three different update methods: "always update", "update on inaccuracy" and "never update". The last option reduces to standard Byzantine agreement. The performance of the three accuracy models presented in this paper are considered with each of these update methods for two different Byzantine fault models. Always update and never update perform very well under one of the fault models each, while they both perform very poorly for the other. Update on inaccuracy, the method followed in this paper, is always close to the best.

6.1 Experimental Setup and Parameters

For the experimental evaluation, we focus on faulty processes that will always try to make the system agree upon an incorrect value. The faulty processes have complete knowledge of the system including the correct value for each iteration. Our simulation uses two models for faulty processes. Model 1 uses a process that will always propose the incorrect value. Model 2 uses a process that looks at the percentage of its own weight to the weight of all processes and proposes the correct value if its percentage is below a threshold and the incorrect value otherwise. There are two types of non-faulty processes used. The first is an accurate non-faulty process that always proposes the correct value ($d = 0.5, \beta = 1$). The second type of non-faulty process chooses the correct value with probability $\beta = 0.5 + d$, where $d \in [0, 0.5]$. The Queen algorithm [3] is used for step 2 in the ABA algorithm and for all simulations, $n = 41$, $f = 10$ and $\epsilon = 0.1$.

6.2 Results

Simulation results for *deterministic accuracy* are shown in Fig. 2. For this experiment, we had one accurate process, and the other non-faulty processes had a value of $d = 0.00001$. We compare the % of accurate decisions made by the algorithm for 100 iterations, with increasing values of $a_0/(p + q)$ i.e. the starting weights of the accurate processes divided by the total weight of processes in the system. The experiments were performed for the two fault models 1 and 2. As can be seen, having an update method performs much better than not having one with model 1 and always updating performs poorly with model 2. Update on error gives a good compromise between the two.

Results for *probabilistic accuracy* are shown in Fig. 3. For this experiment, all non-faulty processes had $d = 0.02$ and all processes start with uniform weights. We compare the % of accurate decisions with increasing number of iterations. Notice that, on the whole, update on inaccuracy performs the best for these graphs. Always updating seems like the natural method to use but in Fig. 3(b) always update performs the worst.

Simulation results for *at-least-one accuracy* are shown in Fig. 4. For this experiment, we had one non-faulty process which is always accurate i.e. $d = 0.5$, and the remaining non-faulty processes had $d = 0.00001$. The processes start with uniform weights. We compare the % of accurate decisions with increasing number of iterations. For model 1 in Fig. 4(a), updating weights increases the accuracy over iterations. With model 2 always update shows the worse performance with update on accurate being the best. Notice how update on inaccuracy is always close to the best.

7 Conclusion and Future Work

We introduce the problem of Accurate Byzantine Agreement with Feedback where in addition to agreeing on the same value, the processes in the system have to agree on

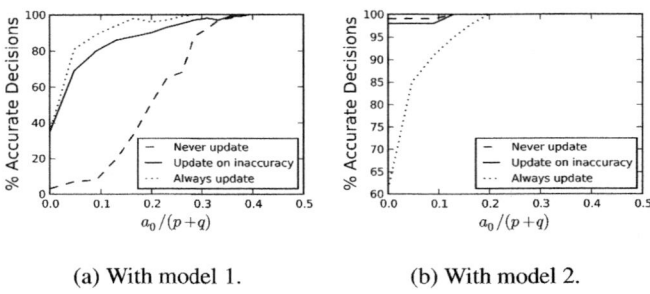

(a) With model 1.   (b) With model 2.

Fig. 2. Deterministic accuracy: Ratio of Accurate Process Weights vs. % Accurate Decisions

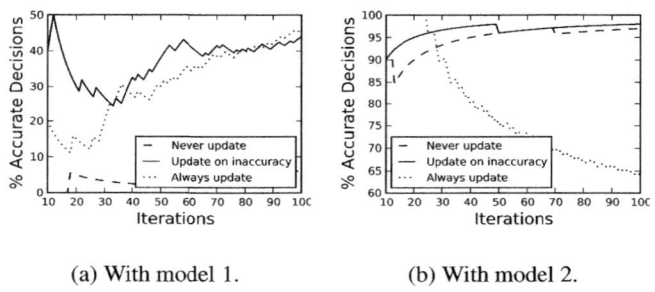

(a) With model 1.   (b) With model 2.

Fig. 3. Probabilistic Accuracy: Iterations vs. % Accurate Decisions, $d = 0.02$

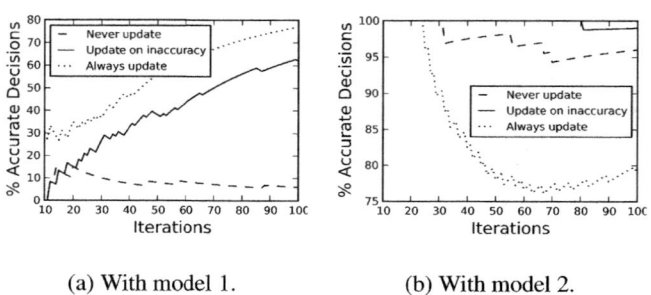

(a) With model 1.   (b) With model 2.

Fig. 4. At-Least One Accuracy: Iterations vs. % Accurate Decisions, One accurate process

the correct value. The notion of correctness is based on the environment or any kind of external feedback common to all the processes in the system. We present an algorithm that solves the problem for various assumptions on the initial accuracy and weight distribution of the processes. We show that if the weight of the accurate processes is greater than 3/4 the weight of the non-faulty processes, then the algorithm always decides on the correct value. We relax this further and show that if a majority of the non-faulty processes are accurate, then for certain assumptions on the faulty and non-faulty processes, the algorithm never makes a mistake. Further, we show that if the probability of accuracy of the non-faulty process is greater than 1/2, then the algorithm's accuracy improves exponentially in the number of mistakes it makes. Finally, we consider the simple assumption that at least one process always proposes the correct value for all iterations and show that the algorithm rarely makes mistakes.

We performed simulations comparing the performance of three different weight update methods: update on inaccuracy, always update and never update (just standard Byzantine agreement). The experiments compared the performance of these solutions under all three accuracy assumptions and the results indicate that while never update and always update perform very well for different fault models, update on inaccuracy performs uniformly well for both fault models.

This problem brings forth further questions. The results in this paper mainly present upper bounds for the problem of accurate Byzantine agreement with feedback. We also need to explore lower bounds for the problem. Also, our results depend on the multiplicative update rule. We wish to explore other update rules, such as additive updates and compare their performance, both theoretically and practically. Designing optimal policies to guarantee maximum probability of correct decisions is also an interesting problem.

References

[1] Abraham, I., Dolev, D., Gonen, R., Halpern, J.: Distributed computing meets game theory: robust mechanisms for rational secret sharing and multiparty computation. In: Proceedings of the Twenty-Fifth Annual ACM Symposium on Principles of Distributed Computing, PODC 2006, pp. 53–62. ACM, New York (2006)

[2] Berman, P., Garay, J.A., Perry, K.J.: Towards optimal distributed consensus. In: Annual IEEE Symposium on Foundations of Computer Science, pp. 410–415 (1989)

[3] Berman, P., Garay, J.A.: Asymptotically Optimal Distributed Consensus (Extended Abstract). In: Ronchi Della Rocca, S., Ausiello, G., Dezani-Ciancaglini, M. (eds.) ICALP 1989. LNCS, vol. 372, pp. 80–94. Springer, Heidelberg (1989)

[4] Bracha, G.: An $O(\log n)$ expected rounds randomized Byzantine generals protocol. Journal of the ACM 34(4), 910–920 (1987)

[5] Castro, M., Liskov, B.: Practical byzantine fault tolerance. In: Third Symposium on Operating Systems Design and Implementation (OSDI), New Orleans, Louisiana. USENIX Association, Co-sponsored by IEEE TCOS and ACM SIGOPS (February 1999)

[6] Clement, A., Marchetti, M., Wong, E., Alvisi, L., Dahlin, M.: Making byzantine fault tolerant systems tolerate byzantine faults. In: 6th USENIX Symposium on Networked Systems Design and Implementation (NSDI) (April 2009)

[7] Cowling, J., Myers, D., Liskov, B., Rodrigues, R., Shrira, L.: Hq replication: A hybrid quorum protocol for byzantine fault tolerance. In: Proceedings of the Seventh Symposium on Operating Systems Design and Implementations (OSDI), Seattle, Washington (November 2006)

[8] Dolev, D., Reischuk, R., Strong, H.R.: Early stopping in byzantine agreement. J. ACM 37(4), 720–741 (1990)

[9] Dolev, D., Strong, H.R.: Authenticated algorithms for byzantine agreement. SIAM J. Comput. 12(4), 656–666 (1983)

[10] Feldman, P., Micali, S.: An optimal probabilistic protocol for synchronous byzantine agreement. SIAM J. Comput. 26(4), 873–933 (1997)

[11] Fitzi, M., Maurer, U.M.: Efficient Byzantine Agreement Secure Against General Adversaries. In: Kutten, S. (ed.) DISC 1998. LNCS, vol. 1499, pp. 134–148. Springer, Heidelberg (1998)

[12] Garay, J.A., Moses, Y.: Fully polynomial byzantine agreement in t + 1 rounds. In: STOC 1993: Proceedings of the Twenty-Fifth Annual ACM Symposium on Theory of Computing, pp. 31–41. ACM, New York (1993)

[13] Garg, V.K., Bridgman, J.: The weighted byzantine agreement problem. In: 25th IEEE International Symposium on Parallel and Distributed Processing, IPDPS 2011 Conference Proceedings, Anchorage, Alaska, USA, May 16-20, pp. 524–531. IEEE (2011)

[14] Hirt, M., Maurer, U.: Complete characterization of adversaries tolerable in secure multiparty computation (extended abstract). In: PODC 1997: Proceedings of the Sixteenth Annual ACM Symposium on Principles of Distributed Computing, pp. 25–34. ACM, New York (1997)

[15] Kale, S.: Efficient algorithms using the multiplicative weights update method. PhD thesis, Princeton, NJ, USA, AAI3286120 (2007)

[16] King, V., Saia, J.: Breaking the o(n2) bit barrier: scalable byzantine agreement with an adaptive adversary. In: Proceeding of the 29th ACM SIGACT-SIGOPS Symposium on Principles of Distributed Computing, PODC 2010, pp. 420–429. ACM, New York (2010)

[17] Lamport, L., Shostak, R., Pease, M.: The byzantine generals problem. ACM Trans. Program. Lang. Syst. 4, 382–401 (1982)

[18] Lee, K.-W., Ewe, H.-T.: Performance study of byzantine agreement protocol with artificial neural network. Inf. Sci. 177(21), 4785–4798 (2007)

[19] Littlestone, N., Warmuth, M.K.: The weighted majority algorithm. Inf. Comput. 108, 212–261 (1994)

[20] Pease, M., Shostak, R., Lamport, L.: Reaching agreements in the presence of faults. Journal of the ACM 27(2), 228–234 (1980)

[21] Rabin, M.O.: Randomized byzantine generals. In: Foundations of Computer Science, pp. 403–409 (1983)

[22] Wang, S.C., Kao, S.H.: A new approach for byzantine agreement. In: Proceedings of the The International Conference on Information Networking, p. 518. IEEE Computer Society, Washington, DC (2001)

Constructing Mid-Points for Two-Party Asynchronous Protocols

Petar Tsankov, Mohammad Torabi-Dashti, and David Basin

ETH Zürich, 8092 Zürich, Switzerland

Abstract. Communication protocols describe the steps that the communication end-points must take in order to achieve a common goal. In practice, networks often contain mid-points, which can relay, redirect, or filter messages exchanged by the end-points. A mid-point can enforce a communication protocol: it forwards the messages that conform to the protocol, and drops them otherwise. Protocol specifications typically define only the end-points' behavior. Implementing a mid-point that enforces a protocol is nontrivial: the mid-point's behavior depends on the end-point's behavior, and also on the behavior of the communication environment in which the protocol executes.

We present a process algebraic framework that takes as input the formal specifications of the protocol and the environment and outputs a specification for a mid-point that enforces the protocol. We prove that the mid-point specifications synthesized by our framework are correct: only messages that could have resulted from correctly executing end-points are forwarded. As an application, we construct a formal model for the mid-point that enforces the TCP three-way handshake protocol.

Keywords: mid-point, specification, synthesis, formal methods, protocol enforcement.

1 Introduction

Context. Communication protocols describe the steps that the communication end-points take in order to achieve a common goal, e.g. to exchange data reliably. In practice, the end-points often communicate over *mid-points*, which relay, redirect, or filter the communication. Firewalls are prominent examples of mid-points. They can not only observe the execution of a protocol between end-points, but also enforce that the protocol is correctly executed. Namely, the mid-point forwards the messages that conform to the protocol, and drops them otherwise. The messages that do not conform to the protocol may have been sent by a faulty end-point or by an adversary, may be the result of communication failures, etc. For example, a mid-point (or firewall) that enforces the TCP protocol should drop ack messages from B to A right after A has sent B a syn message. This is because, according to TCP's three-way handshake, B must reply to A's syn either with a syn&ack or with a rst message.

The behavior of a mid-point that enforces a communication protocol depends on the steps that the end-points must take, and also on the *communication environment* where the protocol should be executed. This is intuitively because the mid-point would observe

A. Fernández Anta, G. Lipari, and M. Roy (Eds.): OPODIS 2011, LNCS 7109, pp. 481–496, 2011.

the actions of the end-points via a "lens", namely the communication channels that connect the end-points to the mid-point. In an asynchronous message-passing environment, for instance, it is possible that an end-point sends message a and then message b, but the mid-point observes the message b before the message a. The mid-point cannot simply dismiss the observed sequence of messages as a violation of the protocol because, depending on the channels' characteristics, the mid-point may observe different events, and events in different orders, compared to the end-points; cf. [2].

Contributions. We present a process algebraic framework[1] for automatically synthesizing formal models for mid-points. The input to the framework is the specification of the end-points of an asynchronous protocol and the characteristics of the channels that connect the end-points to the mid-point. The framework outputs a formal specification for a mid-point that enforces the protocol. Formal specifications for mid-points can in general be used for (model-based) testing, (model-driven) development of mid-points, and formal verification of mid-points. These are all practically important and nontrivial tasks: A case study on three commonly used firewalls (Checkpoint, netfilter/iptables, and ISA Server) shows that different firewall manufacturers implement the mid-point for (enforcing) the TCP protocol differently, and sometimes incorrectly with respect to the TCP specification given in [8] (see [3] for details). A formal specification for TCP mid-points can be used either to avoid or to pinpoint the causes of such discrepancies.

The inputs and the output of our framework are processes specified in the μCRL process algebraic language [12]. The resulting mid-point process can be expanded to a (finite) state machine, if desired. Choosing μCRL for automatically constructing mid-point specifications has two benefits:

1. (Theoretical) The problem of constructing mid-point specifications is reduced to computing parallel compositions in our framework, hence relating the problem to a well-studied body of research. This simplifies the correctness proof for the construction, and also enables us to use bisimulation reductions for minimizing the mid-point processes output by the framework.
2. (Practical) The μCRL process algebra comes with a mature tool support [5,4,6]. This allows us to put the proposed framework immediately into practice: the μCRL toolset has been used for the case study reported in this paper.

We have carried out a case study on constructing a formal model for the mid-point that enforces the TCP three-way handshake protocol [9].

Related work. The closest related work is [3], where the authors give an algorithm for constructing mid-points, assuming that the specifications of the end-points are given as finite-state machines. Our framework is more general and more modular than the algorithm of [3]: (1) end-points are defined as finite-state machines in [3] while in our framework μCRL processes with recursive data types allow for a larger class of end-point specifications, and (2) the algorithm of [3] is tailored for a fixed type of channels while any μCRL process can model the channels in our framework. Thus, our algorithm can be directly applied to settings where different channels have different characteristics.

---

[1] The framework can be downloaded at www.infsec.ethz.ch/research/software

Bhargavan et al. [2] consider a problem which is related to, but nonetheless different from, the mid-point construction problem. In [2], the end-points are assumed to be connected directly via communication channels, and the authors consider the problem of automatically constructing specifications for *monitors* that observe the communication between the end-points. There is a significant difference between monitors and mid-points as the following simple example shows. Suppose that A and B communicate over asynchronous channels. The mid-point, mediating the communication between A and B, knows that if it has not forwarded a message m from A to B, then B could not have received m. However, the monitor, passively observing the communication between A and B, cannot know this: it could be that m has reached B, but m has not reached the monitor due to the asynchronous nature of communication.

Related areas are firewall testing [13,7], the extensive literature on test case generation from Mealy machines (e.g. see [17]), and testing TCP end-point automata [16]. In firewall testing a mid-point is tested. The previous works start with the firewall rules, while our focus is on the interactive nature of stateful firewalls. Test case generation from Mealy machines can be applied to the transition systems produced by our framework for testing mid-points. Testing TCP end-point automata is complementary to our work, as we consider constructing mid-point formal specifications that in turn can be used for testing TCP mid-points.

The remainder of this paper is organized as follows. In Section 2 we give a short introduction to the μCRL process algebra. In Section 3 we describe how we model communication protocols and their environments. In Section 4 we discuss the challenges in constructing mid-point specifications. In Section 5 we give formal definitions and in Section 6 we present our process algebraic framework. In Section 7 we present our case study on the TCP three-way handshake protocol and in Section 8 we draw conclusions. We prove the correctness of our method in Appendix A.

2 The μCRL Process Algebra

For specifying end-points, mid-points, and communication channels, we use the process algebra μCRL [12], which is an extension of the process algebra ACP [1] with abstract data types. Our results however do not depend on this choice in any crucial way, as μCRL is similar to other process calculi such as CSP. In what follows, we provide a brief introduction to μCRL. Its complete syntax and semantics are given in [12].

A μCRL specification consists of data type declarations and process behavior definitions, where processes and actions can be parameterized by data. Data is typed in μCRL and types can be recursive. Each non-empty data type has constructors and possibly non-constructors associated with it. The semantics of non-constructors is given by equations. The presence of a type Bool of Booleans with constants T and F as constructors, and the usual connectives $\land$, $\lor$ and $\neg$ as non-constructors, is always assumed.

A process is specified as a guarded recursive equation that is constructed from a finite set of action labels, process algebraic operators and recursion variables; mutual recursion among processes is allowed. The set of action labels is denoted Act. All members of Act, except for a designated action label τ for silent steps, may be parameterized with data to construct actions. The process algebraic operators $+$ and $\cdot$ denote non-deterministic choice and sequential composition, respectively: The process $p + q$ can behave either as process p or as process q, and the process $p \cdot q$ behaves as process p and when p terminates (if p ever does), it continues as process q. The constant δ denotes a deadlock process, i.e. one that cannot perform any actions. Recursion variables, which can be parameterized with data, are used in the natural way, e.g. $X = a \cdot X$, with $a \in Act$, describes a process that performs action a and then recurs, thereby performing an infinite number of a actions in sequence. A recursive equation is guarded if all its recursion variables are preceded by an action.

The parallel (asynchronous) composition $p\|q$ interleaves the actions of p and q. Moreover, actions from p and q may synchronize, when this is explicitly allowed by the predefined commutative and associative partial function $| : Act \times Act \rightarrow Act$. Two actions can synchronize only if their data parameters are semantically equal. This implies that synchronization can be used to represent data transfer between processes. Encapsulation $\partial_H(p)$, which renames all occurrences of actions from the set H in p to the deadlock action δ, can be used to force actions to communicate. For example, with $a, b, c \in Act$ and $a|b = c$, the process $(a.\delta)\|(b.\delta)$ behaves as $a.b.\delta + b.a.\delta + c.\delta$. Therefore, $\partial_{\{a,b\}}((a.\delta)\|(b.\delta)) = c.\delta$. The operator ρ is used for renaming: $\rho_{a \rightarrow b}(p)$ simultaneously renames all occurrences of action a to action b in process p.

The summation operator $\sum_{d:D} p(d)$, where d is a free variable in process $p(d)$, provides the possibly infinite choice over a data type D. The conditional construct $p \triangleleft b \triangleright q$, with $b : \mathsf{Bool}$, behaves as p if $b = \mathsf{T}$ and as q if $b = \mathsf{F}$. In particular, the construct $\sum_{d:D} p(d) \triangleleft f(d) \triangleright \delta$, with $f : D \rightarrow \mathsf{Bool}$, chooses values of $d \in D$ such that $f(d)$ is true. The operator $\cdot$ has the strongest precedence, the conditional construct binds stronger than $+$, and $+$ binds stronger than $\sum$.

A μCRL specification describes a labelled transition system (LTS) whose states represent process terms and edges are labelled with actions. The μCRL tool set [5,4], together with LTSmin [6] and CADP [10] which act as μCRL's back-ends, features visualization, simulation, symbolic reduction, (distributed) state space generation and reduction, model checking, and theorem proving capabilities for μCRL specifications.

3   Communication Protocols, Environments, and Mid-Points

Below, we fix a data type Msg for messages. Let the two end-points be indexed by $j \in \{1, 2\}$. Given an end-point j, we refer to its partner (the other) end-point by $\bar{j} = 3 - j$.

Communication protocols. Communication protocols typically describe the steps that the communication end-points take to achieve a common goal, e.g. to exchange data reliably. We therefore define a communication protocol Π as a pair (E^1, E^2), where E^j specifies the protocol for end-point j. Note that we are concerned with two-party communication protocols, as opposed to multi-party protocols. The specifications E^j are subject to a number of restrictions defined below. We define two *communication actions* for each end-point:

$$\mathsf{snd} : \{1, 2\} \times Msg$$
$$\mathsf{rcv} : \{1, 2\} \times Msg$$

Intuitively, $\mathsf{snd}(j, m)$ denotes the event of message m being sent to $E^{\bar{j}}$ (via the communication environment, as defined below), and $\mathsf{rcv}(j, m)$ denotes the event of message m

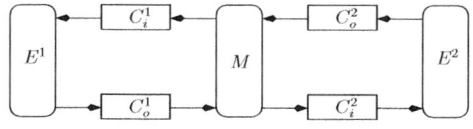

Fig. 1. The general setting: E^1 and E^2 are the end-points and M is the mid-point

with destination E^j being received. We assume that all non-silent actions appearing in E^j are either of the form $\mathrm{snd}(\bar{j}, m)$ or $\mathrm{rcv}(j, m)$, for $j \in \{1, 2\}$ and some $m \in Msg$. All internal actions of E^j are therefore modeled by the silent action τ.

Communication environments. Communication protocols are executed in communication environments. A communication environment is a set of channels $\{C^1, \cdots, C^n\}$, with $n > 0$. A channel's behavior can be formally specified as a μCRL process. Therefore, a communication environment Env is defined as a tuple $(C^1, \cdots, C^n)$, where C^i is the specification of channel i for $1 \leq i \leq n$ (see § 4.1 for examples). The specifications C^i are subject to a number of restrictions defined below. We define two *channel actions* for each channel:

$$\text{in} : \{1, \cdots, n\} \times Msg$$
$$\text{out} : \{1, \cdots, n\} \times Msg$$

Intuitively, $\mathrm{in}(i, m)$ with $1 \leq i \leq n$ and $m \in Msg$ denotes the event of message m being sent to channel i, and $\mathrm{out}(i, m)$ denotes the event of message m being received from channel i. We assume that all non-silent actions appearing in C^i are either of the form $\mathrm{in}(i, m)$ or $\mathrm{out}(i, m)$, for some $m \in Msg$. Any other action of channel i (e.g. dropping or duplicating messages) is therefore modeled as a silent step.

Mid-points. We assume that the mid-point is placed in the communication environment such that all the communication between the end-points passes through the mid-point. See Figure 1.

The communication protocol $\Pi = (E^1, E^2)$ is executed in environment Env by placing the channels C_i^j, C_o^j between E^j and M, as shown in Figure 1. We model the communication environment Env as a quadruple $(C_i^1, C_o^1, C_i^2, C_o^2)$. The subscript i denotes "input" and the subscript o denotes "output". We remark that each of the channels C_i^j, C_o^j may in reality consist of several channels linked together. In our model, say, C_o^1 is therefore the specification of a channel that simulates the behavior of all the channels that are used along the communication path that connects E^1 to M.

Note that the mid-point is assumed to be able to distinguish between messages arriving from different channels. In practice, the modeled environment is an IP network and the mid-point is placed such that it interconnects the networks of E^1 and E^2. The mid-point must be the only entity connecting the two networks to ensure that it can observe all messages exchanged by the end-points. Each network is connected on a different port, hence our assumption is reasonable.

Protocol specifications are usually informal. We however assume that a formal specification for the end-points E^1 and E^2 is available. The characteristics of the communication channels C_i^j, C_o^j, with $j \in \{1, 2\}$, are also assumed to be formally specified. In § 4.1, we give formal specifications for a number of common channel types, such as lossy channels and reliable asynchronous channels. Our goal is to automatically construct a formal specification for the mid-point M that enforces the protocol, given formal specifications for E^j, C_i^j, C_o^j, with $j \in \{1, 2\}$. The notion of *enforcement* is formally defined in § 5.

4   Challenges

In this section, we describe the main aspects that should be considered when constructing formal models for mid-points: *channel fidelity* and *non-determinism*.

4.1   Channels Fidelity

Channels fidelity refers to the fact that the sequence of events executed at the end-point and the sequence of events observed by the mid-point may differ depending on the characteristics of their communication environment. Depending on their properties, channels C_i^j, C_o^j distort the way the mid-point views the actions of E^j.

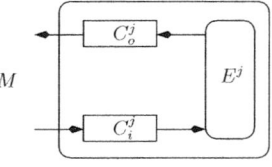

Fig. 2. Mid-point's view of end-point E^j

The mid-point views the actions of E^j via the "lenses" C_i^j and C_o^j; see Figure 2. We illustrate this with an example. Assume that the specification of the end-point E^1 is $E^1 = \text{rcv}(1, x) \cdot \text{snd}(2, y) \cdot \delta + \text{rcv}(1, x) \cdot \delta$. That is, E^1 receives message x and then sends message y, or it receives message x and then stops. Furthermore, assume that the channel C_i^1 is reliable, while the channel C_o^1 is lossy (i.e. it can lose messages). Assume also that the mid-point M sends x to C_i^1. As long as M does not receive message y on C_o^1, it does not know whether E^1 executes $\text{rcv}(1, x) \cdot \text{snd}(2, y) \cdot \delta$ or $\text{rcv}(1, x) \cdot \delta$. This is because message y can be lost by C_o^1 and therefore these two executions of E^1 are indistinguishable to the mid-point.

We remark that given formal specifications of the end-point E^j and the channels C_i^j and C_o^j, we can compute the behavior of the end-point as seen from the point of view of the mid-point; see § 6. As examples, the behavior of reliable, resilient, and lossy channels are formalized in μCRL below. We assume the data structures *Queue* and *Set* are given with their usual operators, which we use to model how channels store the messages passed to them. [2]

Reliable channel. Messages are not lost, duplicated, or reordered in this model. The channel stores messages in a queue. When a message is received, modeled by action $\text{in}(i, m)$, the reliable channel i inserts the message in the queue. When the queue is not empty, the channel removes the first message from the queue and delivers it via action $\text{out}(i, m)$. Below, we omit the name of the channel i from the action labels in and out.

$$C_{reliable}(Q : Queue) = \sum_{m:Msg} \text{in}(m) \cdot C_{reliable}(enqueue(Q, m))$$
$$+ \sum_{m:Msg} \text{out}(m) \cdot C_{reliable}(dequeue(Q)) \vartriangleleft m = head(Q) \vartriangleright \delta$$

---

[2] For a formal specification see www.infsec.ethz.ch/research/software

Models of reliable channels are useful, e.g., when the mid-point is co-located at one of the end-points. For such a mid-point, the sequence of observed events matches the sequence of events executed by the end-point.

Resilient channel. Messages are not lost, but they may be duplicated or reordered in transmission. The channel stores received messages in a set. A message may be delivered multiple times after it is inserted in the channel.

$$C_{resilient}(S : Set) = \sum_{m:Msg} \text{in}(m) \cdot C_{resilient}(S \cup \{m\})$$
$$+ $$
$$\sum_{m:Msg} \text{out}(m) \cdot C_{resilient}(S) \triangleleft m \in S \triangleright \delta$$

In practice, messages can be sent over different routes due to link failures, traffic load balancing, etc. This leads to messages arriving out of order, or multiple times, at the destination.

Lossy channel. Messages are lost and reordered, but are not duplicated. The channel stores messages in a multiset. When a message is in the multiset, it may be delivered or simply removed from the channel buffer.

$$C_{lossy}(S : Set) = \sum_{m:Msg} \text{in}(m) \cdot C_{lossy}(S \cup \{m\})$$
$$+ $$
$$\sum_{m:Msg} \text{out}(m) \cdot C_{lossy}(S \setminus \{m\}) \triangleleft m \in S \triangleright \delta$$
$$+ $$
$$\sum_{m:Msg} \tau \cdot C_{lossy}(S \setminus \{m\}) \triangleleft m \in S \triangleright \delta$$

In practice, channel have finite buffers; when their buffer is full the channels lose messages. Messages are also dropped in case of link failures.

4.2 Non-determinism

Non-determinism in specifications is generally used to allow different alternative behaviors. The alternative behaviors can model, e.g., under-specification (that is, the implementations can follow one or several of the provided alternatives) and abstraction (for instance, probabilistic choices can be modeled as non-deterministic choices).

Since the specifications of the end-points are given in μCRL in our framework, non-determinism in end-points can be naturally expressed using the choice operator $+$. For instance, consider the end-point specification $E^1 = \text{rcv}(1, x) \cdot (\text{snd}(2, y) + \text{snd}(2, z)) \cdot \delta$. That is, E^1 executes $\text{rcv}(1, x)$ and then non-deterministically executes $\text{snd}(2, y)$ or $\text{snd}(2, z)$. The mid-point needs to consider both the executions $\text{rcv}(1, x) \cdot \text{snd}(2, y)$ and $\text{rcv}(1, x) \cdot \text{snd}(2, z)$ as valid, since they comply with the specification of E^1.

5 Formal Definitions

We assume that the protocol specification $\Pi = (E^1, E^2)$, and the communication environment specification $Env = (C_i^1, C_o^1, C_i^2, C_o^2)$ are given in μCRL, and they conform to the restrictions specified in § 3. Our goal here is to define when a mid-point M enforces the protocol described by $\Pi = (E^1, E^2)$, executing in the communication environment described by $Env = (C_i^1, C_o^1, C_i^2, C_o^2)$. We first define how the protocol Π executes in the communication environment Env. We define the set of actions $Act = \{a : \{1, 2\} \times Msg \mid a \in \{\mathsf{snd}, \mathsf{rcv}, \mathsf{in}, \mathsf{out}, \alpha, \beta, \mathsf{com}\}\}$ and the synchronization rules $\mathsf{snd}|\mathsf{in} = \mathsf{com}, \mathsf{out}|\mathsf{rcv} = \mathsf{com}, \alpha|\beta = \mathsf{f}$. We define two processes P and Q that describe how E^1 and E^2 execute in the communication environment:

$$P = \tau_{\{\mathsf{com}\}} \partial_{\{Act\setminus\{\alpha,\beta,\mathsf{com}\}\}} (E^1 \| \rho_{\{\mathsf{out}\to\alpha\}} C_o^1 \| \rho_{\{\mathsf{in}\to\beta\}} C_i^1)$$
$$Q = \tau_{\{\mathsf{com}\}} \partial_{\{Act\setminus\{\alpha,\beta,\mathsf{com}\}\}} (E^2 \| \rho_{\{\mathsf{out}\to\alpha\}} C_o^2 \| \rho_{\{\mathsf{in}\to\beta\}} C_i^2)$$

Note that we rename the actions out and in in C_o^j and C_i^j to α and β, respectively, and force communication between α and β actions in order to link each input channel C_i^j to the output channel $C_o^{\bar{j}}$, for $j \in \{1, 2\}$; see Figure 3. Finally, we define our reference model R:

$$R = \partial_{\{\alpha,\beta\}}(P \| Q)$$

Intuitively, R describes how the mid-point observes the execution of E^1 and E^2 in the communication environment defined by Env.

We now define how arbitrary end-points, constrained by a mid-point M, execute in the communication environment. We assume the extreme case when the end-points arbitrarily execute snd and rcv actions over the set of messages Msg; we model this as $\perp^j = \sum_{m:Msg}(snd(\bar{j}, m) + \mathsf{rcv}(j, m)) \cdot \perp^j$. Let M be the mid-point process such that M executes only $\mathsf{f}(j, m)$ actions, for $j \in \{1, 2\}$ and $m \in Msg$. Action $\mathsf{f}(j, m)$ denotes that the mid-point forwards message m to end-point j. We define processes P' and Q' that describe how the arbitrary end-points execute in the communication environment:

$$P' = \tau_{\{\mathsf{com}\}} \partial_{\{Act\setminus\{\alpha,\beta,\mathsf{com}\}\}} (\perp^1 \| \rho_{\{\mathsf{out}\to\alpha\}} C_o^1 \| \rho_{\{\mathsf{in}\to\beta\}} C_i^1)$$
$$Q' = \tau_{\{\mathsf{com}\}} \partial_{\{Act\setminus\{\alpha,\beta,\mathsf{com}\}\}} (\perp^2 \| \rho_{\{\mathsf{out}\to\alpha\}} C_o^2 \| \rho_{\{\mathsf{in}\to\beta\}} C_i^2)$$

We set the synchronization rules to $\alpha|\mathsf{f} = c_1, \mathsf{f}|\beta = c_2, c_1|\beta = \lambda, \alpha|c_2 = \lambda$, and define our implementation model I:

$$I = \partial_{\{\alpha,\beta,\mathsf{f},c_1,c_2\}}(P' \| M \| Q')$$

Given the synchronization rules, a message delivered by an output channel (action α) is received by an input channel (action β) only after synchronizing with the mid-point (action f).

A symmetric binary relation B over processes is a *bisimulation relation* [15,14] iff $(P, P') \in B$ implies that for any action a and any message m, $P \xrightarrow{a(m)} P_1 \implies P' \xrightarrow{a(m)} P_1'$ with $(P_1, P_1') \in B$. Two processes P and P' are bisimilar, denoted $P \equiv P'$, iff there is bisimulation relation B such that $(P, P') \in B$. The Bisimilarity of two processes intuitively indicates that the two processes are indistinguishable from an observer's point of view. This is the core of our definition of enforcement.

Definition 1 (Enforcement). *Mid-point M enforces the communication protocol described by $\Pi = (E^1, E^2)$ in the communication environment described by $Env = (C_i^1, C_o^1, C_i^2, C_o^2)$ iff $I \equiv \rho_{\mathsf{f}\to\lambda} R$.*

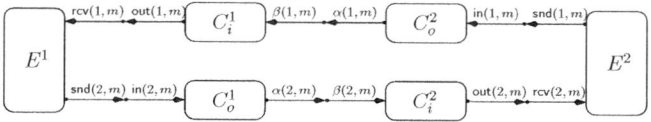

Fig. 3. μCRL action synchronization

Note that we rename action f to λ so that we can compare the implementation and the reference models. The intuition behind this definition is that if the reference and the implementation processes have executed the same protocol steps until some point in time and the reference process can continue the protocol execution with some step s, then the implementation process can also execute s. Conversely, if the implementation process can continue by taking a step s', then the reference process can also take s'.

6   The Framework

In this section we present our framework which computes a formal specification of the mid-point. The framework takes as an input the protocol specification $\Pi = (E^1, E^2)$, and the communication environment specification $Env = (C_i^1, C_o^1, C_i^2, C_o^2)$. Π and Env are both given in μCRL and must conform to the restrictions specified in § 3. In § 4.1 we provided several common channel specifications that can be used as an input to our framework. We remark that our framework is modular and each of the four channels can have a different specification. The mid-point specification computed by our framework enforces the communication protocol in the environment defined by Env. A message from E^j to $E^{\bar{j}}$ is allowed, i.e. forwarded to $E^{\bar{j}}$, if it could have been sent by E^j, and rejected otherwise. An incorrect message could result from a faulty end-point or due to communication channel noise.

We distinguish three steps performed in our framework. The first step (*construction*) takes as inputs the specifications of Π and Env given in μCRL and outputs a specification of M in μCRL. Step two (*minimization*) minimizes the state space of M using a branching bisimilarity algorithm. Optionally, the specification of M can be expanded to a finite state machine using a standard ϵ-removal algorithm in the third step. All three steps are automated using the μCRL toolset.

6.1   Mid-Point Construction

The mid-point construction computes a process that enforces the protocol executed by the two end-points. We define one *enforcement action* for the mid-point:

$$f : \{1, 2\} \times Msg$$

Intuitively, the action $f(j, m)$ denotes the event of message m being forwarded to end-point E^j for some message $m \in Msg$ and $j \in \{1, 2\}$. By forwarding a message m to E^j we mean that the mid-point receives a message on $C_o^{\bar{j}}$ and inserts it in channel C_i^j. To determine what messages should be forwarded by the mid-point, we compute the parallel composition of the input μCRL processes E^j, C_i^j, and C_o^j for $j \in \{1, 2\}$. We link channel C_o^1 to C_i^2 and channel C_o^2 to C_i^1, as illustrated in Figure 3. The channels are linked by renaming the action out in channels C_o^1 and C_o^2 to α, renaming the action in in channels C_i^1 and C_i^2 to β, and forcing communication between α and β actions. Given the synchronization between α and β actions, every message delivered by an output channel is inserted into the corresponding input channel.

We synchronize actions that must happen together. Figure 3 illustrates the actions performed by the end-points and the channel processes. We declare the following synchronization rules:

$$\mathsf{snd} \mid \mathsf{in} = \mathsf{com}$$
$$\mathsf{out} \mid \mathsf{rcv} = \mathsf{com}$$
$$\alpha \mid \beta = \mathsf{f}$$

snd | in enforces that output channel C_o^j receives a message from E^j only when E^j triggers a send message event (action snd); we synchronize these two actions to action com which denotes communication between an end-point and a channel. out | rcv enforces that end-point E^j receives a message from input channel C_i^j only when C_i^j triggers a deliver message event (action out). We also force communication between α and β actions to enforce that an input channel C_i^j gets a message from $C_o^{\bar{j}}$ only when $C_o^{\bar{j}}$ triggers a deliver message event (action α).

The mid-point process is synthesized by computing the parallel composition of the processes $E^1, E^2, C_i^1, C_o^1, C_i^2, C_o^2$ and then hiding all actions that are unobservable to the mid-point. Intuitively, the parallel composition of the input processes gives us a process that describes all possible protocol executions in the given environment. Hiding all actions unobservable by the mid-point gives us the mid-point's point of view of the protocol executions. The mid-point receives messages from C_o^1 and C_o^2, and sends messages to C_i^1 and C_i^2. Therefore, M observes the α and β events that are synchronized to action f and hence we do not hide action f. As an example, action $\mathsf{f}(1, m)$ indicates that upon receiving message m, the mid-point should forward it to end-point E^1. The mid-point cannot observe communication between an end-point and a channel and hence we hide the action com. We compute the mid-point process as follows:

$$M = \partial_{\{\alpha,\beta\}}(\ \tau_{\{\mathsf{com}\}}\partial_{\{Act\setminus\{\alpha,\beta,\mathsf{com}\}\}}(E^1 \| \rho_{\{\mathsf{out}\to\alpha\}}C_o^1(\emptyset) \| \rho_{\{\mathsf{in}\to\beta\}}C_i^1(\emptyset)) \|$$
$$\tau_{\{\mathsf{com}\}}\partial_{\{Act\setminus\{\alpha,\beta,\mathsf{com}\}\}}(E^2 \| \rho_{\{\mathsf{out}\to\alpha\}}C_o^2(\emptyset) \| \rho_{\{\mathsf{in}\to\beta\}}C_i^2(\emptyset)))$$

Theorem 1. M enforces *the communication protocol Π in the communication environment Env.*

Proof. We show that $I \equiv \rho_{\mathsf{f}\to\lambda}R$, where I is the implementation model and R is the reference model, both defined in § 5. According to Theorem 2 (given in Appendix A) $I \equiv \rho_{\mathsf{f}\to\lambda}M$ if $P \preceq P'$ and $Q \preceq Q'$, where P and Q are the processes that describe how E^1 and E^2 execute in Env, and P' and Q' describe how arbitrary end-points execute in Env; these processes are all defined in § 5. By construction, the mid-point process M is equivalent to the reference model R, therefore, $I \equiv \rho_{\mathsf{f}\to\lambda}R$ holds if $P \preceq P'$ and $Q \preceq Q'$. Recall that $P = \tau_G \partial_H(E^1 \| \rho_K C_o^1 \| \rho_L C_i^1)$ and $P' = \tau_G \partial_H(\bot^1 \| \rho_K C_o^1 \| \rho_L C_i^1)$ for $G = \{\mathsf{com}\}, H = \{Act\setminus\{\alpha,\beta,\mathsf{com}\}\}, K = \{\mathsf{out} \to \alpha\}$, and $L = \{\mathsf{in} \to \beta\}$. Using the fact that $(P\|X) \preceq (P'\|X)$ if $P \preceq P'$ (proved in Lemma 1 in Appendix A) and that $E^1 \preceq \perp^1$, we have $P \preceq P'$; analogously $Q \preceq Q'$. ☐

We remark that the computed mid-point is *permissive*: it forwards messages that could have resulted from correctly executing end-points. If the mid-point M receives a message sent by an intruder, and the mid-point cannot distinguish between the intruder's message and the end-point's message, M will forward the message. Constructing a permissive mid-point is the best we can do as we do not want to block legitimate messages and interfere with the protocol execution.

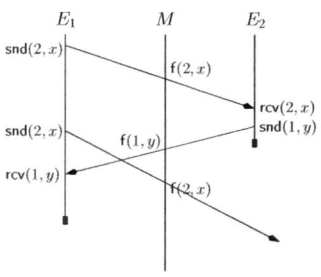

Fig. 4. A permissive mid-point

Note that when the mid-point forwards a message to E^j, there is no guarantee that E^j can receive the message. Using the end-point specifications we can compute a mid-point that blocks messages that cannot be received by the receiving end-point. We illustrate this observation using a simple example. Consider two end-points with specifications $E^1 = \mathrm{snd}(2, x) \cdot (\mathrm{rcv}(1, y) \cdot \delta + E^1)$ and $E^2 = \mathrm{rcv}(2, x) \cdot \mathrm{snd}(1, y) \cdot \delta$. E^1 repeatedly sends x to E^2 until it receives y from E^2, then terminates. E^2 receives x from E^1, sends y, and terminates. An acceptable execution is illustrated in Figure 4. The second x message from E^1 is forwarded to E^2, although E^2 has already terminated after sending message y to E^1.

6.2 State Space Minimization

The mid-point process M has a state space associated to it. The computation of M involves hiding all events that the mid-point cannot observe, which appear as τ events in M. Due to the τ events, the mid-point's state space can be large. We reduce the state space by applying branching bisimulation reduction on the mid-point process. The choice of branching bisimulation reduction is motivated by the fact that the notion of enforcement in our framework is based on bisimulation, and branching bisimulation reduction preserves the branching structure of processes while removing the action τ [11]. Our framework computes a process M', which is branching bisimilar to M. The state space of M' is potentially smaller than the state space of M, and M' is branching bisimilar to M (hence Theorem 1 holds for M' as well).

6.3 The Mid-Point as a State Machine

The μCRL specification of the mid-point is in the form of a linear process equation which can be automatically expanded to a state machine. The state space can be explored by a depth-first search. The generated state space contains τ transitions for all actions unobservable by the mid-point. To eliminate all τ transitions, we apply a standard ϵ-removal algorithm. The output is a state machine that can also be used as a mid-point specification. For example, Figure 5 (in Section 7) illustrates the mid-point state machine for enforcing the TCP three-way handshake protocol, output by our framework. Clearly, this step cannot be completed if the state space of the mid-point is infinite.

7   TCP Case Study

An evaluation on three popular firewalls (Checkpoint, netfilter/iptables, and ISA Server) shows that different firewall manufacturers implement mid-points for the TCP protocol differently and incorrectly [3], i.e. they forward messages that should not be sent by the end-points if they implement the protocol correctly. We performed a case study on the TCP protocol to demonstrate how our framework constructs a specification for a mid-point that enforces the protocol. The mid-point specification synthesized by our framework eliminates any ambiguities concerning which packets should be forwarded by the mid-point.

A formal mid-point specification has several applications in practice. It can be used for model-based testing in order to test an implementation for inconsistencies. The tester can use the mid-point specification to generate test cases and run them against the implementation. Additionally, when the mid-point specification is relatively simple, which is the case of the TCP mid-point, a software engineer can use the formal specification to perform code inspection, i.e. systematically examine the source code of the mid-point using the formal specification as a reference. Another application of our framework is model-driven development for mid-points, e.g., using the formal specification to automatically generate the implementation of a stateful TCP firewall.

Firewalls typically distinguish between internal and external networks. The policy for handling TCP connections initiated from the external network are usually handled differently from TCP connections initiated from the internal network. To reflect this, we take the TCP protocol specification [9] and construct two end-point specifications: one that models the *initiator* role and another that models the *responder* role. Below we give the specification of the two roles in μCRL, where we assume that E^1 represents the initiator role and E^2 the responder role.

Initiator end-point. It is the end-point that initiates a TCP connection. Below we give the μCRL specification for the initiator role:

$$
\begin{aligned}
E^1 = \ &\mathsf{snd}(2,\mathsf{syn}) \cdot \mathsf{rcv}(1,\mathsf{synack}) \cdot \mathsf{snd}(2,\mathsf{ack}) \cdot \\
&(\ \mathsf{rcv}(1,\mathsf{fin}) \cdot \mathsf{snd}(2,\mathsf{ack}) \cdot \mathsf{snd}(2,\mathsf{fin}) \cdot \mathsf{rcv}(1,\mathsf{ack}) \\
&\quad + \\
&\quad \mathsf{snd}(2,\mathsf{fin}) \cdot (\ \mathsf{rcv}(1,\mathsf{ack}) \cdot \mathsf{rcv}(1,\mathsf{fin}) \cdot \mathsf{snd}(2,\mathsf{ack}) \\
&\qquad\qquad + \\
&\qquad\quad \mathsf{rcv}(1,\mathsf{fin}) \cdot \mathsf{snd}(2,\mathsf{ack}) \cdot \mathsf{rcv}(1,\mathsf{ack}))) \cdot \delta
\end{aligned}
$$

Responder end-point. The responder end-point waits for an initiator end-point to open a TCP connection. The actions performed by the responder are symmetric to the initiator actions. We assume that the responder role can initiate a tear-down after it has sent a synack to E^1, i.e. before receiving an ack from E^1.

$$
\begin{aligned}
E^2 = \ &\mathsf{rcv}(2,\mathsf{syn}) \cdot \mathsf{snd}(1,\mathsf{synack}) \cdot (\mathsf{rcv}(2,\mathsf{ack}) \cdot E_T^2 + E_T^2) \\
E_T^2 = \ &\mathsf{rcv}(2,\mathsf{fin}) \cdot \mathsf{snd}(1,\mathsf{ack}) \cdot \mathsf{snd}(1,\mathsf{fin}) \cdot \mathsf{rcv}(2,\mathsf{ack}) \cdot \delta \\
&+ \\
&\mathsf{snd}(1,\mathsf{fin}) \cdot (\ \mathsf{rcv}(2,\mathsf{ack}) \cdot \mathsf{rcv}(2,\mathsf{fin}) \cdot \mathsf{snd}(1,\mathsf{ack}) \\
&\qquad\qquad + \\
&\qquad\quad \mathsf{rcv}(2,\mathsf{fin}) \cdot \mathsf{snd}(1,\mathsf{ack}) \cdot \mathsf{rcv}(2,\mathsf{ack})) \cdot \delta
\end{aligned}
$$

In our case study we assume that the environment can lose and reorder packets, but cannot duplicate messages. For the channel specification we use the μCRL specification of a lossy channel as defined in § 4.1. We compute M using our framework and perform the optional step 3 to expand the state space of M to a state machine, given in Figure 5. The input alphabet to the mid-point automaton is $f(j, m)$, $j \in \{1, 2\}, m \in \{\text{ack}, \text{synack}, \text{syn}, \text{fin}\}$. Action $f(j, m)$ denotes that M receives a message m from end-point $E^{\bar{j}}$ and forwards it to end-point E^j.

Although the end-points have a small number of non-deterministic choices in their specifications, the mid-point process can receive different types of messages in most states, as depicted in Figure 5. This is explained by the effect of the environment, which can reorder and drop messages. For instance, assume M

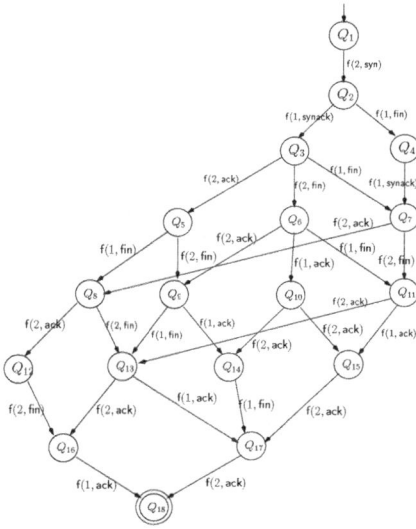

Fig. 5. Mid-point automaton for TCP

is in state Q_2, i.e. it has forwarded the initial syn message to E^2. E^2 replies to the syn with a synack message and afterwards it can send a fin. The network may reorder the two messages. Therefore, M would forward the fin message if it is received before the synack message.

The TCP specification computed by our framework is equivalent to the TCP automaton presented in [3]. The environment models in both case studies exercise the same properties, hence, the mid-point specification is identical, as expected. In contrast to [3] which fixes the behavior of the environment, we can easily modify the channel specifications and compute a mid-point specification for a different environment. For instance, suppose that the mid-point is co-located at one of the end-points, say E^1. To handle this scenario, we set the specifications of C_i^1 and C_o^1 to reliable channels and re-run our framework on the new inputs.

8   Conclusions and Future Work

We give a process algebraic approach to automatically synthesizing a formal specification for a mid-point that enforces a communication protocol. Formal mid-point specifications can be used for model-based testing, for model-driven development, and for formal verification of mid-points. In this paper we have systematically explored the aspects that must be considered when constructing formal models for mid-points. Our approach to handling these challenges can be applied to other related problems; for instance, our framework can be extended to synthesize specifications for passive monitors. Passive monitors are entities that observe messages exchanged over a channel and can be used to check security properties or to guard against network intrusion.

An interesting direction for future work is synthesizing more restrictive mid-points. As we mentioned in § 6, our current framework implementation computes mid-point specifications that are in some cases too permissive. For instance, forwarding a message to an end-point does not guarantee that the receiving end-point can actually receive the message. This may happen, e.g., when an end-point repeatedly re-transmits a message until receiving an acknowledgment from the other end-point or when a channel can duplicate messages. We can modify our framework to compute a mid-point process that forwards a message only if it could have been sent by the source end-point and it can be received by the destination end-point. We remark that such a mid-point achieves more than enforcing the protocol and can be seen as an additional optimization, e.g. to reduce network traffic.

Acknowledgments. The work has been supported by the EU FP7 projects SPACIOS (no. 257876).

References

1. Bergstra, J., Klop, J.: Algebra of communicating processes with abstraction. Theor. Comput. Sci. 37, 77–121 (1985)
2. Bhargavan, K., Chandra, S., McCann, P., Gunter, C.: What packets may come: Automata for network monitoring. In: POPL, pp. 206–219. ACM (2001)
3. von Bidder-Senn, D., Basin, D., Caronni, G.: Midpoints Versus Endpoints: From Protocols to Firewalls. In: Katz, J., Yung, M. (eds.) ACNS 2007. LNCS, vol. 4521, pp. 46–64. Springer, Heidelberg (2007)
4. Blom, S., Calamé, J.R., Lisser, B., Orzan, S., Pang, J., van de Pol, J., Dashti, M.T., Wijs, A.J.: Distributed Analysis with μCRL: A Compendium of Case Studies. In: Grumberg, O., Huth, M. (eds.) TACAS 2007. LNCS, vol. 4424, pp. 683–689. Springer, Heidelberg (2007)
5. Blom, S., Fokkink, W., Groote, J.F., van Langevelde, I., Lisser, B., van de Pol, J.: μCRL: A Toolset for Analysing Algebraic Specifications. In: Berry, G., Comon, H., Finkel, A. (eds.) CAV 2001. LNCS, vol. 2102, pp. 250–254. Springer, Heidelberg (2001)
6. Blom, S., van de Pol, J., Weber, M.: LTSmin: Distributed and Symbolic Reachability. In: Touili, T., Cook, B., Jackson, P. (eds.) CAV 2010. LNCS, vol. 6174, pp. 354–359. Springer, Heidelberg (2010)
7. Brucker, A., Brügger, L., Kearney, P., Wolff, B.: Verified firewall policy transformations for test case generation. In: ICST, pp. 345–354. IEEE Computer Society (2010)
8. Brucker, A.D., Brügger, L., Wolff, B.: Model-based firewall conformance testing. In: 8th International Workshop on Formal Approaches to Testing of Software, Tokyo, Japan, pp. 103–118 (2008)
9. Postel, J. (ed.): Transmission control protocol (1981)
10. Fernandez, J., Garavel, H., Kerbrat, A., Mounier, L., Mateescu, R., Sighireanu, M.: CADP - A Protocol Validation and Verification Toolbox. In: Alur, R., Henzinger, T.A. (eds.) CAV 1996. LNCS, vol. 1102, pp. 437–440. Springer, Heidelberg (1996)
11. van Glabbeek, R.: The linear time – branching time spectrum II. In: Best, E. (ed.) CONCUR 1993. LNCS, vol. 715, pp. 66–81. Springer, Heidelberg (1993)
12. Groote, J., Ponse, A.: The syntax and semantics of μCRL. In: Algebra of Communicating Processes 1994. Workshops in Computing Series, pp. 26–62. Springer, Heidelberg (1995); Also as technical report CS-R9076, CWI, Amsterdam, The Netherlands (December 1990)
13. Mayer, A., Wool, A., Ziskind, E.: Offline firewall analysis. Int. J. Inf. Sec. 5(3), 125–144 (2006)
14. Milner, R.: Communication and concurrency. PHI Series in computer science. Prentice Hall (1989)

15. Park, D.: Concurrency and automata on infinite sequences. In: Deussen, P. (ed.) TCS 1981. LNCS, vol. 104, pp. 167–183. Springer, Heidelberg (1981)
16. Paxson, V.: Automated packet trace analysis of TCP implementations. In: SIGCOMM, pp. 167–179. ACM Press (1997)
17. Utting, M., Legeard, B.: Practical Model-Based Testing: A Tools Approach. Morgan Kaufmann (2007)

A   Proof of Correctness

We start with a definition: A binary relation S over processes is a *simulation* relation iff $(P, P') \in S$ implies that $P \xrightarrow{a(m)} P_1 \implies P' \xrightarrow{a(m)} P_1'$, with $(P_1, P_1') \in S$, for all actions a and messages m. Process P simulates process P', denoted $P' \preceq P$, iff there is a simulation relation S such that $(P, P') \in S$.

Below, we fix

- The reference model: $M = \partial_{\{\alpha, \beta\}}(P\|Q)$ and $\alpha|\beta = f$.
- The implementation model: $I = \partial_{\{\alpha, \beta, f, c_1, c_2\}}(P'\|M\|Q')$ and the synchronization rules $\alpha|f = c_1, f|\beta = c_2, c_1|\beta = \lambda, \alpha|c_2 = \lambda$.

Theorem 2. $I \equiv \rho_{f \to \lambda} M$ if $P \preceq P'$ and $Q \preceq Q'$.

Proof. We define the relation B as $(S, S') \in B$ iff

$$S = \rho_{f \to \lambda} \partial_{\{\alpha, \beta\}}(P\|Q)$$

and $S' = \partial_{\{\alpha, \beta, f, c_1, c_2\}}(P'\|M\|Q')$ for all processes P, P', Q, Q' with $P \preceq P'$ and $Q \preceq Q'$. Below, we show that B is indeed a bisimulation relation. In the following we refer to the assumption $P \preceq P'$ and $Q \preceq Q'$ as the *simulation assumption*. We split the proof into two parts:

- Assume $S \xrightarrow{\lambda} S_1$. We claim $S' \xrightarrow{\lambda} S_1'$ and $(S_1, S_1') \in B$. Notice that in order for S to perform λ, $\partial_{\{\alpha, \beta\}}(P\|Q)$ must execute f, and in turn the processes P and Q must execute α and β respectively (the symmetric case is trivial; hence omitted here). Let $P \xrightarrow{\alpha} P_1$ and $Q \xrightarrow{\beta} Q_1$. Due to the simulation assumption, $P' \xrightarrow{\alpha} P_1'$ and $Q' \xrightarrow{\beta} Q_1'$ and $P_1 \preceq P_1'$ with $Q_1 \preceq Q_1'$. That is,

$$S' = \partial_{\{\alpha, \beta, f, c_1, c_2\}}(\alpha \cdot P_1'\|f \cdot \delta_{\alpha, \beta}(P_1\|Q_1)\|\beta \cdot Q_1')$$

  Given the aforementioned synchronization rules, we have $S' \xrightarrow{\lambda} S_1'$ where $S_1' = \partial_{\{\alpha, \beta, f, c_1, c_2\}}(P_1'\|\delta_{\alpha, \beta}(P_1\|Q_1)\|Q_1')$. It is immediate that $(S_1, S_1') \in B$.
- Assume $S' \xrightarrow{\lambda} S_1'$, with $S_1' = \partial_{\{\alpha, \beta, f, c_1, c_2\}}(P_1'\|\delta_{\alpha, \beta}(P_1\|Q_1)\|Q_1')$ for some P_1, Q_1, P_1' and Q_1'. We claim $S \xrightarrow{\lambda} S_1$ and $(S_1, S_1') \in B$. Notice that in order for S' to perform λ, the following two conditions must be satisfied:
  - The process $\partial_{\{\alpha, \beta\}}(P\|Q)$ must execute f. This implies that $P \xrightarrow{\alpha} P_1$ and $Q \xrightarrow{\beta} Q_1$ (the symmetric case is omitted here). Then it is immediate that $S \xrightarrow{\lambda} \partial_{\{\alpha, \beta\}}(P_1\|Q_1)$.

- Moreover, $P' \xrightarrow{\alpha} P'_1$ and $Q' \xrightarrow{\beta} Q'_1$ (the symmetric case is omitted). Due to the simulation assumption, $P_1 \preceq P'_1$ and $Q_1 \preceq Q'_1$. Now it is immediate that $(\partial_{\{\alpha,\beta\}}(P_1\|Q_1), S'_1) \in B$.

These two points prove our claim.

This completes the proof. □

Lemma 1. $(P\|X) \preceq (P'\|X)$ *for all* X, *if* $P \preceq P'$.

Proof. Let $P \preceq P'$ for some processes P and P'. We define the binary relation S over processes as: $((Q\|Z), (Q'\|Z)) \in S$ for all $Q \preceq Q'$ and any Z. Obviously we have $((P\|X), (P'\|X)) \in S$. Below, we show that S is indeed a simulation relation.

Suppose that $(P\|X) \xrightarrow{a} Y$. It must be that either P or X executed a, or that P and X executed some b and c, respectively, and $b|c = a$. We look at these three cases:

- $(P\|X) \xrightarrow{a} (P_1\|X)$. This implies that $P \xrightarrow{a} P_1$. Given that $P \preceq P'$ it follows that $P' \xrightarrow{a} P'_1$ and $P_1 \preceq P'_1$. It is immediate that $(P'\|X) \xrightarrow{a} (P'_1\|X)$. Clearly, $(P_1\|X, P'_1\|X) \in S$.
- $(P\|X) \xrightarrow{a} (P\|X_1)$. This implies that $X \xrightarrow{a} X_1$. Then, $(P'\|X) \xrightarrow{a} (P'\|X_1)$, and hence $(P\|X_1, P'\|X_1) \in S$.
- $(P\|X) \xrightarrow{a} (P_1\|X_1)$. This implies that $P \xrightarrow{b} P_1$ and $X \xrightarrow{c} X_1$, for some $b, c \in Act$ and $b|c = a$. Given that $P \preceq P'$, it follows that $P' \xrightarrow{b} P'_1$ and $P_1 \preceq P'_1$. It is immediate that $(P'\|X) \xrightarrow{a} (P'_1\|X_1)$. Hence $(P_1\|X_1, P'_1\|X_1) \in S$.

This completes the proof. □

Optimal Instrumentation of Data-flow in Concurrent Data Structures

Samaneh Navabpour[1], Borzoo Bonakdarpour[2], and Sebastian Fischmeister[1]

[1] Department of Electrical and Computer Engineering
University of Waterloo
200 University Avenue West
Waterloo, Ontario, Canada, N2L 3G1
{snavabpo,sfischme}@uwaterloo.ca
[2] School of Computer Science
University of Waterloo
200 University Avenue West
Waterloo, Ontario, Canada, N2L 3G1
borzoo@cs.uwaterloo.ca

Abstract. In this paper, we propose an automated technique for optimal instrumentation of *multi-threaded* programs for debugging and testing of concurrent data structures. We define a notion of *observability* that enables debuggers to trace back and locate errors through data-flow instrumentation. Observability in a concurrent program enables a debugger to extract the value of a set of *desired variables* through instrumenting another (possibly smaller) set of variables. We formulate an optimization problem that aims at minimizing the size of the latter set. In order to cope with the exponential complexity of the problem, we present a SAT-based solution. Our approach is fully implemented and experimental results on popular concurrent data structures (e.g., linked lists and red-black trees) show significant performance improvement in optimally-instrumented programs using our method as compared to ad-hoc over-instrumented programs.

Keywords: Debugging, testing, multi-thread, concurrent programs, instrumentation, optimization.

1 Introduction

Debugging is a systematic process of finding and reducing the number of defects in a computer program. Program debugging is a continual de facto step in the software development process and often requires significant human and computing resources. The debugging process ranges over a variety of techniques such as traditional or breakpoint-style debuggers, event monitoring systems, and static analysis for which different aspects and tools are employed. Incorporating these techniques mostly requires adding extra instructions to the program under scrutiny, called *instrumentation*.

A. Fernández Anta, G. Lipari, and M. Roy (Eds.): OPODIS 2011, LNCS 7109, pp. 497–512, 2011.
© Springer-Verlag Berlin Heidelberg 2011

The main problems associated with instrumenting (and, hence, debugging) programs are increased complexity, the *probe effect*, and non-repeatability. The probe effect refers to the problem that any attempt to observe the behavior of a system may change its behavior. Furthermore, such problems are amplified significantly in the context of concurrent programs. This is due to the fact that instrumenting these programs complicates their inherent non-deterministic nature, causing different executions for the same data and more unpredictable context switches.

Moreover, although there have been significant advances in the multi-core technology, it is currently unclear to what extent software products can be multi-threaded to take advantage of these new chips. Thus, in the presence of challenges in developing, testing, and maintaining scalable multi-threaded programs, having access to effective debugging tools for concurrent programs is highly beneficial. This benefit is even more crucial in the context of concurrent embedded safety-critical applications, where deviation of a mutated program from its specification may result in catastrophic consequences.

In [10], we proposed a notion of *observability* as the ability to test various features of a *sequential program* by observing the program's outcome to check if it conforms to the software's specification. The traditional methods for achieving observability incorporate ad-hoc instrumentation techniques [13,14,12] that cause the observed outcome of the software to be produced by a mutated program which can violate its correctness. Our approach to contain such probe effects in [10] is to introduce minimal instrumentation to sequential programs under debugging.

With this motivation, in this paper, we extend the concept of observability to the context of *concurrent programs*. Our contributions in this paper are as follows:

- We formally define the notion of observability for concurrent programs. This notion has a different nature as compared to sequential programs due to the existence of shared variables and interleaving scenarios. Roughly speaking, observability in a concurrent program enables a debugger to extract the value of a set of *desired variables* through instrumenting another set of variables. We call the latter the set of *naturally observable* variables.
- We formulate an optimization problem to tackle under and over-instrumentation defects. In other words, given a multi-threaded program and a set of desired variables, our goal is to identify the minimum set of naturally observable variables (that will be instrumented for debugging) through which one can extract the value of all desired variables.
- Since the complexity of our optimization problem is exponential, we encode the problem as a propositional satisfiability problem to leverage powerful SAT-solvers to solve our problem.
- Our method is fully implemented in a tool chain. We use LLVM [7] and the method presented in [9] to extract program data-flow dependencies. This is

achieved by implementing a new pass over LLVM that takes the source code and a set of desired variables as input and generates the full set of data-flow dependencies as output. Using the extracted dependencies, we automatically generate a SAT model which is the input to the SMT-solver Yices [1]. The solution to the SAT model is the set of variables that need to be instrumented (the naturally observable variables).

– We conduct experiments on two popular concurrent data structures: *linked lists* and *red-black trees*. We consider different implementations of these data structures with respect to different liveness criteria and synchronization primitives, such as lock-based, software transactional memory (STM), lock-free, and obstruction-free implementations. Our experiments show that our method effectively optimizes instrumentation instructions, resulting in significant performance improvement (in some cases up to 50 times), as compared to ad-hoc over-instrumented programs.

Organization. In Section 2, we present the preliminary concepts. Section 3 is dedicated to define our notion of observability in concurrent programs and the statement of our optimization problem to reduce instrumentation. We describe our approach to solve the optimization problem in Section 4. Section 5 presents the results of our experiments. Finally, we make concluding remarks and discuss future work in Section 6.

2   Preliminaries

2.1   Concurrent Control-Flow Graphs

Intuitively, a *concurrent control-flow graph* (CCFG) [8] is a control-flow graph which incorporates constructs to model concurrency. We use a cobegin/coend construct to express concurrent execution of threads. The cobegin/coend construct contains two or more blocks of code, which may in turn contain other cobegin/coend constructs.

Definition 1. *A Concurrent Control Flow Graph (CCFG) is a directed graph* $G = \langle N, A, n^0 \rangle$ *such that:*

– *N is the set of nodes in G. Each node is a* basic block. *Without loss of generality, we assume that each basic block contains only one instruction.*
– n^0 *is the initial node with indegree 0, which represents the initial basic block of G.*
– *A is a set of arcs* (n, m), *where* $n, m \in N$. *An arc* (n, m) *exists in A, iff the execution of basic block n in a thread thr encoded in G immediately leads to the execution of basic block m in thread thr.*   □

For the sake of clarity, we distinguish different types of basic blocks in a CCFG. These types include *Entry, Exit, Cobegin, Coend, Compute, ThreadEntry*, and *ThreadExit*. Arcs that involve any type of basic block except for *Compute* are

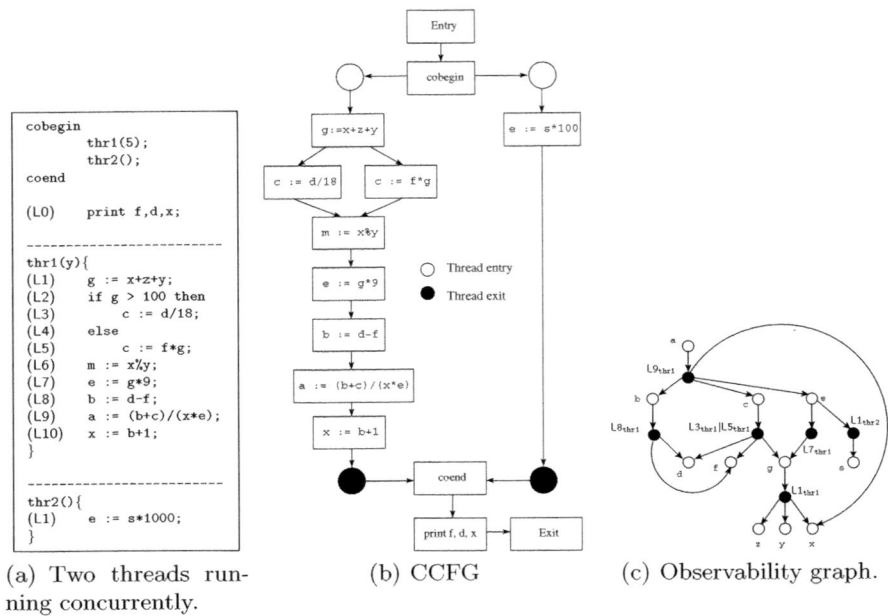

(a) Two threads run-
ning concurrently.

(b) CCFG

(c) Observability graph.

Fig. 1. A C program and its concurrent control-flow graph

trivially added to a CCFG. Arcs which involve *Compute* basic blocks are specified in Definition 1. Figure 1(a) shows an example, where the program consists
of two threads thr1 and thr2 running concurrently. We consider variables with
the same name in different threads as shared variables. For instance, variable
e is a shared variable. The CCFG of the program in Figure 1(a) is shown in
Figure 1(b).

Notation: Let $\pi = n_1, n_2, \ldots, n_k$ be a sequence of nodes of a CCFG G and *thr*
be a thread encoded in G. By $thr(\pi)$, we denote the sequence of nodes where
the nodes of all threads except for *thr* are removed from π.

Definition 2. *Let $G = \langle N, A, n^0 \rangle$ be a CCFG. An execution path π of G between two nodes m and m' in N is a sequence $n_1, n_2, \ldots, n_k$, such that:*

1. *$n_1 = m$ and $n_k = m'$,*
2. *for all threads thr of G, we have: (i) $thr(\pi) = x_1, x_2, \ldots, x_j$ is a total order,
   and (ii) for all i, where $1 \leq i \leq j - 1$, $(x_i, x_{i+1}) \in A$, and*
3. *the causal relation between nodes in π is a partial order.*   □

Intuitively, in Definition 2, Condition 2 requires that the order of basic blocks
of the same thread in π must follow the isolated sequential execution of that
thread. Moreover, Condition 3 expresses that the order of basic blocks of a set
of concurrent threads in π must be in an ordered interleaving fashion.

Notation: To distinguish instructions of different threads, we denote an instruction of a thread thr by line_number$_{\text{thr}}$. When the thread name is irrelevant or clear from the context, we omit it.

For example, an execution path of the CCFG in Figure 1(b) is L1$_{\text{thr1}}$, L2$_{\text{thr1}}$, L3$_{\text{thr1}}$, L6$_{\text{thr1}}$, L1$_{\text{thr2}}$, L7$_{\text{thr1}}$, L8$_{\text{thr1}}$, L9$_{\text{thr1}}$, L10$_{\text{thr1}}$, L0.

2.2   Data-Flow Dependencies in Concurrent Programs

Definition 3. *We say that the value of a variable v depends on the value of variable v' iff $v = F(v', V)$, where F is an arbitrary function and V is the remaining set of F's arguments called* parameters *[2].*   □

In a source code, any instruction that updates the value of a variable creates a dependency. For instance, in function thr1, in Figure 1(a), instruction L9 creates a dependency between variable a and variables b, x, e, and c. We represent a dependency by a tuple $\langle v, n, v' \rangle$, where n is an instruction (i.e., a node in the corresponding CCFG) and v is a variable whose value can be extracted from variable v' via instruction n. In this case, we say that instruction n *defines* the value of variable v. In our example, we have $\langle$a, L9$_{\text{thr1}}$, b$\rangle$ and $\langle$a, L9$_{\text{thr1}}$, c$\rangle$.

Some data dependencies are resolved based upon runtime circumstances, e.g., in conditional statements. For instance, in thr1, the value of variable c at line L9 is defined by either L3 or L5. As a result, it is incorrect to have the dependency $\langle$c, L3$_{\text{thr1}}$, d$\rangle$, as we cannot determine whether this dependency indeed holds at run time. Thus, in order to compute data dependencies statically, we consider a conservative set of instructions that can define the value of a variable at run time. Hence, we require a representation that conveys that c depends on variable d or on variables f and g.

In order to resolve this issue, we combine dependencies that define the same variable into one dependency as follows. When a variable v is defined via instructions n_1 and n_2, where $\langle v, n_1, v_1 \rangle$ and $\langle v, n_2, v_2 \rangle$ are in two separate and mutually exclusive conditional branches, we combine instructions n_1 and n_2 into one instruction $n' = n_1 \mid n_2$. Hence, we replace dependencies $\langle v, n_1, v_1 \rangle$ and $\langle v, n_2, v_2 \rangle$ with $\langle v, n', v_1 \rangle$ and $\langle v, n', v_2 \rangle$. This implies that v depends on v_1 when one of the two instructions n_1 or n_2 execute. The same concept applies for the dependency between v and v_2. In our example, for dependency $\langle$c, L3$_{\text{thr1}}$, d$\rangle$, we have $\langle$c, L3$_{\text{thr1}}$ | L5$_{\text{thr1}}$, d$\rangle$. In other words, c may depend on d if L3$_{\text{thr1}}$ or L5$_{\text{thr1}}$ execute. In this case, parameters of L3$_{\text{thr1}}$ | L5$_{\text{thr1}}$ is the set {d, g, f}.

During program execution, the value of a variable depends on variables used/defined by a sequence of instructions leading to the instruction that defines the variable. For instance, the value of variable a defined by L9$_{\text{thr1}}$ indirectly depends on: (1) variables d and f used by instruction L8$_{\text{thr1}}$ which defines b, (2) variables d, f, and g used by L3$_{\text{thr1}}$ | L5$_{\text{thr1}}$ which define c, (3) variable g used by instruction L7$_{\text{thr1}}$ which defines e, and (4) variables x, z, and y used by L1$_{\text{thr1}}$ which defines g.

```
cobegin
        thr2();
coend
cobegin
        thr1(5);
coend
```

Fig. 2. Two threads running sequentially

Definition 4. *Let G be a CCGF and v be a variable. A* dependency chain *for v is a sequence* $\sigma = \langle v_1, n_1, v_2 \rangle \langle v_2, n_2, v_3 \rangle \ldots \langle v_{k-1}, n_{k-1}, v_k \rangle$ *of dependencies where:*

- $v_1 = v$, *and*
- *the sequence* $n_k, n_{k-1}, \ldots, n_2, n_1$ *is an execution path of G between basic blocks* n_1 *and* n_k *[2].* □

Clearly, determining data dependencies relies on the structure of the source code. For example, in the source code of Figure 1(a) dependency chains $\sigma_1 = \langle a, L9_{thr1}, e \rangle \langle e, L1_{thr2}, s \rangle$ and $\sigma_2 = \langle a, L9_{thr1}, e \rangle \langle e, L7_{thr1}, g \rangle$ are both possible. This is caused by the fact that threads thr1 and thr2 run concurrently. However, if we change the structure as shown in Figure 2, then dependency σ_1 is invalid while σ_2 is still valid.

Typically, one does not need to enumerate all dependency chains of a variable in order to extract the value of that variable. In other words, we only need to identify a subset of all dependency chains that is *maximal*.

Definition 5. *Let* $\mathcal{S}_v$ *be a set of dependency chains for a variable v. We say that* $\mathcal{S}_v$ *is a* maximal dependency set *for v iff for all dependency chains* $\sigma \in \mathcal{S}_v$, *there does not exist a dependency chain* σ', *where* $\sigma\sigma' \in \mathcal{S}_v$. □

For example, $\mathcal{S}_a = \{\langle a, L9_{thr1}, b \rangle, \langle a, L9_{thr1}, b \rangle \langle b, L8_{thr1}, d \rangle, \langle a, L9_{thr1}, x \rangle, \langle a, L9_{thr1}, e \rangle\}$ is *not* a maximal dependency chain of a, as the dependency chain $\langle a, L9_{thr1}, b \rangle$ is a prefix of the dependency chain $\langle a, L9_{thr1}, b \rangle \langle b, L8_{thr1}, d \rangle$. To convert $\mathcal{S}_a$ into a maximal dependency set, we must either remove $\langle a, L9_{thr1}, b \rangle$ or $\langle a, L9_{thr1}, b \rangle \langle b, L8_{thr1}, d \rangle$ from $\mathcal{S}_a$.

Intuitively, a program slice [9] for a variable v is a maximal dependency chain set that covers all dependency chains that start with v.

Definition 6. *Let* $\mathcal{S}_v$ *be a maximal set of dependency chains for a variable v. We say that* $\mathcal{S}_v$ *is the* program slice *for v iff there does not exist a dependency chain* σ *for v, such that* $\sigma\sigma'$ *is not in* $\mathcal{S}_v$ *for some* σ'. □

For example, the program slice for variable a includes all dependency chains for a built from instructions $L9_{thr1}$, $L8_{thr1}$, $L7_{thr1}$, $L5_{thr1}$, $L3_{thr1}$, $L1_{thr1}$, and $L1_{thr2}$.

3   Observability in Concurrent Programs

Definition 7. *A value of a variable v is* naturally observable   *iff   the value is an output or input of the system.*   □

For example, in Figure 1(a), variables d,f,x, and y are naturally observable. Note that *instrumented* variables are considered as program outputs and, hence, naturally observable variables as well.

We now use the notion of program slices in Definition 6 to define what it means for a variable to be observable. Intuitively, a variable is observable if there exists a *sub-slice* of the variable, where each dependency chain in the sub-slice ends with a variable that is naturally observable.

Definition 8. *A sub-slice S' of a slice S is a set of dependency chains, where each chain in S' is a prefix of a chain in S.*   □

To motivate the idea of observability, notice that given the value of naturally observable variables d, f, x, and y in our running example, we cannot extract the value of a using a's program slice. To extract the values of a, we require the values of b, c, e, and x. Variable a does not have a dependency with the value of x printed on line L0, since a uses the value of x before it is redefined at line L10 and printed on L0. On the other hand, a has dependencies with d and f via variables b and c at lines L8, L5, and L3. Moreover, d and f can only lead to extracting the value of b and c at lines L8 and L3. Hence, the value of c is still unknown at L9 since we can not predict if c will be defined via line L3 or L5 at runtime. As a result, we cannot guarantee determining the value of c at line L9 without having the value of g. In addition, based on thr1's code, it is clear that d, f, and y can not be used to extract values of e and x at line L9. Hence, the values of x, e, and c are still required for extracting the value of a. Therefore, no sub-slice of a provides enough information to extract a's value.

We now formally define the constraints that need to be satisfied to extract the value of a variable in concurrent programs.

Definition 9. *A sub-slice S is* complete   *iff*

1. *for each dependency $\langle v, n, v' \rangle$ in a dependency chain of S, there exists a dependency $\langle v, n, v'' \rangle$ in at least one dependency chain of S, for each variable v'' in n's parameter set.*
2. *for every dependency prefix $\langle v, n_{thr}, sv \rangle \langle sv, m_{thr}, v' \rangle$ in S, if there exists another thread thr' running concurrently with thr that contains an instruction of the form:*
   $$(\mathsf{L})\quad sv := F(v'');$$
   i.e., sv is a shared variable also defined by thr', then there must exist a dependency chain $\sigma' \in S$ that contains $\langle sv, \mathsf{L}_{thr'}, v'' \rangle$.   □

For example, the sub-slice $S_a = \{\langle a, L9_{thr1}, b \rangle, \langle a, L9_{thr1}, x \rangle, \langle a, L9_{thr1}, e \rangle\}$ is not complete, since it violates both Conditions 1 and 2 of Definition 9. To satisfy Condition 1, we add dependency $\langle a, L9_{thr1}, c \rangle$ to S_a and to satisfy Condition

2, we add dependency $\langle \mathtt{a}, \mathtt{L1}_{\mathtt{thr2}}, \mathtt{e} \rangle$ to $\mathcal{S}_{\mathtt{a}}$, as $\mathtt{e}$ is defined by $\mathtt{thr2}$ which runs concurrently with $\mathtt{thr1}$.

Definition 10. *A variable v is* observable *iff there exists a complete sub-slice $\mathcal{S}_v$ where:*

- *every dependency chain $\sigma \in \mathcal{S}_v$ ends in a naturally observable variable, and*
- *every shared variable sv in $\mathcal{S}$ is naturally observable.*

We call $\mathcal{S}_v$ an observable sub-slice. $\qquad\qquad\qquad\qquad\qquad\qquad$ □

To clarify the need for Condition 2 in Definition 10, consider Figure 1(a). In order to observe the value of variable $\mathtt{a}$, we require the value of variable $\mathtt{e}$. Variable $\mathtt{e}$ is updated by both lines $\mathtt{L7}_{\mathtt{thr1}}$ and $\mathtt{L1}_{\mathtt{thr2}}$. Since $\mathtt{thr1}$ and $\mathtt{thr2}$ run concurrently, we can not predict which of the two lines $\mathtt{L7}_{\mathtt{thr1}}$ or $\mathtt{L1}_{\mathtt{thr2}}$ is last to update $\mathtt{e}$. Hence, we need to explicitly extract the time at which both lines execute, so one can determine which instruction defines the value of $\mathtt{e}$ used at line $\mathtt{L9}_{\mathtt{thr1}}$. As a result, we need to explicitly make $\mathtt{e}$ naturally observable at both lines $\mathtt{L7}_{\mathtt{thr1}}$ and $\mathtt{L1}_{\mathtt{thr2}}$ to be capable of observing the time at which the instructions execute and consequently extract which instruction was the last to define $\mathtt{e}$.

Problem Statement. As mentioned earlier, in addition to inputs and outputs, we consider *instrumented* variables as naturally observable variables, as their value can be explicitly observed. When program development is divided among multiple development groups, the program may suffer from *under-* or *over-instrumentation* defects caused by developers due to lack of knowledge about other developments.

Our goal is to optimize data-flow instrumentation in concurrent programs to tackle over- and under-instrumentation defects. Informally, given a set of *desired variables* required for debugging, our goal is to find the minimum set of variable in the program that should be made naturally observable (i.e., instrumented), so that the set of desired variables become observable. Formally, we aim at solving the following optimization problem:

Given a concurrent program and a set V of desired variables to be made observable, decide whether there exists a set of variables V', where $|V'| \leq k$ for some positive integer k, such that by making variables in V' naturally observable, there exists an observable sub-slice $\mathcal{S}_v$ for all $v \in V$.

---

4 Approach

In this section, we propose our approach to solve the optimization problem, introduced in Section 3. Our method consists of three steps: (1) extracting program slices of variables required to be observed (i.e., desired variables), (2) building a graph representation of slices, and (3) transforming the optimization problem using the graph built in Step 2 into a satisfiability decision problem. These steps are discussed in Subsections 4.1, 4.2, and 4.3, respectively.

4.1   Extracting Program Slices

Given a concurrent program and a set V of desired variables, we first extract the program slices of V from the Static Single Assignment (SSA) [4] representation of the program by leveraging the slicing algorithm proposed in [9]. Our slicing approach takes the following steps for all $v \in V$:

1. We find the threads, say *thr*, that execute instructions defining v. Then, we extract the dependency chains of v by only using instructions of *thr*; i.e., we do not expand the chains over different threads. We use a reachability algorithm [11] to extract these chains.
2. For every chain σ found in Step 1, we extract the last variable v' of σ. We check if v' is defined by an instruction of a thread, say *thr'*, which does not run concurrently with *thr*. If so, we find dependency chains of v' in *thr'* using the method in Step 1. Subsequently, we append the newly extracted chains to σ and create a new set of chains which we add to the set of dependency chains of v. We repeat this step until no new chains are created.
3. For every chain σ identified in Steps 1 and 2, we extract the instructions, say L, in σ which use a shared variable *sv*. Next, we extract the threads, say *thr*, that execute instruction L. Then, we check if *sv* is defined by instructions executed by a different thread, say *thr'*, that runs concurrently with *thr*. If so, we find the instructions, say L', that define *sv* in *thr'*. We subsequently check if data dependency is possible from the *sv* used in L to *sv* defined in L'. To this end, we perform a lightweight static analysis to take synchronization issues into account. For instance, if L and L' are both protected with the same lock (e.g., a mutex or transaction), then data dependency between L and L' can be eliminated. If a dependency is possible, we apply Steps 1 and 2 to extract the corresponding dependency chains for *sv* in *thr'*. Then, we append *sv*'s dependency chains to σ and create a new set of chains which we add to the set of dependency chains of v. We repeat this step until no new chains are created.
4. Finally, we test whether each dependency chain σ found for v is indeed possible by checking if there exists an execution path in the CCFG of the program that creates σ. If not, we discard σ from the set.

4.2   Building Observability Graph

Let v be a desired variable and $\mathcal{S}_v$ be the program slice for $v \in V$. In order to find the minimum number of variables for instrumentation in a systematic fashion, we build the *observability graph* [10] that encodes program slice $\mathcal{S}_v$. Let $\mathcal{V}_{\mathcal{S}_v}$ be the set of all variables involved in $\mathcal{S}_v$ and $\mathcal{I}_{\mathcal{S}_v}$ be the set of all instructions involved in $\mathcal{S}_v$. We construct the observability graph $\mathcal{G} = \langle V_{\mathcal{G}}, A_{\mathcal{G}} \rangle$ as follows.

- *(Vertices)* $V_{\mathcal{G}} = C_{\mathcal{G}} \cup U_{\mathcal{G}}$, where $C_{\mathcal{G}} = \{c_i \mid i \in \mathcal{I}_{\mathcal{S}_v}\}$ and $U_{\mathcal{G}} = \{u_v \mid v \in \mathcal{V}_{\mathcal{S}_v}\}$. We call the set $C_{\mathcal{G}}$, *context vertices* and the set $U_{\mathcal{G}}$, *variable vertices*.
- *(Arcs)* $A_{\mathcal{G}} = \{(u, c) \mid u \in U_{\mathcal{G}} \wedge c \in C_{\mathcal{G}} \wedge$ variable u is defined by context $c\} \cup \{(c, u) \mid u \in U_{\mathcal{G}} \wedge c \in C_{\mathcal{G}} \wedge$ variable u is used by context $c\}$.

For example, the observability graph (for simplicity constructed from the original source code and not from its SSA mode) of variable a is presented in Figure 1(c). For instance, context vertex $L9_{thr1}$ shows dependency of a to variables (directly) b, c, e and (indirectly) x. Also, Figure 1(c) shows that shared variable e affects the value of a through instruction $L1_{thr2}$ as well.

In the context of an observability graph, notice that a variable vertex v is observable if there exists a context vertex c, such that (1) (v, c) is an arc in the graph, and (2) all variable vertices, say v', are observable, where (c, v') is an arc in the graph. Thus, our objective is to find the minimum number of variable vertices of the graph whose instrumentation makes the root vertex of the graph observable.

4.3   SAT-Based Optimization

In [10], we prove that the optimization problem for observability graphs of sequential programs is NP-complete. Thus, in the context of concurrent programs, the problem involves two exponential blow-ups: one for computing program slices [9] (and, hence, an observability graph), and (2) solving the optimization problem [10]. In order to cope with the second exponential blow-up, we transform our optimization problem into the propositional satisfiability problem (SAT).

Let $\mathcal{G} = \langle V_{\mathcal{G}}, A_{\mathcal{G}} \rangle$ be an observability graph and $V' \subseteq V_{\mathcal{G}}$ represents the set of desired variables. We include the following variables:

- $X = \{x_v \mid v \in V_{\mathcal{G}}\}$: each variable vertex v is mapped to a Boolean variable x_v, where $x_v = true$ if v is observable and $false$ otherwise.
- $Z = \{z_v \mid v \in V_{\mathcal{G}}\}$: each variable vertex v is mapped to a Boolean variable z_v, where $z_v = true$ if v is instrumented and $false$ otherwise.
- $Q = \{q_c \mid c \in (C_{\mathcal{G}} \cup C_H)\}$: each context vertex c is mapped to a Boolean variable q_c, where $q_c = true$ if all variables used by c are observable and $false$ otherwise. In addition, $C_H = \{c_v \mid v \in V_{\mathcal{G}}\}$ contains context vertices for each variable $v \in V_{\mathcal{G}}$ representing $hypothetical$ instructions that would instrument v; i.e., such instrumentations do not exist in the original code and will only be added to the code if v is chosen to be instrumented. Each context vertex $c \in C_H$ is mapped to a Boolean variable q_c: the value of $q_c = true$ if c is added to the code to instrument the corresponding variable and $false$ otherwise.
- $Y = \{y_z \mid z \in Z\}$: for each variable $z \in Z$, we include an integer variable y_z for our optimization objective.

Constraints on variable vertices.   Obviously, every desired variable must be observable. Hence, we add the following constraint for each $v \in V'$:

$$x_v \iff true.$$

Moreover, each variable $x_v \in X$ is $true$ if and only if the value of the variable v is observable via the context vertex that defines v or by instrumenting v:

$$x_v \iff (q_{c_v} \vee q_{c'_v}),$$

where c_v is the context vertex that defines v and c'_v is the instruction that instruments v. Finally, we require that $y_z \in Y$, where $z \in Z$, has value 1 when $z = true$ and has value 0 otherwise:

$$(y_z = 1) \iff z \qquad \text{and} \qquad (y_z = 0) \iff \neg z.$$

Constraints on context vertices. The value of a variable $q_c \in Q$, where $c \in C_{\mathcal{G}}$, must be $true$ if and only if all the variables used by c are observable:

$$q_c \iff \bigwedge_{v \in V_c} x_v,$$

where $V_c = \{u \mid (c, u) \in A_{\mathcal{G}}\}$. On the other hand, if $c_v \in C_H$, then q_c must be $true$ if and only if when v is instrumented; i.e., when $z_v = true$.

$$q_{c_v} \iff z_v$$

In this expression, when $z_v = true$ an instruction is added to the code to instrument v.

Optimization objective. Following the optimization criterion presented in Section 3, we require that the number of variables to be instrumented is not greater than K:

$$\sum_{y \in Y} y \leq K,$$

for some positive integer K specified by the user.

5   Experiments

Our goal in this section is twofold: (1) to demonstrate the effectiveness of our method through measuring the number of instrumentations removed from an *over-instrumented* program after applying our method, and (2) to evaluate the impact of our approach by studying the performance (i.e., execution time) of optimally instrumented programs as compared to their over-instrumented versions. We note that in the over-instrumented versions, any instruction that can change the state of concurrent data structures is instrumented for debugging purposes.

Our approach is implemented in a tool chain consisting three phases:

1. First, we implement a new pass over LLVM [7] that takes a program's source code and the set of desired variables as input. The pass extracts program slices using the method described in 4.1 and the static single assignment (SSA) mode of the program code. We currently do not handle alias and pointer data structures.
2. Given the extracted program slices, we transform the respective optimization problem into a SAT formula in the input language of our SAT-solver using the method described in Section 4.3.
3. We solve the generated SAT model using the Yices SMT-solver [1]. The solution presents the set of variables that need to be instrumented in the source code (set of naturally observable variables).

5.1 Experimental Setup

Our two case studies are concurrent implementations of *linked-lists* and *red-black trees*. We use regular arrays to eliminate possible pointer analysis. In both test cases, *insert*, *delete*, and *search* operations can run concurrently, where a thread inserts data elements (i.e., *producer*) and a thread deletes data elements (i.e., a *consumer*). In addition, we consider the following synchronization methods for each case study:

1. **Lock-based.** These algorithms employ blocking data structures (e.g., semaphores and mutexes) to enforce linearizable insertion and deletion. In particular, we use our mutex-based implementation of concurrent linked lists and the algorithm in [5] for concurrent red-black trees (with 656 lines of code and 57 desired variables).

2. **Non-blocking.** This group of solutions ensures that threads competing for a shared resource do not have their execution indefinitely postponed by mutual exclusion. We use the following implementation for our experiments:
   - We use the *lock-free* algorithm in [6] for concurrent linked lists (with 1180 lines of code and 85 desired variables) implemented by the CAS (compare-and-swap) operation. Lock-free algorithms ensure that if the program threads run sufficiently long at least one of the threads makes progress.
   - We use the *obstruction-free* algorithm in [3] for concurrent linked lists (with 797 lines of code and 71 desired variables) implemented by virtual locks and the CAS operation. Obstruction-free algorithms guarantee that at any point, a single thread executed in isolation (i.e., with all obstructing threads suspended) for a bounded number of steps will complete its operation.
   - Algorithms based on *software transactional memory* (STM) hide synchronization issues from the programmer; i.e., the programming language provides the programmer with atomic constructs in which reading and writing shared variables take place. In particular, we use our own STM-based implementation of concurrent linked lists and the algorithm in [5] for concurrent red-black trees (with 730 lines of code and 56 desired variables).

The set of desired variables in our experiments include the data contained in the linked-list/red-black tree at any point of execution and the temporary variables used in search, addition, and deletion operations. Thus, instrumenting all these variables is likely to result in over-instrumentation.

Parameters that affect the execution time of experiments are: (1) number of producer and consumer threads, (2) number of insert and delete operations, (3) type of data elements (e.g., long, short, int), (4) time consumed by each instrumentation instruction, (5) number of shared variables, and (6) structure of the source code (i.e., synchronization method). In our experiments, we keep the number of producer and consumer threads, type of data elements, and number

Table 1. Detailed numbers for the instrumented version with I/O delay= $100\mu s$ and number of insert operations = 200

Application	Concurrency	Count	Mean	Median	SEM	CI-95	Min	Max
1 Linked-list	Nested-locks	5	14.13	14.01	0.22	0.30	13.65	14.94
2 Linked-list	Lock-free [6]	5	5526.37	5529.79	6.53	9.06	5502.49	5539.07
3 Linked-list	Obstruction-free [3]	5	2686.26	2683.78	8.01	11.10	2663.26	2707.92
4 Linked-list	STM	5	257.13	258.05	1.14	1.58	252.72	258.90
5 Red-black tree	Nested-locks [5]	20	3.95	3.95	0.00	0.00	3.95	3.95
6 Red-Black tree	STM [5]	10	4.46	4.46	0.00	0.00	4.46	4.46

of shared variables as constants. The rest are obviously variables in our experiments. In particular, we incorporate different numbers of insert operations to study the impact of our optimization on long running programs. Different durations of the instrumentation instruction show the impact of our method for different instrumentation technologies. For instance, `printf()` statements normally take $80\mu s$, whereas EEPROM data logs take $1ms$. All experiments in this section are run on workstations ranging from a Core2Duo to a Core I3 quad-core machines with sufficient memory. Each test series is completed on the same machine, hence, the execution-time measurements from one test series are comparable.

We measure the execution time using the well-accepted utility time. Since individual measurements can be inaccurate (due to context switches or I/O operations between the program and time), we repeated each experiment several times and carried out solid statistical analysis. We have collected sufficient data to have representative and robust results. While we cannot provide the key metrics for all individual data points, Table 1 shows the results for the data series with the least number of values—it is the series where several configurations take more than an hour to complete. In Table 1, SEM and CI-95 abbreviate Standard Error of the Mean and 95% Confidence Interval, respectively. We also performed the following consistency checks on the data: measurements must be positive and growing with respect to the number of insert operations and the amount of I/O delay.

Table 2. Reduction in Instrumentation

Application	Concurrency	Original Inst.	Optimized Inst.
1 Linked-list	Nested-locks	43	20
2 Linked-list	Lock-free [6]	49	23
3 Linked-list	Obstruction-free [3]	42	24
4 Linked-list	STM	28	15
5 Red-black tree	Nested-locks [5]	320	205
6 Red-black tree	STM [5]	294	189

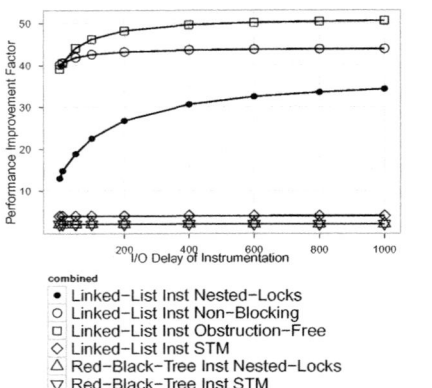

(a) Performance improvement vs. instrumentation instruction I/O delay with constant number of insert operations (= 100)

(b) Performance improvement vs. the number of insert operations with constant instrumentation I/O delay (= $100\mu s$).

Fig. 3. Performance evaluation of instrumentation optimization

5.2   Results and Analysis

Reduction in number of instrumentations.   We apply our method to over-instrumented implementations to optimize the instrumentations of the source code. Table 2 shows that our method achieves a 45% reduction on average across our case studies. Although the set of desired variables is common among different implementations, we observe different reductions in instrumentation, as the amount of reduction depends on the structure of the code. For instance, we do not require instrumentation in the atomic sections of STM-based algorithms, since the changes are local to the threads and do not affect shared variables; i.e., we only need to instrument the values committed into the shared variables by the threads (experiments 4 and 6). On the other hand, in the lock-free and obstruction-free implementations, we require more instrumentation due to lack of synchronization and more possible interleaving scenarios that must be observed. In nested-locks implementations, since the changes carried out in between the locks directly affect shared variables, we need to instrument the code in between the locks. Hence, it requires more instrumentation as compared to STM-based algorithms, but less as compared to non-blocking algorithms, as they have less interleaving scenarios.

Enhancement in performance.   We now compare the performance of over-instrumented test cases against the performance of their optimally instrumented versions in terms of execution time. In the first set of experiments (see Figure 3(a)), we compare the performance of the case studies, where the I/O delay (time consumption) of instrumentation instructions varies from $1\mu s$ to $1ms$, while the number of insert operations is constant ($= 100$). We have collected 1692 data samples from these experiments for statistical soundness (discussed later). Obviously, Figure 3(a) shows that the performance of optimally instrumented implementations is significantly better than the over-instrumented versions. For example, the performance of both red-black tree implementations improve with a factor of two. The reason behind this small improvement is that our method is forced to place 48% of the required instrumentation in loop structures to make desired variables observable. Hence, the effect of the I/O delay of each instrumentation on the performance is multiplied by the loop counts. On the other hand, Figure 3(a) shows a 40 times performance improvement in lock-free and obstruction-free implementations, as the majority of the instrumentations introduced by our method reside outside loop structures. In general, the improvement factor differs from one case study to another, since instrumentation locations tightly depend upon the structure of the source code. In addition, the results show that the improvement factor in performance is insensitive to different durations for instrumentation instructions.

In the second set of experiments, we compare the execution time of the case studies, where the number of concurrent insertions in the linked-list/red-black tree varies from 1 to 1000 (see Figure 3(b)). We collected 966 data samples from these experiments for statistical soundness. Obviously, Figure 3(a) shows that the performance of the optimally instrumented implementations is better than over-instrumented versions. The results show that we achieve a small improvement in the performance of red-black tree implementations due to the same reason discussed in the previous experiment. In addition, the results show that the improvement in performance in each case study is insensitive to the number of insertions, although the improvement factor differs from one case study to another (as the improvement factor depends upon the source code structure).

6   Conclusion and Future Work

In this paper, we introduced an automated technique to optimize instrumentation of *multi-threaded* programs to achieve software *observability*. Intuitively, observability in a concurrent program enables a debugger to extract the value of a set of *desired variables* through instrumenting another (possibly smaller) set of variables, called *naturally observable*. Thus, our optimization method identifies the minimum set of naturally observable variables whose instrumentation makes the value of desired variables extractable. Since our optimization problem is NP-complete, we encoded the problem as a propositional satisfiability problem (SAT) to leverage powerful SAT-solvers to tackle our problem. In our tool chain, we used LLVM and a slicing algorithm to extract program data-flow dependencies. Our experimental results on concurrent linked lists and red-black trees using different concurrency techniques show significant gains (up to 50 times) in performance of optimally-instrumented programs using our method as compared to ad-hoc over-instrumented programs.

For future work, we are considering two main research directions: (1) techniques for solving the optimization problem more efficiently, and (2) extending the concept of observability to other domains. Another direction is to devise probabilistic methods that make desired variables observable with certain probabilities.

Acknowledgement. This research was supported in part by NSERC DG 357121-2008, ORF RE03-045, ORE RE04-036, ORF-RE04-039, ISOP IS09-06-037, APCPJ 386797-09, and CFI 20314 with CMC.

References

1. Yices: An SMT Solver, http://yices.csl.sri.com
2. Ammann, P., Offutt, J.: Introduction to Software Testing. Cambridge University Press, New York (2008)
3. Attiya, H., Hillel, E.: Built-In Coloring for Highly-Concurrent Doubly-Linked Lists. In: Dolev, S. (ed.) DISC 2006. LNCS, vol. 4167, pp. 31–45. Springer, Heidelberg (2006)
4. Cytron, R., Ferrante, J., Rosen, B.K., Wegman, M.N., Zadeck, F.K.: Efficiently Computing Static Single Assignment Form and the Control Dependence Graph. ACM Transactions on Programming Languages and Systems 13(4), 451–490 (1991)
5. Fraser, K., Harris, T.: Concurrent programming without locks. ACM Trans. Comput. Syst. 25(2) (2007)
6. Harris, T.L.: A Pragmatic Implementation of Non-Blocking Linked-Lists. In: Welch, J.L. (ed.) DISC 2001. LNCS, vol. 2180, pp. 300–314. Springer, Heidelberg (2001)
7. Lattner, C., Adve, V.: Llvm: A compilation framework for lifelong program analysis and transformation. In: CGO, p. 75 (2004)
8. Lee, J., Midkiff, S.P., Padua, D.A.: A constant propagation algorithm for explicitly parallel programs. International Journal of Parallel Programming 26(5), 563–589 (1998)
9. Nanda, M.G., Ramesh, S.: Slicing concurrent programs. In: ISSTA, pp. 180–190 (2000)
10. Navabpour, S., Bonakdarpour, B., Fischmeister, S.: Software debugging and testing using the abstract diagnosis theory. In: LCTES, pp. 111–120 (2011)
11. Ottenstein, K.J., Ottenstein, L.M.: The program dependence graph in a software development environment. In: SDE, pp. 177–184 (1984)
12. Schmid, U.: Monitoring of Distributed Real-Time Systems. Real-Time Systems 7(1), 33–56 (1994)
13. Thane, H., Hansson, H.: Towards Systematic Testing of Distributed Real-Time Systems. In: RTSS, pp. 360–369 (1999)
14. Thane, H., Sundmark, D., Huselius, J., Pettersson, A.: Replay Debugging of Real-Time Systems Using Time Machines. In: IPDPS, p. 8 (2003)

Fault-Tolerant Aggregation: Flow-Updating Meets Mass-Distribution*

Paulo Sérgio Almeida[1], Carlos Baquero[1], Martín Farach-Colton[2], Paulo Jesus[1], and Miguel A. Mosteiro[3]

[1] Depto. de Informática (CCTC-DI), Universidade do Minho, Braga, Portugal
{psa,cbm,pcoj}@di.uminho.pt
[2] Dept. of Computer Science, Rutgers University, Piscataway, NJ, USA & Tokutek, Inc.
farach@cs.rutgers.edu
[3] Dept. of Computer Science, Rutgers University, Piscataway, NJ, USA & LADyR, GSyC, Universidad Rey Juan Carlos, Madrid, Spain
mosteiro@cs.rutgers.edu

Abstract. Flow-Updating (FU) is a fault-tolerant technique that has proved to be efficient in practice for the distributed computation of aggregate functions in communication networks where individual processors do not have access to global information. Previous distributed aggregation protocols, based on repeated sharing of input values (or *mass*) among processors, sometimes called Mass-Distribution (MD) protocols, are not resilient to communication failures (or *message loss*) because such failures yield a loss of mass.

In this paper, we present a protocol which we call *Mass-Distribution with Flow-Updating (MDFU)*. We obtain MDFU by applying FU techniques to classic MD. We analyze the convergence time of MDFU showing that stochastic message loss produces low overhead. This is the first convergence proof of an FU-based algorithm. We evaluate MDFU experimentally, comparing it with previous MD and FU protocols, and verifying the behavior predicted by the analysis. Finally, given that MDFU incurs a fixed deviation proportional to the message-loss rate, we adjust the accuracy of MDFU heuristically in a new protocol called *MDFU with Linear Prediction (MDFU-LP)*. The evaluation shows that both MDFU and MDFU-LP behave very well in practice, even under high rates of message loss and even changing the input values dynamically.

Keywords: Aggregate computation, Distributed computing, Radio networks, Communication networks.

---

* This work is supported in part by the Comunidad de Madrid grant S2009TIC-1692, Spanish MICINN grant TIN2008–06735-C02-01, Portuguese FCT grants PTDC/EIA-EIA/104022/2008 and SFRH/BD/33232/2007, and National Science Foundation grant CCF-0937829.

A. Fernández Anta, G. Lipari, and M. Roy (Eds.): OPODIS 2011, LNCS 7109, pp. 513–527, 2011.

1   Introduction

The distributed computation of algebraic aggregate functions is particularly challenging in settings where the processing nodes do not have access to global information such as the input size. A good example of such scenario is Sensor Networks [1, 28] where unreliable sensor nodes are deployed at random and the overall number of nodes that actually start up and sense input values may be unknown. Under such conditions, well-known techniques for distributing information throughout the network such as Broadcast [22] or Gossiping [12] cannot be directly applied, and data collection is only practicable if aggregation is performed. Even more challenging is that loss of messages between nodes or even node crashes are likely in such harsh settings. It has been proved [3] that the problem of aggregating values distributedly in networks where processing nodes may join and leave arbitrarily is intractable. Hence, arbitrary adversarial message loss also yields the problem intractable, but a weaker adversary, for instance a stochastic one as in Dynamic Networks [8], is of interest. In this paper, under a stochastic model of message loss, we study communication networks where each node holds an input value and the average of those values [1] must be obtained by all nodes, none of whom have access to global information of the network, *not even the total number of nodes n.*

A classic distributed technique for aggregation, sometimes called **Mass-Distribution** (MD) [11], works in rounds. In each round, each node shares a fraction of its current average estimation with other nodes, starting from the input values [4, 6, 7, 20, 27, 29, 32, 33]. Details differ from paper to paper but a common problem is that, in the face of message loss, those protocols either do not converge to a correct output or they require some instantaneous failure detector mechanism that updates the topology information at each node in each round. Recently [18, 19], a heuristic termed **Flow-Updating** (FU) addressed the problem assuming stochastic message loss [19], and even assuming that input values change and nodes may fail [18]. The idea underlying FU is to keep track of an aggregate function of all communication for each pair of communicating nodes, since the beginning of the protocol, so that a current value at a node can be re-computed from scratch in each round. Empirical evaluation has shown that FU behaves very well in practice [18, 19], but such protocols have eluded analysis until now.

In this paper, we introduce the concept of FU to MD. First, we present a protocol that we call **Mass-Distribution with Flow-Updating** (MDFU). The main difference with MD is that, instead of computing incrementally, the average is computed from scratch in each round using the initial input value and the accumulated value shared with other nodes so far (which we refer to as either **mass shared**, or **flow passed**). The main difference with FU is that if messages are not lost the algorithm is exactly MD, which facilitates the theoretical analysis of the convergence time under failures parameterized by the failure probability (or **message-loss rate**).

---

[1] Other algebraic aggregate functions can be computed in the same bounds using an average protocol [7, 20].

Our results. We first leverage previous work on bounding the mixing time of Markov chains [30] to show that, for any $0 < \xi < 1$, the convergence time of MDFU under reliable communication is $2\ln(n/\xi)/\Phi(G)^2$, where $\Phi(G)$ is the conductance of the underlying graph characterizing the execution of MDFU on the network. Then, we show that, with probability at least $1 - 1/n$, for a message-loss rate $f < 1/\ln(2\Delta e)^3$, the multiplicative overhead on the convergence time produced by message loss is less than $1/(1 - \sqrt{f\ln(2\Delta e)^3})$, and it is constant for $f \leq 1/(e(2\Delta e)^e)$, where Δ is the maximum number of neighbors of any node. Also, we show that, with probability at least $1 - 1/n$, for any $0 < \xi < 1$, after convergence the expected average estimation at any node is in the interval $[(1 - \xi)(1 - f)\overline{v}, (1 + \xi)\overline{v}]$. This is the first convergence proof for an FU-based algorithm.

In MDFU, if some flow is not received, a node computes the current estimation using the last flow received. Thus, in presence of message loss, nodes do not converge to the average and only some parametric bound can be guaranteed as shown. Aiming to improve the accuracy of MDFU, we present a new heuristic protocol that we call **MDFU with Linear Prediction** (MDFU-LP). The difference with MDFU is that if some flow is not received a node computes the current estimation using an estimation of the flow that should have been received.

We evaluate MDFU and MDFU-LP experimentally and find that the performance of MDFU is comparable to FU and other competing algorithms under reliable communication. In the presence of message loss, the empirical evaluation shows that MDFU behaves as predicted in the analysis converging to the average with a bias proportional to the message-loss rate. This bias is not present in the original FU, which converges to the correct value even under message loss. In a third set of evaluations, we observe that MDFU-LP converges to the correct value even under high message loss rates, with the same speed as under reliable communication. We also test MDFU under changing input values to verify that it tolerates dynamic changes in practice, in contrast to classic MD algorithms, which need to restart the computation each time values are changed.

Roadmap. In Section 2 we formally define the model and the problem, and we give an overview of related work. Section 3 includes the details of MDFU and its analysis, whereas its empirical evaluation is covered in Section 4. In Section 5 we present the details of MDFU-LP and its experimental evaluation. Section 6 evaluates MDFU in a dynamic setting, where input values change over time.

2 Preliminaries

Model. We consider a static connected communication network formed by a set V of n processing **nodes**. We assume that each node has an identifier (ID). Any pair of nodes $i, j \in V$ such that i may send messages to j without relying on other nodes (one hop) are called **neighbors**. We assume that the IDs are assigned so that each node is able to distinguish all its neighbors. The set of ordered pairs of neighbors (or, **edges**) is called E. The network is symmetric, meaning that, for any $i, j \in V$, $(i, j) \in E$ if and only if $(j, i) \in E$. The set of neighbors of a given node i is denoted as N_i and $|N_i|$ is called the **degree** of i. For each pair of nodes $i, j \in V$, the maximum degree between i, j is denoted as $D_{ij} = \max\{|N_i|, |N_j|\}$. The maximum degree throughout the network is denoted as $\Delta = \max_{i \in V} |N_i|$. Each node i knows N_i and D_{ij} for each $j \in N_i$, but does not know the size of the whole network n. The time is slotted in **rounds** and each round is divided in two **phases**. In each round, a node is able to send (resp. receive) one message to (resp. from) all its neighbors (communication phase) and to perform local computations (computation phase). However, for each $(i, j) \in E$ and for each communication phase, a message from i to j is lost independently with probability f. This is a crucial difference with previous work where, although edge-failures are considered, messages are not lost thanks to the availability of some failure detection mechanism. More details are given in the previous work section. Nodes are assumed to be reliable, i.e. they do not fail.

Problem. Each node i holds an input value v_i, for $1 \le i \le n$. The aim is for each node to compute the average $\overline{v} = \sum_{i=1}^{n} v_i / n$ without any global knowledge of the network. We focus on the algorithmic cost of such computation, counting only the number of rounds that the computation takes after simultaneous startup of all nodes, leaving aside medium access issues to other layers. This assumption could be removed as in [11].

Previous Work. Previous work on aggregate computations has been particularly prolific for the area of Radio Networks, including both theoretical and experimental work [9, 13, 14, 15, 16, 17, 20, 21, 23, 24, 25, 26, 35]. Many of those and other aggregation techniques exploit global information of the network [11, 13, 23, 24], or are not resilient to message loss [4, 6, 20].

FU is a recent fault-tolerant approach[18, 19] inspired on the concept of flows (from graph theory). Like common MD techniques, it is based on the execution of an iterative averaging process at all nodes, and all estimates eventually converge to the system-wide average. MD protocols exchange "mass", which lead them to converge to a wrong result in the case of message loss. In contrast, FU does not exchange "mass". Instead it performs idempotent flow exchanges which provide resilience against message loss. In particular, FU keeps the initial input value at each node unchanged (in a sense, always conserving the global mass), exchanging and updating flows between neighbors for them to produce a new estimate. The estimate is computed at each node from the input values and the contribution of the flows. No theoretical bounds on the performance of the algorithm were provided. Empirical evaluation shows that FU performs better than classic MD algorithms, especially in low-degree networks, and it supports high levels of message loss [19]. Moreover, it self-adapts to dynamic changes (i.e. nodes leaving/arriving and input value change) without any restart mechanism (like other approaches), and tolerates node crashes [18].

MD protocols for average computations in arbitrary networks based on gossiping (exchange values in pairs) were studied in [4, 20]. Results in [4] are presented for all gossip-based algorithms by characterizing them by a matrix that models how the algorithm evolves while sharing values in pairs iteratively.

As in our results, the time bounds shown are given as a function of the spectral decomposition of the graph underlying the computation. The work is focused on optimizing distributedly the spectral gap, in order to minimize convergence time. The dynamics of the model are motivated by changes in topology induced by nodes leaving and joining the network. Those changes may be introduced in the probability of establishing communication between any two nodes. However, the delivery of messages has to be reliable to ensure mass conservation. An algorithm called Push-Sum that takes advantage of the broadcast nature of Radio Networks (i.e., it is not restricted to gossip) is included in [20], yielding similar bounds. Chen, Pandurangan, and Hu [6] present an MD algorithm that first builds a forest over the network, where each root collects the information, and then a gossiping algorithm among the roots is used. The authors show a reduction on the energy consumption with respect to the uniform gossip algorithm. On the other hand, the MD algorithm presented in [7] relies on a different randomly chosen local leader in each round to distribute values. The bounds given are also parameterized by the eigen-structure of the underlying graph. This result was extended more recently [5] to networks with a time-varying connection graph, but the protocol requires to update the matrix underlying such graph in each round.

MD protocols have been used also for Distributed Average Consensus [27, 29, 31, 32, 33, 34] within Control Theory, but they do not apply to our model. For example, in [33, 34] the model includes unreliable communication links, but the algorithm requires instantaneous update of the topology information held at each node at the beginning of each round. Others, either rely on similar features [27, 29, 31] or do not consider changes in topology at all [32].

The common problem in all the MD protocols is that they are not resilient to message loss, because it implies a loss of mass. Hence, if messages are lost, they need to restart the computation from scratch. In MDFU, message loss has an impact on convergence time, which we show to be small, but the computation recovers from those losses, yielding the correct value. In fact, it is this characteristic of MDFU and FU in general what makes the technique suitable for dynamic settings in which the input values change with time.

3   MDFU

As in previous work [4, 7, 11, 20], MDFU is based on repeatedly sharing among neighbors a fraction of the average estimated so far. Unlike in those papers, in MDFU the estimation is computed from scratch in each round, as in FU [18, 19]. For that purpose, each node keeps track of the cumulative value passed to each neighbor (or, cumulative flow) since the protocol started. Together with the original input value, those flows allow each node to recompute the average estimation in each round. Should some flow from node i to node j be lost, j temporarily computes the estimation using the last flow received from i. Further details can be found in Algorithm 1.

Recall that the aim is to compute the average $\overline{v} = \sum_{i=1}^{n} v_i/n$ of all input values. Let $e_i(r)$ be the average ***estimate*** of node i in round r, and

---

Algorithm 1. MDFU. Pseudocode for node i. e_i is the estimate of node i. $F_{in}(j)$ is the cumulative inflow from node j. $F_{out}(j)$ is the cumulative outflow to node j.

```
// initialization
1  e_i ← v_i
2  foreach j ∈ N_i do
3      F_in(j) ← 0
4      F_out(j) ← e_i/(2D_ij)

5  foreach round do
       // communication phase
6      foreach j ∈ N_i do
7          Send j message ⟨i, F_out(j)⟩
8      foreach ⟨j, F⟩ received do
9          F_in(j) ← F
       // computation phase
10     e_i ← v_i + ∑_{j∈N_i}(F_in(j) − F_out(j))
11     foreach j ∈ N_i do
12         F_out(j) ← F_out(j) + e_i/(2D_ij)
```

$\varepsilon(r) = \max_i\{|e_i(r) - \overline{v}|/\overline{v}\}$ be the maximum relative ***error*** of the average estimates in round r. We want to bound the number of rounds after which the maximum relative error is below some parametric value ξ.

In each round, a node shares a fraction of its current estimate with each neighbor. Therefore, the execution of each round can be characterized by a ***transition matrix***, denoted as $\mathbf{P} = (p_{ij})$, $\forall i, j \in V$, such that for any round r where messages are not lost

$$p_{ij} = \begin{cases} 1/(2D_{ij}) & \text{if } i \neq j \text{ and } (i,j) \in E, \\ 1 - \sum_{k \in N_i} 1/(2D_{ik}) & \text{if } i = j, \\ 0 & (i,j) \notin E \end{cases}$$

and $\mathbf{e}(r+1) = \mathbf{e}(r)\mathbf{P}$, where $\mathbf{e}(\cdot)$ is the row vector $(e_1(\cdot)e_2(\cdot)\ldots e_n(\cdot))$.

3.1   Convergence Time for $f = 0$

Consider first the case when the communication is reliable, that is $f = 0$. Then, the above characterization is round independent and, given that $\mathbf{P}$ is stochastic, it can be seen as the transition matrix of a time-homogeneous Markov chain $(X_r)_{r=1}^{\infty}$ with finite state space V. Furthermore, $(X_r)_{r=1}^{\infty}$ is irreducible, and aperiodic, then it is ergodic and it has a unique stationary distribution. Given that $\mathbf{P}$ is doubly stochastic such stationary distribution is $\pi_i = 1/n$ for all $i \in V$. Thus, bounding the convergence time of $(X_r)_{r=1}^{\infty}$ we have a bound for the convergence time of MDFU without message loss. The following notation will be useful. Let G be a weighted undirected graph with set of nodes V and where, for each pair $i, j \in V$, the edge (i, j) has weight $\pi_i p_{ij}$. G is called the underlying graph of the Markov chain $(X_r)_{r=1}^{\infty}$. The following quantity characterizes the likelihood that the chain does not stay in a subset of the state space with small stationary probability. Let the **conductance** of graph G be

$$\Phi(G) = \min_{\substack{\emptyset \subset S \subset V \\ \sum_{i \in S} \pi_i \leq 1/2}} \frac{\sum_{i,j \in S} p_{ij} \pi_i}{\sum_{i \in S} \pi_i}.$$

The following theorem shows the convergence time of MDFU with reliable communication parameterized in the conductance of G.

Theorem 1. *For any communication network of n nodes running MDFU, for any $0 < \xi < 1$, and for $r_c = 2\ln(n/\xi)/\Phi(G)^2$, if $f = 0$, it holds that $\varepsilon(r) \leq \xi$ for any round $r \geq r_c$, where $\Phi(G)$ is the conductance of the underlying graph characterizing the execution of MDFU on the network.*

Proof. We want to find a value of r_c such that for all $r \geq r_c$ it holds that $\max_i\{|e_i(r) - \overline{v}|/\overline{v}\} \leq \xi$. Then, we want $\max_i\{|e_i(r)/\sum_{j \in V} v_j - 1/n|\} \leq \xi/n$. Given that $e_i(r) = \sum_{j \in V} v_j (\mathbf{P}^r)_{ji}$, it is enough to have $\max_{j,i \in V}\{|(\mathbf{P}^r)_{ji} - 1/n|\} \leq \xi/n$. On the other hand, given that $p_{ij}\pi_i = p_{ji}\pi_j$ for all $i, j \in V$, the Markov chain is time-reversible. Then, as proved in [30], it is $\max_{i,j \in V} |(\mathbf{P}^r)_{ij} - \pi_j|/\pi_j \leq \lambda_1^r / \min_{j \in V} \pi_j$, where λ_1 is the second largest eigenvalue of $\mathbf{P}$ (all the eigenvalues of $\mathbf{P}$ are positive because $p_{ii} \geq 1/2$ for all $i \in V$). Given that $\pi_i = 1/n$ for all $i \in V$, we have $\max_{i,j \in V} |(\mathbf{P}^r)_{ij} - 1/n| \leq \lambda_1^r$. Thus, from the inequality above, it is enough to have $\lambda_1^r \leq \xi/n$. As proved also in [30], given that $(X_r)_{r=1}^\infty$ is ergodic and time-reversible, it is $\lambda_1 \leq 1 - \Phi(G)^2/2$. Then, it is enough $(1 - \Phi(G)^2/2)^r \leq \xi/n$. Given that $\Phi(G) \leq 1$, using that $1 - x \leq e^{-x}$ for $x < 1$, the claim follows.

3.2   Convergence Time for $f > 0$

Mixing time of a multiple random walk. Recall that we carry out an average computation of n input values where each node i shares a $1/(2D_{ij})$ fraction of its estimate in each round of the computation with each neighboring node j. We have characterized each round of the computation with a transition matrix $\mathbf{P}$ so that in each round r the vector of estimates $\mathbf{e}(r)$ is multiplied by $\mathbf{P}$.

The Markov chain defined in Section 3.1 that models the average computation is also a characterization of a random walk, that is, a stochastic process on the set of nodes V where a particle moves around the network randomly. In our case, for each round, instead of choosing the next node where the particle will be located uniformly among neighbors, the matrix of transition probabilities is $\mathbf{P}$. A state of this process (which of course is also Markovian) is a distribution of the location of the particle over the nodes. The measure of this random walk that becomes relevant in our application is the mixing time, that is, the number of rounds before such distribution will be *close* to uniform. The mixing time of this random walk is the same as the convergence time of the Markov chain $(X_r)_{r=1}^\infty$, setting appropriately for each case the desired maximum deviation with respect to the stationary distribution as follows.

A useful representation of this process in our application is to assume a set S of particles, all of the same value ν, so that at the beginning each node i holds a subset S_i of particles such that $|S_i|\nu = v_i$. In order to analyze the computation along many rounds, we assume that ν is small enough so that particles are not divided. We define the **mixing time** of this multiple random walk as the number of rounds before the distribution of all particles is within ξ/n of the uniform, for $0 < \xi < 1$. Without message loss, it can be seen that the mixing time of the above defined multiple random walk is the same as the convergence time of the Markov chain $(X_r)_{r=1}^{\infty}$ defined in Section 3.1. We consider now the case where messages may be lost.

The following lemma shows that, for $f < 1/\ln(2\Delta e)^3$, the multiplicative overhead on the mixing time produced by message loss is less than $1/(1 - \sqrt{f \ln(2\Delta e)^3})$, and it is constant for $f \leq 1/(e(2\Delta e)^e)$. The proof, left to the full version of this work in [2] for brevity, uses concentration bounds on the delay that any particle may suffer due to message loss.

Lemma 1. *Consider any communication network of n nodes running MDFU, any $0 < f \leq 1/\ln(2\Delta e)^3$, any $0 < \xi < 1$, let $r_c = 2\ln(n/\xi)/\Phi(G)^2$, and let*

$$q = \begin{cases} 1/e & \text{if } f \leq 1/(e(2\Delta e)^e) \\ f\left(\sqrt{4\ln(2\Delta e)^3/f - 3} - 1\right)/2 & \text{otherwise.} \end{cases}$$

Consider a multiple random walk modeling MDFU as described. With probability at least $1 - 1/n$, after $r = r_c/(1-q)$ rounds it holds that $\max_{x \in S, i \in V} |p_x(i) - 1/n| \leq \xi/n$, where $p_x(i)$ is the probability that particle x is located at node i.

The expected number of particles at each node as a function of f. Analyzing a multiple random walk of a set of particles, in Lemma 1 we obtained a bound on the time that any particle takes to converge to a stationary uniform distribution. However, for any probability of message loss $f > 0$ and for any round, there is a positive probability that some particles are located in the edge buffers defined in the proof of such lemma. Hence, the fact that each particle is uniformly distributed over nodes does not imply that the expected average held at the nodes has converged, because only particles located at nodes are uniformly distributed. We bound the expected error in this section. The proof of the following lemma, left to the full version of this work in [2] for brevity, is based on computing the overall expected ratio of particles in nodes with respect to delayed particles.

Lemma 2. *Consider a multiple random walk modeling MDFU under the conditions of Lemma 1. Then, with probability at least $1 - 1/n$, for any round $r \geq r_c/(1-q)$, the expected number of particles $\mathbf{E}(|S_i^{(r)}|)$ in each node i is $(1 - \xi)(1 - f)|S|/n \leq \mathbf{E}(|S_i^{(r)}|) \leq (1 + \xi)|S|/n$.*

Based on the previous lemmata, the following theorem shows the convergence time of MDFU.

Theorem 2. *Consider any communication network of n nodes running MDFU. For any $0 < f \leq 1/\ln(2\Delta e)^3$, let $q = 1/e$ if $f \leq 1/(e(2\Delta e)^e)$, or $q = f\left(\sqrt{4\ln(2\Delta e)^3/f - 3} - 1\right)/2$ otherwise, and let $r_c = 2\ln(n/\xi)/\Phi(G)^2$. Then, with probability at least $1 - 1/n$, for any $0 < \xi < 1$ and any round $r \geq r_c/(1-q)$, the expected average estimation at any node $i \in V$ is $(1-\xi)(1-f)\bar{v} \leq \mathbf{E}(e_i^{(r)}) \leq (1+\xi)\bar{v}$, where $\Phi(G)$ is the conductance of the underlying graph characterizing the execution of MDFU on the network.*

Proof. From Lemmas 1 and 2, we know that, under the conditions of this theorem, for any round $r \geq r_c/(1-q)$ and any node $i \in V$, with probability at least $1 - 1/n$ the expected number of particles (of the multiple random walk modeling MDFU) is $(1-\xi)(1-f)|S|/n \leq \mathbf{E}(|S_i^{(r)}|) \leq (1+\xi)|S|/n$. Then, multiplying by the value of each particle the claim follows.

4   Empirical Evaluation of MDFU

We evalutated MDFU in a synchronous network simulator, using an Erdős–Rényi[10] network with 1000 nodes and 5000 links (giving an average degree of 10). The input values were chosen as when performing node counting [17]; i.e., all values being 0 except a random node with value 1; this scenario is more demanding, leading to slower convergence, than uniformly random input values. The evaluation aimed at: 1) comparing its convergence speed under no loss with competing algorithms; 2) evaluating its behavior under message loss; 3) checking its ability to perform continuous estimation over time-varying input values.

4.1   Convergence Speed against Related Algorithms under No Faults

To evaluate wether MDFU is a practical algorithm in terms of convergence speed, we compared it against three other algorithms: the original Flow-Updating [18, 19](FU), Distributed Random Grouping [7] (DRG), and Push-Synopses [20]. Figure 1 shows the coefficient of variation of the root mean square error as a function of the number of rounds (averaging 30 runs), with CV(RMSE) = $\sqrt{\sum_{i \in V}(e_i - \bar{v})^2/n}/\bar{v}$.

It can be seen that MDFU is competitive, providing approximate estimates slightly faster than FU and DRG and giving reasonably accurate results roughly in line with them. It loses to them for very high precision estimation and to Push-Synopses for all precisions (but both DRG and Push-Synopses are not fault-tolerant).

4.2   Fault Tolerance: Resilience to Message Loss

To evaluate the resilience of MDFU to message loss, we performed simulations using different rates of message loss (0, 1%, 5%, 10%), where each individual

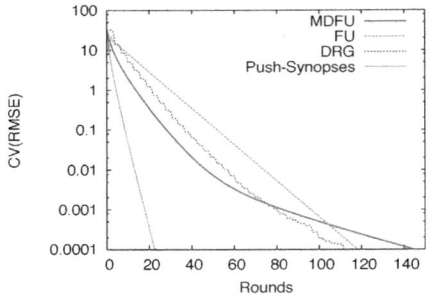

Fig. 1. CV(RMSE) over rounds in a 1000 node 5000 link Erdos-Renyi network

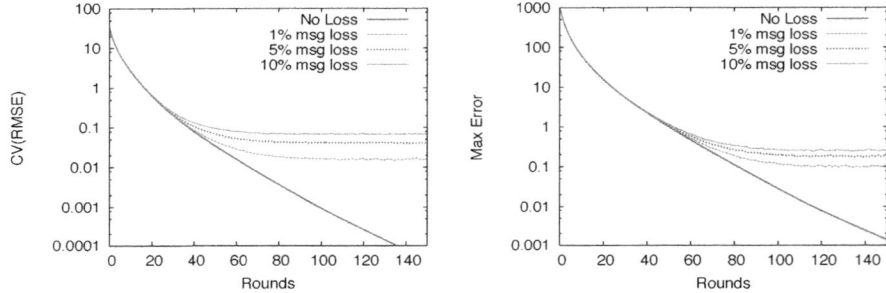

Fig. 2. Coefficient of variation of the RMSE and maximum relative error for MDFU in a 1000 node 5000 link Erdos-Renyi network

message may fail to reach the destination with these given probabilities. We measured the effect of message loss on both the CV(RMSE) and also on the maximum relative error. As can be seen in Figure 2, as long as there is some message loss, they do not tend to zero anymore, but converge to a value that is a function of the message loss rate.

We also measured the behavior of the average of the estimates over the whole network, and observed that there is a deviation from the correct value ($\overline{v}$, the average of the input values) towards lower values. Figure 3 shows the relative deviation from the correct value over time, for different message loss rates. It can be seen that this bias is roughly proportional to the message loss rate (for these small message loss rates).

Relating these results with the theoretical analysis of MDFU, we can see that this bias should not come as a surprise. From Theorem 2, the expected value of the estimation converges to a band between $(1-f)\overline{v}$ and $\overline{v}$. The relative deviation of the lower boundary is thus proportinal to the message loss rate. Figure 3 also shows this boundary for the different message loss rates.

This kind of bias was not present in the original FU, in which the average of the estimates tends to the correct value. In MDFU the message loss rate

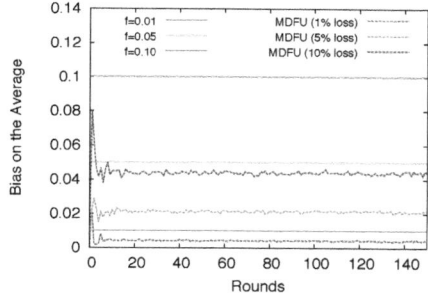

Fig. 3. Bias on the average estimation over rounds in a 1000 node 5000 link Erdos-Renyi network

limits the precision that can be achieved, but it does not impact convergence, contrary to classic mass distribution algorithms where, given message loss, the more rounds pass, the more mass is lost and the more the estimates deviate from the correct value, failing to converge.

5   MDFU with Linear Prediction

The explanation for the behavior of MDFU under message loss lies in that only the estimate converges, but flows keep steadily increasing over time. This can be seen in the formula: $F_{out}(j) \leftarrow F_{out}(j) + e_i / (2D_{ij})$ where the flow sent to some neighbor increases at each round by a value depending on the estimate and their mutual degrees. What happens is that during convergence, the extra flow that each of two nodes send over a link tend to the same value, and the extra outgoing flow cancels out the extra incoming flow. We can say that it is the *velocity* (rate of increase) of flows over a link that converge (to some different value for each link).

This means that, even if the estimate had already converged to the correct value, given a message loss, the extra flow that should have been received is not added to the estimate, implying a discrete deviation from the correct value. This discrete deviation does not converge to zero; thus, we have a bias towards lower values and the relative estimation error is prevented from converging to zero given some message loss rate.

Here we improve MDFU by exploring *velocity convergence*. We keep, for each link, the velocity (rate of increase) of the flow received. If a message is lost, we predict what would have been the flow received, given the stored flow, the velocity and the rounds passed since the last message received over that link, i.e., we perform a *linear prediction* of incoming flow. When a message is received we update the flow and recalculate the velocity. This algorithm is presented in Algorithm 2.

Under no message loss MDFU-LP is the same as MDFU and the theoretical results on convergence speed also apply to MDFU-LP. Under message loss

---

Algorithm 2. MDFU-LP. Pseudocode for node i. e_i is the estimate of node i. $F_{in}(j)$ is the cumulative inflow from node j. $F_{out}(j)$ is the cumulative outflow to node j. $V(j)$ is the velocity of incoming flow from node j. $R(j)$ is the number of rounds since the last message received from node j.

```
    // initialization
1   e_i ← v_i
2   foreach j ∈ N_i do
3       F_in(j) ← 0
4       F_out(j) ← e_i / (2D_ij)
5       V(j) ← 0
6       R(j) ← 1

7   foreach round do
        // communication phase
8       foreach j ∈ N_i do
9           Send j message ⟨i, F_out(j)⟩
        // computation phase
10      foreach ⟨j, F⟩ received do
11          V(j) ← (F − F_in(j))/R(j)
12          R(j) ← 0
13          F_in(j) ← F
14      e_i ← v_i + Σ_{j∈N_i} (F_in(j) + V(j) × R(j) − F_out(j))
15      foreach j ∈ N_i do
16          F_out(j) ← F_out(j) + e_i / (2D_ij)
17          R(j) ← R(j) + 1
```

---

the velocities converge over time and the prediction will be increasingly more accurate. Therefore, message loss should not cause discrete deviations in the estimate, allowing the estimation error to converge to zero.

We have evaluated MDFU-LP for the same network as before, but now with a wide range of message loss rates. We have observed that the behavior under message loss rates below 50% is almost indistinguishable from the behavior under no message loss. Figure 4 shows the CVRMSE and maximum relative error for 0%, 60%, 70%, and 80% message loss rates. It can be seen that even for 60% loss rate, after 60 rounds we have basically the same estimation errors as under no message loss.

6   Continuous Estimation over Time-Varying Input Values

Up to thus point we have considered that the input values v_i are fixed throughout the computation. In most practical situations this will not be the case and input values will change along time. The common approach in MD algorithms is to periodically reset the algorithm and start a new run that freezes the new input values and aggregates the new average. Naturally, resets are inefficient and mechanisms that can adapt the ongoing computation have the potential to adjust the estimates in a much shorter number of rounds.

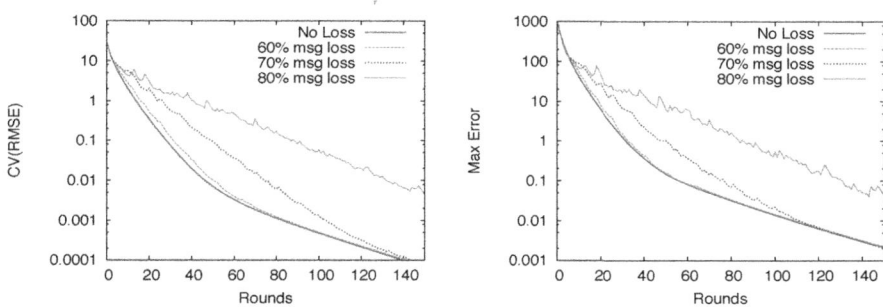

Fig. 4. Coefficient of variation of the RMSE and maximum relative error for MDFU-LP in a 1000 node 5000 link Erdos-Renyi network

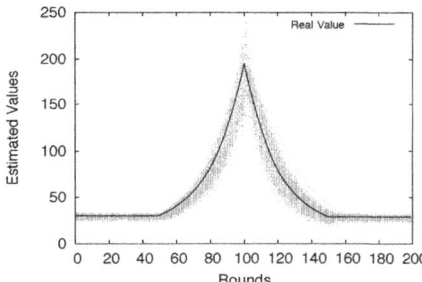

Fig. 5. Estimated value over rounds in a 1000 node 5000 link Erdos-Renyi network, with changes of the initial input value at 50% of the nodes

Without any further modifications, MDFU (and MDFU-LP) share with FU the capability of adapting to input value changes, since v_i is considered in the computation of the local estimate e_i, and this regulates how much the outgoing flows are to be incremented. If v_i decreases, e_i decreases in the same proportion and node i will share less through its flows to the neighbours. The converse occurring when v_i increases. The overall effect is convergence to the new average, even if multiple nodes are having changes in their input values.

In Figure 5 we show an example of how MDFU handles input value changes. In this setting, starting at round 50 and during 50 rounds, we increase by 5% in each round the input value in 500 nodes (a random half of the 1000 nodes). In the following 50 rounds, the same 500 nodes will have its value decreased by 5% per round. Initial input values are chosen uniformly at random (from 25 to 35) and the run is made with message loss at 10%. In Figure 5 one can observe that individual estimates[2] closely follow the global average, with only a slight lag of some rounds.

---

[2] To avoid clutering the graph only shows individual estimate evolution for a random sample of 100 of the 1000 nodes.

526 P.S. Almeida et al.

Notice that the lag could never be zero, since we are updating the new global average (black line) instantaneously and even the fastest theoretical algorithm would need information that takes *diameter* rounds to acquire.

References

1. Akyildiz, I.F., Su, W., Sankarasubramaniam, Y., Cyirci, E.: Wireless sensor networks: A survey. Computer Networks 38(4), 393–422 (2002)
2. Almeida, P., Baquero, C., Farach-Colton, M., Jesus, P., Mosteiro, M.A.: Fault-tolerant aggregation: Flow-updating meets mass-distribution, arXiv:1109.4373v1 (September 2011)
3. Bawa, M., Garcia-Molina, H., Gionis, A., Motwani, R.: Estimating aggregates on a peer-to-peer network. Technical report, Stanford University, Database group (2003)
4. Boyd, S., Ghosh, A., Prabhakar, B., Shah, D.: Randomized gossip algorithms. IEEE/ACM Transactions on Networking 14(SI), 2508–2530 (2006)
5. Chen, J.-Y., Hu, J.: Analysis of distributed random grouping for aggregate computation on wireless sensor networks with randomly changing graphs. IEEE Trans. Parallel Distr. Syst. 19(8), 1136–1149 (2008)
6. Chen, J.-Y., Pandurangan, G., Hu, J.: Brief announcement: locality-based aggregate computation in wireless sensor networks. In: PODC 2009: Proceedings of the 28th ACM Symposium on Principles of Distributed Computing, pp. 298–299. ACM, New York (2009)
7. Chen, J.-Y., Pandurangan, G., Xu, D.: Robust computation of aggregates in wireless sensor networks: distributed randomized algorithms and analysis. IEEE Trans. Parallel Distr. Syst. 17(9), 987–1000 (2006)
8. Clementi, A.E.F., Pasquale, F., Monti, A., Silvestri, R.: Communication in dynamic radio networks. In: Proc. 26th Ann. ACM Symp. on Principles of Distributed Computing, pp. 205–214 (2007)
9. Dimakis, A.G., Sarwate, A.D., Wainwright, M.J.: Geographic gossip: Efficient averaging for sensor networks. IEEE Transactions on Signal Processing 56(3), 1205–1216 (2008)
10. Erdos, P., Renyi, A.: On random graphs–i. Publicationes Matematicae 6, 290–297 (1959)
11. Fernández Anta, A., Mosteiro, M.A., Thraves, C.: An early-stopping protocol for computing aggregate functions in sensor networks. In: Proc. of the IEEE 15th Pacific Rim International Symposium on Dependable Computing, pp. 357–364 (2009)
12. Gasieniec, L.: Randomized gossiping in radio networks. In: Kao, M.-Y. (ed.) Encyclopedia of Algorithms. Springer, Heidelberg (2008)
13. Gupta, I., van Renesse, R., Birman, K.P.: Scalable fault-tolerant aggregation in large process groups. In: DSN, pp. 433–442. IEEE Computer Society (2001)
14. Heidemann, J.S., Silva, F., Intanagonwiwat, C., Govindan, R., Estrin, D., Ganesan, D.: Building efficient wireless sensor networks with low-level naming. In: SOSP, pp. 146–159 (2001)
15. Intanagonwiwat, C., Govindan, R., Estrin, D., Heidemann, J., Silva, F.: Directed diffusion for wireless sensor networking. IEEE/ACM Transactions on Networking 11(1), 2–16 (2003)
16. Intanagonwiwat, C., Estrin, D., Govindan, R., Heidemann, J.S.: Impact of network density on data aggregation in wireless sensor networks. In: ICDCS, pp. 457–458 (2002)

17. Jelasity, M., Montresor, A., Babaoglu, O.: Gossip-based aggregation in large dynamic networks. ACM Transactions on Computer Systems 23(3), 219–252 (2005)
18. Jesus, P., Baquero, C., Almeida, P.S.: Fault-tolerant aggregation for dynamic networks. In: Proc. of the 29th IEEE Symposium on Reliable Distributed Systems, pp. 37–43 (2010)
19. Jesus, P., Baquero, C., Almeida, P.S.: Fault-Tolerant Aggregation by Flow Updating. In: Senivongse, T., Oliveira, R. (eds.) DAIS 2009. LNCS, vol. 5523, pp. 73–86. Springer, Heidelberg (2009)
20. Kempe, D., Dobra, A., Gehrke, J.: Gossip-based computation of aggregate information. In: Proc. of the 44th IEEE Ann. Symp. on Foundations of Computer Science, pp. 482–491 (2003)
21. Kollios, G., Byers, J.W., Considine, J., Hadjieleftheriou, M., Li, F.: Robust aggregation in sensor networks. IEEE Data Engineering Bulletin 28(1), 26–32 (2005)
22. Kowalski, D.R., Pelc, A.: Time complexity of radio broadcasting: adaptiveness vs. obliviousness and randomization vs. determinism. Theoretical Computer Science 333, 355–371 (2005)
23. Krishnamachari, B., Estrin, D., Wicker, S.B.: The impact of data aggregation in wireless sensor networks. In: ICDCS Workshops, pp. 575–578. IEEE Computer Society (2002)
24. Madden, S., Franklin, M.J., Hellerstein, J.M., Hong, W.: Tag: a tiny aggregation service for ad-hoc sensor networks. In: Proc. of the 5th Symp. on Operating Systems Design and Implementation, pp. 131–146 (2002)
25. Madden, S., Szewczyk, R., Franklin, M.J., Culler, D.: Supporting aggregate queries over ad-hoc wireless sensor networks. In: Proceedings of the Fourth IEEE Workshop on Mobile Computing Systems and Applications, p. 49 (2002)
26. Nath, S., Gibbons, P.B., Seshan, S., Anderson, Z.R.: Synopsis diffusion for robust aggregation in sensor networks. In: Proceedings of the 2nd International Conference on Embedded Networked Sensor Systems, pp. 250–262 (2004)
27. Olfati-Saber, R., Murray, R.M.: Consensus problems in networks of agents with switching topology and time-delays. Transactions on Automatic Control 49(9), 1520–1533 (2004)
28. Rentala, P., Musumuri, R., Saxena, U., Gandham, S.: Survey on sensor networks, http://citeseer.nj.nec.com/479874.html
29. Scherber, D.S., Papadopoulos, H.C.: Locally constructed algorithms for distributed computations in ad-hoc networks. In: Proceedings of the 3rd International Symposium on Information Processing in Sensor Networks, pp. 11–19 (2004)
30. Sinclair, A., Jerrum, M.: Approximate counting, uniform generation and rapidly mixing markov chains. Information and Computation 82(1), 93–133 (1989)
31. Spanos, D., Olfati-Saber, R., Murray, R.: Dynamic consensus on mobile networks. In: 16th IFAC World Congress (2005)
32. Xiao, L., Boyd, S.: Fast linear iterations for distributed average. Systems and Control Letters 53, 65–78 (2004)
33. Xiao, L., Boyd, S., Lall, S.: A scheme for robust distributed sensor fusion based on average consensus. In: Proceedings of the 4th International Symposium on Information Processing in Sensor Networks, pp. 63–70 (2005)
34. Xiao, L., Boyd, S., Lall, S.: A Space-Time Diffusion Scheme for Peer-to-Peer Least-Squares Estimation. In: Proceedings of the 5th International Conference on Information Processing in Sensor Networks, pp. 168–176 (2006)
35. Zhao, J., Govindan, R., Estrin, D.: Computing aggregates for monitoring wireless sensor networks. In: Proc. of the 1st IEEE Intl. Workshop on Sensor Network Protocols and Applications, pp. 139–148 (2003)

Provably Good Scheduling of Sporadic Tasks with Resource Sharing on a Two-Type Heterogeneous Multiprocessor Platform

Gurulingesh Raravi[1], Björn Andersson[1,2], and Konstantinos Bletsas[1]

[1] CISTER-ISEP Research Center, Polytechnic Institute of Porto, Portugal
[2] Software Engineering Institute, Carnegie Mellon University, Pittsburgh, USA
{ghri,baa,ksbs}@isep.ipp.pt, baandersson@sei.cmu.edu

Abstract. Consider the problem of scheduling a set of implicit-deadline sporadic tasks to meet all deadlines on a two-type heterogeneous multiprocessor platform where a task may request at most one of $|R|$ shared resources. There are m_1 processors of type-1 and m_2 processors of type-2. Tasks may migrate only when requesting or releasing resources. We present a new algorithm, FF-3C-vpr, which offers a guarantee that if a task set is schedulable to meet deadlines by an optimal task assignment scheme that only allows tasks to migrate when requesting or releasing a resource, then FF-3C-vpr also meets deadlines if given processors $2+3 \cdot \left\lceil \frac{|R|}{\min(m_1, m_2)} \right\rceil$ times as fast. As far as we know, it is the first result for resource sharing on heterogeneous platforms with provable performance.

Keywords: heterogeneous multiprocessor systems, real-time scheduling, resource sharing.

1 Introduction

In heterogeneous multiprocessor platforms (i) not all processors are of the same type and (ii) task execution times depends on the processor type. Many manufacturers offer chips combining different types of processors [1,11,12,13,17]. Clearly, such chips are key components in heterogeneous systems, and such systems are increasingly used in practice. Yet, despite this trend, the state-of-art in real-time scheduling theory for heterogeneous multiprocessors is under-developed. The reasons include (i) processors typically sharing low-level hardware resources (e.g. caches, interconnects), which makes task execution times interdependent and (ii) dispatching limitations (e.g. some processors depend on another processor for dispatching [10]). Such idiosyncratic challenges must be addressed on a case-by-case basis, accounting for the particularities of the architecture. The state-of-art does offer some general ideas on analyzing shared low-level hardware resources [3,15,16,19] and scheduling co-processors [8,9]. Ultimately though, the dependency of the task execution time on the processor-type is what inherently complicates the design of scheduling algorithms for heterogeneous platforms.

A. Fernández Anta, G. Lipari, and M. Roy (Eds.): OPODIS 2011, LNCS 7109, pp. 528–543, 2011.
© Springer-Verlag Berlin Heidelberg 2011

The problem of scheduling independent *implicit-deadline sporadic tasks* (i.e., for each task, its deadline is equal to its minimum inter-arrival time) on heterogeneous multiprocessors has been studied in the past, both for generic [5,6,20] and for two-type [4] platforms but without considering the case when tasks share resources. One might partition tasks to processors and apply a resource-sharing protocol conceived for identical multiprocessors (e.g. D-PCP [18]). However, protocols such as D-PCP are not as effective in minimizing *priority inversion* when used in heterogeneous multiprocessors. For example, a task holding a shared resource may be executing on a processor where it runs slowly – causing large priority inversion to other tasks and poor schedulability. Therefore, a resource-sharing protocol for heterogeneous platforms ought to be cognizant of the execution speed of each task on each processor. It should also provide a finite bound on how much worse it performs, compared to an optimal scheme.

This paper introduces an algorithm, FF-3C-vpr, for scheduling tasks that share resources on a two-type heterogeneous multiprocessor. It offers a guarantee that if a task set can be scheduled to meet deadlines by an optimal scheme that allows a task to migrate only when requesting or releasing a resource then FF-3C-vpr also meets deadlines if given processors $2 + 3 \cdot \left\lceil \frac{|R|}{\min(m_1, m_2)} \right\rceil$ times as fast. Notably this is the first result with provably good performance for resource sharing on heterogeneous multiprocessors – which are increasingly relevant.

In this paper, Section 2 briefs the system model and assumptions. Section 3 gives the main idea of FF-3C-vpr. Section 4 lists notations and results used later. Section 5 discusses virtual processors – integral to our algorithm, presented in Section 6 along with the proof of its performance. Section 7 concludes.

2   System Model and Assumptions

We consider the problem of scheduling implicit-deadline sporadic tasks that share resources on a two-type heterogeneous multiprocessor platform with *restricted migration* (defined later). The system is specified as follows:

- **Processors (Π):** The platform consists of m processors of which $m_1 \geq 1$ processors are of type-1 and $m_2 \geq 1$ processors are of type-2.
- **Shared Resources (R):** A set R of $|R|$ resources that tasks share.
- **Task set (τ):** There are n *implicit-deadline sporadic tasks* – for each task τ_i, its deadline is equal to its minimum inter-arrival time, denoted as T_i.
- **Execution Time and Utilization:** The worst-case execution time of τ_i on a type-z processor ($z \in \{1, 2\}$) is denoted by C_i^z and its utilization by U_i^z.

We make the following assumptions:

- **Sharing the resources:** Each task may request at most one resource from R (known at design time) and at most once by each job of that task.
- **Virtual processors:** *Virtual processors* are logical constructs, used as task assignment targets by our algorithm. A virtual processor vp_i acts equivalent to a (physical) processor of the same type with (scaled) speed $\frac{1}{f}$ – and we

assume that it can be "emulated" on a physical processor of the same type (of speed 1), using no more than $\frac{1}{f}$ of its processing capacity[1].

- **Restricted migration:** A job of a task may only migrate to another processor during execution when it requests a resource; it must then migrate back to the original processor upon releasing the resource. We call this model *restricted migration*. Migrations between processors of any type are allowed.

3   Overview of Our Approach

The key to our approach is to distinguish between three *phases* in the execution of a task and make different scheduling provisions for each of them (Figure 1):

- **Phase-A** of a task spans from its arrival until it requests a shared resource.
- In its **Phase-B**, the task is holding (or waiting for) the shared resource.
- In its **Phase C**, the task has released the resource.

The main structure of our approach is as follows:

1. Split the task execution into phases A, B and C – in essence creating three subtasks out of it. The phase-B and phase-C subtasks of a task "arrive" (i.e. first become ready to execute) at a (respective) fixed time offset to the arrival of the respective phase-A subtask. This ensures that subtasks "inherit" the inter-arrival time of the original task and exhibit no arrival jitter.
2. Use m physical processors to create a set VP of virtual processors, formed by disjoint sets $\mathrm{VP_{AC}}$ and $\mathrm{VP_B}$ (i.e. $\mathrm{VP=VP_{AC}\cup VP_B}$ and $\mathrm{VP_{AC} \cap VP_B = \emptyset}$).
3. Phases A and C of a task are assigned (both) to a virtual processor $vp_j \in \mathrm{VP_{AC}}$. Phase-B of the same task is assigned to a virtual processor $vp_k \in \mathrm{VP_B}$.
4. The phase-A and phase-C subtasks of a task are scheduled using preemptive EDF on their assigned virtual processor in $\mathrm{VP_{AC}}$; the phase-B subtask is scheduled on its assigned virtual processor in $\mathrm{VP_B}$ using non-preemptive EDF – as a way of serializing accesses to shared resources[2].

---

[1] One intuitive way of achieving this is by dividing time to short slots of length S and using $\frac{1}{f} \cdot S$ time units in each slot to serve the workload of vp_i. By selecting S, we can then make the speed of the emulated processor arbitrarily close to $\frac{1}{f}$ (and in practice, S need rarely be impractically short) [14]. In strict terms, a sufficient condition for emulating m_1 type-1 virtual processors from $\mathrm{VP_{AC}}$ onto m_1 type-1 physical processors is: $\displaystyle\sum_{\substack{vp_i \in VP_{AC} \\ vp_i \text{ is type}-1}} V_i < m_1$, where V_i is the speed of virtual processor vp_i (and similarly for type-2 processors in $\mathrm{VP_{AC}}$ and for $\mathrm{VP_B}$ processors). For more details (including how to tradeoff spare processing capacity for longer S), see [14].

[2] Observe that implementing multiple virtual processors on the same physical processor might in practice involve frequent "context-switching" between those. Yet, whenever a physical processor "context-switches" between a phase-B virtual processor and some other virtual processor mapped to it, this does not violate the semantics of non-preemptive scheduling on the phase-B virtual processor because we are only interested (for the purposes of resource access serialization) in ensuring that phase-B subtasks never preempt each other – and this property is not violated.

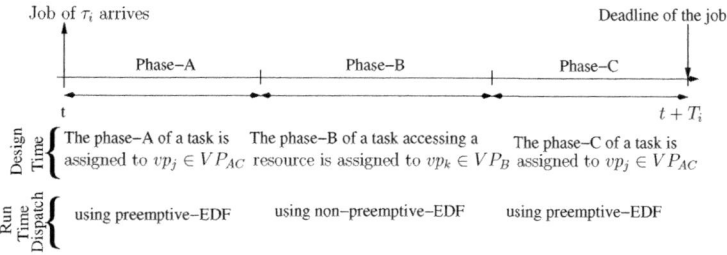

Fig. 1. Three execution phases of a job along with the design time (task assignment) and run time (task dispatching) decisions of FF-3C-vpr

Steps 1-3 are performed at *design time*; step 4 is carried out at *run time*. Despite using virtual processors, our algorithm by-construction ensures that the "restricted migration" assumption is not violated – discussed in Section 5 and 6. Subtasks corresponding to task phases are assigned *constrained* deadlines, i.e. not exceeding their inter-arrival time (inherited from the original task).

4   Few Notations and Useful Results

4.1   Notations

Let $\Pi(m_1, m_2)$ denote a two-type heterogeneous multiprocessor platform having m_1 processors of type-1 and m_2 processors of type-2. Let $\Pi(m_1, m_2) \cdot \langle s_1, s_2 \rangle$ denote a platform in which the speed of a type-1 and type-2 processor is respectively, s_1 and s_2 times the speed of a type-1 and type-2 processor in $\Pi(m_1, m_2)$ platform (where s_1 and s_2 are positive real-numbers, i.e. $s_1 > 0$ and $s_2 > 0$).

Let the predicate $sched(A, \tau, \Pi(m_1, m_2) \cdot \langle s_1, s_2 \rangle)$ signify that a task set τ *meets all its deadlines* if scheduled by an algorithm A on a platform $\Pi(m_1, m_2) \cdot \langle s_1, s_2 \rangle$. The term *meets all its deadlines* in this and other predicates means 'meets deadlines for every possible valid arrival of jobs of tasks in τ'.

We use $sched(\text{nmo}, \tau, \Pi(m_1, m_2) \cdot \langle s_1, s_2 \rangle)$ to signify that there exists a *non-migrative-offline* preemptive schedule which meets all deadlines for the specified system. Here, *non-migrative* schedule refers to a schedule in which all the jobs of a task execute on the same processor to which the task is assigned. In this predicate (and others), the term *offline* means that the schedule (i) can contain inserted idle times and (ii) can be generated using knowledge of future task arrival times (irrespective of whether such knowledge is available in practice).

The predicate $sched(\text{rmo}, \tau, R, \Pi(m_1, m_2) \cdot \langle s_1, s_2 \rangle)$ signifies that there exists a *restricted-migration-offline* preemptive schedule which meets all deadlines for the specified system when tasks share resources from R. As mentioned in Section 2, each task requests at most one resource from R and each job of that task may request that resource at most once during its execution. The term "restricted migration" has the same meaning as discussed in Section 2.

Similarly, $sched(A, \tau, R, \Pi(m_1, m_2) \cdot \langle s_1, s_2 \rangle)$ signifies that τ "sharing the resources" (see Section 2) from R *meets all its deadlines* when scheduled by an algorithm A on $\Pi(m_1, m_2) \cdot \langle s_1, s_2 \rangle$ with "restricted migration" (see Section 2).

Finally, in the above predicates, the suffix -δ (where applicable, i.e. in (sub-)task-partitioned schemes) to a scheduling algorithm (or algorithm class) implies that the schedulability of τ (other than just being established via some exact test) must additionally be ascertainable via a (potentially pessimistic) *density-based* uniprocessor schedulability test. This means that for the sub-set τ' of (sub-)tasks assigned on every type-z processor of speed V, it has to hold that $\sum_{i \in \tau'} \delta_i^z \leq V$, where $\delta_i^z = \frac{C_i^z}{D_i^z}$ is the *density*, C_i^z is the execution time (w.r.t. a processor of speed 1) and D_i^z is the deadline of a task τ_i on a type-z processor.

On a type-z processor: Let $C_{i,1}^z$ denote the execution time of a task τ_i before requesting a resource, i.e. in its phase-A. Let $C_{i,2(k)}^z$ denote the execution time of τ_i while holding resource R^k (where k is the index of the resource used by τ_i), i.e. in its phase-B. Let $C_{i,3}^z$ denote the execution time of a task τ_i after releasing the resource, i.e. in its phase-C. Note that $\forall \tau_i \in \tau$: $C_{i,1}^z + C_{i,2(k)}^z + C_{i,3}^z = C_i^z$.

We derive three new *constrained-deadline* (denoted by D_i^z) *sporadic task* sets (i.e., for each task, its deadline is less than or equal to its minimum inter-arrival time) namely, $TD_A(\tau)$, $TD_{B,R^k}(\tau)$ and $TD_C(\tau)$ from implicit-deadline sporadic task set τ by modifying the parameters of the tasks in τ. Intuitively, (i) a task $\tau_{i(A)} \in TD_A(\tau)$ represents phase-A execution of $\tau_i \in \tau$, (ii) a task $\tau_{i(B)} \in TD_{B,R^k}(\tau)$ represents phase-B execution of $\tau_i \in \tau$, accessing the resource R^k and (iii) a task $\tau_{i(C)} \in TD_C(\tau)$ represents phase-C execution of $\tau_i \in \tau$.

$TD_A(\tau)$, $TD_{B,R^k}(\tau)$ and $TD_C(\tau)$ are defined as follows – for each task $\tau_i \in \tau$:

$$\tau_{i(A)} = \{T_{i(A)} = T_i, \qquad D_{i(A)}^z = \frac{C_{i,1}^z}{C_i^z} \cdot T_i, \qquad C_{i(A)}^z = C_{i,1}^z\}$$

$$\tau_{i(B)} = \{T_{i(B)} = T_i, \qquad D_{i(B)}^z = \frac{C_{i,2(k)}^z}{C_i^z} \cdot T_i, \qquad C_{i(B)}^z = C_{i,2(k)}^z\}$$

$$\tau_{i(C)} = \{T_{i(C)} = T_i, \qquad D_{i(C)}^z = \frac{C_{i,3}^z}{C_i^z} \cdot T_i, \qquad C_{i(C)}^z = C_{i,3}^z\}$$

Observe that $TD_A(\tau)$, $TD_{B,R^k}(\tau)$ and $TD_C(\tau)$ are derived such that the densities of $\tau_{i(A)}$, $\tau_{i(B)}$ and $\tau_{i(C)}$ are equal to the utilization of $\tau_i \in \tau$. For example,

$$\forall \tau_{i(A)} \in TD_A(\tau) : \delta_{i(A)}^z = \frac{C_{i(A)}^z}{D_{i(A)}^z} = \frac{C_{i,1}^z}{\frac{C_{i,1}^z}{C_i^z} \cdot T_i} = \frac{C_i^z}{T_i} = U_i^z \text{ of } \tau_i \in \tau \qquad (1)$$

4.2   Useful Results

Lemma 1 and Lemma 2 (re-)state the speed competitive ratios of FF-3C (which is 2 – see Th. 1 in [4]) and of uniprocessor non-preemptive EDF (at most 3 –see Lem. 1 in [2]). FF-3C is a *non-migrative* scheduling scheme for implicit-deadline sporadic tasks (without resource sharing) on a two-type heterogeneous platform.

Lemma 1. $sched(nmo, \tau, \Pi(m_1, m_2)) \Rightarrow sched(FF\text{-}3C, \tau, \Pi(m_1, m_2) \cdot \langle 2, 2 \rangle)$

Lemma 2. $sched(nmo\text{-}np, \tau, \Pi(1,0)) \Rightarrow sched(nm\text{-}np\text{-}EDF, \tau, \Pi(1,0) \cdot \langle 3,3 \rangle)$

The heterogeneous multiprocessor in Lemma 2 (with only one processor of type-1) is (trivially) a uniprocessor. (Lemma 2 also holds for $\Pi(0,1)$ platform.)

Lemma 3 states that if a task is non-preemptive EDF-schedulable on a uniprocessor, it is also non-preemptive non-migrative (i.e. partitioned) EDF-schedulable on a platform with one more processor.

Lemma 3. $sched(nm\text{-}np\text{-}EDF, \tau, \Pi(1,0)\cdot\langle 3,3 \rangle) \Rightarrow sched(nm\text{-}np\text{-}EDF, \tau, \Pi(1,1)\cdot\langle 3,3 \rangle)$

The intuition behind Lemma 3 is that if the additional (type-2) processor is ignored, τ is schedulable on the original (type-1) processor. (The lemma also holds for platform $\Pi(0,1) \cdot \langle 3,3 \rangle$ in left-hand side predicate.)

Lemma 4. *(Combining Lemma 2 and Lemma 3)*
$sched(nmo\text{-}np, \tau, \Pi(1,0)) \Rightarrow sched(nm\text{-}np\text{-}EDF, \tau, \Pi(1,1) \cdot \langle 3,3 \rangle)$

The following lemma states that if implicit-deadline task set τ is non-migrative offline schedulable on $\Pi(m_1, m_2)$ then constrained-deadline sporadic task set $TD_A(\tau)$ derived from τ (as described in Section 4.1) is also non-migrative schedulable (e.g. under partitioned preemptive EDF) on $\Pi(m_1, m_2)$ *and additionally* this can be established via use of a (potentially pessimistic) *density-based* schedulability test. It is easy to see that the claim holds since the density of a task $\tau_{i(A)}$ in $TD_A(\tau)$ is always equal to the utilization of the corresponding task τ_i in τ.

Lemma 5. $sched(nmo, \tau, \Pi(m_1, m_2)) \Rightarrow sched(nmo\text{-}\delta, TD_A(\tau), \Pi(m_1, m_2))$

Proof. Let us assume that a non-migrative-offline feasible schedule exists for τ on $\Pi(m_1, m_2)$. So, there must exist a schedule in which the following holds:

$$\forall p \in \Pi(m_1, m_2) : \sum_{\tau_i \in \tau[p]} U_i^z \leq 1 \tag{2}$$

where $\tau[p]$ denotes the set of tasks assigned to processor p. Now, we show that there also exists a non-migrative-offline feasible schedule for $TD_A(\tau)$ on $\Pi(m_1, m_2)$. We know that for every task $\tau_i \in \tau$ there exists a task $\tau_{i(A)} \in TD_A(\tau)$. We also know from Expression 1 that $\forall \tau_{i(A)} \in TD_A(\tau) : \delta_{i(A)}^z = U_i^z$ of $\tau_i \in \tau$. Hence, by assigning the tasks in $TD_A(\tau)$ to $\Pi(m_1, m_2)$ in exactly the same way as the tasks in τ are assigned to $\Pi(m_1, m_2)$ (i.e. if $\tau_i \in \tau$ is assigned to processor p then we assign $\tau_{i(A)} \in TD_A(\tau)$ to p), we obtain:

$$\forall p \in \Pi(m_1, m_2) : \sum_{\tau_{i(A)} \in TD_A(\tau)[p]} \delta_{i(A)}^z \leq 1 \tag{3}$$

The above inequality corresponds to density-based schedulability test, on every processor p, for partitioned preemptive EDF (which is a non-migrative algorithm). Thus, $TD_A(\tau)$ is also non-migrative-offline schedulable on $\Pi(m_1, m_2)$.

Lemma 6. *(This largely follows from Lemma 1)*
$sched(nmo\text{-}\delta, TD_A(\tau), \Pi(m_1, m_2)) \Rightarrow sched(FF\text{-}3C\text{-}\delta, TD_A(\tau), \Pi(m_1, m_2) \cdot \langle 2,2 \rangle)$

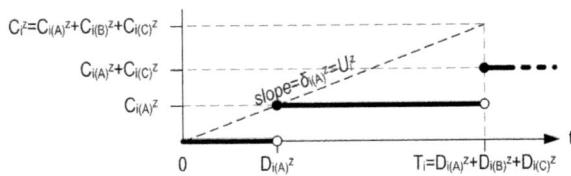

Fig. 2. Assigning phase-C sub-tasks to the same virtual processor as the respective phase-A sub-tasks (earlier assigned using a density-based test) preserves schedulability

Proof. Assume that predicate $sched(\text{nmo-}\delta, TD_A(\tau), \Pi(m_1, m_2))$ holds. Then, since the density of every (sub-)task in $TD_A(\tau)$ is equal to the utilization of the corresponding (original) task in τ, predicate $sched(\text{nmo}, \tau, \Pi(m_1, m_2))$ holds as well. In that case, we know from Lemma 1 that $sched(\text{FF-3C}, \tau, \Pi(m_1, m_2) \cdot \langle 2, 2 \rangle)$ holds. But then, since the density of every (sub-)task in $TD_A(\tau)$ is equal to the utilization of the corresponding (original) task in τ, it follows that: $sched(\text{FF-3C-}\delta, TD_A(\tau), \Pi(m_1, m_2) \cdot \langle 2, 2 \rangle)$.

Finally, a lemma that will be relied upon for assigning phase-C subtasks:

Lemma 7. *If, for a set $TD_A(\tau)[p]$ of phase-A subtasks,*

$$\delta_{TD_A(\tau)[p]} \overset{def}{=} \sum_{\tau_{i(A)} \in TD_A(\tau)[p]} \frac{C_{i(A)}^z}{D_{i(A)}^z} \leq V$$

then $TD_A(\tau)[p] \cup TD_C(\tau)[p]$ (where $TD_C(\tau)[p]$ is the set of the respective phase-C subtasks) is preemptive-EDF schedulable on a type-z (virtual) processor vp_p of speed V.

Proof. That $\delta_{TD_A(\tau)[p]} \leq V$ means that $TD_A(\tau)[p]$ is schedulable under preemptive EDF on vp_p. We now show that the *demand-bound function*[3], dbf(τ', t), of a task set $\tau' = TD_A(\tau)[p] \cup TD_C(\tau)[p]$ is upper bounded at every instant t by $\delta_{TD_A(\tau)[p]} \cdot t$ and hence is *also* schedulable on vp_p under preemptive EDF. Note that, for every phase-A subtask $\tau_{i(A)} \in TD_A(\tau)$ (and respective phase-C subtask $\tau_{i(C)} \in TD_C(\tau)$):

$$\text{dbf}(\{\tau_{i(A)}, \tau_{i(C)}\}, t) \leq \delta_{i(A)}^z \cdot t = \frac{C_{i(A)}^z \cdot t}{D_{i(A)}^z} \quad (4)$$

This is easy to verify because, the maximum "slope" to any point in the graph (Figure 2) of dbf$(\{\tau_{i(A)}, \tau_{i(C)}\}, t)$ from the origin is $\delta_{i(A)}^z = \frac{C_{i(A)}^z}{D_{i(A)}^z}$ (which is equal

---

[3] The *demand bound function* of a task τ_i, $dbf(\tau_i, t)$, is the maximum possible computation demand by jobs of τ_i, that have both release and deadline within any interval of length t. The demand bound function of a task set τ is defined as: $dbf(\tau, t) = \sum_{\tau_i \in \tau} dbf(\tau_i, t)$ [7].

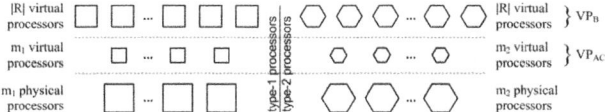

Fig. 3. $m + 2\,|R|$ virtual processors created from m physical processors on a two-type heterogeneous multiprocessor platform ($m = m_1 + m_2$)

to U_i^z of $\tau_i \in \tau$, as per our choice of $D_{i(A)}^z$), at abscissa $t = D_{i(A)}^z$. Summation of Equation 4 over all $\tau_{i(A)} \in TD_A(\tau)[p]$ (and respective $\tau_{i(C)} \in TD_C(\tau)[p]$) yields:

$$\mathrm{dbf}(TD_A(\tau)[p] \cup TD_C(\tau)[p], t) \le t \cdot \sum_{\tau_{i(A)} \in TD_A(\tau)[p]} \delta_{i(A)}^z = \delta_{TD_A(\tau)[p]} \cdot t$$

5   Creating Virtual Processors on a Two-Type Heterogeneous Multiprocessor Platform

We create $m + 2\,|R|$ virtual processors from m physical processors on a two-type heterogeneous multiprocessor platform as shown in Figure 3. The main idea is as follows. We treat physical processors of each type as an identical multiprocessor platform and create a certain number of virtual processors of the corresponding type from this platform. To be precise, m_1 physical processors of type-1 are treated as an identical multiprocessor platform and $m_1 + |R|$ virtual processors (of type-1) are created from them and ordered as shown in the left half of Figure 3 (i.e. left side of the vertical *solid line*). Analogously, m_2 physical processors of type-2 are treated as an identical multiprocessor platform and $m_2 + |R|$ virtual processors (of type-2) are created from them and ordered as shown in the right half of Figure 3 (i.e. right side of the vertical *solid line*). Now, if we look at each row in Figure 3 (separated by *horizontal lines*), it represents a two-type heterogeneous multiprocessor platform (for example, the second row represents a two-type heterogeneous multiprocessor platform with m_1 virtual processors of type-1 and m_2 virtual processors of type-2). Thus, $m + 2\,|R|$ virtual processors are created from m physical processors on a two-type heterogeneous platform. Precisely, we create the virtual processors with following specifications:

- m **virtual processors (denoted as VP$_{AC}$):** m_1 virtual processors of type-1 each of speed $\dfrac{2}{2+3\left\lceil \frac{|R|}{m_1} \right\rceil}$ times the speed of a physical processor of type-1 and m_2 virtual processors of type-2 each of speed $\dfrac{2}{2+3\left\lceil \frac{|R|}{m_2} \right\rceil}$ times the speed of a physical processor of type-2. They are used to schedule phase-A and phase-C of a task execution and are referred to as '*virtual processors in* VP$_{AC}$'.
- $2\,|R|$ **virtual processors (denoted as VP$_B$):** $|R|$ virtual processors of type-1 each of speed $\dfrac{3}{2+3\left\lceil \frac{|R|}{m_1} \right\rceil}$ times the speed of a physical processor of type-1 and $|R|$ virtual processors of type-2 each of speed $\dfrac{3}{2+3\left\lceil \frac{|R|}{m_2} \right\rceil}$ times the

---

Algorithm 1. FF-3C-vpr($\tau, \Pi^2(m_1, m_2), R$): for scheduling tasks that share resources on a two-type heterogeneous multiprocessor platform

---

```
// Lines 1-17 execute offline; line 18 executes at run-time.
```
1 Create $TD_A(\tau)$, $TD_{B,R^k}(\tau)$ and $TD_C(\tau)$ from τ as described in Section 4.1
2 $\{VP_{AC}, VP_B\} := VP\_Create(\Pi^2(m_1, m_2), R)$ `// Create` VP_{AC} `and` VP_B
   `virtual processors and store them in arrays of structures`
3 **for** $i = 1$ **to** $|R|$ **do** //Form $|R|$ pairs from $2|R|$ virtual processors in VP_B
4      $Pair_B[i] := \langle \mathrm{VP_B}[i], \mathrm{VP_B}[|R| + i] \rangle$
5 **end**
6 Assign $TD_A(\tau)$ to virtual processors in $\mathrm{VP_{AC}}$ using FF-3C
7 **for** $i = 1$ **to** n **do**
8   **if** τ_i *requests a resource* **then**
9   let k denote the resource that task τ_i requests
10   **if** $(C^1_{i(B)} \leq C^2_{i(B)})$ **then**
11   assign $\tau_{i(B)}$ to $\mathrm{VP_B}[k]$
12   **else**
13   assign $\tau_{i(B)}$ to $\mathrm{VP_B}[|R| + k]$
14   **end**
15   **end**
16 **end**
17 Assign $TD_C(\tau)$ to virtual processors in $\mathrm{VP_{AC}}$ using the assignment made by FF-3C for phase-A of tasks on line 6, i.e. if $\tau_{i(A)}$ of $TD_A(\tau)$ was assigned to $\mathrm{VP_{AC}}[j]$ processor then assign $\tau_{i(C)}$ of $TD_C(\tau)$ to $\mathrm{VP_{AC}}[j]$ processor
18 Dispatch tasks in (i) $TD_A(\tau)$ with preemptive EDF on $\mathrm{VP_{AC}}$, (ii) $TD_B(\tau)$ with non-preemptive EDF on VP_B and (iii) $TD_C(\tau)$ with preemptive EDF on VP_{AC}

---

speed of a physical processor of type-2. They are used to schedule phase-B of task execution and are referred to as *'virtual processors in* $\mathrm{VP_B}$*'*.

We ensure that no virtual processor is created using two or more physical processors, i.e., the capacity of a virtual processor comes from a single physical processor alone. The pseudo-code for creating virtual processors, referred to as VP_Create in the rest of the paper, can be found in Appendix (Section 8.1). Since VP_Create creates a virtual processor out of the processing capacity of a single respective physical processor, within each of its phases, any job executes on only one physical processor (i.e. does not migrate between different physical processors). However, it can migrate to a different physical processor at the boundaries separating (i) its phase-A and phase-B and (ii) its phase-B and phase-C executions. FF-3C-vpr adheres to the "restricted migration" model by assigning phase-A and phase-C of a task to the same physical processor.

6   FF-3C-vpr and Its Speed Competitive Ratio

6.1   The FF-3C-vpr Algorithm

The pseudo-code of FF-3C-vpr is listed in Algorithm 1. The algorithm works as follows. On line 1, it creates three subsets of tasks, i.e. $TD_A(\tau)$, $TD_{B,R^k}(\tau)$ and

$TD_C(\tau)$ from the given task set τ. On line 2, it creates $m + 2|R|$ virtual processors specified in Section 5 from m physical processors. On lines 3-5, it groups $2|R|$ phase-B virtual processors into $|R|$ *pairs of processors*, each pair containing one processor of each type, i.e. one processor of type-1 and one processor of type-2. Each pair of processors, $Pair_B[k]$ where $k = \{1, \cdots, |R|\}$, is used for scheduling phase-B of tasks that access the resource R^k. At any time instant, only one processor from each heterogeneous pair is used for executing the tasks: this is, in each case, the processor of the type on which the given task executes fastest (termed the *favorite* processor type for that task); the other processor is kept idle during the execution of the task. This technique ensures mutual exclusion for accessing each resource. Moreover, it effectively creates, out of each pair, the equivalent of a hypothetical single virtual processor whereupon every task would execute as fast as on its (respective) favorite processor type. This design choice aims at minimizing blocking times related to resource sharing. On line 6, the algorithm assigns phase-A of a task (in $TD_A(\tau)$) to virtual processors in VP$_{AC}$ using FF-3C [4]. On lines 7-16, it assigns phase-B of a task (in $TD_{B,R^k}(\tau)$) accessing resource R^k to that virtual processor in $Pair_B[k]$ which is of its *favorite* processor type in phase-B. On line 17, it assigns phase-C of a task (in $TD_C(\tau)$) to a virtual processor in VP$_{AC}$ in the same manner as that of assignment of a task in $TD_A(\tau)$ to a virtual processor in VP$_{AC}$ by FF-3C (on line 6). Instead of running FF-3C again on $TD_C(\tau)$ task set, the algorithm makes use of the output of FF-3C (that was run on line 6 to assign tasks in $TD_A(\tau)$ on VP$_{AC}$) to assign $TD_C(\tau)$. Line 17 ensures that phase-C of a task is assigned to that virtual processor in VP$_{AC}$ to which phase-A of the same task has been assigned. Assigning phase-C subtasks on the same virtual processor as its corresponding phase-A subtask (i) does not endanger the schedulability of a previously schedulable virtual processor; intuitively, this is because these two subtasks have precedence constraints – Lemma 7 provides formal proof and (ii) ensures that the "restricted migration" assumption is not violated. On line 18, FF-3C-vpr schedules tasks executing in their phase-A onto VP$_{AC}$ using preemptive EDF, tasks in their phase-B onto VP$_B$ using non-preemptive EDF and tasks in their phase-C onto VP$_{AC}$ using preemptive EDF. Lines 1-17 can be performed at design time and only line 18 has to be performed at run time.

6.2   Time Complexity of FF-3C-vpr

We now show that the time-complexity of FF-3C-vpr is a polynomial function of the number of tasks (n), processors (m) and/or resources $(|R|)$. From FF-3C-vpr pseudo-code (Algorithm 1), we can observe that the time-complexity for:

- creating $TD_A(\tau)$, $TD_{B,R^k}(\tau)$ and $TD_C(\tau)$ subsets (on line 1) is $O(n)$.
- creating the virtual processor subsets, VP$_{AC}$ and VP$_B$ (on line 2) is $O(m)$.
- forming the virtual processor pairs (on lines 3-5) is $O(|R|)$.
- assigning $TD_A(\tau)$ on VP$_{AC}$ using FF-3C (on line 6) is $O(n \cdot \max(m, \log n))$ [4].
- assigning $TD_{B,R^k}(\tau)$ on VP$_B$ (on lines 7-16) is $O(n)$.
- assigning $TD_C(\tau)$ on VP$_{AC}$ (on line 17) is $O(n)$.

Thus the time-complexity of FF-3C-vpr is at most

$$
\left(\underbrace{O(n)}_{\substack{\text{create} \\ \text{subtasks}}} + \underbrace{O(m)}_{\substack{\text{create virtual} \\ \text{processors}}} + \underbrace{O(|R|)}_{\substack{\text{form virtual} \\ \text{processor pairs}}} + \underbrace{O(n \cdot \max(m, \log n)}_{\text{assign } TD_A(\tau)} + \underbrace{O(n)}_{\text{assign } TD_{B,R^k}(\tau)} \right.
$$

$$
\left. + \underbrace{O(n)}_{\text{assign } TD_C(\tau)} \right) = O(\max(n \cdot \max(m, \log n)), |R|) = O(n \cdot \max(m, \log n))
$$

6.3   The Speed Competitive Ratio of FF-3C-vpr Algorithm

Theorem 1. *The speed competitive ratio of FF-3C-vpr is* $2 + 3 \cdot \left\lceil \frac{|R|}{\min(m_1, m_2)} \right\rceil$.

Proof. The proof considers separately the scheduling of each of the three phases and then combines the results. Let us look at phase-A first. Combining Lemma 5 and Lemma 6 and applying the result to virtual processors in VP$_{AC}$ yields:

$$
sched(\text{nmo}, \tau, \Pi(m_1, m_2)) \Rightarrow sched(\text{FF-3C-}\delta, TD_A(\tau), \Pi(m_1, m_2) \cdot \langle 2, 2 \rangle) \quad (5)
$$

Now consider phase-C. Since a task in its phase-A cannot be in its phase-C simultaneously (and vice versa), the respective sub-tasks are not independent. Treating them as such would be potentially pessimistic; conversely, accounting for these precedence constraints during (sub-)task assignment could improve performance. Indeed, our algorithm assigns each phase-C sub-task to the same virtual processor as its respective phase-A sub-task (Algorithm 1, line 17.).

For convenience, let us introduce a notation say, FF-3C-δ+cp for this (sub-) task assignment strategy (using FF-3C-δ to assign phase-A subtasks and "copy-ing" the assignment for respective phase-C subtasks, as done by FF-3C-vpr on line 6). Then, applying Lemma 7 to Equation 5 yields:

$$
sched(\text{nmo}, \tau, \Pi(m_1, m_2)) \Rightarrow
$$
$$
sched(\text{FF-3C-}\delta\text{+cp}, TD_A(\tau) \cup TD_C(\tau), \Pi(m_1, m_2) \cdot \langle 2, 2 \rangle) \quad (6)
$$

Now, let us consider phase-B. Recall that to ensure a task executes at the speed associated with its favorite processor type, when accessing a shared resource R^k, we create two virtual processors (one of each type) for the execution of tasks holding the resource. These are termed *processor pair* $Pair_B[k]$ (see lines 3-5 of the FF-3C-vpr algorithm) and execute in mutual exclusion. We assign the task accessing the resource R^k to $Pair_B[k]$ — whichever processor is favorite to the task in consideration, executes it; the other one sits idle. This minimizes overall blocking time in the system.

If a task set τ in which tasks share a single resource say, R^k, is non-migrative-offline non-preemptive schedulable on $\Pi(m_1, m_2)$ then $TD_{B,R^k}(\tau)$ is also non-migrative-offline non-preemptive schedulable on $\Pi(m_1, m_2)$. $\forall R^k \in R$:

$$
sched(\text{nmo-np}, \tau, \Pi(m_1, m_2)) \Rightarrow sched(\text{nmo-np}, TD_{B,R^k}(\tau), \Pi(m_1, m_2)) \quad (7)
$$

If $TD_{B,R^k}(\tau)$ in which tasks share a single resource R^k is non-migrative-offline non-preemptive schedulable on $\Pi(m_1, m_2)$ then the same task set is also non-migrative-offline non-preemptive schedulable on $\Pi(1,1)$, i.e.: $\forall R^k \in R$:

$$sched(\text{nmo-np}, TD_{B,R^k}(\tau), \Pi(m_1, m_2)) \Rightarrow sched(\text{nmo-np}, TD_{B,R^k}(\tau), \Pi(1,1)) \quad (8)$$

Hence, combining Equations 7 and 8 gives: $\forall R^k \in R$:
$$sched(\text{nmo-np}, \tau, \Pi(m_1, m_2)) \Rightarrow sched(\text{nmo-np}, TD_{B,R^k}(\tau), \Pi(1,1)) \quad (9)$$

Without loss of generality, Lemma 4 can be rewritten as:

$$sched(\text{nmo-np}, \tau, \Pi(1,1)) \Rightarrow sched(\text{nm-np-EDF}, \tau, \Pi(1,1) \cdot \langle 3,3 \rangle) \quad (10)$$

Applying Equation 10 to a task set $TD_{B,R^k}(\tau)$ gives: $\forall R^k \in R$:
$$sched(\text{nmo-np}, TD_{B,R^k}(\tau), \Pi(1,1)) \Rightarrow sched(\text{nm-np-EDF}, TD_{B,R^k}(\tau), \Pi(1,1) \cdot \langle 3,3 \rangle) \quad (11)$$

Combining Equation 9 and 11 and applying the result to VP$_B$ virtual processors: $\forall R^k \in R$,
$$sched(\text{nmo-np}, \tau, \Pi(m_1, m_2)) \Rightarrow sched(\text{nm-np-EDF}, TD_{B,R^k}(\tau), \Pi(1,1) \cdot \langle 3,3 \rangle) \quad (12)$$

Combining the above intermediate results: dividing type-1 and type-2 processor speeds by, respectively, $2 + 3 \left\lceil \frac{|R|}{m_1} \right\rceil$ and $2 + 3 \left\lceil \frac{|R|}{m_2} \right\rceil$ in Equations 6 and 12 gives:

$$sched(\text{nmo}, \tau, \Pi(m_1, m_2) \cdot \left\langle \frac{1}{2 + 3 \left\lceil \frac{|R|}{m_1} \right\rceil}, \frac{1}{2 + 3 \left\lceil \frac{|R|}{m_2} \right\rceil} \right\rangle) \Rightarrow$$
$$sched(\text{FF-3C-}\delta\text{+cp}, TD_A(\tau) \cup TD_C(\tau), \Pi(m_1, m_2) \cdot \left\langle \frac{2}{2 + 3 \left\lceil \frac{|R|}{m_1} \right\rceil}, \frac{2}{2 + 3 \left\lceil \frac{|R|}{m_2} \right\rceil} \right\rangle) \quad (13)$$

$$\forall R^k \in R : sched(\text{nmo-np}, \tau, \Pi(m_1, m_2) \cdot \left\langle \frac{1}{2 + 3 \left\lceil \frac{|R|}{m_1} \right\rceil}, \frac{1}{2 + 3 \left\lceil \frac{|R|}{m_2} \right\rceil} \right\rangle) \Rightarrow$$
$$sched(\text{nm-np-EDF}, TD_{B,R^k}(\tau), \Pi(1,1) \cdot \left\langle \frac{3}{2 + 3 \left\lceil \frac{|R|}{m_1} \right\rceil}, \frac{3}{2 + 3 \left\lceil \frac{|R|}{m_2} \right\rceil} \right\rangle) \quad (14)$$

In the right-hand sides of Equations 13 and 14, the processor specifications match those created by FF-3C-vpr. Note also that under FF-3C-vpr (which only allows "restricted migration"), phase-A and phase-C sub-tasks are assigned to virtual processors in VP_{AC} and phase-B sub-tasks are assigned to virtual processors in VP_B (and $VP_{AC} \cap VP_B = \emptyset$). Hence by combining Equations 13 and 14 we get:

$$sched(\text{rmo}, \tau, R, \Pi(m_1, m_2) \cdot \left\langle \frac{1}{2 + 3 \left\lceil \frac{|R|}{m_1} \right\rceil}, \frac{1}{2 + 3 \left\lceil \frac{|R|}{m_2} \right\rceil} \right\rangle) \Rightarrow$$
$$sched(\text{FF-3C-vpr}, \tau, R, \Pi(m_1, m_2)) \quad (15)$$

We know that higher speed processors do not jeopardize the schedulability of a task set. Hence, we can write:

540   G. Raravi, B. Andersson, and K. Bletsas

$$sched(\text{rmo}, \tau, R, \Pi(m_1, m_2) \cdot \langle \min(s_1, s_2), \min(s_1, s_2) \rangle) \Rightarrow$$
$$sched(\text{rmo}, \tau, R, \Pi(m_1, m_2) \cdot \langle s_1, s_2 \rangle)$$

Substituting $s_1 = \frac{1}{2+3\left\lceil \frac{|R|}{m_1} \right\rceil}$ and $s_2 = \frac{1}{2+3\left\lceil \frac{|R|}{m_2} \right\rceil}$ in the above equation and combining with Equation 15 and rewriting gives:

$$sched(\text{rmo}, \tau, R, \Pi(m_1, m_2) \cdot \left\langle \frac{1}{2 + 3 \cdot \max\left(\left\lceil \frac{|R|}{m_1} \right\rceil, \left\lceil \frac{|R|}{m_2} \right\rceil\right)}, \right.$$
$$\left. \frac{1}{2 + 3 \cdot \max\left(\left\lceil \frac{|R|}{m_1} \right\rceil, \left\lceil \frac{|R|}{m_2} \right\rceil\right)} \right\rangle) \Rightarrow sched(\text{FF-3C-vpr}, \tau, R, \Pi(m_1, m_2)) \qquad (16)$$

Multiplying processor speeds in Equation 16 by $2+3 \cdot \max\left(\left\lceil \frac{|R|}{m_1} \right\rceil, \left\lceil \frac{|R|}{m_2} \right\rceil\right)$:

$$sched(\text{rmo}, \tau, R, \Pi(m_1, m_2)) \Rightarrow sched(\text{FF-3C-vpr}, \tau, R, \Pi(m_1, m_2)) \cdot$$
$$\left\langle 2 + 3 \cdot \max\left(\left\lceil \frac{|R|}{m_1} \right\rceil, \left\lceil \frac{|R|}{m_2} \right\rceil\right), 2 + 3 \cdot \max\left(\left\lceil \frac{|R|}{m_1} \right\rceil, \left\lceil \frac{|R|}{m_2} \right\rceil\right) \right\rangle) \qquad (17)$$

By rewriting the RHS of the above equation, we get:

$$sched(\text{rmo}, \tau, R, \Pi(m_1, m_2)) \Rightarrow sched(\text{FF-3C-vpr}, \tau, R, \Pi(m_1, m_2)) \cdot$$
$$\left\langle 2 + 3 \cdot \left\lceil \frac{|R|}{\min(m_1, m_2)} \right\rceil, 2 + 3 \cdot \left\lceil \frac{|R|}{\min(m_1, m_2)} \right\rceil \right\rangle)$$

7   Conclusions

We proposed a new algorithm, FF-3C-vpr, for scheduling implicit-deadline sporadic tasks with restricted migration to meet all the deadlines on a two-type heterogeneous multiprocessor platform where each task can access at most one shared resource. We showed that FF-3C-vpr has a speed competitive ratio of $2 + 3 \cdot \left\lceil \frac{|R|}{\min(m_1, m_2)} \right\rceil$. If $R=\emptyset$ this becomes 2 and FF-3C-vpr reduces to FF-3C.

Acknowledgments. This work was partially supported by the REHEAT project, ref. FCOMP-01-0124-FEDER-010045, funded by FEDER funds through COMPETE (POFC – Operational Programme Thematic Factors of Competitiveness), National Funds through FCT – Portuguese Foundation for Science and Technology and REJOIN project of FLAD (Luso-American Development Foundation).

References

1. AMD Inc.: The AMD Fusion Family of APUs, http://sites.amd.com/us/fusion/apu/Pages/fusion.aspx
2. Andersson, B., Easwaran, A.: Provably good multiprocessor scheduling with resource sharing. Real-Time System 46(2), 153–159 (2010)

3. Andersson, B., Easwaran, A., Lee, J.: Finding an Upper Bound on the Increase in Execution Time Due to Contention on the Memory Bus in COTS-Based Multicore Systems. In: WiP of 30th IEEE Real-Time Systems Symposium (2009)
4. Andersson, B., Raravi, G., Bletsas, K.: Assigning real-time tasks on heterogeneous multiprocessors with two unrelated types of processors. In: 31st IEEE Real-Time Systems Symposium, pp. 239–248 (2010)
5. Baruah, S.: Task partitioning upon heterogeneous multiprocessor platforms. In: Proceedings of the 10th IEEE International Real-Time and Embedded Technology and Applications Symposium, pp. 536–543 (2004)
6. Baruah, S.: Partitioning real-time tasks among heterogeneous multiprocessors. In: Proc. of the 33rd International Conference on Parallel Processing (2004)
7. Baruah, S., Mok, A., Rosier, L.: Preemptively scheduling hard-real-time sporadic tasks on one processor. In: IEEE Real-Time Systems Symposium (1990)
8. Bletsas, K.: Worst-case and Best-case Timing Analysis for Real-time Embedded Systems with Limited Parallelism. Ph.D. thesis, The University of York (2007)
9. Gai, P., Abeni, L., Buttazzo, G.C.: Multiprocessor DSP scheduling in system-on-a-chip architectures. In: 14th Euromicro Conference on Real-Time Systems (ECRTS 2002), Vienna, Austria, pp. 231–238 (June 2002)
10. Gschwind, M., Hofstee, H.P., Flachs, B., Hopkins, M., Watanabe, Y., Yamazaki, T.: Synergistic Processing in Cell's Multicore Architecture. IEEE Micro 26(2) (2006)
11. IBM Corp.: The Cell Project, http://www.research.ibm.com/cell/
12. IEEE Spectrum: With Denver Project NVIDIA and ARM Join CPU-GPU Integration Race,
    http://spectrum.ieee.org/tech-talk/semiconductors/processors/
    with-denver-project-nvidia-and-arm-join-cpugpu-integration-race
13. Intel Corporation: The 2nd generation Intel Core processor family,
    http://www.intel.com/en_IN/consumer/
    products/processors/core-family.htm
14. Bletsas, K., Andersson, B.: Notional Processors: An Approach for Multiprocessor Scheduling. In: Proceedings of the 15th IEEE International Real-Time and Embedded Technology and Applications Symposium, pp. 3–12 (2009)
15. Li, Y., Suhendra, V., Liang, Y., Mitra, T., Roychoudhury, A.: Timing Analysis of Concurrent Programs Running on Shared Cache Multi-Cores. In: Proceedings of the 30th IEEE Real-Time Systems Symposium, pp. 57–67 (2009)
16. Lv, M., Guan, N., Yi, W., Yu, G.: Combining Abstract Interpretation with Model Checking for Timing Analysis of Multicore Software. In: Proceedings of the 31st IEEE Real-Time Systems Symposium, pp. 339–349 (2010)
17. NVIDIA: Dell and NVIDIA Workstation Solutions,
    http://www.nvidia.com/object/IO_16084.html
18. Rajkumar, R., Sha, L., Lehoczky, J.: Real-Time Synchronization Protocols for Multiprocessors. In: 9th IEEE Real-Time Systems Symposium, pp. 259–269 (1988)
19. Raravi, G., Andersson, B.: Calculating an upper bound on the finishing time of a group of threads executing on a GPU: A preliminary case study. In: 16th IEEE International Conference on Embedded and Real-Time Computing Systems and Applications – WiP Session, pp. 5–8 (2010)
20. Baruah, S.: Feasibility analysis of preemptive real-time systems upon heterogeneous multiprocessor platforms. In: 25th IEEE Real-Time Systems Symposium (2004)

8   Appendix

8.1   Algorithm for Creating Virtual Processors

In our notation, PP denotes the set of physical processors, pp_i denotes the ith physical processor and vp_i denotes the ith virtual processor. The VP_Create pseudo-code for creating the specified virtual processors (in Section 5) is listed in Algorithm 2. It, in turn, uses the sub-routine VP$_{\text{ABC}}$_Create (Algorithm 3).

The VP_Create function on line 2 calls sub-routine VP$_{\text{ABC}}$_Create to create $m_1 + |R|$ virtual processors of type-1 from m_1 physical processors of type-1. The sub-routine first creates m_1 virtual processors (see lines 1-5 in Algorithm 3) from m_1 physical processors and then creates $|R|$ virtual processors (see lines 6-20 in Algorithm 3) from the remaining capacity of type-1 processors. Observe that no virtual processor is created using two physical processors, i.e. the capacity of a virtual processor comes from a single physical processor alone. Similarly, VP_Create() on line 3 creates $m_2 + |R|$ virtual processors of type-2 from m_2 physical processors of type-2.

Since VP$_{\text{ABC}}$_Create creates a virtual processor out of the processing capacity of a single respective physical processor, within each of its phases, any job executes on only one physical processor (i.e. does not migrate between different physical processors). However, it can migrate to a different physical processor at the boundaries separating (i) its phase-A and phase-B and (ii) its phase-B and phase-C. FF-3C-vpr adheres to the "restricted migration" model by assigning phase-A and phase-C of a task to the same physical processor.

The following observations can be made regarding our specification and creation of virtual processors. After creating one VP$_{\text{AC}}$ virtual processor (for phase-A and phase-C) from every physical processor (lines 1-5 in the sub-routine shown in Algorithm 3), let us see (i) how much capacity remains in each of the physical processors and (ii) how many phase-B virtual processors (i.e. virtual processors in VP$_{\text{B}}$) can be created from that capacity. For ease of explanation, consider the case of type-1 processors. After creating one virtual processor, i.e. one in VP$_{\text{AC}}$ (for phase-A and phase-C) of speed $\frac{2}{2+3\left\lceil\frac{|R|}{m_1}\right\rceil}$ (times the speed of a physical processor of type-1) from each physical processor, every physical processor is left with a capacity: $1 - \frac{2}{2+3\left\lceil\frac{|R|}{m_1}\right\rceil} = \frac{3\left\lceil\frac{|R|}{m_1}\right\rceil}{2+3\left\lceil\frac{|R|}{m_1}\right\rceil}$. As per our specification

---

Algorithm 2. VP_$Create$(PP, $|R|$): for creating virtual processors from a two-type heterogeneous platform

---

Input   : PP, $|R|$
Output: VP$_{\text{AC}}$, VP$_{\text{B}}$
// PP denotes the set of physical processors
// $|R|$ denotes the number of shared resources
1   VP$_{\text{AC}}[1,\cdots,m] := \{0,\cdots,0\}$ VP$_{\text{B}}[1,\cdots,2|R|] := \{0,\cdots,0\}$
2   VP$_{\text{ABC}}$_$Create$(PP, VP$_{\text{AC}}$, VP$_{\text{B}}$, 0, 0, 1)
3   VP$_{\text{ABC}}$_$Create$(PP, VP$_{\text{AC}}$, VP$_{\text{B}}$, m_1, $|R|$, 2)
4   **return** VP$_{\text{AC}}$, VP$_{\text{B}}$

Algorithm 3. $\text{VP}_{\text{ABC}}\text{\_}Create(\text{PP}, \text{VP}_{\text{AC}}, \text{VP}_{\text{B}}, lb, si, z)$: for creating phase-AC and phase-B virtual processors

---

    Input : $\text{PP}, \text{VP}_{\text{AC}}, \text{VP}_{\text{B}}, lb, si, z$
    Output: $\text{VP}_{\text{AC}}, \text{VP}_{\text{B}}$
    `// lb denotes the starting index for array` VP_{AC}
    `// si denotes the starting index for array` VP_B
    `// z denotes the processor type`
1 $\text{VP}_{\text{AC}}[lb + 1, \cdots, lb + m_z] := \{0, \cdots, 0\}$ `// initialize the relevant`
      `elements in` VP_{AC} `to zero`
2 **for** $i = 1$ **to** m_z **do**
3 | Create a virtual processor say, vp_i^{ACz} from pp_i of speed $\dfrac{2}{2+3\left\lceil\frac{|R|}{m_z}\right\rceil}$ times the
    | speed of pp_i
4 | $\text{VP}_{\text{AC}}[lb + i] := vp_i^{ACz}$
5 **end**
6 $cnt := 1, flag := 0$
7 **for** $i = 1$ **to** m_z **do**
8 | **for** $j = 1$ to $\left\lceil\frac{|R|}{m_z}\right\rceil$ **do**
9 | | create a virtual processor say, vp_{cnt}^{Bz} from pp_i of speed $\dfrac{3}{2+3\left\lceil\frac{|R|}{m_z}\right\rceil}$ times
    | | the speed of pp_i
10 | | $\text{VP}_{\text{B}}[si + cnt] := vp_{cnt}^{Bz}$
11 | | **if** $(cnt = |R|)$ **then**
12 | | | $flag := 1$
13 | | | break
14 | | **end**
15 | | $cnt := cnt + 1$
16 | **end**
17 | **if** $(flag = 1)$ **then**
18 | | break
19 | **end**
20 **end**

(in Section 5), the phase-B virtual processor must have $\dfrac{3}{2+3\left\lceil\frac{|R|}{m_1}\right\rceil}$ times the speed of a physical processor of type-1. Hence, it is possible to create:

$$\left\lfloor \frac{\frac{3\left\lceil\frac{|R|}{m_1}\right\rceil}{2+3\left\lceil\frac{|R|}{m_1}\right\rceil}}{\frac{3}{2+3\left\lceil\frac{|R|}{m_1}\right\rceil}} \right\rfloor = \left\lfloor\left\lceil\frac{|R|}{m_1}\right\rceil\right\rfloor = \left\lceil\frac{|R|}{m_1}\right\rceil \geq 1$$

phase-B virtual processors from the remaining capacity of every physical processor of type-1. This allows us to successfully create $|R|$ phase-B virtual processors from the remaining capacity of m_1 processors of type-1. Analogous reasoning holds for type-2 processors as well.

A Dynamic Elimination-Combining Stack Algorithm

Gal Bar-Nissan, Danny Hendler, and Adi Suissa*

Department of Computer Science
Ben-Gurion University of the Negev
Be'er Sheva, Israel

Abstract. Two key synchronization paradigms for the construction of scalable concurrent data-structures are *software combining* and *elimination*. Elimination-based concurrent data-structures allow operations with *reverse semantics* (such as *push* and *pop* stack operations) to "collide" and exchange values without having to access a central location. Software combining, on the other hand, is effective when colliding operations have *identical semantics*: when a pair of threads performing operations with identical semantics collide, the task of performing the combined set of operations is delegated to one of the threads and the other thread waits for its operation(s) to be performed. Applying this mechanism iteratively can reduce memory contention and increase throughput.

The most highly scalable prior concurrent stack algorithm is the *elimination-backoff stack* [5]. The elimination-backoff stack provides high parallelism for symmetric workloads in which the numbers of *push* and *pop* operations are roughly equal, but its performance deteriorates when workloads are asymmetric.

We present DECS, a novel Dynamic Elimination-Combining Stack algorithm, that scales well for all workload types. While maintaining the simplicity and low-overhead of the elimination-bakcoff stack, DECS manages to benefit from collisions of both identical- and reverse-semantics operations. Our empirical evaluation shows that DECS scales significantly better than both blocking and non-blocking best prior stack algorithms.

1 Introduction

Concurrent stacks are widely used in parallel applications and operating systems. As shown in [11], LIFO-based scheduling reduces excessive task creation and prevents threads from attempting to dequeue and execute a task which depends on the results of other tasks. A concurrent stack supports the *push* and *pop* operations with linearizable LIFO semantics. *Linearizability* [7], which is the most widely used correctness condition

---

* Supported by a grant from the Lynne and William Frankel Center for Computer Sciences and by the *Israel Science Foundation* (grant number 1227/10).

A. Fernández Anta, G. Lipari, and M. Roy (Eds.): OPODIS 2011, LNCS 7109, pp. 544–561, 2011.

for concurrent objects, guarantees that each operation appears to have an atomic effect at some point between its invocation and response and that operations can be combined in a modular way.

Two key synchronization paradigms for the construction of scalable concurrent data-structures in general, and concurrent stacks in particular, are *software combining* [13,3,4] and *elimination* [1,10]. Elimination-based concurrent data-structures allow operations with reverse semantics (such as *push* and *pop* stack operations) to "collide" and exchange values without having to access a central location. Software combining, on the other hand, is effective when colliding operations have identical semantics: when a pair of threads performing operations with identical semantics collide, the task of performing the combined set of operations is delegated to one of the threads and the other thread waits for its operation(s) to be performed. Applying this mechanism iteratively can reduce memory contention and increase throughput.

The design of efficient stack algorithms poses several challenges. Threads sharing the stack implementation must synchronize to ensure correct linearizable executions. To provide scalability, a stack algorithm must be highly parallel; this means that, under high load, threads must be able to synchronize their operations without accessing a central location in order to avoid sequential bottlenecks. Scalability at high loads should not, however, come at the price of good performance in the more common low contention cases. Hence, another challenge faced by stack algorithms is to ensure low latency of stack operations when only a few threads access the stack simultaneously.

The most highly scalable concurrent stack algorithm known to date is the lock-free *elimination-backoff stack* of Hendler, Shavit and Yerushalmi [5] (henceforth referred to as the HSY stack). It uses a single elimination array as a backoff scheme on a simple lock-free central stack (such as Treiber's stack algorithm [12][1]). If the threads fail on the central stack, they attempt to eliminate on the array, and if they fail in eliminating, they attempt to access the central stack once again and so on. As shown by Michael and Scott [9], the central stack of [12] is highly efficient under low contention. Since threads use the elimination array only when they fail on the central stack, the elimination-backoff stack algorithm enjoys similar low contention efficiency.

The HSY stack scales well under high contention if the workload is symmetric (that is, the numbers of *push* and *pop* operations are roughly equal), since multiple pairs of operations with reverse semantics succeed

---

[1] Treiber's algorithm is a variant of an algorithm previously introduced by IBM [8].

in exchanging values without having to access the central stack. Unfortunately, when workloads are asymmetric, most collisions on the elimination array are between operations with identical semantics. For such workloads, the performance of the HSY stack deteriorates and falls back to the sequential performance of a central stack.

Recent work by Hendler et al. introduced *flat-combining* [2], a synchronization mechanism based on coarse-grained locking in which a single thread holding a lock performs the combined work of other threads. They presented flat-combining based implementations of several concurrent objects, including a flat-combining stack (FC stack). Due to the very low synchronization overhead of flat-combining, the FC stack significantly outperforms other stack implementations (including the elimination-backoff stack) in low and medium concurrency levels. However, since the FC stack is essentially sequential, its performance does not scale and even deteriorates when concurrency levels are high.

Our Contributions: This paper presents DECS, a novel Dynamic Elimination-Combining Stack algorithm, that scales well for all workload types. While maintaining the simplicity and low-overhead of the HSY stack, DECS manages to benefit from collisions of both identical- and reverse-semantics operations.

The idea underlying DECS is simple. Similarly to the HSY stack, DECS uses a contention-reduction layer as a backoff scheme for a central stack. However, whereas the HSY algorithm uses an elimination layer, DECS uses an *elimination-combining layer* on which concurrent operations can dynamically either eliminate or combine, depending on whether their operations have reverse or identical semantics, respectively. As illustrated by Fig. 1-(a), when two identical-semantics operations executing the HSY algorithm collide, both have to retry their operations on the central stack. With DECS (Figure 1-(b)), every collision, regardless of the types of the colliding operations, reduces contention on the central stack and increases parallelism by using either elimination or combining. Since combining is applied iteratively, each colliding operation may attempt to apply the combined operations (multiple *push* or multiple *pop* operations) of multiple threads - its own and (possibly) the operations delegated to it by threads with which it previously collided, threads that are awaiting their response.

We compared DECS with a few prior stack algorithm, including the HSY and the FC stacks. DECS outperforms the HSY stack on all workload types and all concurrency levels; specifically, for asymmetric workloads, DECS provides up to 3 times the throughput of the HSY stack.

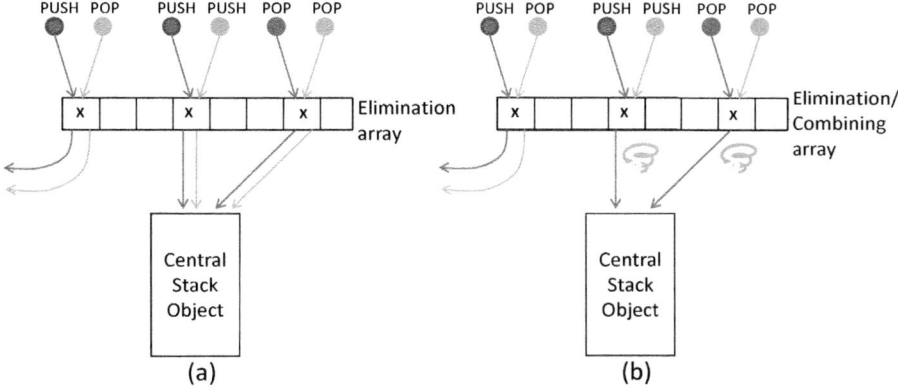

Fig. 1. Collision-attempt scenarios: (a) Collision scenarios in the elimination-backoff stack; (b) Collision scenarios in DECS

The FC stack outperforms DECS in low and medium levels of concurrency. The performance of the FC stack deteriorates quickly, however, as the level of concurrency increases. DECS, on the other hand, continues to scale on all workload types and outperforms the FC stack in high concurrency levels by a wide margin, providing up to 4 times its throughput.

For some applications, a nonblocking [6] stack may be preferable to a blocking one because nonblocking implementations are more robust in the face of thread failures. Whereas the elimination-backoff stack is lock-free, both the FC and the DECS stacks are blocking. We present NB-DECS, a lock-free [6] variant of DECS that allows threads that delegated their operations to a combining thread and have waited for too long to cancel their "combining contracts" and retry their operations. The performance of NB-DECS is slightly better than that of the HSY stack when workloads are symmetric and for *pop*-dominated workloads, but it provides significantly higher throughput for *push-dominated* asymmetric workloads.

The remainder of this paper is organized as follows. We describe the DECS algorithm in Section 2 and report on its performance evaluation in Section 3. A high-level description of the NB-DECS algorithm is provided in Section 4. We conclude the paper in Section 5 with a short discussion of our results. Correctness proofs and the pseudo-code of the NB-DECS algorithm appear in the full paper.

2   The Dynamic Elimination-Combining Algorithm

In this section we describe the DECS algorithm. Figure 2-(a) presents the data-structures and shared variables used by DECS. Similarly to the

HSY stack, DECS uses two global arrays - *location* and *collision* - which comprise its elimination-combining layer. Each entry of the *location* array corresponds to a thread $t \in \{1..N\}$ and is either *NULL* or stores a pointer to a *multiOp* structure described shortly. Each non-empty entry of the *collision* array stores the ID of a thread waiting for another thread to collide with it. DECS also uses a *CentralStack* structure, which is a singly-linked-list of *Cell* structures - each comprising an opaque *data* field and a *next* pointer.

Push or *pop* operations that access the elimination-combining layer may be combined. Thus, in general, operations that are applied to the central stack or to the elimination-combining layer are *multi-ops*; that is, they are either *multi-pop* or *multi-push* operations which represent the combination of multiple *pop* or *push* operations, respectively. A multi-op is performed by a *delegate thread*, attempting to perform its own operation and (possibly) also those of one or more *waiting threads*. The *length* of a multi-op is the number of operations it consists of (which is the number of corresponding waiting threads plus 1). Each multi-op is represented by a *multiOp* structure (see Fig. 2-(a)), consisting of a thread identifier *id*, the operations type (*PUSH* or *POP*) and a *Cell* structure (containing the thread data in case of a multi-push or empty in case of a multi-pop). The

---

(a) Data Structures and Shared Variables

```
1  define Cell: struct {data: Data, next: Cell };
2  define multiOp: struct {id,op,length,cStatus: int, cell: Cell, next,last: multiOp
   };
3  global CentralStack: Cell;
4  global collision: array of [1,...,N] of int init EMPTY;
5  global location: array of [1,...,N] of multiOp init null;
```

(b) **Data pop()**	(c) push(Data: *data*)
6 **multiOp** mOp = initMultiOp();	14 **multiOp** mOp = initMultiOp(*data*);
7 **while true do**	15 **while true do**
8 **if** *cMultiPop(mOp)* **then**	16 **if** *cMultiPush(mOp)* **then**
9 **return** *mOp.cell.data*;	17 **return**;
10 **else if** *collide(mOp)* **then**	18 **else if** *collide(mOp)* **then**
11 **return** *mOp.cell.data*;	19 **return**;
12 **end**	20 **end**
13 **end**	21 **end**

Fig. 2. (a): Data structures, (b) and (c): DECS Push and Pop functions

next field points to the structure of the next operation of the multiOp (if any). Thus, each multiOp is represented by a *multiOp list* of structures, the first of which represents the operation of the delegate thread. The *last* field points to the last structure in the multiOp list and the *length* field stores the multi-op's length. The *cStatus* field is used for synchronization between a delegate thread and the threads awaiting it and is described later in this section.

Figures 2-(b) and 2-(c) present the code performed by a thread when it applies a *push* or a *pop* operation to the DECS stack. A *pop* (*push*) operation starts by initializing a *multiOp* record in line **6** (line **14**). It then attempts to apply the *pop* (*push*) operation to the central stack in line **8** (line **16**). If this attempt fails, the thread then attempts to apply its operation to the elimination-combining layer in line **10** (line **18**). A thread continues these attempts repeatedly until it succeeds. A *pop* operation returns the data stored at the cell that it received either from the central stack (line **9**) or by way of elimination (line **11**).

Central Stack Functions. Figures 3-(a) and 3-(b) respectively present the pseudo-code of the cMultiPop and cMultiPush functions applied to the central stack.

The cMultiPop function receives as its input a pointer to the first *multiOp* record in a multi-op list of *pop* operations to be applied to the central stack. It first reads the central stack pointer (line **22**). If the stack is empty, then all the *pop* operations in the list are linearized in line **22** and will return an *empty* indication. In lines **25–30**, an EMPTY_CELL is assigned as the response of all these operations and the *cStatus* fields of all the *multiOp* structures is set to FINISHED in order to signal all waiting threads that their response is ready. The cMultiPop function then returns *true* indicating to the delegate thread that its operation was applied.

If the stack is not empty, the number m of items that should be popped from the central stack is computed (lines **31–35**); this is the minimum between the length of the multi-pop operation and the central stack's size. The $nTop$ pointer is set accordingly and a CAS is applied to the central stack attempting to atomically pop m items (line **36**). If the CAS fails, *false* is returned (line **50**) indicating that cMultiPop failed and that the multi-pop should be next applied to the elimination-combining layer.

If the CAS succeeds, then all the multi-pop operations are linearized when it occurs. The cMultiPop function proceeds by iterating over the multi-op list (lines **39–48**). It assigns the m cells that were popped from the central stack to the first m pop operations (line **43**) and assigns an

(a) **boolean cMultiPop(multiOp: *mOp*)**	
22 *top = CentralStack*;	**36** **if** *CAS(&CentralStack, top, nTop)* **then**
23 **if** *top =* **null then**	**37** *mOp.cell = top*;
24 **repeat**	**38** *top = top.next*;
25 *mOp.cell =* EMPTY_CELL;	**39** **while** *mOp.next ≠* **null do**
26 *mOp.cStatus =* FINISHED;	**40** **if** *top =* **null then**
27 *mOp=mOp.next*;	**41** *mOp.next.cell =* EMPTY_CELL;
28 **until** *mOp =* **null** ;	**42** **else**
29 **return true;**	**43** *mOp.next.cell = top*;
30 **end**	**44** *top = top.next*;
31 *nTop = top.next*;	**45** **end**
32 *m = 1*;	**46** *mOp.next.cStatus =* FINISHED;
33 **while** *nTop ≠* **null** ∧ *m < mOp.length* **do**	**47** *mOp.next=mOp.next.next*;
34 *nTop = nTop.next, m++*;	**48** **end**
35 **end**	**49** **return true;**
	50 **else return false;**

(b) **boolean cMultiPush(multiOp: *mOp*)**
51 *top = CentralStack*;
52 *mOp.last.cell.next=top*;
53 **if** *CAS(&CentralStack, top, mOp.cell)* **then**
54   **while** *mOp.next ≠* **null do**
55   *mOp.next.cStatus =* FINISHED;
56   *mOp.next = mOp.next.next*;
57   **end**
58   **return true;**
59 **else**
60   **return false;**
61 **end**

Fig. 3. (a) and (b): Central stack operations

EMPTY_CELL to the rest of the pop operations, if any (line **41**). It then sets the *cStatus* of all these operations to FINISHED (line **46**), signalling all waiting threads that their response is ready. The cMultiPop function then returns *true*, indicating that it was successful (line **49**).

The cMultiPush function receives as its input a pointer to the first *multiOp* record in a multi-op list of *push* operations to be applied to the central stack. It sets the *next* pointer of the last cell to point to the top of the central stack (line **52**) and applies a *CAS* operation in an attempt to atomically chain the list to the central stack (line **53**). If the CAS succeeds, then all the *push* operations in the list are linearized when it occurs. In this case, the `cMultiPush` function proceeds by iterating over the multi-op list and setting the *cStatus* of the *push* operations to FINISHED (lines **54–57**). It then returns *true* in line **58**, indicating its success. If the CAS fails, `cMultiPush` returns *false* (line **60**) indicating that the multi-push should now be applied to the elimination-combining layer.

Elimination-Combining Layer Functions. The `collide` function, presented in Fig. 4, implements the elimination-combining backoff algorithm performed after a multi-op fails on the central stack.[2] It receives as its input a pointer to the first *multiOp* record in a multi-op list. A delegate thread executing the function first *registers* by writing to its entry in the *location* array (line **62**) a pointer to its *multiOp* structure, thus advertising itself to other threads that may access the elimination-combining layer . It then chooses randomly and uniformly an index into the collision array (line **63**) and repeatedly attempts to swap the value in the corresponding entry with its own ID by using *CAS* (lines **64–66**).

A thread that initiates a collision is called an *active collider* and a thread that discovers it was collided with is called a *passive collider*. If the value read from the collision array entry is not null (line **68**), then it is a value written there by another registered thread that may await a collision. The delegate thread (now acting as an active collider) proceeds by reading a pointer to the other thread's multiOp structure *oInfo* (line **69**) and then verifies that the other thread may still be collided with (line **70**).[3]

If the tests of line **70** succeed, the delegate thread attempts to *deregister* by CAS-ing its *location* entry back to *NULL* (line **71**). If the CAS is successful, the thread calls the `activeCollide` function (line **72**) in an attempt to either combine or eliminate its operations with those of the other thread. If the CAS fails, however, this indicates that some other thread was quicker and already collided with the current thread; in this case, the current thread becomes a passive thread and executes the `passiveCollide` function (line **74**).

If the tests of line **70** fail, the thread attempts to become a passive collider and waits for a short period of time in line **78** to allow other threads to collide with it. It then tries to deregister by CAS-ing its

---

[2] This function is similar to the `LesOP` function of the HSY stack and is described for the sake of presentation completeness.

[3] Some of the tests of line **70** are required because *location* array entries are not re-initialized when operations terminate (for optimization reasons) and thus may contain outdated values.

(c) **boolean** `collide`(multiOp: mOp)

```
62  location[id] = mOp;
63  index = randomIndex();
64  him = collision[index];
65  while CAS(&collision[index], him, id)=false do
66  |   him = collision[index];
67  end
68  if him ≠ EMPTY then
69  |   oInfo = location[him];
70  |   if oInfo ≠ NULL ∧ oInfo.id ≠ id ∧ oInfo.id=him then
71  |   |   if CAS(&location[id], mOp, NULL)=true then
72  |   |   |   return activeCollide(mOp, oInfo);
73  |   |   else
74  |   |   |   return passiveCollide(mOp);
75  |   |   end
76  |   end
77  end
78  wait();
79  if CAS(&location[id], mOp, NULL)=false then
80  |   return passiveCollide(mOp);
81  end
82  return false;
```

Fig. 4. The `collide` function

entry in the *location* array to *NULL*. If the CAS fails - implying that an active collider succeeded in initiating a collision with the delegate thread - the delegate thread, now a passive collider, calls the `passiveCollide` (line **80**) function in an attempt to finalize the collision. If the CAS succeeds, the thread returns *false* indicating that the operation failed on the elimination-combining layer and should be retried on the central stack.

The `activeCollide` function (Figure 5-(a)) is called by an active collider in order to attempt to combine or eliminate its operations with those of a passive collider. It receives as its input pointers to the *multiOp* structures of both threads. The active collider first attempts to swap the passive collider's *multiOp* pointer with a pointer to its own *multiOp* structure by performing a *CAS* on the *location* array in line **83**. If the CAS fails then the passive collider is no longer eligible for collision and the function returns *false* (line **92**), indicating that the executing thread must retry its multi-op on the central stack. If the *CAS* succeeds, then the collision took place. The active collider now compares the type of its multi-op with that of the passive collider (line **84**) and calls either the `combine` or the `multiEliminate` function, depending on whether the multi-ops have identical or reverse semantics, respectively (lines **84–89**).

(a) boolean activeCollide
(multiOp: *aInf, pInf*)

```
83 if CAS(&location[pInf.id], pInf,
       aInf) then
84  |   if aInf.op = pInf.op then
85  |   |    combine(aInf,pInf);
86  |   |    return false;
87  |   else
88  |   |    multiEliminate(aInf,pInf);
89  |   |    return true;
90  |   end
91 else
92  |   return false;
93 end
```

(b) boolean passiveCollide
(multiOp: *pInf*)

```
94  aInf = location[pInf.id];
95  location[pInf.id] = null;
96 if pInf.op ≠ aInf.op then
97  |    if pInf.op = POP then
98  |    |    pInf.cell = aInf.cell;
99  |    end
100 |    return true;
101 else
102 |    await(pInf.cStatus ≠ INIT);
103 |    if pInf.cStatus = FINISHED
    |    then
104 |    |    return true;
105 |    else
106 |    |    pInf.cStatus = INIT;
107 |    |    return false;
108 |    end
109 end
```

(c) combine(multiOp: *aInf, pInf*)

```
110 if aInf.op = PUSH then
111  |    aInf.last.cell.next = pInf.cell;
112 end
113 aInf.last.next = pInf;
114 aInf.last = pInf.last;
115 aInf.length = aInf.length +
     pInf.length;
```

(d) multiEliminate(multiOp:
**   *aInf, pInf*)**

```
116 aCurr = aInf;
117 pCurr = pInf;
118 repeat
119  |    if aInf.op = POP then
120  |    |    aCurr.cell = pCurr.cell;
121  |    else
122  |    |    pCurr.cell = aCurr.cell;
123  |    end
124  |    aCurr.cStatus = FINISHED;
125  |    pCurr.cStatus = FINISHED;
126  |    aInf.length = aInf.length - 1;
127  |    pInf.length = pInf.length - 1;
128  |    aCurr = aCurr.next;
129  |    pCurr = pCurr.next;
130 until aCurr = null ∨ pCurr =
     null ;
131 if aCurr ≠ null then
132  |    aCurr.length = aInf.length;
133  |    aCurr.last = aInf.last;
134  |    aCurr.cStatus = RETRY;
135 else if pCurr ≠ null then
136  |    pCurr.length = pInf.length;
137  |    pCurr.last = pInf.last;
138  |    pCurr.cStatus = RETRY;
139 end
```

Fig. 5. (a) The `activeCollide`, (b) `passiveCollide`, (c) `combine` and (d) `multiEliminate` functions

Observe that `activeCollide` returns *true* in case of elimination and *false* in case of combining. The reason is the following: in the first case it is guaranteed that the executing thread's operation was matched with a reverse-semantics operation and so was completed, whereas in the latter case the operations of the passive collider are delegated to the active collider which must now access the central stack again.

The `passiveCollide` function (Figure 5-(b)) is called by a passive collider after it identifies that it was collided with. The passive collider first reads the multi-op pointer written to its entry in the *location* array by the active collider and initializes its entry in preparation for future operations (lines **94**–**95**). If the multi-ops of the colliding threads-pair are of reverse semantics (line **96**) then the function returns *true* in line **100** because, in this case, it is guaranteed that the colliding delegate threads exchange values. Specifically, if the passive thread's multi-op type is *pop*, the thread copies the cell communicated to it by the active collider (line **98**).

If both multi-ops are of identical semantics, then the passive collider's operations were delegated to the active thread and the executing thread ceases to be a delegate thread. In this case, the thread waits until it is signalled (by writing to the *cStatus* field of its *multiOp* structure) how to proceed. There are two possibilities: (1) *cStatus = FINISHED* holds in line **103**. In this case, the thread's operation response is ready and it returns *true* in line **104**. (2) *cStatus = RETRY* holds (line **105**) indicating that the executing thread became a delegate thread once again. This occurs if a thread to which the current thread's operation was delegated eliminated with a multi-op that had a shorter list than its own and the first operation in the "residue" is the current thread's operation. In this case, the thread changes the value of its *cStatus* back to *INIT* (line **106**) and returns *false*, indicating that the operation should be retried on the central stack.

The `combine` function (Figure 5-(c)) is called by an active collider when the operations of both colliders have identical semantics. It receives as its input pointers to the *multiOp* structures of the two colliders. It delegates the operations of the passive collider to the active one by concatenating the *multiOp* list of the passive collider to that of the active collider, and by updating the *last* and *length* fields of its *multiOp* record accordingly (lines **113**–**115**). In addition, if the type of both multi-ops is *push*, then their cell-lists are also concatenated (line **111**); this allows the delegate thread to push all its operations to the central stack by using a single *CAS* operation.

The `multiEliminate` function (Figure 5-(d)) is called by an active collider when the operations of both colliders have reverse semantics. It receives as input pointers to the *multiOp* records of the active and passive colliders. In the loop of lines **118**–**130**, as many pairs of reverse-semantics operations as possible are matched until at least one of the operation lists is exhausted. All matched operations are signalled by writing the value *FINISHED* to the *cStatus* field of their *multiOp* structure, indicating that they can terminate (lines **124–125**). Note that both lists contain at least one operation, thus at least a single pair of operations are matched. If the lengths of the multi-ops are unequal, then a "residue" sublist remains. In this case, the *length* and *last* fields of the *multiOp* structure belonging to the first waiting thread in the residue sub-list are set. Then that thread is signalled by writing the value *RETRY* to the *cStatus* field of its *multiOp* structure (in line **134** or line **138**). This makes the signaled thread a delegate thread once again and it will retry its multi-op on the central stack.

3   DECS Performance Evaluation

We conducted our performance evaluation on a Sun SPARC T5240 machine, comprising two UltraSPARC T2 plus (Niagara II) chips, running the Solaris 10 operating system. Each chip contains 8 cores and each core multiplexes 8 hardware threads, for a total of 64 hardware threads per chip. According to common practice, we ran our experiments on a single chip to avoid communication via the L2 cache. The algorithms we evaluated are implemented in C++ and the code was compiled using GCC with the -O3 flag for all algorithms.

We compare DECS with the Treiber stack[4] and with the most effective known stack implementations: the HSY elimination-backoff stack, and a flat-combining based stack.[5],[6]

In our experiments, threads repeatedly apply operations to the stack for a fixed duration of one second and we measure the resulting *throughput* - the total number of operations applied to the stack - varying the number of threads from 1 to 128. Each data point is the average of three runs. We measure throughput on both symmetric (push and pop operations are equally likely) and asymmetric workloads. Stacks are pre-populated with

---

[4] We evaluated two variants of the Treiber algorithm - with and without exponential backoff. The variant using exponential backoff performed consistently better and is the version we compare with.

[5] We downloaded the most updated flat-combining code from https://github.com/mit-carbon/Flat-Combining.

[6] The Treiber, HSY and DECS algorithms need to cope with the "ABA problem" [8], since they use dynamic-memory structures that may need to be recycled and perform CAS operations on pointers to these structures. We implemented the simplest and most common ABA-prevention technique that includes a tag with the target memory locations so that both the memory location and the tag are manipulated together atomically, and the tag is incremented with each update of the target memory location [8].

enough cells so that pop operations do not operate on an empty stack also in asymmetric workloads.

Figures 6-(a) through (c) compare the throughput of the algorithms we evaluate in symmetric (50% push, 50% pop), moderately-asymmetric (25% push, 75% pop) and fully-asymmetric (0% push, 100% pop) workloads, respectively. It can be seen that the DECS stack outperforms both the Treiber stack and the HSY stack for all workload types and all concurrency levels.

Symmetric Workloads

We first analyze performance on a symmetric workload, which is the optimal workload for the HSY stack. As shown in Fig. 6-(a), even here the HSY stack is outperformed by DECS by a margin of up to 31% (when the number of threads is 64). This is because, even in symmetric workloads, there is a non-negligible fraction of collisions between operations of identical semantics from which DECS benefits but the HSY stack does not. Both DECS and the HSY stack scale up until concurrency level 64 - the number of hardware threads. When the number of software threads exceeds the number of hardware threads, the HSY stack more-or-less maintains its throughput whereas DECS slightly declines but remains significantly superior to the HSY stack.

The FC stack incurs the highest overhead in the lack of contention (concurrency level 1) because the single running thread still needs to capture the FC lock. Due to its low synchronization overhead it then exhibits a steep increase in its throughput and reaches its peak throughput at 24 threads, where it outperforms DECS by approximately 33%. The FC stack does not continue to scale beyond this point, however, and its throughput rapidly deteriorates as the level of concurrency rises. For concurrency levels higher than 40, its performance falls below that of DECS and it is increasingly outperformed by DECS as the level of concurrency is increased: for 64 threads, DECS provides roughly 33% higher throughput, and for 128 threads DECS outperforms FC by a factor of 4. For concurrency levels higher than 96, the throughput of the FC stack is even lower than that of the Treiber algorithm. The reason for this performance deterioration is clear: the FC algorithm is essentially sequential, since a single thread performs the combined work of other threads. The Treiber algorithm exhibits the worst performance since it is sequential and incurs significant synchronization overhead. It scales moderately until concurrency level 16 and then more-or-less maintains its throughput.

Figure 6-(d) provides more insights into the behavior of the DECS and HSY stacks in symmetric workloads. The HSY curve shows the percentage

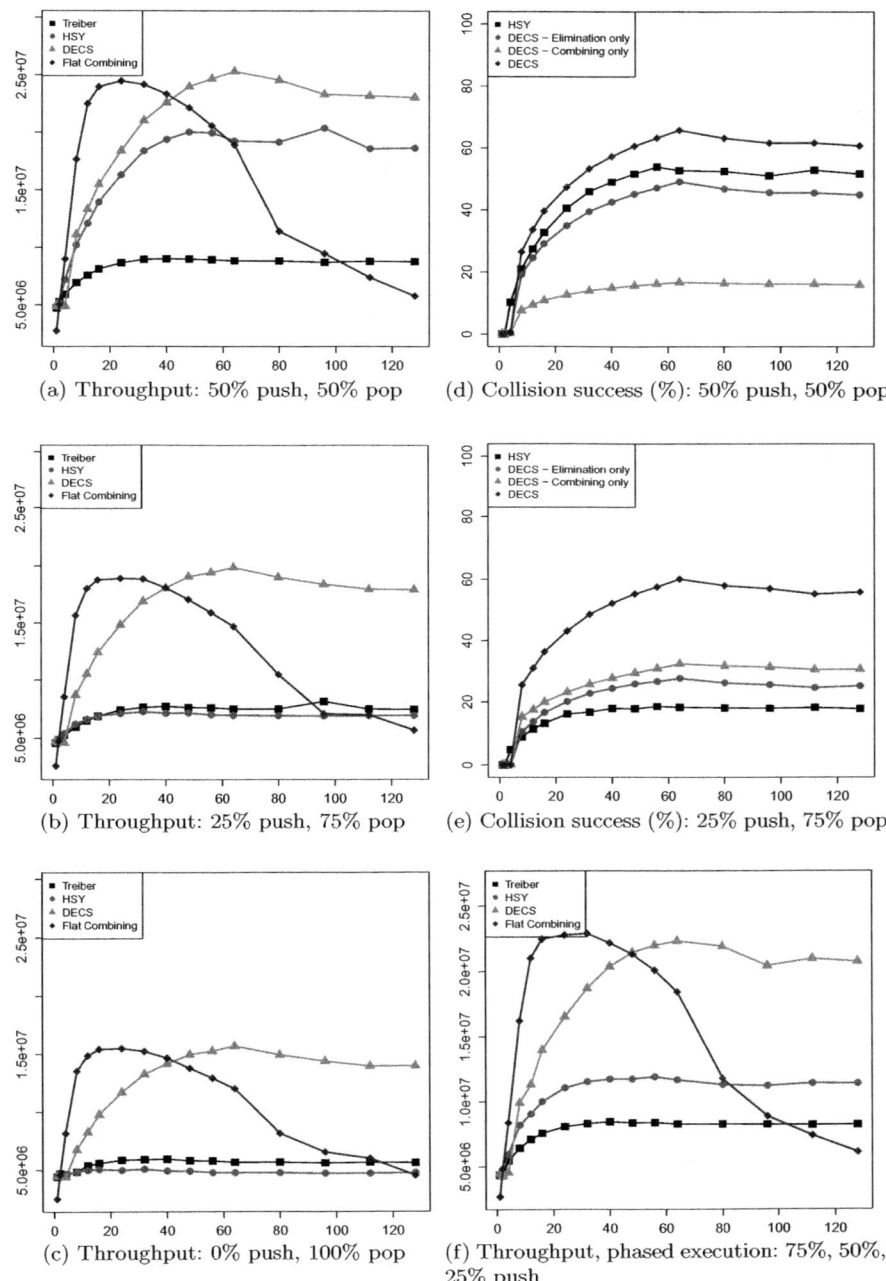

Fig. 6. Throughput and collision success rates. X-axis: threads #; Y-axis in (a)-(c), (f): throughput.

of operations completed by elimination. The DECS curve shows the percentage of operations not applied directly to the central stack. These are the operations completed by either elimination or combining.[7] The curves titled "Elimination only" and "Combining only" show a finer partition of the DECS operations according to whether they completed through elimination or combining. It can be seen that the overall percentage of operations not completed on the central stack is higher for DECS than for the HSY stack by up to 30% (for 64 threads), thus reducing the load on the central stack and allowing DECS to perform better than the HSY stack.

Asymmetric Workloads

Figures 6-(b) and 6-(c) compare throughput on moderately- and fully-asymmetric workloads, respectively. The relative performance of DECS, the FC and the Treiber stacks is roughly the same as for the symmetric workload; nevertheless, DECS performance decreases because, as can be seen in Fig. 6-(e), the ratio of DECS operations that complete via elimination is significantly reduced for the 25% push workload. This ratio drops to 0 for the 0% push workload. This reduction in elimination is mostly compensated by a corresponding increase in the ratio of DECS operations that complete by combining.

The performance of the HSY stack, however, deteriorates for asymmetric workloads because, unlike DECS, it cannot benefit from collisions between operations with identical semantics. When the workload is moderately asymmetric (Figure 6-(b)), the HSY stack scales up to 32 threads but then its performance deteriorates and falls even below that of the Treiber algorithm for 48 threads or more. In these levels of concurrency, the low percentage of successful collisions makes the elimination layer counter-effective. The throughput of the DECS algorithm exceeds that of the HSY stack by a factor of up to 3. The picture is even worse for the HSY algorithm for fully asymmetric workloads (Figure 6-(c)), where it performs almost consistently worse than the Treiber algorithm. In these workloads, DECS' throughput exceeds that of the HSY algorithm significantly in all concurrency levels 8 or higher; the performance gap increases with concurrency up until 64 threads and DECS provides about 3 times the throughput for all concurrency levels 64 or higher.

---

[7] Whenever a muli-op is applied to the central stack, the operation of the delegate thread is regarded as applied directly to the central stack and those of the waiting threads are counted as completed by combining. Similarly, when two multi-ops of reverse semantics collide, the operations of the delegate threads are counted as completed by elimination and those of the waiting threads as completed by combining.

The asymmetric workloads we considered above serve for highlighting the performance tradeoffs between the algorithms we evaluate, as a function of the ratio of *push* and *pop* operations in the workload. A more realistic scenario is when the stack goes through phases, in each of which it is initially filled with items, then it is accessed by a symmetric workload and finally it is being emptied. Figure 6-(f) compares throughput for a phased execution that lasts 3 seconds. In the first second, workload consists of 75% push operations. In the following second, workload is symmetric. Finally, in the third second, the workload consists of 75% pop operations. It can be seen that the performance tradeoffs between the evaluated algorithms are the same as for the non-phased workloads. DECS is consistently better than the HSY stack by a margin of up to 94% (for 80 threads). It is outperformed by FC by a margin of up to 86% (for 12 threads) but, due to its better scalability, provides 21% higher performance for 64 threads and outperforms FC by a factor of approximately 3.4 for 128 threads. The picture is similar for fully asymmetric phased workloads.

4   The Nonblocking DECS Algorithm

For some applications, a nonblocking stack may be preferable to a blocking one because it is more robust in the face of thread failures. The HSY stack is nonblocking - specifically lock-free [6] - and hence guarantees global progress as long as some threads do not fail-stop. In contrast, both the FC and the DECS stacks are blocking. In this section, we provide a high-level description of NB-DECS, a lock-free variant of our DECS algorithm that allows threads that delegated their operations to another thread and have waited for too long to cancel their "combining contracts" and retry their operations. A full description of the NB-DECS algorithm appears in the full paper, where we also present a comparative evaluation of this new algorithm.

Recall that waiting threads await a signal from their delegate thread in the `passiveCollide` function (line **102** in Fig. 5-(b)). In the DECS algorithm, a thread awaits until the delegate thread writes to the *cStatus* field of its *multiOp* structure but may wait indefinitely. In NB-DECS, when a thread concludes that it waited "long enough" it attempts to *invalidate* its *multiOp* structure. To prevent race conditions, invalidation is done by applying *test-and-set* to a new *invalid* field added to the *multiOp* structure. A delegate thread, on the other hand, must take care not to assign a cell of a valid *push* operation to an invalid multi-op structure of a *pop* operation.

This raises the following complications which NB-DECS must handle. (1) A delegate thread may pop invalid cells from the central stack. Therefore, in order not to assign an invalid cell to a *pop* operation, the delegate thread must apply *test-and-set* to each popped cell to verify that it is still valid (and if so to ensure it remains valid), which hurts performance; (2) A delegate thread performing a pop multi-op must deal with situations in which some of its waiting threads invalidated their multi-op structures. If the delegate thread were to pop from the central stack more cells than can be assigned to valid multi-op structures in its list, linearizability would be violated. Consequently, unlike in DECS, the delegate thread must pop items from the central stack *one by one*, which also hurts the performance of NB-DECS as compared with DECS; (3) The `multiEliminate` function, called by an active delegate thread when it collides with a thread with reverse semantics, must also verify that valid cells are only assigned to valid pop multi-ops. Once again, *test-and-set* is used to prevent race conditions.

Due to the extra synchronization introduced in NB-DECS for allowing threads to invalidate operations that are pending for too long, the throughput of NB-DECS is, in general, significantly lower than that of the (blocking) DECS stack. However, NB-DECS provides better performance than the HSY stack for all workload types, providing up to 2 times the throughput for push-dominated workloads.

5   Discussion

We present DECS, a novel Dynamic Elimination-Combining Stack algorithm. Our empirical evaluation shows that DECS scales significantly better than (both blocking and nonblocking) best known stack algorithms for all workload types, providing throughput that is significantly superior to that of both the elimination-backoff stack and the flat-combining stack for high concurrency levels. We also present NB-DECS - a lock-free variant of DECS. NB-DECS provides lower throughput than (the blocking) DECS due to the extra synchronization required for satisfying lock-freedom but may be preferable for some applications since it is more robust to thread failures. NB-DECS outperforms the elimination-backoff stack, the most scalable prior lock-free stack on almost all workload types. The key feature that makes DECS highly effective is the use of a dynamic elimination-combining layer as a backoff scheme for a central data-structure. We believe that this idea may be useful for obtaining high-performance implementations of additional concurrent data-structures.

Acknowledgements. We thank Yehuda Afek and Nir Shavit for allowing us to use their Sun SPARC T5240 machine.

References

1. Afek, Y., Korland, G., Natanzon, M., Shavit, N.: Scalable Producer-Consumer Pools Based on Elimination-Diffraction Trees. In: D'Ambra, P., Guarracino, M., Talia, D. (eds.) Euro-Par 2010, Part II. LNCS, vol. 6272, pp. 151–162. Springer, Heidelberg (2010)
2. Hendler, D., Incze, I., Shavit, N., Tzafrir, M.: Flat combining and the synchronization-parallelism tradeoff. In: SPAA, pp. 355–364 (2010)
3. Hendler, D., Kutten, S.: Bounded-wait combining: constructing robust and high-throughput shared objects. Distributed Computing 21(6), 405–431 (2009)
4. Hendler, D., Kutten, S., Michalak, E.: An Adaptive Technique for Constructing Robust and High-Throughput Shared Objects. In: Lu, C., Masuzawa, T., Mosbah, M. (eds.) OPODIS 2010. LNCS, vol. 6490, pp. 318–332. Springer, Heidelberg (2010)
5. Hendler, D., Shavit, N., Yerushalmi, L.: A scalable lock-free stack algorithm. J. Parallel Distrib. Comput. 70(1), 1–12 (2010)
6. Herlihy, M.: Wait-free synchronization. ACM Transactions On Programming Languages and Systems 13(1), 123–149 (1991)
7. Herlihy, M.P., Wing, J.M.: Linearizability: a correctness condition for concurrent objects. ACM Transactions on Programming Languages and Systems (TOPLAS) 12(3), 463–492 (1990)
8. IBM. IBM System/370 Extended Architecture, Principles of Operation, publication no. SA22-7085 (1983)
9. Michael, M.M., Scott, M.L.: Nonblocking algorithms and preemption-safe locking on multiprogrammed shared — memory multiprocessors. Journal of Parallel and Distributed Computing 51(1), 1–26 (1998)
10. Shavit, N., Touitou, D.: Elimination trees and the construction of pools and stacks. Theory of Computing Systems (30), 645–670 (1997)
11. Taura, K., Matsuoka, S., Yonezawa, A.: An efficient implementation scheme of concurrent object-oriented languages on stock multicomputers. In: Principles Practice of Parallel Programming, pp. 218–228 (1993)
12. Treiber, R.K.: Systems programming: Coping with parallelism. Technical Report RJ 5118, IBM Almaden Research Center (April 1986)
13. Yew, P., Tzeng, N., Lawrie, D.: Distributing hot-spot addressing in large-scale multiprocessors. IEEE Transactions on Computers C-36(4), 388–395 (1987)

Author Index

GPSR Compliance

*The European Union's (EU) General Product Safety Regulation (GPSR)
is a set of rules that requires consumer products to be safe and our
obligations to ensure this.*

*If you have any concerns about our products, you can contact us on
ProductSafety@springernature.com*

In case Publisher is established outside the EU, the EU authorized
representative is:

Springer Nature Customer Service Center GmbH
Europaplatz 3
69115 Heidelberg, Germany

Batch number: 09490872

Printed by Printforce, the Netherlands